# Introduction to
# Engineering Thermodynamics

Introduction to
# Engineering Thermodynamics

**Robert C. Fellinger**
**William J. Cook**

*Iowa State University*

wcb
Wm. C. Brown Publishers
Dubuque, Iowa

**Book Team**





Library of Congress Catalog Card Number: 83–072444

ISBN 0–697–08606–2



Printed in the United States of America
10 9 8 7 6 5 4 3 2 1

# Contents

# 3

## The First Law of Thermodynamics 70



# 4

## Characteristics of Pure Substances 104



# 5

## Properties and Processes for Perfect Gases 128

# 6

## The Second Law of Thermodynamics 165



# 7

## Entropy, A Consequence of the Second Law 191



# 8

## Second-Law Analysis 248

# 9

## Vapor Cycles for Power and Refrigeration 292



# 10

## Air-Standard Cycles 339

# 11

## Perfect Gas Mixtures, Psychrometry, and Air Conditioning Processes 381



# 12

## Combustion Process Fundamentals 423

# 13

## Some Aspects and Applications of One-Dimensional Compressible Flow 472



# 14

## Compressors, Fans, Pumps, and Turbines 506

# Preface

Thermodynamics is a science integral to many disciplines and of interest in numerous professional careers. Topics in thermodynamics of particular interest to engineers form **engineering thermodynamics,** an engineering science that is included in the curricula of several branches of engineering. This book has been prepared for students whose curricula include a single course or a two-term sequence in engineering thermodynamics. The text is written and arranged so that the fundamentals of the subject and selected engineering applications can be covered in a three-credit semester course. It is also suitable for a more extensive two-term sequence of courses in which additional applications are covered.

The traditional classical approach to thermodynamics has been taken in writing this text. The authors feel that this approach best serves the student who takes a single or introductory course in engineering thermodynamics. A knowledge of mathematics through integral calculus and courses in university chemistry and physics should be prerequisites for courses taught from this text. Both the SI and English systems of units have been included because the authors believe that it is important for today's engineers to be conversant in both systems. Throughout the book numerous example problems illustrate the material covered. These are viewed as an integral part of the text material. Further, the authors are of the firm opinion that effective study of engineering thermodynamics must include the solving of problems that illustrate and emphasize each new concept as it is presented. To this end, the book contains more than 500 problems of varying difficulty and complexity.

The first seven chapters cover the fundamentals of engineering thermodynamics in a concise manner. The remaining seven chapters are devoted, for the most part, to engineering applications. The first chapter introduces the subject to the student through discussion of simplified versions of the first and second laws of thermodynamics. This chapter emphasizes the relation between thermodynamics and energy, introduces the concept of the grade or quality of energy, and includes selected information on energy resources and utilization, particularly as these relate to the energy picture in the United States.

Basic definitions and concepts—properties, processes, cycles, heat and work interactions, mass conservation—are presented and discussed in Chapter 2. The first law is formally introduced and the related equations are developed in Chapter 3. Stored energy is shown to be a consequence of the first law. The general open-system energy equation is developed. Subsequent emphasis is placed on energy equations for closed and steady-flow open systems. The first law is introduced prior to discussion of the relationships among the various thermodynamic properties of substances. These relationships are discussed in Chapters 4 and 5. Where necessary prior to Chapter 4, numerical values of properties are given. This approach allows early focusing on the nature of the first law without dealing with the complexities related to the evaluation of properties of particular substances. Chapter 4 deals with the properties of pure substances that undergo phase changes. Tables of properties for such substances are introduced and discussed. Chapter 5 covers properties and processes for perfect (ideal) gases. This order of presentation allows perfect gases to be treated as special cases of general pure substances.

The second law is introduced in Chapter 6 through the Kelvin-Planck and Clausius statements. The reversible process is presented in relation to the second law. The chapter deals mainly with the second law for cyclic processes. The property of entropy is shown to be a consequence of the second law in Chapter 7. The second law as it applies to noncyclic processes is concentrated on in this chapter. The chapter closes with a section on the usefulness of heat energy for producing work, that is, the grade of heat energy.

Chapters 1 through 7 are intended to be studied in the order in which they are presented. The remaining chapters are, for the most part, independent, and most need not be covered in their entirety to maintain course continuity. Chapter 8 is a chapter on the basics of second-law analysis, a topic of current interest in relation to the optimum utilization of energy. A section on cogeneration is included among the topics on vapor cycles for power and refrigeration in Chapter 9. Chapter 10 discusses air-standard gas power cycles and includes an introduction to combined power cycles. Perfect gas mixtures, psychrometrics, and air-conditioning processes are presented in Chapter 11. A portion of this material provides background for Chapter 12, which covers the fundamentals of combustion processes. The material in Chapter 13 deals with one-dimensional compressible flow. One topic in this chapter (isentropic stagnation states) supports Chapter 14, a chapter on compressors, pumps, fans, and turbines. The applications chapters provide an opportunity to select topics that best fit the needs and interests of particular branches of engineering. For example, students majoring in aerospace engineering would likely find Chapter 13 and parts of Chapter 10 more relevant to their interests than Chapter 9. Although some applications are included in the illustrative material in the first seven chapters, additional applications are important not only for their intrinsic value, but also in demonstrating the relevance of the fundamentals of engineering thermodynamics.

Included in the appendices are English and SI abridged tables of properties for water. Some instructors will find these adequate. The authors recommend that each student have a more complete table, such as the **Steam Tables** by Keenan, Keyes, Hill, and Moore, as part of their professional library; English units are preferred as a majority of the problems presented here that require steam tables involve English units.

The authors gratefully acknowledge the contributions to this book by the late Dr. H. J. Steover. His detailed review of the first draft produced many suggestions on the manner of presentation. The authors thank reviewers Larry Witte, University of Houston; Adrian Bejan, University of Colorado; John W. Cipolla, Jr., Northeastern University; Warren M. Heffington, Texas A & M University; Dean S. Schupe, University of Cincinnati; and Satish Ramadhyani, Purdue University for their expert criticism—and praise. Also acknowledged are the helpful comments from numerous students who have used various versions of this text at Iowa State University. Finally, the authors wish to express their gratitude to their families for their support and patience while the authors were engaged in the enjoyable but time-consuming task of writing this book.

Although diligent efforts have been made by the authors, editor, and publisher to produce an errorfree book, it is recognized that this may not have been achieved. Comments, criticism, and suggestions on the manner of presentation and content of this text are welcome.

**Robert C. Fellinger**
**William J. Cook**

# Symbols

See Table 2.1 and Appendix A for additional unit symbols.

|  |  |
|---|---|
| $A$ | area; closed-system availability |
| $a$ | acceleration; closed-system availability per unit mass |
| $B$ | British thermal unit; open-system availability |
| $b$ | open-system availability per unit mass |
| $C$ | constant (as in $pv^n = C$); Celsius temperature |
| $c$ | speed of sound (sonic speed) |
| $C_p$ | specific heat, $(\partial h/\partial T)_p$ |
| $C_v$ | specific heat, $(\partial u/\partial T)_v$ |
| $d$ | exact differential |
| $\delta$ | inexact differential |
| $E$ | stored energy |
| $e$ | stored energy per unit mass (specific stored energy) |
| $F$ | force; Fahrenheit temperature |
| $F_g$ | weight (body force due to gravity) |
| $g$ | local gravitational acceleration |
| $g_0$ | gravitational constant in Newton's second law, $F = ma/g_0$ |
| $H$ | enthalpy |
| $h$ | enthalpy per unit mass; head (column height) |
| $\bar{h}$ | enthalpy per unit mole |
| $HP$ | horsepower |
| $\Delta H_f$ | enthalpy of formation |
| $\Delta H_C$ | enthalpy of combustion |
| $\Delta H_R$ | enthalpy of reaction |
| $I$ | irreversibility |
| $J_e$ | Joule's equivalent |
| $K$ | Kelvin temperature |
| $KE$ | kinetic energy |
| $kmol$ | kilogram mole |
| $kW$ | kilowatt |
| $kWh$ | kilowatt-hour |
| $L$ | length |
| $lb\ mol$ | pound mole |
| $M$ | molecular weight (molar mass) |
| $m$ | mass |
| $\dot{m}$ | mass rate of flow |
| $m_f$ | mass fraction |
| $N$ | number of constituents |
| $n$ | number of moles; polytropic exponent |
| $n_i$ | number of inbound streams |
| $n_o$ | number of outbound streams |
| $N_m$ | Mach number |
| $p$ | pressure |
| $PE$ | mechanical potential energy |
| $p_i$ | ice-point pressure; partial-pressure of $i$th constituent |
| $P_r$ | reduced pressure, $p/p_{cp}$ |
| $p_s$ | steam-point pressure |
| $Q$ | heat transfer |
| $q$ | heat transfer per unit mass |
| $\dot{Q}$ | heat-transfer rate |
| $Q_a$ | available portion of heat energy |
| $Q_u$ | unavailable portion of heat energy |
| $R$ | gas constant; Rankine temperature |
| $\bar{R}$ | universal gas constant |
| $r_c$ | compression ratio |
| $r_{co}$ | cutoff ratio |
| $r_p$ | pressure ratio |
| $S$ | entropy; air supplied |
| $s$ | entropy per unit mass |
| $T$ | absolute temperature |
| $t$ | temperature, Fahrenheit or Celsius |
| $T_r$ | reduced temperature, $T/T_{cp}$ |
| $U$ | internal energy |
| $u$ | internal energy per unit mass |
| $\Delta U_f$ | internal energy of formation |
| $\Delta U_c$ | internal energy of combustion |
| $\Delta U_r$ | internal energy of reaction |
| $V$ | volume; velocity |
| $V'$ | partial volume; tangential component of velocity |
| $\dot{V}$ | volume rate of flow |
| $v$ | specific volume |
| $W$ | work |

$w$ work per unit mass
$\dot{W}$ power
$W$ work
$w_s$ specific weight
$x$ quality; mole fraction; distance
$Z$ compressibility factor
$z$ elevation (height)
$\oint$ cyclic integral (integral around a cycle)

## Greek Letters

$\beta$ angle
$\beta_R$ refrigerator coefficient of performance
$\beta_P$ heat pump coefficient of performance
$\gamma$ specific heat ratio, $C_p/C_v$
$\Delta$ change in a quantity; final property value minus initial property value for a state change
$\epsilon$ second-law effectiveness; regenerator effectiveness
$\eta$ efficiency
$\theta$ angle; temperature; absolute thermodynamic temperature
$\theta_i$ ice-point temperature
$\theta_s$ steam-point temperature
$\rho$ density (mass per unit volume)
$\Sigma$ summation
$\tau$ time
T torque
$\phi$ relative humidity
$\omega$ humidity ratio (specific humidity); angular velocity

## Subscripts

$a$ average; actual; air
$c$ compressor
$C$ Celsius temperature; Carnot
$cp$ critical point (critical state)
$cr$ isentropic critical state (Mach number equals unity)
$da$ dry air
$F$ Fahrenheit temperature
$f$ saturated liquid; final state
$fg$ change in a property between saturated-liquid state and saturated-vapor state at constant pressure
$g$ saturated vapor
$H$ high temperature
$I$ due to internal irreversibilities
$i$ saturated solid; initial state; summation index; ice point; inbound section; $i$th constituent
$if$ change in a property between saturated-solid state and saturated-liquid state at constant pressure
$L$ low temperature
$m$ mixture
$o$ open system; atmosphere (environment); outbound section
$pr$ products
$r$ reservoir
$re$ reactants
$rev$ reversible
$s$ isentropic; steam point; open system in Chapter 8
$\sigma$ system
$t$ turbine; isentropic stagnation state
$tp$ triple point
$u$ unavailable; useful
$w$ water

## Superscripts

$0$ standard state
property per unit mole
$*$ state where Mach number equals unity; perfect gas behavior in Section 5.7
$\cdot$ time rate

# Introduction to
# Engineering Thermodynamics

# 1

# Thermodynamics: A Study of Energy

*The study of engineering thermodynamics will provide fundamental knowledge related to energy and should enhance your understanding of energy's role in today's society. It will acquaint you with some aspects of what has been done in the past, as well as with the current state of energy engineering. Advances in energy engineering are products of the human mind. Our energy future, the solution of existing problems, and the development of new energy sources and systems, is a challenge for each new generation of engineers. And the challenge is not only technical. Engineering decisions are increasingly tempered by the attitudes of society. Concerns about acid rain, the handling of nuclear waste, and increased carbon dioxide in the atmosphere as a result of fuel combustion are but a few of the challenges that must be met. Our planet has limited natural resources. How these resources are utilized to serve humanity best depends heavily on the ingenuity of engineers.*

*At this point you may wonder, "What is engineering thermodynamics?" For the moment we shall say it is a study of energy that is focused on engineering systems. (A more exact description will be developed in the next chapter with the introduction of some of the terminology of thermodynamics.) But what is energy? Again, we have yet to reach a point where this word can be assigned a precise thermodynamic definition. Each of you has some idea of what energy is, since you have encountered the term in previous courses, as well as in everyday life. We shall rely on that general knowledge, along with some additional comments, in this first chapter. Let us describe energy as follows.*

*Energy is a nonmaterial agent stored in many forms. It can be converted from one form to another. It may be transferred or transported from one body or location to another. It is utilized each day in many different forms for innumerable purposes. Changes in our lifestyle and standard of living have been, and continue to be, highly dependent on our ability to utilize energy for greater productivity, creature comfort, and, to a degree, personal freedom. Quantitative evaluation of energy stored, converted, and transferred is the primary goal of engineering thermodynamics.*

## 1–1 Some Aspects of Engineering Thermodynamics

Central to our study of engineering thermodynamics are two laws, which will be presented and discussed in subsequent chapters. The first relates to conservation of energy. For the present let us use a rather loose statement related to the first law and note that while energy may exist in a multiplicity of forms and may undergo changes in form, it is conserved. The concept of such conservation will be illustrated by examples.

The second of these laws, among its many ramifications, establishes limits on the amount of heat that can be converted to work by devices that promote energy transformations by means of what we shall call cycles. Here we have used three terms—heat, work, and cycle—which have not yet been defined. At this point let us think of heat and work as ways in which energy can be transferred from one region to another. Their differences lie in what causes the energy transfer to take place. In the case of heat, the transfer is caused by a difference in temperature between the regions. Each of you has encountered the term "work" in your study of physics and engineering mechanics. The form of work most familiar to you at this time no doubt is that associated with a force acting through some distance. Although we shall treat other forms of work, let us use that concept for the moment. The important point to keep in mind is that heat and work are mechanisms for energy transfer. The terms are never used in conjunction with any form of stored energy.

Finally, what is a cycle? Pending the thermodynamic definition of the term, think of a cycle as a series of operations or processes that are repeated in a regular, periodic manner. Two examples of devices that are identified with cyclic operation are steam power plants for electric generation and the home refrigerator-freezer. In contrast to the latter, an electric toaster, a gas or electric stove, or a lighting fixture does not operate on a cycle in the thermodynamic sense. An illustration of a cycle will be presented later.

To illustrate some energy conversions, let us use the simple energy triangle shown in Figure 1.1. The four blocks identify different forms of energy and the arrows indicate energy conversions from one form to another. In each case energy is conserved. However, the conversions may not be complete. We shall see, for example, that thermal energy cannot be completely converted to mechanical energy on a continuous basis. Stored energy forms *(S)* include, but are not limited to, chemical energy in fuels that is released during combustion, nuclear energy released during fission, energy associated with the velocity of a wind, and energy in water behind the dam of a hydroelectric generating station. The thermal energy block *(T)* represents both energy that can be stored—as, for example, in a region at high temperature—and energy that can be transferred as heat to or from a region. The ways in which thermal energy may be stored, transferred, or converted to another energy form are fundamental topics in engineering thermodynamics. Mechanical energy *(M)* is typified by the work at the shaft of an engine, turbine, pump, compressor, or similar device. The electric block *(E)* represents electric energy delivered by a generator, battery, or similar device, or, conversely, electric energy as might be supplied to a motor. Electric energy warrants a place in the energy picture because, at this time, it is the most efficient form in which we can transport usable energy in large quantities over long distances.

Let us now consider some conversions that are not cyclic in nature. Windmills, which were a common sight on farms a number of years ago, fall into this class. Stored energy associated with the wind velocity was converted to mechanical energy as the wind passed over the blades of the mill fixed to a shaft. This mechanical energy typically was used to power a water-well pump. In terms of Figure 1.1, the windmill was an $S$ to $M$ converter,

**Figure 1.1** The Energy Triangle

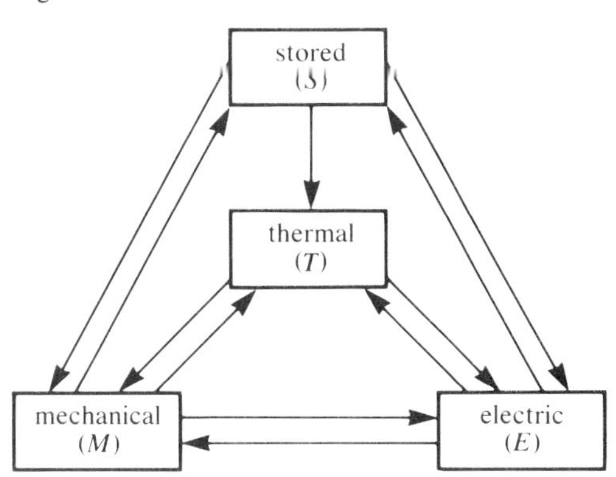


though some thermal energy was lost to the atmosphere through the dissipation of heat resulting from friction effects. Currently attention is being redirected to wind energy (which is now included as a part of solar energy) for electric generation. The flat blades of the old windmill have been replaced by a variety of propellers of advanced aerodynamic design. The windmill has become the wind turbine that supplies mechanical energy to drive electric generators capable of producing of the order of 2500 kW (kilowatts) of electric power. These units follow the $S \rightarrow M \rightarrow E$ path on the energy triangle. With blades longer than the wing span of a Boeing 747 aircraft, such units represent one method of using a renewable energy source, the wind, rather than nonrenewable fuels for electric generation.

Hydroelectric generating stations also follow the $S \rightarrow M \rightarrow E$ conversion sequence. Stored energy is represented by the height (head) of water behind a dam and above the inlet to the hydraulic turbine. As water passes through the turbine, mechanical power at the turbine shaft drives an electric generator. The reversed conversion $E \rightarrow M \rightarrow S$ in which electrically driven pumps return water to a lake behind a dam is of interest because, as yet, economical methods of storing large quantities of electric energy have not been developed.

Batteries and fuel cells are converters in which stored energy is converted directly to electric energy. In a primary (galvanic or voltaic) battery the $S \rightarrow E$ conversion is one of chemically stored energy being converted directly to electric energy. Since primary cells—for example, dry cells—would require replacement of their exhausted chemical constituents to return them to their original condition, they typically are discarded after the component materials have been exhausted. Secondary batteries, such as storage batteries used in automobiles, can be recharged $(E \rightarrow S)$ by supplying a reversed electric current to the unit.

The preceding examples using the energy triangle did not involve what we are, for the present, calling thermal energy, nor did they include cyclic devices in which thermal energy is converted to mechanical or electric work forms. The interrelationships of thermal energy—in a variety of forms—and work are primary considerations in engineering thermodynamics. With this in mind let us next consider an energy conversion system that involves

**Figure 1.2** Simplified Steam–Electric Power Plant

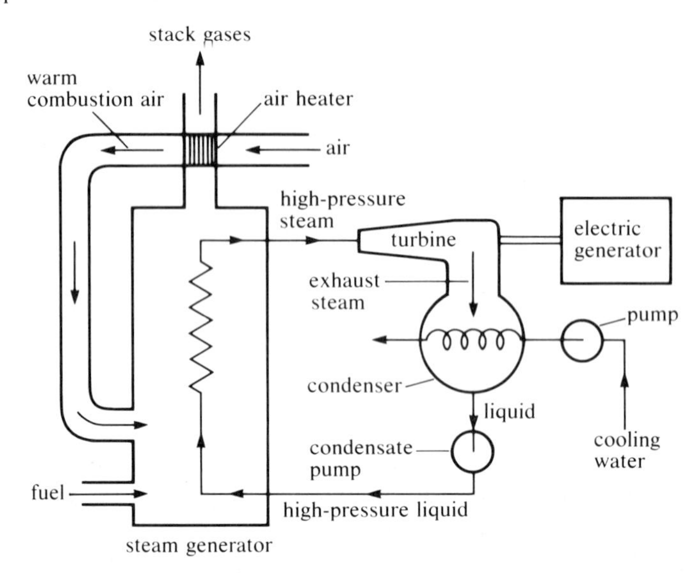


thermal and mechanical energies but does not operate in a cycle. Any gun, from a hunting rifle to a 16-inch cannon, follows the $S \rightarrow T \rightarrow M$ conversions. Stored chemical energy in the gunpowder is converted to thermal energy in high-temperature gases produced when the weapon is fired. As these gases expand in the barrel, thermal energy is converted to mechanical energy (work), which accelerates the projectile (shell) through the gun barrel. While some of the work is dissipated as heat (thermal energy) as the projectile moves down the barrel, most of it is converted to a form of stored mechanical energy—the kinetic energy associated with the mass and muzzle velocity of the projectile. Finally, the expanding gases at a lower temperature leave the barrel and discharge into the atmosphere. An accounting of all energy forms for the sequence must satisfy the energy conservation concept.

Devices that operate in a cyclic manner and convert thermal energy to mechanical energy are called heat engines. An example of a heat engine is shown in Figure 1.2, which presents a simplified diagram of a steam-electric power plant that uses coal as a fuel. The generation of electric energy in such a plant follows the $S \rightarrow T \rightarrow M \rightarrow E$ lines on the energy triangle. The series of processes that the water enclosed in the system undergoes constitutes a cycle. Combustion of the coal releases the stored chemical energy *(S)* in the fuel and produces high-temperature gaseous combustion products in the steam generator. More than 90 percent of this thermal energy is transferred either to the water, which is converted to steam in the steam generator, or to the combustion air as it passes through the air heater. The remainder is discharged to the atmosphere via the stack gases. The high-pressure steam delivered by the steam generator expands through the turbine to a low pressure and produces work. About one-third of the thermal energy supplied to generate the steam is converted to work by the

**Figure 1.3** Compressed-Air Energy-Storage System

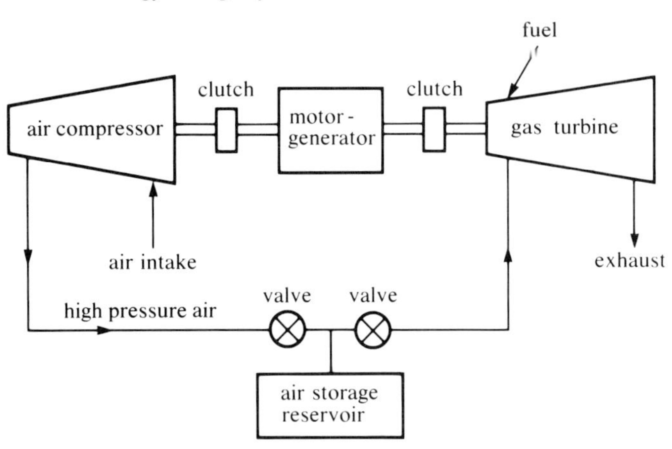


turbine. The mechanical energy delivered by the turbine drives the generator, producing the desired electric energy. This completes the $S \rightarrow T \rightarrow M \rightarrow E$ conversions. The cycle continues as some of the thermal energy in the steam exhausted to the condenser is transferred to the condenser cooling water, thereby condensing the steam to liquid. The cooling water and the condensing steam are separated by metallic surfaces and do not mix. The liquid is returned to the steam generator at high pressure by the condensate pump to be converted to steam again and delivered to the turbine. The thermal energy given up by the condensing steam to the cooling water is in turn rejected to the atmosphere. As the electric energy is used, it too eventually is rejected to the atmosphere. The end result is that, in time, all the stored energy originally released by burning the fuel finds its way to the atmosphere.

Earlier we noted that current technology does not provide a method of storing large amounts of electric energy. During any day the demand on an electric generating station undergoes continuous change. Residential and commercial demands decrease during the night. The same is true for industrial plants operating one or two shifts. To utilize the capacity of a generating station more fully and to provide power for peak load periods, methods to utilize $E \rightarrow M \rightarrow S$ conversions are of interest. Currently, pumped hydro storage is the common technique that achieves these conversions. Excess electric energy from a fuel-fired generating station drives pumps that discharge water into a holding pond behind a dam. This yields stored energy that can then be converted to electric energy in hydroelectric generation as discussed earlier. Reversible motor-generators, that is, units with the capability of serving either as motors or as generators, are used in such pumped hydropower systems. Compressed-air energy storage (CAES), somewhat similar to pumped hydro storage, is another method of storing energy from off-peak electric generation. This method is described in Figure 1.3. During off-peak hours the reversible motor-generator drives compressors that discharge high-pressure air into storage reservoirs such as underground caverns, thus following $E \rightarrow M \rightarrow S$ conversions. When electric demand is high, compressed air from storage is supplied to combustion (gas) turbines. Fuel and the air are burned in the combustion chambers of the unit, producing high-pressure and high-temperature gases that expand through the turbines, converting thermal energy to mechanical energy, which, in turn, drives the reversible motor-generator to produce electric energy. When operating in the storage mode, the conversions are $E \rightarrow M \rightarrow S$. During electric generation the conversions follow $S \rightarrow T \rightarrow M \rightarrow E$ where part of the stored energy is that in the fuel supplied. Both pumped hydro and CAES reduce the size (capacity) required in a conventional generating station.

To this point we have discussed the fact that although energy may exist in various forms, and may undergo changes in form, it is conserved. For the purposes of this chapter we have accepted this as a statement of the first law. Now let us consider the second law. Although this law has many far-reaching aspects, the feature that will serve our discussion here deals with the extent to which thermal energy can be converted to work by a heat engine. We have noted that a heat engine is a cyclic device that receives thermal energy and converts a portion of it to work. A parameter that describes the extent to which this conversion takes place is thermal efficiency, which is defined as

$$\text{Heat engine thermal efficiency} = \frac{\text{Work produced}}{\text{Thermal energy supplied}}$$

As noted earlier, the thermal efficiency of a heat engine in the form of a steam-electric power plant is about 33 percent. Thus about 67 percent of the thermal energy is not converted to work. The question immediately arises: Why does such a heat engine not convert a larger fraction of the thermal energy supplied to work? The answer to this question comes from the second law. Our study of this law later in the text will provide a simple expression for the upper limit of thermal efficiency for heat engines. For our purposes this expression is written as

$$\text{Maximum heat engine thermal efficiency} = \frac{t_{\text{source}} - t_{\text{surr}}}{t_{\text{source}} + 460}$$

where $t_{\text{source}}$ is the temperature of the region from which the thermal energy is received and $t_{\text{surr}}$ is the temperature of the heat engine surroundings, that is, the atmospheric temperature. The temperatures are in degrees Fahrenheit (F). For illustrative purposes we will use 40 F as a representative value for the temperature of the surroundings. Any real heat engine will have a thermal efficiency less than the maximum value. However, this expression leads to some important points regarding the conversion of thermal energy to work. Note first that the maximum thermal efficiency is not unity since $t_{\text{source}}$ is a finite number. (Practical considerations limit $t_{\text{source}}$ to values in the range of about 1500 to 2500 F.) Hence a heat engine cannot convert to work all of the thermal energy supplied to it. To pursue this further, let us consider a source of thermal energy at 1540 F and let this source supply 100 thermal

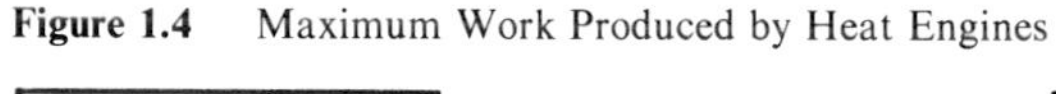

**Figure 1.4** Maximum Work Produced by Heat Engines

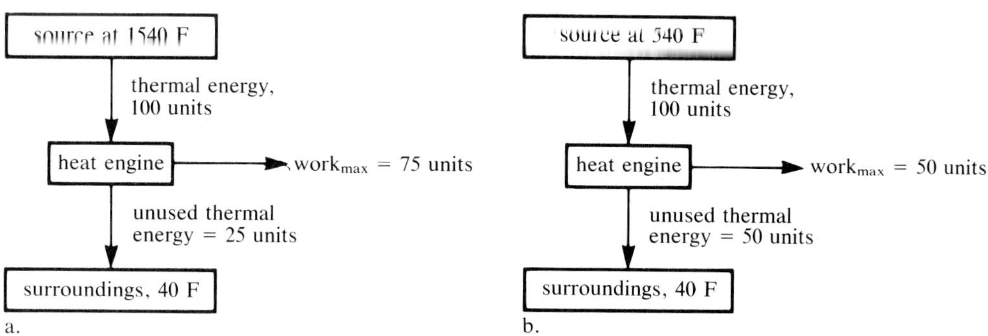


energy units to a heat engine, as depicted in Figure 1.4a. From the above expression we can compute the maximum thermal efficiency for the heat engine as 0.75. The maximum work that the engine could produce is then 75 units. In this section we shall term the maximum work the "work potential." Thus the work potential of the 100 units of thermal energy from the source at 1540 F is 75 units. Consider now the similar case shown in Figure 1.4b in which 100 units of thermal energy are delivered to the heat engine from the source at 540 F. The work potential calculated in the same manner as that for Figure 1.4a is 50 units. In each case energy is conserved; the thermal energy supplied is equal to the work produced plus the unused thermal energy, as shown in the figure. Although the same amount of energy is supplied to each heat engine, the work potential is quite different. We describe this result by saying that a given amount of thermal energy from a high-temperature source has a higher quality or grade than does the same amount of thermal energy from a source at a lower temperature. From this discussion it follows that when a given amount of thermal energy flows (without producing work) from a region at one temperature to a region at a lower temperature, it is degraded—that is, its work potential is decreased. This points out that the usefulness of an amount of thermal energy for producing work depends not only on the quantity of energy, but also on the temperature of the source from which it is supplied. And thus we recognize that all energy is not of the same quality or grade and that degradation of thermal energy is a natural process in that thermal energy flows to regions of lower temperature.

The above discussion has set forth only a few of the concepts of the two laws of thermodynamics that are fundamental to the understanding of energy engineering and to the numerous related topics of engineering thermodynamics. The remainder of this chapter discusses some aspects of the supply, demand, and use patterns related to the energy picture in the United States. This material is intended to serve as background for the energy-related topics in our study of engineering thermodynamics.

**Figure 1.5** U.S. Energy Demand by Fuel Consumed

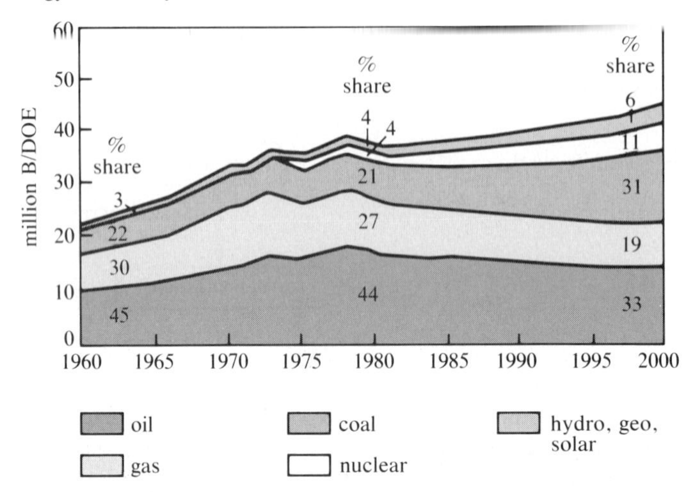


(Exxon Company, U.S.A., A Division of Exxon Corporation.)

## 1–2 Energy Sources: Supply and Demand

Currently more than 90 percent of the energy consumed in the United States comes from three nonrenewable resources—oil, natural gas, and coal. As shown in Figure 1.5, the figure exceeds 95 percent when nuclear fuels are included. The ordinate on Figure 1.5, million B/DOE, is millions of barrels per day of oil equivalent, with one barrel equivalent to 5.55 million British thermal units. The British thermal unit (Btu) is approximately the energy required to raise the temperature of one pound of water one degree Fahrenheit. Consumption through 1980 is well documented. Forecasts of future demands are subject to many variables, ranging from public attitudes and laws to costs and the national economy. The uncertainties associated with nuclear fuels for electric generation are one example. The growth potential for renewable resources such as hydro and geothermal is limited by the availability of suitable locations. The future of solar energy is dependent on technological developments needed to make this source cost effective.

Next let us review the reserves and production of the three major energy sources shown in Figure 1.5. Again, forecasting reserves is subject to uncertainties. For this reason we shall include only proved reserves for all areas except the centrally planned economies (CPE) such as the USSR, Eastern Europe, the People's Republic of China, and other Communist-bloc nations. For the last group, proved and less certain figures are given. *Proved reserves* are the amounts of hydrocarbon fuels in known underground reservoirs that are estimated with reasonable certainty to be recoverable in the future under existing economic conditions using current technology. The uncertainty associated with proved reserves is of the order of plus or minus 10 percent. Of course, the only exact figures are after-the-fact knowledge, that is,

**Table 1.1** Crude Oil Reserves and Production (millions of barrels)

| | Reserves, January 1, 1982 | Production, 1981 |
|---|---|---|
| **United States** | 29,783 | 3,132 |
| **Canada** | 7,300 | 469 |
| **Latin America** | 84,982 | 2,179 |
| **Western Europe** | 24,634 | 937 |
| **Africa** | 56,172 | 1,658 |
| **Middle East** | 362,840 | 5,729 |
| **Far East** | 19,151 | 1,014 |
| **CPE** | 85,845 | 5,328 |
| **Total world** | 670,709 | 20,446 |

(Exxon Company, U.S.A., a Division of Exxon Corporation)

the amount recovered from an oil field until it stops producing. That end point, too, is a moving target. During the 1930s about 20 percent of the crude oil in place could be brought to the surface. At present the recovery is near 40 percent as a result of improved technology and the price–demand relationship. Increasing prices have justified higher cost recovery methods, such as water, steam, and carbon dioxide injections to augment conventional free-flowing and pumping recovery.

Reserve and production figures are prepared regularly by major energy companies, industry groups, and governmental agencies. A typical compilation for crude oil is given in Table 1.1. The unit, a barrel, is equal to 42 U.S. gallons. While reserve and production figures are continually changing, inspection of the data in Table 1.1 clearly shows the serious problem the United States would face if oil imports were to cease. In 1980, for example, approximately 60 percent of the U.S. oil supply was from domestic production and 40 percent from imports. A modest decline in this foreign dependence has been experienced more recently, however, it is apparent that the United States will continue to rely heavily on imported oil.

Production of natural gas in the United States has been declining since 1973 when it peaked at 22 trillion cubic feet per year. As shown in Figure 1.6, the U.S. supply of this clean-burning fuel is primarily from domestic production. The ordinate on Figure 1.6, TCF/yr, is trillions of cubic feet per year. Approximately one-third of the U.S. natural gas consumption is for residential and commercial heating. Regulated pricing and environmental requirements in the past have encouraged the use of natural gas as a fuel for electric generation. It is to be hoped that this use will decline as utilities and heavy industries switch to coal, a much more abundant fuel. Production of synthetic gas from coal, a fairly common practice prior to 1940, is presently receiving increased attention.

**Figure 1.6** U.S. Gas Supply

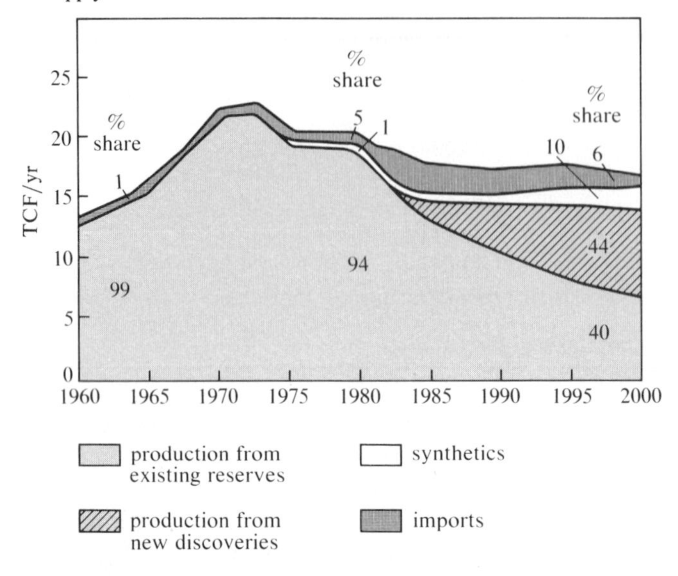


(Exxon Company, U.S.A., A Division of Exxon Corporation.)

Worldwide reserves and production of natural gas are shown in Table 1.2. Comparison of these data with Figure 1.6 indicates the need for new discoveries. Although gas from Alaska's North Slope is expected to provide about 5 percent of the U.S. production by 1990, new discoveries are imperative if we are to continue to use natural gas in the residential, commercial, and industrial sectors. While one may be inclined to think of natural gas only as a fuel, it is also a primary feed stock in the production of petrochemicals, fertilizers, and synthetic materials.

Coal is America's most abundant fuel source. It has been observed that the United States has more coal than the Middle East has oil. While current domestic reserves of crude oil and natural gas indicate supplies in the decade range, domestic coal reserves are adequate for a 200 or 300 years' supply at the current rate of consumption. Recognizing the folly of looking centuries ahead in a society experiencing rapid technological change, perhaps a better perspective is that approximately 90 percent of the United States' fossil fuel reserves are in the form of coal. At this time the major fraction of coal use in the United States is for fuel to generate steam for the production of electricity. Such centralized use in large generating systems is virtually mandated by air-quality regulations since coal contains sulfur and its products of combustion include sulfur oxides and particulates, which must be reduced to regulated levels before discharge to the atmosphere. At present this can be accomplished economically only in large-scale facilities.

**Table 1.2** Natural Gas Reserves and Production (billions of cubic feet)

| | Reserves, January 1, 1982 | Production, 1981 |
|---|---|---|
| **United States** | 198,000 | 19,596 |
| **Canada** | 89,900 | 2,623 |
| **Latin America** | 176,323 | 3,227 |
| **Western Europe** | 150,650 | 6,906 |
| **Africa** | 211,667 | 1,947 |
| **Middle East** | 762,490 | 1,542 |
| **Far East** | 127,616 | 2,765 |
| **CPE** | 1,194,700 | 19,210 |
| **Total world** | 2,911,346 | 57,816 |

(Exxon Company, U.S.A., a Division of Exxon Corporation)

**Table 1.3** Coal Reserves and Production* (millions of tons)

| | Reserves, 1980 | Production, 1980 |
|---|---|---|
| **United States** | 245,585 | 829 |
| **Canada** | 6,497 | 40 |
| **Latin America** | 5,718 | 17 |
| **Western Europe** | 155,122 | 555 |
| **Africa** | 37,549 | 133 |
| **Middle East** | — | — |
| **Far East** | 87,179 | 338 |
| **CPE** | 451,127 | 2,232 |
| **Total world** | 988,777 | 4,144 |

*Including lignite

(Exxon Company, U.S.A., a Division of Exxon Corporation)

The higher sulfur content of the product as well as the costs of deep mining are reducing the use of coal from the midwestern, southern, and Appalachian states. To meet air-quality standards, increasing amounts of surface-mined, low-sulfur western coal are being used. This coal, found in a band reaching roughly from Montana to Arizona, is often blended with eastern coal to meet emission standards. While western coals generally produce, on combustion, less energy per pound, this is offset by lower production costs as well as by the desired lower sulfur content. Past use and a forecast of future production of coal in the United States are shown in Figure 1.7.

The convenience of having electric energy available at the flip of a switch has become a part of our life-style. Some of the problems associated with its generation are no doubt familiar. Whether the fuel is fossil—coal, oil, gas—or nuclear, electric generation continues to experience many challenges. Aside from the technical aspects of power production, one

**Figure 1.7** U.S. Coal Supply

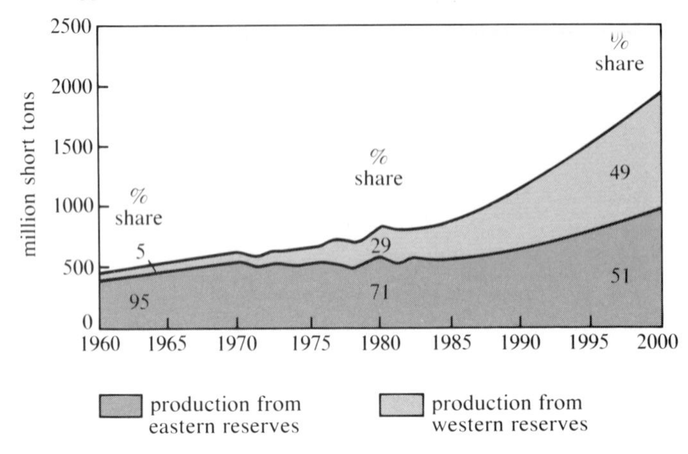


(Exxon Company, U.S.A., A Division of Exxon Corporation.)

must immediately recognize that the bulk of electric generation is from nonrenewable energy sources. The customer desires low-cost power, but the economics of the industry are such that this desire is not likely to be met in the foreseeable future. Escalating costs for fuel, equipment, and operation translate to higher rate structures. As noted above, demands for a cleaner atmosphere have necessitated the use of equipment such as stack gas scrubbers and precipitators to reduce undesirable stack emissions from coal-burning plants. At this time about 20 percent of the cost of a generating station goes into emission-control equipment. Concerns—some real, some imaginary—about the safety of nuclear generating stations have essentially halted new construction of such plants. Some years ago the Atomic Energy Commission forecast that approximately 1000 nuclear generating units would be operating in the United States by the year 2000. With that forecast came concern that known reserves of uranium would be expended in a few decades. The nuclear picture has changed markedly. As this is being written no new reactors for power generation have been ordered since 1978. Many stations have been canceled or construction deferred or placed on a "stretch-out" schedule. The U.S. Department of Energy in 1982 estimated that about 170, not 1000, nuclear reactors for electric generation would be in operation by the end of this century. The lack of a comprehensive program for reprocessing spent fuel rods and disposing of nuclear waste materials presents a major problem for the industry as well as a significant public concern.

While it appears that the electric generating industry will continue to be the largest user of coal, the production of synthetic gas and liquid fuels from coal well may become an important future use of this resource. Changes in governmental policies, a poor worldwide economy, and abundant oil and gas supplies during the early 1980s have curtailed activity in the synthetic fuels area. With changes in these factors and escalating costs for crude oil and natural gas, synthetic fuels derived from coal are likely to become economically viable.

**Table 1.4** Representative Higher Heating Values of Some Fuels

| | | Btu/lb | Btu/ft³* | Btu/U.S. Gallon |
|---|---|---|---|---|
| Hydrogen, $H_2$ | | 61,005 | 325 | |
| Natural gas (Louisiana) | | 21,840 | 1,002 | |
| Commercial propane | | 21,560 | | 91,500 |
| Commercial butane | | 21,180 | | 102,760 |
| Automotive gasoline | | 20,260 | | 124,800 |
| Aviation motor gasoline | | 20,460 | | 119,770 |
| Jet fuel | | 20,070 | | 129,000 |
| Methanol | | 10,264 | | 68,150 |
| Ethanol | | 13,161 | | 87,170 |
| No. 2 fuel oil | | 19,460 | | 138,750 |
| No. 6 fuel oil | | 18,300 | | |
| Coal, Pennsylvania anthracite | 0.60% sulfur | 12,925 | | |
| Coal, Pennsylvania bituminous | 1.82% sulfur | 13,325 | | |
| Coal, Illinois bituminous | 2.70% sulfur | 11,910 | | |
| Coal, Montana sub-bituminous | 0.43% sulfur | 11,140 | | |
| Coal, Wyoming sub-bituminous | 0.30% sulfur | 9,345 | | |
| Lignite, North Dakota | 0.40% sulfur | 7,225 | | |
| Wood, white pine | Dried to 8% moisture | 8,900 | | |
| Wood, elm | Dried to 8% moisture | 8,810 | | |
| Wood, hickory | Dried to 8% moisture | 8,670 | | |
| Wood, black oak | Dried to 8% moisture | 8,180 | | |

*At 60 F and 14.7 psi (pounds per square inch) absolute pressure.

## 1-3 Comparison of Some Energy Sources

A goal of energy engineering is to obtain, in usable form, the maximum energy that a fuel or other source can provide. Conversions from stored energy to a desired form must satisfy the laws of thermodynamics described earlier in a general context. The energy potentially available from combustible fuels such as petroleum products, coal, gases, and wood is expressed as the heating value of the fuel. The higher heating value (HHV) is the stored energy that is released when a fuel is burned completely (all carbon burning to carbon dioxide, all hydrogen to water, etc.), the products of combustion are cooled back to the fuel-oxidizer temperature, and all water formed during combustion is condensed to liquid. The lower heating value (LHV) is similar except the water formed by combustion of hydrogen in the fuel is not condensed to liquid. Higher heating values of a sample of fuels are given in Table 1.4. These figures are in British thermal units per pound (Btu/lb) or per unit volume. In reviewing the data in Table 1.4, it should be recognized that most of the materials listed (e.g., natural gas, petroleum products, coal, and wood) vary somewhat in chemical composition. The values presented are representative and can be used for relative comparisons. Precise information for a specific lot of any fuel is determined by specified test procedures established by the American Society for Testing and Materials (ASTM).

Table 1.5 Sample Solar Radiation Data for Ames, Iowa (42.0 north latitude)

| | (Btu/ft² day) |
|---|---|
| January | 641 |
| February | 932 |
| March | 1204 |
| April | 1484 |
| May | 1767 |
| June | 1992 |
| July | 1973 |
| August | 1694 |
| September | 1352 |
| October | 1009 |
| November | 689 |
| December | 527 |

Nuclear fuel pellets are enriched with the isotope uranium–235 to produce the fissile fuel. Typically the fuel pellets contain about 3 percent of this isotope. Fission of 1 gram of U–235 results in the release of approximately 82,000,000 Btu of thermal energy. For comparison, this amount of energy is equivalent to the higher heating value of more than three U.S. tons of coal. Stated in another way, when one ton of U–235 is fissioned, the energy released is equivalent to the higher heating value of approximately three million tons of coal.

Increasing attention is being directed to the use of solar energy, a source not limited by finite resources of the earth. Table 1.5 gives *monthly averaged daily total solar radiation* on a horizontal surface for Ames, Iowa. The values vary seasonally with the sun's angle and the latitude of a particular location.

Additional factors that affect solar radiation are elevation above sea level, atmospheric constituents such as smog, and the general clearness of the atmosphere. Recalling that the sample data for Table 1.5 are for a horizontal flat plate, it is evident a collector normal to the sun's rays and tracking the sun during the day would yield higher solar radiation values. The fraction of solar radiation that can be collected depends on the type of collector used.

## 1–4 Energy Conversion and Utilization

In Section 1–2 primary energy sources were discussed. Next let us look at how and where energy is used. In reviewing energy use or demand, electric energy as well as the primary sources considered in Section 1–2 are included. Energy demand is frequently expressed in terms of three large market categories: residential–commercial, transportation, and industrial. A fourth, smaller category is the nonenergy use of oil, gas, and metallurgical coal as feed stock or raw material rather than as fuel. Each sector is influenced by the national and world economies, prices of the energy form supplied to the ultimate consumer, and energy conservation. Thus the forecasts for the period 1980 to 2000 shown in Figure 1.8 must be recognized as estimates for that period. Energy losses in the generation and transmission of electric energy have been apportioned to each sector. These energy losses amount to more than 20 percent of the U.S. energy supply requirements.

**Figure 1.8** U.S. Energy Demand by Consuming Sector

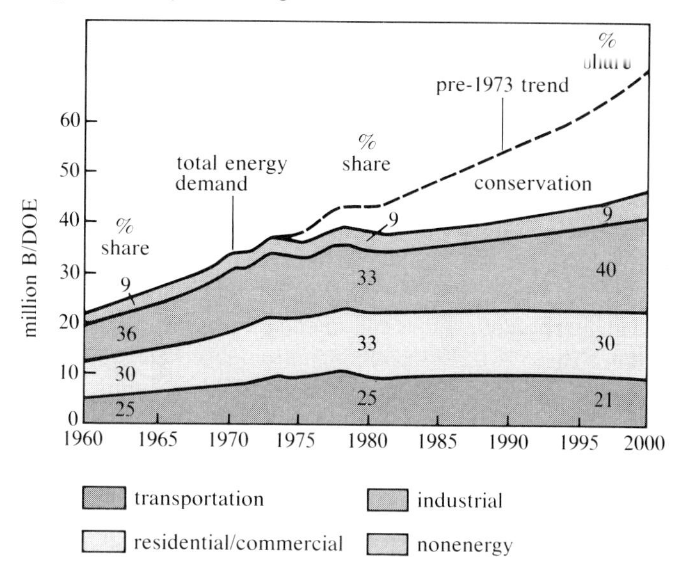


(Exxon Company, U.S.A., A Division of Exxon Corporation.)

Conservation efforts are predicted to have a significant effect on energy demand in the United States. This is indicated in Figure 1.8 by the departure of the projected total demand from the pre-1973 trend.

Energy demand for the residential–commercial sector includes that for homes, apartments, theaters, hotels, motels, schools, hospitals, shopping centers, office buildings, and similar establishments. Most energy supplied is electric, natural gas, or oil. As about half of the energy demand typically is for heating and cooling these spaces, energy conservation measures for the building shell are desirable first steps in reducing consumption. The demand by the residential–commercial sector is expected to decline slightly as a result of energy conservation and a lower growth rate in the establishment of new households.

Energy use for transportation accounts for approximately one-fourth of all U.S. energy demand. With more than 95 percent of this energy supplied by petroleum products, the transportation sector is one in which more efficient use of energy directly results in decreased demand for crude oil. New automotive vehicle designs are decreasing vehicle weights, thereby requiring less engine power and yielding higher fuel economy. Americans' desires for the personal freedom afforded by driving where they want and when they want would appear to preclude a significant decrease in the demand for automotive fuels. While pricing has some effect on fuels for transportation, to effect any significant reduction in demand, say in a national emergency, probably would require rationing. Long-distance passenger travel by common carrier has shifted from railways to aircraft, a change that has increased energy use per passenger-mile. Transportation accounts for more than one-half of the total petroleum requirements in the United States. Thus a more energy-efficient transportation scenario along with some alternative energy sources would result in a marked decrease in our dependence on imported petroleum.

The industrial sector energy demand covers the primary energy sources for essentially all manufacturing and mining industries. It includes products for immediate use, ranging from computer chips to aircraft, from kitchen utensils to tanks, and from those used in the mining and processing of ores to intermediate products such as sheet aluminum, structural steel, and copper wire. As in the other sectors, electric energy is included in the demand. The production of process steam, the firing of cement kilns, and the operation of smelters are other uses of energy in the industrial sector. Approximately 40 percent of the total industrial energy use is for the generation of process steam and 25 percent is for direct process heating. Energy conservation for these two uses is receiving increased attention. One technique that promises to afford more efficient use of energy sources in the industrial sector is the utilization of cogeneration systems. This concept, discussed in Chapter 9, couples simultaneous generation of electricity and steam or other process heating demands for industrial operations. With approximately 20 percent of the total industrial use of energy going for electric drive applications, the increasing employment of cogeneration systems in the paper, petroleum, petrochemical, and steel industries is a viable energy conservation step.

In this and the two preceding sections we have reviewed energy sources and resources, energy supply and demand, and some energy utilization patterns for the United States. This picture is not at all typical of that for the rest of the world. Energy use per capita in the United States is about six times the world average. With but 6 percent of the world's population, the United States consumes about one-third of the energy used worldwide.

The trend in the United States is toward a diversity of energy sources. While energy supplies for the near term appear to be adequate, the long-term supply is far less certain. New technologies, most requiring long lead times, will develop and influence energy supply and demand in the United States and elsewhere. The topics studied in engineering thermodynamics provide a foundation for dealing with both current and emerging energy technologies.

## 1–5 Summary

In this chapter we have used simplified terminology to introduce the subject of engineering thermodynamics. The study of energy is central to engineering thermodynamics. Energy is a nonmaterial agent that can be stored in many forms and, within certain limits, converted from one form to another. Quantitative evaluation of energy stored, converted, and transferred is the primary goal of engineering thermodynamics.

Two basic laws of thermodynamics will be addressed in subsequent chapters. The first of these is stated in simple terms: Although energy may exist in a multiplicity of forms and may undergo changes in form, energy is conserved. Several examples of conversions between stored, mechanical, thermal, and electric energy have been discussed. In each example, energy conservation had to be satisfied. The second basic law, among its many ramifications, establishes limits on the fraction of the thermal energy supplied to a continuously operating system that can be converted to work. This law also has led to the fact that the usefulness of an amount of thermal energy for producing work depends not only on the quantity of thermal energy, but also on the temperature of the source from which it is supplied.

Currently about 90 percent of the energy consumed in the United States comes from three nonrenewable resources—oil (44 percent), natural gas (25 percent), and coal (23 percent). Approximately 60 percent of the present U.S. oil supply comes from domestic production and 40 percent from imports. Although the United States has limited oil and natural gas resources, it has abundant coal reserves. Energy demand is frequently expressed in terms of three large market categories: residential–commercial, transportation, and industrial. These categories constitute about 90 percent of the energy demand in the United States, with the remainder going to nonenergy use of fuels as industrial feed stocks and raw materials. Energy use in the three major categories is approximately equal at the present time. Energy use per capita in the United States is about six times the world average; with 6 percent of the world's population, the United States consumes about one-third of the energy used. Projections of energy demand in the United States, while subject to measures of uncertainty, forecast a modest increase in energy use during the remainder of this century. The recent energy-demand trend has been influenced significantly by energy conservation.

## Selected References

Beckman, W. A., S. A. Klein, and J. A. Duffie. *Solar Heating Design.* New York: Wiley-Interscience, 1977.

Dixon, J. R. *Thermodynamics I: An Introduction to Energy.* Englewood Cliffs, N.J.: Prentice-Hall, 1975.

*Energy Consumption Measurement.* Washington, D.C.: National Academy of Sciences, 1977.

*Energy Outlook 1980–2000.* Exxon Company, U.S.A., 1980.

*Exploring Energy Choices.* The Ford Foundation, 1974.

Fowler, John M. *Energy and the Environment.* New York: McGraw-Hill, 1975.

Fraas, A. P. *Engineering Evaluation of Energy Systems.* New York: McGraw-Hill, 1982.

Fryling, G. R., ed. *Combustion Engineering.* New York: Combustion Engineering, Inc. 1966.

Goss, W. P., and J. G. McGowan. Energy requirements for passenger ground transportation systems. New York: American Society of Mechanical Engineers, paper number 73–1CT–24.

Harrah, B. *Alternate Sources of Energy, A Bibliography of Solar, Geothermal, Wind and Tidal Energy and Environmental Architecture.* Metuchen, N.J.: Scarecrow Press, 1975.

Hartnett, J. P. *Alternative Energy Sources.* New York: Academic Press, 1976.

Hirst, E., and J. C. Moyers. Efficiency of energy use in the United States. *Science* 179 (March 1973).

Loveland, W. D., B. I. Spinard, and C. H. Wang, eds. *Proceedings.* Conference on Magnitude and Deployment of Energy Resources, Office of Energy Research and Development, Oregon State University, 1975.

*Patterns of Energy Consumption in the United States.* Stanford Research Institute, 1972.

Penner, S. S. *Non-nuclear Energy Technologies.* Reading, Mass.: Addison-Wesley, 1975.

*Science.* 184 (4134) (April 19, 1974). Published by American Association for the Advancement of Science.

*Statistical Year Book.* Edison Electric Institute (published annually).

*Steam/Its Generation and Use.* New York: Babcock & Wilcox Co. 1972.

Thorndike, E. *Energy and Environment: A Primer for Scientists and Engineers.* Reading, Mass.: Addison-Wesley, 1976.

## Problems

1.1 Describe an energy conversion that follows the electric-to-thermal path in Figure 1.1. Identify a mechanical-to-thermal energy conversion.

1.2 Using local cost data, determine the cost of 1,000,000 Btu of energy if it is supplied as (a) number 2 fuel oil, (b) natural gas, (c) coal, (d) local wood supply for stove or fireplace, and (e) electricity. Indicate assumptions made in using rate schedules for electricity and natural gas.

1.3 Suppose there were no petroleum fuels available; how might the energy requirements be satisfied by other means for (a) residential heating, (b) industrial furnaces, and (c) transportation?

1.4 Suppose there were no natural gas or petroleum fuels available; how might the energy requirements for (a) residential and commercial heating, (b) transportation, and (c) steel making be satisfied?

1.5 Imagine a situation in which no fossil fuels are available. How might the energy for (a) the residential–commercial sector, (b) transportation, (c) electric generation, and (d) the industrial sector be supplied?

# 2

# Some Basic Definitions and Concepts

*The purpose of this chapter is to set forth some of the basic definitions and concepts of thermodynamics. Central to this purpose are the terminology and the language of the subject. As with any science, thermodynamics has a unique language. Words in common usage and broad context become more specific and well may be defined in a manner different from that in a dictionary. For example, in Chapter 1 the word energy was used frequently, yet energy was not a defined quantity. This is but one case where some reliance has been placed on previous usage without a precise definition of a term. The language of thermodynamics will unfold as we proceed through this text. A clear understanding of this language by the student is an obvious first requisite in developing a sound, basic grasp of the subject. Where appropriate, operational definitions are used. An* ***operational definition*** *is both qualitative and quantitative in that it not only describes the term, but also provides a description of the manner in which a numerical evaluation of that term can be accomplished.*

## 2–1 Thermodynamics

In a broad sense thermodynamics is concerned with energy, energy transfers, and energy transformations. However, although this is indicative of the general nature of thermodynamics, it does not suffice as a definition because the terms used are undefined. Thermodynamics is basic to the understanding of energy-related concepts in science and engineering. A subject so broad quite logically develops into specialized treatments for a variety of disciplines. We shall define *engineering thermodynamics* as that subject dealing with systems, properties of systems, and interactions between systems. Definitions of the terms *system, property,* and *interaction* follow in this and subsequent chapters.

In this text fundamentals of engineering thermodynamics are developed first, with as much rigor and logic as necessary brevity permits. Once they are established, we proceed to apply these fundamentals to selected engineering applications. Such applications include, but are not limited to, refrigerators, heat pumps, air conditioning systems, some aspects of combustion, gas compressors, internal combustion engines, gas turbines, jet engines, and cycles for electric power generation. But only after the foundation has been built and the fundamental concepts established, can one proceed to the analysis of these energy-conversion devices and systems.

The study of thermodynamics can be undertaken on a macroscopic scale, a microscopic scale, or both. The macroscopic approach is a "large-scale" approach that does not require knowledge of the behavior of individual particles (molecules). The microscopic approach is based on the statistical behavior of large numbers of such particles. It requires the assumption of a molecular model for matter and can provide information not discernible through

**Figure 2.1** Examples of Systems. *a.* A closed system. *b.* An open system (control volume). *c.* An open system.

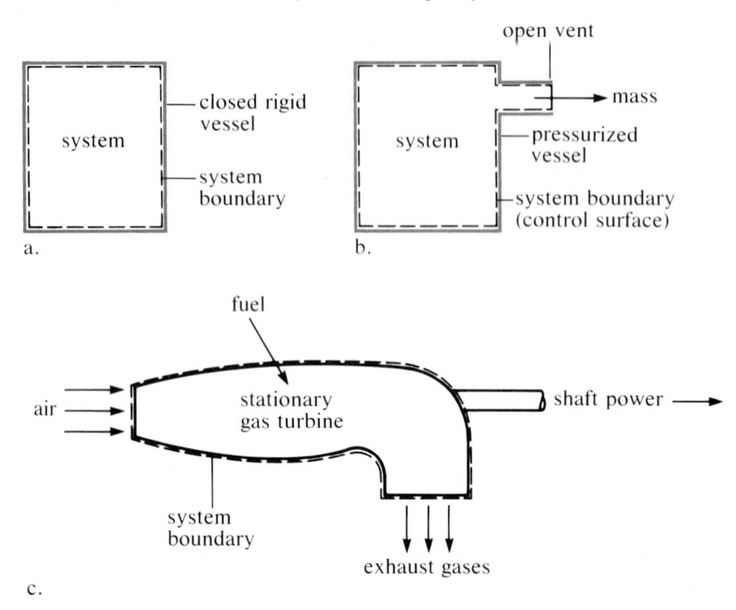


the macroscopic approach. In this introductory text we shall be considering large quantities of matter with substances considered on a continuum basis, that is, on the macroscopic scale. This approach, also called the classical approach, lends itself well to the analysis of typical engineering systems.

## 2–2 Thermodynamic Systems

A *thermodynamic system* is a region or fixed quantity of matter defined by its boundary. The boundary, either real or imaginary, rigid or deformable, defines the extent of the system. A *closed system* is one which has no mass crossing its boundaries. Therefore, the boundary encloses a fixed quantity of matter. An *open system*, conversely, is one for which mass does cross the system boundary. For an open system, the time rate at which mass enters the system may be the same as that at which mass leaves the system or the mass flow rates for the inbound and outbound streams may differ. In the case in which all inbound and outbound mass flow rates are zero, we obviously have a closed system. The term *control volume* is synonymous with open system and *control surface* is used interchangeably with open-system boundary.

The *surroundings* (environment) of a thermodynamic system are those regions external to a system that influence or are affected by the system. An *isolated system,* as the name implies, is one that has no interactions with its surroundings; that is, the behavior of the system is not influenced by anything outside its boundary. From this we see that an isolated system is necessarily a closed system.

Figure 2.1 shows three examples of systems. All of the boundaries are three dimensional. In Figure 2.1a the system boundary encloses only the substance in the vessel. The boundary in Figure 2.1b consists of the inner surface of the vessel and an imaginary surface at the vent across which mass flows. In Figure 2.1c the system boundary is an imaginary surface that encloses the turbine. Fuel, air, and exhaust gases cross this boundary and power is transmitted across it by means of a rotating shaft. The choice of the boundary was arbitrary in each case. Note that in defining a system we are not restricted by the complexity of the region or substance under study. The system can be as simple as a quantity of a substance in a tank or as complex as a gas turbine. Definition of a system by selection of boundaries is an essential first step in the application of the concepts of thermodynamics.

## 2–3 The Working Fluid

The particular substance within the boundary of a system that undergoes changes and in doing so influences the system's behavior is referred to as the *working fluid* or the *working medium*. Substances such as air, water, common gases, and refrigerants are examples. In general, substances existing as solids, liquids, gases, and mixtures thereof are of interest as working media in engineering thermodynamics. With the single exception of chemically reactive systems, which we address in the treatment of combustion processes, the working media considered in this text will be pure substances. A *pure substance* is one that has a homogenous and fixed chemical composition. It need not be a single substance. For example, a homogenous mixture of gases qualifies as a pure substance provided there is no chemical reaction of the gases. A substance—water, for example—that can exist in a solid, liquid, or gaseous form is viewed as a pure substance if in the various forms the chemical composition does not change, that is, the water molecule remains $H_2O$.

A pure substance may exist as a solid, a liquid, or a vapor (gas). These different forms are known as *phases* of a pure substance. Usually a phase is separated from other phases of a system by a distinct bounding surface as, for example, ice (solid) in water (liquid). Depending on the pressure and temperature, a pure substance may have more than one solid phase. However, for our purposes it will be adequate to assume that pure substances have only one solid phase, one liquid phase, and one vapor (gas) phase.

## 2–4 Properties and States

*Properties* of thermodynamic systems are quantities that describe the *state* (condition) of a system. Some properties—such as pressure, temperature, specific volume (or its reciprocal, density), and electrical resistivity—are directly observable, that is, measurable with laboratory instruments. Other properties we shall encounter are not directly measurable but can be evaluated on the basis of their definition and observable quantities. They include stored energy, a consequence of the first law of thermodynamics, and entropy, a consequence of the second law of thermodynamics. These laws and related properties will be discussed in detail later in this text.

The state of a thermodynamic system is specified by the values of its properties. A change of state requires a change in at least one property of the system. Equilibrium states are of particular interest. A thermodynamic system is said to be in an *equilibrium state* when a spontaneous change in state cannot occur. Thus, if the system is isolated from its surroundings and does not spontaneously progress to a new state, it is at an equilibrium state. An equilibrium state is one for which there are no unbalanced potentials within the system or between the system and its surroundings. A system in equilibrium can undergo a state change only when one or more external influences are imposed on it.

Most properties of a system fall into one of two categories: *intensive* or *extensive*. Consider at a given instant a region within a system that is not necessarily in equilibrium. As the region size approaches a small value, all *intensive properties* approach respective finite limits for the region. Thus an intensive property is single valued at a particular point in a system. Pressure, temperature, and specific volume—or its reciprocal, density—are examples of intensive properties. Some other properties we will encounter, such as internal energy per unit mass, are also intensive properties. The adjective *specific*, when indicating per unit mass, implies an intensive property. Specific volume $v$, defined as volume $V$ divided by mass $m$ ($v = V/m$), is an example. An *extensive* property is one that depends on the extent—mass or volume—of a system. Characteristic of an extensive property is that the property is the sum of the values for all parts of the system. For $n$ parts of any system the extensive property mass $m$ is

$$m = \sum_{i=1}^{i=n} m_i$$

where $m_i$ is the value of the extensive property $m$ for any one part of the system. As another illustration, consider a system whose boundary encompasses three separate and distinct parts; $a$, $b$, and $c$. The volume $V$ of the system is an extensive property having the value $V = V_a + V_b + V_c$.

In a mathematical sense intensive properties are point (state) functions and their differentials are *exact differentials*, that is, the values of their integrals depend on the end states only. Let $Z$ be a property of a system. At a given equilibrium state (1) (i.e., in a condition where all properties are fixed), the property $Z$ will be single valued. Now consider a change in the state of the system to a new equilibrium state (2). For the arbitrary property $Z$ we can write

$$\int_1^2 dZ = Z_2 - Z_1$$

where $Z_2$ is the value of the property at state (2) and $Z_1$ is the value at state (1). The manner in which the system was caused to change state from (1) to (2) does not affect the magnitude of the change in the property $Z$.

As noted earlier, the state of a system is described by values of the properties of the system. Since numerous system properties can be identified, a question arises regarding the number of properties required to fix a state. Experience indicates that, in the absence of electrical, magnetic, gravitational, motion, and surface tension effects, an equilibrium state of a pure substance is fixed by specifying values for two *independent intensive* properties. A substance under these conditions is referred to here as a *simple substance*, and the related system will be termed a *simple system*. Equilibrium for a simple system is easily described; the value of any given intensive property is the same at all points throughout the system, a condition that does not exist when a simple system is not in equilibrium. Examples of intensive properties that are applicable at this point in our development for specifying states of simple systems are pressure, temperature, and specific volume. Others will be introduced in succeeding chapters. Note that two *independent* intensive properties must be specified to fix the state of a simple system. Not all intensive properties are always independent. For instance, the temperature at which water boils depends on the pressure. We say that pressure and temperature are dependent properties for this condition. However, pressure and temperature are always independent for any single phase of a simple substance. Pressure and specific volume are always independent, regardless of the phases present. Once values for two independent intensive properties have been specified, all other intensive properties are fixed for a simple system and the state of the system is completely determined. In engineering thermodynamics a large class of problems involves substances that can be treated as simple substances. Chapters 4 and 5 deal to a large extent with the relations between the various properties of pure substances. Prior to the study of these chapters, numerical values for properties will be given where needed.

## 2–5 Process, Path, Cycle for a Closed System

When a change in the state of a system occurs the system is said to undergo a *process*. When a process for a closed system proceeds in such a manner that the value of each property at a given location within the system differs by no more than an infinitesimal amount from the value of the corresponding property at any other location in the system, the process is termed quasi-static. Thus a *quasi-static process* is one in which departures from equilibrium during the process, both within the system and between the system and its surroundings, are infinitesimal.

It should be noted that not all closed-system processes occur in a quasi-static manner. The quasi-static process is very useful for purposes of visualization and serves as a reasonable approximation for many processes that occur in certain closed systems.

For a simple system undergoing a quasi-static process, the series of equilibrium states through which the system passes can be plotted on a diagram with two independent properties, say pressure and specific volume, as the coordinates. Figure 2.2a shows a piston-cylinder assembly that contains a unit mass of a substance. Thus the volume exhibited by the substance is numerically equal to the specific volume $v$. (The system boundary is defined as the inner surface of the cylinder and the left-hand face of the piston.) The series of states the system passes through as it proceeds from a given state (1) to a given state (2) (for example, those along lines $B$ or $C$ of Figure 2.2a) is termed the *path* for a process. In the figure, $B$ and $C$ are only two of the many paths that could be followed during a quasi-static process between states (1) and (2).

A *cycle* is a continuous series of processes that returns the system to its original state. Referring again to Figure 2.2a, one cycle illustrated is that from (1) to (2) along path $B$ and (2) back to (1) along path $A$. Another cycle is (1) to (2) along path $C$ and (2) to (1) along path $A$. Remembering properties are single valued at a given state, we can write, using specific volume as an example,

$$\oint dv = \underset{\text{B}}{\int_2^1 dv} + \underset{\text{A}}{\int_2^1 dv} = 0$$

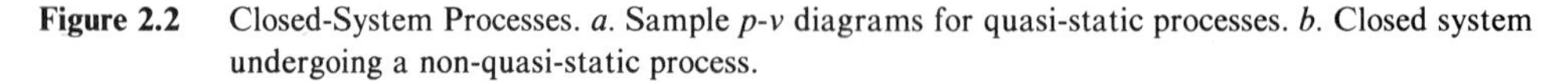

Figure 2.2 Closed-System Processes. *a.* Sample *p-v* diagrams for quasi-static processes. *b.* Closed system undergoing a non-quasi-static process.

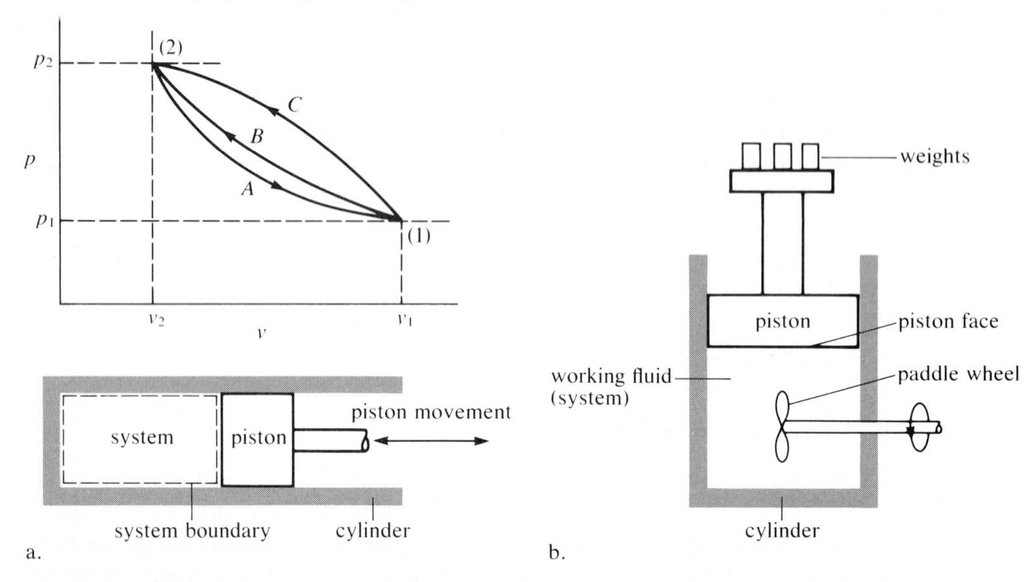


where the integral sign $\oint$ indicates integration around the cycle. This provides an alternate definition of a property as being a characteristic of a system whose cyclic integral equals zero.

For processes that are not quasi-static, the path cannot be plotted since, at any time during such a process, properties generally are not single valued because of nonequilibrium conditions within the system. Consider the apparatus in Figure 2.2b in which the violent stirring action of a paddle wheel results in a change in state of the working fluid and an upward movement of the piston. Since such a process taking place during a certain time period would result, at minimum, in finite departures from equilibrium of properties such as pressure, temperature, and velocity, the process would not be quasi-static. For such a process, only initial and final equilibrium states could be identified. Due to finite departure from equilibrium during the process, properties would not be single valued and a path for such a process could not be defined.

Prior to a more detailed discussion of properties, it is necessary to discuss the dimensions and units of quantities that are encountered in thermodynamics. The next section addresses this subject.

## 2–6 Dimensions and Units

*Dimensions* are names that are used to characterize distinct, unique, physical quantities. Examples of dimensions and the symbols used here to represent them are mass $M$, length $L$, and time $\tau$. A group of dimensions from which the dimensions of all other quantities can be formed is called a *primary group*. The group of dimensions given in the foregoing example serves as a primary group for the discussion in this section. The dimensions of several other quantities of interest here are formed from this primary group. Thus, in terms of primary dimensions, the dimensions of density (mass per unit volume) are $M/L^3$ and the dimensions of velocity are $L/\tau$. A *unit* is an arbitrary magnitude of a dimension that is used to measure that dimension. For example, the dimension length can be measured in inches, feet, meters, light-years, and so on. Similarly, the dimension mass can be measured in such units as the pound mass, the slug, and the kilogram. The dimensions of force are related to the primary dimensions by Newton's second law, which states that the rate of change of linear momentum is proportional to the force applied, or

$$\mathbf{F} \; \alpha \; \frac{d(m\mathbf{V})}{d\tau} \tag{2.1}$$

From the right-hand side of eq. (2.1) we see that, in terms of the primary dimensions, the dimensions of force are $ML/\tau^2$. The unit of force is not usually formed in terms of the units of this dimension group, but is established through a definition. The proportionality factor in eq. (2.1) depends on the definition of the force unit. For a fixed mass $m$ undergoing linear motion, eq. (2.1) can be written as

$$F = \frac{m}{g_0} \, a \tag{2.2}$$

where $a$ is the acceleration experienced by the mass when it is subjected to a force $F$. The proportionality factor is $1/g_0$. The value of $g_0$ depends on the system of units chosen. Two systems of interest are the system widely used in the United States, which will be referred to as the English system, and the system chosen for worldwide scientific and engineering work, the System Internationale, which is usually abbreviated as SI. Some mechanical quantities and the units for each are given in Table 2.1. The standards for these units are well established and are accepted throughout the world.

**English System** The pound force, lbf, is the unit of force in the English system. This unit is defined as follows:

*The pound force is that force which gives to a mass of 1 pound an acceleration of 32.17 ft/sec².*

The value of $g_0$ for this system is, from eq. (2.2) and the force definition,

$$g_0 = \frac{m}{F}\, a = \frac{1 \text{ lbm}}{1 \text{ lbf}} \left| \frac{32.17 \text{ ft}}{\text{sec}^2} \right. = 32.17 \, \frac{\text{lbm}}{\text{lbf}} \, \frac{\text{ft}}{\text{sec}^2} \tag{2.2a}$$

**System Internationale** The newton (N) is the SI force unit and is defined as follows:

*The newton is that force which gives to a mass of 1 kilogram an acceleration of 1 m/s².*

The value for the constant $g_0$ for this system is

$$g_0 = \frac{m}{F}\, a = \frac{1 \text{ kg}}{1 \text{ N}} \left| \frac{1 \text{ m}}{\text{s}^2} \right. = 1 \, \frac{\text{kg}}{\text{N}} \, \frac{\text{m}}{\text{s}^2} \tag{2.2b}$$

**Table 2.1** Mechanical Quantities and Units for the SI and English System of Units

| | | English System $g_0 = 32.17$ lbm ft/(lbf sec²) | | SI $g_0 = 1$ kg m/(N s²) | |
|---|---|---|---|---|---|
| **Quantity** | **Dimension** | **Unit Name** | **Symbol or Units** | **Unit Name** | **Symbol or Units** |
| mass* | $M$ | pound mass | lbm | kilogram | kg |
| length* | $L$ | foot | ft | meter | m |
| time* | $\tau$ | second | sec | second | s |
| force | $ML/\tau^2$ | pound force† | lbf | newton† | N |
| torque | $ML^2/\tau^2$ | | ft lbf | | N m |
| acceleration | $L/\tau^2$ | | ft/sec² | | m/s² |
| velocity | $L/\tau$ | | ft/sec | | m/s |
| frequency | $1/\tau$ | hertz | Hz = 1/sec | hertz | Hz = 1/s |
| pressure | $M/\tau^2 L$ | | lbf/ft² | pascal | Pa = 1 N/m² |
| energy | $ML^2/\tau^2$ | | ft lbf | joule | J = 1 N m |
| power | $ML^2/\tau^3$ | | ft lbf/sec | watt | W = 1 J/s |

*These quantities are taken as primary quantities for the English system as well as the SI in this text.

†See definitions in text.

It can be seen that $g_0$ in eq. (2.2) is different for the two systems of units.

Appendix A lists values for certain physical constants in both the SI and English systems. Conversion factors that are useful in working between the two systems are also given, as are prefixes that are used with the SI.

**Weight, Specific Weight, and Density** The weight of an object is the force due to the local acceleration of gravity. A freely falling object (an object falling in a vacuum) experiences an acceleration of gravity $g$. The average value on the earth's surface is

$$g = 32.17 \text{ ft/sec}^2 = 9.81 \text{ m/s}^2$$

Weight $F_g$ is related to mass by

$$F_g = \frac{ma}{g_0} = \frac{mg}{g_0} \tag{2.2c}$$

The weight acts downward through the mass center. If the object is to remain stationary, an upward-acting force equal to the weight must be applied through the mass center.

By dividing through eq. (2.2c) by $V$, the volume, the relation of specific weight $w_s$ and density $\rho$ is obtained.

$$w_s \equiv \frac{F_g}{V} = \frac{m\,g}{V\,g_0} = \rho\,\frac{g}{g_0} \tag{2.3}$$

with $m/V \equiv \rho$ being the density (mass per unit volume) and $w_s$ the specific weight (weight per unit volume). Note that the relation between specific weight and density is dependent on the local gravitational acceleration.

---

## Example 2.a

The gravitational acceleration on the surface of Mars is 12.95 ft/sec². A block of material with a fixed volume of 1.10 ft³ weighed on a spring scale in a standard gravity field (32.17 ft/sec²) has a weight of 43.74 lbf.

(a) Calculate the mass of the block.
(b) What is the mass of the block on the surface of Mars?
(c) If the block were weighed, using a spring scale calibrated at standard gravity, what weight would the spring scale indicate on the surface of Mars?
(d) Determine the density and the specific volume of the block.
(e) Determine the specific weight of the block on the surface of Mars.

Give answers in the English system of units.

---

**Solution:**

(a) From eq. (2.2c) we write Newton's law as

$$F_g = \text{weight} = \frac{m}{g_0} g$$

Solving for the mass $m$, and using $g_0$ from eq. (2.2a),

$$m = \frac{F_g \, g_0}{g} = \frac{43.74 \text{ lbf}}{} \left| \frac{32.17 \text{ lbm ft}}{\text{lbf sec}^2} \right| \frac{\text{sec}^2}{32.17 \text{ ft}}$$

$$= 43.74 \text{ lbm}$$

(b) Mass represents a fixed quantity of matter and is independent of local gravity. Thus the mass of block is 43.74 lbm regardless of its location.

(c) Using eq. (2.2c)

$$\text{Weight} = F_g = \frac{43.74 \text{ lbm}}{} \left| \frac{\text{lbf sec}^2}{32.17 \text{ lbm ft}} \right| \frac{12.95 \text{ ft}}{\text{sec}^2} = 17.61 \text{ lbf}$$

(d) From the definition of density,

$$\rho = m/V = 43.74 \text{ lbm}/1.10 \text{ ft}^3 = 39.76 \text{ lbm/ft}^3$$

$$v = 1/\rho = 0.02515 \text{ ft}^3/\text{lbm}$$

or

$$v = V/m = 1.1 \text{ ft}^3/43.74 \text{ lbm} = 0.02515 \text{ ft}^3/\text{lbm}$$

Note that both $\rho$ and $v$ are independent of the local acceleration of gravity.

(e) From eq. (2.3)

$$w_s = \rho \frac{g}{g_0} = (39.76)(12.95)/(32.17) = 16.00 \text{ lbf/ft}^3$$

---

## Example 2.b

(a) A body with a volume of 0.037 m³ has a density of 6500 kg/m³. What is its weight at a location where $g = 9.68$ m/s²?

(b) The body undergoes linear motion in a horizontal plane. A net force of 1500 N acts parallel to the plane on the body. What is the acceleration experienced by the body?

**Solution:**

(a) From eqs. (2.2c) and (2.2b)

$$F_g = \frac{mg}{g_0} \quad \text{and} \quad g_0 = 1 \text{ kg m/(N s}^2)$$

$$m = \rho V = (6500 \text{ kg/m}^3)(0.037 \text{ m}^3) = 240.5 \text{ kg}$$

Thus

$$F_g = \frac{(240 \text{ kg})(9.68 \text{ m/s}^2)}{1 \text{ (kg m)/(N s}^2)} = 2328 \text{ N}$$

(b) Using eq. (2.2)

$$F = \frac{ma}{g_0}$$

and solving for the acceleration $a$ we obtain

$$a = \frac{F g_0}{m} = \frac{(1500\ \text{N})}{(240.5\ \text{kg})} \frac{1\ \text{kg m}}{(\text{N s}^2)}$$
$$= 6.24\ \text{m/s}^2$$

Note that the local acceleration of gravity does not play a role in part (b).

---

## 2–7 Pressure

*Pressure,* a directly observable characteristic of a system and hence a property, is defined as force per unit area. As noted earlier, one approach to engineering thermodynamics is to view systems from the macroscopic viewpoint; that is, individual particles and molecules are not considered and the fluid is considered to be a continuum with the smallest dimension of the system several orders of magnitude greater than the mean free path of the molecules. (The mean free path is the average distance a molecule travels before it collides with another molecule.) The microscopic view, that is, consideration of individual particles within a system, is useful in discussing pressure. In a typical system molecular activity is continuous, with molecules impacting on the boundaries of the system. The differential normal force due to these collisions with a differential area of the system boundary indicates the nature of pressure, which then can be defined as

$$P = \frac{dF}{dA}$$

where $dA$ is much greater than the square of the mean free path. As the number of molecules impacting on a given area increases, the pressure undergoes a proportionate change. As a lower limit we have the case where there are no molecules in a system, thus establishing a condition of zero pressure or a perfect vacuum.

Figure 2.3 illustrates the terminology commonly associated with pressures, with an absolute pressure of zero identical to a perfect vacuum. As can be seen from this illustration, an absolute pressure greater than the barometric (atmospheric) pressure is equal to the gage pressure plus the barometric pressure in consistent units. Note that gage pressures are always measured relative to atmospheric pressure. Absolute pressures less than atmospheric are evaluated by subtracting the vacuum from the barometric pressure.

A variety of instruments is available to measure pressure. For engineering applications in which pressure is constant or varies slowly with time, pressures are typically measured by Bourdon gages or manometers. The Bourdon gage operates by means of a change in shape of a curved tube element caused by a change in pressure. A mechanical linkage transmits motion to a pointer that indicates gage pressure on a graduated scale marked on a circular clocklike face. Proper calibration is necessary to ensure correct pressure readings.

In the SI the standard pressure unit is the pascal (Pa), which is defined as 1 $\text{N/m}^2$. In the English system the units of pressure are either $\text{lbf/ft}^2$ or $\text{lbf/in}^2$, which are abbreviated,

**Figure 2.3** Pressure Terminology

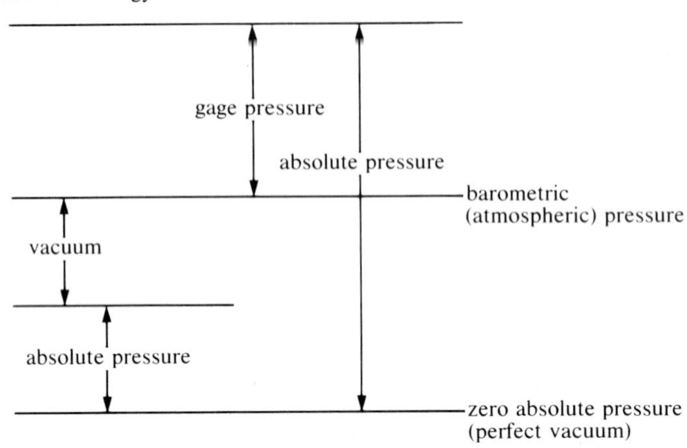


respectively, as psf and psi. The standard sea-level atmospheric pressure in English units is

$$14.696 \text{ lbf/in}^2 = 2116.2 \text{ lbf/ft}^2$$

whereas in the SI the value is

$$1.01325 \times 10^5 \text{Pa} = 101.325 \text{ kPa}$$

The unit denoted by kPa is the kilopascal ($10^3$ Pa) and is frequently used as a convenient unit of pressure in the SI. The above pressures are, of course, absolute pressures. In practice both gage and absolute pressures are encountered. In this text the notation psia and psfa will be used to denote absolute pressures in the English system. Similarly, psig and psfg will be used to note corresponding gage pressures. The pascal unit will be understood to indicate absolute pressure unless the symbol is followed by a gage notation. For example, 10 kPa indicates an absolute pressure whereas 50 kPa gage designates a gage pressure.

Small pressure differences can be measured by manometers. Figure 2.4a shows a manometer that is used to determine the difference in the time–steady pressure of a fluid flowing in a horizontal duct and the pressure of the atmosphere. The difference in pressure is related to the column height of the manometer fluid. This is shown in the following analysis. Consider in Figure 2.4b the differential length $dz$ of the column of the manometer fluid (at rest) between (1) and (2) in Figure 2.4a. The absolute pressure differs by an amount $dp$ between $z$ and $z + dz$. In summing the forces in the vertical direction ($z$ is positive upward) and recognizing that this sum must be zero for the fluid at rest, we obtain

$$pA - (p + dp)A - w_s A dz = 0$$

**Figure 2.4** Simple Manometer. a. Manometer arrangement. b. Differential element of manometer fluid

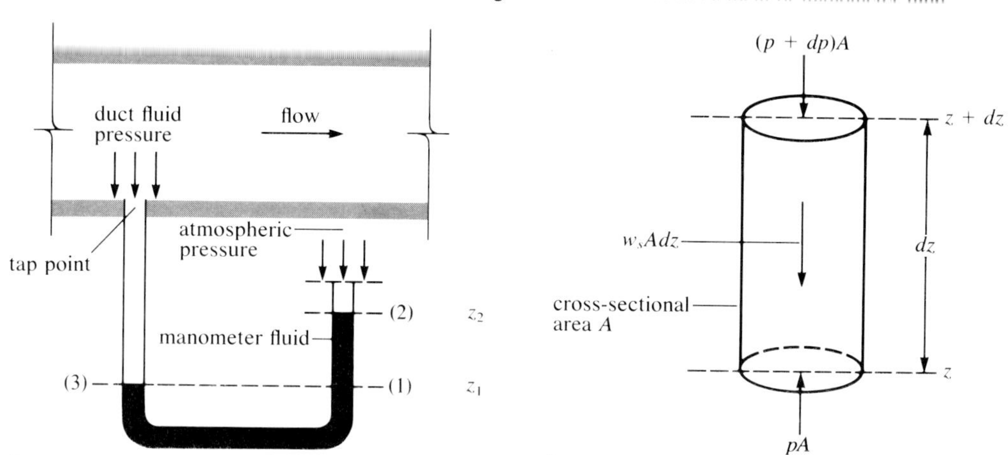


where the third term is the weight of the column of fluid of height $dz$. This expression reduces to

$$dp = -w_s dz \tag{2.4}$$

This is the *hydrostatic equation* and is the pressure–elevation relation for any fluid at rest. A positive $dz$ results in a negative $dp$. Thus it is seen that $p$ decreases with increasing elevation in a fluid at rest. The above expression can be integrated for the manometer as follows:

$$\int_1^2 dp = p_2 - p_1 = -\int_1^2 w_s dz = -w_s\,(z_2 - z_1) \tag{2.4a}$$

where $w_s$ is assumed constant. The elevation change $z_2 - z_1$ is sometimes termed the head $h$, a positive number. Thus

$$p_1 - p_2 = w_s h \tag{2.4b}$$

Equations (2.4) show that in a given fluid at rest, pressure changes only with elevation. Hence in Figure 2.4a the pressure at point (3) is the same at point (1). When $w_s$ for the manometer fluid is much greater than $w_s$ for the fluid flowing, the pressure at (3) is essentially the fluid pressure in the duct at the tap point shown in Figure 2.4a. Thus the absolute fluid pressure is

$$p_{\text{fluid}} = p_3 = p_1 = p_2 + w_s\,(z_2 - z_1)$$

where $p_2$ is the atmospheric pressure.

Let us apply eq. (2.4b) to two common manometer fluids, water and mercury (specific gravity* = 13.6), to determine the pressure equivalent of a given head. Consider first quantities measured in the English system of units. Assume standard acceleration of gravity, $g = 32.17\ \text{ft/sec}^2$. The density of water at room temperature is 62.4 lbm/ft³. From eq. (2.3) the specific weight is then 62.4 lbf/ft³. We then have for the pressure change $\Delta p$ corresponding to a head of 1 inch of water:

$$\Delta p = \frac{62.4\ \text{lbf}}{\text{ft}^3}\left|\frac{1\ \text{in}}{}\right|\frac{\text{ft}}{12\ \text{in}} = 5.20\ \text{lbf/ft}^2$$

$$= \frac{5.20\ \text{lbf}}{\text{ft}^2}\left|\frac{\text{ft}^2}{144\ \text{in}^2}\right. = 0.03611\ \text{lbf/in}^2$$

For mercury, in the same gravitational field as the water, we have the density

$$\rho_{\text{Hg}} = 13.6\ (62.4\ \text{lbm/ft}^3) = 848.6\ \text{lbm/ft}^3$$

and the specific weight

$$w_s = 848.6\ \text{lbf/ft}^3$$

Thus for a head of 1 inch of mercury

$$\Delta p = \frac{848.6\ \text{lbf}}{\text{ft}^3}\left|\frac{1\ \text{in}}{}\right|\frac{\text{ft}}{12\ \text{in}} = 70.72\ \text{lbf/ft}^2$$

$$= 0.491\ \text{lbf/in}^2$$

Next consider head and other pertinent quantities measured in SI units. Let the head be measured in centimeters. The density of water at room temperature is 1 kg/liter. Thus the specific weight is

$$w_s = \frac{1\ \text{kg}}{\text{liter}}\left|\frac{\text{liter}}{10^{-3}\text{m}^3}\right|\frac{9.81\ \text{m/s}^2}{1\ \text{kg m/(N s}^2)}$$

$$= 9810\ \text{N/m}^3$$

Therefore, for a head of 1 cm of water,

$$\Delta p = \frac{9810\ \text{N}}{\text{m}^3}\left|\frac{1\ \text{cm}}{}\right|\frac{\text{m}}{100\ \text{cm}}$$

$$= 98.1\ \text{N/m}^2 = 98.1\ \text{Pa}$$

*The specific gravity of a substance is defined as the ratio of its density to that of water at specified values of temperature and pressure. These values are usually 60 F and 1 atmosphere pressure.

For a head of 1 cm of mercury,

$$\Delta p = 1334 \text{ Pa}$$

## Example 2.c

A certain fluid has a density of 37.56 lbm/ft³. What is the pressure difference indicated by a manometer column height of 27.92 inches of the fluid in a region where the local gravitational acceleration is 7.876 ft/sec²?

**Solution:**

From eq. (2.3)

$$w_s = \rho \frac{g}{g_0} = 37.56 \frac{7.876}{32.17} = 9.196 \text{ lbf/ft}^3$$

And from eq. (2.4b)

$$\Delta p = hw_s = \frac{27.92}{12}(9.196) = 21.40 \text{ lbf/ft}^2$$
$$= \frac{21.40}{144} = 0.1486 \text{ lbf/in}^2$$

### 2–8 The Work Interaction

One of the fundamental concepts that we encounter in the study of thermodynamics is work. This section deals with this concept as it applies to engineering thermodynamics.

From elementary physics we recall that mechanical work is defined as the product of a force $F$ and a distance $x$, where the force $F$ acts in the direction of the displacement $x$. For cases where this force does not act in the direction of the displacement, and also varies with the displacement, the work on a body undergoing a displacement from $x_1$ to $x_2$ as shown in the sketch is

$$\text{Work} = \int_{x_1}^{x_2} F \cos\theta \, dx \tag{2.5}$$

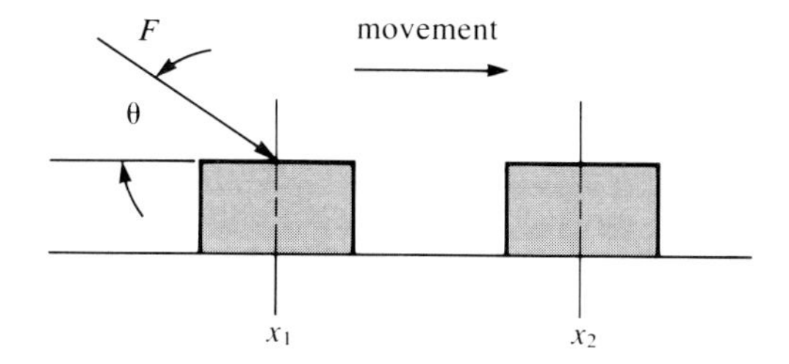

where $\theta$ is the angle between the line of action of the force and the displacement. In thermodynamics this definition is, of course, retained, but it is necessary that we broaden our view of work in relation to a thermodynamic system and its surroundings. The following view is taken. *Work* is an interaction between a system and its surroundings; work is done by a system if the sole effect external to the system could be the raising of a weight. The magnitude of the work is the product of the weight and the vertical distance it rises. The definition does not require that a weight actually be raised. Rather it states that work is done by a system during an interaction between a system and its surroundings if the only effect external to the system *could be* the lifting of a weight. This means of identifying work may seem odd when it is first encountered. Nonetheless it clearly defines the work interaction and thus permits us to distinguish the work interaction from other interactions that can occur between a system and its surroundings—for example, the heat interaction, which is discussed in Section 2–13. From the definition it is apparent that work is done by a system only when the surroundings are affected. Work is always associated with an interaction across the boundary between a system and its surroundings and has no meaning with regard to a state of a single system. While the definition indicates evaluation of the amount of work in terms of a mechanical system, the weight and a lifting mechanism, it is broad enough to cover all types of work.

We shall use the sign convention that *work done by a system is positive* and *work done on a system is negative.* Work is denoted by the symbol $W$ and, of course, has units of the product of force and length. In the English system of units, work has the unit foot-pounds force (ft lbf). In SI work has the units Newton meter (N m). It is often convenient to compute the work per unit mass of working fluid. This quantity is denoted by the symbol $w$, and is defined as $w = W/m$.

The magnitude of work is independent of time. Regardless of the time required to raise a given weight through a specified elevation change, the same amount of work is done. However, the time rate at which work is done is an important quantity called *power.* Power will be denoted by the symbol $\dot{W}$ and has units of work per unit time. Thus,

$$\dot{W} = \frac{dW}{d\tau} \tag{2.5a}$$

where $\dot{W}$ can be time dependent. Familiar engineering power units are the horsepower and the kilowatt. Horsepower (HP) is the English system unit for power and is equal to 33,000 ft lbf/min. The SI power unit is the watt (W), which is equal to 1 J/s = 1 N m/s. The kilowatt (kW) is $10^3$ W. The amount of work done in a time interval $\tau_2 - \tau_1$ is related to power by

$$W = \int_{\tau_1}^{\tau_2} \dot{W}\, d\tau \tag{2.5b}$$

Where a force acts through a distance as illustrated in the previous sketch, the power can be written as

$$\dot{W} = \frac{dW}{d\tau} = -F\cos\theta\,\frac{dx}{d\tau} = -F\cos\theta\,V \tag{2.5c}$$

This equation is consistent with the sign convention adopted for work. Work is positive, that is, work is done by the body, when $F\cos\theta$ and the velocity $V$ have opposite signs.

There are numerous ways in which work can be done on or by a system. These are called work modes. All work modes are related in some form to a force acting through a distance in accordance with eq. (2.5). The movement of an electric charge through an electric field involves a force acting through a distance. To move a conductor through a magnetic field requires that a force act through a distance. Similarly, a force acting through a distance to overcome surface tension effects when a film of liquid is stretched results in work being performed on the film. In this text we will not consider systems that are influenced by electric or magnetic fields or surface tension because, for many applications of engineering thermodynamics, these effects are negligible. We will consider other work modes that are frequently encountered in engineering thermodynamics. Important among these is mechanical work. This work is associated with movement of the whole system, as in the case of a moving closed system, or with movement of a portion of the boundary of a system, either closed or open. Four types of mechanical work are commonly encountered. These are:

1. Work involved in an elevation change of the system.
2. Work related to a change in velocity of a system.
3. Work associated with compression or expansion processes in a system.
4. Shaft or rotor work.

Each of these will be discussed separately. However, it should be recognized that some of the work modes may occur simultaneously, and these may be interrelated. For example, consider again the apparatus in Figure 2.2b. Let the cylinder contain a gas and let only work interactions take place. As shaft work is done on the gas by rotating the paddle wheel, the gas will expand and expansion work will be done in moving the piston upward. Thus the work forms in items 3 and 4 are involved and occur simultaneously.

To illustrate the work involved with an elevation change of a system, consider a mass $m$ that is acted on by a force equal to its weight that causes its elevation to change from elevation $z_1$ to elevation $z_2$. No other interactions take place. The mass is initially and finally at rest. The work is

$${}_1W_2 = -\int_{z_1}^{z_2} F_g dz = -\int_{z_1}^{z_2} m\frac{g}{g_0}\,dz = -m\frac{g}{g_0}(z_2 - z_1) \tag{2.5d}$$

To obtain the final expression, it has been assumed that the elevation change is small enough that the local acceleration of gravity can be assumed constant. In the above expression the term ${}_1W_2$ denotes the work for the process between states (1) and (2). This notation is used to point out clearly that work is associated with a process and that it is not identified with a state. The negative sign is consistent with the selected sign convention for work. If the mass is raised, $z_2$ is larger than $z_1$ and the work is negative, indicating work has been done on the mass by some part of its surroundings to elevate it. The quantity $m(g/g_0)z$ is called mechanical *potential energy,* or simply potential energy. We note that the work associated with an elevation change is then the negative of the potential energy change. Rearrangement of the above expression yields the work per unit mass:

$$ {}_1w_2 = \frac{{}_1W_2}{m} = -\frac{g}{g_0}(z_2 - z_1) \tag{2.5e} $$

In either of these work expressions the reference for elevation $z$ is arbitrary. We note that the work required to move a mass through a given elevation change depends on the elevation change only and is therefore independent of the path involved during the elevation change.

An expression for the work involved in changing only the velocity of a system can be developed by considering a change in velocity of a mass $m$ moving in a horizontal plane. To change the velocity of the mass, a force must be applied through a distance. Thus the work is

$$ {}_1W_2 = -\int_1^2 F dx = -\int_1^2 \frac{ma}{g_0} dx = -\int_1^2 \frac{m}{g_0}\frac{dV}{d\tau} dx $$
$$ = -\int_1^2 \frac{mVdV}{g_0} $$

where $dx/d\tau$ is, by definition, the velocity $V$. Thus

$$ {}_1W_2 = -\frac{m}{g_0}\left(\frac{V_2^2 - V_1^2}{2}\right) \tag{2.5f} $$

The work per unit mass is

$$ {}_1w_2 = \frac{{}_1W_2}{m} = -\left(\frac{V_2^2 - V_1^2}{2g_0}\right) \tag{2.5g} $$

These equations yield work values with signs that are consistent with the sign convention adopted for work. The quantity $mV^2/2g_0$ is defined as the translational *kinetic energy.* The only requirement for the velocity reference is that it be a nonaccelerating reference. The above equations show that the work associated with changing the velocity of a mass from $V_1$ to $V_2$ is the same regardless of how (by what path) the change takes place. Hence this work is independent of the path.

The type of work in item 3 of our list is of interest for both closed and open systems. At this point it is best illustrated for a closed system. Here, compression and expansion work occurs when forces act normal to a moving boundary, as in the piston–cylinder system il-

**Figure 2.5** Two Forms of Mechanical Work Associated with Moving Boundaries. *a*. Compression–expansion work for a closed system. *b*. Rotor work on a closed system.

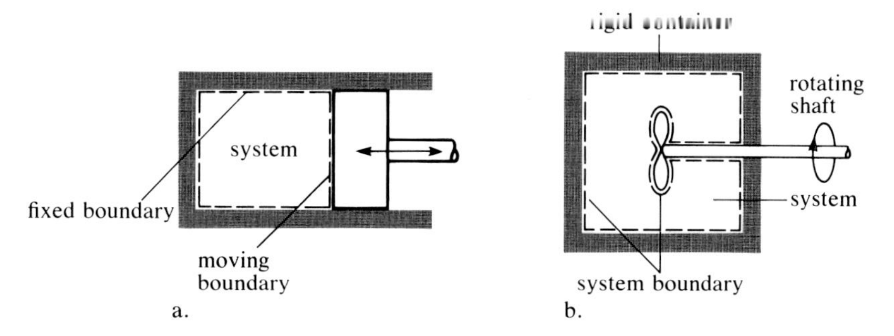


lustrated in Figure 2.5a. As the piston moves leftward and compresses the contents of the system (a gas, for example), the piston does work on the system. When the piston moves to the right, the system expands, doing work on the piston. The quasi-static frictionless process of compression or expansion in a closed system is of special interest and is dealt with in detail in the next section.

Shaft work and shaft power, item 4 of our list, are frequently encountered in engineering thermodynamics. When the boundary of a system, either closed or open, passes through a rotating shaft on which a torque acts, work is transmitted across the boundary. The shaft work is given by

$$W_s = -\int_{\theta_1}^{\theta_2} \mathrm{T}\, d\theta \tag{2.5h}$$

where $d\theta$ is the angle through which the torque T acts. T is the torque that is exerted by the surroundings on the rotating part of the system boundary (the cross section of the shaft through which the boundary passes) and is taken as positive in the positive $\theta$ direction. According to eq. (2.5a) and (2.5h), shaft power is given by

$$\dot{W}_s = \frac{dW_s}{d\tau} = -\mathrm{T}\frac{d\theta}{d\tau} = -\mathrm{T}\omega \tag{2.5i}$$

where $\omega$ is the angular speed of the shaft.

The boundary need not pass through a rotating shaft for shaft work to be involved in a process. Figure 2.5b illustrates this. Note that the system boundary has been chosen to be coincident with the surface of the rotor and does not include the rotor. As the rotor turns as a result of the application of a torque, work is done on the system (a liquid, for example) by forces that are both normal and tangential to the moving boundary. Equations (2.5h) and (2.5i) still apply, with the torque T taken as that applied to the shaft. The stirring action in Figure 2.5b produces significant viscous effects within the system. In fact, the very nature of the process is associated with frictional effects in the fluid due to the relative motion of adjacent fluid elements. The work for such a process can only be negative, that is, the system illustrated cannot produce work. This is in contrast to the system in Figure 2.5a, for which the work can be either positive or negative. This illustrates one difference between stirring work and compression and expansion work for a closed system.

**Figure 2.6** Electric Current Crossing a System Boundary

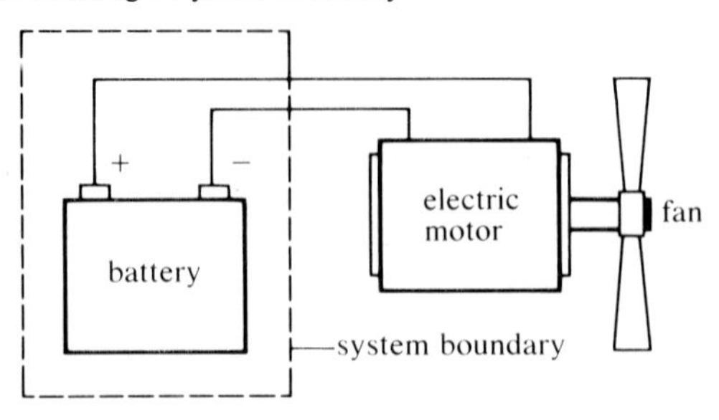


An interaction frequently encountered is that of the flow of electric current across the boundary of a system. This is illustrated in Figure 2.6, in which the system, a battery, supplies electric current to drive an electric motor, which in turn drives a fan. The question that arises is: Does the flow of electric current across a boundary constitute work? From the definition of work, the question that must be addressed is: Could the sole external effect of the current flowing across the boundary be the lifting of a weight? If this is answered in the affirmative, then the interaction is the work interaction. First notice that in real operation the motor will have some losses (bearing friction and electrical losses, for example) and, as a result, not all of the energy supplied by the battery will be delivered by the motor to drive the fan. Thus, if we were to attempt to replace the fan by a pulley and rope to lift a weight, the sole external effect would not be the lifting of a weight. However, let us consider a motor in which friction losses and electrical losses have been made vanishingly small. In concept it is possible to accomplish this. In this limiting case, the sole effect external to the system could be the raising of a weight. Thus we conclude that electric current crossing a boundary constitutes a work interaction.

## Example 2.d

A mass of 1.40 kg undergoes a change in elevation from 1.9 to 0.8 m. The local acceleration of gravity is 9.7 m/s$^2$. What is the change in its potential energy and what is the work associated with this elevation change?

**Solution:**

The potential energy change is

$$\Delta PE = (z_2 - z_1)\frac{mg}{g_0} = \frac{(0.8 - 1.9)\ \text{m}}{} \left| \frac{1.4\ \text{kg}}{} \right| \frac{9.7\ \text{m}}{\text{s}^2} \left| \frac{\text{N s}^2}{1\ \text{kg m}} \right.$$

$$= -14.9\ \text{N m}$$

According to eq. (2.5d) the work is $-\Delta PE$. Thus

$$_1W_2 = +14.9 \text{ N m}$$

The sign indicates that the system (the mass) does work on its surroundings during the noted elevation change.

---

## Example 2.e

A body with a mass of 10 lbm moving in a horizontal plane undergoes a change in velocity from $V_1 = 23$ ft/sec to $V_2 = 90$ ft/sec. Find the kinetic energy change and the related work.

**Solution:**

From the definition of kinetic energy,

$$\Delta KE = \frac{m}{g_0}\,\frac{V_2^2 - V_1^2}{2}$$

$$= \frac{10 \text{ lbm}}{2} \left| \frac{\text{lbf sec}^2}{32.17 \text{ lbm ft}} \right| \frac{(90^2 - 23^2)\text{ft}^2}{\text{sec}^2}$$

$$= 1177 \text{ ft lbf}$$

From eq. (2.5f) $W = -1177$ ft lbf. It should be understood that this work is *only* that required to impart the noted velocity change. It does not include, for example, the work required to overcome friction that might exist between the plane and the body.

---

## Example 2.f

A shaft transmitting work across the boundary of a system experiences a resisting torque from the surroundings of 68 N m while rotating at 1000 revolutions per minute. Evaluate the shaft power.

**Solution:**

From the discussion in this section, $\dot{W}_s = -T\omega$. Since the torque resists the shaft motion, its sign is opposite to that of $\omega$. Thus,

$$\dot{W}_s = -\frac{-68 \text{ N m}}{} \left| \frac{(2\pi)\ 1000 \text{ rad}}{\text{min}} \right| \frac{\text{min}}{60 \text{ s}} = 7120 \text{ N m/s}$$

$$= 7120 \text{ J/s} = 7120 \text{ W} = 7.120 \text{ kW}$$

The positive sign indicates that work is done by the system on its surroundings.

**Figure 2.7** A Closed System Undergoing a Frictionless Quasi-Static Process

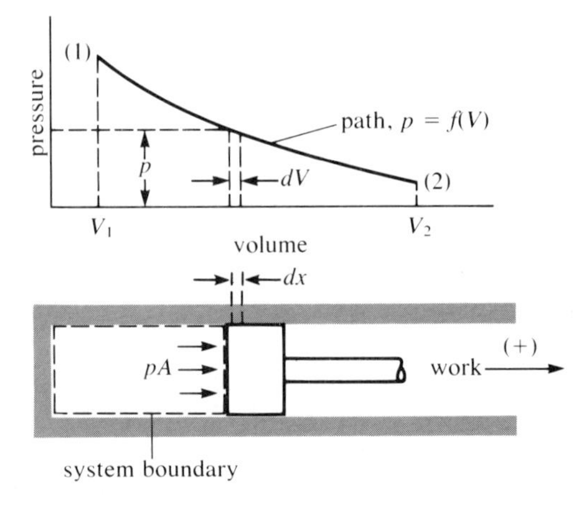


## 2–9 Work for a Closed System Undergoing a Frictionless, Quasi-Static Process

Let us now consider in more detail compression and expansion processes for a closed system as illustrated in Figure 2.5a. Suppose that the working fluid undergoes a quasi-static frictionless expansion process. As noted earlier, a quasi-static process is one in which only infinitesimal departures from equilibrium occur. Consider the closed system in Figure 2.7 with the inner cylinder walls and piston face as the boundaries. For purposes of illustration let us take as the system $m$ mass units of the working fluid. For the quasi-static process, the pressure on the fixed boundaries and that on the moving boundary (piston face) will not differ by more than an infinitesimal amount at any instant. Each property will be uniform throughout the system or differ infinitesimally with location as the process, an expansion, takes place.

The system is not subject to magnetic, electrical, gravitational, motion, or surface effects; that is, we are dealing with a simple system. For a portion of the process in which the volume change is $dV$ and the piston moves to the right a distance $dx$, we can write, using the definition of mechanical work,

$$\delta W = F dx = (pA)\frac{dV}{A} = p dV \tag{2.6}$$

where $p$ is the absolute pressure of the system acting on the piston face, $A$ is the area of the piston face, and $\delta W$ is the differential work. The notation $\delta$ is used to characterize an inexact differential. The integral of an inexact differential depends on the path as well as the end states. For a known path, that is, for a given functional relation of $p$ and $V$ between states (1) and (2), the work done by the system on the piston face is

$$_1W_2 = \int_1^2 p dV \tag{2.6a}$$

Inspection of the $p$–$V$ diagram of Figure 2.7 shows the work for the process considered to be a path function; that is, the work is dependent on the path connecting states (1) and (2). Referring to eq. (2.6a) and Figure 2.7, it is evident that the work is equal to the area under the path to the abscissa for closed systems undergoing frictionless quasi-static processes. For such processes, eq. (2.6) also tells us that this form of work will be zero when no change in volume occurs. The sign on $dV$ in eq. (2.6) yields the sign on the work since $p$, the absolute pressure, is always positive. Therefore, the work will be positive for expansion processes and negative for compression processes.

Equation (2.6a) yields the work for a system containing $m$ mass units. In many cases there is interest in the work per unit mass, $w$. Division of eq. (2.6a) by the mass $m$ yields

$$_1w_2 = \frac{_1W_2}{m} = \frac{\int_1^2 p dV}{m} = \int_1^2 p dv \tag{2.6b}$$

where $dv$ has been recognized as $dV/m$ from the definition of specific volume $v = V/m$ and the fact that $m$ is a constant for a closed system.

It is important to note that the processes for which eqs. (2.6) are applicable do not involve any internal frictional effects; that is, such systems do not involve stirring work as illustrated in Figure 2.5b. Thus it is clear that eqs. (2.6) are restricted to closed-system quasi-static frictionless processes.

In eqs. (2.6) the pressure $p$ is the absolute pressure. Also, these equations give the work done by the system on the piston face. This will equal the useful work delivered to the surroundings when there are no friction losses as the piston moves relative to the cylinder and there is no opposing pressure acting on the back side of the piston. If a constant surroundings pressure, say the atmospheric pressure $p_0$, acts on the outside of the piston, the net useful work delivered in the absence of friction will be the piston face work less the amount of work done in pushing back the atmosphere. Letting $A$ be the area of the piston face, the work done on the atmosphere becomes

$$W_{\text{atmos}} = p_0 (A) (x_2 - x_1) = p_0 (V_2 - V_1) \quad \textbf{(2.6c)}$$

or per unit mass

$$w_{\text{atmos}} = p_0(v_2 - v_1) \quad \textbf{(2.6d)}$$

Note, however, that when the system undergoes a cycle, the cyclic integral of the specific volume, a property, will be equal to zero, with the result that the *net* work on the atmosphere is zero for a cycle. In subsequent chapters we focus on the work done by the system without regard for work done on or by the atmosphere.

## Example 2.g

A closed system consisting of 0.2 pound mass of working fluid undergoes a frictionless quasi-static process in the piston–cylinder assembly shown in the figure from an initial state (1) at 70 psia and a volume of 1.20 ft³ to a final state (2) where the pressure is 20 psia. The path for the process is $pv^n$ = constant, with $n$ having a value of 1.35 (refer to Figure 2.7 for the $p$-$V$ diagram). The surrounding atmospheric pressure is 14.30 psia. As the system expands from (1) to (2), the frictionless mechanism shown causes the weight to rise a distance of 120 feet. The local acceleration of gravity is 31.8 ft/sec².

(a) Calculate the work done by the system.
(b) What is the mass of the weight lifted?

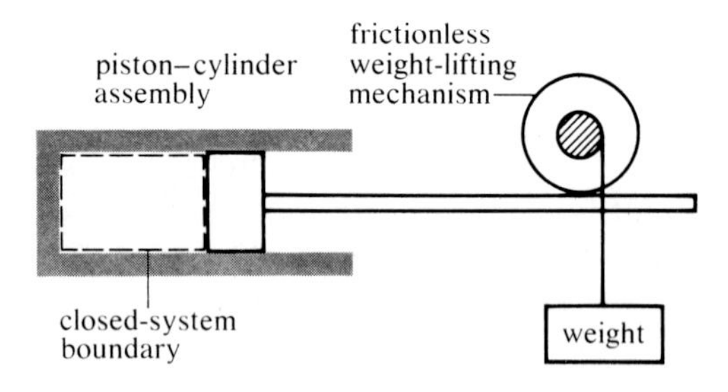


**Solution:**

(a) Initial calculation basis: 1 pound mass of the working fluid.

$$dw = pdv \quad \text{and} \quad p_1 v_1^n = p_2 v_2^n = pv^n = C$$

At the initial state

$$v_1 = \frac{V_1}{m} = \frac{1.20}{0.2} = 6.00 \text{ ft}^3/\text{lbm}$$

At the final state, using the path equation,

$$v_2 = \left(\frac{p_1}{p_2}\right)^{1/n} v_1 = \left(\frac{70}{20}\right)^{1/1.35} 6.00 = 15.18 \text{ ft}^3/\text{lbm}$$

From the path equation, $p = C/v^n$ at any state during the process. Thus the work equation becomes

$$_1w_2 = \int_1^2 pdv = C\int_1^2 \frac{dv}{v^n}$$

which integrates to

$$_1w_2 = C\left[\frac{v^{-n+1}}{-n+1}\right]_{v_1}^{v_2}$$

When substituting the specific volume limits, it is convenient to select the value of the constant $C$ containing the limit. Thus

$$\begin{aligned} _1w_2 &= \frac{(p_2v_2^n)(v_2^{-n+1}) - (p_1v_1^n)(v_1^{-n+1})}{1-n} \\ &= \frac{p_2v_2 - p_1v_1}{1-n} \\ &= \frac{(20)(144)(15.18) - (70)(144)(6.00)}{1-1.35} \\ &= 47{,}890 \text{ ft lbf/lbm} \end{aligned}$$

Finally, taking note of the mass of the system,

$$_1W_2 = (0.2)(47{,}890) = 9578 \text{ ft lbf}$$

(b) The work calculated in part (a) is done by the system on the surroundings in two ways: The weight is lifted and the atmosphere is pushed back, or

$$_1W_2 = W_{\text{atmos}} + W_{wt}$$

Using eq. (2.6d), the work done on the atmosphere is

$$\begin{aligned} W_{atmos} &= m\, w_{\text{atmos}} = 0.2\,(14.30)(144)(15.18 - 6.0) \\ &= 3781 \text{ ft lbf} \end{aligned}$$

The work done on the weight in lifting it is equal to its change in potential energy

$$W_{wt} = {}_1W_2 - W_{\text{atmos}} = m\frac{g}{g_0}(z_2 - z_1)$$

or

$$9578 - 3781 = m\frac{31.8}{32.17}(120)$$

Solving for $m$, we obtain 49.9 lbm.

---

**Figure 2.8** The Zeroth Law

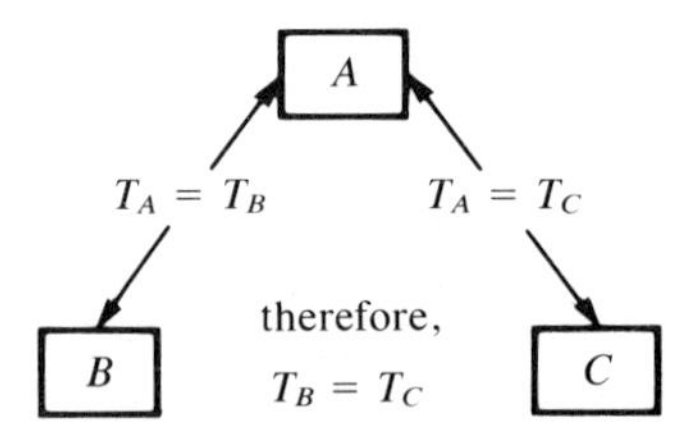


## 2–10 Equality of Temperature and the Zeroth Law

Efforts to establish temperature concepts have spanned several centuries, with Galileo (1564–1642) one of the early investigators. He and other pioneers devised many instruments in their attempts to measure a quantity that had not yet been defined. The human sensing of "hot" and "cold" was inadequate by the subjective nature of the terms. Seasonal changes in the weather, while obvious to an individual, provided little insight into this undefined concept. This inadequacy is easily experienced at the present time. Think of being outdoors on a winter day. Touching a steel post or rail produces an entirely different sensation than that experienced when touching a piece of wood at the same time and in the same location.

In this book we first define *equality of temperature.* Consider a closed system $A$ and any other closed system $B$ and allow the two systems to interact without either system undergoing a work interaction and while isolated from all other bodies. If no change in the properties of either system takes place, then we say the two systems have equal temperature, or *equality of temperature* of the two systems is said to exist. Equality of temperature also can be used in defining *thermal equilibrium;* namely, two systems are in thermal equilibrium with each other when they meet the equality of temperature criterion.

Now, with equality of temperature established, we can proceed to the zeroth law and the concept of a thermometer. Consider the three separate and distinct closed systems $A$, $B$, and $C$ shown in Figure 2.8. The size and contents of the systems are immaterial, however, each individual system must be at an equilibrium state. In the absence of work let system $A$ interact (communicate) with system $B$. If there is no change in any property of either system, then, by the definition of equality of temperature, we establish that the two systems are at equal temperature. Next repeat the process using systems $A$ and $C$, with $A$ at its same state fixed by its properties. If, by observation, it is established that equality of temperature exists between $A$ and $C$, we can then say, without further investigation, that systems $B$ and $C$ are at equal temperature. A generalization of the foregoing observations is the *zeroth law*, which states that two systems, each having equality of temperature with a third system, have equality of temperature with each other. An analogous statement of the zeroth law is that two systems, each in thermal equilibrium with a third system, are in thermal equilibrium with each other.

The concept of a thermometer as an instrument utilizing thermal equilibrium is thus established. In the specific example described above, system $A$ has served as a thermometer in that it indicates that systems $B$ and $C$ are at equal temperature, even though $B$ and $C$ have not, as a pair, been subjected to an equality-of-temperature test.

## 2–11 Thermometers and Temperature Scales

This section is devoted to a discussion of the ideas behind thermometers and temperature scales. In the next section we define a scale of temperature by an operational definition. We shall not, however, explicitly define temperature.

A number of different thermometers and temperature scales have been used. Each type of thermometer has a particular thermometric substance and we rely on a change in a thermometric property of that substance to indicate a change in temperature. The absolute thermodynamic scale of temperature that will be introduced following the second law of thermodynamics and some of the corollaries of that law is not dependent on the behavior of any thermometric substance.

The sealed-stem thermometer was invented in 1653 by the Grand Duke Ferdinand II of Tuscany, one of the founders of the Florentine Academy of Experiment. Present-day examples of that thermometer are the alcohol-in-glass and the mercury-in-glass thermometers. The utility of any thermometer requires that readings taken with the instrument be reproducible. This demand, in turn, requires what are known as fixed points. Fixed points commonly used include the fusion, or melting, point and the boiling point of various pure substances at a specified pressure. Some of the earliest attempts to establish fixed points were, in light of present knowledge, crude and humorous. Summer heat and winter cold were proposed. Gabriel Fahrenheit (1686–1736) once assigned a numerical value of 96 to human body temperature and to the coldest temperature he could produce repeatedly, the freezing point of a saturated salt–$H_2O$ system, he assigned a value of 0. The melting point of butter, certainly not a fixed point, nevertheless was used as a fixed point by some of the pioneers of thermometry.

To reveal the arbitrary nature of temperature scales, let us consider the three different thermometers shown in Figure 2.9. Thermometer $a$ is a closed-stem, mercury-in-glass thermometer, $b$ is a constant-volume gas thermometer of the type first used by Amonton in the year 1700, and $c$ is simply a bar of copper of fixed mass and isotropic behavior. For thermometer $a$ the thermometric substance is mercury and the thermometric property is the coefficient of volumetric expansion of mercury. The stem bore has a constant cross-sectional area. For thermometer $b$ the gas is the thermometric substance and pressure is the thermometric property. Thermometer $c$ has copper as the thermometric substance and, arbitrarily, the length $L$ is chosen as the thermometric property.

**Figure 2.9** Three Thermometers

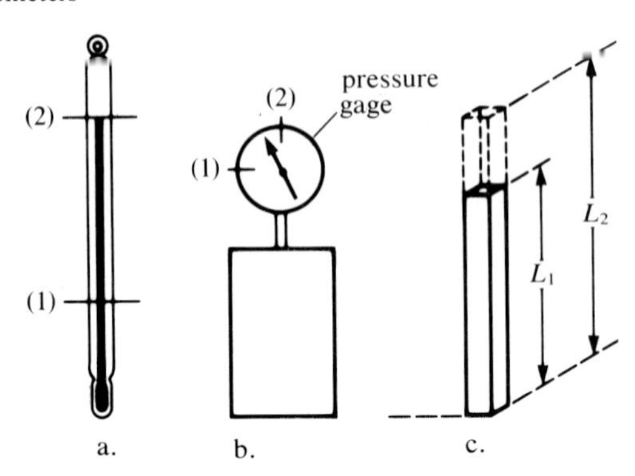


Consider two fixed points, a water–ice system at a pressure of 1 standard atmosphere (the ice point) and boiling water at 1 standard atmosphere (the steam point). Each of the thermometers is placed in the water–ice bath, remaining until thermal equilibrium (equality of temperature) is reached. At thermal equilibrium with the water–ice system we establish a column height of the mercury for thermometer *a,* the location of the gage pointer for *b,* and the length of *c.* For each thermometer the ice point is marked as "1" in Figure 2.9. Note that no numerical values are assigned, as yet, to the ice point for any of the thermometers. The procedure is next repeated for each of the thermometers using the boiling water. When the thermometric substance is at equal temperature with the boiling water, we can mark point "2" for each thermometer and call it the steam point.

With these fixed points established, the arbitrary nature of temperature scales can be illustrated. Any two numbers of nonidentical value could be assigned to the two fixed points for any of the three thermometers. For example, thermometer *a* could have a value of $-25$ for the ice point and 200 for the steam point; thermometer *b,* 100 for the ice point and 700 for the steam point; thermometer *c,* 2.173 for the ice point and 3.465 for the steam point. Each of these numerical values would be reproducible for the particular thermometer at the designated fixed points. Celsius (1701–1744), who proposed the centigrade scale in 1742, once considered a scale assigning the steam point a value of zero and the ice point a value of 100. The unrelated nature of temperature scales is not confined to values assigned to fixed points but, additionally, extends to the graduations between fixed points. The mercury-in-glass thermometer, for example, might have equal linear graduations between the ice point and the steam point. These linear intervals could be extended beyond either bound of the ice point–steam point range. The constant-volume gas thermometer could have graduations of equal arc interval between the fixed points with extrapolations of equal interval above the steam point and below the ice point. The intervals, for any one of the thermometers, could be any nonlinear choice. Common usage today associates a hotter body with a greater temperature and a colder body with a lower temperature. Reference to Celsius' scale having a steam point of zero and an ice point of 100 indicates the arbitrary relationship of hotness and coldness to numerical values of temperature.

Suppose we take the mercury-in-glass thermometer and the constant-volume gas thermometer and assign a value of 25 to the ice point and a value of 150 to the steam point of

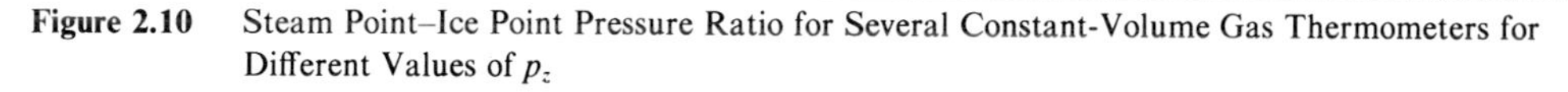

**Figure 2.10** Steam Point–Ice Point Pressure Ratio for Several Constant-Volume Gas Thermometers for Different Values of $p_z$

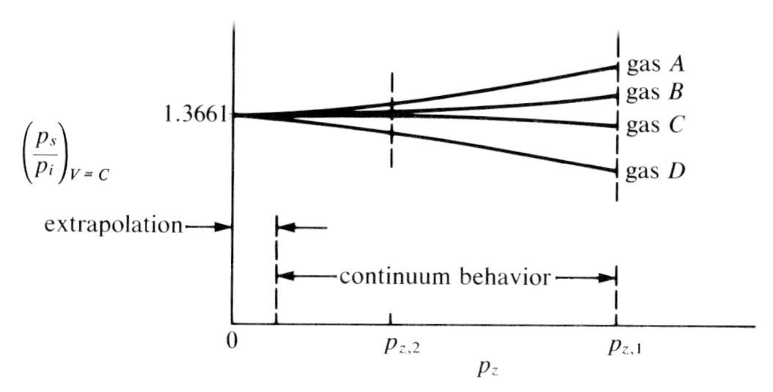


each. Continuing, let us divide the steam point–ice point interval of the mercury-in-glass thermometer into 125 equal increments. In a like manner let us divide the steam point–ice point arc on the constant-volume gas thermometer pressure gage into 125 equal arc increments. We then have two thermometers with the same numerical values for the ice point and the steam point and assigned numerical values between the given fixed points. At some fixed point between the ice point and the steam point, say the boiling point of ethyl alcohol at 1 standard atmosphere, temperature readings of that fixed point are taken with each of the thermometers. Will each thermometer yield the same numerical value for the temperature of this intermediate fixed point? Would other thermometers having the same ice point and steam point values and equal divisions yield the same numerical value for the intermediate fixed point? For each question the answer is the same—no. Why? Because the relation between temperature and the thermometric property is not the same for all thermometric substances. From this simple illustration emerges the need for standardization, that is, selection as a standard of one thermometer with a defined relation between temperature and thermometric property for temperatures ranging between selected fixed points. The International Practical Temperature Scale of 1968 provides the necessary standardization.

## 2–12 Gas Thermometry

In the preceding section a constant-volume gas thermometer (*b*, Figure 2.9) was discussed. Now we shall discuss in more detail this particular type of thermometer and temperature scales that are commonly used. Consider several constant-volume gas thermometers of identical volume, each containing a different gas. Each of the thermometers is filled with the exact amount of gas to yield identical pressure readings $p_{z,1}$ at a specific fixed point, say the melting point of zinc, at 1 standard atmosphere. Next each thermometer is placed in boiling water at a pressure of 1 atmosphere. For each of the thermometers, the pressure $p_s$ is read when thermal equilibrium is achieved between the gas in the thermometer and the boiling water. Next the thermometers are placed in an ice-water bath, which is at 1 atmosphere, and, after thermal equilibrium has been established, pressures $p_i$ are read for each thermometer. The ratio $p_s/p_i$ is then plotted for each of the gas thermometers as an ordinate value with $p_z$ as the abscissa. The result for $p_z = p_{z,1}$ is shown in Figure 2.10.

**Figure 2.11** The Constant–Volume Gas Thermometer Scale of Temperature

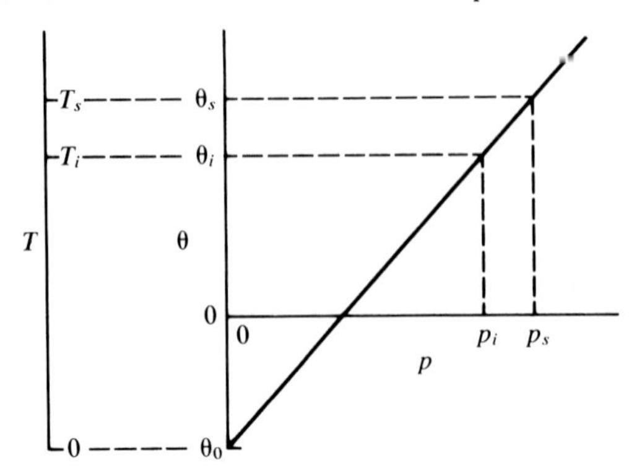


Next a portion of the gas is released from each thermometer so that each indicates a new value $p_{z,2}$ when in equilibrium with the melting zinc. The experiments are repeated and values of $p_s/p_i$ are determined for each thermometer. These data are plotted in Figure 2.10 at $p_{z,2}$. The experiments are repeated for a series of values of $p_z$ to the minimum $p_z$ where each gas still behaves as a continuum. When the $p_s/p_i$ locus for each gas is extrapolated to the ordinate, it is found that these lines intersect the ordinate at the single value $p_s/p_i = 1.3661$.

This important fact permits us to define a temperature scale based on the behavior of gases at low pressures. Further, when such a scale is established, we can identify an absolute scale of temperature. Consider again the constant-volume gas thermometer containing any gas at low pressure. Let $\theta_i$ and $\theta_s$ be the temperatures on a constant-volume gas thermometer scale of temperature corresponding to the ice point and the steam point respectively. The *constant-volume gas thermometer scale of temperature* is defined by a straight line fitted through the steam and ice points on a graph of $\theta$ versus the pressure exhibited by a constant-volume gas thermometer. This is illustrated in the right-hand side of Figure 2.11.

The equation for the straight line is

$$\theta = \theta_0 + \left(\frac{\theta_s - \theta_i}{p_s - p_i}\right) p \tag{2.7}$$

where $\theta_0$ is the intercept on the $\theta$ axis at $p = 0$. Multiplying and dividing the term in the parentheses by $p_i$ yields

$$\theta = \frac{\theta_s - \theta_i}{\left[\dfrac{p_s}{p_i} - \dfrac{p_i}{p_i}\right]}\left[\frac{p}{p_i}\right] + \theta_0 = \frac{\theta_s - \theta_i}{\left[\dfrac{p_s}{p_i} - 1\right]}\left[\frac{p}{p_i}\right] + \theta_0 \tag{2.7a}$$

where $p_s/p_i = 1.3661$ from Figure 2.10. Equation (2.7a) constitutes an operational definition of a scale of temperature. The *absolute* scale of temperature $T$ has the value of zero at the point on the $\theta$ scale where $p = 0$, that is, at $\theta = \theta_0$. This scale is shown on the left-hand side of Figure 2.11. It has only positive values in the same sense that an absolute pressure scale has only positive values. The relation between the absolute $T$ scale and the $\theta$ scale (not an absolute scale) is

$$T = \theta - \theta_0 \tag{2.8}$$

It is seen in Figure 2.11 that the interval $T_s - T_i$ is equal to the interval $\theta_s - \theta_i$. Equations (2.7a) and (2.8) form an operational definition of absolute temperature.

To illustrate the foregoing discussion, let us consider the Fahrenheit scale (not an absolute scale) for which the steam point $\theta_s$ is 212 F and the ice point $\theta_i$ is 32 F. From eq. (2.7a) $\theta_0$ can be determined by substitution at, say, the ice point:

$$32 = \frac{212 - 32}{1.3661 - 1}\left[\frac{p_i}{p_i}\right] + \theta_0 \quad \text{or} \quad \theta_0 = -459.67 \text{ F}$$

The absolute Fahrenheit scale, called the Rankine scale, is given by eq. (2.8) as

$$T = \theta - (-459.67) \doteq \theta + 460$$

or

$$T_R = t_F + 460 \tag{2.8a}$$

where $t_F$ rather than $\theta$ is used to denote temperatures on the Fahrenheit scale and $T_R$ denotes temperatures on the absolute Rankine scale.

Another commonly used absolute scale of temperature, the Kelvin scale, is related to the Celsius scale, which has 0 C for the ice point and 100 C for the steam point. From eq. (2.7a), substitution of $\theta_i = 0$ and $\theta_s = 100$ and evaluation at the ice point yields

$$0 = \frac{100 - 0}{1.3661 - 1}\frac{p_i}{p_i} + \theta_0 \quad \text{or} \quad \theta_0 = -273.15 \text{ C}$$

The absolute Kelvin scale is given by eq. (2.8) as

$$T = \theta - (-273.15) \simeq \theta + 273$$

or

$$T_K = t_C + 273 \tag{2.8b}$$

where $t_C$ is the temperature on the Celsius scale.

The relations between temperatures on the various scales are as follows:

$$\begin{aligned} T_R &= 1.8\ T_K \\ t_C &= (t_F - 32)(5/9) \\ t_F &= (t_C)(9/5) + 32 \end{aligned} \tag{2.8c}$$

**Figure 2.12** Heat-Transfer Modes. *a.* Conduction. *b.* Convection. *c.* Radiation.

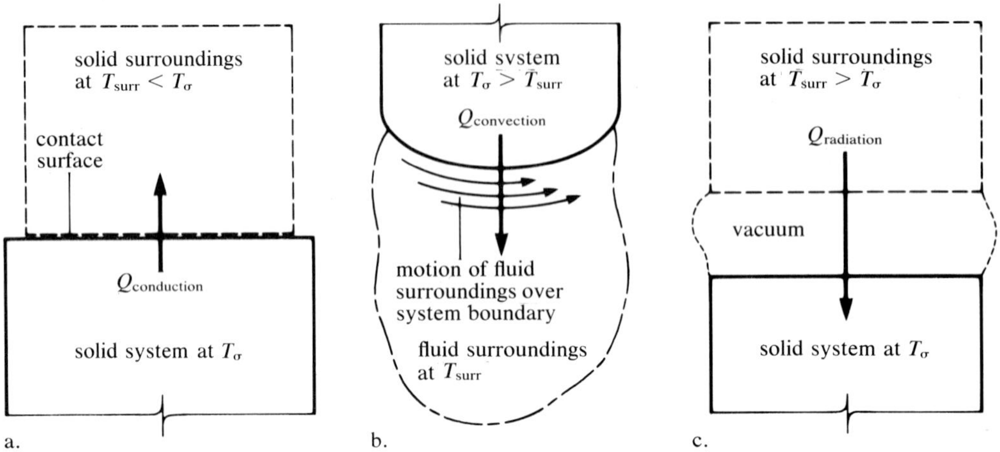

## 2–13 The Heat Interaction

When two systems are not in thermal equilibrium, and therefore do not have equality of temperature, an interaction will occur when the systems are brought into thermal communication. This interaction, which is attributable solely to a difference in temperature of the two systems, is defined as the *heat interaction.* As the interaction takes place, each system undergoes a process. Thus each system will experience a change in state. An energy transfer called *heat transfer* occurs during this interaction. It can be seen, therefore, that heat is never associated with a single system. Further, it is not associated with a state, but with a process, and thus is not a property of a system. The definition follows the reasoning expressed by Poincare in the 1890s. Today we find such loose usage of the word "heat" as "heat rises" or a body has some "heat content."

A sign convection for heat interactions must be selected. We shall designate the heat interaction or heat transfer $Q$ as *positive* when heat transfer is to a system and *negative* when it is from a system. Using the temperature scales previously established, the heat interaction will be positive for a system if the system temperature is less than that of its surroundings. Consider a system $A$ and its surroundings, system $B$. For all cases where the temperature of system $A$ is less than that of system $B$, the heat transfer for system $A$ will be positive and that for system $B$, negative. The equation for this heat interaction is, then, $Q_A = -Q_B$, with $Q_B$ being negative when $Q_A$ is positive. We shall use the notation $Q/m = q$ for heat transfer per unit mass of working fluid. The symbol $\dot{Q}$ will denote heat transfer rate, that is, heat transfer per unit time.

It is useful to examine the ways in which heat transfer can occur. Heat transfer is a boundary phenomenon and can occur at a boundary by one or more heat-transfer modes. These are *conduction, convection,* and *radiation.* Conduction requires physical contact between media at the boundary. This is illustrated for the case of a solid part of a system in contact with a solid part of its surroundings in Figure 2.12a. The symbol $\sigma$ is used to denote the system. Convection heat transfer is promoted by movement of a liquid or a gas over a surface.

Figure 2.12b illustrates heat transfer from a solid system to liquid or gaseous surroundings promoted by motion of the liquid or gas at a temperature lower than that of the system. In the strictest sense, convection heat transfer at the boundary may occur by conduction since, for most flows, fluid particles very close to a surface cling to that surface and remain motionless, thus promoting heat transfer by conduction. Nevertheless, heat transfer in such cases is said to occur by convection. Radiation heat transfer occurs through propagation of electromagnetic waves and requires no intervening medium. As an illustration, Figure 2.12c shows heat transfer to a system occurring by radiation from a solid in its surroundings. The system and the solid are separated by a vacuum. It should be noted that the three modes of heat transfer occur simultaneously in many heat interactions.

An *adiabatic* process is one in which there is no heat interaction, that is, $Q \equiv 0$. Perfect thermal insulation at the system boundary would be required for a process to be adiabatic if the system and its surroundings were at different temperatures.

In the English system of units, the British thermal unit (Btu, denoted by B in this text), is the unit used to measure the heat energy transferred in the heat interaction. The corresponding SI unit is the Joule (J).

## Example 2.h

In the diagram the boundary for system $A$ encloses the oil. System $B$ is the ice and the water and the enclosing uninsulated rigid container. The outer boundary of system $A$ is adiabatic. The two systems are initially in thermal equilibrium. The motor drives the paddle wheel for a period of time, resulting in an increase in the temperature of the oil. Later it is observed that some of the ice has melted. Describe the work and heat interactions for the composite system (system $A$ and system $B$) and for the separate systems.

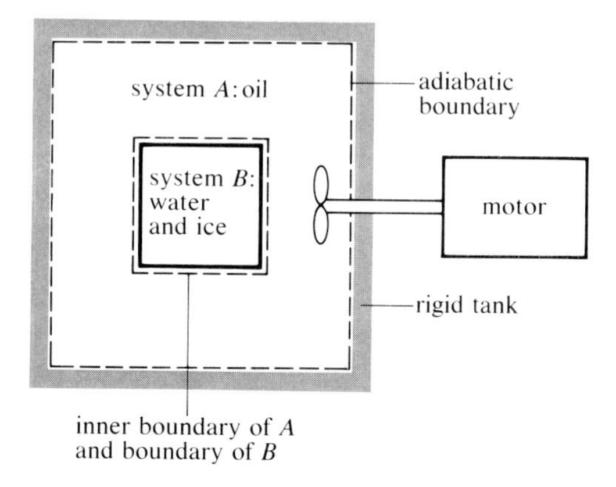


**Solution:**

The composite system boundary is the outer boundary of $A$ and thus is adiabatic; hence $Q$ for the composite system is zero. The work for this system is that done by the motor in turning the paddle wheel and is negative. The process of melting the ice occurs inside the composite system boundary and, therefore, is not a heat interaction for the composite system since it is not an interaction that resulted from a temperature difference across the system boundary. The work for system $A$ is the work done on it by the paddle wheel. Its outer boundary is adiabatic but the heat interaction occurs across its inner boundary since the temperature of $A$ was greater than the temperature of $B$ and the inner container is uninsulated. $Q_A$ is negative. For system $B$ the work is zero (it has no moving boundaries) and $Q_B$ is positive. For this example, $Q_A = -Q_B$.

---

**Figure 2.13** Conservation of Mass for an Open System. *a.* Mass flow for an open system. *b.* Mass flow across an open-system boundary.

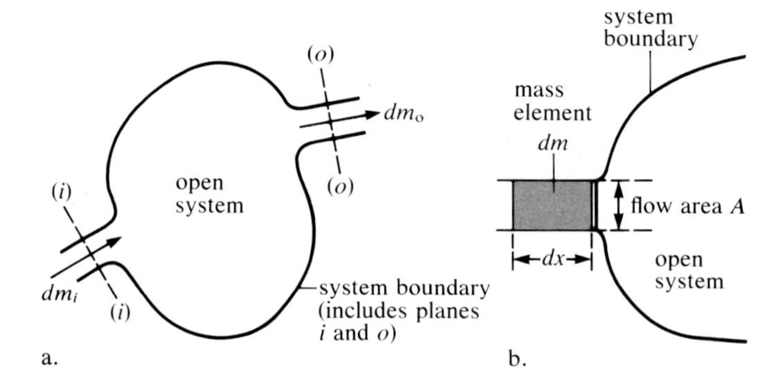


## 2–14 Conservation of Mass

The principle of conservation of mass states that mass cannot be created or destroyed.† We are interested in this fundamental concept in relation to thermodynamic systems. Application of this principle to a closed system results in the simple statement that the mass within the boundary is a constant, since for a closed system no mass crosses the boundary.

Application of the principle to an open system must account for the flow of mass across the system boundary, and the possible accumulation or depletion of mass within the system boundary.

Consider the open system shown in Figure 2.13a. The flow is not necessarily time steady. During a time interval $d\tau$ an element of mass $dm_i$ flows into the system by crossing the boundary at *(i)*. During the same interval an element of mass $dm_o$ flows out of the system

†Actually this statement has to be modified if the conversion between mass and energy through the well-known expression $E = mc^2$ is considered. However, such conversion is not addressed in this book.

at *(o)*. The mass increments $dm_i$ and $dm_o$ are not necessarily equal. Thus there can be an incremental change in the mass within the system. This is given by

$$dm_\sigma = dm_i - dm_o \tag{2.9}$$

where the symbol $\sigma$ is used to denote the system. An expression involving time rates can be formed as follows:

$$\frac{dm_\sigma}{d\tau} = \frac{dm_i}{d\tau} - \frac{dm_o}{d\tau}$$

or

$$\dot{m}_\sigma = \dot{m}_i - \dot{m}_o \tag{2.10}$$

The terms $\dot{m}_i$ and $\dot{m}_o$ are always positive and represent, respectively, the instantaneous mass rate of flow in at *(i)* and out at *(o)*. The term $\dot{m}_\sigma$ represents the corresponding instantaneous time rate of mass accumulation within the system. If $\dot{m}_i > \dot{m}_o$, the accumulation is positive. If $\dot{m}_i < \dot{m}_o$, the accumulation is negative, that is, a depletion of mass within the system is taking place at the instant in time considered. If the system has several $(n_i)$ inbound streams and several $(n_o)$ outbound streams, the mass conservation statement for the open system becomes

$$\dot{m}_\sigma = \sum_{i=1}^{n_i} \dot{m}_i - \sum_{o=1}^{n_o} \dot{m}_o \tag{2.11}$$

Equation (2.11) is generally applicable to open-system flows that are not time steady and will be referred to in subsequent discussion.

An important relation between mass rate of flow, area, specific volume, and velocity at a section of an open-system boundary across which mass flows is called the *continuity equation*. This relation can be developed as follows: Figure 2.13b shows an element of mass $dm$ that will move across the system boundary in time $d\tau$. The element has a volume $Adx$, which, when divided by the specific volume $v$, equals $dm$. The instantaneous mass rate of flow across the boundary is then

$$\frac{dm}{d\tau} = \frac{dx}{d\tau}\frac{A}{v}$$

This yields the continuity equation

$$\dot{m} = \frac{VA}{v} \tag{2.12}$$

where $V$ is the fluid velocity, which is assumed to be uniform across the area $A$. This velocity, often referred to as the bulk velocity, is the average velocity at the section where the specific volume is $v$ and the area $A$ that yields the mass rate of flow $\dot{m}$.‡ In all cases the direction of $V$ is normal to the plane establishing the section area $A$. Note that the product $VA$ is the *volume rate* of flow across the boundary. The continuity equation, (2.12), applies to any section of an open system across which either inbound or outbound mass flow occurs.

## Example 2.i

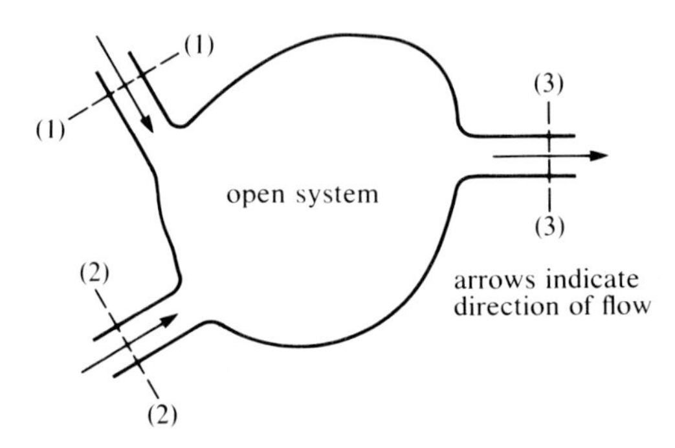


The following information is known at a particular instant in time for the flows at the numbered sections of the open system shown in the illustration.

| (1) | (2) | (3) |
|---|---|---|
| $\dot{m}_1 = 5.3$ lbm/sec | $V_2 = 38{,}000$ ft/hr | volume rate of flow |
| $v_1 = 9.6$ ft³/lbm | $A_2 = 0.40$ ft² | $= 2000$ ft³/min |
| $A_1 = 1.4$ ft² | $v_2 = 0.90$ ft³/lbm | $v_3 = 3.2$ ft³/lbm |

Find

(a) the flow velocity at (1) and (b) the rate of mass accumulation for the system.

**Solution:**

(a) Equation (2.12) applies. Solving for $V_1$

$$V_1 = \frac{\dot{m}_1 v_1}{A_1} = \frac{5.3\ \text{lbm}}{\text{sec}}\left|\frac{9.6\ \text{ft}^3}{\text{lbm}}\right|\frac{}{1.4\ \text{ft}^2} = 50.1\ \text{ft/sec}$$

‡A more general expression of the continuity equation is

$$\dot{m} = \int_A \frac{V dA}{v}$$

with velocity and specific volumes dependent on their location in the cross section, that is, $v = f(A)$ and $V = f(A)$.

(b) Equations (2.11) and (2.12) apply as follows:

$$\dot{m}_\sigma = \sum_{i=1}^{n_i} \dot{m}_i - \sum_{o=1}^{n_o} \dot{m}_o = \dot{m}_1 + \dot{m}_2 - \dot{m}_3$$

Using eq. (2.12) we have

$$\dot{m}_\sigma = \dot{m}_1 + \frac{V_2 A_2}{v_2} - \frac{V_3 A_3}{v_3}$$

where $V_3A_3$ is the volume rate of flow at (3). Choosing lbm/sec as the unit for the mass flow rates, substitution into the above equation yields

$$\dot{m}_\sigma = 5.3\,\frac{\text{lbm}}{\text{sec}} + \frac{38{,}000\ \text{ft}}{3600\ \text{sec}}\left|\frac{0.40\ \text{ft}^2}{}\right|\frac{\text{lbm}}{0.90\ \text{ft}^3}$$

$$-\frac{2{,}000\ \text{ft}^3}{60\ \text{sec}}\left|\frac{\text{lbm}}{3.2\ \text{ft}^3}\right. = -0.43\ \text{lbm/sec}$$

This answer indicates that at the instant in time for which the data are given, the mass within the system boundary is decreasing at the rate of 0.43 lbm/sec.

---

## 2–15 Steady Flow

In many applications of engineering thermodynamics, open systems that behave in a *steady-flow* manner can be identified. Steady flow is said to exist in an open system when all properties at each point within the system and on the system boundaries remain constant with respect to time. The properties referred to include intensive thermodynamic properties as well as those associated with the flow, such as velocity and elevation. In particular, the following conditions exist for steady flow.

1. At any given section of the boundary where working fluid enters the system, its chemical composition, state, and flow properties are uniform across the section and do not change with time. At other entering sections the composition, state, and flow properties can be different, but they do not vary with time. Similar statements apply to sections of the boundary where the working fluid leaves the system. One important consequence of this is obtained by application of the continuity expression, eq. (2.12). At any section of the boundary across which mass flows, the mass rate of flow $\dot{m}$ is constant since $V$, $A$, and $v$ in the continuity equation are not time dependent.

2. The properties of the working fluid may differ from point to point within a steady-flow system, but at any given point the properties do not vary with time. In general, properties of the working fluid change as it passes through a steady-flow system. The purpose of many steady-flow systems is to alter one or more of the properties of the working fluid.
3. The summation of the mass rates of flow for all streams entering the system is equal to the summation of the mass rates of flow for all streams leaving the system. This can be established by examining eq. (2.11) for steady flow. For steady flow $\dot{m}_\sigma = 0$, that is, the amount of mass within the steady open system boundary is neither increasing nor decreasing, although the identity of the mass inside the boundary does change with time. From this result eq. (2.11) becomes

$$\sum_{i=1}^{n_i} \dot{m}_i = \sum_{o=1}^{n_o} \dot{m}_o \tag{2.13}$$

where $n_i$ is the number of inbound streams and $n_o$ is the number of outbound streams.
4. Work and heat interactions with the surroundings take place at constant rates, that is, at rates that do not vary with time.

Numerous examples of steady flow will be encountered in our study of engineering thermodynamics. This occurs because many systems are designed to operate in a continuous manner; after a transient period of start-up, the conditions of steady flow are reached. Consider, for example, a stationary gas turbine, such as that described in Figure 2.1c, which is designed to produce mechanical power at a fixed level. After start-up, a transient period exists during which the internal parts eventually reach steady operating temperatures; constant rates of flow for the fuel, air, and exhaust gases are established; and the design power level is reached. The system then operates in a steady-flow manner.

Occasionally we encounter systems that are, in the strictest sense, not steady-flow systems, but that can with good accuracy be treated as such. For example, consider a reciprocating gas compressor, a device that employs mechanical elements such as pistons, cylinders, and valves to compress the flowing gas. The flow through such a device during normal operation may be pulsating flow and therefore not steady. However, if the properties at points within such a system vary with time in a periodic manner (i.e., states are reproduced at fixed time intervals), then the system may be treated as a steady-flow system, provided time-average values are used for all of the pertinent quantities.

---

## Example 2.j

Let the open system described in the figure for Example 2.i be a steady-flow system. Find the volume rate of flow at (3) if the conditions at the numbered sections are as follows:

| (1) | (2) | (3) |
|---|---|---|
| $\dot{m} = 12.0 \text{ kg/s}$ | $V_2 = 3.2 \text{ m/s}$ | $v_3 = 0.20 \text{ m}^3/\text{kg}$ |
| $v_1 = 0.6 \text{ m}^3/\text{kg}$ | $A_2 = 0.04 \text{ m}^2$ | |
| $A_1 = 0.13 \text{ m}^2$ | $\rho_2 = 17.8 \text{ kg/m}^3$ | |

---

**Solution:**

Since the flow is steady, eq. (2.13) applies and

$$\dot{m}_1 + \dot{m}_2 = \dot{m}_3$$

In applying eq. (2.12) to section (2) we get

$$\dot{m}_3 = \dot{m}_1 + \rho_2 V_2 A_2 = 12 \text{ kg/s} + \frac{17.8 \text{ kg}}{\text{m}^3} \left| \frac{3.2 \text{ m}}{\text{s}} \right| 0.04 \text{ m}^2$$

$$= 2.28 \text{ kg/s}.$$

The volume rate of flow at (3) is, from eq. (2.12),

$$V_3 A_3 = \dot{m} v_3 = [2.28 \text{ kg/s}][0.20 \text{ m}^3/\text{kg}] = 0.456 \text{ m}^3/\text{s}$$

---

### 2–16 Work for Frictionless Steady-Flow Processes

In Section 2–9 the work for frictionless quasi-static closed-system compression and expansion processes was discussed. It is useful at this point to consider, as a parallel to this closed-system work, the work for *frictionless steady-flow* systems.

Consider a steady-flow frictionless system with one inbound stream and one outbound stream as illustrated in Figure 2.14a. Consider also an element of mass flowing through the system. In general the element will simultaneously undergo an elevation change, a velocity change, and an expansion or a compression process. It is assumed that as the element flows through the system, it passes through equilibrium states described by a well-defined continuous path such as that illustrated in Figure 2.14b. Also shaft work or its equivalent in general will be involved.

There are several ways to develop the expression for the work for such a process. However, these are relatively complicated and it will serve our purposes at this point to focus on the

**Figure 2.14** The Frictionless Steady-Flow Process. *a.* Frictionless steady-flow system. *b.* A path for a frictionless steady-flow process.

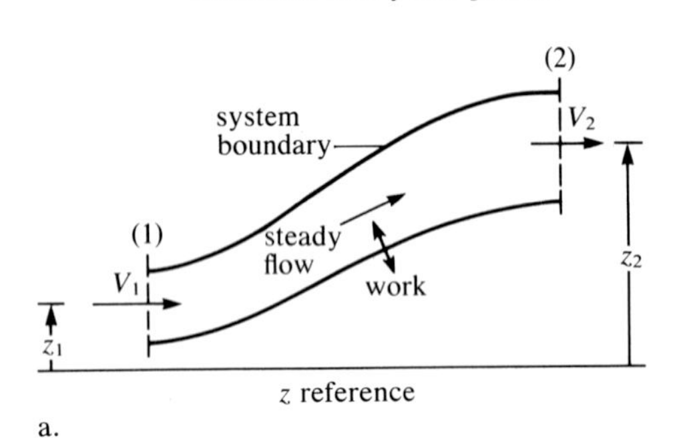

a.

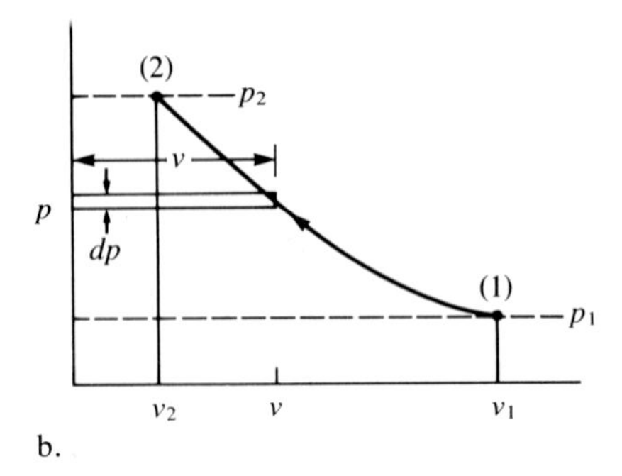

b.

result. (A formal development based on concepts from the second law of thermodynamics will be presented in Section 7–5.) The expression obtained is

$$_1w_2 = -\int_1^2 v\,dp - \frac{V_2^2 - V_1^2}{2g_0} - \frac{g}{g_0}(z_2 - z_1) \tag{2.14}$$

where $_1w_2$ is the work per unit mass for the process. It is evident from examination of the integral in eq. (2.14) and the *p-v* diagram in Figure 2.14b that the absolute value of the integral is represented by the area to the left of the path that describes the variation of $p$ with $v$ for the process. The second and third terms on the right side of eq. (2.14) represent respectively the changes in kinetic and potential energy per unit mass of the working fluid as it passes through the steady-flow system. Only in the absence of kinetic and potential energy changes does the integral in eq. (2.14) represent all of the work.

Equation (2.14) is valid for any steady-flow system involving a frictionless process and one inbound and one outbound stream, including those processes with heat transfer. The work is typically shaft work. The important case of zero work is illustrated in an example at the end of this section.

For *any* steady-flow system with one inbound and one outbound stream, the power $\dot{W}$ for the system is given by the product of the mass rate of flow $\dot{m}$ and the work per unit mass. Thus

$$\dot{W} = \dot{m}w$$

## Example 2.k

A nozzle is a varying area passage used to increase the velocity of a fluid passing through it. Suppose a nozzle, operating in a frictionless, steady-flow manner, has the following characteristics:

| | **Inlet** | **Outlet** |
|---|---|---|
| pressure | 120 psia | 15 psia |
| specific volume | 3.0 ft³/lbm | |
| velocity | 450 ft/sec | |

During the expansion through the nozzle the pressure–specific volume relation $pv^{1.372} =$ constant is maintained. Changes in mechanical potential energy may be neglected, as is frequently done, because the magnitude of such changes is trivial in comparison with other terms. Additionally, there is no work during a nozzle process. On the basis of these conditions, calculate the mass rate of flow, in lbm/min, if the cross-sectional area at the outlet is 1.749 in².

---

**Solution:**

From the continuity equation, eq. (2.12), the mass rate of flow at the outlet, section 2, will be

$$\dot{m} = \frac{A_2 V_2}{v_2}$$

which, with $A_2$ known, requires evaluation of the velocity and specific volume at the outlet. In utilizing the $p$-$v$ relation for the process, we have

$$v_2 = \left(\frac{p_1}{p_2}\right)^{1/1.372} (v_1) = 13.65 \text{ ft}^3/\text{lbm}$$

Applying eq. (2.14) to this problem, we get

$$0 = \int_1^2 v\,dp + \int_1^2 \frac{V\,dV}{g_0}$$

with the limits 1 and 2 indicating inlet and outlet respectively. Noting that $p_1 v_1{}^n = p_2 v_2{}^n = pv^n = \text{constant} = C$ with $n = 1.372$ for the process we have

$$v = \left(\frac{C}{p}\right)^{1/n}$$

and the steady-flow equation becomes

$$0 = C^{1/n} \int_1^2 \frac{dp}{p^{1/n}} + \int_1^2 \frac{V\,dV}{g_0}$$

Integrating, the preceding equation becomes

$$0 = C^{1/n} \left[\frac{p^{-(1/n)+1}}{-\dfrac{1}{n} + 1}\right]_1^2 + \frac{V_2^2 - V_1^2}{2g_0}$$

When substituting limits for the first term choose values of the constant $C$ that contain the limit. Thus,

$$0 = \frac{p_2^{1/n} v_2 p_2^{-(1/n)+1} - p_1^{1/n} v_1 p_1^{-(1/n)+1}}{-\dfrac{1}{n} + 1} + \frac{V_2^2 - V_1^2}{2g_0}$$

or

$$0 = \frac{n}{n-1}[p_2 v_2 - p_1 v_1] + \frac{V_2^2 - V_1^2}{2g_0}$$

Substituting numerical values

$$0 = \frac{1.372}{0.372}[(15)(144)(13.65) - (120)(144)(3)] + \frac{V_2^2 - 450^2}{(2)(32.17)}$$

which yields

$$V_2 = 2347 \text{ ft/sec}$$

From the continuity equation written at section 2

$$\dot{m} = \frac{1.749}{144}\bigg|\frac{2347}{13.65}\bigg|\frac{60}{\ }$$

$$= 125.3 \text{ lbm/min}$$

---

## Example 2.1

A working fluid enters a turbine at 1380 kPa, specific volume 0.106 m³/kg, with negligible velocity. Expansion through the turbine—a prime mover or power-producing unit—is frictionless and steady flow with $pv^{1.312}$ = constant = $C$. At the turbine outlet (exhaust) the pressure is 103.4 kPa and the area 0.128 m². The change in mechanical potential energy through the turbine is negligible. Determine the power, in kilowatts, delivered by the unit when the flow rate is 45,450 kg/hr.

**Solution:**

Let us assume a basis of 1 kg for initial calculations. On this basis, using eq. (2.14),

$$-w = \int_1^2 v\,dp + (V_2^2 - V_1^2)/2g_0$$

with 1 and 2 as the inlet and outlet conditions. Following the integration procedures of the preceding example we get

$$-w = \frac{n}{n-1}\,[p_2v_2 - p_1v_1] + \frac{V_2^2}{2g_0}$$

for the work per kilogram. The specific volume $v_2$ is

$$v_2 = \left(\frac{1380}{103.4}\right)^{1/1.312}(0.106) = 0.764 \text{ m}^3/\text{kg}$$

The exhaust velocity $V_2$ is determined using the continuity equation at that section

$$V_2 = \dot{m}v_2/A_2$$

$$= \frac{45{,}450 \text{ kg}}{3600 \text{ s}}\left|\frac{0.764 \text{ m}^3}{\text{kg}}\right|\frac{}{0.128 \text{ m}^2} = 73.36 \text{ m/s}$$

Thus the work per kilogram is

$$_1w_2 = -\frac{1.312}{1.312-1}\left[\frac{103.4\times10^3 \text{ N}}{\text{m}^2}\left|\frac{0.764 \text{ m}^3}{\text{kg}}\right.\right.$$

$$\left.-\frac{1380\times10^3 \text{ N}}{\text{m}^2}\left|\frac{0.106 \text{ m}^3}{\text{kg}}\right.\right]$$

$$-\frac{(73.36)^2 \text{ m}^2}{2 \quad \text{s}^2}\left|\frac{\text{N s}^2}{1 \text{ kg m}}\right.$$

$$= 280{,}240 \text{ N m/kg} = 280.24 \text{ kJ/kg}$$

The power delivered is

$$\dot{W} = {}_1w_2\dot{m} = \frac{280.24 \text{ kJ}}{\text{kg}}\left|\frac{45{,}450 \text{ kg}}{\text{hr}}\right|\frac{\text{hr}}{3600 \text{ s}}$$

$$= 3538 \text{ kJ/s} = 3538 \text{ kW}$$

## 2-17 Summary

Engineering thermodynamics deals with systems, properties of systems, and interactions between systems and their surroundings. A system is a region or a fixed quantity of matter separated from its surroundings by a boundary. Mass does not cross the boundary of a closed system; hence the mass of a closed system is of fixed identity. An open system is a region that has mass crossing its boundary. An isolated system is one that does not interact in any way with its surroundings.

A pure substance is matter that is homogeneous and of fixed chemical composition, and it may exist in one or more of the following phases: solid, liquid, or vapor (gas). Properties of a system are quantities that describe the state (condition) of a system. An equilibrium state exists when a spontaneous change in state cannot occur. An extensive property is a quantity that depends on the extent—mass or volume—of a system. Intensive properties are single valued at a particular point in a system. Unique values for intensive properties of a system can be specified only when the system is in an equilibrium state. A simple system is one whose working fluid is a pure substance and is not influenced by motion, surface tension, or gravitational or electric and magnetic field effects. Two independent intensive properties establish the state of a simple system.

When a change in state of a system takes place, a process is said to have occurred. A quasi-static process is one in which departures from equilibrium are infinitesimal. The series of states through which a system passes during a process is called a path.

Work is an interaction between a system and its surroundings. Work is done by a system if the sole effect external to the system could be the lifting of a weight. Mechanical work is attributable to the action of a force on a moving boundary of a system. Work done by a system is positive; work done on a system is negative. For a frictionless quasi-static process occurring in a simple closed system, the work per unit mass is given by

$$ {}_1w_2 = \int_1^2 p\,dv $$

Work for this process is path dependent, that is, it depends on how $p$ varies with $v$.

Two bodies are said to have equality of temperature if all properties of each body remain unchanged when the bodies, collectively isolated from their surroundings, are brought into thermal communication with each other. Two systems are in thermal equilibrium with each other when they have equality of temperature. The zeroth law states that two systems that have equality of temperature with a third have equality with each other. The constant-volume gas thermometer scale of temperature provides an operational definition of temperature. Absolute temperature based on this scale is

$$ T = \frac{\theta_s - \theta_i}{(p_s/p_i) - 1}(p/p_i) $$

where the subscripts $s$ and $i$ denote the steam and ice points respectively.

An interaction between a system and its surroundings that is due solely to a difference in temperature between the system and its surroundings is the heat interaction. Heat transfer occurs during the heat interaction. Heat transfer $Q$ is positive when heat transfer is to the system and negative when it is from the system. Since heat is an interaction between a system and its surroundings, it is not a property of a system.

The principle of mass conservation applied to an open system is written as

$$\dot{m}_\sigma = \sum_{i=1}^{n_i} \dot{m}_i - \sum_{o=1}^{n_o} \dot{m}_o$$

where $\dot{m}$ is the mass rate of flow, the subscript $i$ represents the inbound streams, the subscript $o$ denotes the outbound streams, and $\dot{m}_\sigma$ denotes the time rate of change of mass within the system boundary. At any given section of the boundary where mass crosses

$$\dot{m} = \frac{VA}{v}$$

where $A$ is the cross-sectional area, $V$ is the average velocity across the section, and $v$ is the specific volume of the working fluid. Steady flow exists in an open system when all properties at each point within the system and on the system boundary are independent of time. The statement of mass conservation for a steady-flow system is

$$\sum_{i=1}^{n_i} \dot{m}_i = \sum_{o=1}^{n_o} \dot{m}_o$$

For frictionless steady flow through an open system with one inbound and one outbound stream, the work per unit mass is

$${}_1w_2 = -\int_1^2 v\,dp - \frac{V_2^2 - V_1^2}{2g_0} - \frac{g}{g_0}(z_2 - z_1)$$

## Selected References

Carli, A., and A. Favaro. *Bibliografia Galileiana,* 1896.
Fahrenheit, Gabriel D. *Philosophical Transactions* (German), 1724.
Jones, J. B., and G. A. Hawkins. *Engineering Thermodynamics.* New York: Wiley, 1960.
Keenan, J. H. *Thermodynamics.* New York: Wiley, 1941 (Reprint by M.I.T. Press).
Keenan, J. H. "Adventure in Science." *Mechanical Engineering,* May 1958, p. 79–83.
Poincaré, Henri. *La Valeur de la Science,* 1904.
Poincaré, Henri. *Science et Methode,* 1908.
Stoever, H. J. *Engineering Thermodynamics.* New York: Wiley, 1951.
Zemansky, M. W., and R. H. Dittman. *Heat and Thermodynamics,* 6th ed. New York: McGraw-Hill, 1981.
Zemansky, M. W., M. M. Abbott, and H. C. VanNess. *Basic Engineering Thermodynamics,* 2nd ed. New York: McGraw-Hill, 1975.

## Problems

2.1 One of the systems of units used in the United States and elsewhere employs the following units. Force, the pound force; mass, the slug; length, the foot; time, the second. These quantities are related by the following definition: 1 pound force is the force that gives to a mass of 1 slug an acceleration of 1.0 foot per second squared. Determine $g_0$ for this system of units. How many pounds mass are there in a slug?

2.2 A body with a mass of 23.00 lbm is in a location where the acceleration of gravity is 29.36 ft/sec². Calculate the force required to give the body a horizontal acceleration of 50 ft/sec².

2.3 The acceleration of gravity at the surface of the earth's moon is 1.58 m/s². An astronaut is to carry equipment that will subject him to a force (load) not to exceed 200 N. What is the maximum mass of the equipment to remain within the specified load limit?

2.4 In a certain location a mass of 165 pounds is weighed using a spring scale that has been calibrated in a standard gravity location. The spring scale reads 158 pounds when the mass is placed on it. Determine the local acceleration of gravity.

2.5 A spring with a modulus of 9 N/cm has one end fixed and its axis vertical. A mass of 4.5 kg is attached to the bottom of the spring. If the local acceleration of gravity is 4.2 m/s², what is the spring extension?

2.6 A certain material has a density of 50 lbm/ft³. If the material is a liquid, what manometer column height of the liquid would correspond to a pressure difference of 8.50 psi in a region where the local acceleration of gravity is 12.95 ft/sec²?

2.7 One cubic foot of a material weighs (on a spring scale) 42 pounds in a region where the local acceleration of gravity is 27.00 ft/sec². Calculate the density of the material in lbm/ft³. If the material is a liquid, what manometer column height of the liquid would be equivalent to a pressure of 4.13 psi in a region where local gravity is 5.19 ft/sec²?

2.8 Assume liquid $H_2O$ has a density of 1000 kg/m³. What is the pressure equivalent, in kPa, of a manometer column of water 10 cm high in a location having standard, sea-level gravitational acceleration?

2.9 Repeat problem 2.8 for liquid mercury with a specific gravity of 13.60.

2.10 Assuming the barometric pressure is 14.53 psia and the local acceleration of gravity is 32.17 ft/sec², calculate the absolute pressure in psia of the following:

(a) A vacuum of 10 inches of mercury
(b) A gage pressure of 153.0 psig
(c) A gage pressure equivalent to a manometer column height of 3.54 inches of liquid water

2.11 A cylindrical tank with a vertical axis is filled to a depth of 15 meters with a liquid having a specific gravity of 0.943. A pressure gage is attached to the bottom of the tank. What gage pressure, kPa, will the gage show? What additional information would be needed to calculate the absolute pressure at the bottom of the tank? The top of the tank is open to the atmosphere.

2.12 Seawater has an average density of 64.0 lbm/ft³. A submersible vessel is to be designed for operation at depths of up to 600 feet in seawater. Assuming a design barometric pressure of 29.92 inches Hg, what is the maximum pressure to which the submersible vessel will be subjected?

2.13 A truck has a gross weight of 73,000 lbf. The truck is moving up a hill that has a rise of 5 feet per 100 feet of highway. Resistances due to friction, drag, and so on are 20 lbf/ton. If the truck is to maintain a constant speed of 45 mph, what horsepower is required?

2.14 An exploratory oil well 10 cm in diameter is drilled to a depth of 3660 meters through earth having an average density of 2355 kg/m³. Calculate the work required to bring the drilled earth to the surface. State assumptions made in solving the problem.

2.15 Suppose the pressure and temperature of the atmosphere at sea level are 14.696 psia and 70 F respectively, that the temperature at an altitude of 30,000 feet is − 48 F, and that the temperature varies linearly with altitude. If the density of the atmosphere is

$$\rho = 2.702 \frac{p}{T}$$

where $p$ is the pressure in psia and $T$ the Rankine temperature, what is the atmospheric pressure at an altitude of 30,000 feet above sea level?

2.16 A closed system undergoes a cycle consisting of three frictionless, quasi-static processes. The first process takes place at a constant pressure of 100 kPa from an initial volume of 0.670 m³ to a final volume of 0.142 m³. The next process takes place at constant volume to a pressure of 500 kPa. The cycle is closed with a process that has a straight-line path on $p$-$v$ coordinates. Calculate the net work for the cycle.

2.17 A closed system undergoes a quasi-static and frictionless process along the path $p = 3.75v + 10$ where $p$ is in psia and $v$ is in ft³/lbm. Initially the system is at 15 psia. During the frictionless, quasi-static process the system does work in the amount of 53,760 ft lbf/lbm. Determine the pressure at the final state.

2.18 A closed system consisting of 1 kg of a substance undergoes a frictionless, quasi-static process along the path $p = 96 + 12.30v^2$ where $p$ is in kPa and $v$ is in m³/kg. For an initial pressure of 200 kPa and a final pressure of 105 kPa (a) sketch the $p$-$v$ diagram and (b) calculate work for the process.

2.19 A closed system undergoes a process along the path $pv^{1.376} = C$ from an initial state at 15 psia, 12.78 ft³/lbm, to a final state where the pressure is 127 psia. Calculate the work for the frictionless, quasi-static process. Determine the slope of the path on $p$-$v$ coordinates at a state where the pressure is 57 psia.

2.20 A closed system of 0.123 kg undergoes a frictionless, quasi-static process along the path $pV = C$ from 100 kPa to a final state where the volume is twice the initial volume. During the process 8140 Nm of work is done by the system. What is the final specific volume of the system?

2.21 A closed system initially at 15 psia and 13.08 ft³/lbm undergoes a frictionless, quasi-static process. During the process 16,600 ft lbf/lbm of work is done on the system. The process takes place along the path $pv^n = C$ where the exponent $n$ is 1.45. Determine the final pressure and the final specific volume for the system.

2.22 A piston–cylinder assembly is arranged so that the centerline of the piston is vertical. The piston has a diameter of 0.25 m and the cylinder volume occupied by a gas is initially 0.050 m³. The gas expands in a quasi-static constant pressure process to lift the piston 0.20 m. The pressure in the cylinder is 125 kPa and the surrounding atmospheric pressure is 95 kPa. The local acceleration of gravity is 9.75 m/s². What is the mass of the piston? What fraction of the work produced by the expanding gas goes into lifting the weight?

2.23 Is it possible to have a process of closed system during which the $\int pdv = 0$ but the work is not zero? Explain. It is possible to generate a series of equilibrium states that can be plotted on $p$-$v$ coordinates with a smooth curve drawn through the several points. Could the $\int pdv$ be evaluated between the initial and final points on such a curve? Describe a situation satisfying these conditions but for which the $\int pdv$ would not be equal to work.

2.24 On a certain constant volume gas thermometer temperature scale the ice point is $-22°$ and the steam point $89°$. What is the steam-point temperature on an absolute gas scale having the same number of degrees between the steam point and ice point?

2.25 A Fahrenheit and a Celsius thermometer are placed in a constant-temperature bath. After reaching thermal equilibrium with the bath, both thermometers indicate the same numerical value. What is the bath temperature on (a) the Fahrenheit scale and (b) the Celsius scale? Would it be possible to achieve identical numerical readings with Rankine and Kelvin scale thermometers?

2.26 It is proposed that a constant volume gas thermometer temperature scale be established on which the boiling point of benzene at 1 atmosphere would be set equal to 100 Z (Z being the new temperature unit) and the temperature of boiling sulfur dioxide at 1 atmosphere be set equal to $-50$ Z. The following are known:

Boiling point of benzene at 1 atmosphere = 80.1 C.
Boiling point of sulfur dioxide at 1 atmosphere = $-3.78$ C.

(a) What constant would have to be added to Z scale temperatures to produce an absolute gas scale having the same fixed point interval as the Z scale?
(b) What is the boiling point of $H_2O$ at 1 atmosphere expressed on the Z scale?

2.27 Steam flows through a turbine at the rate of 900,000 lb/hr. At the turbine exhaust the steam velocity is 400 ft/sec and its specific volume 398.0 $ft^3$/lbm. The exhaust is circular in cross section. Calculate the required exhaust diameter in inches for steady flow.

2.28 Water is supplied to a pump at 20 cm Hg vacuum through a 15.2-cm inside-diameter line with a velocity of 183 m/min. The pump discharges the water at 136 kPa gage. The pump outlet, also 15.2 cm in diameter, is 0.6 m above the inlet (centerline to centerline). Assuming that the pump operates in a frictionless steady-flow manner, calculate the power in kilowatts required to drive the pump. The density of water is 1000 $kg/m^3$.

2.29 In a refinery iso-octane (specific gravity 0.702) enters a pump at the rate of 500 gals/min. The pump discharge line is located 36 ft above the pump inlet. Through the pump the pressure increases 12.05 psi. Assuming frictionless operation, calculate the power in kilowatts required to drive the pump.

2.30 Two streams enter a vessel, mix, and leave as a single stream in steady flow. The first entering stream flows through a line of 23 cm inside diameter with a velocity of 150 m/s and a density of 1.60 $kg/m^3$. The second entering stream enters through a line of 0.032 $m^2$ area at the rate of 27,300 kg/hr and has a specific volume of 0.521 $m^3$/kg. At the outlet the velocity is 107 m/s and the specific volume 0.44 $m^3$/kg. Calculate (a) the velocity of the second entering stream and (b) the mass rate of flow at the outlet.

2.31 An air compressor operates in steady flow taking in 500 $ft^3$/min at 14 psia. Discharge from the frictionless compressor is at 120 psia. Intake and discharge velocities can be neglected. If the compressor operates with $pv = C$, what horsepower is required to drive the compressor?

2.32 Solve problem 2.31 for compression following the path $pv^{1.30} = C$. Compare the result with that for problem 2.31 utilizing areas on a $p$-$v$ diagram.

2.33 A nozzle is a varying area passage used to produce an increase in velocity as the pressure decreases during steady flow through the passage. Suppose a gas entered a nozzle at 680 kPa, specific volume 0.25 $m^3/kg$, with a velocity of 61 m/s, and expanded to 136 kPa at the nozzle outlet. The expansion is frictionless, steady flow with $pv^{1.67} = C$. What outlet area would be required for a flow rate of 0.62 kg/s?

2.34 A certain substance undergoes a frictionless, steady-flow expansion through a nozzle. At the nozzle inlet section the velocity is 600 ft/sec, the pressure is 100 psia, and the specific volume is 2.736 $ft^3/lbm$. During the expansion through the nozzle the quantity $pv^{1.400}$ remains constant. At the nozzle outlet section the velocity is 1906 ft/sec. Calculate the pressure in psia at the nozzle outlet.

2.35 A turbine operates in steady flow with fluid entering the unit at 1.379 MPa, specific volume 0.312 $m^3/kg$. At the turbine exhaust the pressure is 105 kPa and the velocity is 150 m/s. Expansion through the machine follows $pv^{1.40} = C$. Assuming the turbine delivers 746 kW, calculate (a) the mass rate of flow, kg/hr and (b) the exhaust area, $m^2$. Assume frictionless operation. The inlet velocity is negligible.

2.36 The working fluid enters a steady-flow device with a velocity of 100 ft/s, specific volume 5.27 $ft^3/lbm$, and pressure 100 psia. During the frictionless process 30,000 ft lbf/lbm of work is done by the device. The fluid leaves the unit at a velocity of 300 ft/sec. Assuming the process follows $pv^{1.40} = C$, calculate the final pressure, psia.

2.37 A fan takes in air from the atmosphere at 14.0 psia, 69 F with a density of 0.0715 $lbm/ft^3$, at the rate of 6,500 $ft^3/min$. The fan discharges into a duct of 2.87 $ft^2$ cross-sectional area at a gage pressure equivalent to 3.75 in. $H_2O$. Calculate the horsepower required (a) to produce the kinetic energy change and (b) the pressure rise. Assume the specific volume of the air remains constant and that the process is frictionless.

# 3

# The First Law of Thermodynamics

*In this chapter we shall build on the concepts developed in the preceding chapter by stating the first law of thermodynamics and then establishing an operational definition of stored energy. A definition of the term "energy" is presented. The general energy equations for closed and open systems, applicable to processes for all substances including those in chemically reactive systems, follow from the first law. The parameters of thermal efficiency, associated with conversion of heat to work, and coefficient performance, associated with refrigeration and heat pump cycles, are introduced. Finally, a sampling of other statements of the first law is presented with commentary on the variety of presentations.*

## 3–1 The First Law of Thermodynamics

In the preceding chapter the terms "cycle," "work," and "heat" were defined. Thus, in a form similar to that used by Clausius in 1850, we can state the first law of thermodynamics using these terms as follows: *When a closed system undergoes a cycle, the net work and the net heat are proportional.* The sign conventions adopted for heat and work apply to this statement. If, for example, the net work for a cycle is positive, then according to the first law, the net heat transfer is also positive. Similarly, if the net work is negative, then the net heat transfer is negative. Cycles that have net work done by the system (positive work) are power cycles. Refrigeration and heat pump cycles are examples with negative net work quantities. Net work must be supplied to such cycles.

A mathematical statement of the first law is

$$\oint \delta Q \propto \oint \delta W \tag{3.1}$$

for all closed systems. The value of the proportionality factor depends only on the units of heat and work.

It should be noted that the first law of thermodynamics is empirical in character. Numerous experiments have been conducted to verify the first law by either direct or indirect means, and all such experiments have firmly supported it. This law has been accepted as a far-reaching fundamental principle and cannot be deduced or established from other fundamental principles.

As an illustration of eq. (3.1), consider the apparatus and the closed system cycle described in Figure 3.1a. The system boundaries are the surfaces that enclose the working fluid, a gas, in the cylinder. The system is initially in thermal equilibrium with its surroundings. The piston travel is limited by the stops shown. For simplicity, let the piston be weightless. The cycle begins with the piston down and the system at state (1) on the $p$-$v$ diagram. Let

**Figure 3.1** Cycles for Closed Systems. *a.* Closed-system processes. *b.* Steady-flow processes.

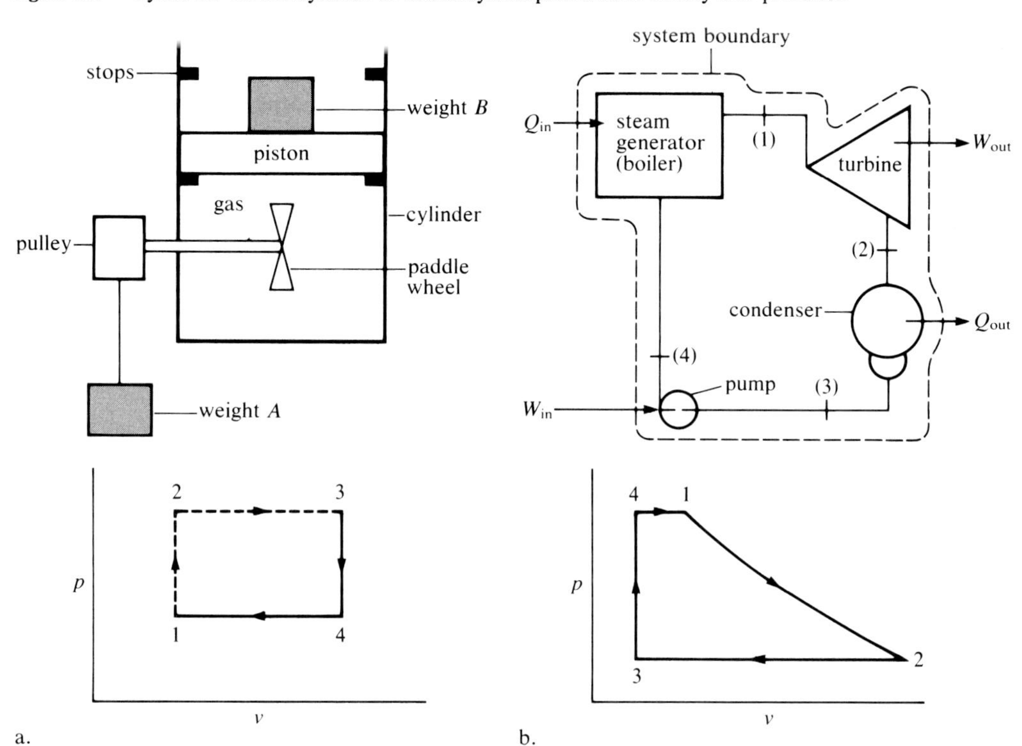


the system be initially thermally insulated and allow weight $A$ to fall. This will cause the paddle wheel to do work on the working fluid, thereby increasing its pressure and temperature. The pressure will increase at constant volume (1) to (2) on the $p$-$v$ diagram. At state (2) a force equal to $p_2$ times the piston face area will just balance the piston and weight $B$. (We temporarily interrupt the fall of weight $A$ to allow the equilibrium state (2) to be established.) As previously noted, during the period in which paddle wheel work is being done on the system, the system is not in a state of equilibrium. This is indicated by the dashed path from (1) to (2). As weight $A$ continues to fall, the system volume increases along the path from (2) to (3), again in a nonequilibrium process, causing the piston to reach the upper stop, at which point the falling of weight $A$ is terminated and weight $B$ is moved horizontally off the piston. During the process (2) to (3) work is done on the system by the paddle wheel and the system does work in lifting the piston and weight $B$ and by pushing back the surrounding atmosphere. The cycle is completed by removing the insulation from the system and allowing heat transfer to take place to the lower temperature surroundings while the paddle wheel is inactive. This will occur first along the constant volume path (3) to (4) and then along the constant pressure path (4) to (1), during which work is done on the system by the surrounding atmosphere as the volume decreases. These two processes are essentially quasi-static and frictionless; the system traces a series of equilibrium states as indicated by the solid lines joining the states. The cycle could be repeated after placing another weight $B$ on the piston.

Let us now examine the cycle in light of the first law. Equation (3.1) states that the algebraic sums of work and heat are proportional for the cycle. Thus the sum of the negative work of the paddle wheel performed during the process (1) to (2) to (3), the positive work done in lifting weight $B$ and pushing back the atmosphere, and the negative work involved with the process (4) to (1) must be proportional to the heat transfer for the process (3) to (4) to (1), which as noted above is negative. (Recall that the process (1) to (2) to (3) is adiabatic.) Thus the net work is negative. Note that the processes in which the paddle wheel is active involve friction in the working fluid due to the relative motion of adjacent fluid elements, while the heat transfer processes are essentially quasi-static and frictionless. This cycle has little practical value. However, an important point of this illustration is that the first law is independent of the nature of the processes that make up the cycle. The first law is also independent of the working fluid under consideration.

Although eq. (3.1) is stated for a closed system, the cyclic processes inside the closed-system boundary may be either closed or steady-flow. This is illustrated in Figure 3.1b, which shows four components of a simple power plant cycle that uses water as the working fluid. (See Figure 1.2 also.) Each component operates in a steady-flow manner. The system chosen has only heat and work crossing its boundaries and the working fluid undergoes a cycle as shown in the simplified $p$-$v$ diagram in the figure. While we have not yet discussed property relationships for substances, we can trace the path for the working fluid as it undergoes the cycle. High-pressure steam at (1) expands through the turbine to a low pressure at (2) producing work, after which the expanded steam is condensed to a liquid by cooling it in the constant-pressure process (2) to (3). The specific volume decreases greatly during this process. Liquid, for which specific volume does not change greatly with pressure, is pumped from (3) to the steam-generator pressure at (4), after which it is vaporized at constant pressure by heat-transfer processes and delivered to the turbine at (1) completing the cycle. Work is required to pump the liquid from $p_3$ to $p_4$. We assume that the processes (1) to (2) and (3) to (4) are adiabatic. Only heat transfer is involved in the processes (2) to (3) and (4) to (1). Under these conditions the first law requires that the algebraic sum of the turbine work and the pump work be proportional to the algebraic sum of the heat transfer for the processes (4) to (1) (a positive value) and (2) to (3) (a negative value). Since a power plant produces work, the net heat transfer will be positive, that is, $Q_{in}$ is greater in magnitude than $Q_{out}$.

## 3–2 The Proportionality of Heat and Work Units

The possibility that heat and work might be different forms of the same "thing" perplexed scientists for more than a century. This confusion is understandable when one recognizes that the "caloric theory of heat" was in vogue until approximately the middle of the 19th century.

Caloric was thought to be a weightless substance that passed from high-temperature bodies, for example, flames, and penetrated bodies at lower temperatures. Bodies at higher temperatures supposedly contained more caloric than those at lower temperatures. When two bodies at different temperatures were in contact, the higher-temperature body was presumed to lose caloric to the lower-temperature body until each reached some intermediate temperature.

John Locke, in 1722, observed that "the axle trees of carts and coaches are often made hot, sometimes to a degree that it sets them on fire, by the rubbing of the naves of the wheels upon them." This observation was made with regard to systems that were initially at the same temperature, and that were not subjected to caloric "penetration" from a flame or high-temperature body, but were affected only by the nave (hub) of the wheel rubbing on the axle.

Count von Rumford (Sir Benjamin Thompson) also was intrigued by friction effects and noted "there seems to be an inexhaustible source of heat from friction in boring brass cannons, thus heat cannot be a material substance." In his efforts to relate work and heat, he noted, in 1798, that 2 horses (the power source for the boring bars) boiled off 19 pounds of water in a period of 2 hours and 20 minutes. His paper "Enquiry Concerning the Source of Heat which is Excited by Friction" was presented in 1798.

James Prescott Joule was the person who, in the 1840s, convincingly disproved the caloric theory and established the proportionality of work and heat units. He used a number of different experimental systems in his studies and recognized the need for extremely accurate instrumentation. The unit of heat had for some time been defined as either the calorie (the amount of heat necessary to change the temperature of 1 gram of water from 14.5 to 15.5 C at constant atmospheric pressure) or the British thermal unit, B (the amount of heat required to raise the temperature of 1 pound mass of water from 63 to 64 F at constant atmospheric pressure). These definitions are often called the calorimetric definitions of the two heat units. The unit of work was defined on a mechanical basis as presented in Chapter 2.

With these independent definitions for heat and work units it is evident that any evaluation of their relation was subject to experimental error, with the accepted value for the proportionality factor changing as experimental techniques became more sophisticated. Joule's original work in determining the proportionality factor was not precise by today's standards. He eventually selected a technique using paddle wheels rotating in water. His work quantities were established in terms of a falling weight and the distance through which it moved. Water temperatures were observed using thermometers of 0.005 F interval. On the basis of experiments of this nature, Joule adopted, in 1847, the relation Work = 781.5 ft lbf = 1 British thermal unit = Heat. The water temperature rise in these experiments was of the order of 0.5 F.

Succeeding investigators used more refined techniques with the result that the proportionality factor became more accurate. However, this continual change was undesirable. As a result, the First International Steam Tables Conference (1929) defined the relation of heat and work units as 778.26 ft lbf/B. More recently the proportionality factor was established as

$$J_e = 778.169 \text{ ft lbf/B}$$

In SI units

$$J_e = 1 \text{ N m/J}$$

where $J_e$ is known as Joule's equivalent. On this basis the proportionality of eq. (3.1) becomes the equality

$$\oint \delta Q = \oint \frac{\delta W}{J_e} \tag{3.1a}$$

Of course, the units on both sides of this equation must be the same. This requirement can be fulfilled by use of $J_e$ or by using the same units for both heat and work. The proportional factor $J_e$ will not be written in equations in subsequent sections of this book. In any case where homogeneity of units is uncertain, the student should use a unit equation to establish consistency of units.

### 3–3 Change in Stored Energy, a Consequence of the First Law

From the first law, as given in eq. (3.1a), we can investigate the relation of heat and work for closed-system processes. Figure 3.2 shows three processes for a closed system. Paths between the arbitrary states (1) and (2) are shown as broken lines since the processes are not necessarily frictionless and quasi-static. The coordinates are not identified because this section is not limited to pure substance. Thus more than two independent properties may be required to establish states (1) and (2).

By the first law we can write for the cycle (1) → (2) → (1) with processes $A$ and $C$:

$$\oint \delta Q = \underset{A}{\int_1^2 \delta Q} + \underset{C}{\int_2^1 \delta Q} = \underset{A}{\int_1^2 \delta W} + \underset{C}{\int_2^1 \delta W} = \oint \delta W$$

or

$$\underset{A}{\int_1^2 (\delta Q - \delta W)} = \underset{C}{\int_1^2 (\delta Q - \delta W)} \tag{a}$$

In like manner the cycle (1) → (2) → (1) with processes $B$ and $C$ yields

$$\underset{B}{\int_1^2 \delta Q} + \underset{C}{\int_2^1 \delta Q} = \underset{B}{\int_1^2 \delta W} + \underset{C}{\int_2^1 \delta W}$$

or

$$\underset{B}{\int_1^2 (\delta Q - \delta W)} = \underset{C}{\int_1^2 (\delta Q - \delta W)} \tag{b}$$

**Figure 3.2** Processes and Cycles for a Closed System

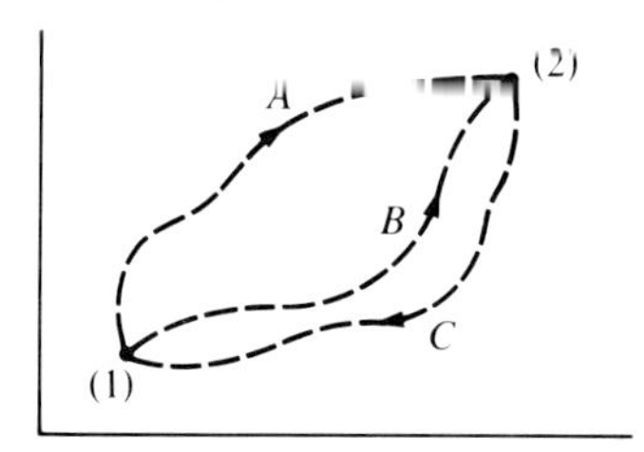


From eqs. (a) and (b) we observe that the integral from (1) to (2) for process $A$ is identical to the integral from (1) to (2) for process $B$. Since $A$ and $B$ represent arbitrary processes, we conclude that we would obtain the same value for the integral regardless of the process between states (1) and (2). Thus

$$\int_1^2 (\delta Q - \delta W) = \text{constant} \tag{c}$$

for *all* closed-system processes between states (1) and (2) including those subject to gravity, motion, and electric, magnetic, and surface effects, as well as systems undergoing chemical reactions. The only restrictions on eq. (c) are the limitation to closed systems and that states (1) and (2) must be equilibrium states.

Recalling that a property is a characteristic of a system, changes of which are independent of the path and depend only on the end states, it is apparent that the constant of eq. (c) can be defined as the *change* of a property. That change we shall define as the change in the property *stored energy E,*

$$\int_1^2 (\delta Q - \delta W) \equiv E_2 - E_1$$

or

$${}_1Q_2 - {}_1W_2 = E_2 - E_1 \tag{3.2}$$

This equation is limited to closed-system processes and constitutes the operational definition of stored energy.

Equation (3.2) is the basis for another mathematical form of the first law for a closed system. A differential change in stored energy is

$$\delta Q - \delta W = dE$$

Let this change take place in a time interval $d\tau$. Thus the time-rate expression for a closed system is

$$\frac{\delta Q}{d\tau} - \frac{\delta W}{d\tau} = \frac{dE}{d\tau}$$

or

$$\dot{Q} - \dot{W} = \dot{E} \tag{3.2a}$$

This equation must be satisfied at any instant in time. Also, $\dot{Q}$ and $\dot{W}$ must be interpreted as net values. $\dot{E}$ represents the instantaneous time rate of change of stored energy within the boundary of the system. For time-steady processes in a closed system, $\dot{E}$ is zero. Therefore, $\dot{Q}_{net} = \dot{W}_{net}$.

*Energy* is a general term that includes stored energy in its various forms, heat and work. Previously it was emphasized that heat and work are interactions between systems and are never associated with the state of a system. We now can say that heat is energy transferred to or from a system due to a temperature difference between a system and its surroundings. In like manner, work is an energy transfer to or from a system with mechanical work being an energy transfer due to a force acting through a distance. Equation (3.2) is the most general energy equation for closed systems. It is applicable to processes involving all forms of work and to substances that can be treated as pure substances, as well as to those that cannot be so treated.

In view of the fact that the term *energy* includes all forms of stored energy and heat and work, it is evident that we can use the same unit for the different forms of energy. The common energy units in the English system are Btu (B) or the foot pound force (ft lbf), while in the SI system the common energy unit is the kilojoule (1000 N m = 1000 J = 1 kJ). The corresponding specific energy units (energy per unit mass) are the B/lbm, the ft lbf/lbm, and the kJ/kg. Time rates associated with energy are expressed in energy units per unit time (e.g., B/sec, ft lbf/sec, kJ/s) or in terms of the horsepower (HP) or the kilowatt (kW). Appendix A provides a list of conversion factors that relate the various energy units.

## Example 3.a

During the start-up process of a DC electric motor the following values were measured at a certain instant in time: input voltage and current, 24 V and 10 A; torque applied to output shaft, 21.1 in lbf; shaft speed, 920 revolutions per minute. Assume that the motor is thermally insulated from the surrounding air and determine the time rate of change of stored energy within the motor.

**Solution:**

Equation (3.2a) is applicable at any instant in time regardless of whether or not the system to which it is applied is time steady. Since the motor is in a start-up process, we interpret the process as an unsteady one. Pertinent values are given at an instant in time and eq. (3.2a) can be applied to evaluate $\dot{E}$. From the discussion in Section 2–8 related to electrical work we conclude that the flow of electric current to the motor constitutes power delivered to the motor, that is, $\dot{W}_{\text{elect}}$ is negative. Since $\dot{Q}$ is assumed to be zero

$$\dot{E} = -\dot{W}_{\text{net}} = -(\dot{W}_{\text{elect}} + \dot{W}_s)$$

Since DC electric power is the product of voltage and current

$$\dot{W}_{\text{elect}} = -(24)(10) = -240 \text{ watts}$$

The shaft power is, from eq. (2.5i), the product of torque and angular speed

$$\dot{W}_s = -\text{T}\omega = -\left(\frac{-21.1}{12}\text{ ft lbf}\right)(920)(2\pi)/\text{min}$$

$$= 10{,}164 \left.\frac{\text{ft lbf}}{\text{min}}\right|\left.\frac{1.356 \text{ J}}{\text{ft lbf}}\right|\frac{\text{min}}{60 \text{ s}}$$

$$= 229.7 \text{ W}$$

Therefore

$$\dot{E} = -(-240 + 229.7) = +10.3 \text{ W}$$

$$= 10.3 \text{ W}\left(\frac{3.413 \text{ B}}{\text{W hr}}\right) = 35.2 \text{ B/hr}$$

When time-steady operation is achieved, eq. (3.2a) still applies. However, $\dot{E}$ is then zero and the power quantities will not vary with time. Losses in the conversion of electric power to shaft power will be manifest in heat transfer $\dot{Q}$. Thus we see that with such losses the motor cannot reach time-steady operation if it is perfectly thermally insulated.

---

## 3–4 Stored Energy and Internal Energy

Stored energy represents all of the energy stored in a system at a given state. Energy can be stored in a system in various forms. In this text we shall limit stored energy to mechanical potential and kinetic energies and *internal energy U*. The internal energy is that stored in a system when the system is not subject to gravitational, motion, magnetic, electrical, or surface tension effects. Internal energy is a property and as such is single valued at a particular state of a given system. For the given conditions, stored energy is written as

$$E = U + KE + PE$$

Utilizing the expressions for potential and kinetic energies developed earlier

$$E = U + m\frac{V^2}{2g_0} + mz\frac{g}{g_0}$$

and on a unit mass basis

$$e = u + \frac{V^2}{2g_0} + z\frac{g}{g_0}$$

where $e = E/m$ and $u = U/m$. The first law, eq. (3.2), written in differential form becomes

$$\delta Q - \delta W = dU + m\,d\left(\frac{V^2}{2g_0}\right) + m\,d\left(\frac{g}{g_0}z\right) \tag{3.2b}$$

Introducing the heat transfer per unit mass $q = Q/m$ and the work per unit mass $w = W/m$, the above equation on a unit mass basis becomes

$$\delta q - \delta w = du + d\left(\frac{V^2}{2g_0}\right) + d\left(\frac{g}{g_0}z\right) \tag{3.2c}$$

The first law, stated in mathematical form as eq. (3.2) or eqs. (3.2b) and (3.2c), does not establish absolute values for stored energy or internal energy. Rather, it provides only the method to evaluate changes in these quantities. To assign a numerical value to the internal energy at any state we can write

$$u = (u - u_0) + u_0$$

where $u_0$ is an arbitrary constant at a fixed reference state. The change $u - u_0$ can then be evaluated as

$$\int(\delta q - \delta w) = u - u_0$$

for a simple closed system.

Most of the closed systems treated in this text will be stationary, that is, not subject to changes in elevation or velocity. With these added limitations the differential forms of the energy equation, eqs. (3.2b) and (3.2c), integrate to

$$_1Q_2 = U_2 - U_1 + {_1W_2} \tag{3.2d}$$

$$_1q_2 = u_2 - u_1 + {_1w_2} \tag{3.2e}$$

Internal energy is related to the molecular nature of matter. We recall that a pure substance is matter that is invariant in chemical composition. For pure substances, changes in internal energy are related only to changes in the kinetic and potential energies of the molecules of the substance. Changes in the structure of the molecules of a substance, brought about by a chemical reaction for example, also result in changes in internal energy. In such cases the substance no longer can be treated as a pure substance. However, regardless of the means by which changes in internal energy have taken place, $\Delta U$ can be determined according to eq. (3.2d) by measuring heat and work for the process. Internal energy $U$ is an extensive property while specific internal energy $u = U/m$ is an intensive property.

## Example 3.b

For a certain closed system process $Q$ and $W$ were measured and found to be $-3050$ B and $-1.90$ kWh respectively. What is the change in the internal energy of the system for the process if the system is stationary?

**Solution:**

Applying eq. (3.2d) we obtain

$$\begin{aligned} \Delta U = U_2 - U_1 &= {}_1Q_2 - {}_1W_2 \\ &= -3050\ \text{B} - (-1.9\ \text{kWh})\left(\frac{3413\ \text{B}}{\text{kWh}}\right) \\ &= +\ 3435\ \text{B} = 3435\ \text{B}\left(\frac{1.055\ \text{kJ}}{\text{B}}\right) = 3624\ \text{kJ} \end{aligned}$$

Note that no knowledge of the nature of the process or the substance was necessary to obtain the solution.

### 3–5 Heat Interactions; Path Functions

At this point we can show that heat interactions are not properties, but rather are path dependent. Consider a closed stationary system undergoing a frictionless quasi-static process. Equation (3.2d) is applicable.

$${}_1Q_2 = {}_1W_2 + U_2 - U_1$$

Previously the work for such a process was shown to be a path function. Since internal energy is a property, the change of internal energy between any given states (1) and (2) is constant and independent of the path. Therefore, the heat transfer ${}_1Q_2$ must also be path dependent. Generally we see from eq. (3.2d) that heat transfer is path dependent whenever work is path dependent. For an adiabatic process between two given states, this equation shows that the work is equal to the negative of the internal energy change and hence is independent of the adiabatic path between the two states, that is, the work is the same for all adiabatic paths between the two states.

In review let us again note that since work and heat interactions are path dependent, their differentials are inexact differentials. These are denoted by $\delta Q$ and $\delta W$ and the quantities for a process are noted ${}_1Q_2$ and ${}_1W_2$. This notation is used to denote path dependency as contrasted with properties such as $U$, changes of which are independent of the path and depend only on the end states for the process.

The following example is an illustration of an application of the first law to a cycle and the processes that make up the cycle.

## Example 3.c

The working fluid of a certain stationary closed system undergoes a cycle that involves four processes. The table lists each of the processes and known values for $Q$, $W$, and $\Delta U$. Also given are values for $U_1$ and $U_4$. Using the given information, complete the table and the internal energy list.

| Process | $Q$, kJ | $W$, kJ | $\Delta U$, kJ |
|---|---|---|---|
| 1–2 | | 37.7 | 626.9 |
| 2–3 | 0 | −256.3 | |
| 3–4 | −920.2 | | |
| 4–1 | | 42.7 | |

$U_1 = 213.8$ kJ
$U_2 =$ ________
$U_3 =$ ________
$U_4 = 256.5$ kJ

**Solution:**

Each of the processes is governed by the first law for a process as written in eq. (3.2d) and the complete cycle must satisfy $\oint \delta W = \oint \delta Q$, eq. (3.1a). For the process 1–2,

$${}_1Q_2 = {}_1W_2 + {}_1\Delta U_2 = {}_1W_2 + U_2 - U_1$$

Substituting known values,

$${}_1Q_2 = +37.7 + 626.9 = +664.6 \text{ kJ}$$

Also,

$$U_2 = U_1 + {}_1\Delta U_2 = 213.8 + 626.9 = 840.7 \text{ kJ}$$

For the process 2–3,

$${}_2Q_3 = {}_2W_3 + U_3 - U_2$$

$$0 = -256.3 + U_3 - 840.7$$

Thus,

$$U_3 = 1097 \text{ kJ}$$

$$_2\Delta U_3 = U_3 - U_2 = 1097 - 840.7 = 256.3 \text{ kJ}$$

For the process 3–4

$$_3Q_4 = {_3W_4} + U_4 - U_3$$

$$-920.2 = {_3W_4} + 256.5 - 1097$$

Thus

$$_3W_4 = -79.7 \text{ kJ}$$

$$_3\Delta U_4 = U_4 - U_3 = 256.5 - 1097 = -840.6 \text{ kJ}$$

The value for $_4Q_1$ can be found using eq. (3.1a) as follows:

$$\oint \delta W = W_{\text{net}} = {_1W_2} + {_2W_3} + {_3W_4} + {_4W_1} = \oint \delta Q$$
$$= Q_{\text{net}} = {_1Q_2} + {_2Q_3} + {_3Q_4} + {_4Q_1}$$

Substituting values

$$W_{\text{net}} = +37.7 - 256.3 - 79.7 + 42.7 = -255.6 \text{ kJ}$$
$$= 664.6 + 0 - 920.2 + {_4Q_1}$$

From the above equation

$$_4Q_1 = 0$$

Also,

$$_4\Delta U_1 = U_1 - U_4 = 213.8 - 256.5 = -42.7 \text{ kJ}$$

The value for $_4Q_1$ can also be computed by

$$_4Q_1 = {_4W_1} + U_1 - U_4 = +42.7 + 213.8 - 256.5 = 0$$

## 3–6 Enthalpy, a Composite Property

Several thermodynamic properties are composite properties, that is, they are a combination of properties. The property of this nature most frequently encountered in engineering thermodynamics is *enthalpy* defined as

$$h \equiv u + pv$$

Since $u$, $p$, and $v$ are intensive properties, enthalpy is obviously an intensive property. This property for many years was called "heat content." Even today, terminology such as "latent heat of vaporization" is used in describing the enthalpy change, at constant pressure, for a phase change from liquid to vapor. The word "enthalpy" stems from the Greek word *enthalpo,* which means "heat within."

As noted earlier, internal energy $u$ can be assigned a numerical value only when an arbitrary reference state is selected. It follows that enthalpy, in like manner, must be evaluated with respect to a reference state. Fortunately most problems in engineering thermodynamics involve changes in these properties, and for such changes the reference state value cancels out of energy equations. At a given state we can write

$$h = (h - h_0) + h_0$$

where $h_0$ is the arbitrarily chosen constant value of enthalpy at the reference state.

We will find in the next section that enthalpy appears in a natural way in the first law applied to open systems, but one example indicating the utility of this property is its role in the analysis of a constant-pressure frictionless quasi-static closed-system process. For such a process the energy equation is

$$\delta q = du + \delta w = du + pdv$$

The differential of enthalpy is

$$dh = du + pdv + vdp$$

For any constant-pressure process $vdp$ is equal to zero. Thus for constant-pressure processes with the given limitations we obtain from the above equations

$$[dh = \delta q]_{p = \text{const}}$$

or

$$[h_2 - h_1 = {}_1q_2]_{p = \text{const}}$$

regardless of the working fluid undergoing the process.

Figure 3.3 A Generalized Open System

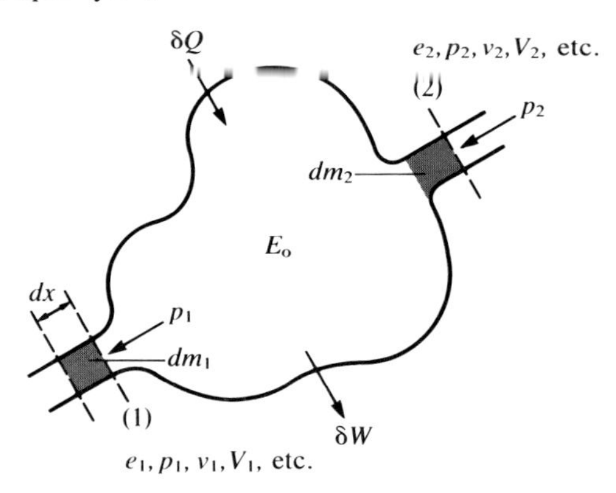


## 3–7 The General Energy Equation

The first law for a closed-system process between an initial state $i$ and a final state $f$ is

$$ {}_iQ_f - {}_iW_f = E_f - E_i \tag{3.2} $$

Next we develop an energy equation for open systems (those having mass crossing their boundaries) using eq. (3.2). A generalized open system (control volume) is shown in Figure 3.3. The boundaries of the open system (control surfaces) are section (1) (inbound flow), section (2) (outbound flow), and the system surface between the two sections. Figure 3.3 shows the open system at the beginning of the process we shall study. To utilize eq. (3.2), we must describe a closed system. At the beginning, the closed system includes all mass within the open-system boundaries, including $dm_2$, plus the differential mass $dm_1$, which will enter the open system in a time interval $d\tau$. At the end of the time interval $d\tau$ the closed system includes the same mass but the differential mass $dm_2$ has passed section (2) in leaving the open system. To aid in visualizing this closed system, the fluid immediately upstream from $dm_1$ can be thought of as an imaginary piston that will force the mass $dm_1$ past the open-system boundary, section (1). Another imaginary piston, initially having its face at section (2), is, in like manner, the fluid immediately downstream from section (2), and is acted on by the fluid moving out of the system. Thus the closed-system process we shall consider is one in which the differential mass $dm_1$ enters the open system at section (1) and, during the same time interval $d\tau$, the differential mass $dm_2$ passes section (2). In general the mass $dm_1$ is not equal to the mass $dm_2$. The stored energy within the *open* system is $E_o$ at the beginning of the process and $E_o + dE_o$ at the completion of the process. The specific stored energy of the inbound mass $dm_1$ is $e_1$ as it passes section (1). The specific stored energy of the element $dm_2$ is $e_2$ as it passes section (2), leaving the open system. During the process heat transfer $\delta Q$ may occur between the open system and its surroundings. Work interactions also occur. However, we shall identify these later.

Now let us apply eq. (3.2) in differential form to the pseudo-closed system process. For the process of the above-described closed system taking place in a time interval $d\tau$, eq. (3.2) becomes

$$\delta Q = (E_o + dE_o + e_2\, dm_2) - (E_o + e_1\, dm_1) + \delta W \tag{a}$$

where $dE_o$ represents the change in stored energy of the open system and $e_1$ and $e_2$ are, respectively, the specific stored energy of the inbound fluid at section (1) and the outbound fluid at section (2). Possible work quantities include shaft work, the energy necessary to move $dm_1$ past section (1), the energy involved in the passage of $dm_2$ past section (2), and other forms of work, each of which will be identified in the following. The sign convention for work, established earlier, assigns positive values to work done by a system and negative values to work done on a system.

For the inbound mass $dm_1$, the imaginary "piston" of fluid immediately upstream from it exerts a force $p_1A_1$ as $dm_1$ passes section (1). The energy, sometimes called flow energy, is a work quantity for the pseudo-closed system. Its magnitude is

$$\delta W_1 = Fdx = p_1A_1dx = p_1A_1\frac{dV_1}{A_1} = p_1v_1dm_1$$

where $dV_1$ is the volume of the differential mass $dm_1$ evaluated at section (1). The work quantity will be negative since it represents "work" done on the fluid element. In like manner, the fluid element $dm_2$, as it passes section (2), moves other fluid downstream and does "work" on it equal to

$$\delta W_2 = p_2v_2dm_2$$

This expression is derived in a manner identical to that applied to the inbound mass $dm_1$. By the sign convention for work, $\delta W_2$ will be positive. Other possible work interactions for the open system are shaft work $\delta W_s$, work due to the movement of the open-system boundary normal to itself (piston face work), and work associated with electric current crossing the boundary. The last two are collectively denoted by the term $\delta W_b$.* Including these quantities, eq. (a) becomes

$$\delta Q = dE_o + (e_2 + p_2v_2)dm_2 - (e_1 + p_1v_1)dm_1 + \delta W_s + \delta W_b \tag{b}$$

*The term $W_b$, of course, could include other forms of work such as those associated with surface tension and electric and magnetic field effects. As noted earlier, these are not discussed in this text.

with the stored energy $e$ including internal energy, kinetic energy, and potential energy. Thus eq. (b) becomes

$$\delta Q = dE_o + \left(u_2 + p_2 v_2 + \frac{V_2^2}{2g_0} + \frac{g}{g_0} z_2\right) dm_2 - \left(u_1 + p_1 v_1 + \frac{V_1^2}{2g_0} + \frac{g}{g_0} z_1\right) dm_1 + \delta W_s + \delta W_b \tag{c}$$

With enthalpy defined as $u + pv$, eq. (c) becomes

$$\delta Q = dE_o + \left(h_2 + \frac{V_2^2}{2g_0} + \frac{g}{g_0} z_2\right) dm_2 - \left(h_1 + \frac{V_1^2}{2g_0} + \frac{g}{g_0} z_1\right) dm_1 + \delta W_s + \delta W_b \tag{3.3}$$

While eq. (3.3) was developed for only one inbound and one outbound stream, the equation may be extended to any number of streams by expanding the notation 1 to include a summation of all streams entering the open system, and, in like manner, expanding the notation 2 to include all leaving streams. We shall designate eq. (3.3) as the general energy equation. It must be developed further to provide a time-rate equation or integrated. Applications are illustrated in the following.

Since the derivation of eq. (3.3) proceeded from the energy equation for a closed system, that equation should, under closed-system conditions, reduce to eq. (3.2). For a closed system no mass crosses the system boundaries, and hence $dm_1$ and $dm_2$ are zero. The remaining terms are

$$\delta Q = dE + \delta W_s + \delta W_b$$

The work terms are combined in the closed-system energy equation. For a process between an initial state $i$ and a final state $f$ we obtain

$${}_iQ_f = {}_iW_f + E_f - E_i$$

where ${}_iW_f$ designates all forms of work. This equation is identical to eq. (3.2). The work associated with the expansion or contraction of a system (piston face work), which is included in $W_b$, is most frequently encountered for closed systems.

## Example 3.d

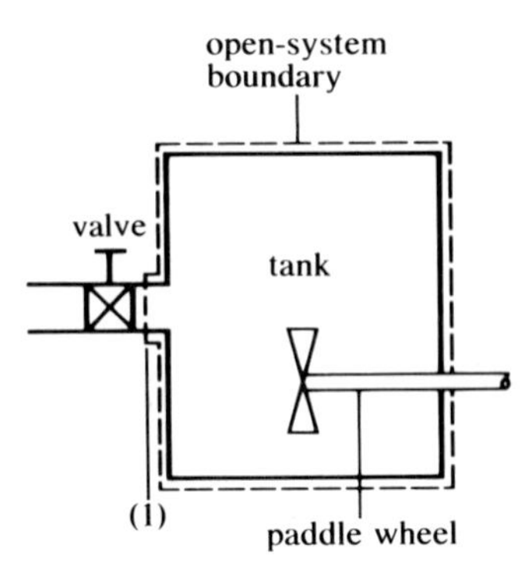


A rigid tank with a volume of 75 ft³ contains a working fluid at 60 psia and 500 F. At this state the specific volume is 9.403 ft³/lbm and the enthalpy is 1283 B/lbm. A line connected to the tank carries the same fluid at a pressure of 100 psia and 600 F, a state where the enthalpy is 1329 B/lbm. A valve in the line connected to the tank is opened, admitting 20 lbm of the working fluid into the tank. During the process 1000 B of work is done on the contents of the tank by a paddle wheel (rotor). The fluid in the tank at the completion of the process is at a pressure of 75 psia and an enthalpy of 693 B/lbm. No fluid leaves the tank. Calculate the heat transfer for the process.

**Solution:**

In view of the given conditions, the boundary shown in the figure is chosen. It is evident that an open system is involved. The process, therefore, must be analyzed using the general energy equation, eq. (3.3). No mass flows out of the system and only shaft work is involved. Thus eq. (3.3) reduces to

$$\delta Q = dE_o - \left(h_1 + \frac{V_1^2}{2g_0} + \frac{g}{g_0} z_1\right)dm_1 + \delta W_s$$

With no information given regarding the velocity of the entering fluid or elevation effects, we assume $V_1$ is small and take $z_1 = 0$. In addition, for a stationary closed system $dE_o$ is $dU_o$. Therefore

$$\delta Q = dU_o - h_1 dm_1 + \delta W_s$$

We now integrate the above equation between the initial condition $i$ and the final condition $f$ of the system. Integration produces

$$\begin{aligned} Q &= U_f - U_i - h_1(m_f - m_i) + W_s \\ &= m_f u_f - m_i u_i - h_1(m_f - m_i) + W_s \end{aligned}$$

From the given data

$$m_i = \frac{V_i}{v_i} = \frac{75}{9.403} = 7.976 \text{ lbm and } m_f = m_i + 20 = 27.98 \text{ lbm}$$

At the end of the process,

$$v_f = \frac{V_f}{m_f} = \frac{75}{27.976} = 2.681 \text{ ft}^3/\text{lbm}$$

The initial and final internal energies in the tank are

$$u_i = h_i - \frac{p_i v_i}{J_e} = 1283 - \frac{(60)(144)(9.403)}{778} = 1178.6 \text{ B/lbm}$$

and

$$u_f = 693 - \frac{(75)(144)(2.6809)}{778} = 655.79 \text{ B/lbm}$$

Substituting the above values into the energy equation

$$\begin{aligned} Q &= (27.98)\,(655.79) - (7.9762)(1178.6) - (1329)(20) \\ &\quad + (-1000) \\ &= 18{,}634 \text{ B} \end{aligned}$$

---

## 3–8 The Energy Equation for Steady-Flow Processes

The conditions for steady flow were presented in Section 2–15. The steady-flow energy equation follows from the general energy equation, eq. (3.3), when the conditions of steady flow are satisfied. The requirement of constant properties at any location within a steady-flow system demands that $dE_o$ be zero. The added requirement of mass rate of flow into a steady-flow system being equal to the mass rate of flow leaving the system requires that $dm_1$ and $dm_2$ in eq. (3.3) be identical. The flow for many devices encountered in engineering is steady. The nature of such flows eliminates work due to expansion or contraction of the system. For steady flow we will drop the subscripts on work and combine shaft work and the work associated with electric current crossing a boundary into a single term, $\delta W$. The former is identified by a shaft acted on by a force or a torque crossing a boundary and the latter by electric current crossing a boundary. When the flow of electric current is not mentioned, we shall interpret the term to mean shaft work. Under the above-noted conditions eq. (3.3) becomes

$$\delta Q = \left(h_2 + \frac{V_2^2}{2g_0} + \frac{g}{g_0} z_2\right)dm - \left(h_1 + \frac{V_1^2}{2g_0} + \frac{g}{g_0} z_1\right)dm + \delta W$$

Since the quantities $\delta Q$, $dm$, and $\delta W$ are associated with the time interval $d\tau$, the above equation can be written as

$$\frac{\delta Q}{d\tau} = (h_2 + \frac{V_2^2}{2g_0} + \frac{g}{g_0} z_2) \frac{dm}{d\tau} - (h_1 + \frac{V_1^2}{2g_0} + \frac{g}{g_0} z_1)\frac{dm}{d\tau} + \frac{\delta W}{d\tau}$$

or rearranging

$$\dot{m}(h_1 + \frac{V_1^2}{2g_0} + \frac{g}{g_0} z_1) + \dot{Q} = \dot{W} + \dot{m}(h_2 + \frac{V_2^2}{2g_0} + \frac{g}{g_0} z_2) \quad \textbf{(3.4)}$$

where $\delta Q/d\tau = \dot{Q}$, $dm/d\tau = \dot{m}$, and $\delta W/d\tau = \dot{W}$. Equation (3.4) is the steady-flow equation written on a time-rate basis. In steady flow where more than one inbound or outbound stream crosses the system boundary, eq. (3.4) becomes

$$\Sigma\left[\dot{m}[h + \frac{V^2}{2g_0} + \frac{g}{g_0} z]\right]_{\text{in}} + \dot{Q} = \Sigma\left[\dot{m}[h + \frac{V^2}{2g_0} + \frac{g}{g_0} z]\right]_{\text{out}} + \dot{W} \quad \textbf{(3.4a)}$$

For steady-flow systems in which there is a single inbound stream and a single outbound stream, the steady-flow equation can be written on a unit mass basis. The expression for this case can be formed by dividing eq. (3.4) by $\dot{m}$. The result is

$$h_1 + \frac{V_1^2}{2g_0} + \frac{g}{g_0} z_1 + {}_1q_2 = h_2 + \frac{V_2^2}{2g_0} + \frac{g}{g_0} z_2 + {}_1w_2 \quad \textbf{(3.4b)}$$

where ${}_1q_2 = \dot{Q}/\dot{m}$ and ${}_1w_2 = \dot{W}/\dot{m}$.

---

## Example 3.e

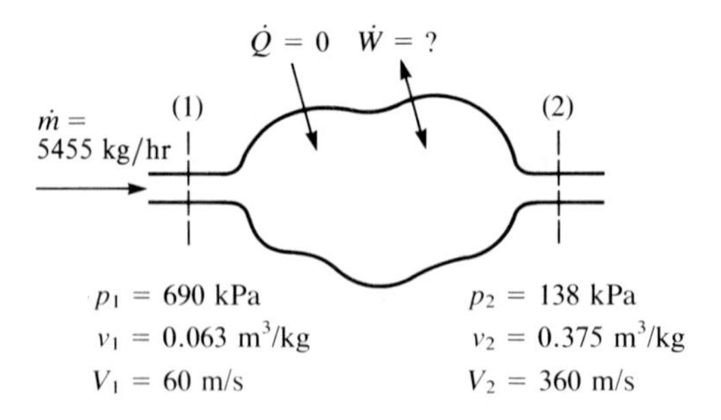


The unit shown operates in a steady-flow manner. The process is adiabatic and changes in potential energy are negligible. Assuming the internal energy of the working fluid decreases 58 kJ/kg between sections (1) and (2), calculate the power supplied to or delivered by the unit and the area at section (2).

**Solution:**

Under the given conditions the steady-flow energy equation on a unit mass basis, eq. (3.4b), reduces to

$$ {}_1w_2 = h_1 - h_2 + \frac{V_1^2 - V_2^2}{2g_0} $$

or, from the definition of enthalpy,

$$ {}_1w_2 = (u_1 + p_1v_1) - (u_2 + p_2v_2) + \frac{V_1^2 - V_2^2}{2g_0} $$

Substituting the given values for Sections (1) and (2) we have

$$ \begin{aligned} {}_1w_2 &= +58\ \text{kJ/kg} + \frac{690\ \text{kN}}{\text{m}^2}\bigg|\frac{0.063\ \text{m}^3}{\text{kg}} - \frac{138\ \text{kN}}{\text{m}^2}\bigg|\frac{0.375\ \text{m}^3}{\text{kg}} \\ &\quad + \frac{(60\ \text{m/s})^2 - (360\ \text{m/s})^2}{2}\bigg|\frac{\text{N s}^2}{1\ \text{kg m}}\bigg|\frac{\text{kJ}}{10^3\ \text{N m}} \\ &= -13.28\ \text{kJ/kg} \end{aligned} $$

The power is

$$ \begin{aligned} \dot{W} &= \dot{m}_1w_2 = (5455\ \text{kg/hr})(-13.28\ \text{kJ/kg})(1\ \text{hr}/3600\ \text{s}) \\ &= -20.12\ \text{kJ/s} = -20.12\ \text{kW} \end{aligned} $$

The negative sign indicates that power is supplied to the system. The area at the outlet, Section (2), is determined by

$$ \begin{aligned} A_2 &= \frac{\dot{m}v_2}{V_2} = \frac{5455\ \text{kg}}{\text{hr}}\bigg|\frac{0.375\ \text{m}^3}{\text{kg}}\bigg|\frac{\text{s}}{360\ \text{m}}\bigg|\frac{\text{hr}}{3600\ \text{s}} \\ &= 1.58 \times 10^{-3}\ \text{m}^2 \end{aligned} $$

where it has been recognized that $\dot{m}_1 = \dot{m}_2$ since the flow is steady and there are one inbound stream and one outbound stream.

## Example 3.f

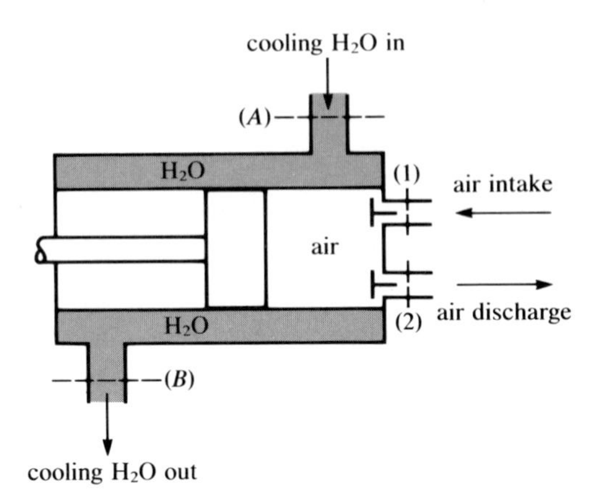


The following data were obtained for a water-jacketed air compressor operating under steady-flow conditions:

| | Air | | Water | |
|---|---|---|---|---|
| | **Intake (1)** | **Discharge (2)** | **Intake (A)** | **Discharge (B)** |
| **Flow** | 500 ft³/min | | 800 lbm/hr | |
| **Pressure, psia** | 14.00 | 95.00 | | — |
| **Enthalpy, B/lbm** | 143.2 | 181.1 | 50.00 | 120.00 |
| **Specific volume, ft³/lbm** | 13.76 | 2.653 | | — |
| **Area, in²** | 20.00 | 2.000 | | — |

The power supplied to drive the compressor is 55.36 HP. Calculate the heat transfer from the unit and the volume rate of flow of the air leaving the unit as it operates under the stated steady-flow conditions.

### Solution:

This problem illustrates the necessity to define system boundaries before proceeding with the problem. The circulating water and the air could be treated separately by defining systems for each. However, we shall select a boundary that encloses the compressor, across which water and air flow and heat and work interactions take place. Because the air compressor is a reciprocating device, the flow of the air through the system will be pulsating in nature and therefore not truly steady. However, as discussed in Section 2–15, such processes can be treated with good accuracy as steady-flow provided time-averaged flow quantities are used. Interpreting the given values to be such, we note that for the system defined above the applicable energy equation is eq. (3.4a), which becomes

$$\dot{m}_1 h_1 + \frac{\dot{m}_1 V_1^2}{2g_0} + \dot{m}_A h_A + \dot{Q} = \dot{m}_2 h_2 + \frac{\dot{m}_2 V_2^2}{2g_0} + \dot{m}_B h_B + \dot{W}$$

where the kinetic energy change of the water and potential energy changes of both the water and air are assumed to be negligible. For steady flow $\dot{m}_1 = \dot{m}_2 = \dot{m}_a$ and $\dot{m}_A = \dot{m}_B = \dot{m}_w$. Under these conditions the steady-flow equation becomes

$$\dot{Q} = \dot{m}_a(h_2 - h_1) + \dot{m}_w(h_B - h_A) + \frac{\dot{m}_a(V_2^2 - V_1^2)}{2g_0} + \dot{W}$$

The mass rate of flow of the air can be calculated from the given data at section (1) as

$$\dot{m}_1 = \frac{\dot{V}_1}{v_1} = \frac{500 \text{ ft}^3}{\text{min}} \left| \frac{\text{lbm}}{13.76 \text{ ft}^3} \right. = 36.34 \text{ lbm/min}$$

The continuity equation, written at section (1), yields the velocity $V_1$ as

$$V_1 = \frac{36.34 \text{ lbm}}{\text{min}} \left| \frac{13.76 \text{ ft}^3}{\text{lbm}} \right| \frac{}{20 \text{ in}^2} \left| \frac{144 \text{ in}^2}{\text{ft}^2} \right| \frac{\text{min}}{60 \text{ sec}}$$

$$= 60.00 \text{ ft/sec}$$

The mass rate of flow at section (2) is the same as that at section (1) for the steady-flow process. Thus,

$$\dot{m}_1 = \dot{m}_2 = \dot{m}_a = \frac{A_1 V_1}{v_1} = \frac{A_2 V_2}{v_2}$$

and the right-hand equality yields $V_2$ as

$$V_2 = \left(\frac{A_1}{A_2}\right)\left(\frac{v_2}{v_1}\right)(V_1)$$

$$= \frac{20.00 \text{ in}^2}{2.000 \text{ in}^2} \left| \frac{2.653 \text{ ft}^3}{\text{lbm}} \right| \frac{\text{lbm}}{13.76 \text{ ft}^3} \left| \frac{60.00 \text{ ft}}{\text{sec}} \right.$$

$$= 115.7 \text{ ft/sec}$$

Upon substituting these values into the steady-flow equation we have

$$\dot{Q} = 36.34(181.1 - 143.2) + \frac{810}{60}(120.0 - 50.00)$$

$$+ \frac{36.34(115.7^2 - 60.00^2)}{(2)(32.17)(778)} + \frac{-(55.36)(2545)}{60}$$

$$= -18.92 \text{ B/min}$$

Under steady-flow conditions the volume rate of flow of the air leaving the compressor is calculated as

$$(\dot{m}v)_2 = A_2V_2 = \text{volume rate of flow at (2)}$$

$$= \frac{2.000}{144} \bigg| \; 115.7 = 1.6069 \text{ ft}^3/\text{sec} = 96.4 \text{ ft}^3/\text{min}$$

**Figure 3.4** A Heat Engine

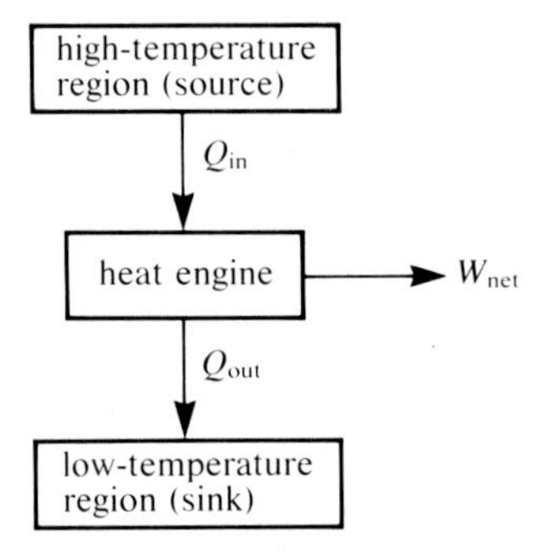


## 3–9 Heat Engines, Refrigerators, and Heat Pumps; Parameters Identifying Their Effectiveness

In the discussion of heat engine, refrigeration, and heat pump cycles it is useful to introduce the additional subscripts "in" and "out" in conjunction with work and heat-transfer quantities. Work or heat-transfer terms with either of these subscripts indicate absolute values and therefore such quantities have magnitude only. This notation is useful in denoting the directions of work and heat transfer with respect to the working fluid. Where confusion might arise with regard to work and heat-transfer terms, absolute signs will be used.

For purposes of thermodynamic analysis we define a *heat engine* as a system of one or more components that operates in a continuous and cyclic manner and converts to work a portion of the heat transferred to the working fluid. A heat engine receives heat from a high-temperature region (a source) and converts only part of the heat to work. The remainder of the energy supplied is rejected by the heat engine to a low-temperature part of its surroundings (a sink). The simple power plant shown in Figure 3.1b is one example of a heat engine. Heat from a high-temperature source is supplied to generate the steam in the process (4) to (1). In process (2) to (3) heat is rejected by the condensing exhaust steam to the cooling water that passes through the condenser. The working fluid undergoes a cycle as shown in the *p-v* diagram in the figure and net work is produced.

The general description of a heat engine is given in Figure 3.4. A heat engine is characterized by heat exchange in the direction shown with two regions at different temperatures, the working fluid undergoing a cycle, and the production of work.

Thermal efficiency for a heat engine is defined as the ratio of the net work output to the total heat supplied from the high-temperature source and is denoted by the symbol $\eta$. Thus

$$\eta = \frac{W_{net}}{Q_{in}} \tag{3.5}$$

and since thermal efficiency is dimensionless, $W_{net}$ and $Q_{in}$ must be in the same units. The first law relates net heat and net work as

$$\oint \delta Q = \oint \delta W \quad \text{or} \quad Q_{net} = Q_{in} - Q_{out} = W_{net}$$

and alternate expressions for thermal efficiency are

$$\eta = \frac{Q_{net}}{Q_{in}} = \frac{Q_{in} - Q_{out}}{Q_{in}} \tag{3.5a}$$

Equations (3.5) and (3.5a) may also be written on the basis of energy per unit mass for the heat and work quantities or in terms of time rates for these quantities.

One of the important aspects of heat engines with which thermodynamics deals is the upper limit of heat engine thermal efficiency. We have implied from the above and previous discussion that $\eta_{max}$ is less than unity. We shall delay further discussion of this matter until the basis for establishing $\eta_{max}$, the second law of thermodynamics, is introduced.

Engines such as the gas turbine illustrated in Figure 2.1 and reciprocating internal combustion engines are work-producing devices that, in the strictest sense, are not heat engines because their working fluids do not undergo cycles. Air and fuel are mixed, combustion takes place, work is produced, and combustion products are exhausted to the surroundings. No cycle for the working fluids can be identified. However, we can still compute the thermal efficiency for such engines by application of eq. (3.5) by replacing $Q_{in}$ with an appropriate quantity that measures the energy released in the burning of the fuel. Note that eq. (3.5a) is not applicable to such engines because a cycle of the working fluid cannot be identified. Caution in the choice of a basis for calculations must be exercised to ensure that $\eta$ is dimensionless.

A *refrigerator* is a device that operates in a cyclic manner, and, by energy input (usually work), continually transfers heat from one region to another region at a higher temperature. We shall consider refrigerators that operate by only work input because such devices account for a large fraction of today's refrigerating capacity.‡ Figure 3.5 illustrates a simple refrigerator that uses a working fluid that undergoes changes in phase during its cycle. In such a system the evaporator pressure is lower than the condenser pressure. As high-pressure liquid at (1) flows through the expansion valve, part of it flashes to vapor. Liquid in the evaporator

‡The absorption refrigeration cycle operates mainly from heat input and requires only a relatively small shaft work input. The Servel refrigerator requires no work input and operates solely on heat as an energy source. The reader interested in these cycles should consult the references at the end of Chapter 9.

**Figure 3.5** A Simplified Refrigeration Cycle

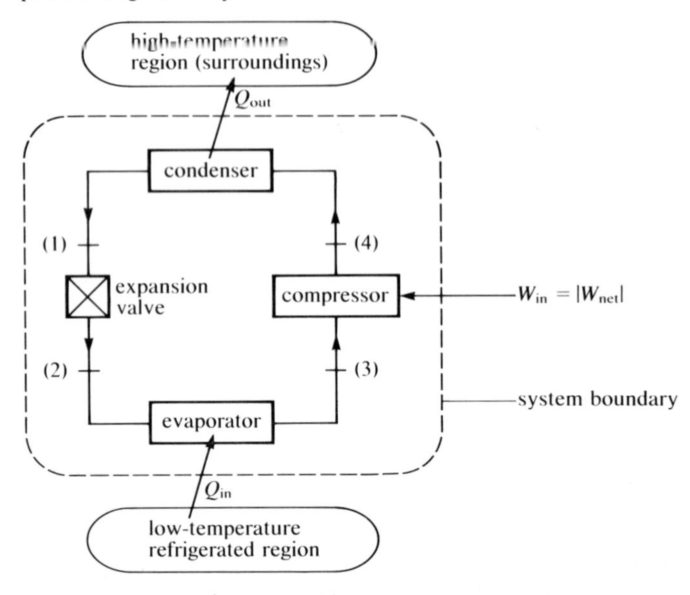


undergoes a phase change to vapor brought about by heat transfer from the refrigerated region that, in practice, surrounds the evaporator. The compressor typically draws only vapor from the evaporator at (3) and compresses it to the condenser pressure at (4). In the condenser the vapor is condensed to liquid by heat rejection in the process (4) to (1) to the high-temperature region, usually the surrounding atmosphere. Only heat and work cross the boundary and each component of the system operates in a steady-flow manner.

The optimum refrigerator produces the maximum refrigerating effect, $Q_{in}$, with minimum work input. The *coefficient of performance* of a refrigerator $\beta_R$ is defined as

$$\beta_R \equiv \frac{\text{refrigerating effect}}{\text{net work supplied}} = \frac{Q_{in}}{|W_{net}|} = \frac{Q_{in}}{Q_{out} - Q_{in}} \tag{3.6}$$

As with thermal efficiencies of heat engines, the upper bound for coefficient of performance is a consequence of the second law and cannot be established at this point.

Capacities (refrigerating effect) for refrigeration systems are commonly expressed in tons of refrigerating capacity. A ton of refrigerating capacity is defined as

$$\begin{aligned} 1 \text{ ton} &= 200 \frac{\text{B}}{\text{min}} = 12{,}000 \frac{\text{B}}{\text{hr}} \\ &= 210.9 \frac{\text{kJ}}{\text{min}} = 12{,}658 \frac{\text{kJ}}{\text{hr}} \end{aligned}$$

The name of this unit is based on the fact that this refrigerating effect will freeze approximately 1 ton (2000 lbm) of water at 32 F to ice at 32 F in a period of 24 hours.

Heat pumps are similar to refrigerators with one important exception. The $Q_{out}$ quantity, representing energy for heating purposes, is the desired effect. Thus the coefficient of performance for a heat pump is

$$\beta_P \equiv \frac{\text{heating effect}}{\text{net work supplied}} = \frac{Q_{out}}{|W_{net}|} = \frac{Q_{out}}{Q_{out} - Q_{in}} \tag{3.7}$$

Sources of low-temperature energy $Q_{in}$ for heat pumps include the atmosphere and the earth. The cycle described in Figure 3.5 could serve as a heat pump cycle if the low-temperature region were replaced by the source region for the heat pump.

The coefficient of performance expressions in eqs. (3.6) and (3.7) may also be written on the basis of energy per unit mass for the heat and work quantities or in terms of time rates for these quantities.

---

## Example 3.g

Consider again the cycle described in Example 3.c. Let the cycle operate between two regions of different temperature. Determine the value for the appropriate performance parameter (thermal efficiency or coefficient of performance) for the cycle.

**Solution:**

It must be determined whether the cycle is for a heat engine or a refrigerator or a heat pump. This can be determined by examining the sign on the net work. Noting that the sign is negative, we conclude that the cycle is not one for a heat engine (heat engines produce work). Recognizing that both refrigerators and heat pumps require that net work be supplied to the cycle, we conclude that the cycle could apply to either. If the cycle is treated as a refrigerator, the coefficient of performance is given by eq. (3.6).

$$\beta_R = \frac{Q_{in}}{|W_{net}|} = \frac{{}_1Q_2}{|W_{net}|} = \frac{664.6}{255.6} = 2.60$$

where ${}_1Q_2$ is chosen for $Q_{in}$ because $Q_{in}$ is always positive with respect to the working fluid. Treating the cycle as a heat pump cycle and using eq. (3.7),

$$\beta_P = \frac{Q_{out}}{|W_{net}|} = \frac{|{}_3Q_4|}{|W_{net}|} = \frac{920.2}{255.6} = 3.60$$

where ${}_3Q_4$ is recognized as $Q_{out}$ by its negative sign. This quantity is negative with respect to the working fluid but is positive when viewed from the region receiving the heat.

---

## 3–10 Commentary on Some Other Statements of the First Law

In 1850 Rudolf Clausius stated the first law in a form that differs little from present, rigorous statements of that law. Clausius' statement was: "For a cyclic process in any closed system, the net work is proportional to the net heat." Unfortunately some of the later authors did not follow his example. Frequently the first law has been couched as a generalized statement of conservation of energy, for example, "Energy can neither be created nor destroyed." Such statements usually have two significant faults. First, the statement appears in a text before energy is defined (in some cases it never is defined), and second, the broad generalization exceeds the limit of finite systems that can be subjected to experimental verification.

A limited, random sample of first-law statements includes

1. "When work is expended in producing heat, the quantity of heat generated is proportional to the work done, and conversely, when heat is employed to do work, a quantity of heat precisely equivalent to the work done disappears." This statement, from Goodenough's *Principles of Thermodynamics* (1911), one of the earliest texts addressed to engineering thermodynamics, leaves something to be desired when compared with the Clausius statement of the first law.
2. "When transfers take place between energy in the thermal and energy in mechanical forms, the amount of work performed is proportional to the amount of thermal energy which disappears and, if the change takes place in the opposite direction, the amount of thermal energy generated is equivalent to the amount of mechanical energy expended" (Young & Young, 1941). As a minimum, this statement is incomplete in that no reference to the type of system (closed or open) is included.
3. Keenan, in his classic text *Thermodynamics,* also published in 1941, states the first law as: "If any system is carried through a cycle (the end state being precisely the same as the initial state) then the summation of the work delivered to the surroundings is proportional to the summation of the heat taken from the surroundings." Note that a closed system is implied by the relation of the initial and final states.
4. Stoever, in *Essentials of Engineering Thermodynamics* (1953), notes that "the First Law is the Principle of the Conservation of Energy applied to finite systems," and then states the first law in this manner: "If any system undergoes a process during which energy is added to or removed from it (in the form of work or heat), none of the energy added is destroyed within the system and none of the energy removed is created within the system." Note that this statement is addressed to a process, and not necessarily to a cycle.
5. Zemansky, Abbot, and Van Ness (1975) have approached the first law in a procession of steps, defining heat thus: "When a system whose surroundings are at a different temperature undergoes a process during which work may be done, the energy transferred by nonmechanical means, equal to the difference between the internal energy change and the work, is called heat." This approach obviously requires prior definitions of work and internal energy. The authors proceed through a logic sequence, which yields: "The energy $E$ of any system and its surroundings, considered together, is conserved, thus, $\Delta E_{\text{total}} = 0$."

This limited sample indicates the variety of approaches to, and statements of, the first law. A study of the logic followed in reaching various statements of the law is, of itself, a most enlightening experience in thermodynamics. The only general criterion that must be satisfied in the several approaches to statements of the first law is that the law must be expressed in terms of finite systems that, by their nature, allow experimental verification of the law.

## 3–11 Summary

The first law of thermodynamics states that the cyclic integral of heat is proportional to the cyclic integral of work for all closed systems. When heat and work are measured in the same units, the first law is written as

$$\oint \delta Q = \oint \delta W$$

This expression is independent of the working fluid and the nature of the processes that make up the cycle. The first law is empirical. It cannot be deduced or established from other fundamental principles. The property $E$, stored energy, is a consequence of the first law. For any process (1) to (2) in a closed system, heat, work, and stored energy are related by

$${}_1Q_2 - {}_1W_2 = E_2 - E_1$$

This expression constitutes an operational definition of stored energy. On a time-rate basis the relation between heat, work, and stored energy for a closed system is

$$\dot{Q} = \dot{W} + \dot{E}$$

Energy is a general term that includes stored energy in its various forms and heat and work. Stored energy represents all of the energy of a system at a given state. Here we limit stored energy to internal energy, and mechanical energy to kinetic and potential energies. Thus $E$ is written as

$$E = U + m\frac{V^2}{2g_0} + m\frac{g}{g_0}z$$

where $U$, the internal energy, includes forms of stored energy that are independent of electric, magnetic, surface tension, gravity, and motion effects. For a stationary closed system, the first law becomes

$${}_1Q_2 = {}_1W_2 + U_2 - U_1$$

The general energy equation applicable to both closed and open systems is

$$\delta Q = dE_o + (h_2 + \frac{V_2^2}{2g_0} + \frac{g}{g_0} z_2)dm_2 - (h_1 + \frac{V_1^2}{2g_0} + \frac{g}{g_0} z_1)dm_1 + \delta W_s + \delta W_b$$

where $\delta W_s$ represents shaft work and $\delta W_b$ represents other forms of work. This expression may be extended to more than one inbound and one outbound stream. For a steady-flow system with multiple inbound and multiple outbound streams, the general energy equation becomes

$$\Sigma\left(\dot{m}\ (h + \frac{V^2}{2g_0} + \frac{g}{g_0} z)\right)_{in} + \dot{Q} = \Sigma\left(\dot{m}\ (h + \frac{V^2}{2g_0} + \frac{g}{g_0} z)\right)_{out} + \dot{W}$$

For steady-flow systems that involve one inbound and one outbound stream, the energy equation may be written on a unit mass basis as

$$h_1 + \frac{V_1^2}{2g_0} + \frac{g}{g_0} z_1 + {}_1q_2 = h_2 + \frac{V_2^2}{2g_0} + \frac{g}{g_0} z_2 + {}_1w_2$$

Frequently the first law is called the conservation of energy principle. Many of its applications require application of the conservation of mass principle also.

A heat engine is a system that operates in a continuous and cyclic manner and converts to work a portion of the heat transferred to its working fluid. Heat engine thermal efficiency is defined as

$$\eta = \frac{W_{net}}{Q_{in}}$$

Application of the first law to the heat engine cycle ($W_{net} = Q_{net}$) yields

$$\eta = \frac{Q_{net}}{Q_{in}} = \frac{Q_{in} - Q_{out}}{Q_{in}}$$

A refrigerator is a device that operates in a cyclic manner and, by work input, continually transfers heat from one region to another region at a higher temperature. The coefficient of performance of a refrigerator is defined as

$$\beta_R = \frac{Q_{in}}{|W_{net}|} = \frac{Q_{in}}{Q_{out} - Q_{in}}$$

where $Q_{in}$ is the desired refrigerating effect. A heat pump is similar to a refrigerator except that $Q_{out}$, energy used for heating purposes, is the desired effect. The heat-pump coefficient of performance is defined as

$$\beta_P = \frac{Q_{out}}{|W_{net}|}$$

## Selected References

Clausius, R. *The Mechanical Theory of Heat*. English translation by W. R. Browne, 1879.
Goodenough, G. A. *Principles of Thermodynamics*. New York: Henry Holt, 1911.
Joule, J. P. "Scientific Papers of J. P. Joule." Physical Society of London, 1885–1887.
Keenan, J. H. "Should the Btu Be Abandoned." *Heat Power News & Views*, vol. 10, no. 40, December 1955.
Keenan, J. H. *Thermodynamics*. New York: Wiley, 1941.
Obert, E. F. *Thermodynamics*. New York: McGraw-Hill, 1948.
Roller, D. *The Early Development of the Concepts of Temperature and Heat—The Rise and Decline of the Caloric Theory*. Cambridge, Mass.: Harvard University Press, 1950.
Spalding, D. B., and E. H. Cole. *Engineering Thermodynamics*. New York: McGraw-Hill, 1958.
Stoever, H. J. *Essentials of Engineering Thermodynamics*. New York: Wiley, 1953.
Thompson, Sir Benjamin (Count von Rumford). "Enquiry Concerning the Source of Heat which is Excited by Friction." The Royal Society, 1798.
Van Wylen, G. J., and R. E. Sonntag. *Fundamentals of Classical Thermodynamics*, 2nd ed. New York: Wiley, 1973.
Young, V. W., and G. A. Young. *Elementary Engineering Thermodynamics*. New York: McGraw-Hill, 1941.
Zemansky, M. W., and H. C. Van Ness. *Basic Engineering Thermodynamics*. New York: McGraw-Hill, 1966.

## Problems

3.1 A closed system undergoes a cycle consisting of two processes. During the first process the system does work in the amount of 42.2 kJ while heat transfer to the system is 28.5 kJ. During the second process, returning the system to its initial state, 26.4 kJ of heat is transferred from the system. What is the work for the second process? What are the net work and net heat transfer for the cycle?

3.2 A closed system undergoes a cycle consisting of three frictionless quasi-static processes. The system, initially at 15 psia, specific volume 2 ft$^3$/lbm, goes to 60 psia in a constant-volume process. During the second process pressure varies linearly with volume to a state at 15 psia, 14 ft$^3$/lbm. The cycle is closed with a constant-pressure process. During the first process the internal energy increases by 200 B/lbm. During the constant-pressure process the enthalpy decreases by 280 B/lbm. Calculate (a) work for the second process, (b) heat transfer for the second process, and (c) the change in internal energy and the heat transfer for the constant-pressure process.

3.3 A revolving flywheel, having stored energy due to its mass and angular speed, is brought to rest by a piston compressing gas in a cylinder. Neglecting all friction effects and assuming no heat transfer through the cylinder or piston, compare the change in stored energy of the flywheel with the change in stored energy of:

(a) The gas contained by the piston and cylinder.
(b) The flywheel–piston–cylinder combination.

3.4 A household refrigerator–freezer combination is powered by a 110-V electric motor. The unit is placed in a perfectly insulated room. If the unit operates with the refrigerator door open and the freezer door closed, will the room temperature increase, decrease, or remain constant? Repeat with the freezer door open and refrigerator door closed. Repeat with both doors closed. State any assumptions made in arriving at the answers.

3.5 A 1-kg copper bar at 27 C is put into a rigid, insulated closed vessel containing 1 kg of a gas at 530 C. The isolated system proceeds to a condition where the two components are at equal temperature. Comment on the stored energy change of the gas, the copper, and the composite system, justifying your conclusions on the basis of the first law.

3.6 A spring, with its axis vertical, rests on a desk. A book is placed on top of the spring, decreasing the spring length. Neglecting any heat transfer, what will be the stored energy change and the work (positive, zero, negative) for (a) the book, (b) the spring, (c) the desk, and (d) the book–spring–desk combination? State any assumptions made.

3.7 The following data are for a closed-system cycle of frictionless, quasi-static processes:

| **Process** | **$W$, ft lbf** | **$Q$, B** | **$\Delta E$** | **Remarks** |
|---|---|---|---|---|
| 1–2 | +70,500 | | | Constant pressure |
| 2–3 | | −10 | | Straight line on $p$ versus $v$ |
| 3–1 | | | 0 | $pV =$ constant |

Additional data are

$$p_1 = p_2 = 70 \text{ psia}$$
$$V_1 = 3.00 \text{ ft}^3$$
$$p_3 = 10 \text{ psia}$$

Sketch the cycle on $p$-$V$ coordinates and calculate (a) work for the process from (2) to (3), (b) heat transfer for the process from (1) to (2), and (c) the stored energy change $E_2$–$E_1$.

3.8 A closed system undergoes a frictionless, quasi-static cycle and 0.85 m³ of the working fluid is initially at a pressure of 105 kPa. It undergoes a constant-volume process to a state (2), which is at 485 kPa. Expansion from state (2) to (3) follows the path $pV = C$. The cycle is completed by a constant-pressure process. Calculate the net heat transfer for the cycle.

3.9 For a closed system we know $\delta Q = \delta W + dE$. By an example show that this statement is not valid for an open system.

3.10 One pound of gas undergoes a frictionless, quasi-static cycle consisting of the following processes: expansion from 75 psia along the path $pv^{1.40} = C$ to 15 psia and a volume of 10 ft³; compression from 15 psia along the path $pv = C$ to a volume equal to that at the initial state; constant-volume process to close the cycle. For the closed system draw the $p$-$v$ diagram and calculate the net heat transfer for the cycle.

3.11 A battery is charged by applying a current of 50 amperes at 12 volts for a period of 30 minutes. During the charging process heat transfer from the battery is at the rate of 7.56 kJ/min. Calculate the change in stored energy of the battery during the charging process.

3.12 Benzene (specific gravity = 0.90) flows adiabatically through a pump at the rate of 1.80 $ft^3$/sec. The inlet pressure is 10 in Hg vacuum. The outlet, located 3 feet above the inlet, is at 24 psia. The inlet cross-sectional area is 0.36 $ft^2$ and the outlet area is 0.18 $ft^2$. Calculate the power required to drive the pump, the enthalpy change, B/lbm, across the pump and the internal energy change, B/lbm, across the pump. Assume the benzene is incompressible.

3.13 A machine operates adiabatically in a steady-flow manner with working fluid entering the machine at 690 kPa, specific volume 0.0625 $m^3$/kg, with a velocity of 60 m/s. At the outlet the pressure is 138 kPa, the specific volume is 0.375 $m^3$/kg, and the velocity is 365 m/s. Through the machine the internal energy decreases by 26.4 kJ/kg. For a flow rate of 5440 kg/hr determine the power in kilowatts and the outlet cross-sectional area in square meters.

3.14 A 115-volt direct-current electric motor runs at 3600 revolutions per minute delivering a torque of 220 in-lbf. The motor draws 83 amperes. Calculate the heat transfer when the motor is in steady-state operation at the given conditions.

3.15 Prove for all substances that the heat transfer for any constant-pressure, frictionless, and quasi-static process is equal to the enthalpy change.

3.16 A closed system undergoes a frictionless, quasi-static process along the path $p = 3.75v + 10$ where $p$ is in psia and $v$ in $ft^3$/lbm. The initial pressure is 15 psia. During the process 53,760 ft lbf/lbm of work is done by the system and the enthalpy increases by 400 B/lbm. Sketch the path on $p$-$v$ coordinates. Calculate the final pressure and specific volume, the internal energy change, and the heat transfer for the process.

3.17 Under what conditions will the $\int v dp$ be equal to the enthalpy change in a steady-flow process?

3.18 A certain substance undergoes a frictionless, adiabatic expansion through a nozzle (no work or potential energy change). At the nozzle inlet the velocity is 183 m/s, the pressure is 690 kPa, and the specific volume is 0.171 $m^3$/kg. Expansion through the nozzle follows $pv^{1.40}$ = constant. At the nozzle outlet the velocity is 580 m/s. Calculate (a) the enthalpy change, $h_2 - h_1$, and (b) the pressure at the nozzle outlet.

3.19 A refrigerant enters a condenser as a vapor at 80 C and leaves as a liquid at 35 C. The mass flow rate of the refrigerant is 0.20 kg/s. The condenser is water cooled and has its outer surface insulated. The cooling water enters at 10 C and leaves at 20 C. The initial and final values of the refrigerant enthalpies are 233 kJ/kg and 70 kJ/kg respectively. The enthalpy of the water increases by 42 kJ/kg. Assume no changes in kinetic or potential energies and that steady flow prevails. Find the mass flow rate of the water.

3.20 A frictionless air compressor operates under steady-flow conditions with $pv^{1.30}$ = constant. Air enters the compressor at the rate of 500 $ft^3$/min at 14 psia and 14 $ft^3$/lbm. The air leaves the compressor at 95 psia with negligible velocity. The inlet area is 20 $in^2$. In passing through the compressor the enthalpy of the air increases by 38 B/lbm. Calculate (a) the horsepower to drive the compressor and (b) the rate of heat transfer from the air, B/min.

3.21 During an adiabatic, frictionless, steady-flow process 1 kg of a gas expands from an initial state at 700 kPa, 0.312 $m^3$/kg, along the path $pv^{1.40}$ = constant. The gas does 40.68 kJ of work during the expansion. Calculate the final specific volume and the enthalpy change for the process. Assume that the changes in potential and kinetic energies are negligible.

3.22 Liquid $H_2O$ flows steadily downward through a vertical, converging tube. The enthalpy at a point 500 feet above a reference plane is 100.0 B/lbm; its enthalpy at a point 100 feet above the reference plane is 99.5 B/lbm. Assuming adiabatic flow, determine the velocity of the water stream at the lower elevation if its velocity at the upper level is 30 ft/sec. Assume standard sea-level acceleration of gravity. Would your result be affected if the process took place in a region where the local acceleration of gravity was 31.56 ft/sec$^2$?

3.23 A closed system of 0.270 lbm of a gas undergoes a frictionless, quasi-static process. Initially the gas is at 30 psia, 140 F, and occupies a volume of 2.00 ft$^3$. The process takes place along the path $pV$ = constant until the volume is doubled. During the process 28.5 B of heat transfer from the systems takes place. Calculate the enthalpy change for the process.

3.24 A certain substance undergoes a frictionless, adiabatic expansion through a nozzle. At the inlet to the nozzle the pressure is 690 kPa, the specific volume is 0.171 m$^3$/kg, and the velocity is 183 m/s. During the process the quantity $pv^{1.67}$ remains constant. At the nozzle outlet the pressure is 138 kPa and the area 11.5 cm$^2$. Assuming the potential energy change to be negligible, calculate the mass rate of flow through the nozzle in kilograms per hour.

3.25 Suppose the internal energy of a substance is expressed as

$$u = 831.0 + 0.617\, pv$$

where $u$ is in B/lbm, $p$ in psia, and $v$ in ft$^3$/lbm. A closed system of the substance undergoes a constant volume process that is adiabatic. Assuming the initial pressure is 10 psia, the initial specific volume is 3.00 ft$^3$/lbm, and the final pressure is 20 psia, calculate the work and the enthalpy change for the process.

3.26 For a certain substance the relation between pressure, specific volume, and absolute temperature is $pv = 53T$, where $p$ is in lbf/ft$^2$, $v$ is in ft$^3$/lbm, and $T$ is in Rankine units. The substance expands adiabatically through a turbine between the following states:

| | Inlet | Outlet |
|---|---|---|
| temperature | 1500 R | 800 R |
| pressure | — | 20 psia |
| velocity | negligible | 400 ft/sec. |

Elevation change is negligible. For the substance the internal energy is $u = 0.18T$, B/lbm, where $T$ is in Rankine units.

(a) Calculate the turbine work, B/lbm.
(b) For a flow rate of 10,000 lbm/min, calculate turbine power in horsepower.
(c) For a flow rate of 10,000 lbm/min, calculate the area of the turbine exhaust (outlet) in square feet.

3.27 Consider the cycle described in Figure 3.1a. Is this a heat engine cycle, a refrigeration cycle, a heat pump cycle, or none of these? Comment on the practicality of the cycle.

3.28 In Chapter 6 we will find that the second law of thermodynamics precludes the complete conversion of heat to work in a cyclic process. However, it is possible to convert work completely to heat in a cyclic process. Devise a cycle that will accomplish this. Is the conversion of work to heat desirable?

3.29 In a steam-electric generating station the heat rejected ($\dot{Q}_{out}$) is $10^9$ B/hr. Assuming the thermal efficiency is 38.5 percent, calculate the net power, in kilowatts, generated.

3.30 An automotive engine burns a petroleum fuel with a higher heating value of 19,000 B/lbm and a density of 57.4 lbm/ft³. Assuming $Q_{in}$ to be the mass rate of flow of the fuel multiplied by the higher heating value, calculate the fuel consumption in gallons per hour if the engine has a thermal efficiency of 30 percent when delivering 210 HP. Also calculate the specific fuel consumption, which is defined as the ratio of fuel use rate to power output.

3.31 A closed system of 1 kg of a substance undergoes the cycle shown below. Assuming each process is frictionless and quasi-static, calculate:

(a) Work for each process.
(b) If the thermal efficiency of the cycle is 30 percent, the heat transfer *from* the substance, in kJ/kg, for one cycle.

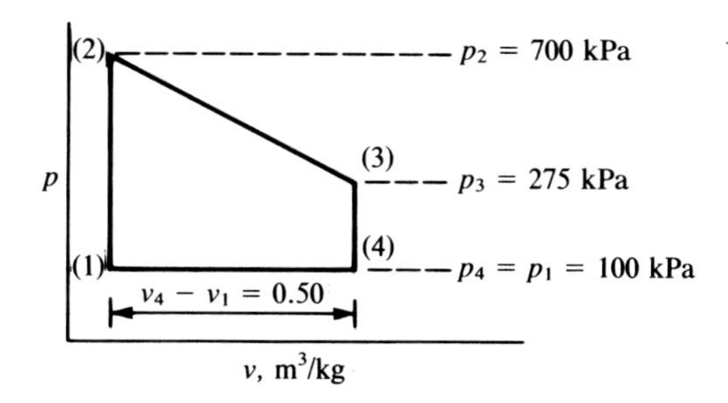


3.32 A refrigeration unit has a coefficient of performance of 3.20 and a refrigerating capacity of 3 tons. The unit is electrically driven using power that costs 4.2 cents per kilowatt-hour. Determine the cost of operating the unit continually for one 24-hour day.

3.33 A small, window air conditioner has a refrigerating capacity of 6,000 B/hr. The unit is powered by a 0.5-HP electric motor. Determine the coefficient of performance and the energy required, in kilowatt-hours, to operate the unit for 6 hours.

3.34 Write a formal definition of a heat pump.

3.35 A heat pump uses the atmosphere as an energy source. The electrically powered unit delivers 75,000 kJ/hr to the heated space and has a coefficient of performance of 2.15. Electric energy cost is 4.1 cents per kilowatt-hour. Calculate the energy cost to operate the unit for one hour. Compare this cost with that for the same heating effect produced using resistance heating only.

3.36 A heat pump with a coefficient of performance of 2.07 has a heating capacity of 50,000 B/hr. This capacity is augmented with resistance heating for midwinter operation. On a particular day 100,000 B/hr is required to maintain a temperature of 68 F in the heated space. Assuming electric energy is available at 4.5 cents per kilowatt-hour, calculate the energy cost, per hour, to (a) operate the heat pump and (b) operate the resistance heating.

# 4

# Characteristics of Pure Substances

*In Chapters 2 and 3 several properties of thermodynamic substances were defined. Energy equations for systems in which property changes occur were developed from the first law. The properties considered included pressure, temperature, specific volume, internal energy, and enthalpy. Processes for closed and open systems were considered, but methods to evaluate properties and property changes for specific substances are yet to be addressed. In this chapter we look in more detail at properties for pure substances, with emphasis on those substances that may exist in one or more phases in engineering applications.*

## 4–1 Phases and Phase Changes of Pure Substances

Pure substances may exist as solids, liquids, or vapors (gases). As noted in Chapter 2, these physical forms are known as phases of the substance. While several different solid phases of a pure substance may be identified, most substances encountered in thermodynamic systems have only one solid phase at the temperatures and pressures normally encountered.

In the earlier discussion of a pure substance, we noted that in the absence of gravity, motion, magnetic, electrical, and surface effects, an equilibrium state is fixed by two independent intensive properties. Experience has shown that pressure and temperature are dependent properties during a phase change of a pure substance. To illustrate, consider a system that is initially liquid $H_2O$. When a phase change from liquid to vapor occurs at a fixed pressure, the temperature will remain constant at a unique value that depends only on the pressure. During this phase change an additional independent intensive property must be known to establish a state and the proportion of each phase that exists at that state. In general, for two-phase systems the phase proportions and the state at either a known pressure or known temperature are fixed by either pressure or temperature *and* one additional independent property, such as specific volume, specific internal energy, or specific enthalpy. In single-phase systems pressure and temperature are always independent, and thus are adequate to fix the state of such systems.

Water, being economical and readily available, is encountered as the working fluid in many engineering systems. We shall use it as the example substance in this chapter. Figure 4.1 shows a *phase diagram* for $H_2O$. In this diagram the solid, liquid, and vapor regions are separated by the fusion, sublimation, and vapor-pressure curves. Only one solid phase is shown since pressures of the order of 30,000 psia (207 MPa) and greater are necessary to produce other solid phases of $H_2O$. The solid and liquid regions as well as the slopes of the lines bounding them are exaggerated in the figure. The triple point, the unique pressure–temperature combination at which solid, liquid, and vapor coexist, is indicated as the

**Figure 4.1** Phase Diagram for $H_2O$

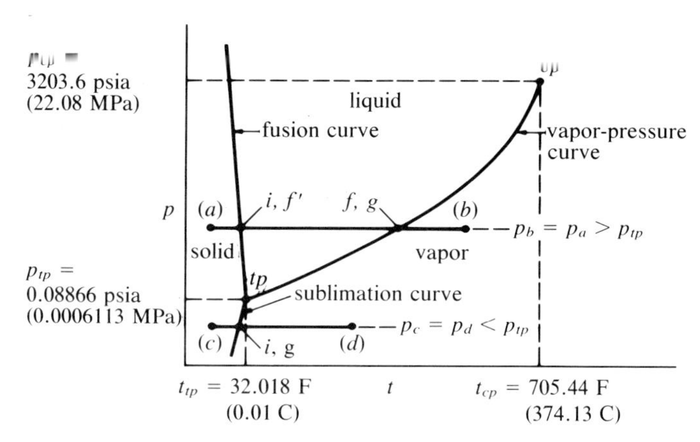


point *tp* on the diagram. On other property diagrams with coordinates such as pressure and specific volume a triple line appears. The triple point for $H_2O$ has, since 1954, been used as a standard point on the International Practical Temperature Scale, with the triple point fixed at 0.01 C. The distinction between liquid and vapor phases vanishes for states at pressures and temperatures greater than those at the *critical point.* This point is denoted as *cp* in Figure 4.1. Values for the critical temperature and critical pressure for several pure substances are given in Table 4.1.

In Section 3.6 it was shown for a quasi-static frictionless closed-system constant-pressure process that the heat transfer is equal to the change in enthalpy. Consider such a process occurring between (a) and (b) in Figure 4.1. The process takes place at a pressure greater than the triple-point pressure. As heat is transferred to the system at (a), the temperature will increase until the fusion curve is reached. At this point an interruption in temperature increase of the solid will be observed, and as heat transfer continues, a phase change from solid to liquid will take place at constant temperature. The heat transfer necessary to complete the solid-to-liquid phase change is the *enthalpy of melting* or, for processes proceeding in the opposite direction, the *enthalpy of fusion.* Following the phase change the temperature will again increase as heat is transferred to the $H_2O$, which is now in the liquid phase. It will be observed that in the liquid region the temperature increases almost linearly with the heat transferred until the vapor pressure curve is reached. Again the phenomenon of heat transfer producing no change in temperature will be observed. At a given pressure, a precise amount of energy, the *enthalpy of vaporization,* must be transferred to the $H_2O$ before a temperature change is observed. Following vaporization, further heat transfer will produce an increase in temperature as the system progresses in the vapor region along the constant pressure path to state (b).

Consider now the constant-pressure process (c) to (d) that takes place at a pressure less than the triple-point pressure in Figure 4.1. This process exhibits one significant difference when compared with process (a) to (b) in that the liquid phase never occurs. When the system reaches the sublimation curve, an interruption in temperature increase will be observed as heat transfer continues. The phase change in this instance is from solid to vapor.

**Table 4.1**

| | | Triple-Point Data | | | |
|---|---|---|---|---|---|
| | | Pressure | | Temperature | |
| Substance | Formula | psia | kPa | F | C |
| ammonia | $NH_3$ | 0.88 | 7.1 | −108 | − 77.8 |
| carbon dioxide | $CO_2$ | 75 | 517 | − 70 | − 56.7 |
| helium | He | 0.747 | 5.15 | −456 | −271 |
| hydrogen | $H_2$ | 0.994 | 6.85 | −434 | −259 |
| nitrogen | $N_2$ | 1.86 | 12.8 | −346 | −210 |
| oxygen | $O_2$ | 0.039 | 0.27 | −361 | −218 |
| water | $H_2O$ | 0.08866 | 0.6113 | 32.018 | 0.010 |
| | | **Critical-Point Data** | | | |
| | | **Pressure** | | **Temperature** | |
| | | **psia** | **kPa** | **F** | **C** |
| ammonia | $NH_3$ | 1639 | 11,300 | 270 | 132.2 |
| carbon dioxide | $CO_2$ | 1071 | 7384 | 88 | 31.1 |
| helium | He | 34 | 234 | −450 | −267.8 |
| hydrogen | $H_2$ | 188 | 1296 | −400 | −240.0 |
| nitrogen | $N_2$ | 493 | 3399 | −233 | −147.2 |
| oxygen | $O_2$ | 731 | 5040 | −182 | −118.9 |
| water | $H_2O$ | 3204 | 22,080 | 705.44 | 374.136 |

Such a process is termed *sublimation*, with the requisite heat transfer at constant pressure being the *enthalpy of sublimation.*

Figure 4.2 shows enthalpy versus temperature diagrams for both of the constant-pressure processes identified in Figure 4.1. The states in Figure 4.2 correspond to those in Figure 4.1. The various enthalpy changes identified with the phase changes are shown in the figure. The figure clearly indicates that the temperature remains constant during a constant pressure phase change. Temperature change occurs with heat transfer only when a single phase is present. As alluded to above, processes may proceed in both directions along the constant pressure paths shown. It is evident from examination of Figure 4.2 that enthalpy is a very descriptive property for systems undergoing phase changes.

## 4–2 Terminology for Two-Phase Systems of Pure Substances

The vapor-pressure curve, as illustrated in Figure 4.1 for $H_2O$, extends from the triple point to the critical point. The equation for this curve relating pressure and temperature during a phase change from liquid to vapor is the vapor-pressure equation. Similarly definite relations between pressure and temperature exist along the fusion and sublimation curves in Figure 4.1. The relations for each of these curves differ markedly between different pure substances, as indicated by the values of pressure and temperature at the critical and triple points listed in Table 4.1. The temperature at which a constant-pressure phase change occurs

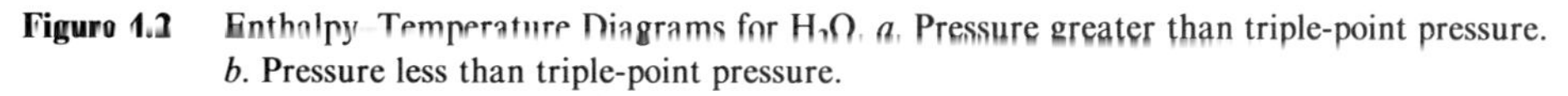

**Figure 4.2** Enthalpy-Temperature Diagrams for $H_2O$. *a.* Pressure greater than triple-point pressure. *b.* Pressure less than triple-point pressure.

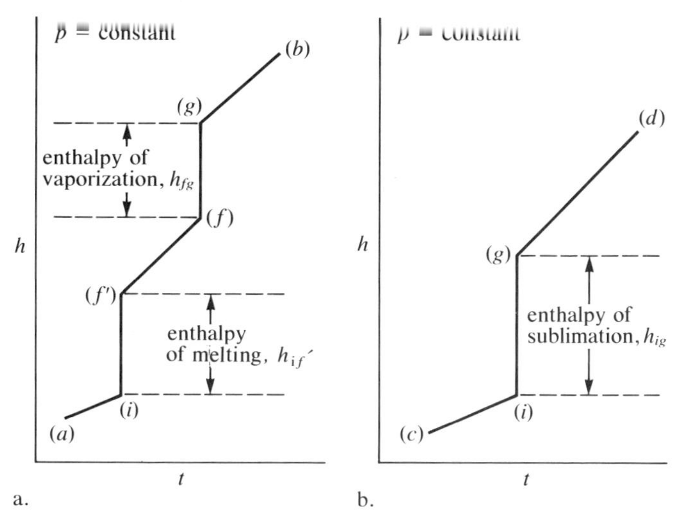


is the *saturation temperature.* In like manner, the pressure at which a constant-temperature phase change occurs is the *saturation pressure.* Thus the vapor-pressure curve, the fusion curve, and the sublimation curve in Figure 4.1 are curves of saturation pressure versus saturation temperature for the respective phase changes associated with the curves.

A liquid existing at a temperature less than the saturation temperature for the given pressure is said to be *subcooled* or *compressed* liquid. For example, the saturation temperature for $H_2O$ at a pressure of 1 standard atmosphere is 212 F. Liquid at 70 F and 1 standard atmosphere is then subcooled or compressed liquid. *Saturated liquid* is liquid existing at a saturation pressure and the corresponding saturation temperature, as state $(f)$ in Figures 4.1 and 4.2a. The vapor at these conditions is *saturated vapor* shown as $(g)$ in Figures 4.1 and 4.2a. Vapor at a temperature greater than the saturation temperature for a particular pressure is *superheated vapor,* shown as state $(b)$ in Figures 4.1 and 4.2a. A system consisting of saturated liquid and saturated vapor in equilibrium is a *mixture.* Any state between $(f)$ and $(g)$ on Figure 4.2a is a mixture.

Similar terminology may be applied to systems of solid and liquid in equilibrium and to solid and vapor in equilibrium at states where the pressure is less than the triple-point pressure.

## 4–3 The p-v-t Relation for Liquids and Vapors

The relation between $p$, $v$, and $t$ for a substance is called its equation of state. For many substances the mathematical relation of pressure, specific volume, and temperature is so complex that its use for ordinary engineering calculations is impractical. Examples of such substances include $H_2O$ and most refrigerants. Because of this complexity, tables of properties are prepared for the substances. These tables enable one to determine the properties

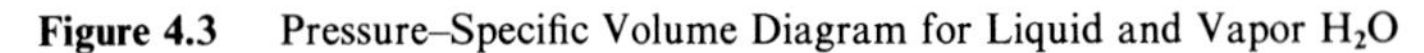

**Figure 4.3** Pressure–Specific Volume Diagram for Liquid and Vapor $H_2O$

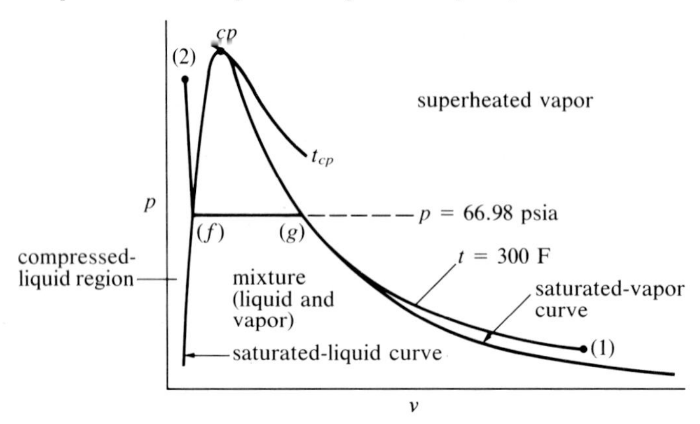


of a pure substance if two independent properties are known. The most widely used tables of properties are those for $H_2O$. For this reason properties from the *Steam Tables* (*English units*) prepared by Keenan et al.* will be used to illustrate the procedures employed in treating substances having complex *p-v-t* relations. Appendices B and C present abridged *Steam Tables;* SI units in Appendix B and English units in Appendix C.

Tables 1 and 2 of *Steam Tables* contain essentially the same data. However, Table 1 is arranged in terms of saturation temperatures and Table 2 has saturation pressure as the primary entry. The interdependency of Tables 1 and 2 is established by the vapor-pressure equation. (Values from Tables 1 and 2 have been combined into a single saturation table in the abridged tables given in Appendices B and C.) From either Table 1 or Table 2 values of $v_f$ and $v_g$ can be obtained for a given saturation temperature (Table 1) or saturation pressure (Table 2). The subscripts *f, fg,* and *g* indicate, respectively, saturated liquid, the change in going from saturated liquid to saturated vapor, and saturated vapor. This information can be used to prepare a *p-v* diagram or a *t-v* diagram for $H_2O$. Consider the *p-v* diagram for $H_2O$ in Figure 4.3. The saturated-liquid line, terminating at the critical point, is developed using $v_f$ data from Table 2. The region to the left of this line has been exaggerated to facilitate illustration of compressed-liquid states. In like manner $v_g$ data from Table 2 are used to develop the saturated-vapor curve, which approaches infinite specific volume as the pressure approaches zero.

Three regions are apparent once the saturation lines have been developed. That to the left of the saturated-liquid line contains compressed-liquid states. Between the saturated-liquid and saturated-vapor lines we have mixtures of saturated liquid and saturated vapor. To the right of the saturated vapor line are superheated-vapor states.

Table 3 of the *Steam Tables* (Tables B–2 and C–2 in Appendices B and C) contains data for states in the superheat region where pressure and temperature are independent properties. Isotherms (constant-temperature lines) can be plotted on *p-v* coordinates by selecting a given temperature, say 300 F, and noting the specific volumes at a series of pressure and

*Keenan, J. H., F. G. Keyes, P. G. Hill, and J. G. Moore. *Steam Tables* (*English Units*). New York: Wiley, 1969.

**Figure 4.4** The *p-v-t* Surface for a Substance that Contracts on Solidification

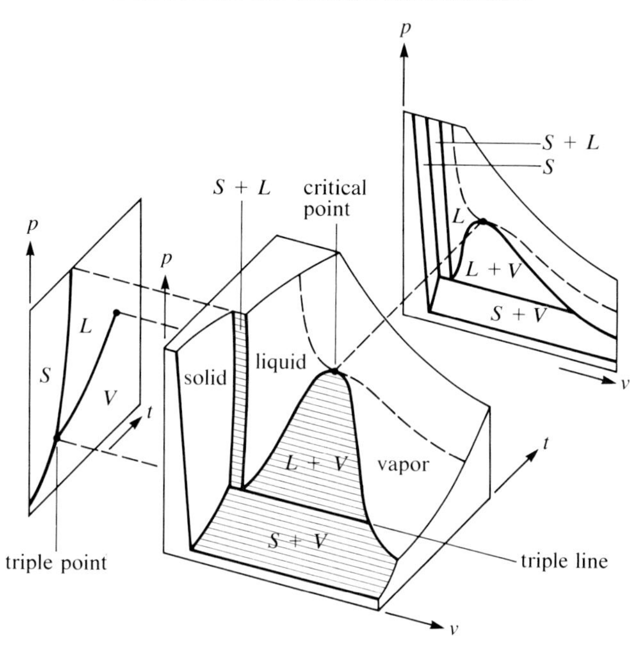


the selected temperature. These data, from Table 3, establish the isotherm (1) to ($g$) in Figure 4.3. Through the mixture region from ($g$) to ($f$) the isotherm is coincident with its saturation pressure since $p$ and $t$ are dependent. In the compressed-liquid region the isotherm ($f$) to (2) has a negative slope of extremely large magnitude.

An exact plot of the subcooled liquid region isotherm ($f$) to (2) in Figure 4.3 can be developed by using the data of Table 4 (Tables B–4 and C–4 in Appendices B and C). Inspection of this table reveals minute changes in properties with pressure for a fixed temperature. For example, at 300 F the specific volume changes from 0.017416 to 0.017343 ft$^3$/lbm (0.4 percent change) when the pressure increases from 500 to 1500 psia. At moderate pressures the effect of pressure on other thermodynamic properties of the subcooled liquid at a given temperature is small. For this reason the properties of subcooled liquids can, with good accuracy, be assumed to depend only on *temperature* when pressures are less than about 500 psia. Using this approximation, specific volumes, internal energies, and enthalpies are taken as $v_f$, $u_f$, and $h_f$ from Table 1 at the *temperature* of the compressed liquid.

## 4–4 The p-v-t Surface

A graphical representation of the equation of state for a substance is the *p-v-t* surface. Figure 4.4 shows a *p-v-t* surface for a substance that contracts during solidification. The *p, v,* and *t* coordinates are mutually perpendicular and each point on the surface represents a possible equilibrium state for the substance. Quasi-static processes (processes passing through equilibrium states) trace paths on the surface. The areas of the surfaces that represent single phases are generally double-curved surfaces. The areas that are shown by multiple lines

**Figure 4.5** The *p-v-t* Surface for $H_2O$

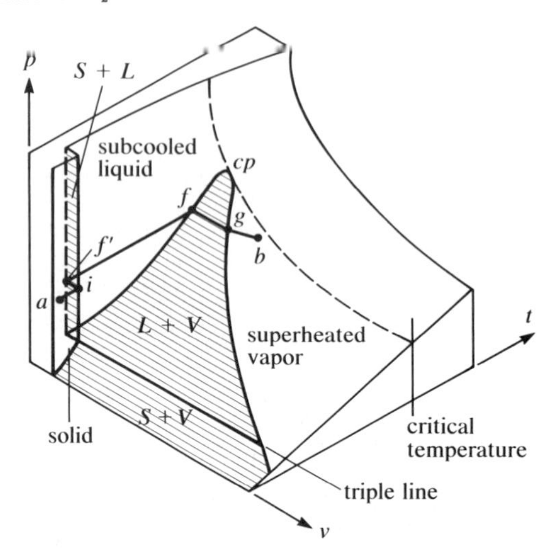


(Courtesy of Prof. Kenneth R. Jolls, Department of Chemical Engineering, Iowa State University of Science and Technology.)

represent two-phase regions. These lines are straight lines and are parallel to the specific volume axis, that is, they are formed by the intersection of planes of constant pressure and the corresponding saturation temperature. The abbreviations denote the phases present for the mixtures in these regions. Note in Figure 4.4 that contraction on solidification (freezing) is indicated by the decrease in specific volume in proceeding at constant pressure from the single-phase liquid region to the single-phase solid region. Projections of the surface onto *p-v* and *p-t* planes are shown in the diagram. Note that the triple line appears as a point in the *p-t* plane and that the critical point is projected on to both planes. It is easy to see from the *p-v-t* surface that any constant-pressure process for which $p > p_{cp}$ cannot involve a phase change from liquid to vapor. Such a process starting in the liquid region and proceeding to a high temperature traces a constant pressure path along the curved surface above the critical point.

Most substances contract on freezing. However, water is a notable exception. The *p-v-t* surface for water is shown in Figure 4.5. As in Figure 4.4, the lined surfaces indicate two-phase regions. The solid–liquid region is a hidden surface in the diagram. States (*a*), (*i*), (*f'*), (*f*), (*g*), and (*b*) in Figures 4.1 and 4.2a are duplicated in Figure 4.5. The freezing process occurs from *f'*, a hidden point, to *i*, which, along with *a*, is on the surface representing the solid phase. This process involves an increase in specific volume, indicating an expansion on freezing. The specific volume scale for the liquid and solid phases has been expanded for clarity because of the small magnitude of the specific volumes of these phases. Figure 4.1 is produced by projecting the *p-v-t* surface in Figure 4.5 onto the *p-t* plane. In like manner, Figure 4.6 shows a projection of the *p-v-t* surface onto the *p-v* plane. We can now identify Figure 4.3 as a projection of the subcooled liquid, liquid and vapor mixture, and the superheated vapor regions of the *p-v-t* surface onto the *p-v* plane. These regions are those most frequently encountered. Therefore, *p-v* diagrams such as Figure 4.3 will suffice for many applications.

Figure 4.6 Projection of the $p$-$v$-$t$ Surface for $H_2O$ onto the $p$-$v$ Plane

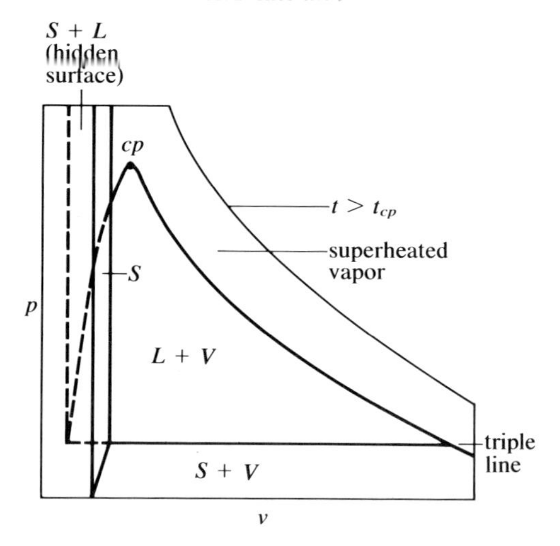


## 4–5 Properties of Equilibrium Two-Phase Mixtures

In engineering applications the most common two phase mixtures of a pure substance are those of saturated liquid and saturated vapor. These mixtures, as noted earlier, will be at a fixed temperature for any one pressure of a pure substance. The *quality* of such mixtures is, by definition,

$$x \equiv \frac{\text{mass vapor}}{\text{mass of liquid} + \text{mass of vapor}} = \frac{m_g}{m_f + m_g} \tag{4.1}$$

Thus the quality $x$ is the mass fraction of saturated vapor in a liquid–vapor equilibrium mixture. Utilizing quality, the properties of such mixtures are calculated.

Values for the properties specific volume, internal energy, and enthalpy at states in the liquid–vapor mixture region are found by adding the contribution of the liquid and of the vapor. For a unit mass of mixture

$$a = (1 - x)a_f + xa_g \tag{4.1a}$$

where $a$ is any specific property (specific volume, specific internal energy, or specific enthalpy). The product $xa_g$ is the contribution to the mixture specific property $a$ made by $x$ mass units of vapor ($x \leq 1$), and $(1 - x)a_f$ is the contribution made by $(1 - x)$ mass units of liquid. By introducing the quantity $a_{fg} = a_g - a_f$, eq. (4.1a) can be written as

$$a = a_f + xa_{fg} \tag{4.1b}$$

Equations (4.1a) and (4.1b) yield identical results. However, eq. (4.1a) is usually more convenient for specific volume calculations, while eq. (4.1b) is somewhat more convenient to use for the evaluation of other specific properties.

As an example, suppose we have a mixture of $H_2O$ at 80 psia, quality 59 percent. The specific enthalpy of the mixture is given by

$$h = h_f + xh_{fg}$$

The values for $h$ tabulated in the *Steam Tables* are consistent with the table reference value for internal energy, $u_f = 0$ at the triple point, and the definition of enthalpy, $h = u + pv$.

From Table 2 of the *Steam Tables* the values of $h_f$ and $h_{fg}$ are obtained and we have

$$\begin{aligned} h &= 282.21 + (0.59)(901.4) \\ &= 814.0 \text{ B/lbm of mixture} \end{aligned}$$

The given state is in the mixture region; thus the temperature is the saturation temperature for 80 psia. From Table 2, the mixture temperature is 312.07 F.

Calculation of the specific volume of liquid–vapor mixtures can be carried out in two ways, depending on the accuracy desired. Consider an $H_2O$ mixture at 210 F, quality 93 percent. The exact specific volume of the mixture is

$$v = (1 - x)v_f + xv_g$$

At 210 F, from Table 1, values of $v_f$ and $v_g$ are obtained and the specific volume is

$$\begin{aligned} v &= (0.07)(0.016702) + (0.93)27.82 \\ &= 0.00117 + 25.87 \\ &= 25.87 \text{ ft}^3\text{/lbm of mixture} \end{aligned}$$

For these conditions it is apparent the volume of the liquid portion of the mixture is very small and the specific volume can be taken as

$$v \cong xv_g$$

The accuracy of this approximation decreases as the mixture pressure or temperature increases. Low quality also will increase the error of this approximation. Ignoring the contribution of the liquid in the calculation of other mixture properties such as internal energy and enthalpy does not yield acceptable results since the saturated liquid and saturated vapor values are frequently of the same order of magnitude.

The methods used in the preceding paragraph to calculate properties of liquid–vapor mixtures of $H_2O$ are applicable to other mixtures such as refrigerants. Abridged tables of properties for two refrigerants, ammonia and dichlorodifluormethane (Freon–12), are given in Appendices D and E.

The properties of liquid–solid or solid–vapor equilibrium mixtures are treated in the same manner as liquid-vapor mixtures. For $H_2O$ the saturated-solid and saturated-vapor properties for temperatures at or below the triple point are given in Table 6 of the *Steam Tables*. Corresponding abridged tables are given in Appendices B and C.

## Example 4.a

The following table gives several states for 1 pound of $H_2O$. Complete the table. Show each state on a $p$-$v$ diagram.

| State | Pressure, Psia | Temperature, F | Specific Volume, $v$, ft³/lbm | Enthalpy, $h$, B/lbm | Internal Energy, $u$, B/lbm |
|---|---|---|---|---|---|
| a | 14.7 | 60 | | | |
| b | 500 | 800 | | | |
| c | 100 | | | 975 | |
| d | 5000 | 150 | | | |
| e | 1250 | | | | 732.0 |

**Solution:**

(a) The temperature is less than the saturation temperature for the given pressure, 212 F, therefore the $H_2O$ is subcooled. At the indicated pressure and temperature the effect of pressure on properties of subcooled liquid is negligible. Therefore the properties are taken as those with the $f$ subscript at the given temperature, 60 F. From Table 1

$$v = v_{f_{60}} = 0.016035 \text{ ft}^3/\text{lbm}$$

$$h = h_{f_{60}} = 28.08 \text{ B/lbm}$$

$$u = u_{f_{60}} = 28.08 \text{ B/lbm}$$

(b) From Table 2 we read the saturation temperature at 500 psia as 467.13 F. Since the given temperature is larger than this value, state $b$ lies in the superheat region. From the superheat table, Table 3

$$v = 1.4407 \text{ ft}^3/\text{lbm}$$

$$u = 1278.8 \text{ B/lbm}$$

$$h = 1412.1 \text{ B/lbm}$$

(c) State $c$ can be located by examining the enthalpy values in Table 2 at 100 psia. The given enthalpy is less than $h_g$ but greater than $h_f$, therefore, a mixture is indicated and

$$h = 975.0 = h_f + xh_{fg}$$
$$= 298.61 + x(889.2)$$

which yields the quality

$$x = \frac{975.0 - 298.61}{889.2} = 0.7607$$

The mixture temperature will be the saturation temperature for 100.0 psia. From Table 2

$$t = 327.86 \text{ F}$$

Since the pressure is moderate and the quality high, the specific volume can be approximated as

$$\begin{aligned} v &\cong x\, v_g \\ &= (0.7607)4.434 \\ &= 3.373 \text{ ft}^3/\text{lbm} \end{aligned}$$

Had the exact equation been used, the specific volume would have been

$$\begin{aligned} v &= v_f + x\, v_{fg} \\ &= 0.017736 + 0.7607(4.434 - 0.017736) \\ &= 3.377 \text{ ft}^3/\text{lbm} \end{aligned}$$

Internal energy is given by

$$\begin{aligned} u &= u_f + x\, u_{fg} = 298.28 + 0.7607(807.5) \\ &= 912.6 \text{ B/lbm} \end{aligned}$$

(d) The pressure 5000 psia is much greater than the saturation pressure at 150 F (3.722 psia). Therefore state $d$ is in the subcooled liquid region. Because of the high pressure it is appropriate to use the compressed liquid table, Table 4, to find the properties. From that table

$$v = 0.016104 \text{ ft}^3/\text{lbm}$$

$$h = 130.17 \text{ B/lbm}$$

$$u = 115.27 \text{ B/lbm}$$

Had the effect of pressure been neglected, values from Table 1 would be

$$v = v_{f_{150}} = 0.016343 \text{ ft}^3/\text{lbm}$$

$$h = h_{f_{150}} = 117.96 \text{ B/lbm}$$

$$u = u_{f_{150}} = 117.95 \text{ B/lbm}$$

(e) A check of Table 2 at 1250 psia indicates a mixture since the given internal energy is greater than $u_f$ but less than $u_g$. For the mixture the temperature is 572.56 F. The quality is determined by

$$732.0 = 573.4 + x(528.3)$$

$$x = 0.300$$

In this problem $(1 - x)v_f$ is not insignificant. The specific volume and enthalpy are calculated as

$$v = 0.02250 + (0.300)(0.3454 - 0.02250)$$
$$= 0.1194 \text{ ft}^3/\text{lbm}$$

$$h = 578.6 + 0.300(603.0)$$
$$= 759.5 \text{ B/lbm}$$

Each of the states ($a$) through ($e$) is shown on the $p$-$v$ diagram below. Values of $p$ and $t$ at each state are noted on the figure. Scales have been distorted to display the states.

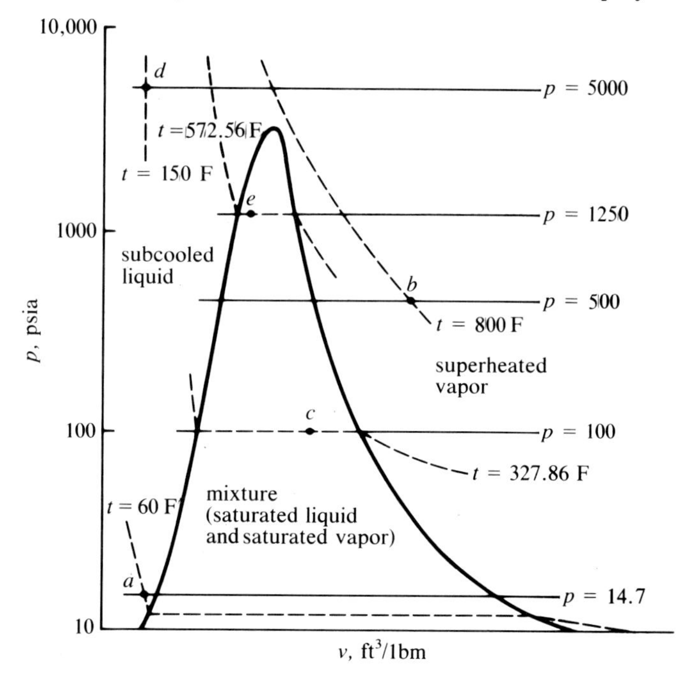

## Example 4.b

A rigid 0.150-m$^3$ tank contains $H_2O$ at $t_1 = 320$ C and $p_1 = 2050$ kPa. The temperature of the $H_2O$ is decreased to 150 C. Find the internal energy change for the $H_2O$ and the final pressure. Use the property tables in Appendix B.

**Solution:**

Neither the tank volume nor the mass changes during the process. Therefore, the final specific volume is equal to the initial specific volume. Inspection of Table B–1, the saturation table, reveals that $p_1 = 2.05$ MPa is less than $p_{sat}$ at 320 C. Thus state (1) is in the superheat region and from Table B–2

$$v_1 = 127.44 \times 10^{-3} \text{ m}^3/\text{kg} = v_2$$

$$u_1 = 2806.9 \text{ kJ/kg}$$

State (2) lies in the mixture region since $v_1 = 0.12744$ m$^3$/kg is between $v_f = 0.0010905$ m$^3$/kg and $v_g = 0.3928$ m$^3$/kg at 150 C. Thus $p_2 = p_{sat} = 475.8$ kPa. The quality at (2) is found from

$$v_2 = (1 - x_2)v_f + x_2 v_g$$

$$0.12744 = (1 - x_2)0.0010905 + x_2\, 0.3928$$

which yields $x_2 = 0.324$. Ignoring the product $(1 - x_2)v_f$ in the above equation produces $x_2 = 0.323$. Continuing with values from Table B–1

$$\begin{aligned} u_2 &= u_f + x_2 u_{fg} \\ &= 631.68 + (0.324)(2559.5 - 631.68) \\ &= 1256.3 \text{ kJ/kg} \end{aligned}$$

The specific internal energy change is, then,

$$u_2 - u_1 = 1256.3 - 2806.9 = -1550.6 \text{ kJ/kg}$$

The mass is determined as

$$m = V/v = 0.150/0.12744 = 1.177 \text{ kg}$$

and the internal energy change is

$$\begin{aligned} U_2 - U_1 &= m(u_2 - u_1) = 1.177\,(-1550.6) \\ &= -1825 \text{ kJ} \end{aligned}$$

## Example 4.c

Ammonia initially at a temperature of −6.7 C and a quality of 0.87 undergoes a process to state (2) where $t_2$ is 18 C greater than the critical temperature and $p_2$ is 9650 kPa less than the critical pressure. Find $p_1$ and the change in specific volume and enthalpy for the process. Use the ammonia property tables (Appendix D) and give final answers in SI units.

**Solution:**

The values for $p_2$ and $t_2$ are found using the critical point data for ammonia from Table 4.1.

$$t_2 = 132.2 + 18 = 150.2 \text{ C}$$

$$p_2 = 11{,}300 - 9650 = 1650 \text{ kPa}$$

It is noted that the values in the tables in Appendix D are given in English units. Hence, to determine property values, it is necessary to convert the given properties into English units. This is accomplished by use of the relation between temperature on the Fahrenheit and Celsius scales given in Section 2–11 and the pressure-conversion factor 6.893 kPa/psia from Appendix A. The converted values are shown in parentheses in the following table.

| | **State (1), $x_1$ = 0.87** | **State (2)** |
|---|---|---|
| $t$, C | −6.7 | 150.2 |
| $t$, F | (19.9) | (302.4) |
| $p$, kPa | (332.3) | 1650 |
| $p$, psia | 48.21 | (239.4) |

The value for $p_1$ in psia units was read as the saturation pressure at $t$ = 20 F from Table D–1. Using Table D–1 and Appendix A,

$$\begin{aligned} h_1 &= (1 - x_1)h_f + x_1 hg \\ &= (1 - 0.87)64.7 + 0.87(617.8) = 545.90 \text{ B/lbm} \\ &= (545.90 \text{ B/lbm})[(2.326 \text{ kJ/kg})/(1 \text{ B/lbm})] \\ &= 1270 \text{ kJ/kg} \end{aligned}$$

Similarly

$$\begin{aligned} v_1 &= (1 - 0.87)(0.02474) + 0.87(5.910) = 5.145 \text{ ft}^3/\text{lbm} \\ &= (5.145 \text{ ft}^3/\text{lbm})(1 \text{ m}^3/\text{kg}/16.02 \text{ ft}^3/\text{lbm}) \\ &= 0.3213 \text{ m}^3/\text{kg} \end{aligned}$$

State (2) is found in the superheat region. From Table D–3, at $p$ = 240 psia and $t$ = 300 F,

$$v_2 = 1.894 \text{ ft}^3/\text{lbm} = 0.1182 \text{ m}^3/\text{kg}$$

$$h_2 = 762.0 \text{ B/lbm} = 1772 \text{ kJ/kg}$$

Thus

$$h_2 - h_1 = 1772 - 1270 = 502 \text{ kJ/kg}$$

$$v_2 - v_1 = 0.1182 - 0.321 = -0.203 \text{ m}^3/\text{kg}$$

States (1) and (2) are shown on the following *p*-*v* diagram. The scales are distorted to display the states in relation to the critical and triple-point values.

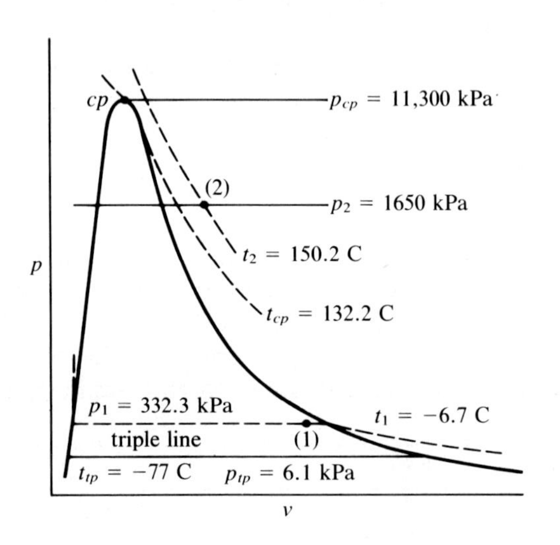


## Example 4.d

A throttling process is a steady-flow process that takes place in such a manner that no work or heat transfer occurs and changes in potential and kinetic energies are negligible. A partially closed valve, an orifice, or a similar restriction that produces a pressure drop is present in the flow. Consider a throttling process for a refrigerant, Freon 12, across a valve as shown in the figure. Describe the state at the exit section.

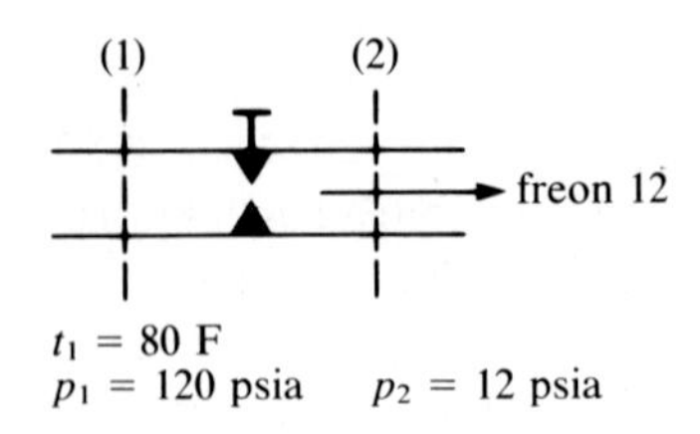

**Solution:**

Noting that work, heat transfer, and changes in kinetic and potential energies are zero, the steady-flow energy equation reduces to

$$h_1 = h_2$$

From Table E–1 for Freon 12, Appendix E, we note that state (1) is a subcooled liquid state since $t_1 = 80$ F is less than the saturation temperature corresponding to $p_1 = 120$ psia, 93.29 F. Consequently, $h_1$ is taken as $h_f$ at 80 F. From Table E–2, $h_1 = 26.365$ B/lbm. The independent properties $p_2$ and $h_2 = h_1$ establish state (2). Examination of Table E–2 at $p_2 = 12$ psia reveals that $h_2 = 26.365$ is between $h_f$ and $h_g$. Therefore, state (2) lies in the mixture region and $t_2$ is the saturation temperature at 12 psia, $-30$ F. The quality at state (2) is determined from the following equation.

$$h_2 = h_f + x_2\, h_{fg}$$

$$26.365 = 2.1120 + x_2(74.015 - 2.1120)$$

From this expression, $x_2 = 0.337$, indicating that 33.7 percent of the mixture at (2) is vapor.

## 4–6 Specific Heats for Pure Substances

For a pure substance existing in any one phase the constant-pressure specific heat $C_p$ and the constant-volume specific heat $C_v$ are defined as

$$C_p \equiv \left(\frac{\partial h}{\partial T}\right)_p \quad \text{and} \quad C_v \equiv \left(\frac{\partial u}{\partial T}\right)_v \tag{4.2}$$

These specific heats are properties and, as such, are single valued at a given state of the substance. Typically

$$C_p = f(T, P)$$

and

$$C_v = f(T, v)$$

The constant-pressure specific heats for liquid $H_2O$, given on pages 120 and 121 of the *Steam Tables,* illustrate the state dependency. For $H_2O$ in the vapor phase, similar data appear on page 122 of the *Steam Tables*. The specific heat at a given state is called the *instantaneous specific heat* (either $C_p$ or $C_v$).

In practice, constant-pressure specific heats of a pure substance frequently are expressed as a pure temperature function at a known, constant pressure. From this information $C_p$ at other pressures and temperatures can be calculated. The values of $C_v$ can then be determined if the $p$-$v$-$T$ relation (equation of state) is known.

## 4–7 Summary

The relationships among the properties of a pure substance, particularly one undergoing phase changes, are complex. As a result, tables of properties have been prepared for a number of substances encountered in engineering.

When two phases of a pure substance coexist in an equilibrium mixture, the mixture pressure and the mixture temperature have a definite relationship. These values of pressure and temperature are called the *saturation pressure* and the *saturation temperature* and the two phases are called *saturated phases*. The term *quality, x,* denotes the ratio of the mass of the vapor to the mass of the vapor and liquid in a mixture of saturated liquid and saturated vapor. Quality ranges from zero (all saturated liquid) to unity (all saturated vapor). The expressions

$$a = a_f + xa_{fg}$$

$$a = (1 - x)\, a_f + xa_g$$

where $a$ is $h$, $u$, or $v$, permit these properties to be found for liquid–vapor mixtures. The subscripts $f$ and $g$ refer to saturated liquid and saturated vapor, respectively, and subscript $fg$ denotes $a_g - a_f$.

Liquid that exists at a temperature lower than the saturation temperature corresponding to its pressure is called *subcooled* or *compressed* liquid. Vapor existing at a temperature higher than the saturation temperature corresponding to its pressure is called *superheated vapor.*

A pure substance exhibits a *triple point*—a condition under which three phases (solid, liquid and gas, or vapor) can coexist in equilibrium. At low pressure $v_g$ and $v_f$ differ by large values. As pressure increases, $v_g - v_f = v_{fg}$ decreases, until a state is reached where $v_{fg} = 0$. This state is called the critical point. At a pressure and temperature greater than those at the critical point, liquid and vapor phases are not distinguished.

Specific heats for pure substances are defined as

$$C_v \equiv \left(\frac{\partial u}{\partial T}\right)_v \quad C_p \equiv \left(\frac{\partial h}{\partial T}\right)_p$$

and are thermodynamic properties.

### Selected References

Ellenwood, F. O., and C. O. Mackey. *Thermodynamic Charts.* New York: Wiley, 1944.

Gibbs, J. W. *The Collected Works of J. W. Gibbs.* New Haven, Conn.: Yale University Press, 1928.

Hougen, O. A., K. M. Watson, and R. A. Ragatz. *Chemical Process Principles, Part I,* 2nd ed. New York: Wiley, 1958.

Jones, J. B., and G. A. Hawkins. *Engineering Thermodynamics.* New York: Wiley, 1960.

Moore, W. J. *Physical Chemistry,* 3rd ed. Englewood Cliffs, N.J.: Prentice-Hall, 1964.

Potter, J. H. *Steam Charts, ASME SI–10.* New York: American Society of Mechanical Engineers, 1976.

Stoever, H. J. *Engineering Thermodynamics.* New York: Wiley, 1951.

Wheeler, L. P. *Josiah Willard Gibbs, The History of a Great Mind,* 2nd ed. New Haven, Conn.: Yale University Press, 1952.

## Problems

4.1 Consider a system of 1 lbm of dry air at 14.7 psia and 70 F. Assume the air is composed of only nitrogen, oxygen, and argon. At 1 atmosphere the boiling point of argon is −303 F, that of nitrogen −320.4 F, and that of oxygen −297.3 F. Is the air at 14.7 psia and 70 F a pure substance? Would the air be a pure substance at −200 F? What would be the minimum temperature at which this air at 14.7 psia could be considered a pure substance in the gas phase? What would be the maximum temperature at which this air could be considered a pure substance in the liquid phase?

4.2 A mixture of 1 mole of oxygen and 0.5 mole of hydrogen, each in the gas phase, is contained in a rigid, perfectly insulated tank. The mixture is ignited by a spark of negligible energy and, after a period of several minutes, all the hydrogen has burned to $H_2O$. Is the initial mixture a pure substance? At any time during the combustion process would the system be a pure substance? At the completion of the process would the system be a pure substance? What would be the internal energy change for the process?

4.3 A rigid tank is divided into two parts by a partition. One side contains oxygen at 130 kPa and the other nitrogen at 300 kPa, each at 30 C. The partition is broken and the two gases mix until uniform pressure, temperature, and composition are reached. For the system bounded by the tank walls, is the system a pure substance (a) at the initial state? (b) at the final state?

4.4 Suppose a container holds water at 70 F and ice at 20 F. Is the ice–water combination a pure substance? Would the system be a pure substance when half of the ice had melted?

4.5 Plot a phase ($p$-$t$) diagram for $H_2O$ that has 0 F as the lowest temperature and the critical temperature the highest temperature. Identify regions on the diagram where

(a) Only liquid exists
(b) Only solid exists
(c) Only vapor exists
(d) Solid to vapor phase changes can occur
(e) There is no differentiation between liquid and vapor

4.6 Plot the constant specific volume line $v = 4.434$ ft³/lbm and the constant enthalpy lines $h = 100$ B/lbm, $h = 900$ B/lbm, and $h = 1100$ B/lbm on the phase diagram.

4.7 Assuming the saturation curve for solid–liquid $H_2O$ to be a linear relation of pressure and temperature and the melting point of $H_2O$ to be 19.77 F at 2000 atmospheres, what is the melting-point temperature for $H_2O$ (a) at 1000 psia? (b) at 10,000 psia? What phase or phases would exist if $H_2O$ were at 10,000 psia and 10 F?

4.8 The triple point of carbon dioxide is at 517 kPa and −56.7 C. Suppose $CO_2$ gas initially at 16 C and 500 kPa is cooled at constant pressure to −70 C. Will any $CO_2$ liquid be formed during the process?

4.9 Describe the conditions under which oxygen can undergo sublimation.

4.10 On log-log paper plot the following constant-temperature lines (isotherms) on *p-v* coordinates for $H_2O$.

(a) 210 F from saturated vapor to a pressure of 0.200 psia
(b) 300 F from quality 50 percent to pressure of 1.00 psia
(c) 700 F from saturated vapor to a pressure of 14.696 psia
(d) 720 F from a pressure of 3300 psia to a pressure of 10 psia

4.11 Calculate the specific volume of each of the following:

(a) $H_2O$ at 10 psia, 50 percent quality
(b) $H_2O$ at 100 psia, 50 percent quality
(c) Ammonia at 30 F, 80 percent quality
(d) Refrigerant F-12 at 0 F, 95 percent quality

4.12 A 5 $ft^3$ tank contains 2 lbm of $H_2O$ at 300 F. Calculate the mass of liquid, the mass of vapor, the volume occupied by the liquid, and the volume occupied by the vapor.

4.13 Two kilograms of $H_2O$ are trapped in a rigid 0.62-$m^3$ tank. The temperature of the $H_2O$ is 150 C. What is the pressure? Calculate the mass of the liquid and the mass of the vapor and the volume occupied by each phase.

4.14 A rigid tank contains 2 lbm of saturated water and 3 lbm of saturated steam at 40 psia. What is the volume of the tank? If the temperature is raised to 500 F, what will the pressure be?

4.15 Determine the temperature of $H_2O$ at 150 psia if its specific volume is (a) 2.000 $ft^3/lbm$ and (b) 4.000 $ft^3/lbm$.

4.16 Determine the pressure of steam at 300 F if its specific volume is 0.500 $ft^3/lbm$, (b) 2.000 $ft^3/lbm$, and (c) 20.30 $ft^3/lbm$.

4.17 What is the specific volume of $H_2O$ (a) at 1500 psia and 1000 F and (b) at 1500 psia and 500 F?

4.18 A rigid tank contains steam at 200 psia and 800 F. If the steam is cooled to 400 F, what will its pressure be?

4.19 A rigid tank contains 3 lbm of saturated water and 1 lbm of saturated steam at 60 psia. Heat is transferred to the $H_2O$ until all the liquid is evaporated. What is the final pressure?

4.20 Complete the following table for 1 pound of $H_2O$:

| *p*, psia | *t*, F | *v*, $ft^3/lbm$ | *h*, B/lbm | *u*, B/lbm |
|---|---|---|---|---|
| 70 | 500 | | | |
| 20 | | | 930.0 | |
| | 300 | 5.600 | | |
| 14.696 | 70 | | | |
| | 10 | | −50.00 | |
| 5000 | 200 | | | |
| 5000 | 800 | | | |

4.21 Complete the following table for $H_2O$. Use Appendix B.

| $p$, kPa | $t$, C | $v$, m³/kg | $h$, kJ/kg | $u$, kJ/kg |
|---|---|---|---|---|
| 360 | 180 | | | |
| 150 | | | 2000 | |
| | 150 | 0.205 | | |
| 101.3 | 45 | | | |
| 2050 | | | | 2807 |

4.22 Throttling is a steady-flow process that takes place in such a manner that no heat or work interactions occur and the changes in potential and kinetic energies are negligible. Suppose $H_2O$ is throttled across an insulated valve from an initial pressure of 50 psia to a final pressure of 15 psia. Assuming the $H_2O$ entering the valve is saturated liquid, determine the mass proportions of liquid and vapor downstream from the valve.

4.23 Initially at a pressure of 3400 kPa and a quality of 95.1 percent, $H_2O$ is throttled to a pressure of 100 kPa. Find the final temperature and the specific volume change. Refer to the previous problem for a description of the throttling process.

4.24 Refrigerant F-12 enters the expansion valve of a refrigeration unit as subcooled liquid at 80 F and 200 psia. The refrigerant is throttled across the valve to a temperature of −5 F. If the mass rate of flow into the valve is 12 lbm/min, what is the volume rate of flow in ft³/min, immediately downstream from the expansion valve?

4.25 Steam at 100 psia is throttled to 15 psia. If the temperature of the steam at 15 psia is 260 F, what is the quality of the 100 psia steam?

4.26 Complete the following table for $H_2O$.

| $m$, lbm | $V$, ft³ | $v$, ft³/lbm | $p$, psia | $t$, F | $h$, B/lbm |
|---|---|---|---|---|---|
| 3.00 | 54.37 | | 20 | | |
| 1.00 | | | 15 | 96.0 | |
| | 0.9374 | | | 600.0 | 1076.0 |
| 1.00 | | | 4000 | 500.0 | |
| | 0.6560 | | 5000 | | 1363.4 |
| 4.736 | | | 440 | 454.14 | |
| | 8000 | 4000 | | −15.0 | |
| 6.157 | | 0.16941 | 1000 | | |

4.27 Calculate the mass proportion of saturated $H_2O$ liquid and vapor at 101.3 kPa that will, when heated at constant volume, pass through the critical state. If the initial volume of the system is $1.34 \times 10^{-3}$ m$^3$, how much heat must be transferred to the system to bring it to the critical state?

4.28 One pound mass of $H_2O$, initially at 20 psia and specific volume 15.74 ft$^3$/lbm, is heated at constant pressure until the temperature reaches 340 F. Calculate the heat transfer assuming the process takes place in a frictionless, quasi-static manner in a closed system. Calculate the enthalpy change for the process and compare it with the heat transfer.

4.29 A mixture of ice, water, and water vapor is in equilibrium at the triple point in a closed rigid container. It is slowly heated until the pressure is 200 psia, at which state only saturated vapor is present in the container. The mass of the initial mixture is 1 lbm. What are the proportions of liquid and vapor at the instant the last of the ice melts? Find the heat transfer required to bring this liquid-vapor mixture to the final state.

4.30 On Cartesian coordinates plot an enthalpy–pressure diagram for $H_2O$. Show the following on the diagram:

(a) Saturated liquid and saturated vapor lines
(b) Liquid–vapor mixture region
(c) Subcooled liquid region
(d) Triple line
(e) Solid–vapor mixture region
(f) Liquid–solid mixture region

Neglect the effect of pressure on enthalpy of liquid at the freezing point.

4.31 A closed system of saturated steam ($x = 100$ percent) occupies a volume of 0.32 m$^3$ at 120 C. The system is cooled at constant volume to $p = 100$ kPa, then heated at constant pressure back to the initial temperature. Each process is frictionless and quasi-static. Sketch the two paths on $p$-$v$ coordinates showing saturation lines and constant-property lines fixing the states. Determine the heat transfer for each process.

4.32 A closed system of 1 lbm of $H_2O$ undergoes the following cycle in which all processes are frictionless and quasi-static:

1–2 Expansion along $pv =$ constant from 100 psia and specific volume 3.989 ft$^3$/lbm to 5 psia
2–3 Constant pressure until $v_3 = v_1$
3–1 Constant volume

For the cycle:

(a) Sketch a $p$-$v$ diagram showing states, saturation lines, and the three paths.
(b) Calculate the thermal efficiency of the cycle.

4.33 A closed system of 1 kg of $H_2O$ undergoes the following frictionless, quasi-static heat engine cycle:

1–2 The steam at 100 kPa, quality 63.6 percent, is heated at constant volume to the saturated-vapor line.
2–3 The path is a straight line from state (2) to state (3), which is at 100 kPa and 160 C.
3–1 Constant pressure.

For this cycle:

(a) Sketch a *p-v* diagram for the cycle. Show saturation lines.
(b) Calculate heat transfer for each process.
(c) Calculate cycle thermal efficiency.

4.34 A turbine operates adiabatically in a steady-flow manner. Steam enters the turbine at 600 psia, 900 F, with a velocity of 200 ft/sec. At the turbine exhaust the pressure is 2.00 psia, the quality is 91.5 percent, and the velocity is 500 ft/sec. The turbine power output is 500 kW. Calculate the mass of steam flowing through the turbine per hour.

4.35 Refer to problem 4.34. The steam leaving the turbine enters a condenser. The heat transferred from the condensing steam is equal to that transferred to the circulating water passing through the tubes of the condenser. The condensing process takes place at constant pressure with the condensate leaving the condenser subcooled at 100 F. The temperature of the circulating water increases from 72 F to 90 F as it passes through the condenser. Calculate the circulating water flow rate in gallons per minute assuming negligible kinetic and potential energy changes for the circulating water.

4.36 Using finite differences calculate $(\partial h/\partial T)_p$ for $H_2O$ at 60 psia and 400 F. Compare your result with $C_p$, at that state, as given in the *Steam Tables*.

4.37 A certain substance has the following constant-pressure specific heat at 14.7 psia:

$$C_p = 0.250 + 2.60 \times 10^{-8} T^2 \text{ B/(lbm R)}$$

Assuming the substance undergoes a constant pressure process at 14.7 psia from an initial temperature of 530 R to a final temperature of 1500 R, calculate:

(a) The instantaneous $C_p$'s at 530 R and 1500 R.
(b) The mean $C_p$ between 530 R and 1500 R.
(c) The enthalpy change during the process.
(d) The enthalpy change if a constant $C_p$ equal to that at 530 R is used. Compare the results for parts (c) and (d).

4.38 A closed system of 2 pounds mass of $H_2O$, initially at 80 F and 14.7 psia, is in a rigid container. When 778.16 ft lbf of work is done on the water by a paddle wheel and 10 B of heat is transferred to the system during the process, it is observed that the final temperature is 90 F. From these data determine the mean $C_v$ of $H_2O$ at the given specific volume.

4.39 Neglecting the effect of pressure on enthalpy, calculate the approximate $C_p$ of liquid $H_2O$ at 60 F, 212 F, and 300 F.

4.40 A closed system undergoes a constant-volume process for which the mean specific heat is 0.43 B/(lbm R). The initial temperature is 460 R and the final temperature is 660 R. Between 460 R and 560 R the mean specific heat is 0.35 B/(lbm R). Assuming that the specific heat at this particular specific volume varies linearly with the Rankine temperature, calculate the instantaneous $C_v$ of the substance at 635 R. A sketch of $C_v$ versus $T$ may be of help in setting up the solution.

4.41 The illustration shows a superheating calorimeter that can be used to determine the state of liquid–vapor mixtures. The mixture is throttled between (1) and (2), then superheated by an electrical resistance unit between (2) and (3). The flow rate is determined by condensing and weighing the fluid leaving the superheating section. During a particular test it is found that 20 pounds of $H_2O$ per hour pass through the calorimeter. Input to the resistance heater of 1200 W yields steam at section (3) having a pressure of 15.0 psia and a temperature of 260 F. If the steam entering the calorimeter (section [1]) is at 180 psia, what is the quality of the steam at (1)?

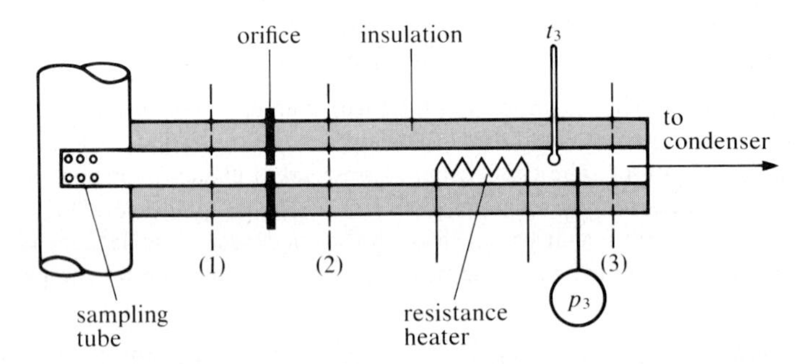


4.42 As shown in the following illustration, steam at 1200 psia and 1000 F (section [1]) is throttled to 825 psia before entering a turbine. At a point in the turbine where the pressure is 84 psia (section [3]), steam is extracted from the turbine at the rate of 7600 lbm/hr. Flow from the extraction point to the turbine exhaust is at the rate of 82,000 lbm/hr. Neglecting potential and kinetic energies, and assuming adiabatic conditions, calculate the power, in kilowatts, delivered by the turbine.

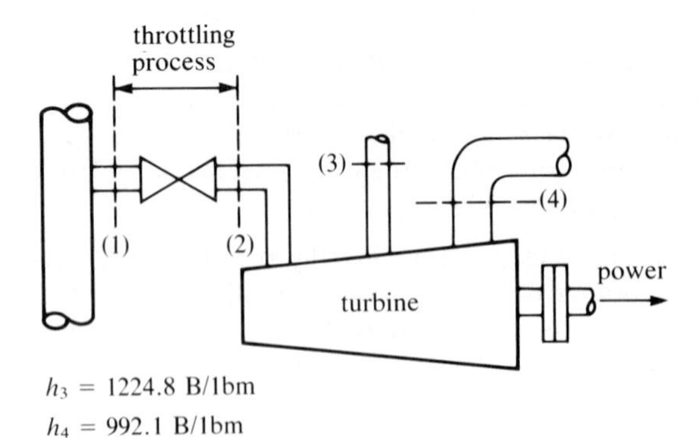


$h_3 = 1224.8$ B/lbm

$h_4 = 992.1$ B/lbm

4.43 As shown in the following illustration, two streams of $H_2O$ mix in a steady-flow process to produce one leaving stream. The process is adiabatic and there is no work.

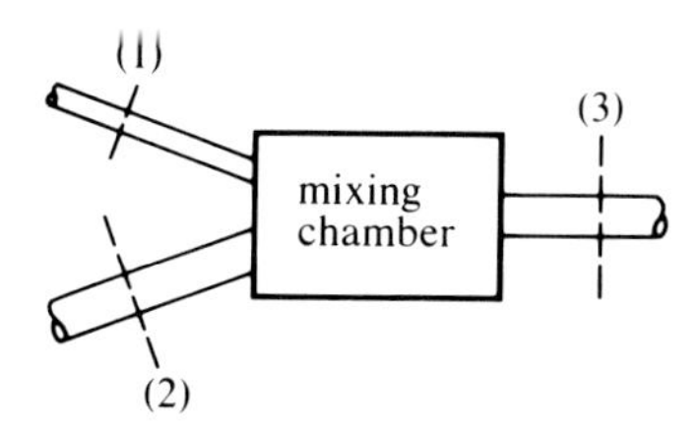


| | Section 1 | Section 2 | Section 3 |
|---|---|---|---|
| **Pressure, psia** | 15.00 | 15.00 | 15.00 |
| **Temperature, F** | 70 | — | — |
| **Specific volume, ft³/lbm** | — | 21.36 | — |
| **Velocity, ft/sec** | negligible | 500 | 300 |
| **Area, ft²** | — | 0.1452 | — |
| **Flow rate, lbm/min** | — | — | 1250 |

Calculate the mass rate of flow at section (1), the enthalpy per lbm at section (3), and the volume rate of flow at sections (2) and (3).

# 5

# Properties and Processes for Perfect Gases

*The properties of the gaseous phase of pure substances are frequently of interest in engineering thermodynamics. This chapter is devoted, for the most part, to the study of properties and processes for gases that behave as* ***perfect gases.***

## 5–1 Definition and Description of Perfect Gases

As mentioned in Chapter 4, the equation relating pressure, temperature, and specific volume of a pure substance is called the equation of state. A *perfect gas* is a gas having the equation of state

$$pv = RT \tag{5.1}$$

and whose internal energy $u$ depends only on temperature, that is,

$$u = u(T) \tag{5.2}$$

In this two-part definition the temperature $T$ is on the absolute gas scale of temperature. The *gas constant* $R$ has different values for gases of different molecular weights.

Generally all gases tend to behave as perfect gases as the pressure approaches zero. Perfect gas behavior is also exhibited by many gases at finite pressures, provided the temperature is significantly higher than the critical temperature. This will be discussed further in Section 5–7.

The equation of state for a system of $m$ mass units of a perfect gas is obtained by multiplying eq. (5.1) by $m$ to yield

$$p(mv) = mRT \tag{5.1a}$$

or

$$pV = mRT \tag{5.1b}$$

The equation of state for perfect gases may also be written on a mole basis. A mole of a substance is related to the molecular weight $M$ of the substance. Molecular weights of several substances are given in column 3 of Table 5.1. The mass associated with 1 mole depends on

**Table 5.1**

| Substance | Formula | Molecular Weight, $M$ | Volume of $M$ lbm at 14.7 psia and 32 F, $ft^3$ | Gas Constants: SI Units $\bar{R} = 8314$ N m/(kmol K) | English Units $\bar{R} = 1545$ lbf ft/(lbmol R) |
|---|---|---|---|---|---|
| | | | | $R$, N m/(kg K) | $R$, lbf ft/(lbm R) |
| air, dry | | 28.97 | 358 | 287.0 | 53.35 |
| argon | Ar | 39.95 | — | 208.1 | 38.69 |
| ammonia | $NH_3$ | 17.03 | 354 | 488.2 | 90.74 |
| carbon dioxide | $CO_2$ | 44.01 | 356 | 188.9 | 35.11 |
| carbon monoxide | CO | 28.01 | 358 | 296.8 | 55.17 |
| helium | He | 4.003 | 359 | 2077 | 386.1 |
| hydrogen | $H_2$ | 2.016 | 359 | 4124 | 766.5 |
| nitrogen | $N_2$ | 28.01 | 358 | 296.8 | 55.16 |
| oxygen | $O_2$ | 32.00 | 358 | 259.8 | 48.29 |
| sulfur dioxide | $SO_2$ | 64.06 | 350 | 129.8 | 24.12 |

(Courtesy of Prof. Kenneth R. Jolls, Department of Chemical Engineering, Iowa State University of Science and Technology)

the system of units used. In the English system $M$ pounds mass of a substance is defined as a pound mole (lb mol). In the SI system $M$ kilograms constitutes 1 kilomole (kmol). An exact definition of the kilomole is as follows.

*The kilomole is the amount of substance of a system that contains as many elementary entities as there are carbon atoms in 12 kg of carbon 12. The elementary entities must be specified and may be atoms, molecules, ions, electrons, other particles, or specified groups of such particles.*

The number of atoms in 12 kg of carbon 12 is Avogadro's number, $6.024 \times 10^{26}$/kmol. In this text the term *mole* with no prefix or modifier designates either a lb mol or a kmol.

The number of moles $n$ in $m$ mass units of a substance is

$$n = m/M \tag{5.3}$$

Equation (5.1a), written for $M$ mass units (1 mole), becomes

$$p(Mv) = MRT \tag{5.1c}$$

where the product $Mv$ is the volume occupied by 1 mole. It has been determined by experiment that $Mv$ is approximately the same for many gases when measured at a fixed temperature and pressure. Sample data for such experiments, given in column 4 of Table 5.1, indicate the approximate constancy of the volume occupied by 1 pound mole of the gases at the given temperature and pressure. Ammonia, carbon dioxide, and sulfur dioxide are at states closer to their saturated vapor lines, and as a result exhibit values of $Mv$ that differ from those for the other gases by a small amount. While dry air (no water vapor present) is a mixture of several gases, it is treated as a pure substance when its composition does not change.

Letting $\bar{v}$ represent $Mv$, the volume per mole of a perfect gas, eq. (5.1c), becomes

$$p\bar{v} = MRT$$

With $\bar{v}$ a constant at a fixed pressure and temperature, as suggested by Table 5.1, it is apparent that the quantity $MR$ is a constant. The product $MR$ is called the *universal gas constant*, $\bar{R}$.

Thus the gas constant $R$ for any perfect gas is

$$R = \frac{\bar{R}}{M} \tag{5.1d}$$

The equation of state on a unit-mole basis now can be expressed as

$$p\bar{v} = \bar{R}T \tag{5.1e}$$

or, for $n$ moles,

$$p(n\bar{v}) = n\bar{R}T$$

Since $\bar{v}$ is the volume per mole, the product $n\bar{v}$ is the volume $V$. Thus for $n$ moles of a perfect gas

$$pV = n\bar{R}T \tag{5.1f}$$

The universal gas constant and values for $R$ for several gases are given in both systems of units in Table 5.1.

Figure 5.1 shows the $p$-$\bar{v}$-$T$ surface for perfect gases. Equation (5.1e) is applicable for interpretation of the figure. Recalling that $\bar{R}$ is the universal gas constant, we see from this equation that the single $p$-$\bar{v}$-$T$ surface applies to all perfect gases. Lines on the surface for constant pressure, specific volume, and temperature are evident from inspection of the figure. From eq. (5.1e) it can be seen that for constant $\bar{v}$, $p$ is proportional to $T$. Similarly, for constant pressure, $\bar{v}$ is proportional to $T$. For constant temperature, $p\bar{v}$ is a constant. Therefore, lines of constant temperature are hyperbolas in the $p$-$\bar{v}$ plane. Surfaces based on eq. (5.1) differ as different gases are considered since $R$ as given by eq. (5.1d) varies inversely with molecular weight. Nonetheless, the interpretation of the $p$-$\bar{v}$-$T$ surface given above also applies to the $p$-$v$-$T$ surface.

**Figure 5.1** Computer-Generated $p$-$\bar{v}$-$T$ Surface for a Perfect Gas

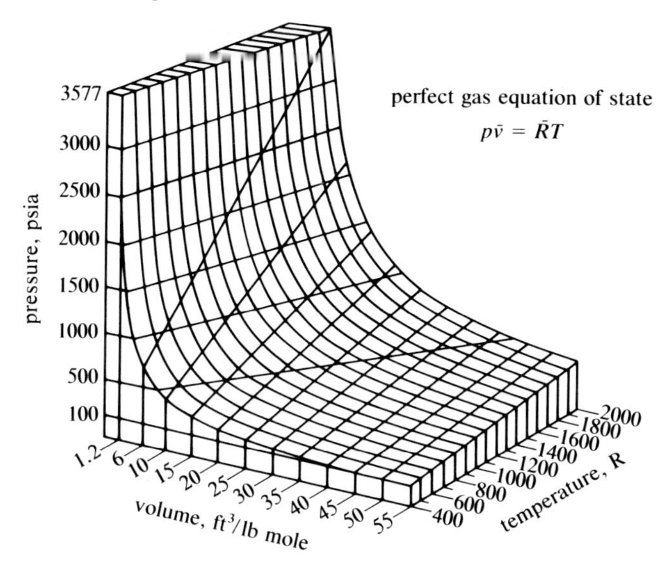


(Courtesy of Prof. Kenneth R. Jolls, Department of Chemical Engineering, Iowa State University of Science & Technology)

## Example 5.a

A mixture of perfect gases having a molecular weight of 36.14 occupies a volume of 4.50 m³ at a temperature of 30 C and a pressure of 120.0 kPa. Find the mass of the mixture and the number of moles.

**Solution:**

Using eq. (5.1b)

$$m = (pV)/(RT)$$

where

$$\begin{aligned} R = \overline{R}/M &= 8314\ [\text{N m}/(\text{kmol K})]/[36.14\ \text{kg/kmol}] \\ &= 230.0\ \text{N m}/(\text{kg K}) \end{aligned}$$

we obtain

$$m = \begin{array}{c|c|c|c|c} 120\ \text{kPa} & 10^3\ \text{N/m}^2 & 4.5\ \text{m}^3 & \text{kg K} & \\ \hline & \text{kPa} & & 230.0\ \text{N m} & (273 + 30)\text{K} \end{array}$$

$$= 7.749\ \text{kg}$$

From eq. (5.3)

$$n = m/M = 7.749\ \text{kg}/(36.14\ \text{kg/kmol}) = 0.2144\ \text{kmol}$$

Alternately, the number of moles can be found from eq. (5.1f) as

$$n = \frac{pV}{RT} = \frac{120\ \text{kPa}}{} \left| \frac{10^3\ \text{N/m}^2}{\text{kPa}} \right| \frac{\text{kmol K}}{8314\ \text{N m}} \left| \frac{4.5\ \text{m}^3}{(273+30)\text{K}} \right.$$

$$= 0.2144\ \text{kmol}$$

and $m$ can be found from eq. (5.3)

$$m = nM = (0.2144\ \text{kmol})(36.14\ \text{kg/kmol}) = 7.749\ \text{kg}$$

---

## Example 5.b

A rigid tank has a volume of 13.6 ft³ and contains 2.390 lbm of helium at a temperature of 175 F. In a certain process the temperature of the helium remains constant while 1.53 lbm of helium is released from the tank. Determine the initial and final pressures in the tank.

**Solution:**

From eq. (5.1b) and Table 5.1, the initial pressure is

$$p_1 = m_1RT_1/V_1 = \frac{2.390\ \text{lbm}}{} \left| \frac{386.1\ \text{lbf ft}}{\text{lbm R}} \right| \frac{(175 + 460)\ \text{R}}{13.6\ \text{ft}^3}$$

$$= 43{,}086\ \text{lbf/ft}^2\ \text{absolute}$$

$$= (43{,}086\ \text{lbf/ft}^2)/(144\ \text{in}^2/\text{ft}^2) = 299.2\ \text{psia}$$

The final pressure may be written as

$$p_2 = m_2RT_2/V_2$$

Dividing this by the expression for $p_1$ and using the given information we obtain

$$p_2 = p_1\frac{m_2}{m_1} = 299.2\,\frac{(2.390 - 1.53)}{2.390} = 107.7\ \text{psia}$$

---

For steady-flow open systems the mass rate of flow for each entering stream has a fixed value and the same is true for each leaving stream. The volume rate of flow $VA$ and the mass rate of flow $\dot{m}$ at any section are related according to the continuity equation, eq. (2.12). For perfect gases it is useful to combine eqs. (2.12) and (5.1) to yield

$$\dot{m} = (VA)/v = \frac{(VA)p}{RT}$$

or

$$p\dot{V} = \dot{m}RT \tag{5.4}$$

where $\dot{V} = VA$. A similar expression can be obtained for a section having a mass rate of flow of $\dot{n}$ moles per unit time. Dividing the continuity equation by the molecular weight $M$ and combining with eq. (5.1e) we obtain

$$\frac{\dot{m}}{M} = \dot{n} = \frac{VA}{Mv} = \frac{\dot{V}}{\bar{v}} = \frac{\dot{V}p}{\bar{R}T}$$

or

$$p\dot{V} = \dot{n}\bar{R}T \tag{5.5}$$

## Example 5.c

Carbon dioxide flows in a pipe that has a diameter of 15 cm. At a certain section the velocity is 60 m/s and the pressure and temperature are respectively 275 kPa and 500 K. Find the mass flow rate in kmol/s.

**Solution:**

From eq. (5.5)

$$\dot{n} = \frac{p\dot{V}}{\bar{R}T} = \frac{pVA}{\bar{R}T}$$

$$= \frac{275 \times 10^3 \text{ N}}{\text{m}^2} \left| \frac{60 \text{ m}}{\text{s}} \right| \frac{\pi\,(0.15)^2 \text{ m}^2}{4} \left| \frac{\text{kmol K}}{8314 \text{ N m}} \right| \frac{}{500 \text{ K}}$$

$$= 0.07014 \text{ kmol/s}$$

### Example 5.d

A steady-flow system has one entering and one leaving stream. Nitrogen enters the system at 70 psia, 700 F, at the rate of 1000 ft³/min. At the leaving section the pressure is 15 psia and the temperature 120 F. Calculate the volume rate of flow at the leaving section.

**Solution:**

The volume and mass rates of flow are related by eq. (5.4) as

$$p_1 \dot{V}_1 = \dot{m}_1 R T_1$$

at the inlet and

$$p_2 \dot{V}_2 = \dot{m}_2 R T_2$$

at the exit. In steady flow $\dot{m}_1 = \dot{m}_2$ and $R$ is constant for the perfect gas, nitrogen. Thus

$$\frac{p_2 \dot{V}_2}{p_1 \dot{V}_1} = \frac{\dot{m}_2 R T_2}{\dot{m}_1 R T_1}$$

which reduces to

$$\frac{p_2 \dot{V}_2}{p_1 \dot{V}_1} = \frac{T_2}{T_1}$$

with $\dot{V}_2$ becoming

$$\begin{aligned} \dot{V}_2 &= \dot{V}_1 \frac{p_1}{p_2} \frac{T_2}{T_1} \\ &= 1000 \frac{(70)(120 + 460)}{(15)(700 + 460)} \\ &= 2333 \text{ ft}^3/\text{min} \end{aligned}$$

## 5–2 Internal Energy and Enthalpy Changes for Perfect Gases

The constant-volume specific heat $C_v$ has been defined as

$$C_v \equiv \left(\frac{\partial u}{\partial T}\right)_v \tag{4.2}$$

**Table 5.2** Tabulated Values for the Internal Energy and Enthalpy of Dry Air*

| Temperature | | Internal Energy | | Enthalpy | |
|---|---|---|---|---|---|
| K | R | kJ/kg | B/lbm | kJ/kg | B/lbm |
| 250 | 450 | 178.52 | 76.75 | 250.28 | 107.60 |
| 300 | 540 | 214.36 | 92.16 | 300.47 | 129.18 |
| 350 | 630 | 250.32 | 107.62 | 350.78 | 150.81 |
| 400 | 720 | 286.49 | 123.17 | 401.31 | 172.53 |
| 450 | 810 | 320.40 | 137.75 | 449.57 | 193.28 |
| 500 | 900 | 359.87 | 154.72 | 503.39 | 216.42 |
| 550 | 990 | 397.25 | 170.79 | 555.12 | 238.66 |
| 600 | 1080 | 435.17 | 187.09 | 607.39 | 261.13 |
| 650 | 1170 | 473.66 | 203.64 | 660.23 | 283.85 |
| 700 | 1260 | 512.75 | 220.45 | 713.68 | 306.83 |
| 750 | 1350 | 552.44 | 237.51 | 767.72 | 330.06 |
| 800 | 1440 | 592.73 | 254.84 | 822.36 | 353.56 |
| 850 | 1530 | 633.73 | 272.46 | 877.71 | 377.35 |
| 900 | 1620 | 674.88 | 290.21 | 933.34 | 401.27 |

*Based on data from Hilsenrath, J., et al. *Tables of Thermal Properties of Gases.* National Bureau of Standards Circular 564, November 1, 1955.

By definition (eq. [5.2]), the internal energy of a perfect gas depends only on temperature. The combination of these two definitions permits us to say

$$C_v = \frac{du}{dT} = \frac{d(u(T))}{dT} = C_v(T)$$

for perfect gases. This expression can be written as

$$du = C_v(T)dT$$

or

$$u_2 - u_1 = \int_{T_1}^{T_2} C_v(T)dT \tag{5.6}$$

which is the basic equation for calculating the change in internal energy for perfect gases.

Three procedures are available to evaluate changes in internal energy for perfect gases. First, if the relation between $C_v$ and temperature is known, eq. (5.6), integrated between the initial and final temperatures for the process, yields the change in internal energy. The second procedure is to use a table of properties for the gas involved. The internal energy values tabulated in such tables are obtained by integrating eq. (5.6) for the particular gas from a reference temperature to the tabulated temperatures. Table 5.2, which lists values of internal energy for air for several values of temperature, is an example of such a table. The third method is to assume that $C_v$ is constant. The choice of which method to use depends on the

**Table 5.3** Values of $C_p$, $C_v$, and $\gamma$ at Various Temperatures for Several Perfect Gases.*

| Temperature | | $C_p$ | | $C_v$ | | | $C_p$ | | $C_v$ | | |
|---|---|---|---|---|---|---|---|---|---|---|---|
| K | R | kJ/(kg K) | B/(lbm R) | kJ/(kg K) | B/(lbm R) | $\gamma$ | kJ/(kg K) | B/(lbm R) | kJ/(kg K) | B/(lbm R) | $\gamma$ |
| | | | | **Air** | | | | | **Argon, Ar** | | |
| 300 | 540 | 1.005 | 0.2399 | 0.7177 | 0.1714 | 1.400 | 0.5204 | 0.1243 | 0.3122 | 0.0746 | 1.667 |
| 400 | 720 | 1.013 | 0.2420 | 0.7263 | 0.1734 | 1.395 | 0.5204 | 0.1243 | 0.3122 | 0.0746 | 1.667 |
| 500 | 900 | 1.029 | 0.2458 | 0.7424 | 0.1773 | 1.386 | 0.5204 | 0.1243 | 0.3122 | 0.0746 | 1.667 |
| 600 | 1080 | 1.051 | 0.2510 | 0.7639 | 0.1824 | 1.376 | 0.5204 | 0.1243 | 0.3122 | 0.0746 | 1.667 |
| 700 | 1260 | 1.075 | 0.2567 | 0.7878 | 0.1881 | 1.365 | 0.5204 | 0.1243 | 0.3122 | 0.0746 | 1.667 |
| 800 | 1440 | 1.099 | 0.2623 | 0.8116 | 0.1938 | 1.354 | 0.5204 | 0.1243 | 0.3122 | 0.0746 | 1.667 |
| | | | | **Carbon Dioxide, $CO_2$** | | | | | **Hydrogen, $H_2$** | | |
| 300 | 540 | 0.8457 | 0.2019 | 0.6567 | 0.1568 | 1.288 | 14.31 | 3.416 | 10.18 | 2.432 | 1.406 |
| 400 | 720 | 0.9390 | 0.2242 | 0.7501 | 0.1791 | 1.252 | 14.48 | 3.457 | 10.35 | 2.472 | 1.399 |
| 500 | 900 | 1.014 | 0.2421 | 0.8250 | 0.1970 | 1.229 | 14.51 | 3.467 | 10.39 | 2.481 | 1.397 |
| 600 | 1080 | 1.075 | 0.2568 | 0.8863 | 0.2116 | 1.213 | 14.55 | 3.474 | 10.42 | 2.489 | 1.396 |
| 700 | 1260 | 1.126 | 0.2689 | 0.9372 | 0.2238 | 1.201 | 14.60 | 3.487 | 10.48 | 2.502 | 1.393 |
| 800 | 1440 | 1.169 | 0.2791 | 0.9797 | 0.2340 | 1.193 | 14.69 | 3.509 | 10.57 | 2.524 | 1.390 |
| | | | | **Nitrogen, $N_2$** | | | | | **Oxygen, $O_2$** | | |
| 300 | 540 | 1.040 | 0.2482 | 0.7428 | 0.1774 | 1.400 | 0.9183 | 0.2193 | 0.6585 | 0.1572 | 1.395 |
| 400 | 720 | 1.044 | 0.2493 | 0.7472 | 0.1784 | 1.397 | 0.9409 | 0.2247 | 0.6810 | 0.1626 | 1.382 |
| 500 | 900 | 1.056 | 0.2521 | 0.7591 | 0.1813 | 1.391 | 0.9716 | 0.2320 | 0.7118 | 0.1700 | 1.365 |
| 600 | 1080 | 1.075 | 0.2566 | 0.7780 | 0.1858 | 1.382 | 1.003 | 0.2395 | 0.7431 | 0.1774 | 1.350 |
| 700 | 1260 | 1.098 | 0.2621 | 0.8010 | 0.1913 | 1.371 | 1.031 | 0.2461 | 0.7709 | 0.1841 | 1.337 |
| 800 | 1440 | 1.122 | 0.2679 | 0.8252 | 0.1971 | 1.360 | 1.054 | 0.2518 | 0.7945 | 0.1897 | 1.327 |

*Based on data from Hilsenrath, J., et al. *Tables of Thermal Properties of Gases.* National Bureau of Standards Circular 564, November 1, 1955.

temperature range for the process, the particular gas, the availability of tables of properties, and the accuracy desired. Table 5.3 lists $C_v$ values for a range of temperatures for several gases. The data in this table indicate that the approximation of constant $C_v$ for a small range of temperature, for example, 100 K, involves only a small error. However, as the temperature range increases, relatively large changes in $C_v$ may occur. Of the gases listed, carbon dioxide exhibits the largest $C_v$ change with temperature while argon exhibits no change at all. For large temperature changes, such as those encountered in certain combustion processes, the variation in specific heats must be considered if high accuracy is desired.

The constant-pressure specific heat $C_p$ is, by definition,

$$C_p \equiv \left(\frac{\partial h}{\partial T}\right)_p \tag{4.2}$$

and enthalpy has been defined as

$$h \equiv u + pv$$

For a perfect gas

$$pv = RT$$

Thus the enthalpy of a perfect gas may be expressed as

$$h = u + RT$$

With $R$ constant and $u$ a pure temperature function for a perfect gas, it follows that the enthalpy of a perfect gas depends only on its temperature. Using the same logic as that for the internal energy we get

$$dh = C_p(T)dT$$

$$h_2 - h_1 = \int_{T_1}^{T_2} C_p\,(T)dT \tag{5.7}$$

for perfect gases. The three alternate methods described for calculation of internal energy changes are also applicable to enthalpy changes of a perfect gas. The only difference is the use of $C_p$ in place of $C_v$ in determining the change of enthalpy of a perfect gas. Table 5.2 tabulates typical values of enthalpy for air obtained by integration of eq. (5.7), and Table 5.3 gives values of $C_p$ for a range of temperature for several perfect gases.

For constant specific heats, eqs. (5.6) and (5.7) become

$$u_2 - u_1 = C_v\,(T_2 - T_1) \tag{5.6a}$$

and

$$h_2 - h_1 = C_p\,(T_2 - T_1) \tag{5.7a}$$

## 5–3 Specific Heat Relations for Perfect Gases

In the preceding section expressions for internal energy and enthalpy differentials for a perfect gas were developed. In combining these with the differential of $h = u + RT$, we get

$$dh = du + RdT = C_p dT = C_v dT + RdT$$

which, at a given temperature, yields the exact relation

$$C_p = C_v + R \tag{5.8}$$

for all perfect gases.

The ratio of the specific heats is, by definition,

$$\gamma \equiv \frac{C_p}{C_v} \tag{5.9}$$

With $C_p$ and $C_v$ dependent on temperature only, it follows that their ratio will depend only on temperature. Values of $\gamma$ are listed at various temperatures for several gases in Table 5.3. Simultaneous solutions of eqs. (5.8) and (5.9) yield

$$C_v = \frac{R}{\gamma - 1} \quad \text{and} \quad C_p = \frac{\gamma R}{\gamma - 1} \tag{5.10a,b}$$

at a given temperature for a perfect gas.

**Figure 5.2** $C_p$ Versus Temperature for a Perfect Gas

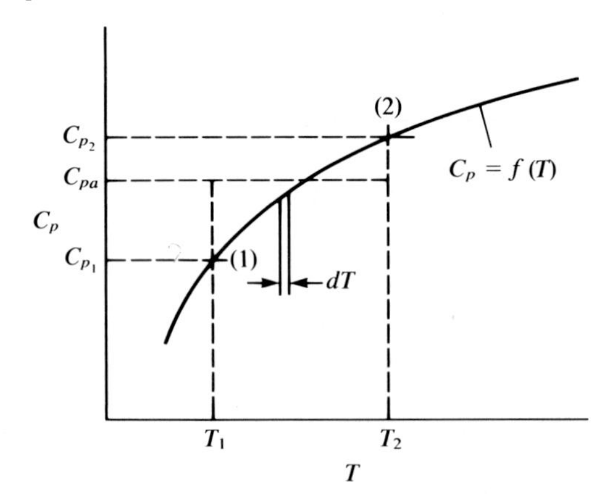


Specific heat values at a given state are called *instantaneous specific heats.* We now introduce the term *average specific heat.* Consider the variation of $C_p$ with temperature. A generalized illustration of this variation is shown in Figure 5.2.

The change in enthalpy of a perfect gas as its temperature changes from $T_1$ to $T_2$ is, by eq. (5.7),

$$h_2 - h_1 = \int_{T_1}^{T_2} C_p dT \tag{5.7}$$

Inspection of the integral reveals it to be the area beneath the $C_p$ curve between $T_1$ and $T_2$. The average specific heat of the gas between $T_1$ and $T_2$, $C_{p_a}$, is the ordinate value that, when multiplied by $T_2 - T_1$, will yield the enthalpy change $h_2 - h_1$. By definition

$$\underset{T_1 \to T_2}{C_{p_a}} \equiv \frac{h_2 - h_1}{T_2 - T_1} = \frac{\int_{T_1}^{T_2} C_p dT}{T_2 - T_1} \tag{5.11}$$

A similar expression can be developed for $C_{v_a}$.

Specific heat data on a mole basis are frequently given in the literature and are often used in engineering applications. Perfect gas specific heats on the basis of a mole of substance are

$$\overline{C}_v = \frac{d\overline{u}}{dT} \tag{5.12}$$

$$\overline{C}_p = \frac{d\overline{h}}{dT} \tag{5.13}$$

**Table 5.4** $\overline{C}_p$–$T$ Relations for Several Perfect Gases*

| Gas | $\overline{C}_p$, B/(lb mol R), $T$ in Rankine | Range, R |
|---|---|---|
| air, and nitrogen, $N_2$ | $9.47 - \frac{3.47(10^3)}{T} + \frac{1.16(10^6)}{T^2}$ | 540–9000 |
| oxygen, $O_2$ | $11.515 - \frac{172}{T^{1/2}} + \frac{1530}{T}$ | 540–5000 |
| hydrogen, $H_2$ | $5.76 + \frac{0.578\ T}{1000} + \frac{20}{T^{1/2}}$ | 540–4000 |
| carbon dioxide, $CO_2$ | $16.2 - \frac{6.53(10^3)}{T} + \frac{1.41(10^6)}{T^2}$ | 540–6300 |
| water vapor, $H_2O$ | $19.86 - \frac{597}{T^{1/2}} + \frac{7500}{T}$ | 540–5400 |

*Sweigart, R. L., and M. W. Beardsley. *Empirical Specific Heat Equation Based Upon Spectroscopic Data.* Georgia School of Technology Engineering Experiment Station Bulletin 2, 1938.

where the bar indicates that the quantities are on the basis of a mole. Changes in internal energy and enthalpy are obtained by integrating these expressions. The relations between specific heats on a mass basis and on a mole basis are

$$\overline{C}_v = C_v M, \quad \overline{C}_p = C_p M \tag{5.14}$$

Therefore, from eq. (5.8)

$$\overline{C}_p = \overline{C}_v + MR$$

$$\overline{C}_p = \overline{C}_v + \overline{R}$$

Table 5.4 gives equations that relate $\overline{C}_p$ to $T$ for several common gases. These gases, as well as many others, behave as perfect gases provided the pressure is low.

While our focus in this section has been on specific heats for perfect gases, it is instructive to examine the conditions under which both pressure and temperature influence specific heats of gases. When substances in the gaseous phase behave as perfect gases, pressure has no influence on specific heats. High pressures produce departures from perfect gas behavior and consequently specific heats vary with both pressure and temperature. Figure 5.3 shows the variation of $C_p$ with temperature for air at five values of pressure ranging from 0.1 to 100 atmospheres. Perfect gas behavior for air is expected at the two lower values of pressure. High values of pressure have the greatest influence in the lower range of temperature. For many engineering applications the influence of pressure on specific heats can be neglected unless very large values of pressure are involved. More will be said in Section 5–7 regarding departure from perfect gas behavior.

**Figure 5.3** $C_p$ Versus $T$ for Air at Five Values of Pressure

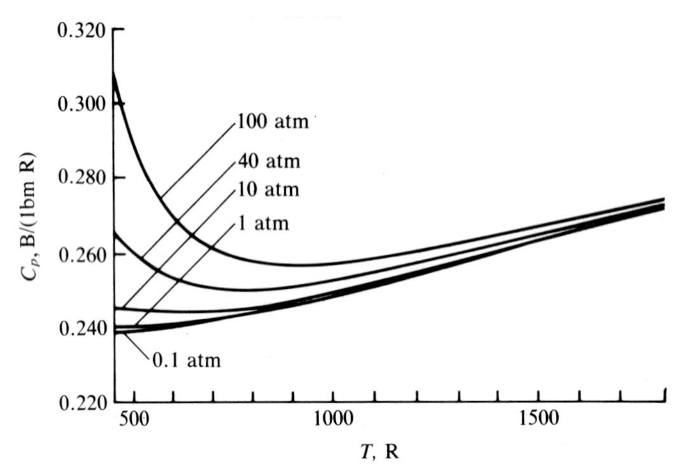


(Based on data from NBS Circular 564, 1955)

**Table 5.5** Approximate Values for $C_p$, $C_v$, and $\gamma$ at Moderate Temperatures for Several Perfect Gases.* $\overline{R}$ = 8314 N m/(kmol K) = 1545 lbf ft/(lb mol R)

| | | $C_p$ | | $C_v$ | | $\gamma$ |
|---|---|---|---|---|---|---|
| **Gas** | **M** | **B/(lbm R)** | **kJ/(kg K)** | **B/(lbm R)** | **kJ/(kg K)** | |
| air, dry | 28.97 | 0.240 | 1.00 | 0.172 | 0.718 | 1.400 |
| argon | 39.95 | 0.124 | 0.520 | 0.0746 | 0.312 | 1.667 |
| carbon dioxide | 44.01 | 0.202 | 0.846 | 0.157 | 0.657 | 1.288 |
| carbon monoxide | 28.01 | 0.249 | 1.04 | 0.178 | 0.743 | 1.399 |
| helium | 4.004 | 1.24 | 5.19 | 0.744 | 3.12 | 1.667 |
| hydrogen | 2.016 | 3.42 | 14.3 | 2.43 | 10.2 | 1.405 |
| nitrogen | 28.01 | 0.248 | 1.04 | 0.177 | 0.743 | 1.400 |
| oxygen | 32.00 | 0.219 | 0.918 | 0.157 | 0.659 | 1.395 |

*Based on data from Hilsenrath, J., et al. *Tables of Thermal Properties of Gases.* National Bureau of Standards Circular 564, November 1, 1955.

In a number of engineering applications the simplifying assumption of constant specific heats yields sufficiently accurate results. Table 5.5 gives approximate specific heat data for several perfect gases at moderate temperatures that are adequate for use in conjunction with the constant specific heat assumption. In view of the resulting simplicity, we shall assume constant specific heats for the solution of perfect gas problems unless another approach is specified.

## Example 5.e

Determine the enthalpy change for air undergoing a process from an initial pressure and temperature of 10 psia and 540 R to a final state where the pressure and temperature are 100 psia and 1530 R by the following methods: (a) integration to account for variable specific heat; (b) use of the table of properties; (c) use of average specific heat; and (d) assumption of constant specific heat.

**Solution:**

Since we assume that air behaves as a perfect gas, the enthalpy change depends only on the temperature change. Therefore, pressure does not enter into the solution.

(a) Using eqs. (5.7) and (5.14) and the expression for $\overline{C}_p$ for air from Table 5.4

$$h_2 - h_1 = \int_{T_1}^{T_2} \frac{\overline{C}_p(T)}{M} \, dT$$

$$= \int_{540}^{1530} (1/28.97)[9.47 - (3.47 \times 10^{-3}/T) + (1.16 \times 10^6/T^2)]dT$$

$$= 246.9 \text{ B/lbm}$$

(b) From Table 5.2, $h_2 - h_1 = 377.35 - 129.18 = 248.17$ B/lbm

(c) $C_{p_a}$ is given by eq. (5.11). The integral in that equation has been evaluated in part (a). Hence

$$C_{p_a} = (246.9)/(1530 - 540) = 0.2494 \text{ B/(lbm R)}$$

and

$$h_2 - h_1 = C_{p_a}(T_2 - T_1) = 246.9 \text{ B/lbm}$$

(d) Using eq. (5.7a) and $C_p$ for air from Table 5.5

$$h_2 - h_1 = C_p(T_2 - T_1) = 0.24(1530 - 540) = 238 \text{ B/lbm}$$

The difference in the results for parts (a) and (b) is attributable to the small differences in the $C_p$ versus $T$ expressions associated with each method. The answer obtained assuming constant $C_p$ differs from that obtained using variable $C_p$ by about 4 percent. Note that the temperature range is 990 R.

## Example 5.f

A closed system of 0.45 kg of helium initially at 140 kPa and 335 K is compressed during an isothermal, frictionless, quasi-static process to a pressure of 600 kPa. Calculate the heat transfer for the process.

**Solution:**

For a closed system

$$_1q_2 = u_2 - u_1 + {_1w_2}$$

with $u_2 - u_1 = 0$ for a perfect gas undergoing an isothermal process. Thus

$$_1q_2 = {_1w_2} = \int_1^2 p dv$$

for the frictionless, quasi-static process. For the perfect gas

$$p_1 v_1 = RT_1 \quad \text{and} \quad p_2 v_2 = RT_2$$

or, at constant temperature,

$$p_1 v_1 = p_2 v_2 = pv = C = RT$$

The heat transfer is determined as

$$_1q_2 = {_1w_2} = \int_1^2 p dv = \int_1^2 \frac{RT}{v} dv$$

$$= RT \, ln \frac{v_2}{v_1}$$

For constant-temperature processes of a perfect gas

$$\left(\frac{v_2}{v_1}\right)_T = \left(\frac{p_1}{p_2}\right)_T$$

Thus

$$_1q_2 = RT \, ln \frac{p_1}{p_2}$$

$$= (2077)(335) \, ln \, (140/600) = -1.013 \times 10^6 \text{ N m/kg}$$

$$= -1013 \text{ kJ/kg}$$

The negative sign indicates heat transfer from the air.

## Example 5.g

Air enters a long, horizontal pipe at 65 psia, 80 F, at the rate of 40 lbm/min. The pipe cross-sectional area is constant and equal to 3.000 in². The air leaves the pipe at 15 psia and 150 F. Calculate the rate of heat transfer for the process.

**Solution:**

With no work done, the steady-flow equation becomes

$$h_1 + \frac{V_1^2}{2g_0} + {}_1q_2 = h_2 + \frac{V_2^2}{2g_0}$$

The enthalpy change of the perfect gas $h_2 - h_1$ is equal to $C_p(T_2 - T_1)$ and, on the basis selected, the steady-flow equation is

$${}_1q_2 = C_p\,(T_2 - T_1) + \frac{V_2^2 - V_1^2}{2g_0}$$

Velocities can be determined using the continuity equation at the entering and leaving sections.

$$V_1 = \frac{\dot{m}v_1}{A_1} \text{ and } v_1 = \frac{RT_1}{P_1} = \frac{(53.3)(540)}{(65)(144)} = 3.075 \text{ ft}^3/\text{lbm}$$

$$V_1 = \frac{40}{60}\left|\frac{3.075}{}\right|\frac{144}{3.000} = 98.4 \text{ ft/sec}$$

In like manner,

$$v_2 = \frac{(53.3)(610)}{(15)(144)} = 15.05 \text{ ft}^3/\text{lbm}$$

and

$$V_2 = \frac{40}{60}\left|\frac{15.05}{}\right|\frac{144}{3.000} = 481.6 \text{ ft/sec}$$

Substituting into the steady-flow equation

$$\begin{aligned} {}_1q_2 &= 0.24(610 - 540) + \frac{481.6^2 - 98.4^2}{(2)(32.17)(778)} \\ &= 21.24 \text{ B/lbm} \end{aligned}$$

and

$${}_1\dot{Q}_2 = (40)(21.24) = 849.6 \text{ B/min}$$

## 5–4 Closed-System, Frictionless, Quasi-static, Adiabatic Processes for Perfect Gases

For all stationary closed systems undergoing *adiabatic* processes,

$$_1Q_2 = 0 = U_2 - U_1 + {_1W_2}$$

from which it is evident that any work done by the system will result in a decrease in its internal energy. The above equation, applied to a system of a unit mass of a perfect gas, becomes, in differential form,

$$0 = du + \delta w = C_v dT + \delta w \tag{a}$$

The additional restriction of a frictionless, quasi-static process allows $\delta w$ to be replaced by $pdv$, or

$$C_v dT + pdv = 0 \tag{b}$$

With $p$, $v$, and $T$ related by the equation of state, the preceding equation can be written in terms of any pair of these properties. Let us select $p$ and $v$ as the variables, eliminating $T$ in the manner

$$pdv + vdp = RdT$$

or

$$dT = \frac{pdv + vdp}{R}$$

Substituting this expression for $dT$ into eq. (b) yields

$$\frac{C_v}{R}(pdv + vdp) + pdv = 0$$

Rearranging, we obtain

$$(C_v + R)pdv + C_v vdp = 0$$

Recalling $C_v + R = C_p$ for a perfect gas, it follows that

$$\frac{C_p}{C_v} pdv + vdp = \gamma\, pdv + vdp = 0$$

where $\gamma = C_p/C_v$ has been introduced.

For the case of constant specific heats $\gamma$ will also be constant. Separating variables and integrating we obtain

$$ln\, C = \gamma\, ln\, v + ln\, p$$

which yields

$$pv^{\gamma} = \text{constant} = C \tag{5.16}$$

as the path equation for a closed system of a perfect gas with constant specific heats undergoing a process that is frictionless, quasi-static, and adiabatic. The volume $V$ of the closed system may be used in place of the specific volume $v$ in eq. (5.16). This result can be obtained by multiplying eq. (5.16) by the constant $m^{\gamma}$, where $m$ is mass, to yield

$$p(vm)^{\gamma} = pV^{\gamma} = Cm^{\gamma} = \text{constant}$$

The $T$-$v$ and $p$-$T$ relations for such processes are developed from the path equation and the equation of state. To illustrate, consider the $T$ and $v$ relation derived as follows:

$$p_1v_1 = RT_1 \quad \text{and} \quad p_2v_2 = RT_2$$

$$p_1{v_1}^{\gamma} = p_2{v_2}^{\gamma}$$

or

$$\frac{p_1{v_1}^{\gamma}}{\dfrac{p_1v_1}{T_1}} = \frac{p_2{v_2}^{\gamma}}{\dfrac{p_2v_2}{T_2}}$$

which reduces to

$$T_1{v_1}^{\gamma-1} = T_2{v_2}^{\gamma-1}$$

The $p$-$T$ relation is developed in a similar manner. The resultant property relations are

$$\frac{T_2}{T_1} = \left(\frac{p_2}{p_1}\right)^{(\gamma-1)/\gamma} = \left(\frac{v_1}{v_2}\right)^{\gamma-1} \tag{5.17}$$

## Example 5.h

A closed system of 3 lbm of argon undergoes a frictionless, quasi-static adiabatic expansion from 100 psia, 300 F, to 16 psia. Calculate the work done during the expansion.

**Solution:**

The work could be calculated as $\int_1^2 pdV$ with $pV^{\gamma} = C$. Since the process is adiabatic, a simpler method is to use the closed-system energy equation which reduces to

$${}_1W_2 = m\,{}_1w_2 = -m(u_2 - u_1) = -mC_v(T_2 - T_1)$$

For the process

$$T_2 = T_1 \left(\frac{p_2}{p_1}\right)^{(\gamma-1)/\gamma}$$

$$= 760 \left(\frac{16}{100}\right)^{0.667/1.667} = 365 \text{ R}$$

$${}_1W_2 = -(3)(0.0746)(365 - 760)$$
$$= 88.4 \text{ B} = 68{,}780 \text{ ft lbf}$$

The identical result would be obtained if $\int_1^2 pdV$ had been used.

---

## 5-5 Steady-Flow, Frictionless, Adiabatic Processes for Perfect Gases

An equation for frictionless, steady-flow processes was derived in Chapter 2. This equation, eq. (2.14), in differential form is

$$-\delta w = vdp + d(KE) + d(PE)$$

While a heat-transfer term does not appear in the equation, it is valid for all values of heat transfer, including zero heat transfer. The differential form of the steady-flow energy equation, eq. (3.4b), is

$$\delta q - \delta w = dh + d(KE) + d(PE)$$

For a frictionless, adiabatic steady-flow process the two equations above combine to yield

$$vdp = dh \tag{c}$$

With $dh$ equal to $du + pdv + vdp$, eq. (c) becomes

$$0 = du + pdv$$

or

$$0 = C_v dT + pdv \tag{d}$$

for a perfect gas. Comparison of eq. (d) with eq. (b) in Section 5–4 reveals that they are identical. It follows, then, that eq. (5.16) and the property relations given in eq. (5.17) are also valid for adiabatic, frictionless, steady-flow processes for perfect gases with constant specific heats.

## Example 5.i

Helium enters a nozzle at 670 kPa, 225 C, with a velocity of 150 m/s. Expansion through the nozzle is steady, frictionless and adiabatic. If the pressure at the exit plane of the nozzle is 130 kPa, what is the mass rate of flow per unit area at the exit?

**Solution:**

The continuity equation written at the exit plane is

$$\left(\frac{\dot{m}}{A}\right)_2 = \left(\frac{V}{v}\right)_2$$

Determination of the unknowns $V_2$ and $v_2$ will require solutions of the steady-flow equation and the equation of state. With no shaft work done during a nozzle process, the steady-flow energy equation reduces to

$$h_1 + \frac{V_1^2}{2g_0} = h_2 + \frac{V_2^2}{2g_0}$$

For the perfect gas, helium, the expression for $V_2$ becomes

$$V_2 = [2g_0(h_1 - h_2) + V_1^2]^{1/2} = [2g_0C_p(T_1 - T_2) + V_1^2]^{1/2}$$

The exit temperature is

$$T_2 = T_1\left(\frac{p_2}{p_1}\right)^{(\gamma-1)/\gamma} = 498\left(\frac{130}{670}\right)^{0.667/1.667} = 258.4 \text{ K}$$

Substituting into the expression for $V_2$

$$V_2 = \left[\left(2\ \frac{\text{kg m}}{\text{N s}^2}\right)\left(\frac{5190 \text{ J}}{\text{kg K}}\right)[(498 - 258.4)\text{ K}] - (150 \text{ m/s})^2\right]^{1/2}$$
$$= 1570 \text{ m/s}$$

At the exit plane

$$v_2 = \frac{RT_2}{p_2} = \frac{(2077)\ (258.4)}{130{,}000} = 4.128 \text{ m}^3/\text{kg}$$

The exit plane mass rate of flow per unit area then is

$$\left(\frac{\dot{m}}{A}\right)_2 = (1570)/(4.128) = 380 \text{ kg/(s m}^2)$$

## 5-6 The Polytropic Process

A process following the path

$$pv^n = \text{constant} = C \tag{5.18}$$

with $n$ constant during the process is called a *polytropic* process. This expression describes a path that is composed of a continuous series of equilibrium states through which the working fluid passes in undergoing a process along the path. It is applicable to frictionless, quasi-static, closed-system processes or to frictionless processes in steady-flow systems with one inbound and one outbound stream, and is useful in describing paths for several processes of interest. The working fluid need not be a perfect gas and heat transfer may occur.

For a perfect gas undergoing a polytropic process, properties can be related in the manner used to derive eq. (5.17). The resulting relations are

$$\frac{T_2}{T_1} = \left(\frac{p_2}{p_1}\right)^{(n-1)/n} = \left(\frac{v_1}{v_2}\right)^{n-1} \tag{5.19}$$

Note the $p$-$T$ and $v$-$T$ relations are valid only for polytropic processes of a perfect gas.

The work for a closed system undergoing a polytropic process is obtained using the path equation and $\int_1^2 p\,dv$ as

$$_1w_2 = \int_1^2 p\,dv = C\int_1^2 \frac{dv}{v^n} = \frac{p_2v_2 - p_1v_1}{1-n} \tag{5.20}$$

For steady-flow polytropic processes with negligible potential and kinetic energy changes, the work, from eq. (2.14), is

$$_1w_2 = -\int_1^2 v\,dp = \frac{n}{1-n}(p_2v_2 - p_1v_1) \tag{5.21}$$

Equations (5.20) and (5.21) are not limited to perfect gases. If the working fluid is a perfect gas, the $pv$ terms in eqs. (5.20) and (5.21) can be replaced by $RT$. Neither equation is applicable when $n = 1$.

For perfect gases with constant specific heats the closed-system energy equation $_1q_2 = u_2 - u_1 + {}_1w_2$ becomes

$$_1q_2 = C_v(T_2 - T_1) + \frac{p_2v_2 - p_1v_1}{1-n}$$

**Figure 5.4** Paths for Perfect Gas Polytropic Processes

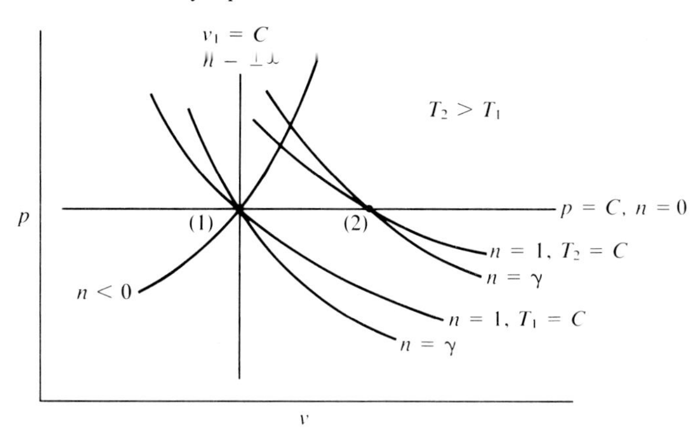


for polytropic processes. Using eq. (5.10a) to replace $C_v$ and substituting $RT$ for the $pv$ terms we have for $n \neq 1$

$$
\begin{aligned}
{}_1q_2 &= \frac{R}{\gamma - 1}(T_2 - T_1) + \frac{R}{1 - n}(T_2 - T_1) \\
&= \frac{(n - \gamma)}{(\gamma - 1)(n - 1)} R(T_2 - T_1) \qquad \textbf{(5.22)}
\end{aligned}
$$

Inspection of eq. (5.22) reveals, for expansions, that heat transfer is positive when $n$ is less than $\gamma$, zero when $n = \gamma$, and negative when $n$ is greater than $\gamma$. A similar analysis is applicable to compression processes.

Additional insight is gained by differentiating eq. (5.18), remembering $n$ is a constant. The result is

$$
\frac{dp}{dv} = -n\frac{p}{v} \qquad \textbf{(5.23)}
$$

At a given state $p$ and $v$ are fixed; hence the slope of the path through the given state on $p$-$v$ coordinates depends solely on $n$. Figure 5.4 shows common processes for perfect gases in terms of $n$. From Figure 5.4 and eq. (5.22) one can determine the sign of work, internal energy change, and heat transfer for closed, perfect-gas systems undergoing polytropic processes if $\gamma$ and $n$ are known.

## Example 5.j

A mixture of perfect gases with a molecular weight of 38 and $C_v = 0.165$ B/(lbm $R$) undergoes a polytropic process in a steady-flow system according to the relation $pv^{1.36} = C$. At the system inlet, section (1), the volume rate of flow is 120 ft³/sec, $p_1 = 90$ psia, and $t_1 = 1200$ F. At the system exit, section (2), the pressure is 15 psia. Kinetic and potential energy changes are negligible. Find (a) the power and (b) the heat tranfer rate for the process.

**Solution:**

(a) The power is given by $\dot{W} = \dot{m}_1w_2$, where

$$\dot{m} = \frac{V_1A_1}{v_1} = \frac{(120)(38)(100)(144)}{(1545)(1200 + 460)} = 25.6 \text{ lbm/sec}$$

and according to eqs. (5.21) and (5.1)

$${}_1w_2 = \frac{n}{(1-n)}(p_2v_2 - p_1v_1) = \frac{n}{(1-n)}R(T_2 - T_1)$$

$T_2$ is obtained from eq. (5.19) as

$$T_2 = (1660)\left(\frac{15}{90}\right)^{0.36/1.36} = 1033\ R$$

Therefore,

$${}_1w_2 = \frac{(1.36)(1545)(1033 - 1660)}{(1 - 1.36)(38)(778)} = 123.8 \text{ B/lbm}$$

$$\dot{W} = 25.6(123.8)(3600)/2545$$
$$= 4480 \text{ HP}$$

(b) The heat transfer rate expression is $\dot{Q} = \dot{m}_1q_2$, where ${}_1q_2$ is given by the steady-flow energy equation which, for the stated conditions, reduces to

$${}_1q_2 = {}_1w_2 + h_2 - h_1 = {}_1w_2 + C_p(T_2 - T_1)$$

From eq. (5.8)

$$C_p = C_v + R = 0.165 + \frac{1545}{778(38)} = 0.217 \text{ B/(lbm R)}$$

Thus,

$${}_1q_2 = 123.8 + 0.217(1033 - 1660) = -12.3 \text{ B/lbm}$$

$$\dot{Q} = 25.6(-12.3) = -315 \text{ B/sec}$$

## 5–7 Departure of Gases from Perfect Gas Behavior

Values for the properties of the gaseous phase of pure substances are most accurately determined from a table of properties for the substance. However, tables of properties are not always available. Also, perfect gas behavior is not exhibited for a given gas at all values of pressure and temperature at which a substance exists in the gaseous phase. As a result, the expression $pv = RT$ may not be sufficiently accurate or its accuracy may be unknown for some applications. Therefore, it is of practical interest to examine the departure of gases from perfect gas behavior.

A number of modifications to the perfect gas equation of state have been proposed to describe the $p$-$v$-$T$ relation for real gases. The simplest approach, the one to be discussed here, is the use of the compressibility factor $Z$, which is defined as

$$Z = \frac{pv}{RT} \tag{5.24}$$

Clearly $Z$ is unity for a perfect gas. For real gases $Z$ differs from unity and its value depends on both pressure and temperature. Charts of $Z$ as a function of pressure and temperature are available for a large number of pure substances.

One approach that relates $Z$ to pressure and temperature that is reasonably accurate and is applicable to a broad range of pure substances is called the *law of corresponding states.* This law assumes that

$$Z = f(P_r, T_r)$$

where

$$P_r = p/p_{cp} = \text{reduced pressure}$$

$$T_r = T/T_{cp} = \text{reduced temperature}$$

In these expressions $p_{cp}$ and $T_{cp}$ are, respectively, the absolute pressure and temperature at the critical point. Table 4.1 lists critical-point data for several substances. Figure 5.5 is a graph of $Z$ versus $P_r$ for various values of $T_r$. Inspection of this figure clearly reveals the conditions for which $Z = 1$. It is seen that perfect gas behavior exists at least for $P_r << 1$ $(p << p_{cp})$ and $T_r >> 1$ $(T >> T_{cp})$. Departure from perfect gas behavior is directly related to departure of $Z$ from unity. Hence knowledge of the critical pressure and temperature and the use of Figure 5.5 readily permit evaluation of the accuracy of the perfect gas equation of state for a given application.

Other properties for real gases can also be formulated in terms of reduced pressure and reduced temperature. Figure 5.6 presents the quantity $(\bar{h}^* - \bar{h})/T_{cp}$ in terms of $P_r$ and $T_r$; $\bar{h}^*$ is the enthalpy for a perfect gas and $\bar{h}$ is the real gas enthalpy at the same temperature. When $Z$ departs from unity, internal energy and enthalpy depend on both pressure and temperature. The method by which changes in enthalpy of real gases are determined by use of Figure 5.6 can be illustrated by use of the $p$-$v$ diagram in Figure 5.7. Pressure and temperature values at states (1) and (2) are assumed to be known, as well as $T_{cp}$ and $p_{cp}$ for the substance under consideration. Therefore, $P_r$ and $T_r$ can be determined for each of the states.

**Figure 5.5** Compressibility Factor for Gases and Vapors

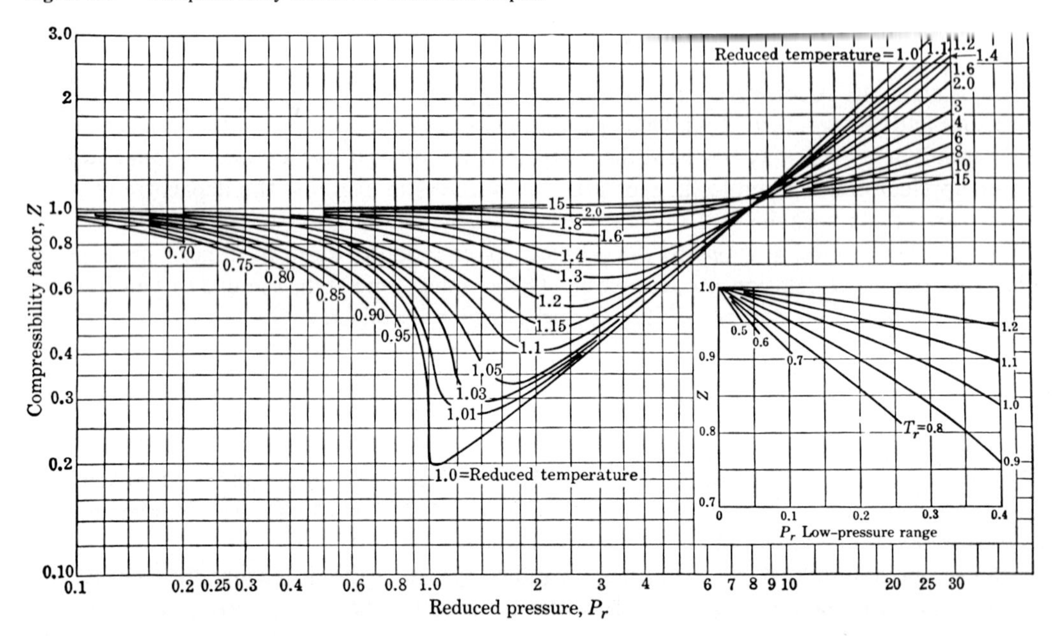


(Reprinted by permission from Hougan and Watson, *Chemical Process Principles*, Part II, Wiley, New York)

States (1*) and (2*) are established by the intersection of lines of constant temperature through states (1) and (2), respectively, with the constant-pressure line $p_1^* = p_2^*$, where the * superscript indicates perfect gas behavior, that is, $P_r = p/p_{cp}$ approaches zero. The enthalpy change between states (1) and (2) can now be evaluated along the paths shown in Figure 5.7 as

$$\bar{h}_2 - \bar{h}_1 = (\bar{h}_2 - \bar{h}_2^*) + (\bar{h}_2^* - \bar{h}_1^*) + (\bar{h}_1^* - \bar{h}_1) \tag{5.25}$$

where $\bar{h}_2 - \bar{h}_2^*$ and $\bar{h}_1^* - \bar{h}_1$ are determined from Figure 5.6 using the known values of $P_r$ and $T_r$ at each state. $\bar{h}_2^* - \bar{h}_1^*$ depends on temperature only since perfect gas behavior is exhibited at the low pressure $p^*$. Its value is given by

$$\bar{h}_2^* - \bar{h}_1^* = \int_1^2 C_p\,(T)\,dT$$

The following examples illustrate the use of Figures 5.5 to 5.7 to determine properties for real gases.

**Figure 5.6** Enthalpy Correction for Gases and Vapors

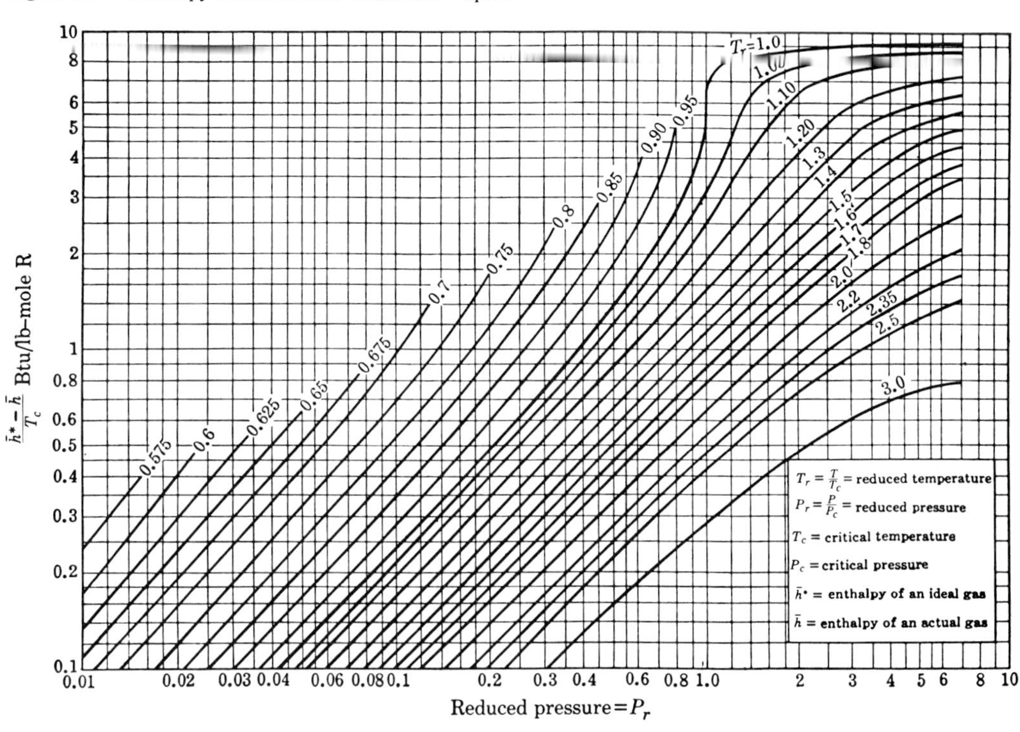


(Reprinted by permission from Hougan and Watson, *Chemical Process Principles*, Part II, Wiley, New York)

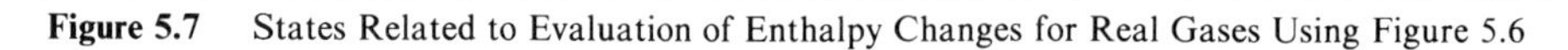

**Figure 5.7** States Related to Evaluation of Enthalpy Changes for Real Gases Using Figure 5.6

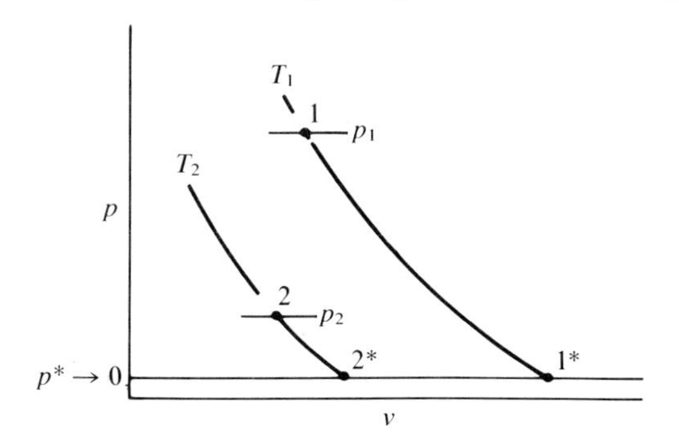

## Example 5.k

Determine the specific volume of oxygen at a gage pressure of 13.8 MPa and a temperature of 0 C by (a) assuming perfect gas behavior and (b) use of the compressibility factor.

**Solution:**

(a) Using $R$ for oxygen from Table 5.1

$$v = RT/p = \frac{259.8\ \text{N m}}{\text{kg K}}\left|\frac{(273 + 0)\ \text{K}}{(13{,}800 + 101.3)\ \text{kPa}}\right|\frac{\text{kPa}}{1000\ \text{N/m}^2}$$

$$= 5.1 \times 10^{-3}\ \text{m}^3/\text{kg}$$

(b) From Table 4.1, the pressure and temperature at the critical point for oxygen are 5040 kPa and −118.9 C. Therefore

$$T_r = T/T_{cp} = 273/(273 - 118.9) = 1.77$$

$$P_r = p/p_{cp} = (13{,}800 + 101.3)/5040 = 2.76$$

From Figure 5.5, $Z = 0.91$. Thus

$$v = Z(RT/p) = 0.91(5.10 \times 10^{-3}) = 4.63 \times 10^{-3}\ \text{m}^3/\text{kg}$$

A difference of about 9 percent is introduced by neglecting the compressibility factor. Note that the accuracy of $pv = RT$ at the given state is unknown until the value of $Z$ is determined.

## Example 5.l

Using the compressibility factor and enthalpy correction charts determine (a) the initial and final specific volumes, (b) the enthalpy change, and (c) the internal energy change when 1 lbm of $H_2O$ undergoes a process from an initial state at 740 F and 3200 psia to a final state at 400 psia and 480 F. Compare the results with those obtained from the *Steam Tables*.

**Solution:**

(a) At the initial state

$$T_{r_1} = \frac{T_1}{T_{cp}} = \frac{(740 + 460)}{(705.4 + 460)} = 1.03$$

and

$$P_{r_1} = \frac{p_1}{p_{cp}} = \frac{3200}{3204} = 1.00$$

At the final state

$$T_{r_2} = \frac{T_2}{T_{cp}} = \frac{(480 + 460)}{(705.4 + 460)} = 0.807$$

and

$$P_{r_2} = \frac{p_2}{p_{cp}} = \frac{400}{3204} = 0.125$$

From Figure 5.5 at the determined reduced pressures and temperatures the compressibility factors are

$$Z_1 = 0.53 \quad \text{and} \quad Z_2 = 0.91$$

yielding the specific volumes

$$v_1 = \frac{Z_1 R T_1}{p_1} = (0.53)\left(\frac{1545}{18}\right)\left(\frac{1200}{(3200)(144)}\right) = 0.118 \text{ ft}^3/\text{lbm}$$

and

$$v_2 = \frac{Z_2 R T_2}{p_2} = (0.91)\left(\frac{1545}{18}\right)\left(\frac{940}{(400)(144)}\right) = 1.28 \text{ ft}^3/\text{lbm}$$

From the *Steam Tables*

$$v_1 = 0.1224 \text{ ft}^3/\text{lbm} \quad \text{and} \quad v_2 = 1.242 \text{ ft}^3/\text{lbm}$$

The deviation from the *Steam Tables* values would have been significantly greater if perfect gas behavior had been assumed.

(b) The enthalpy correction chart (Figure 5.6) corrects for the effect of pressure on enthalpy at a given temperature. Using the previously calculated reduced properties, we obtain the following from Figure 5.6.

$$\left(\frac{\bar{h}^*-\bar{h}}{T_{cp}}\right)_1 = 3.8 \text{ B/(lb mol R)} \quad \left(\frac{\bar{h}^*-\bar{h}}{T_{cp}}\right)_2 = 0.67 \text{ B/(lbmol R)}$$

The enthalpy change is given by eq. (5.25) as

$$\bar{h}_2 - \bar{h}_1 = (\bar{h}_2 - \bar{h}_2^*) + (\bar{h}_2^* - \bar{h}_1^*) + (\bar{h}_1^* - \bar{h}_1)$$

Therefore

$$\bar{h}_2 - \bar{h}_2^* = (-0.67)(T_{cp}) = (-0.67)(1165) = -780.6 \text{ B/lb mol}$$

and

$$\bar{h}_1^* - \bar{h}_1 = (3.8)(1165) = 4427 \text{ B/lb mol}$$

The variable specific heat equation for $H_2O$ at low pressure from Table 5.4 is

$$\bar{C}_p = 19.86 - \frac{597}{\sqrt{T}} + \frac{7500}{T} \text{ B/(lb mol R)}$$

Therefore

$$\bar{h}_2^* - \bar{h}_1^* = \int_{1200}^{940} \left(19.86 - (597/T^{1/2}) + \frac{7500}{T}\right) dT$$
$$= -2241 \text{ B/lb mol}$$

Summing the changes

$$\bar{h}_2 - \bar{h}_1 = -780.6 - 2241 + 4427 = 1405.4 \text{ B/lb mol}$$

$$h_2 - h_1 = 78.1 \text{ B/lbm}$$

From the *Steam Tables*

$$h_2 - h_1 = 1231.5 - 1150.4 = 81.1 \text{ B/lbm}$$

While there is a difference between $\Delta h$ obtained from the *Steam Tables* and $\Delta h$ obtained using the enthalpy correction chart, a very large error results from assuming perfect gas behavior. Such an assumption yields

$$h_2 - h_1 = (\bar{h}_2^* - \bar{h}_1^*)/M = -2241/18 = -124.5 \text{ B/lbm}$$

Also, uncertainty in reading values from the enthalpy correction chart contributes a corresponding uncertainty to final answers obtained using eq. (5.25).

(c) The internal energy change is calculated using the definition of enthalpy

$$\begin{aligned} u_2 - u_1 &= (h_2 - p_2 v_2) - (h_1 - p_1 v_1) \\ &= (h_2 - h_1) - (p_2 v_2 - p_1 v_1) \\ &= 78.1 - \frac{144}{778}\Big((400)(1.28) - (3200)(0.118)\Big) \\ &= 53.2 \text{ B/lbm} \end{aligned}$$

From the *Steam Tables*

$$u_2 - u_1 = 1139.6 - 1077.9 = 61.7 \text{ B/lbm}$$

While this example provides only a limited sample of the use of the compressibility and enthalpy correction charts, it does show that their use yields reasonably accurate results for applications where more precise thermodynamic data are not available.

---

## 5–8 Summary

Substances in the gaseous phase that have an equation of state given by

$$pv = RT$$

and whose internal energy $u$ depends on temperature only are called perfect gases. The gas constant $R$ is given by

$$R = \frac{\overline{R}}{M}$$

where $\overline{R}$ is the universal gas constant and $M$ is the molecular weight of the gas. For $m$ mass units of a perfect gas an equivalent equation of state is

$$pV = mRT$$

A mole is the amount of a substance that has a mass equal to $M$ mass units. Therefore, the number of moles $n$ in a system of mass $m$ is

$$n = \frac{m}{M}$$

The equation of state for a perfect gas can also be written on the basis of a mole of gas as

$$p\bar{v} = \bar{R}T$$

where $\bar{v}$ is the volume per mole of gas. For $n$ moles of a perfect gas

$$pV = n\bar{R}T$$

Specific heats for perfect gases are functions of temperature only and are written as

$$C_v = \frac{du}{dT} \quad C_p = \frac{dh}{dT}$$

Changes in internal energy and enthalpy for perfect gases are given by

$$u_2 - u_1 = \int_{T_1}^{T_2} C_v dT \quad h_2 - h_1 = \int_{T_1}^{T_2} C_p dT$$

When constant specific heats are assumed these changes become

$$u_2 - u_1 = C_v(T_2 - T_1) \quad h_2 - h_1 = C_p (T_2 - T_1)$$

Relations between the specific heats and their ratio $\gamma = C_p/C_v$ are

$$C_p - C_v = R \quad C_v = \frac{R}{\gamma - 1} \quad C_p = \frac{\gamma R}{\gamma - 1}$$

Specific heats on a mole basis are related to those on a unit mass basis by

$$\bar{C}_v = C_v M \quad \bar{C}_p = C_p M$$

The path for a polytropic process is

$$pv^n = \text{constant}$$

where $n$ is a constant for any given polytropic process. Properties for perfect gases undergoing polytropic processes are related by the equations

$$\frac{T_2}{T_1} = \left[\frac{p_2}{p_1}\right]^{(n-1)/n} = \left[\frac{v_1}{v_2}\right]^{n-1}$$

For a closed-system polytropic process the work is given by

$${}_1w_2 = \int_1^2 p dv = \frac{p_2 v_2 - p_1 v_1}{1 - n}$$

For steady-flow polytropic processes with negligible potential and kinetic energy changes the work is

$$_1w_2 = -\int_1^2 v\,dp = \frac{n}{1-n}(p_2v_2 - p_1v_1)$$

For a frictionless quasi-static *adiabatic* process of a perfect gas with constant specific heats or for a frictionless *adiabatic* steady-flow process of a perfect gas with constant specific heats the equation that describes the path is

$$pv^{\gamma} = \text{constant}$$

For this process, the above equations for a polytropic process for a perfect gas apply when $n$ is assigned the value $\gamma = C_p/C_v$.

Departure from perfect gas behavior can be determined by examining $Z$, which is defined as

$$Z = \frac{pv}{RT}$$

For perfect gases $Z$ is unity. $Z$ for real gases can be determined from a compressibility chart if the reduced temperature and reduced pressure defined as

$$T_r = \frac{T}{T_{cp}} \quad P_r = \frac{p}{p_{cp}}$$

are known. Charts for correcting perfect gas enthalpy changes for real gas effects are also prepared in terms of $P_r$ and $T_r$. Results obtained in this manner, although approximate, are useful when a table of properties is not available for the gas under consideration.

## Selected References

Holman, J. P. *Thermodynamics*, 2nd ed. New York: McGraw-Hill, 1974.

Lee, J. F., and F. W. Sears. *Thermodynamics*, 2nd ed. Reading, Mass.: Addison-Wesley, 1963.

McBride, B. J., et al. *Thermodynamic Properties to 6000 K for 210 Substances Involving the First 18 Elements*. NASA SP-3001, 1963.

Reid, R. C., and T. K. Sherwood. *The Properties of Gases and Liquids*. New York: McGraw-Hill, 1958.

Reynolds, W. C., and H. C. Perkins. *Engineering Thermodynamics*. New York: McGraw-Hill, 1970.

Touloukian, Y. S., and T. Makita. *Thermophysical Properties of Matter: Specific Heat*, Vol. 2. New York: Plenum, 1970.

Van Wylen, G. J., and R. E. Sonntag. *Fundamentals of Classical Thermodynamics*, 2nd ed. New York: Wiley, 1973.

Wark, K. *Thermodynamics*, 2nd ed. New York: McGraw-Hill, 1971.

## Problems

In solving the following problems assume constant specific heats from Table 5.5 unless otherwise specified.

5.1 Using the perfect gas equation of state, compute the volume exhibited at 32 F and 14.7 psia by $M$ lbm of each of the gases listed in Table 5.1. $M$ is molecular weight. Compare your results with the corresponding experimental volumes listed in the table.

5.2 A tank is to hold 23 kg of compressed air at a pressure of 1380 kPa and a temperature of 50 C. What is the required tank volume? If the temperature of the air decreases to $-23$ C, what will the pressure be?

5.3 A spherical balloon has a weight when deflated of 200 lbf and a radius of 20 feet when inflated. The atmosphere is at 70 F and 14.50 psia. Assuming the balloon is filled with hydrogen at 70 F and 14.60 psia, determine the force required to anchor it to the ground. Compare the result with that obtained if the balloon is filled with helium at 70 F and 14.60 psia. The local $g$ is 31.0 ft/sec$^2$.

5.4 An automobile tire contains 750 in$^3$ of air at 40 F and 28 psi gage. The barometric pressure is 14.29 psia. If the temperature of the air increases to 140 F, (a) what will its pressure be and (b) what mass of air must be released from the tire to reduce the pressure back to 28 psi gage? Assume the tire does not stretch and that the temperature of the air remains at 140 F for part (b).

5.5 Assuming perfect gas behavior, calculate the volume, at 101.3 kPa and 25 C, occupied by 1 kmol of $N_2$, CO, and $CO_2$. Calculate the specific volume in *m*$^3$/kg for each gas at the given pressure and temperature.

5.6 Air at 14.07 psia and 60 F enters a steady-flow apparatus at the rate of 1200 ft$^3$/min. The air leaves the apparatus at 175 psia and 230 F. Calculate the volume rate of flow at the exit section.

5.7 At low pressures and ambient temperature $H_2O$ is sometimes treated as a perfect gas. Calculate the specific volume of $H_2O$ at 100 F and pressures of 0.20 and 0.60 psia using the perfect gas equation of state. Compare the results with *Steam Tables* values.

5.8 The specific volume of ammonia is 1.320 m$^3$/kg at 103.4 kPa and 10 C. Using the perfect gas equation of state, determine the gas constant $R$ from these data. Compare the result with that obtained when $R$ is calculated from $\overline{R}/M$.

5.9 A rigid container has a volume of 12.15 ft$^3$. Calculate the pound moles of $O_2$, $N_2$, $CH_4$, $H_2$, and $H_2O$, each at 14.696 psia and 300 F, to fill the container, assuming each gas to be a perfect gas. Find the corresponding mass of each gas. Compare the result for $H_2O$ with that found using the *Steam Tables*.

5.10 A pressure vessel with a volume of 0.425 m$^3$ contains 0.0168 kmol of a perfect gas at 96.5 kPa and 21 C. If 0.0439 kmol of gas is added to the contents of the vessel and the temperature after adding the gas is 93 C, what will the pressure be?

5.11 A rigid tank contains 2 lbm of $N_2$ at 20 psia and 500 R. Paddle-wheel work is done on the gas until the temperature reaches 620 R. There is no heat transfer during the process. Calculate the change in internal energy of the $N_2$. Suppose the system were heated (no work) from the same initial to final states. How would the heat transfer for the second process compare with the work for the first?

5.12 One kilogram of air, initially at 105 kPa and 350 K, undergoes a process to a final state at 250 kPa and 850 K. Calculate the internal energy change using data from Table 5.2 and compare the result with that obtained using constant specific heat.

5.13 A closed system of 1 kg of air undergoes a constant-pressure, frictionless quasi-static process during which the temperature of the air increases from 300 to 450 K. Determine the heat transfer for the process using data from Table 5.2; compare the result with that obtained using constant specific heat.

5.14 A closed system of 3 lbm of $CO_2$ expands during an adiabatic process. If the initial temperature is 300 F and if 9600 ft lbf of work is done during the process, what is the final temperature?

5.15 Nitrogen flows adiabatically through a nozzle. At the nozzle inlet the temperature is 150 C and the velocity 150 m/s. At the outlet the temperature is 50 C and the pressure 97.3 kPa. If the nozzle outlet area is 0.0010 m$^2$, what is the mass rate of flow through the nozzle?

5.16 Forty pounds mass per minute of air at 65 psia and 80 F enter a pipe having a constant cross-sectional area of 3 in$^2$. At a downstream section of the pipe the air is at 15 psia and the velocity is six times that entering the pipe. The flow is steady. What is the temperature of the air at the downstream section?

5.17 An air compressor takes in 750 ft$^3$/min at 14 psia and 60 F. Heat transfer from the air is at the rate of 3200 B/min. If the air leaves the compressor at 70 psia and 165 F, how much power, in kilowatts, is required to drive the compressor? Assume kinetic energies are negligible.

5.18 Helium enters a nozzle at 50 psia and 70 F with negligible velocity. It expands adiabatically through the nozzle and leaves with a velocity of 1500 ft/sec. What is the temperature of the helium at the nozzle outlet?

5.19 A small air turbine operates adiabatically. The mass flow rate is 7.0 kg of air per minute and the power delivered is 3.8 kW. If air enters the turbine at 340 kPa and 21 C and the kinetic energies are negligible, at what temperature does the air leave the turbine?

5.20 A stream of air at 14.5 psia and 90 F flows at the rate of 500 lbm/min through a duct having a cross-sectional area of 0.500 ft$^2$. A second stream of air at 14.5 psia and 220 F flows at the rate of 6950 ft$^3$/min through a duct with a cross-sectional area of 0.750 ft$^2$. The two streams mix adiabatically to produce one leaving stream at 14.5 psia. The velocity of the leaving stream is 70 ft/sec. Calculate (a) the temperature of the leaving stream and (b) the cross-sectional area of the outlet duct.

5.21 Under what conditions will the heat transfer be equal to the enthalpy change for (a) a closed-system process and (b) a frictionless steady-flow process?

5.22 A closed system of helium is initially at 100 kPa, 20 C, and occupies a volume of 0.80 m$^3$. The helium is compressed isothermally to a pressure of 1545 kPa. What is the heat transfer for the process?

5.23 During a frictionless, steady-flow process, air at 70 F and 14.5 psia is compressed isothermally. If the heat transfer from the air is 69.5 B/lbm, what is the final pressure? Assume the kinetic energy change to be negligible.

5.24 One pound mole of $O_2$, initially at 70 F and 10 psia, is heated at constant volume to 2000 F. Using variable specific heat data from Table 5.4, calculate the heat transfer for the process.

5.25 Develop an expression for the average specific heat $C_v$ for a perfect gas.

5.26 Using information from Table 5.4, calculate for air (a) the instantaneous $C_v$ of air at 100 and 1000 F and (b) $C_v$ average between 100 and 1000 F.

5.27 One and a half kilograms of an unknown perfect gas occupy a volume of 0.354 $m^3$ at 345 kPa. The molecular weight of the gas is 32.00. The gas undergoes a process to a final state at 105 kPa and 15.5 C. If $C_p$ for the gas is 0.913 kJ/(kg K), what are the internal energy and enthalpy changes for the process?

5.28 An unknown perfect gas undergoes a constant volume process during which the internal energy increases 136.1 kJ/kg. For the process the pressure ratio $p_2/p_1$ is 1.345. If the molecular weight of the gas is 44 and its specific heat ratio $\gamma$ is 1.37, what are the initial and final temperatures for the process?

5.29 Starting with eq. (5.6) show that the internal energy change of a perfect gas with constant specific heats is

$$u_2 - u_1 = \frac{p_2 v_2 - p_1 v_1}{\gamma - 1}$$

and the enthalpy change is

$$h_2 - h_1 = \frac{\gamma}{\gamma - 1}(p_2 v_2 - p_1 v_1)$$

5.30 An unknown perfect gas has a specific heat ratio of 1.53 and a molecular weight of 19. The gas undergoes a constant-pressure closed system process at 27.00 psia during which 57 B of heat transfer to the gas occurs. Calculate the work for the frictionless, quasi-static process. The mass of the system is unknown.

5.31 One kilogram of a perfect gas is heated at constant volume from 21 to 138 C. The internal energy change during the process is 82.92 kJ/kg. Assuming the specific volume of the gas is 0.3121 $m^3$/kg and the specific heat ratio is 1.45, calculate (a) the initial pressure and (b) the enthalpy change during the process.

5.32 Starting with the closed-system energy equation for a frictionless, quasi-static process, show that

$$\delta q = \frac{C_v}{R} v dp + \frac{C_p}{R} p dv$$

for a perfect gas.

5.33 A closed system of 2 lbm of carbon dioxide is compressed from 15 psia, 70 F, to 90 psia in a frictionless, quasi-static process. Calculate the work assuming the process is (a) adiabatic and (b) isothermal. Sketch a *p-v* diagram for the processes.

5.34 A closed system of 1 kg of air undergoes a cycle consisting of the following frictionless quasi-static processes: (a) The air is first heated at constant volume from 40 to 260 C. (b) It is next expanded isothermally back to its initial pressure. (c) Finally it is cooled at constant pressure back to its initial state. Calculate the thermal efficiency of a heat engine operating on this cycle.

5.35 Develop the following expression and list its restrictions. Start with $pv^\gamma = C$.

$$T_2/T_1 = (p_2/p_1)^{(\gamma-1)/\gamma}$$

5.36 A perfect gas undergoes a frictionless, adiabatic expansion through a nozzle. The gas enters the nozzle at 60 psia and 970 R with negligible velocity. It leaves the nozzle at 15 psia, 630 R, and a velocity of 1850 ft/sec. Determine the cross-sectional area of the nozzle outlet for a mass flow rate of 1.320 lbm/sec.

5.37 The contents of a closed rigid tank are cooled from state (1) where $p_1 = 90$ psia and $t_1 = 1500$ F to state (2) where $p_2 = 40$ psia. Find the heat transfer per unit mass if the substance in the tank is (a) nitrogen (b) $H_2O$.

5.38 A compressor takes in air at 14.70 psia and 60 F and discharges it at 75 psia. Assume the entering and leaving velocities are negligible. If the compressor operation is frictionless and adiabatic and if 25 HP is required to drive the compressor, what is the inlet volume rate in cubic feet per minute?

5.39 The following frictionless and quasi-static processes take place in a closed system:

1–2: Constant pressure process, $p_1 = 19$ psia,

$v_1 = 15$ ft³/lbm, $v_2 = 24.2$ ft³/lbm

2–3: Constant volume process, $p_3 = 30$ psia

Find the temperature at each state and the heat transfer for each process if the substance undergoing the processes is (a) carbon dioxide (b) $H_2O$.

5.40 A rigid, perfectly insulated tank has a volume of 0.283 m³. Initially the tank is completely evacuated. A pipeline with a valve connects to the tank. If air in the line is at 136 kPa and 22 C, and if the valve is opened slowly, allowing air to flow into the tank, what will the temperature in the tank be when the pressure reaches 136 kPa?

5.41 A steady-flow polytropic process with $n = 1.3$ takes place between state (1) where $p_1 = 2.75$ *MPa*, $t_1 = 420$ C and state (2) where $p_2 = 0.36$ *MPa*. The kinetic energy at (1) is equal to that at (2). Find the work and heat transfer per unit mass if the working fluid is (a) air (b) $H_2O$.

5.42 During a polytropic process 1 pound mass of helium is compressed from 15 psia and 70 F to 100 psia along the path $pv^{1.537} =$ constant. The process is for a closed system. Calculate the heat transfer.

5.43 A closed system of 1 lbm of oxygen initially at 20 psia and 80 F undergoes a frictionless quasi-static process along the path $pv^{1.25} =$ constant. Assuming the final pressure is 50 psia, determine the heat transfer for the process.

5.44 A closed system of 1 lbm of air undergoes a polytropic process. Initially the air is at 100 psia and 1510 R. At the final state the temperature is 1150 R. During the process 7.15 B/lbm heat transfer to the system takes place. Calculate the initial and final specific volumes.

5.45 Air undergoes a frictionless, steady-flow process in a manner such that $pv^{1.31} =$ constant. At the inlet section the pressure is 100 kPa and the temperature is 27 C. At the discharge the pressure is 700 kPa. Both potential and kinetic energies may be neglected. Calculate the heat transfer and work per kilogram of air.

5.46 A steady-flow polytropic expansion process follows the path $pv^{1.5} = C$ from state (1) where $p_1 = 90$ psia, $t_1 = 1000$ F, and $V_1 = 100$ ft/sec to state (2) where $p_2 = 15$ psia and $V_2 = 500$ ft/sec. The cross-sectional area at (1) is 1.0 ft². The working fluid is a perfect gas with $C_p = 0.220$ B/(lbm R) and $C_v = 0.160$ B/(lbm R). Determine (a) the work per unit mass, (b) the power, and (c) the heat transfer rate for the process.

5.47 A closed system of 1 lbm of air undergoes a polytropic process along the path $pv^{1.32}$ = constant. The air is initially at 15 psia and 70 F. During the process 15 B of heat is transferred *from* the air.

(a) Sketch a path on the $p$-$v$ diagram showing the constant temperature lines $T_1$ and $T_2$.
(b) Calculate the pressure and temperature at the final state.

5.48 A closed system of 1 lbm of air undergoes a cycle consisting of the following frictionless quasi-static process:

$1 \rightarrow 2$ Constant-pressure expansion from −50 to 0 F
$2 \rightarrow 3$ Adiabatic compression to 168 F
$3 \rightarrow 4$ Constant-pressure compression
$4 \rightarrow 1$ Adiabatic expansion

For the cycle calculate the appropriate performance parameter, that is, the thermal efficiency or coefficient of performance.

5.49 A closed system of 0.10 kg of air undergoes the following frictionless, quasi-static cycle:

$1 \rightarrow 2$ Constant volume heating from 80 to 650 C
$2 \rightarrow 3$ Adiabatic expansion to $p_3 = p_1$
$3 \rightarrow 1$ Constant pressure

The cycle is repeated 400 times per minute.

(a) Sketch the $p$-$v$ diagram for the cycle.
(b) Calculate the cycle thermal efficiency.
(c) Calculate the net power output.

5.50 A closed system of 1 lbm of air undergoes the following frictionless, quasi-static cycle:

$1 \rightarrow 2$ Constant volume from 40 to 440 F
$2 \rightarrow 3$ Polytropic expansion along $pv^{1.25} = C$ to $p_1$
$3 \rightarrow 1$ Constant pressure

Calculate the cycle thermal efficiency.

5.51 Compute the volume that 10 lbm of ammonia ($NH_3$) occupies at a pressure of 280 psia and a temperature of 380 F by (a) assuming perfect gas behavior, (b) use of the compressibility factor, and (c) use of the table of properties for ammonia, Appendix D. Compare your results.

5.52 Carbon dioxide undergoes a steady-flow process with no work and negligible kinetic and potential energy changes from 800 psia and 365 F to 107 psia and 80 F. Using the enthalpy correction chart and the expression for $C_p$ from Table 5.4, estimate the heat transfer for the process. Compare your answer with that obtained by neglecting the enthalpy correction.

5.53 A vessel with a volume of 0.0175 m³ contains 5.80 kg of nitrogen at a pressure of 68.9 MPa. Using the compressibility chart, estimate the temperature of the nitrogen.

# 6

# The Second Law of Thermodynamics

*In the preceding chapters the first law, as well as properties, processes, and cycles for different pure substances, were presented. First-law energy equations were formulated for processes and cycles. Power cycles involving conversion of heat to work were encountered, and refrigeration and heat pump cycles that promote the flow of heat energy from one region to another at a higher temperature were discussed. Heat engine thermal efficiency and coefficients of performance for refrigerators and heat pumps were defined and first law concepts have been discussed in relation to these parameters. However, the first law does not impose any limits on heat engine thermal efficiency or on coefficients of performance for refrigerators or heat pumps. While the first law must be satisfied for any process or cycle, it places no restrictions on the direction of the flow of heat or work; for example, for a cycle the cyclic integral of heat is still equal to the cyclic integral of work if the signs are changed on all of the heat and work quantities. Further, the first law does not provide information as to whether a particular process or cycle is possible. Consider, for example, the possibility of constructing a heat engine that would receive heat from the atmosphere at a fixed temperature, convert part of the heat to work, and reject the remainder back to the same atmosphere. A first-law analysis would tell us that the heat engine would produce work in an amount equal to the heat received less the heat rejected. Yet experience has shown that it is impossible to construct such a heat engine. Consider next a hot object and a cold object that are brought into contact while isolated from all other bodies. Heat transfer will occur until both objects are at the same temperature. From the first law, the energy given up by the hot object is equal to that received by the cold object. The first law is not violated if the reverse process, the transfer of heat to restore the original temperatures of the objects, is assumed to take place. Yet we know that such a process does not occur.*

*From the foregoing discussion we see that the first law has limitations. Certain events known to occur cannot be explained by the first law alone. Others that are not possible cannot be rejected by a first-law analysis only. It is the second law of thermodynamics that provides the additional foundation necessary to explain and describe fully the behavior of many systems and processes encountered in engineering thermodynamics. The second law and its associated concepts establish upper bounds for heat engine thermal efficiency, and the coefficients of performance for refrigerators and heat pumps. The second law provides the basis for determining whether certain processes can occur and serves as the foundation for numerous other concepts in thermodynamics. This chapter will deal with the second law as it is related to cyclic processes. Chapter 7 will extend the second law to noncyclic processes.*

*The second law, like the first law, is empirical. It is based solely on observations of nature and cannot be derived or deduced from any other principle or law. This law, among its many ramifications, is an extremely important and useful concept in the evaluation of energy-conversion systems and in dealing with the energy problems facing humankind.*

## 6–1 Statements of the Second Law of Thermodynamics

A number of concepts related to the second law involve heat reservoirs. A *heat reservoir* is a region that has a uniform and constant temperature. It is of such nature and extent that finite amounts of heat transferred to or from it will not change its temperature. When heat is transferred from a reservoir, the reservoir is frequently referred to as a heat "source." When heat is transferred to a reservoir, the reservoir is often termed a heat "sink." The atmosphere is a reservoir that can serve as a source or a sink, depending on the direction of heat transfer. A region maintained at a constant temperature by the combustion of a fuel, for example, can serve as an energy reservoir for a heat engine or as a source of heat energy for other uses.

Two equivalent classical statements of the second law have been formulated—the Clausius statement and the Kelvin-Planck statement. Clausius, in 1850, stated: "The effects of a direct heat interaction between two bodies cannot be completely undone." Current statements of the Clausius concept are perhaps clearer but retain the basic concept he addressed. The variation we shall use is:

> *It is impossible for a cyclic device to convey heat from a reservoir at a given temperature to another at a higher temperature without energy input.*

This statement addresses the direction of heat transfer. It is a matter of experience that in natural (spontaneous) heat transfer processes, heat energy flows from a hot region to a cold region and never in the opposite direction. To cause heat to flow from a region or a reservoir at a given temperature to another at a higher temperature, we must employ a device of some sort to promote the heat transfer. A cyclic device that conveys heat from a low temperature region to a higher temperature region is by definition a refrigerator. In simple terms the Clausius statement declares that in order for a refrigerator to operate, there must be an energy input (either heat or work or both) from the surroundings. A schematic diagram of a violation of the Clausius statement is shown in Figure 6.1a. As noted in section 3–9, we shall focus attention in this text on refrigerators that operate by work input.

The Kelvin-Planck statement of the second law is as follows:

> *No device operating in a cyclic manner can exchange heat with one or more reservoirs at the same temperature and produce a net work output.*

Reflection on this statement leads to the conclusion that complete and continuous conversion of heat to work is not possible. Figure 6.1b illustrates schematically a "perfect heat engine," one that would convert all of the heat supplied to it to work and thus would violate the Kelvin-Planck statement.

The Kelvin-Planck statement and the Clausius statement are equivalent. This equivalence can be established by assuming a violation of one and then showing that the assumed violation results in a violation of the other. The procedure, when applied to each of the second-law statements, establishes their equivalence. To illustrate this logic let us *assume* that the

**Figure 6.1** Violators of the Second Law. *a.* Clausius violator. *b.* Kelvin-Planck violator.

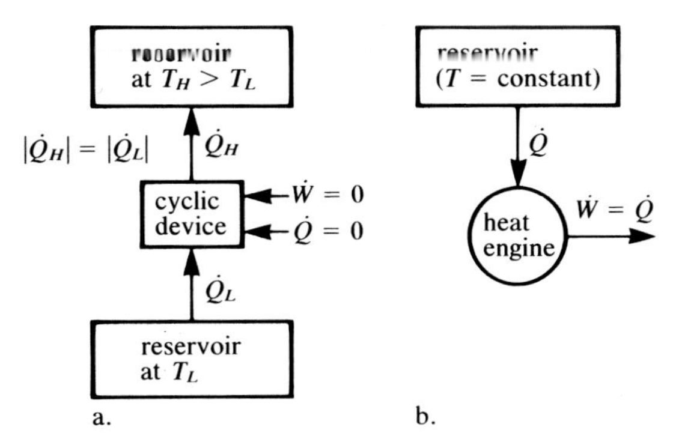


**Figure 6.2** Assumed Violation of Kelvin-Planck Statement

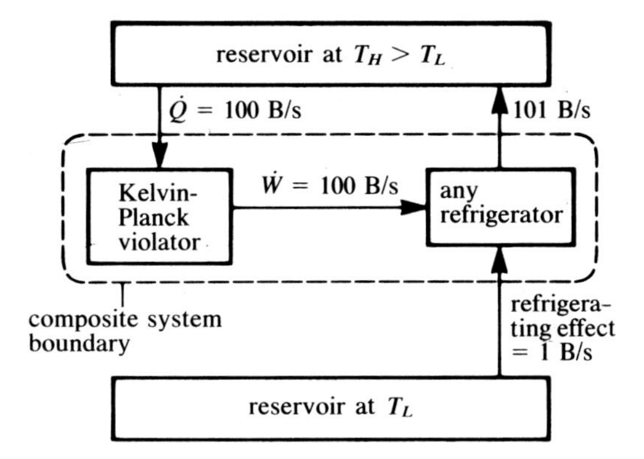


Kelvin-Planck statement can be violated. While any value of $\dot{Q}$ will yield the same result, suppose we let $\dot{Q}$ for the Kelvin-Planck violator be 100 B/sec. The heat engine would then, by the first law, deliver work at the rate of 100 B/sec as indicated in Figure 6.2. This power can be used to drive any refrigerator. (Recall that a refrigerator is a cyclic device, requires energy input, and continually transfers heat from one region to another region at a higher temperature.) If the refrigerator produces a refrigerating effect of only 1 B/sec, it must, by the first law, reject 101 B/sec to the reservoir at $T_H$. The result is a self-contained system defined by the boundary shown in Figure 6.2. While no work is supplied to the cyclic composite system, heat is transferred continuously at a net rate of 1 B/sec from the reservoir at $T_L$ to that at $T_H$. This violates the Clausius statement of the second law. What assumption has been made? Only one, namely, that the Kelvin-Planck statement of the second law could be violated. The conclusion we draw from this exercise is that a violation of the Kelvin-Planck statement results in a violation of the Clausius statement of the second law. The final step in proving the equivalence of the two statements is to assume the Clausius statement can be violated, and then show that such an assumption will result in a violation of the Kelvin-Planck statement.

## 6–2 The Reversible Process

A process closely related to the second law is the reversible process. A system undergoes a *reversible process* if, upon completion of the process, the system and its surroundings can be restored to their initial states. One means to determine whether the system and its surroundings can be returned to their initial states is to apply the second law. The method for this application will be illustrated later. In concept, reversible processes can occur in steady-flow systems as well as in closed systems. Let us first consider a closed system. The following are the conditions under which a closed system process is reversible.

1. The system does not depart internally from equilibrium by more than an infinitesimal amount.
2. The system does not depart from equilibrium with its surroundings by more than an infinitesimal amount.
3. No dissipative effects occur within the system or in its surroundings.

In conditions 1 and 2 we refer at this point to equilibrium in terms of mechanical and thermal equilibrium. These forms of equilibrium constitute the main forms of interest in dealing with simple substances. Thermal equilibrium requires that the temperature be the same everywhere. For mechanical equilibrium to exist, no unbalanced forces are permitted. Mechanical and thermal equilibrium for a simple substance in a closed system exists when the pressure and the temperature are uniform throughout the system. Generally, heat and work interactions are involved in closed system processes. Departure from either thermal or mechanical equilibrium by more than an infinitesimal amount within the system or between the system and its surroundings during a process involving either of these interactions will render the process irreversible. The conditions for a steady-flow process to be reversible are the same as those for a closed system, but an interpretation is required. If an element of mass passing through a steady-flow system undergoes a process that fulfills the conditions for a closed system reversible process, then the steady-flow process is reversible.

Dissipative effects include, but are not limited to, mechanical friction, viscous effects in fluids (e.g., turbulence and shear), and Joulean heating (the flow of electric current through a resistor). Dissipative effects result in the conversion of energy in the form of work, or other energy forms that theoretically could be completely converted to work, into heat or an increase in internal energy. If work is used to accomplish any portion of a process that could be accomplished by heat transfer, the process will be irreversible. An example of this is a paddle-wheel stirring process in a perfectly insulated rigid container. For such a process, the work done on the system is equal to the internal energy increase. The same change in state (and therefore the same internal energy change) could occur by heat transfer to the uninsulated system. Therefore, the stirring process is not reversible.

An irreversible process is, quite simply, one that is not reversible. In the strictest sense, reversible processes do not occur. One significant reason for this is that dissipative effects can never be completely eliminated. Nonetheless, the concept of reversibility is a necessary prerequisite to the development of several fundamental principles that follow from the second law. In a practical sense the reversible process is an optimum process. Efficiencies of numerous devices relate actual performance to the performance the same equipment would exhibit if it could operate reversibly.

Let us now examine the concept of reversibility in more detail. This concept applies to closed as well as open systems. While the three conditions described outline in a general way the reversibility requirements, it is useful to consider certain specific occurrences that render processes irreversible. When these have been identified, we can more readily recognize a process as reversible if these events are not associated with the process. It has been noted that tests to determine whether a process is reversible or irreversible are based on the second law. Examples of such tests are presented below as we consider the various occurrences that produce irreversibilities. Since the Clausius and Kelvin-Planck statements are entirely equivalent, either statement may be used. In some cases one statement will produce the desired result in fewer steps than the other. The establishment of irreversibility via second-law statements follows a consistent logic pattern and usually requires that the "reverse" of the process in question be part of a cycle.

*Unrestrained Expansion* The expansion of a substance in an unrestrained manner is irreversible. In an unrestrained expansion, condition 1 is not fulfilled in that the system departs internally from at least mechanical equilibrium, that is, the pressure in the system departs from uniformity by more than an infinitesimal amount. As an example consider a rigid container with a partition dividing its volume into two parts. Let a gas occupy one of the parts and let the other be evacuated. If the partition is suddenly removed, the gas will expand rapidly in an unrestrained manner to fill the volume completely. During this expansion a nonuniform pressure distribution exists as a result of the presence of time-dependent pressure waves. Expansion of the gas in a restrained manner could occur in a piston–cylinder assembly, with a slowly moving piston providing the restraint. A substance flowing through a throttling valve to a lower pressure also experiences an unrestrained expansion, and, as the next example shows, is irreversible.

## Example 6.a

A steady-flow throttling process involves an unrestrained expansion of the flowing substance. Show that the process is irreversible.

**Solution:**

*Step 1.* Define the system and write the energy equation for the process noting any interactions with the surroundings during the process: The system is defined by sections (1) and (2) and the inner surface of the insulated pipe. During the throttling process there is no heat transfer, no work, and negligible changes in kinetic and potential energy. Thus the steady-flow energy equation reduces to $h_1 = h_2$. States (1) and (2), having the same enthalpy, are shown on the *p-v* diagram.

*Step 2. Assume* the process is reversible. The reversed process $(2 \rightarrow 1)$ would restore the system to state (1) with no heat transfer or work and negligible changes in kinetic and potential energy.

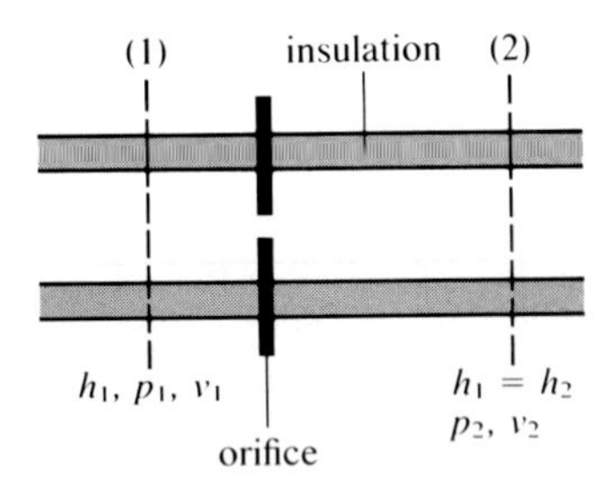


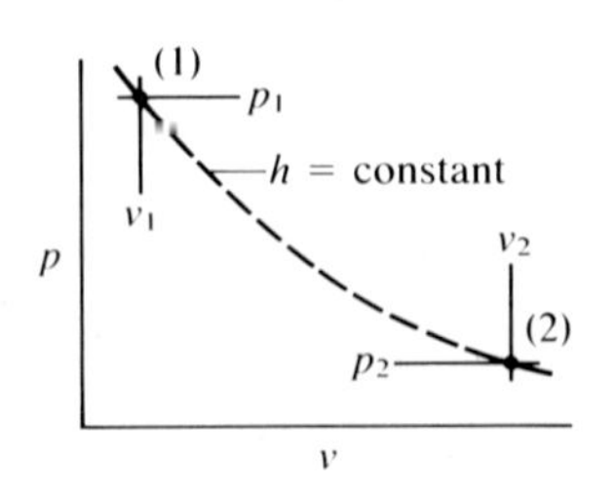


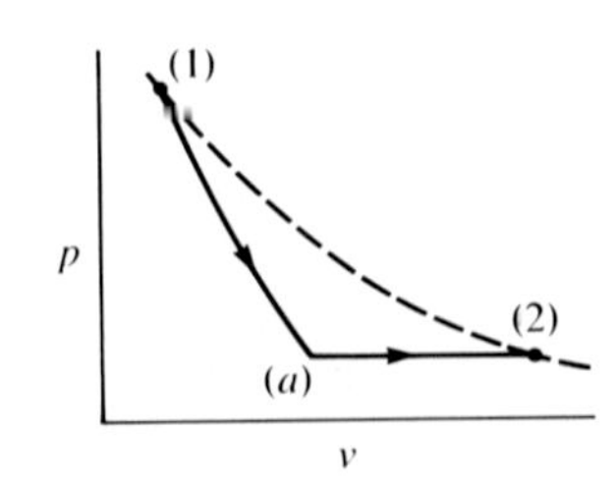


*Step 3*. Make the assumed reverse process part of a cycle. One such cycle would consist of the reversed process, expanding the working fluid reversibly and adiabatically through a turbine from (1) to ($a$), then heating at constant pressure from ($a$) to (2) with the heat transfer from a single reservoir at a temperature greater than $T_2$. Velocities at states (2), (1), and ($a$) could be negligible. (Note that $h_a$ is less than $h_1$ by the work per unit mass done by the turbine. Therefore, the heat transfer, $h_2 - h_a$, is positive.) This cycle would have net work output equal to the turbine work while exchanging heat with the single reservoir at a temperature greater than $T_2$. Thus the cycle would violate the Kelvin-Planck statement of the second law. The only assumption is that made in step 2, namely, that the throttling process was reversible. Since that assumption yielded a violation of the second law, we conclude that it is erroneous and that throttling processes are irreversible.

---

*Heat Transfer Across a Finite Temperature Difference* Direct heat transfer between two systems for which a finite temperature difference exists is irreversible. Let heat transfer occur between a reservoir and a system whose temperature is lower than that of the reservoir by a finite amount. Because of the temperature difference this heat transfer takes place in a natural way and produces no effects in other parts of the system's surrounding. For this process, the system is not in thermal equilibrium with a part of its surroundings, the reservoir, and therefore condition 2 is not fulfilled. From this we conclude the heat-transfer process is not reversible. It is easy to prove by means of the second law that heat transfer across a finite temperature difference is irreversible. Assume that the heat transfer between the reservoir and the above system is reversible. The reversed process involves transfer of heat from the system to the higher-temperature reservoir with no other effects. We immediately recognize a violation of the Clausius statement; hence we prove that the heat transfer process is irreversible.

An important concept that is useful in describing certain ideal processes is that of reversible heat transfer. In view of condition 2 we conclude that reversible heat transfer can occur between a system and its surroundings only when the system and its surroundings depart from temperature equilibrium by no more than an infinitesimal amount. Regardless of the mode of heat transfer (conduction, convection, or radiation), the rate of the transfer depends on the temperatures of the bodies between which heat transfer occurs, with heat-transfer rates increasing as the temperature difference increases. Further, the heat-transfer rate through a surface is proportional to the area normal to the direction of heat flow. As the temperature difference approaches an infinitesimal, $dT$, an infinitesimal heat-transfer rate is also approached when the heat-transfer area is fixed. Thus we see that the reversible transfer of a finite quantity of heat would require either an infinite time or an infinite heat-transfer area. While reversible heat transfer is impossible in practice, such a process may be viewed as a limiting case of real heat transfer. We shall see that the concept of reversible heat transfer plays an important role in a number of aspects of the second law.

*Friction* An important dissipative effect included in condition 3 is friction. Any process involving any form of friction is irreversible. The following example involves friction internal to a fluid.

## Example 6.b

A viscous fluid in a rigid container has work done on it by a paddle wheel. During the process there is no heat transfer. Show by application of the second law that the process is irreversible.

**Solution:**

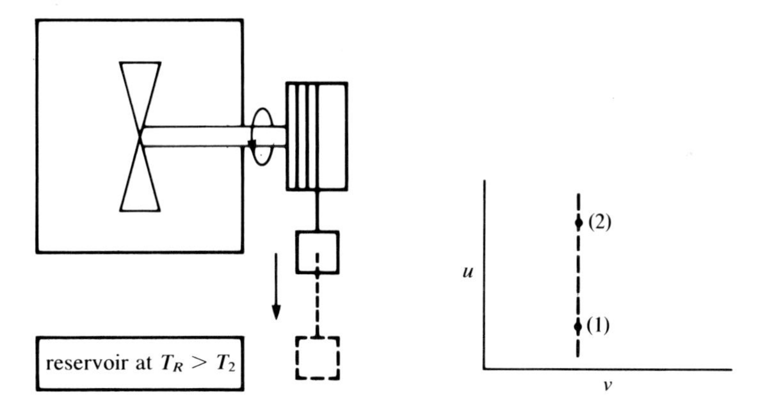


*Step 1.* For the closed system the energy equation for the given process is

$$ {}_1Q_2 = U_2 - U_1 + {}_1W_2 = 0 $$

With ${}_1W_2$ less than zero, $U_2$ must be greater than $U_1$ to satisfy the first law. The initial and final equilibrium states are shown on the *u-v* diagram.

*Step 2. Assume* the process is reversible. During the reversed process ($2 \rightarrow 1$) the internal energy would decrease and work in the amount of

$$ {}_2W_1 = U_2 - U_1 $$

would be done by the system.

*Step 3.* A cycle could then be devised that would consist of the reversed process and a process that would return the system to state (2), during which heat could be transferred from a single heat reservoir whose temperature was greater than $T_2$. The amount of heat transfer from the reservoir would be

$$_1Q_2 = U_2 - U_1$$

This two-process cycle would have net work output while exchanging heat with a single heat reservoir. Such a cycle violates the Kelvin-Planck statement of the second law. The only assumption made was that the original process was reversible. Since that assumption resulted in a violation of the second law, the assumption is incorrect. Thus we have proved that stirring (paddle-wheel) work, or any other action that produces fluid friction, causes a process to be irreversible.

---

*Other Occurrences* A number of other occurrences that render processes irreversible will be mentioned briefly. Mixing of dissimilar substances is irreversible. Consider an adiabatic closed system consisting of two gases at the same pressure and temperature separated by a partition. On removal of the partition a nonequilibrium condition exists within the system in which a difference in concentration of species is present. In time mixing of the gases will occur due to the migration of each gas into the total system volume. Equilibrium is established when the concentration of species is uniform. The two gases cannot be separated unless work is supplied from the surrounding. Phenomena such as shock waves; inelastic deformation of solids; most chemical reactions, including combustion processes; and mixing of streams or amounts of a substance initially at different states are also irreversible.

At this point formal determination of the irreversibility can be made only by direct application of the second law. In Chapter 7 the inequality of Clausius and the entropy principle are presented. These two concepts provide additional methods for determining irreversibility of a process. Methods to quantify the irreversibility of a process will be developed in Chapter 8.

We previously used the terms *quasi-static* and *frictionless* in relation to a closed-system process that at all times is essentially in equilibrium and involves no frictional effects. We have also described steady-flow frictionless processes. We now broaden our description of these processes and will subsequently use the term *reversible* in referring to them. Thus, for example, the work for a reversible closed system process is given by

$$_1W_2 = \int_1^2 p\,dV \qquad \textbf{(2.6b)}$$

and the work for a steady-flow reversible process of a single stream is

$$_1w_2 = -\int_1^2 v\,dp - \frac{V_2^2 - V_1^2}{2g_0} - \frac{g(z_2 - z_1)}{g_0} \qquad \textbf{(2.14)}$$

Processes will be considered for which a system undergoes a reversible process internally, but is externally irreversible. Such processes are termed *internally reversible.* In other words, an internally reversible process is one in which the system behaves internally exactly as it would during a completely reversible process but irreversibilities occur in the surroundings. Use of the term *reversible* implies that both the system and its surroundings satisfy the conditions for a reversible process. The difference between a reversible process and an internally reversible process for a system thus depends only on events taking place in the surroundings during the process. Equations that describe the behavior of reversible processes in terms of only the properties of the system are also applicable to internally reversible processes. Thus the equations for reversible work listed above are also applicable to internally reversible processes.

## 6–3 The Carnot Cycle, a Reversible Cycle

A *reversible cycle* is one in which each process is reversible. While the individual processes comprising the cycle may be either closed system or steady flow, the cycle itself will operate within a closed boundary. In a steady-flow cycle properties of the working fluid at all points in the system do not change with time.

The Carnot heat engine cycle is of particular interest because it is a simple reversible cycle that operates between a single heat source at a fixed temperature (the high-temperature reservoir) and a single heat sink (the low-temperature reservoir) at a constant temperature less than the source temperature. This cycle consists of four reversible processes:

$1 \rightarrow 2$. Reversible heat transfer from a source at $T_H$ to the working fluid at constant temperature.

$2 \rightarrow 3$. Reversible and adiabatic expansion until the temperature of the working fluid decreases to $T_L$.

$3 \rightarrow 4$. Reversible heat transfer from the working fluid at constant temperature to a sink at $T_L$.

$4 \rightarrow 1$. Reversible and adiabatic compression of the working fluid from $T_L$ to $T_H$.

From the conditions for a reversible process given in Section 6–2, it is apparent that the temperature of the working fluid during the isothermal heat-transfer processes, processes $1 \rightarrow 2$ and $3 \rightarrow 4$, can differ only infinitesimally from the reservoir temperatures. For the heat-reception process $1 \rightarrow 2$, the temperature of the working fluid is $dT$ less than $T_H$, and for the heat-rejection process $3 \rightarrow 4$ the working fluid temperature is $dT$ above $T_L$. These differences, of course, are necessary for heat transfer to occur. Note that no particular working fluid is prescribed for the Carnot cycle.

The Carnot cycle described is only one of numerous reversible cycles that can be devised to operate between two reservoirs. Other cycles that can, in theory, operate reversibly between two fixed-temperature regions include the Stirling and Ericsson cycles. For an explanation of these cycles see the works referenced at the end of this chapter.

Figure 6.3 Carnot Heat Engine Cycle. *a.* Perfect gas cycle. *b.* Energy interactions.

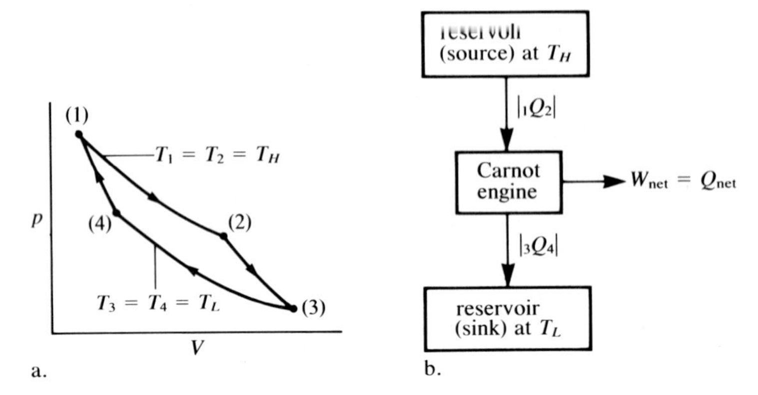


## 6–4 The Perfect Gas Carnot Heat Engine Cycle

In this section we shall develop an expression for the thermal efficiency of a closed-system Carnot heat engine cycle using a perfect gas as the working fluid. Suppose that a closed system of a perfect gas, as defined by eqs. (5.1) and (5.2), undergoes a Carnot heat engine cycle. The processes for this cycle as described in the preceding section can be plotted on a pressure-volume diagram, as shown in Figure 6.3a. Process (1) to (2) is the isothermal process in which heat is transferred reversibly to the gas from the reservoir at $T_H$. Process (2) to (3) is the reversible adiabatic expansion process. The path for this process is given by $pV^\gamma = C$. (Note that this path equation, as related to a closed system, was derived for a frictionless quasi-static adiabatic process; see Section 5–4. We now refer to the process as a reversible adiabatic process.) The process (3) to (4) is the isothermal process in which heat is transferred reversibly to the reservoir at $T_L$. The cycle is completed by the reversible adiabatic process (4) to (1) for which $pV^\gamma = C$. Figure 6.3b indicates the energy interactions between the Carnot engine and its surroundings. Thermal efficiency for a work-producing cycle is defined as

$$\eta \equiv \frac{W_{net}}{Q_{in}} = \frac{Q_{in} - Q_{out}}{Q_{in}} \tag{3.5}$$

With work involved in all four of the processes, as evidenced by the areas below the paths, and heat transfer involved in only two, it is apparent that the thermal efficiency determination will be simpler if the heat-transfer quantities are used. For the reversible isothermal process of a perfect gas at $T_1 = T_2 = T_H$, $U_2 = U_1$, and the closed-system energy equation becomes

$$_1Q_2 = {_1W_2} = \int_1^2 p\,dV$$

For the isothermal process at $T_H$

$$p_1V_1 = p_2V_2 = PV = mRT_H = \text{constant}$$

Thus the heat transfer for the process at $T_H$ is

$$ {}_1Q_2 = mRT_H \int_1^2 \frac{dV}{V} = mRT_H \ln \frac{V_2}{V_1} = Q_{in} $$

Similarly, the heat transfer for the process taking place at constant temperature $T_L$ is

$$ {}_3Q_4 = mRT_L \int_3^4 \frac{dV}{V} = mRT_L \ln \frac{V_4}{V_3} $$

Inspection of this equation indicates the heat transfer will be from the system (negative) since $V_4$ is less than $V_3$. The absolute value of the heat transfer can be expressed as

$$ |{}_3Q_4| = mRT_L \ln \frac{V_3}{V_4} = Q_{out} $$

Upon substituting values for the heat-transfer quantities into eq. (3.5) we have

$$ \eta_C = \frac{mRT_H \ln \dfrac{V_2}{V_1} - mRT_L \ln \dfrac{V_3}{V_4}}{mRT_H \ln \dfrac{V_2}{V_1}} \qquad \textbf{(a)} $$

The processes $2 \rightarrow 3$ and $4 \rightarrow 1$ are reversible and adiabatic. With the assumption of constant specific heats, eq. (5.17) in the form

$$ TV^{\gamma-1} = \text{constant} $$

applies. Thus

$$ T_L V_4^{\gamma-1} = T_H V_1^{\gamma-1} $$

$$ T_L V_3^{\gamma-1} = T_H V_2^{\gamma-1} $$

Dividing these equalities we get

$$ \left(\frac{V_3}{V_4}\right)^{\gamma-1} = \left(\frac{V_2}{V_1}\right)^{\gamma-1} $$

or

$$ \frac{V_3}{V_4} = \frac{V_2}{V_1} $$

Noting this equality, eq. (a) reduces to

$$\eta_C = \frac{T_H - T_L}{T_H} \tag{6.1}$$

where $\eta_C$ is the thermal efficiency of the closed-system Carnot heat engine cycle using a perfect gas with constant specific heats as the working fluid. It is interesting to note that the thermal efficiency for this cycle depends only on the absolute temperatures $T_H$ and $T_L$.

## 6–5 Perfect Gas Carnot Refrigerator and Heat Pump Cycles

The Carnot cycle, a reversible cycle, can be operated in the reversed direction, with each process the reverse of the corresponding process of the Carnot heat engine cycle. Using the same notation as that of Figure 6.3 for the reversed Carnot cycle, the pressure–volume and energy interaction diagrams become those in Figure 6.4. Inspection of Figure 6.4 reveals the only change to be the direction of the cycle—now counterclockwise on the *p-V* diagram—and the direction of each energy interaction. The change in direction of the heat-transfer quantities requires that the signs on $dT$ change. Thus, for process $4 \rightarrow 3$, the heat-reception process, the temperature of the working fluid is $dT$ below $T_L$, and for the heat-rejection process $2 \rightarrow 1$, the working fluid temperature is $dT$ above $T_H$. From Figure 6.4b it is apparent that the reversed cycle is either a refrigerator or a heat pump cycle.

The coefficient of performance of a refrigeration cycle has been defined as

$$\beta_R = \frac{\text{refrigerating effect}}{\text{net work supplied}} \tag{3.6}$$

Following the same procedures as those used in Section 6–4, the coefficient of performance of a closed-system Carnot refrigerator having a perfect gas with constant specific heats as the working fluid is

$$\beta_{R,C} = \frac{|{}_4Q_3|}{|{}_2Q_1| - |{}_4Q_3|} = \frac{T_L}{T_H - T_L} \tag{6.2}$$

Similar analysis of a closed-system Carnot heat pump having a perfect gas with constant specific heats as the working fluid reveals its coefficient of performance to be, by eq. (3.7),

$$\beta_{P,C} = \frac{\text{heating effect}}{\text{net work supplied}} = \frac{|{}_2Q_1|}{|{}_2Q_1| - |{}_4Q_3|} = \frac{T_H}{T_H - T_L} \tag{6.3}$$

In eqs. (6.2) and (6.3), as well as eq. (6.1), temperatures are on the perfect gas absolute scale, that is, either Rankine or Kelvin degrees.

**Figure 6.4** Reversed Carnot Cycle. *a.* Perfect gas cycle. *b.* Energy interactions.

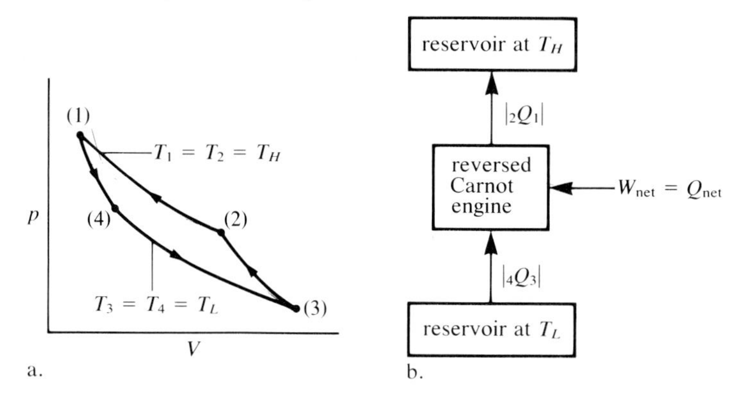


## Example 6.c

A Carnot cycle operates between heat reservoirs at 660 R and 440 R with a maximum pressure of 100.0 psia and a minimum pressure of 15.00 psia. See the $p$-$V$ diagram in Figure 6.3a. The working fluid is 1 lbm of nitrogen. Calculate the heat-transfer quantities at the high and low temperatures. From these data determine the thermal efficiency of a Carnot heat engine cycle, the coefficient of performance of a Carnot refrigerator, and the coefficient of performance of a Carnot heat pump, each operating with nitrogen and between the given temperature limits. Compare the results with those obtained using eqs. (6.1)–(6.3).

**Solution:**

For the heat engine cycle the maximum pressure, 100 psia, is at state (1) and the minimum pressure occurs at state (3). With the process between (2) and (3) reversible and adiabatic, from eq. (5.17) we have

$$\left(\frac{p_2}{p_3}\right) = \left(\frac{T_2}{T_3}\right)^{\gamma/(\gamma-1)} = \left(\frac{T_H}{T_L}\right)^{\gamma/(\gamma-1)}$$

and

$$p_2 = \left(\frac{660}{440}\right)^{1.40/0.4} (15) = 62.00 \text{ psia}$$

The pressure at state (4) can be determined in like manner as

$$p_4 = 100\left(\frac{440}{660}\right)^{3.50} = 24.19 \text{ psia}$$

For the process at $T_H$

$$\frac{v_2}{v_1} = \frac{p_1}{p_2}$$

Thus

$$
\begin{aligned}
{}_1q_2 &= RT_H \ln \frac{v_2}{v_1} \\
&= (55.1)(660) \ln \frac{100.0}{62.00} \\
&= 17{,}384 \text{ ft lbf/lbm}
\end{aligned}
$$

Heat transfer for the isothermal process from (3) to (4) is

$$
\begin{aligned}
{}_3q_4 &= (55.1)(440) \ln \frac{15.00}{24.19} \\
&= -11{,}586 \text{ ft lbf/lbm}
\end{aligned}
$$

The heat engine thermal efficiency is

$$\eta_C = \frac{17{,}384 - 11{,}586}{17{,}384} = 33.3\%$$

By eq. (6.1) the thermal efficiency is

$$\eta_C = \frac{T_H - T_L}{T_H} = \frac{660 - 440}{660} = 33.3\%$$

For the Carnot refrigerator

$$\beta_{R,C} = \frac{11{,}586}{17{,}384 - 11{,}586} = 2.00$$

and by eq. (6.2)

$$\beta_{R,C} = \frac{440}{660 - 440} = 2.00$$

The coefficient of performance of the Carnot heat pump is

$$\beta_{P,C} = \frac{17{,}384}{17{,}384 - 11{,}586} = 3.00$$

or, by eq. (6.3)

$$\beta_{P,C} = \frac{660}{660 - 440} = 3.00$$

**Figure 6.5** Two Heat Engines Operating Between Heat Reservoirs at the Same $T_H$ and the Same $T_L$. *a*. Reversible engine. *b*. Any engine.

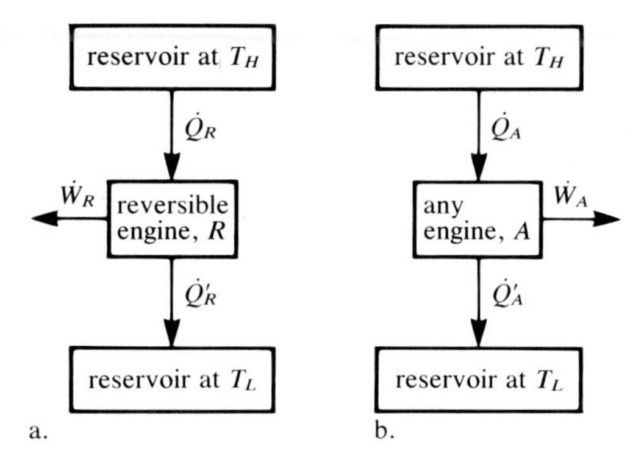


The preceding example illustrates an important point regarding reversible cycles operating between two heat reservoirs at different fixed temperatures. Compare the magnitudes of the heat-transfer quantities for each reservoir when the cycle operates as a heat engine, as a refrigerator, and as a heat pump. Note that the corresponding values are identical in magnitude and differ only by the sign (direction) of the heat transfer. The same observation can be made for the net work of the cycles. These conditions are met *only* by a reversible cycle operating between two fixed-temperature heat reservoirs. They illustrate well the concepts associated with the definition of a reversible process.

## 6–6 Carnot Corollaries for Heat Engine Cycles

An important set of corollaries to the second law, known as the Carnot corollaries, will be discussed in this and the following section. The Carnot corollaries (Carnot principles) deal with heat engines, refrigerators, and heat pumps. The first two corollaries, applicable to heat engine cycles, are

**Corollary 1.** *It is impossible to devise a heat engine that will operate between two heat reservoirs and have a thermal efficiency greater than that of a heat engine operating on a reversible cycle between the same two reservoirs.*

**Corollary 2.** *The thermal efficiency of all heat engines operating on reversible cycles between the same heat reservoirs will be identical, regardless of the cycle or the working fluid.*

Proof of these corollaries is established using the second law. For a reversible engine $R$ operating between a heat source at $T_H$ and a sink at $T_L$, let $\dot{Q}_R$ be the rate at which heat is transferred to the working fluid of the engine from the source at $T_H$, $\dot{W}_R$ the net power produced, and $\dot{Q}'_R$ the rate of heat transfer from the working fluid to the sink at $T_L$. For any other heat engine $A$ operating between the same source and sink temperatures, $\dot{Q}_A$, $\dot{W}_A$, and $\dot{Q}'_A$ are analogous quantities. Figure 6.5 shows the energy interactions for the two engines. The rate quantities for heat and work in the figure and in this discussion are absolute quantities. The arrows indicate their directions relative to the working fluid and the reservoirs. For engine $R$ the thermal efficiency is

$$\eta_R \equiv \frac{\dot{W}_R}{\dot{Q}_R} = \frac{\dot{Q}_R - \dot{Q}'_R}{\dot{Q}_R}$$

and for engine $A$

$$\eta_A \equiv \frac{\dot{W}_A}{\dot{Q}_A} = \frac{\dot{Q}_A - \dot{Q}'_A}{\dot{Q}_A}$$

In proving corollary 1 we can set the rate of heat transfer to the working fluids of the two engines equal, the rate of heat transfer from the working fluids equal, or the net power quantities equal. Note that these are not assumptions. For example, if $\dot{W}_R$ is set equal to $\dot{W}_A$, we are merely comparing two engines delivering equal power. Let us select that condition of $\dot{W}_R$ equal to $\dot{W}_A$ for the proof.

As is true for many proofs relating to the second law, the logical pattern is to assume the opposite of the corollary to be proved. Thus we make the single assumption

$$\frac{\dot{W}_A}{\dot{Q}_A} = \eta_A > \eta_R = \frac{\dot{W}_R}{\dot{Q}_R}$$

in proving corollary 1. With equal power delivered by each engine and $\eta_A$ greater than $\eta_R$, it follows that $\dot{Q}_R$ will be greater than $\dot{Q}_A$. Therefore, by the first law we can relate the rates at which heat is rejected from the engines as follows.

$$\dot{Q}_R - \dot{W}_R = \dot{Q}'_R > \dot{Q}'_A = \dot{Q}_A - \dot{W}_A$$

Noting that $R$ operates on a reversible cycle, we can reverse it with the result that the direction, but not the magnitude, of each energy interaction of the engine will be reversed. Thus engine $A$, delivering $\dot{W}_A$, could be used to power the reversed engine $R$, which now operates as a refrigerator or a heat pump. Such a composite system is shown in Figure 6.6. Inspection of this figure reveals no work interaction between the composite system and its surroundings. However, earlier steps have shown $\dot{Q}_R$ greater than $\dot{Q}_A$ and $\dot{Q}'_R$ greater than $\dot{Q}'_A$. Thus a net heat transfer from $T_L$ to $T_H$ of magnitude $\dot{Q}_R - \dot{Q}_A$ would take place without work being supplied to the cyclic composite system. Such an occurrence is a direct violation of the Clausius statement of the second law. One assumption, namely, $\eta_A$ greater than $\eta_R$, has been made. This assumption has produced a violation of the second law, and hence it is not valid. The result is verification of corollary 1.

By similar logic corollary 2 is proved. Two reversible engines, say $R_1$ and $R_2$, are considered. An assumption that the thermal efficiency of $R_1$ is greater than than of $R_2$ will result in a violation of the second law. Assumption that the efficiency of $R_2$ is greater than $R_1$ also results in a violation of the second law. The only possible conclusion is that the reversible engines $R_1$ and $R_2$ must have equal thermal efficiencies. This is true regardless of the nature of the cycles or of the working fluids of the reversible engines.

**Figure 6.6** The Composite Device

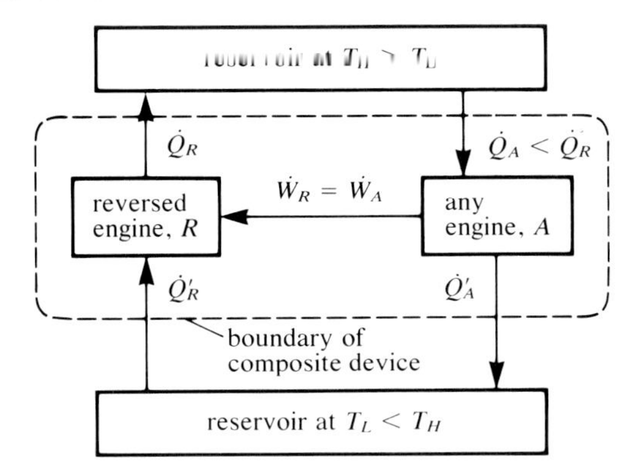


## 6–7 Some Consequences Resulting from Corollaries 1 and 2

From corollary 1 we have seen that no heat engine can have a thermal efficiency greater than that of a reversible heat engine when the engines operate between the same reservoirs. From this we conclude that the thermal efficiency of a heat engine operating on a reversible cycle is the maximum possible. Further, corollary 2 states that *all* reversible heat engines operating between the same temperature limits have the same thermal efficiency. Thus an expression for the thermal efficiency obtained for any reversible heat engine is applicable to all reversible heat engines. In Section 6–4 an expression, eq. (6.1), was developed for the reversible closed-system Carnot heat engine cycle using a perfect gas with constant specific heats as the working fluid. Thus, on the basis of corollaries 1 and 2, we can now state that the maximum thermal efficiency for any heat engine operating between a heat source at $T_H$ and a heat sink at $T_L$ will be

$$\eta_{\text{max}} = \eta_{\text{rev}} = \frac{T_H - T_L}{T_H} \tag{6.4}$$

It is useful to examine heat engine operation in light of eq. (6.4). The temperature $T_L$ of the heat sink for practical heat engine operation is that of the atmosphere. Let us assume the value for $T_L$ to be 537 R (298 K). A reversible heat engine receiving heat from a reservoir at $T_H$ and rejecting heat to the atmosphere at $T_L$ would convert to work the fraction

$$W_{\text{max}}/Q_{\text{in}} = W_{\text{rev}}/Q_{\text{in}} = \left(\frac{T_H - T_L}{T_H}\right)$$

of the heat transferred from the reservoir at $T_H$. By eq. (6.4) no heat engine could convert a greater fraction of $Q_{\text{in}}$ to work when operating between the same heat reservoir temperatures. The ratio $W_{\text{max}}/Q_{\text{in}}$ and the ratio of the heat rejected to the reservoir at $T_L$ to $Q_{\text{in}}$ are shown for several values of $T_H$ in Table 6.1. The results, which exhibit the same trend as that for actual power cycles, indicate the desirability of high source temperatures for heat engines.

**Table 6.1** Maximum Conversion of Heat to Work for Several Heat-Source Temperatures and a Sink at 537 R (298 K)

| Source Temperature, $T_H$ | | $W_{max}/Q_{in}$ | $Q_{out}/Q_{in}$ |
|---|---|---|---|
| R | K | | |
| 3000 | 1667 | 0.8210 | 0.1790 |
| 2000 | 1111 | 0.7315 | 0.2685 |
| 1500 | 833 | 0.6420 | 0.3580 |
| 1000 | 556 | 0.4630 | 0.5370 |
| 750 | 417 | 0.2840 | 0.7160 |
| 650 | 361 | 0.1738 | 0.8262 |
| 600 | 333 | 0.1050 | 0.8950 |
| 550 | 306 | 0.0236 | 0.9763 |
| 537 | 298 | 0.0000 | 1.0000 |

In real heat engines the maximum temperatures are limited by properties of the materials available for continuous high-temperature operation, not by our ability to create high-temperature heat sources. The fact that all heat engines reject to a lower-temperature reservoir a portion of the heat transferred to the working fluid is, by the second law, inescapable. Table 6.1 shows the minimum heat rejection for given source and sink temperatures. For example, each unit of heat transferred from a 1500 R reservoir to the working fluid of a reversible heat engine will result in heat rejection in the amount of 0.358 unit to the 537 R sink. Real heat engines have lower thermal efficiencies than comparable reversible engines and therefore produce less work per unit heat supplied, with the result that more heat is rejected to the environment.

## 6–8 Carnot Corollaries for Refrigerators and Heat Pumps

Four additional corollaries for refrigerators and heat pumps parallel those presented for reversible power cycles. For refrigeration cycles these are:

**Corollary 3.** *Of all refrigerators operating between a refrigerated region at* $T_L$ *and a higher-temperature reservoir at* $T_H$*, none will have a higher coefficient of performance than a refrigerator operating on a reversible cycle.*

**Corollary 4.** *All reversible refrigerators operating between a refrigerated region at* $T_L$ *and a higher-temperature reservoir at* $T_H$ *will have the same coefficient of performance regardless of the cycle or working fluid.*

As noted previously, heat pumps are identical to refrigerators with one exception. A refrigerator is a device to produce cooling of the lower-temperature region while the desired effect for a heat pump is the heating effect at the high-temperature region. The corollaries for a heat pump are:

**Corollary 5.** *Of all heat pumps receiving heat from a reservoir at* $T_L$ *and supplying heat to a region at a higher temperature* $T_H$*, none will have a higher coefficient of performance than a heat pump operating on a reversible cycle.*

**Corollary 6.** *All reversible heat pumps receiving heat from a reservoir at* $T_L$ *and supplying heat to a region at a higher temperature* $T_H$ *will have the same coefficient of performance regardless of the cycle or working fluid.*

The proof of corollaries 3–6 is established by the second law following logic similar to that used in proving corollaries 1 and 2. The coefficients of performance for Carnot refrigerators and heat pumps having a perfect gas as the working fluid were established in Section 6–5. Thus, using arguments similar to those used in Section 6.7 in arriving at eq. (6.4), we can state for any refrigerator that

$$\beta_{R,\max} = \beta_{R,\text{rev}} = \frac{T_L}{T_H - T_L} \tag{6.5}$$

Similarly, for any heat pump,

$$\beta_{P,\max} = \beta_{P,\text{rev}} = \frac{T_H}{T_H - T_L} \tag{6.6}$$

The expressions for $\eta_{\max}$, $\beta_{R,\max}$, and $\beta_{P,\max}$ in eqs. (6.4), (6.5), and (6.6) provide the sought-after limits on the performance of heat engines, refrigerators, and heat pumps. It is important to remember that these expressions were developed from second-law concepts.

Trends for the performance of refrigerators can be assessed by examining eq. (6.5). Since refrigerators are usually used to maintain regions at subatmospheric temperatures, let us assume as a typical atmospheric temperature the value 537 R for $T_H$. A reversible refrigerator receiving heat from a refrigerated region at $T_L$ and rejecting heat to the atmosphere at $T_H$ would, by the definition of $\beta_R$ ($\beta_R = Q_{\text{in}}/W_{\text{net,in}}$) and eq. (6.5), require a power input of

$$\dot{W}_{\text{net,in}} = \frac{T_H - T_L}{T_L} \dot{Q}_{\text{in}}$$

where $\dot{Q}_{\text{in}}$ is the refrigerating effect. The power given by the above expression is the minimum possible for a given $\dot{Q}_{\text{in}}$. The ratio $\dot{W}_{\min}/\dot{Q}_{\text{in}}$, which is the reciprocal of $\beta_{R,\max}$, is listed for several values of $T_L$ in Table 6.2. This ratio is the minimum power requirement per unit of refrigerating effect. The values clearly show a trend to larger power requirements per unit of refrigerating effect as $T_L$ decreases. Real refrigerators exhibit a similar trend.

For a heat pump $\dot{Q}_{\text{out}}$ is the desired effect. From eq. (6.6) the ratio

$$\frac{\dot{W}_{\text{net,in}}}{\dot{Q}_{\text{out}}} = \frac{1}{\beta_{P,\max}} = \frac{T_H - T_L}{T_H}$$

is the minimum power input per unit heating effect for a heat pump. Values of the ratio of minimum power per unit heating effect are listed in Table 6.2. $T_H$ has been assumed to be 537 R. This ratio is seen to increase for decreasing $T_L$. A similar trend exists for real heat pumps, and the power requirement per unit heating effect would, of course, be larger at a given temperature than the corresponding value computed for a reversible heat pump.

**Table 6.2** Ratios for Various Values of Cold-Region Temperatures for Reversible Refrigerators and Heat Pumps with Heat Rejection to a Reservoir at 537 R (298 K)

| Cold Region Temperature $T_L$, R | $\dot{W}_{min}/\dot{Q}_{in}$ | $\dot{W}_{min}/\dot{Q}_{out}$ |
|---|---|---|
| 530 | 0.0132 | 0.0130 |
| 500 | 0.0740 | 0.0689 |
| 460 | 0.1674 | 0.1434 |
| 440 | 0.2205 | 0.1806 |

## 6–9 The Absolute Thermodynamic Scale of Temperature

The possibility of defining a temperature scale that is independent of the behavior of any substance is suggested by the second Carnot corollary, which states that the thermal efficiency of all reversible heat engines operating between the same temperature limits is the same. It is important to note that this statement is independent of the reversible heat engine working fluid and that no reference is made to any particular scale of temperature. (The temperature scale we have used up to this point is the constant-volume gas thermometer scale of temperature defined by eq. [2.7a].) Further, it is evident from our previous discussion of the Carnot cycle that the thermal efficiency of a reversible heat engine changes if the temperature of either of the reservoirs changes. This is true regardless of the scale on which the temperatures of the reservoirs are measured. To investigate this approach to defining a scale of temperature, let us consider a reversible heat engine operating between a high-temperature reservoir and another at a lower fixed temperature. For convenience let the temperature of the latter be the triple point of water. We will use the symbol $\theta$ to represent temperature on the scale under investigation. The thermal efficiency of the reversible heat engine is

$$\eta = \frac{W_{net}}{Q_{in}} = 1 - \frac{Q_{out}}{Q_{in}}$$

Since $\eta$ for the reversible heat engine depends on temperature only, we conclude that the ratio $Q_{out}/Q_{in}$ depends only on the temperatures of the reservoirs. We are at liberty within certain limits to define this temperature dependence as we choose. Following the approach conventionally taken, we define the *absolute thermodynamic scale of temperature* as

$$\frac{Q_{in}}{Q_{out}} = \frac{\theta_H}{\theta_L} \tag{6.7}$$

where $Q_{in}$ is the heat transfer to the working fluid of the reversible engine from a heat reservoir at $\theta_H$ and $Q_{out}$ is the heat transfer from the working fluid of the reversible engine to the reservoir at $\theta_L$. Consider now the expression for the thermal efficiency of a reversible heat engine written in terms of temperatures on the constant-volume gas thermometer scale of temperature. From eq. (6.4) we write

$$\eta_{rev} = 1 - \frac{Q_{out}}{Q_{in}} = 1 - \frac{T_L}{T_H}$$

or

$$\frac{Q_{\text{in}}}{Q_{\text{out}}} = \frac{T_H}{T_L} \tag{a}$$

Thus from eqs. (6.7) and (a)

$$\frac{\theta_H}{\theta_L} = \frac{T_H}{T_L}$$

Recall that we chose the temperature of the low-temperature reservoir to be the triple point of water. To establish numerical values for $\theta$, the thermodynamic temperature, it is convenient to set $\theta$ at the triple point of water equal to $T$ on the absolute gas scale of temperature at the triple point, that is, we set $\theta_L = \theta_{tp} = T_{tp}$. When this is done, the absolute thermodynamic temperature scale and the absolute perfect gas temperature scale are identical, that is, $\theta = T$. With the relation between $\theta$ and $T$ established, we shall henceforth use $T$ to indicate temperature on an absolute thermodynamic scale. We therefore replace the perfect gas scale with the thermodynamic scale. This does not change any of the previous expressions in which $T$ appears. The use of the thermodynamic scale simply broadens our concept of the temperature scale.

While the discussion above was restricted to reversible heat engine cycles operating between reservoirs, that is, regions that have fixed and uniform temperatures, the concept of a reversible heat engine can be extended to operation between heat sources and heat sinks that have temperatures that change. Under such conditions the reversible heat engine would receive an infinitesimal amount of heat $\delta Q_H$, produce work $\delta W_{\text{net}}$, and reject heat $\delta Q_L$ while executing a cycle. During the cycle the temperature of either the source or the sink, or both, could change by $dT$. From the definition of thermodynamic temperature, the heat-transfer quantities are related by

$$\frac{|\delta Q_H|}{|\delta Q_L|} = \frac{T_H}{T_L} \tag{6.8}$$

Of course, this equation also applies to reversible heat engine operation for fixed $T_H$ and $T_L$.

## 6–10 Summary

The second law of thermodynamics is an empirical principle that is independent of other principles and laws. Two equivalent statements of the second law are the Clausius and Kelvin-Planck statements. These statements are, respectively, as follows:

> *It is impossible for a cyclic device to convey heat from a region at a given temperature to another at a higher temperature without energy input.*

> *No device operating in a cyclic manner can exchange heat with one or more reservoirs at the same temperature and produce a net work output.*

A system is said to undergo a reversible process if, upon completion of the process, the system and its surroundings can be restored to their initial states. A process is reversible if the system does not depart internally from equilibrium or from equilibrium with its surroundings by more than an infinitesimal amount and if no dissipative effects occur. Specific occurrences that render processes irreversible include friction of any kind, unrestrained expansion, heat transfer across a finite temperature difference, mixing of dissimilar substances, and spontaneous chemical reactions. Second-law concepts can be applied to prove that these occurrences cause processes to be irreversible. Reversible heat transfer occurs when heat transfer takes place across an infinitesimal temperature difference. Processes for closed systems previously identified as quasi-static and frictionless and frictionless steady-flow processes are now termed reversible processes. Although strictly reversible processes do not occur, the concept of reversibility is a useful tool for describing ideal processes.

An internally reversible process is one for which the process within the system's boundaries is reversible but irreversibilities occur in its surroundings. Expressions for reversible processes that involve only the properties of the system are also valid for internally reversible processes.

A reversible cycle is one in which each process is reversible. The Carnot cycle is a simple reversible cycle that operates between two reservoirs that have different fixed temperatures. A Carnot heat engine receives heat from a high-temperature reservoir at $T_H$ and rejects heat to a lower-temperature reservoir at $T_L$ and produces work. Operated in the reverse direction, a Carnot heat engine becomes a Carnot refrigerator or heat pump.

Six corollaries to the second law provide limits on the performance parameters for heat engines, refrigerators, and heat pumps. The following statements can be made for these devices operating between a reservoir at $T_H$ and a reservoir at $T_L$. Each temperature is on an absolute thermodynamic scale.

The maximum thermal efficiency of any heat engine is

$$\eta_{\max} = \frac{T_H - T_L}{T_H}$$

The maximum coefficient of performance of any refrigerator is

$$\beta_{R,\max} = \frac{T_L}{T_H - T_L}$$

The maximum coefficient of performance for any heat pump is

$$\beta_{P,\max} = \frac{T_H}{T_H - T_L}$$

The absolute thermodynamic scale of temperature is independent of the physical properties of any substance. Numerically it is identical to the constant-volume gas thermometer absolute scale of temperature.

## Selected References

Holman, J. P. *Thermodynamics,* 3rd ed. New York: McGraw-Hill, 1980.
Jones, J. B., and G. A. Hawkins. *Engineering Thermodynamics*. New York: Wiley, 1960.
Keenan, J. H. *Thermodynamics*. New York: Wiley, 1941.
Stoever, H. J. *Engineering Thermodynamics*. New York: Wiley, 1951.
Wark, K. *Thermodynamics,* 3rd ed. New York: McGraw-Hill, 1977.
Zemanskey, M. W., M. M. Abbott, and H. C. Van Ness. *Basic Engineering Thermodynamics,* 2nd ed. New York: McGraw-Hill, 1975.
Zemanskey, M. W., and R. H. Dittman. *Heat and Thermodynamics,* 6th ed. New York: McGraw-Hill, 1981.

# Problems

6.1 Assume a device that violates the Kelvin-Planck statement of the second law is supplied with heat at the rate of 100 B/sec from a heat reservoir at $T_H$. The device is used to supply power to a refrigerator that transfers 50 B/sec of heat from a reservoir at $T_L < T_H$. Show numerically that a violation of the Clausius statement of the second law would result.

6.2 Suppose an apparatus could continually transfer heat from a reservoir at $T_L$ to a higher-temperature reservoir at $T_H$ at the rate of 200 kJ/min with no power supplied to the apparatus. A heat engine having a thermal efficiency of 30 percent is available to operate between the two heat reservoirs. Show numerically that a violation of the Kelvin-Planck statement of the second law could result.

6.3 Each of the following either (a) violates the first law, (b) violates the second law, or (c) is correct. For each indicate which law (if any) is violated. For those violating either law describe the violation.

(1) During an isothermal process a closed system of a perfect gas does 77816.9 ft lbf of work. Heat transfer of 100 B to the system takes place during the process.
(2) A gas expands adiabatically through a nozzle. The enthalpy and velocity leaving the nozzle are each greater than those entering the nozzle.
(3) Cooling water leaves a steam condenser at 150 F. The steam entering the condenser is at 1.50 psia, quality 91.7 percent.
(4) A rigid container holds 1 lbm of air, initially at 500 R and 15 psia. Paddle-wheel work is done on the air until the pressure reaches 30 psia. During the process 50 B of heat is transferred to the system from a reservoir at 600 R.

6.4 Suppose the second law was stated as: "It is impossible for a self-contained machine (unaided by any external agency) to convey heat from a region at a lower temperature to one at a higher temperature."

(a) Describe a device that would be capable of violating this statement.
(b) What modification of the statement is required to make it a valid statement of the second law?

6.5 Suppose a refrigeration unit is supplied with power from a battery. As the unit operates, heat is transferred from the contents of the refrigerator and rejected to the atmosphere. The temperature of the atmosphere is greater than that within the refrigerated region. During the process the battery exchanges heat with the atmosphere. Does this violate the second law? State the basis for your conclusion.

6.6 A closed system of a perfect gas expands isothermally, doing work in the amount of 250 B. During the expansion 250 B of heat transfer to the gas occurs, thus $Q = W$ for the process. Is this complete conversion of heat to work possible? Justify your answer.

6.7 For each of the following determine whether or not the process is reversible. Test for reversibility by making the reversed process a part of a cycle and determine whether or not the cycle results in a violation of the Kelvin-Planck statement of the second law. (Neglect heat transfer for the given process unless the contrary is specified.)

(a) A rigid, perfectly insulated container is divided into two equal volumes by a partition. Gas on one side of the partition is at a pressure of 1 atmosphere. The other side is evacuated. A small hole is made in the partition and, after a time, the pressure is the same on both sides of the partition.
(b) A weight slides down an inclined friction plane.
(c) A block of copper is cooled from 90 to 30 C by immersing it in a large tank of water.

6.8 Consider the cycle described in Figure 3.1a and Section 3–1. This cycle communicates thermally with a single heat reservoir, the surrounding atmosphere. In view of this and in light of the second law, what can be said regarding the net work for the cycle? Is your answer consistent with the remarks related to the net work in Section 3–1 for the cycle?

6.9 A closed system undergoes a Carnot heat engine cycle. The working fluid is at 690 kPa and occupies a volume of 0.085 $m^3$ at the beginning of the isothermal expansion. Assuming the pressure and volume at the end of the adiabatic expansion are 103 kPa and 0.340 $m^3$, respectively, calculate the thermal efficiency if the working fluid is (a) air and (b) helium. Note that the mass of the system is not specified.

6.10 A closed system of air undergoes a Carnot heat engine cycle. At the beginning of the isothermal expansion the air is at 150 psia, 680 R, and occupies a volume of 3.75 $ft^3$. The isothermal expansion continues until the volume is 8.25 $ft^3$. Assuming the minimum temperature of the air during the cycle is 520 R, calculate (a) the heat transfer to the air per cycle, (b) the heat transfer from the air per cycle, and (c) the cycle thermal efficiency.

6.11 A Carnot heat engine operates with 1 lbm of $H_2O$ as the working fluid. At the beginning of the isothermal expansion the temperature is 220 F and the $H_2O$ is saturated liquid. At the end of the heat addition the $H_2O$ is saturated vapor. At the end of the adiabatic expansion the $H_2O$ is at 60 F and its quality is 0.8282. At the beginning of the adiabatic compression the internal energy is 160.10 B/lbm. For the closed system calculate the heat transfer to and from the $H_2O$ and the cycle thermal efficiency. Compare the thermal efficiency with that of the Carnot cycle in the above problem. On the basis of fundamental concepts established to this point, is a general expression for the thermal efficiency of Carnot power cycles justified?

6.12 A closed system of 1 lbm of $CO_2$ undergoes a Carnot refrigeration cycle. Determine the coefficient of performance when the pressure and temperature at the start of the isothermal expansion are 276 kPa and −18 C and the pressure and temperature at the beginning of the isothermal heat rejection are 550 kPa and 66 C.

6.13 Consider a Carnot refrigerator that uses a perfect gas as the working fluid. The refrigerator rejects heat to the atmosphere at 70 F. The closed system is to produce a refrigerating effect of 200 B/min. Work supplied to the refrigerator is electricity at a cost of 8.50 cents/kWh. Determine the cost of power per hour if the refrigerated region is at (a) 0 F, (b) −50 F, (c) −100 F, (d) −250 F, and (e) −350 F. Plot the cost per hour versus temperature of the refrigerated region.

6.14 A Carnot refrigerator operates on closed-system processes using a perfect gas as the working fluid. The refrigerated region is at a constant temperature of 250 K and heat is rejected to the atmosphere at 300 K. Using the notation of Figure 6.4, develop expressions for the refrigerating effect, the heat rejected to the atmosphere, and the net work for the cycle. From these quantities show that the coefficient of performance of the perfect-gas Carnot refrigerator is independent of the perfect gas used, the mass of the gas, and the pressures at any point in the cycle. Compare your result with eq. (6.2).

6.15 A closed-system Carnot heat-pump cycle has a perfect gas as the working fluid. Heat transfer to the working fluid is from the atmosphere at $T_L$. The heating effect is delivered to a constant temperature region at $T_H$. Show that the coefficient of performance for the heat pump is

$$\beta_{P,C} = \frac{T_H}{T_H - T_L}$$

for all perfect gases.

6.16 Heat engine $I$ operates on an irreversible cycle and heat engine $R$ on a reversible cycle. When each is operated between a fixed-temperature reservoir at $T_H$ and a fixed-temperature sink at $T_L$ the thermal efficiency of engine $R$ is 60 percent and, it is claimed, the thermal efficiency of engine $I$ is 70 percent. If each engine rejects 300 B/min to the sink, prove numerically that engine $I$ cannot have the claimed thermal efficiency.

6.17 A reversible engine ($R_1$) is supplied with heat at the rate of 500 kJ/min from a source at 1000 K. The engine rejects heat to a finite body at 750 K. A second reversible engine ($R_2$) is supplied with heat from the finite body at a rate that maintains the body at 750 K. Engine $R_2$ rejects heat at a constant temperature of 300 K. Calculate the net work delivered by the two engines. Now consider a single reversible engine ($R_3$) supplied with heat at the rate of 500 kJ/min from a reservoir at 1000 K and rejecting heat to a sink at 300 K. Determine the net work for engine $R_3$ and compare it with the total net work for engines $R_1$ and $R_2$.

6.18 Suppose that it were possible to build a reversible heat engine. If 50,000 B/hr of heat were supplied to the engine from a source at 500 F, and if the engine delivered 10 HP, what would be the temperature of the sink to which the engine rejected heat?

6.19 A heat engine is to be supplied with 50 kJ/min of heat from a constant-temperature reservoir at 238 C. The engine is to reject 30 kJ/min of heat to a sink at $-18$ C. Is this possible? Justify your answer with appropriate calculations.

6.20 The incident solar radiation (energy from the sun) on a particular day is at the rate of 100 B/hr for each square foot of area. Consider a solar collector with an area of 50 $ft^2$ that concentrates the energy it collects in such a manner that a region (an energy reservoir) can be maintained at 400 F. Assume that one half of the energy incident on the collector goes to this reservoir, which, in turn, serves as a source for a heat engine. If the temperature of the atmosphere is 70 F, what is the maximum power, in kilowatts, that the system can produce?

6.21 A heat reservoir is at 600 F. The atmosphere is at 77 F. For heat transfer of 1000 B from the reservoir to a heat engine, what is the maximum net work that could be produced?

6.22 An inventor claims he can build a refrigerating unit that will be driven by a 0.560-kW motor, will remove 10,550 kJ/hr from a region at 4.5 C, and will reject heat to a region at 32 C. Is this possible? Justify your answer by appropriate calculations.

6.23 Consider a heat engine–refrigerator combination in which the net power output of the engine is used to drive the refrigerator. The heat engine, supplied with heat at the rate of 100 B/min from a reservoir at 1540 F, delivers 1.4 HP while rejecting heat to a sink at 140 F. The refrigeration unit has a capacity of 1 ton (12,000 B/hr), and operates between a refrigerated region of 0 F and a reservoir at 200 F. Is this combination possible? Show appropriate calculations to justify your conclusion.

6.24 A heat pump uses the atmosphere at −10 C as an energy source. The unit is to maintain a temperature of 21 C within a building. Heat loss from the building is at the rate of 50,000 kJ/hr. If the cost of electric energy to power the unit is 7.95 cents/kWh, what is the minimum cost of energy to operate such a unit for 24 hours?

6.25 Mercury freezes at −38 F and a certain form a sulfur melts at 235 F, each at atmospheric pressure. Suppose an absolute thermodynamic scale of temperature is established having the difference in temperature of these two fixed points set equal to 100 degrees. What temperature on this scale would be equivalent to 25 C?

*Note for problems 6.26 and 6.27:* If a reversible engine received only infinitesimal amounts of heat, $\delta Q$, each time it went through one cycle, it could operate between a source and sink whose temperatures change. This would be true because the heat reservoirs would be at essentially constant temperature while a single cycle was performed by the engine. Refer to eq. (6.8).

6.26 Five kilograms of helium, initially at 207 kPa and −46 C, are contained in a rigid vessel. The helium serves as a receiver (sink) for a heat engine supplied with heat from the atmosphere, which is at 21 C. What is the maximum net work the engine can deliver?

6.27 One pound mass of steam initially at 100 percent quality is maintained in a closed system at a constant pressure of one atmosphere and serves as the source of heat for a heat engine. The engine rejects heat to the atmosphere, which is at 70 F. Calculate the maximum work that the engine could deliver. Assume $C_p$ for liquid $H_2O$ to be constant and equal to 1.00 B/(lbm R).

# 7

# Entropy, A Consequence of the Second Law

*The first law of thermodynamics is the basis for the definition of the property we have called* ***stored energy.*** *This definition led to the properties of internal energy and enthalpy. First-law statements for noncyclic processes were shown to include these properties. In a like manner, there exists a property called* ***entropy*** *that is a direct consequence of the second law. Compared with stored energy, entropy is a somewhat more abstract property in that it does not lend itself to direct physical description. Nonetheless, this property permits us to extend the second law from statements related to cyclic processes that were the subject of the last chapter to noncyclic processes. As we shall see, this produces a number of far-reaching results that have wide application in engineering thermodynamics. The property entropy ranks equal in importance to stored energy in the study of thermodynamics. In fact, thermodynamics has been described by some as the science that deals with energy and entropy.*

## 7–1 The Inequality of Clausius

A corollary of the second law known as the inequality of Clausius provides the basis for the discussion in this chapter. This corollary is stated as follows:

> *When a closed system undergoes a cycle, the integral of* $\delta Q/T$ *for the cycle will be less than or equal to zero.*

In equation form the inequality of Clausius is

$$\oint \frac{\delta Q}{T} \le 0 \tag{7.1}$$

This expression can be developed in the following manner. Referring to Figure 7.1, let us focus attention on the closed system, which undergoes a cycle that may be either internally reversible or internally irreversible. During a portion of the cycle the system receives heat $\delta Q$ at a typical location where the temperature is $T$. Also, work $\delta W$ is done by the system. Since the system executes a cycle, the direction of $\delta Q$ on some other parts of the boundary or at other times in the cycle will be opposite to that shown and $\delta W$ may be negative for other parts of the cycle. Since we are concerned with events occurring on and inside the

**Figure 7.1** The Development of the Inequality of Clausius

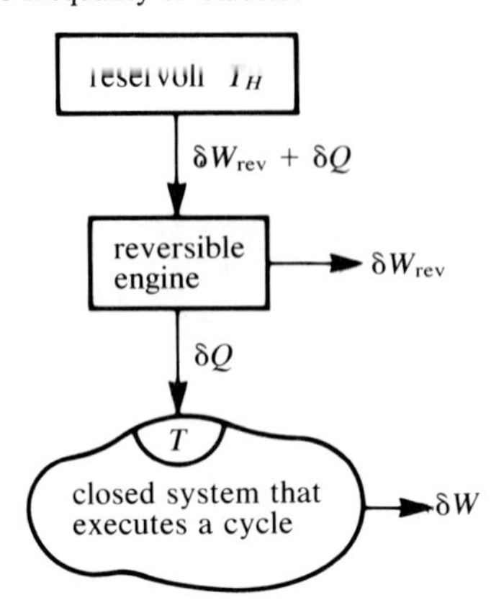


closed-system boundary, we eliminate sources of irreversibility outside the system by requiring that $\delta Q$ be supplied by a reversible engine operating between a reservoir at $T_H$ and a part of the system at temperature $T$ as shown in the figure. We require that the reversible engine execute a whole number of cycles for each cycle completed by the closed system. Work $\delta W_{rev}$ is produced as heat $\delta Q$ is rejected to the closed system by the reversible engine. If the direction of $\delta Q$ were opposite to that shown, the reversible engine would be operated as a refrigerator and require work $\delta W_{rev}$. We would actually employ a multiplicity of reversible devices interacting with the reservoir at $T_H$, each delivering heat to or receiving heat from a different region of the system. For simplicity, we show a single reversible engine in Figure 7.1. We see for this arrangement that any irreversibilities that occur take place within the closed system.

Now consider the composite system composed of the reversible engine and the closed system. From the Kelvin-Planck statement we recognize that since the cyclic composite device interacts with a reservoir at a fixed temperature, the net work for the device cannot be positive. Therefore, we can write

$$\oint \delta W_{rev} + \oint \delta W \leq 0 \tag{a}$$

where the cyclic integral of the reversible work represents the net work for all of the reversible devices involved in delivering heat to or receiving heat from the original closed system during each cycle of that system. Further, from eq. (6.8) we can state for the reversible engine in Figure 7.1 that

$$\frac{\delta Q_H}{T_H} = \frac{\delta Q + \delta W_{rev}}{T_H} = \frac{\delta Q}{T}$$

Note that the above expression is also true if the reversible engine operates as a refrigerator. From this equation we solve for $\delta W_{rev}$ and get

$$\delta W_{rev} = \frac{T_H}{T}\delta Q - \delta Q$$

Introducing $\delta W_{rev}$ into eq. (a), we obtain

$$\oint \frac{T_H}{T}\delta Q - \oint \delta Q + \oint \delta W \leq 0$$

From the first law the sum of the second and third terms on the left is zero. Also, since $T_H$ is a fixed value on the absolute thermodynamic scale, the remaining term can be divided by $T_H$. The final expression is then eq. (7.1). Thus we have proved the inequality of Clausius.

We note that eq. (7.1) involves only the quantities $\delta Q$ and $T$, both of which are related to the original closed system. The manner in which $\delta Q$ is supplied to the system does not influence the behavior of the system. The system will behave the same when the heat-transfer process external to the system is reversible as when it is irreversible. Therefore, we conclude that eq. (7.1), which is applicable to the system, is valid regardless of whether the processes in the surroundings are reversible or irreversible.

The equal sign in eq. (7.1) applies to reversible cycles. Therefore

$$\oint \left(\frac{\delta Q}{T}\right)_{rev} = 0 \tag{7.2}$$

Equation (7.2) is known as the equality of Clausius. This equation can be proved by assuming that it is false and examining the consequences. Let us employ a familiar reversible cycle, the Carnot cycle, in the proof. Assuming that eq. (7.2) is not true, we use the description and notation for the Carnot cycle from Section 6–3 to obtain

$$\oint \left(\frac{\delta Q}{T}\right)_{rev} = \left(\int_1^2 \frac{\delta Q}{T_H} + \int_2^3 \frac{\delta Q}{T} + \int_3^4 \frac{\delta Q}{T_L} + \int_4^1 \frac{\delta Q}{T}\right)_{rev} \neq 0$$

The processes 2 to 3 and 4 to 1 are adiabatic as well as reversible. Also, $T_H$ and $T_L$ are constants. Therefore, the above equation reduces to

$$\frac{{}_1Q_2}{T_H} \neq \frac{-{}_3Q_4}{T_L}$$

or

$$\frac{Q_{in}}{T_H} \neq \frac{Q_{out}}{T_L} \tag{b}$$

**Figure 7.2** Reversible Paths Between Two Equilibrium States

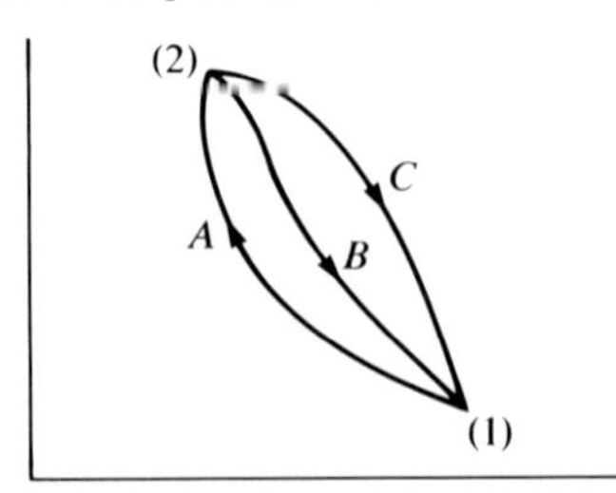


From the expression for the thermal efficiency for reversible cycles, eq. (6.4), we can write

$$\eta_{\text{rev}} = 1 - \frac{Q_{\text{out}}}{Q_{\text{in}}} = 1 - \frac{T_L}{T_H}$$

or

$$\frac{Q_{\text{in}}}{T_H} = \frac{Q_{\text{out}}}{T_L} \tag{c}$$

It is evident that eqs. (b) and (c) cannot both be true. Since eq. (c) is well established by the second law, eq. (b) is a violation of the second law, a result we cannot accept. The same result is obtained if any other reversible cycle is considered in place of the Carnot cycle. Therefore, we conclude that the equal sign in eq. (7.1) holds for reversible cycles, that is, eq. (7.2) is a valid expression.

## 7–2 Entropy Change of a Pure Substance

The equality of Clausius, eq. (7.2), is the basis for the definition of entropy change. We will now develop the equation for entropy change. Figure 7.2 shows three arbitrary reversible paths, *A, B,* and *C,* between any two equilibrium states (1) and (2) for a pure substance in a closed system. Coordinates of the figure are not specified, however, it is assumed that adequate information necessary to fix the states is known. Reversible cycles comprised of paths *A* and *B* or *A* and *C* represent two reversible cycles that include states (1) and (2). Applying eq. (7.2) to the cycle of paths *A* and *B* yields

$$\underset{A}{\int_1^2 \left(\frac{\delta Q}{T}\right)} + \underset{B}{\int_2^1 \left(\frac{\delta Q}{T}\right)} = 0 \tag{a}$$

and, for the cycle of paths *A* and *C,*

$$\underset{A}{\int_1^2 \left(\frac{\delta Q}{T}\right)} + \underset{C}{\int_2^1 \left(\frac{\delta Q}{T}\right)} = 0 \tag{b}$$

Subtracting (b) from (a) yields

$$\underset{B}{\int_2^1 \left(\frac{\delta Q}{T}\right)} \quad \underset{C}{\int_2^1 \left(\frac{\delta Q}{T}\right)} \quad 0 \tag{c}$$

Rearranging (c) we find

$$\underset{B}{\int_1^2 \left(\frac{\delta Q}{T}\right)} = \underset{C}{\int_1^2 \left(\frac{\delta Q}{T}\right)} \tag{d}$$

Noting that (1) and (2) are any two equilibrium states for a closed system or fixed mass of a pure substance, and that $B$ and $C$ are any two reversible paths connecting these states, we conclude that

$$\int_1^2 \left(\frac{\delta Q}{T}\right)_{\text{rev}} = \text{constant} \tag{e}$$

where the subscript rev indicates any reversible process between states (1) and (2). Recalling that a change in a property depends only on the end states, we define the constant of eq. (e) as the change in the extensive property *entropy* between the two states. Using the symbol $S$ for entropy, we write

$$S_2 - S_1 = \int_1^2 \left(\frac{\delta Q}{T}\right)_{\text{rev}} \tag{7.3}$$

or

$$dS = \left(\frac{\delta Q}{T}\right)_{\text{rev}}$$

The extensive property entropy $S$ becomes the intensive property $s$ when entropy per unit mass is considered ($s = S/\text{m}$). Dividing eq. (7.3) by mass we obtain

$$s_2 - s_1 = \int_1^2 \left(\frac{\delta q}{T}\right)_{\text{rev}} \tag{7.3a}$$

and

$$ds = \left(\frac{\delta q}{T}\right)_{\text{rev}} \tag{7.3b}$$

Two important conclusions follow directly from the definition of the change in entropy. For a closed system of a pure substance undergoing a reversible process, the heat transfer is

$$({}_1q_2)_{\text{rev}} = \int_1^2 T\,ds \tag{7.4}$$

and for a reversible isothermal process

$$({}_1q_2)_{\text{rev,isothermal}} = T(s_2 - s_1) \tag{7.4a}$$

The second observation is that all closed-system processes of a pure substance that take place reversibly and adiabatically produce no change in entropy of the system. Such processes are called *isentropic* processes. Conversely, *isentropic* denotes adiabatic and reversible when used to describe processes in this book.

## Example 7.a

One kilogram of a perfect gas initially at 300 K and 125 kPa undergoes a process to a final state at 450 K and 690 kPa. The gas has specific heats as follows: $C_p = 0.8500$ kJ/(kg K) and $C_v = 0.6300$ kJ/(kg K). Calculate the entropy change of the gas for the following reversible paths:

(a) Constant pressure $p_1$ and constant temperature $T_2$
(b) Constant volume $v_1$ and constant pressure $p_2$
(c) Constant temperature $T_1$ and constant volume $v_2$

**Solution:**

Since entropy is a property its change is independent of the type of process that causes the system to change from state (1) to state (2). The process could be closed, steady flow; reversible or irreversible. However, according to eq. (7.3) or (7.3a) the change must be computed by integrating along a *reversible* path. For convenience we choose closed system reversible paths. The expression for reversible heat transfer is obtained from the differential form of the closed-system energy equation in which the work becomes that for a closed-system reversible process, *pdv*. The three reversible paths are shown in the following illustration. The entropy change will be computed using eq. (7.3a).

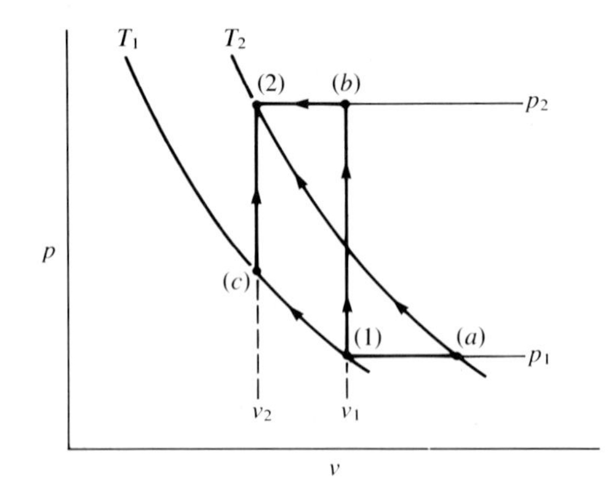

(a) The reversible paths $1 \rightarrow a$ and $a \rightarrow 2$ are specified. For the perfect gas constant pressure process, the closed system energy equation $\delta q = \delta w + du$ becomes

$$(\delta q)_{\text{rev}} = du + pdv = dh = C_p dT$$

and, at constant temperature,

$$(\delta q)_{\text{rev}} = pdv$$

Thus

$$s_a - s_1 = \int_1^a \left(\frac{\delta q}{T}\right)_{\text{rev}} = \int_1^a \left(\frac{C_p dT}{T}\right) = C_p \ln \frac{T_a}{T_1} = C_p \ln \frac{T_2}{T_1}$$
$$= 0.8500 \ln \frac{450}{300} = 0.3446 \text{ kJ/(kg K)}$$

and,

$$s_2 - s_a = \int_a^2 \left(\frac{\delta q}{T}\right)_{\text{rev}} = \int_a^2 \frac{pdv}{T} = R\int_a^2 \frac{dv}{v} = R \ln \frac{v_2}{v_a}$$

For the constant temperature process $a \rightarrow 2$,

$$\frac{v_2}{v_a} = \frac{p_a}{p_2} = \frac{p_1}{p_2}$$

Using the expression $C_p - C_v = R = 0.8500 - 0.6300 = 0.2200$ kJ/(kg K)

$$s_2 - s_a = 0.2200 \ln \frac{125}{690} = -0.3758 \text{ kJ/(kg K)}$$

The entropy change is

$$s_2 - s_1 = (s_2 - s_a) + (s_a - s_1) = -\ 0.03120 \text{ kJ/(kg K)}$$

(b) The paths $1 \rightarrow b$ and $b \rightarrow 2$ are specified. For the perfect gas at constant volume

$$(\delta q)_{\text{rev}} = du = C_v dT$$

and at constant pressure

$$(\delta q)_{\text{rev}} = du + pdv = dh = C_p dT$$

These equations yield, respectively,

$$s_b - s_1 = \int_1^b \left(\frac{\delta q}{T}\right)_{\text{rev}} = \int_1^b \frac{C_v dT}{T} = C_v \ln \frac{T_b}{T_1}$$

and

$$s_2 - s_b = \int_b^2 \left(\frac{\delta q}{T}\right)_{rev} = \int_b^2 \frac{C_p dT}{T} = C_p \, ln \, \frac{T_2}{T_b}$$

The unknown $T_b$ can be determined as

$$\frac{T_b}{T_1} = \frac{p_b}{p_1}$$

since states (1) and ($b$) have the same specific volume. Therefore,

$$T_b = \left(\frac{690}{125}\right)(300) = 1656 \text{ K}$$

Thus,

$$s_b - s_1 = 0.6300 \, ln \, \frac{1656}{300} = 1.0763 \text{ kJ/(kg K)}$$

and

$$s_2 - s_b = 0.8500 \, ln \, \frac{450}{1656} = -\ 1.1075 \text{ kJ/(kg K)}$$

Summing the changes

$$s_2 - s_1 = (s_2 - s_b) + (s_b - s_1) = -\ 0.0312 \text{ kJ/(kg K)}$$

(c) For the constant-temperature path

$$s_c - s_1 = \int_1^c \left(\frac{\delta q}{T}\right)_{rev} = R \, ln \, \frac{v_c}{v_1} = R \, ln \, \frac{v_2}{v_1} = R \, ln \, \frac{T_2 p_1}{p_2 T_1}$$
$$= 0.2200 \, ln \, \frac{(450)(125)}{(690)(300)} = -0.2866 \text{ kJ/(kg K)}$$

and for the constant-volume path at $v_2$

$$s_2 - s_c = \int_c^2 \left(\frac{\delta q}{T}\right)_{rev} = C_v \, ln \, \frac{T_2}{T_c} = C_v \, ln \, \frac{T_2}{T_1}$$

Following these reversible paths

$$s_2 - s_1 = (s_2 - s_c) + (s_c - s_1) = -\ 0.03120 \text{ kJ/(kg K)}$$

We see that for the three reversible paths the entropy change is $-0.0312$ kJ/(kg K). This result is as expected since the reversible requirement for integration of eqs. (7.3) applies to any reversible path.

**Figure 7.3** Some Processes for a Change of Equilibrium States

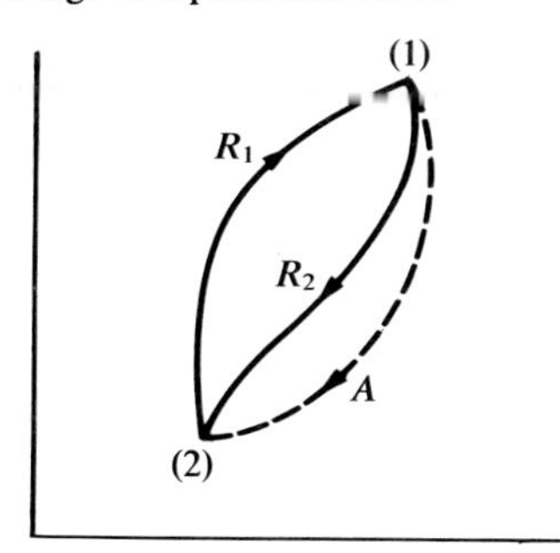


## 7–3 Entropy Change and Irreversibilities

We saw in Section 7–2 that the equality of Clausius leads to the definition of the property entropy and to a means of calculating changes in entropy for a process. The inequality of Clausius, eq. (7.1), yields an expression that provides additional insight regarding the second law for a process. We shall develop that expression with the aid of Figure 7.3. In this figure states (1) and (2) are any two equilibrium states of a system. Paths $R_1$ and $R_2$ depict any two reversible paths connecting the two states. The broken line $A$ is used to indicate any other process (reversible or irreversible) between the two states. The coordinates of Figure 7.3 are undefined; however, they must be sufficient to establish the equilibrium states and the reversible paths.

For the cycle consisting of processes $R_1$ and $A$ the inequality of Clausius, eq. (7.1), yields

$$\underset{R_1,\,A}{\oint}\frac{\delta Q}{T} = \underset{R_1}{\int_2^1}\frac{\delta Q}{T} + \underset{A}{\int_1^2}\frac{\delta Q}{T} \leq 0 \qquad \text{(a)}$$

The equality of Clausius, applied to the cycle along paths $R_1$ and $R_2$, can be expressed as

$$\underset{R_1,R_2}{\oint}\frac{\delta Q}{T} = \underset{R_1}{\int_2^1}\frac{\delta Q}{T} + \underset{R_2}{\int_1^2}\frac{\delta Q}{T} = 0 \qquad \text{(b)}$$

Subtracting (b) from (a), we have

$$\underset{A}{\int_1^2}\frac{\delta Q}{T} - \underset{R_2}{\int_1^2}\frac{\delta Q}{T} \leq 0$$

which rearranges to

$$\underset{A}{\int_1^2}\frac{\delta Q}{T} \leq \underset{R_2}{\int_1^2}\frac{\delta Q}{T} \qquad \text{(c)}$$

With the right side of eq. (c) equal to the entropy change between states (1) and (2), we have for process $A$ (any process)

$$S_2 - S_1 \geq \int_1^2 \frac{\delta Q}{T} \tag{7.5}$$

or in differential form

$$dS \geq \frac{\delta Q}{T} \tag{7.5a}$$

From the definition of entropy change, we see that the equal sign applies when process $A$ is reversible. For all cases where process $A$ is irreversible, that is, all real processes, the inequality sign applies. Hence

$$dS > \frac{\delta Q}{T} \tag{7.5b}$$

for such processes. Care must be taken in interpreting $\delta Q$ in the above equations. This is illustrated in the following example.

## Example 7.b

Consider a unit mass of a perfect gas, $CO_2$, that occupies one half of the volume of a perfectly insulated rigid container as shown in the diagram. The remaining volume is evacuated. Let the partition be removed to allow the gas to expand to fill the container. Calculate the entropy change of the gas and reconcile the result with eq. (7.5).

**Solution:**

It is recognized that the process is irreversible. Nonetheless, we can compute the change in entropy from eq. (7.3a), provided we first establish the final state (2). Since $Q$ and $W$ are zero for the process, the closed-system energy equation results in

$$\delta Q - \delta W = dU = 0$$

Because we are dealing with a perfect gas and $dU = 0$, $T_1 = T_2$. States (1) and (2) are shown on the $p$-$v$ diagram. With state (1) known, state (2) is established by $T_2 = T_1$ and $v_2 = 2v_1$. For computing the entropy change by eq. (7.3a)

$$s_2 - s_1 = \int_1^2 \left(\frac{\delta q}{T}\right)_{\text{rev}}$$

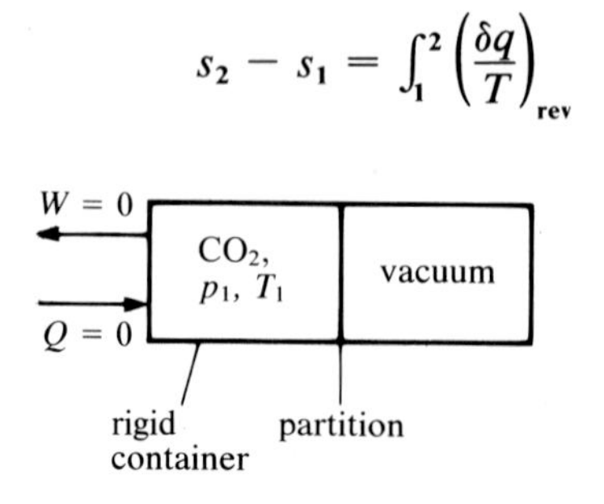


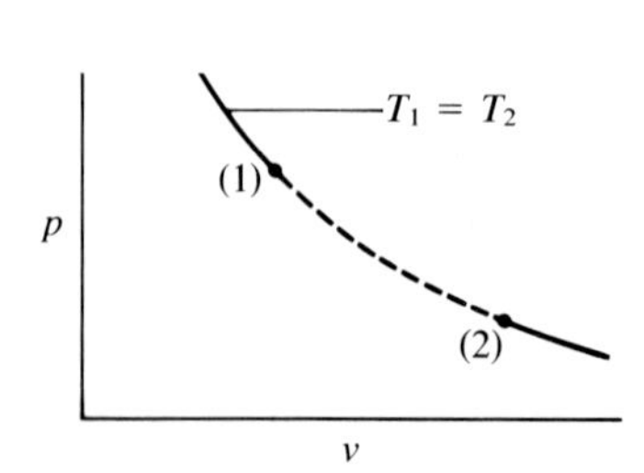

we must construct a reversible path between (1) and (2). Let us choose a reversible isothermal path. For such a path the first law yields for the perfect gas

$$\delta q_{rev} - \delta w_{rev} = du = 0$$

Thus

$$\delta q_{rev} = \delta w_{rev} = pdv$$

In eq. (7.3a), $\delta q_{rev}$ is not $\delta q$ for the actual process. (The actual $q$ for the real process [1] to [2] is zero.) It is the differential heat-transfer quantity that is related to the isothermal reversible path that we have constructed to compute the entropy change. Continuing, we substitute for $\delta q_{rev}$ into eq. (7.3a) to obtain

$$\begin{aligned} s_2 - s_1 &= \int_1^2 \frac{pdv}{T} = R\int_1^2 \frac{dv}{v} = R\,ln\,\frac{v_2}{v_1} \\ &= 188.9\,ln\,2 = 130.9\ \mathrm{J/(kg\ K)} \end{aligned}$$

To apply eq. (7.5) to the process, we note that since the process is irreversible, the inequality sign holds. When we examine irreversible processes, $\delta Q$ in eq. (7.5) is the *actual* heat transfer for the real process. Since in this case the real heat transfer is zero, eq. (7.5) written for a unit mass yields

$$s_2 - s_1 > 0$$

This tells us only that the entropy $s_2$ is greater than $s_1$, which is in agreement with the numerical result obtained by eq. (7.3a).

---

While this example has treated only the adiabatic case, it illustrates that for the inequality sign in eq. (7.5), the heat transfer is always that for the real process. The equality in eq. (7.5) applies only to reversible processes and thus this equation reduces to eq. (7.3). This case will always be identified by the subscript *rev*. It is also important to note that for both the inequality and equality signs in eq. (7.5), the possible changes are positive and negative. However, where the inequality is applicable, the equations must be satisfied in that the left side must be algebraically greater than the right side. (Recall that on an algebraic scale $+1$ is greater than $-2$.)

An alternate way of expressing eq. (7.5b) is

$$dS = \frac{\delta Q}{T} + dS_I \tag{7.6}$$

or

$$S_2 - S_1 = \int_1^2 \frac{\delta Q}{T} + \Delta S_I \tag{7.6a}$$

where $\Delta S_I$ is either a positive quantity or zero. This term represents the entropy increase due to irreversibilities within the system that are present during the process (1) to (2). In eq. (7.6) $\delta Q$ is the actual heat transfer for the process in question. For a reversible process $\Delta S_I$ is zero and eq. (7.6a) reduces to eq. (7.3) and eq. (7.6) reduces to $dS = (\delta Q/T)_{rev}$. Equations (7.6) show that for both reversible and irreversible processes, the entropy of a system can decrease only when heat is transferred from the system. However, the entropy of a system can increase by means of heat transfer to the system and by the presence of internal irreversibilities. Further, it is evident from eq. (7.6a) that for an irreversible adiabatic process occurring in a closed system $S_2$ will be greater than $S_1$.

It is also important to note that although the terms on the right side of eq. (7.6a) can be interpreted qualitatively, they cannot be directly evaluated for many irreversible (real) processes. Nonetheless, the entropy change between any two equilibrium states is always given by eqs. (7.3).

While the notation regarding reversibility in this and the preceding section implies that the reversible processes considered must be completely reversible, that is, reversible both externally and internally, and while the expressions are indeed valid for this condition, they are also correct if the requirement of only internal reversibility is imposed. That this is true follows from the fact that $\Delta S_I$ represents the increase in entropy due to internal irreversibilities. When $\Delta S_I$ is zero, the process is at least internally reversible. Hence for entropy expressions involving the subscript *rev* to apply, the process must be at least internally reversible.

## 7–4 Entropy Change for an Open System

In Section 3–7 the general energy equation for open systems was developed. The development of an equation for the entropy change of such systems follows a similar approach. Figure 7.4a shows the open system at the beginning of a process. The process takes place during a time interval $d\tau$ during which the differential mass $dm_1$ enters the open system at section (1) with specific entropy $s_1$ and a differential mass $dm_2$ leaves the open system at section (2) with specific entropy $s_2$. For the general open system, $dm_2$ and $dm_1$ need not be equal. Recalling that the temperature can vary with location in an open system, consider a location on the open system boundary where the local thermodynamic temperature is $T$ and the boundary area is of magnitude $dA$. During the process heat transfer $\delta Q$ through the

**Figure 7.4** Entropy Change for an Open System: a. System at time $\tau$. b. System at time $\tau + d\tau$

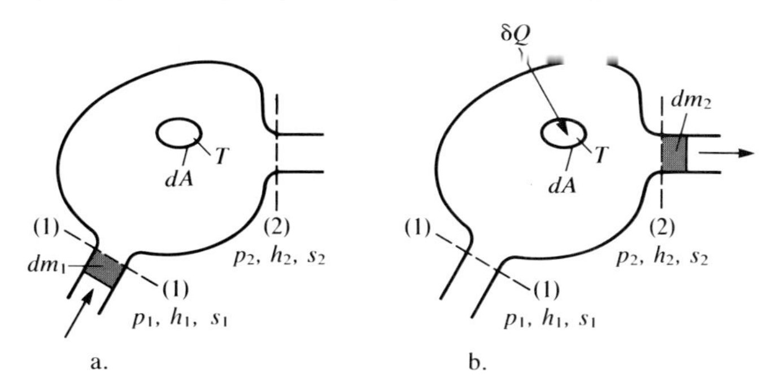


boundary area $dA$ takes place to the open system at the local temperature $T$. The entropy change of the open system can now be formulated by starting with eq. (7.6)

$$dS = \frac{\delta Q}{T} + dS_I$$

which as written is applicable only to closed systems. However, from the above discussion, we see that the expression for entropy change for an open system will include additional terms related to the mass flow across the boundary, that is, the entropy of the open system is increased by $s_1 dm_1$ and decreased by $s_2 dm_2$ in time $d\tau$. Further, for the open system we interpret the term $\delta Q/T$ according to the heat transfer described in relation to Figure 7.4 as

$$\int_A \frac{(\delta Q/dA)}{T}\, dA = \int_A \left(\frac{Q'}{T}\right) dA$$

where $Q'$ is the heat transfer per unit area, for example, B/ft$^2$, and $\delta Q = Q' dA$. The integral over the surface is necessary since, as noted, the term $\delta Q/T$ will vary with area. We now write the entropy change $dS_o$ for the open system as

$$dS_o = s_1 dm_1 - s_2 dm_2 + \int_A \left(\frac{Q'}{T}\right) dA + dS_I \tag{7.7}$$

For a reversible process, $dS_I$ is zero and eq. (7.7) can be written as

$$\left(dS_o = s_1 dm_1 - s_2 dm_2 + \frac{\delta Q}{T}\right)_{\text{rev}}$$

In arriving at eq. (7.7) only one inbound stream and one outbound stream were considered. For multiple streams eq. (7.7) becomes

$$dS_o = \Sigma(s\,dm)_{\text{in}} - \Sigma(s\,dm)_{\text{out}} + \int_A \left(\frac{Q'}{T}\right) dA + dS_I$$

This equation can be written in terms of time rates by dividing it by the differential time $d\tau$ over which the change $dS_o$ takes place. The result is

$$\dot{S}_o = \Sigma(\dot{m}s)_{\text{in}} - \Sigma(\dot{m}s)_{\text{out}} + \int_A\left(\frac{\dot{Q}'}{T}\right) dA + \dot{S}_I \tag{7.7a}$$

where $\dot{Q}'$ is the heat-transfer rate per unit area, for example, B/(ft$^2$ sec). In this equation $\dot{S}_o$ is the time rate of change of entropy within the system boundary and $\dot{S}_I$ is the rate at which entropy increases within the system due to internal irreversibilities. The latter term is always positive for irreversible processes. Each of the terms in the equation may be time dependent. However, the equation must be satisfied at any instant in time.

During a steady-flow process of an open system the state of the working fluid at any point within the open system does not change with time. Thus $\dot{S}_o$ in eq. (7.7a) will be zero for steady-flow processes. The remaining terms will be independent of time. However, $\dot{Q}'$ and $T$ may vary with area. Therefore, we can write for steady-flow processes

$$\Sigma(\dot{m}s)_{\text{out}} - \Sigma(\dot{m}s)_{\text{in}} = \int_A\left(\frac{\dot{Q}'}{T}\right) dA + \dot{S}_I \tag{7.7b}$$

For a single-stream steady-flow process $\dot{m}$ is a constant. Equation (7.7b) divided by $\dot{m}$ becomes

$$s_2 - s_1 = \int_A\left(\frac{q'}{T}\right) dA + \Delta s_I$$

where $q'$ is $\dot{Q}'/\dot{m}$. Each term in this equation has entropy per unit mass as its units, for example, kJ/(kg K). Recognizing the term $q'dA$ as $dq$, we can write the above equation as

$$s_2 - s_1 = \int_1^2 \frac{\delta q}{T} + \Delta s_I \tag{7.7c}$$

where $\Delta s_I$ represents the increase in entropy due to internal irreversibilities. With this term zero for a reversible process, we obtain

$$s_2 - s_1 = \int_1^2 \left(\frac{\delta q}{T}\right)_{\text{rev}} \tag{7.7d}$$

It is evident from eq. (7.7d) that reversible adiabatic single-stream steady-flow processes are isentropic, that is, the entering and leaving specific entropies are equal. Further, the heat transfer for a reversible single-stream steady-flow process is

$$({}_1q_2)_{\text{rev}} = \int_1^2 T ds \tag{7.7e}$$

and for a reversible isothermal process the heat transfer is

$$({}_1q_2)_{\text{rev isothermal}} = T(s_2 - s_1) \tag{7.7f}$$

Equations (7.7e) and (7.7f) are identical to eqs. (7.4) and (7.4a), respectively, which, as noted in Section 7.2, are applicable respectively to reversible processes and reversible isothermal processes in closed systems.

## 7–5 The *Tds* Equations

In previous discussion it was established that for any two equilibrium states of a given pure substance, the change in entropy is independent of the actual process, whether it be a closed or single-stream steady-flow process, reversible or irreversible. This is true because entropy is a property. Therefore, entropy changes can be evaluated utilizing a reversible path for a closed system. For such a path we can write on a unit mass basis

$$\delta q_{\text{rev}} = du + dw_{\text{rev}}$$

By eq. (7.3b), $\delta q_{\text{rev}} = Tds$, and $\delta w_{\text{rev}} = pdv$ by eq. (2.6b). Making these substitutions we obtain

$$Tds = du + pdv \tag{7.8}$$

This equation, considered a *property relation,* is valid for all pure substances.

Enthalpy, defined as

$$h \equiv u + pv$$

has the differential form

$$dh = du + pdv + vdp$$

Rearranging the enthalpy differential we have

$$du + pdv = dh - vdp$$

Substitution of this into eq. (7.8) gives the second *Tds* equation,

$$Tds = dh - vdp \tag{7.9}$$

which also is applicable to all pure substances. Equations (7.8) and (7.9) can be applied to any fixed mass undergoing either a closed-system or a steady-flow process.

In Section 2–16 the equation

$$ {}_1w_2 = -\int_1^2 vdp - \frac{V_2^2 - V_1^2}{2g_0} - \frac{g}{g_0}(z_2 - z_1) \tag{2.14} $$

which applies to steady-flow reversible processes, was presented without derivation. This relation was developed using the concepts of mechanics developed a number of years before the first and second laws of thermodynamics were formulated. Let us now see how it can be derived quite simply from thermodynamic concepts. The steady-flow energy equation on the basis of a unit mass written in differential form is, from eq. (3.4b),

$$ \delta q - dh = \frac{VdV}{g_0} + \frac{g}{g_0}dz + \delta w \tag{a} $$

For a reversible process,

$$ \delta q = Tds \tag{b} $$

and eq. (7.9) can be written as

$$ dh = Tds + vdp \tag{c} $$

Substituting (b) and (c) into (a), we get

$$ Tds - Tds - vdp = \frac{VdV}{g_0} + \frac{g}{g_0}dz + \delta w \tag{d} $$

Rearrangement of eq. (d) and integration between states (1) and (2) yields

$$ {}_1w_2 = -\int_1^2 vdp - \frac{V_2^2 - V_1^2}{2g_0} - \frac{g}{g_0}(z_2 - z_1) $$

which is eq. (2.14). This is but one example of the utility of entropy and the *Tds* equations.

The *Tds* equations provide a means for determining the change in entropy between equilibrium states of a pure substance. Solving eqs. (7.8) and (7.9) for *ds* we get

$$ ds = \frac{du}{T} + \frac{p}{T}dv \tag{7.8a} $$

and

$$ ds = \frac{dh}{T} - \frac{v}{T}dp \tag{7.9a} $$

Integration of either equation to calculate an entropy change will, in general, require an equation of state ($p$-$v$-$T$ relation). Additionally, eq. (7.8a) requires knowledge of the relation of internal energy and temperature along a known constant-specific-volume line and eq. (7.9a) requires knowledge of the enthalpy–temperature relation along a constant-pressure line. Such relations are very complex for substances other than perfect gases. In practice, tables of properties are used to evaluate entropy changes for such substances. This topic is treated in the next section. We devote the remainder of this section to the application of eqs. (7.8a) and (7.9a) to perfect gases.

By substituting perfect gas relations into eqs. (7.8a) and (7.9a) we obtain

$$ds = \frac{C_v dT}{T} + R\frac{dv}{v} \tag{7.8b}$$

and

$$ds = \frac{C_p dT}{T} - R\frac{dp}{p} \tag{7.9b}$$

Integration of terms containing the independent variable $T$ may proceed in two ways. For a perfect gas having variable specific heats the expression for $C_p$ or $C_v$ as a function of temperature must be known. For processes of a perfect gas in which constant specific heats are assumed, the equations integrate to

$$s_2 - s_1 = C_v \ln\frac{T_2}{T_1} + R \ln\frac{v_2}{v_1} \tag{7.8c}$$

and

$$s_2 - s_1 = C_p \ln\frac{T_2}{T_1} - R \ln\frac{p_2}{p_1} \tag{7.9c}$$

Note that $R$ and the specific heats must have the same units when these equations are used. A third method, which takes into account the variations of $C_p$ with temperature, involves the use of tables such as the *Gas Tables* by Keenan et al.* We shall not use such tables in this text. Specific heats of perfect gases will be assumed constant unless a variable specific heat, such as those in Table 5.4, is specified.

Evaluation of entropy changes for perfect gases by the above equations requires only that the appropriate end-state thermodynamic properties and the gas properties be known. The explicit construction of a reversible path is not involved as it was in Example 7.a, in which

---

*Keenan, J. H., J. Chao, and J. Kaye. *Gas Tables,* 2nd ed. New York: Wiley, 1980.

integration of eq. (7.3a) was performed to determine an entropy change. This results from the fact that the reversible path requirement has been satisfied in the development of the above equations. Henceforth we will determine entropy changes for perfect gases by use of eqs. (7.8c) and (7.9c) rather than by use of the reversible-path integration method. Equations (7.8c) and (7.9c) are chosen since we assume constant specific heats unless otherwise stated.

Let us next examine the adiabatic reversible (isentropic) process for perfect gases with constant specific heats. Expressions for this process can be developed by setting the entropy change in eqs. (7.8c) and (7.9c) equal to zero. From eq. (7.9c) we obtain

$$\ln \frac{T_2}{T_1} = \frac{R}{C_p} \ln \frac{p_2}{p_1} = \frac{\gamma - 1}{\gamma} \ln \frac{p_2}{p_1}$$

where $(\gamma - 1)/\gamma$ has been substituted from eq. (5.10b) for $R/C_p$. The above equation yields

$$\frac{T_2}{T_1} = \left[\frac{p_2}{p_1}\right]^{(\gamma-1)/\gamma} \tag{7.10a}$$

The following equation is obtained in a similar manner from eq. (7.8c).

$$\frac{T_2}{T_1} = \left[\frac{v_1}{v_2}\right]^{\gamma-1} \tag{7.10b}$$

By substitution of eq. (7.10b) into (7.10a) we obtain

$$\frac{p_2}{p_1} = \left[\frac{v_1}{v_2}\right]^{\gamma} \tag{7.10c}$$

or

$$pv^\gamma = \text{constant} \tag{7.10d}$$

Thus we conclude that an isentropic process for a perfect gas with constant specific heats is described on *p-v* coordinates by eq. (7.10d). Equations (7.10a) and (7.10b) relate pressure, temperature, and specific volume between any two states along the isentropic path given by eq. (7.10d). Equations (7.10a) through (7.10d) are identical to eqs. (5.16) and (5.17). From this we see that new meaning is attached to the process described by $pv^\gamma$ = constant.

## Example 7.c

A rigid, perfectly insulated tank with a volume of 47.8 $ft^3$ contains air at 20.74 psia and 520 R. Work is done on the air by a paddle wheel until the pressure in the tank reaches 35.27 psia. Calculate (a) the final temperature of the air and (b) the entropy change of the air.

**Solution:**

(a) For the rigid tank

$$p_2 V_2 = mRT_2 \quad \text{and} \quad p_1 V_1 = mRT_1$$

With mass and volume constant

$$\frac{T_2}{T_1} = \frac{p_2}{p_1}$$

and

$$T_2 = \frac{35.27}{20.74}(520) = 884 \text{ R}$$

(b) With pressures and temperatures known, the entropy change is calculated most simply using eq. (7.8c). On a unit mass basis

$$\begin{aligned} s_2 - s_1 &= C_v \ln \frac{T_2}{T_1} = 0.172 \ln \frac{884}{520} \\ &= 0.09127 \text{ B/(lbm R)} \end{aligned}$$

The mass of air in the tank is calculated using the equation of state at (1)

$$m = \frac{(20.74)(144)(47.8)}{(53.3)(520)} = 5.151 \text{ lbm}$$

The entropy change for the system is

$$S_2 - S_1 = m(s_2 - s_1) = (5.151)(0.09127) = 0.4701 \text{ B/R}$$

Note that the process is adiabatic; however, it is not reversible. Thus the $\int \delta Q/T$ for the actual process would not yield the entropy change of the air.

## Example 7.d

A gas that can be treated as air is supplied to a gas turbine at 150 psia and 2000 F at the rate of 30,000 lbm/hr. The gas enters the turbine at a velocity of 400 ft/sec and expands through the turbine isentropically to an exhaust pressure of 15 psia. The exhaust velocity is 700 ft/sec. Calculate (a) the work per unit mass of the gas, (b) the power delivered by the unit, and (c) the exhaust area.

**Solution:**

(a) Since the process is for a perfect gas and is isentropic, the path is $pv^\gamma$ = constant. (We assume constant specific heats.) The *p-v* diagram shows the path and the given values of pressure and temperature. The turbine is a steady-flow device; hence the applicable energy equation is

$$h_1 + \frac{V_1^2}{2g_0} + {}_1q_2 = h_2 + \frac{V_2^2}{2g_0} + {}_1w_2$$

Since the process is adiabatic and the working fluid is a perfect gas,

$${}_1w_2 = C_p(T_1 - T_2) - \frac{V_2^2 - V_1^2}{2g_0}$$

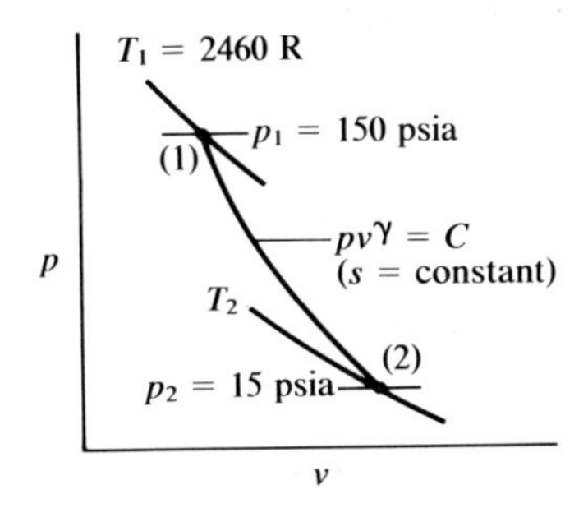


For the isentropic process $T_2$ is related to the given pressure and temperature values by eq. (7.10a). Thus

$$T_2 = 2460\left(\frac{15}{150}\right)^{0.4/1.4} = 1274 \text{ R}$$

Therefore

$${}_1w_2 = 0.24(2460 - 1274) - \frac{700^2 - 400^2}{(2)(32.2)(778)} = 278 \text{ B/lbm}$$

${}_1w_2$ can also be obtained by integration of eq. (2.14).

(b) The power is

$$\dot{W} = \dot{m}_1 w_2 = (30{,}000)\,\frac{270}{2545} = 3277 \text{ HP}$$

(c) From the continuity equation and the perfect gas equation of state,

$$A_2 = \dot{m}\frac{v_2}{V_2} = \frac{30{,}000}{3600}\,\Big|\,\frac{1545}{29}\,\Big|\,\frac{1274}{15}\,\Big|\,\frac{}{144}\,\Big|\,\frac{}{700} = 0.379 \text{ ft}^2$$

---

## 7–6 Entropy for Solid, Liquid, and Vapor Phases of Pure Substances

The equality of Clausius permits a definition of change in entropy, but it does not establish a value for the entropy of a pure substance at a given state. This parallels the situation for internal energy of a pure substance. Recognizing entropy as a property, and thus single valued at a given state of a pure substance, we can write

$$s = (s - s_r) + s_r$$

with $s_r$ an arbitrary constant at a reference state of the pure substance. The difference $s - s_r$ then can be evaluated as the $\int \delta Q/T$ along any reversible path or paths from the reference state to any other state where the entropy has the value $s$. Such methods are used in establishing entropy values in tables of properties. In the *Steam Tables*, for example, the reference is established by setting the entropy of saturated liquid at the triple point equal to zero.

While the $Tds$ relations do not yield explicit equations for entropy changes for general pure substances, as they did for perfect gases, these relations are useful in relating properties of pure substances. Consider, for example, a pure substance undergoing a constant-pressure phase change. For such a process eq. (7.9a) becomes

$$ds = \frac{dh}{T}$$

With temperature remaining constant during a constant-pressure phase change, this equation can be used to relate entropy and enthalpy changes for such processes. For example, consider a constant-pressure process during which saturated liquid undergoes a phase change to saturated vapor. Letting $f$ indicate the saturated-liquid state and $g$ the saturated-vapor state, the above equation becomes

$$\int_f^g ds = \int_f^g \frac{dh}{T}$$

or

$$s_g - s_f = \frac{h_g - h_f}{T}$$

where $T$ is the absolute thermodynamic temperature. Following the notation of the *Steam Tables*, $s_g - s_f = s_{fg}$ and $h_g - h_f = h_{fg}$ or

$$s_{fg} = \frac{h_{fg}}{T} \tag{7.11}$$

The property entropy permits us to extend with relative ease our treatment of processes for general pure substances to additional reversible processes, particularly isentropic processes. To this point we have not dealt with this topic since, as alluded to above, manageable expressions for such processes exist only for perfect gases. The following examples illustrate some of the uses of entropy for general pure substances. While $H_2O$ has been selected as the working fluid for the examples, the procedures employed are applicable to refrigerants such as ammonia and the freon series as well as other substances for which tables of properties are available. Entropy values are tabulated in tables of properties in the same manner as are internal and enthalpy values. In treating entropy in two-phase regions, eqs. (4.1a) and (4.1b) are applicable.

## Example 7.e

At −10 F the enthalpy of sublimation (solid to vapor) of $H_2O$ is 1220.4 B/lbm. Calculate the entropy change during sublimation at the given temperature. Compare your result with $s_{ig}$ at −10 F in Table 6 of the *Steam Tables.*

**Solution:**

Applying eq. (7.11) to the phase change

$$s_{ig} = \frac{h_{ig}}{T} = \frac{1220.4}{459.67 + (-10)}$$
$$= 2.714 \text{ B/(lbm R)}$$

From Table 6 of the *Steam Tables*, we read $s_{ig} = 2.714$ B/(lbm R).

## Example 7.f

A turbine is supplied with steam at 400 psia, 600 F, at the rate of 50,000 lb/hr. The steam enters the unit at a velocity of 400 ft/sec. Expansion through the turbine is reversible and adiabatic to an exhaust pressure of 1.80 psia. The exhaust velocity is 480 ft/sec. Calculate

(a) The work per pound of steam
(b) The horsepower delivered by the unit
(c) The exhaust area in square feet

**Solution:**

(a) The turbine is a steady-flow machine, and hence the applicable energy equation is

$$h_1 + \frac{V_1^2}{2g_0} + {}_1q_2 = h_2 + \frac{V_2^2}{2g_0} + {}_1w_2$$

or, since $q = 0$,

$${}_1w_2 = (h_1 - h_2) - \frac{V_2^2 - V_1^2}{2g_0}$$

From Table 3 of the *Steam Tables,* properties at the inlet are

$$h_1 = 1306.6 \text{ B/lbm} \quad \text{and} \quad s_1 = 1.5892 \text{ B/(lbm R)}$$

At 1.80 psia, $s_f = 0.16840$ and $s_g = 1.9286$. The entropy will remain constant at $s_1 = 1.5892$ during the reversible adiabatic expansion. Comparison of this value with $s_f$ and $s_g$ at 1.80 psia indicates that the exhaust is in the mixture region, and hence the quality at the exhaust must be determined before the enthalpy and specific volume can be established. For the expansion, using eq. (4.1b)

$$s_1 = 1.5892 = s_2 = 0.16840 + (x_2)\ 1.7602$$

and

$$x_2 = \frac{1.5892 - 0.16840}{1.7602} = 0.8123$$

At the exhaust

$$h_2 = 90.18 + (0.8123)(1024.3) = 922.2 \text{ B/lbm}$$

and

$$v_2 = (x_2)(v_{g_2}) = (0.8123)(191.8) = 155.8 \text{ ft}^3/\text{lbm}$$

From the steady-flow equation

$$\begin{aligned} {}_1w_2 &= (1306.6 - 922.21) - \frac{480^2 - 400^2}{2g_0} \\ &= 384.4 - 1.41 \\ &= 383 \text{ B/lbm} \end{aligned}$$

(b) The power is

$$\begin{aligned} \dot{W} &= \frac{(383)(50{,}000)}{2545} \\ &= 7526 \text{ HP} \end{aligned}$$

(c) The continuity equation is used to calculate the exhaust area

$$A_2 = \frac{\dot{m}v_2}{V_2} = \frac{50{,}000}{3600}\left|\frac{155.8}{480}\right. = 4.51 \text{ ft}^2$$

---

## Example 7.g

A closed system of 2.30 kg of $H_2O$, initially at 100 kPa and 300 C, is compressed isothermally and reversibly until the quality is 85.0 percent. Show the process on a *p-v* diagram and find the work for the process.

**Solution:**

The path for the process is shown on the diagram. For a closed system undergoing a reversible process the work on a unit mass basis is

$${}_1w_2 = \int_1^2 p\,dv$$

This integral cannot easily be evaluated because the relation between $p$ and $v$ cannot be expressed mathematically. (The integral could be evaluated graphically.) However, the work can be evaluated by use of the energy equation for a closed system

$${}_1q_2 = {}_1w_2 + u_2 - u_1$$

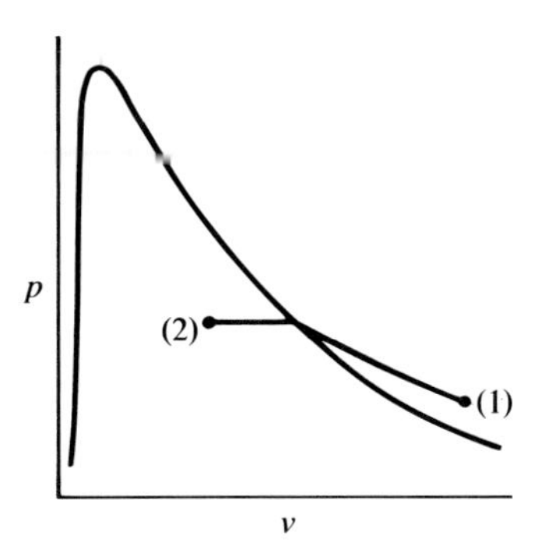


From the definition of entropy change

$$ {}_1q_2 = T(s_2 - s_1) $$

and the energy equation can be expressed as

$$ {}_1w_2 = T(s_2 - s_1) - (u_2 - u_1) $$

At the initial state, from the *Steam Tables,* SI units, Appendix B,

$$ s_1 = 8.2158 \text{ B/(kg K)} \quad \text{and} \quad u_1 = 2810.4 \text{ kJ/kg} $$

At a quality of 0.850 and a temperature of 300 K, the values of $u$ and $s$ for the liquid–vapor mixture are, respectively,

$$ \begin{aligned} u_2 &= u_f + xu_{fg} = 1332 + (0.850)(2563.0 - 1332.0) \\ &= 2378.4 \text{ kJ/kg} \end{aligned} $$

$$ \begin{aligned} s_2 &= 3.2534 + (0.850)(5.7045 - 3.2534) \\ &= 5.3368 \text{ kJ/(kg}\cdot\text{K)} \end{aligned} $$

Therefore, the work is

$$ \begin{aligned} {}_1w_2 &= (300 + 273)(5.3368 - 8.2158) - (2378.4 - 2810.4) \\ &= -1218 \text{ kJ/kg} \end{aligned} $$

For 2.30 kg

$$ {}_1W_2 = -2802 \text{ kJ} $$

**Figure 7.5** Temperature–Entropy Diagrams. *a.* Area on the *T-s* diagram represents heat transfer for a reversible process. *b. T-s* diagram for the Carnot cycle.

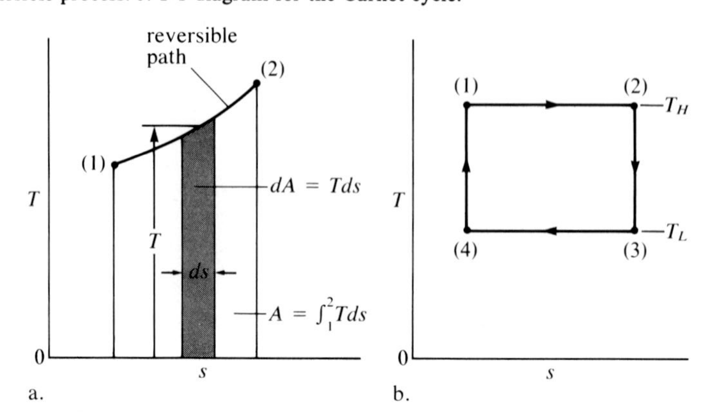


## 7–7 Entropy as a Coordinate on Property Diagrams

Since entropy is a property it can be used as one coordinate on property diagrams. Temperature–entropy diagrams are particularly useful in describing heat transfer for reversible processes. For a unit mass of a pure substance undergoing a reversible process in either a closed or steady-flow system, it has been shown by eqs. (7.4) and (7.7e) that

$$({}_1q_2)_{\text{rev}} = \int_1^2 T ds \tag{7.12}$$

The integral in this equation is the area between the path and the abscissa, $s$, on a $T$-$s$ diagram as illustrated in Figure 7.5a. States (1) and (2) on this diagram can be fixed by specifying, for example, values for temperature and entropy. From eq. (7.12) we see that the noted area represents the heat transfer for the reversible process. This result is similar to that for $p$-$v$ diagrams on which, as discussed in Chapter 2, areas are related to work for reversible processes. Since in eq. (7.12) $T$ is the absolute thermodynamic temperature, we see that heat transfer for reversible processes is positive when entropy increases and negative when entropy decreases. Further, eq. (7.12) is applicable to all pure substances. Therefore, the interpretation of heat transfer for a reversible process as an area on the $T$-$s$ diagram applies to all pure substances. Last, as discussed in Section 7–3, eq. (7.12) is applicable to internally reversible processes as well as completely reversible processes.

It is now possible to describe the familiar Carnot cycle on $T$-$s$ coordinates. Utilizing the description of the Carnot heat engine cycle in Section 6–3, we obtain the $T$-$s$ diagram in Figure 7.5b. The process (1) to (2) is the reversible heat-reception process, with the area below the path representing the heat transfer from the reservoir at $T_H$ per unit mass of working fluid. The process (2) to (3) is the adiabatic and reversible (isentropic) expansion. Process (3) to (4) is the process of reversible heat rejection to the reservoir at $T_L$. The area below the path represents the heat transfer. The isentropic process (4) to (1) completes the cycle. From the first law and eq. (7.12) we can write

$$\oint \delta q = \oint T ds = \text{enclosed area} = (T_H - T_L)(s_2 - s_1) = \oint \delta w$$

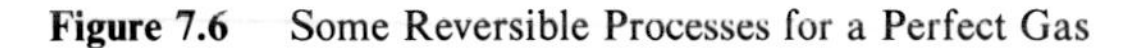

**Figure 7.6** Some Reversible Processes for a Perfect Gas

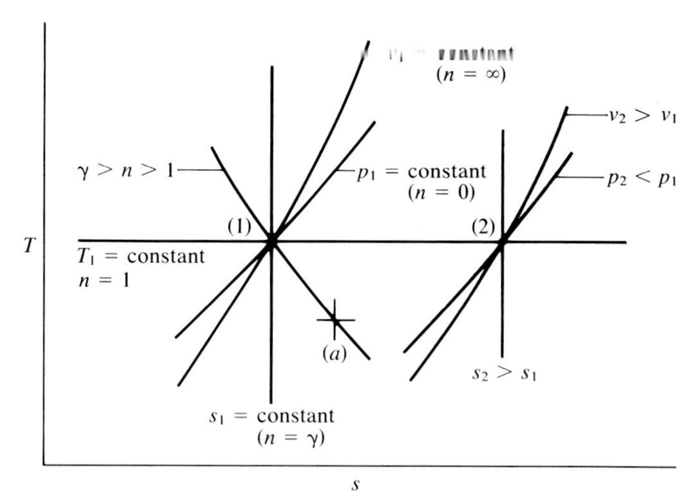


Thus the enclosed area is, on a unit mass basis, both the net heat transfer and the net work. While this result has been obtained for the Carnot cycle, reflection on eq. (7.12) and the first law for a cycle leads to the conclusion that the enclosed area on a *T-s* diagram for a cycle represents both the net heat and the net work, provided that the cycle is at least internally reversible.

Let us now direct our attention to the temperature–entropy diagram for perfect gases. We will use the reversible polytropic process $pv^n = C$ for a perfect gas, which was discussed in Section 5–6 and described in Figure 5.4 as an aid in illustrating different processes. Paths are presented on the *T-s* diagram in Figure 7.6 for various values of $n$ as was done for Figure 5.4. Isothermal paths, $n = 1$, are, of course, horizontal lines on the *T-s* diagram. Recalling that for perfect gases internal energy and enthalpy are functions of temperature only, constant-temperature paths are also paths of constant internal energy and enthalpy. Reversible adiabatic paths, $n = \gamma$, are paths of constant entropy and are therefore vertical lines. Paths marked $n = 1$ and $n = \gamma$ passing through state (1) are shown in Figure 7.6.

The nature of constant-volume lines ($n = \infty$) and constant-pressure lines ($n = 0$) are established by the *Tds* relations, eqs. (7.8b) and (7.9b). At constant volume

$$\left(\frac{\partial T}{\partial s}\right)_v = \frac{T}{C_v}$$

by eq. (7.8b). At constant pressure

$$\left(\frac{\partial T}{\partial s}\right)_p = \frac{T}{C_p}$$

by eq. (7.9b). With $C_p$ greater than $C_v$ for all perfect gases, it is evident that constant-pressure lines have a lesser slope than constant-volume lines. Paths of constant pressure and constant specific volume through state (1) are shown in Figure 7.6. State (2) in Figure 7.6 lies on an isothermal path through state (1), but at a greater entropy. The specific volume at state (2) is greater than that at state (1). Proof of this statement is evident when eq. (7.8c) is applied to the isothermal path between states (1) and (2). Consideration of eq. (7.9c) along an isotherm indicates the direction of increase of constant-pressure lines. Application of this equation between states (1) and (2) shows that $p_2$ is less than $p_1$. Knowledge of the direction of increase of constant-property paths as described in Figure 7.6 is useful in constructing *T-s* diagrams from *p-v* diagrams, and vice versa.

The sign of the heat transfer for a general polytropic process for a perfect gas depends on the relative magnitudes of $n$ for the process and $\gamma$. For example, the reversible process from (1) to (a) with $\gamma \leq n \leq 1$ in Figure 7.6 is an expansion, that is, the specific volume increases. With entropy increasing and $Tds = \delta q$, it is apparent that the heat transfer is positive. If we consider the process (*a*) to (1), a compression process, the heat transfer is negative. Cases with $n > \gamma$ or $n < 1$ can be analyzed in a similar manner.

## Example 7.h

Sketch on *p-v* and *T-s* coordinates the following internally reversible perfect gas cycle that is executed in a piston cylinder assembly.

$1 \rightarrow 2$ Constant-pressure process, $T_2 > T_1$
$2 \rightarrow 3$ Isentropic process, $T_3 < T_1$
$3 \rightarrow 4$ Isothermal process $v_4 = v_1$
$4 \rightarrow 1$ Constant-volume process

**Solution:**

The *T-s* diagram can be completed using the description of the various processes given in Figure 7.6. Both the *T-s* and *p-v* diagrams for the cycle are shown below. The *p-v* diagram is based on previous discussion of processes for perfect gases. Each diagram contributes to the description of the cycle. For example, from the *p-v* diagram we can see that $p_3$ is the lowest pressure in the cycle and from the *T-s* diagram we can see that $T_2$ is the highest temperature. From the *T-s* diagram and eq. (7.12) we see that heat is transferred to the working fluid during processes (4) to (1) and (1) to (2) and that heat is transferred from the working fluid only during process (3) to (4). We also see that $q_{net} = \oint \delta q = \oint Tds$ is positive, that is, the enclosed area on the *T-s* diagram is positive. By the first law, this area is equal to the enclosed area on the *p-v* diagram. Recalling that the cycle is executed in a closed system, we relate the areas as follows:

$$w_{net} = \oint \delta w = \oint pdv = q_{net} = \oint Tds$$

Of course the heat and the work must be measured in the same units.

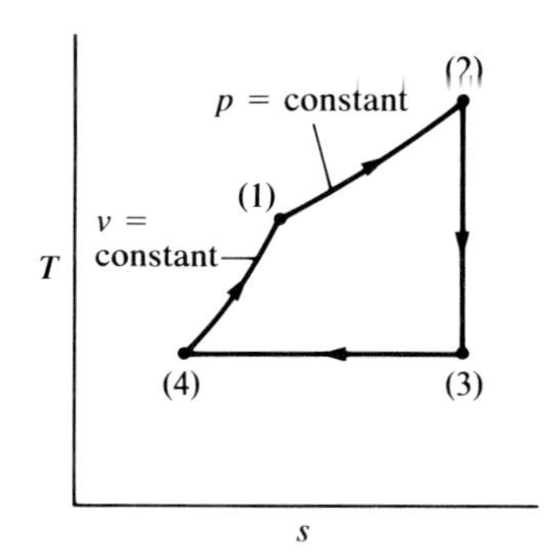


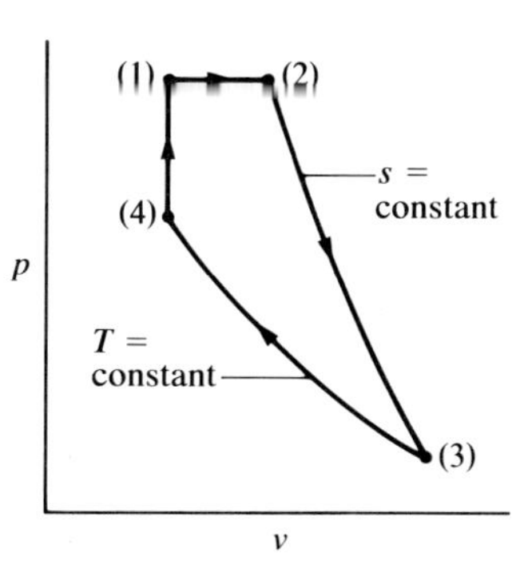


---

Temperature–entropy diagrams for substances that exist in more than one phase such as $H_2O$ and refrigerants differ significantly from those for perfect gases. Figure 7.7a shows a typical $T$-$s$ diagram for the liquid–vapor phases of a pure substance. The area to the left of the saturated-liquid line is the subcooled (compressed) liquid region, and the region to the right of the saturated-vapor curve is the superheated-vapor region. The region between the saturation curves and above the triple line is the mixture region where saturated liquid and saturated vapor coexist. The demarcation between liquid and vapor phases vanishes at temperatures greater than the critical temperature. The arrows on the constant-property paths indicate the direction in which that property increases. The scale in the subcooled-liquid region has been distorted to display the constant-pressure line in that region. The segment of the constant-pressure line in the saturated-mixture region is coincident with the corresponding saturation temperature at that pressure. In the superheated-vapor region the constant-volume line through a given state has a greater slope than does the constant-pressure line passing through that state, as indicated by the intersecting lines for these properties in the superheat region.

Figure 7.7b is a temperature–entropy diagram for the liquid and vapor phases of water. Lines of constant quality are shown in the mixture region. Also shown are selected constant-pressure paths. Lines of constant pressure include the critical pressure, $p_{cp}$ = 3203.6 psia, and a supercritical pressure, $p$ = 5000 psia. While superheated vapor for $H_2O$ does not behave as a perfect gas in a large part of the region of interest, its behavior can be described by perfect gas equations provided the pressure is very low. Perfect gas behavior is indicated on the $T$-$s$ diagram when constant-enthalpy lines are parallel to constant-temperature lines, that is, when enthalpy depends on temperature only. Perfect gas behavior is displayed for $H_2O$ in the vicinity of $h$ = 1160 B/lbm and $p$ = 0.5 psia in Figure 7.7b, for example. More detailed temperature-entropy diagrams for $H_2O$ are included in Appendices B and C.

The liquid region on a temperature–entropy diagram for $H_2O$ warrants added attention. This region lies to the left of the saturated-liquid line. The fact that it is a narrow band is established readily using data from the compressed-liquid table at a given temperature. For example, at 200 F and saturation the specific entropy of $H_2O$ is 0.29400 B/(lbm R). At the same temperature of 200 F the specific entropy is 0.29341 B/(lbm R) at 500 psia, 0.29281 B/(lbm R) at 1000 psia, and 0.29162 B/(lbm R) at 2000 psia. These values illustrate the very small effect of pressure on entropy of the liquid at a given temperature. The change in

**Figure 7.7** The Temperature–Entropy Diagram for Liquid and Vapor Phases of a Pure Substance. *a.* A typical *T-s* diagram. Arrows on constant property lines indicate the direction of increase of the property. *b.* The temperature–entropy diagram for $H_2O$.

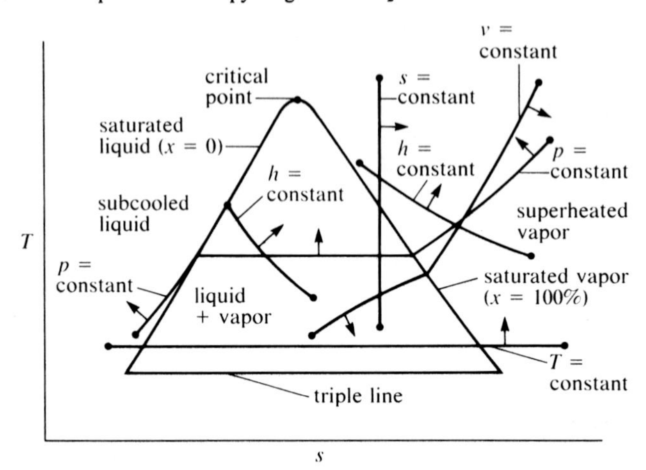


a.

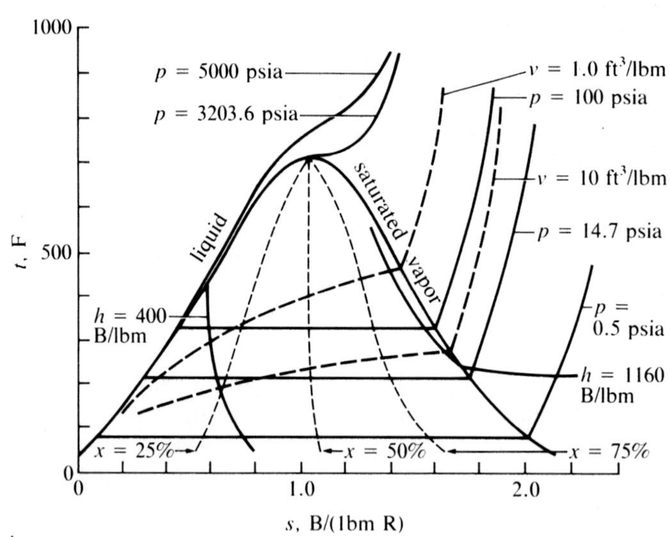


b.

**Figure 7.8** *T-s* Diagram for the Liquid Phase

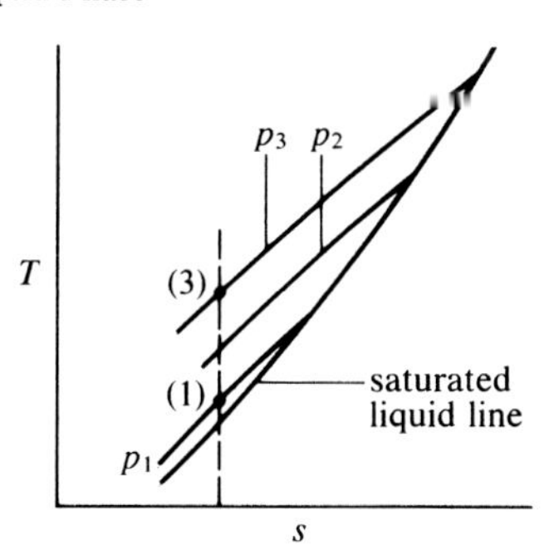


entropy with temperature along a constant-pressure line is significant. The trend can be established by observing specific entropies at a given pressure and a series of temperatures in the compressed-liquid table. Figure 7.8 (not to scale) shows the progression of constant-pressure lines from the subcooled region to the saturated-liquid line. For pressures less than the critical pressure, these lines intercept the saturated-liquid line at the saturation temperature for a given pressure. Recalling that saturation pressure increases with saturation temperature, it is evident that $p_3$ is the highest pressure and $p_1$ the lowest pressure shown in Figure 7.8. The liquid region is of particular interest when evaluating the thermodynamics of pumps. Consider a pump that operates reversibly and adiabatically in steady flow with negligible velocities and with the suction and discharge at the same elevation. The isentropic increase in pressure would follow a path such as $1 \rightarrow 3$ on Figure 7.8. For the stated conditions the steady-flow energy equation reduces to

$$-w = h_3 - h_1 \tag{a}$$

and eq. (7.9) becomes

$$Tds = dh - vdp = 0 \tag{b}$$

Assuming the fluid to be incompressible (constant density) in the liquid phase, eq. (b) yields

$$\int_1^3 dh = \int_1^3 vdp$$

which integrates to

$$h_3 - h_1 = v(p_3 - p_1) \tag{c}$$

Appropriate conversions must be made to ensure that units are consistent. Substituting (c) into (a), we have

$$-w = h_3 - h_1 = v(p_3 - p_1)$$

for the particular process. The assumption of constant specific volume is reasonable for moderate pressure increases. Errors resulting from this assumption increase as the pressure rise increases.

## Example 7.i

The internally reversible cycle described below has $H_2O$ as the working fluid and is executed in a piston–cylinder assembly. Only the liquid and vapor phases are present during the cycle. Sketch the cycle on *T-s* and *p-v* coordinates.

$1 \rightarrow 2$ $x_1 = 50$ percent; constant-pressure process, $T_2 > T_1$
$2 \rightarrow 3$ Isentropic process, $T_3 < T_1$; state (3) is superheated vapor
$3 \rightarrow 4$ Isothermal process, $v_4 = v_1$
$4 \rightarrow 1$ Constant-volume process

**Solution:**

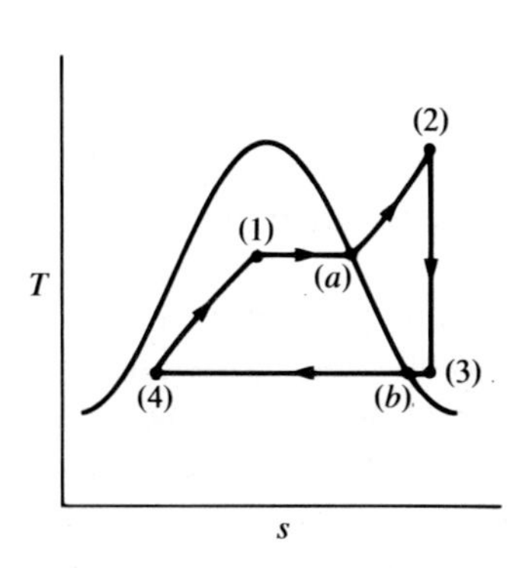


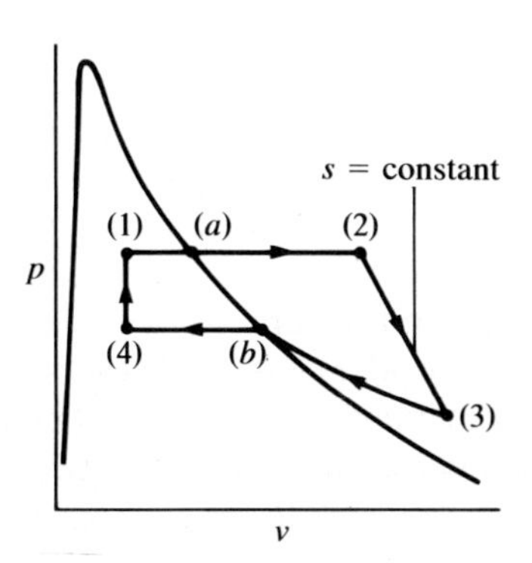


The cycle is shown in the above *T-s* and *p-v* diagrams. The *T-s* diagram is completed with the aid of the descriptions of the various processes given in Figure 7.7a. The process (1) to (*a*) is both constant pressure and temperature since that segment of the constant-pressure process takes place in the mixture region. For the same reason, the segment (*b*) to (4) of the constant-temperature process is constant pressure as well. The isentropic path in the *p-v* diagram deserves attention since the isentropic process has not been described on *p-v* coordinates for the vapor phase of a general pure substance. It can be established that the path for this process is that shown by noting on the *T-s* diagram in Figure 7.7a that the specific volume increases as the temperature decreases when entropy is constant. Thus process (2) to (3) is an isentropic expansion as shown. It is also of interest to observe that at a given state the slope of a constant-entropy line on *p-v* coordinates for any pure substance is always more negative than the slope of a constant-temperature line through the same state. This can be observed at state (3) on the *p-v* diagram. Note that although this cycle and the perfect gas cycle in Example 7.h are composed of the same types of processes, the diagrams differ significantly between the two examples. However, the comments in Example 7.h, related to the interpretation of areas on the diagrams as heat and work, apply to the present example also.

Figure 7.9 Mollier Diagram for $H_2O$

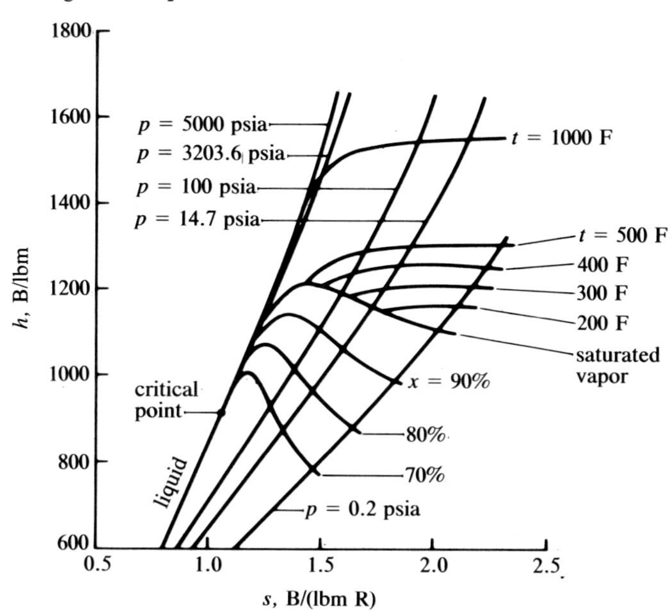


Another property diagram having entropy as a coordinate is the enthalpy–entropy or *h-s* diagram. These diagrams are usually called Mollier diagrams, in recognition of Richard Mollier who first suggested them. Such diagrams are useful for approximate calculations in a number of processes of interest to engineers. Expansions through nozzles and turbines typically are assumed to take place adiabatically. For an ideal nozzle or turbine the process is reversible as well. Thus these ideal processes are isentropic and follow a vertical line on a Mollier diagram.

A skeleton *h-s* diagram for $H_2O$ is given in Figure 7.9. Directions of increasing pressure and temperature are evident from the diagram. The region above the saturated vapor curve is the superheated vapor region, and the region below it is the liquid-vapor mixture region. As was true for the *T-s* diagram, lines of constant temperature are coincident with lines of constant pressure in the mixture region. The area under a path on an *h-s* diagram, that is, the integral of ***hds***, has no significance. More complete Mollier diagrams for $H_2O$ are included in Appendices B and C.

## 7–8 The Entropy Principle

A very important consequence of the second law that provides another example of the usefulness of the property entropy results from consideration of adiabatic processes. For an adiabatic closed system, eq. (7.5a) requires that

$$(dS \geq 0)_{Q=0,\ \text{closed}} \tag{7.13a}$$

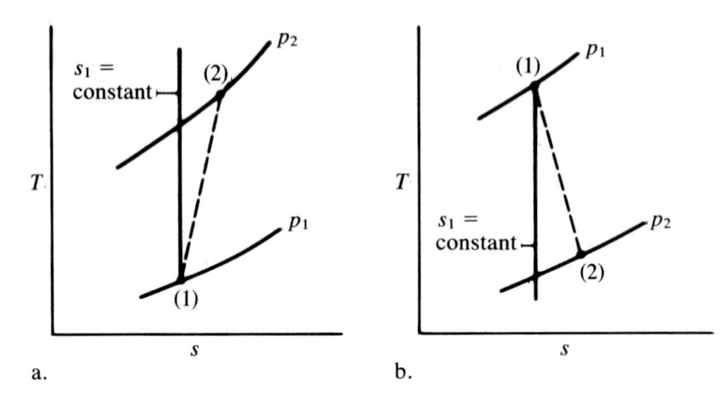


**Figure 7.10** Possible End States for Adiabatic Compression and Expansion Processes Occurring in a Closed or Single-Stream Steady-Flow System. *a.* Adiabatic compression processes. *b.* Adiabatic expansion processes

or

$$(S_2 - S_1 \geq 0)_{Q = 0, \text{ closed}} \tag{7.13b}$$

Now consider an adiabatic steady-flow system with one inbound and one outbound stream. The entropy change for a process in this system is obtained from application of eq. (7.7c). With $\Delta s_I$ in that equation zero or a positive value, we find

$$(s_2 - s_1 \geq 0)_{q = 0, \text{ steady-flow}} \tag{7.14a}$$

or

$$(ds \geq 0)_{q = 0, \text{ steady-flow}} \tag{7.14b}$$

Thus for a substance undergoing an adiabatic process in a closed system, or for mass passing through a single-stream steady-flow system, entropy either will increase or remain constant. A decrease in entropy for an adiabatic process in either system would result in a violation of the second law. This result can be written in a single equation as

$$(dS \geq 0)_{\text{adiabatic, closed or single-stream steady-flow}} \tag{7.15}$$

The equality sign in eq. (7.15) applies to adiabatic internally reversible (isentropic) processes and the inequality sign applies to adiabatic internally irreversible processes. Thus an increase in entropy for a working fluid in either of the adiabatic systems considered tells us that the process is internally irreversible. Equation (7.15) is a statement of what is called the entropy principle or the principle of increase of entropy.

Figure 7.10 provides an illustration of the results obtained from eq. (7.15) for two processes frequently encountered in engineering thermodynamics. The processes could occur in either a closed or a single-stream steady-flow system. Figure 7.10a is a $T$-$s$ diagram for an adiabatic compression process taking place between a fixed state (1) and a given pressure,

$p_2$. Possible states (2) lie on the constant-pressure line $p_2$ at values of $s_2 \geq s_1$. States (2) with entropy values of less than $s_1$ are prohibited. In a similar manner, for an adiabatic expansion process from a fixed state (1) to a pressure $p_2$, possible states (2) lie on the constant-pressure line $p_2$ at $s_2 \geq s_1$. (See Figure 7.10b.) States (2) shown in each part of Figure 7.10 denote possible states for the respective processes. Since in each case $s_2 > s_1$, the processes between states (1) and (2) are irreversible and thus are shown as dashed lines.

## Example 7.j

Steam initially at 0.60 MPa and 100 percent quality undergoes a steady-flow throttling process to a pressure of 0.10 MPa. Determine the temperature change and the entropy change for the process and show the process on *h-s* and *T-s* coordinates.

**Solution:**

Recalling for the throttling process that $q$, $w$, $\Delta KE$, and $\Delta PE$ are zero, the steady-flow energy equation reduces to $h_1 = h_2$. From Appendix B, Table B–1, we obtain $h_1$ as 2756.8 kJ/kg, $t_1$ as 158.85 C, and $s_1$ as 6.7600 kJ/(kg K). State (2) is established by $h_2 = h_1$ and $p_2 = 0.10$ MPa. Process (1) to (2) is shown as the dashed line on the following *h-s* and *T-s* diagrams, since as noted previously, a throttling process is irreversible. From Appendix B, Table B–2, we obtain $t_2 = 140$ C and $s_2 = 7.5659$ kJ/(kg K). Thus

$$t_2 - t_1 = 140 - 158.85 = -18.85 \text{ C}$$

$$s_2 - s_1 = 7.5659 - 6.7600 = 0.8059 \text{ kJ/(kg K)}$$

The increase in entropy is expected because the process is irreversible and adiabatic, and according to eq. (7.15), $s_2$ must be greater than $s_1$. Heat transfer cannot be interpreted as an area on the *T-s* diagram because the process is not reversible.

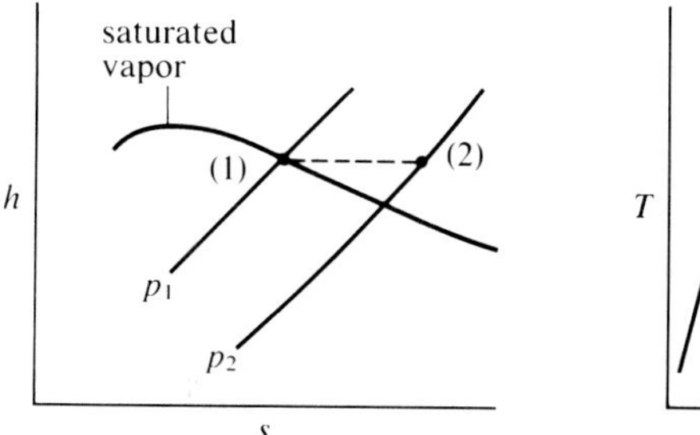


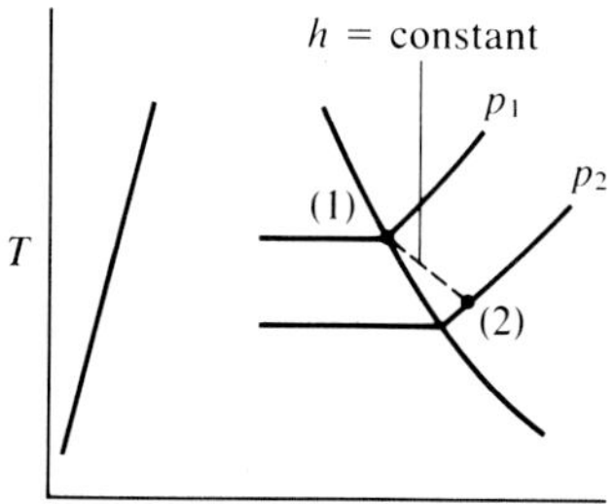

## Example 7.k

Helium enters a convergent–divergent nozzle at 2000 K and 1400 kPa with a negligible velocity. The helium expands adiabatically through the nozzle in a steady-flow process. The pressure at the flow exit plane is 150 kPa. Determine the maximum possible exit velocity.

**Solution:**

With no work involved for the adiabatic process, the steady-flow energy equation reduces to

$$h_1 = h_2 + \frac{V_2^2}{2g_0}$$

Using the expression for perfect gas enthalpy change and solving for $V_2$ we obtain

$$V_2 = [2g_0C_p(T_1 - T_2)]^{1/2}$$

With $T_1$ given, it is evident that $V_2$ is a maximum when $T_2$ is the lowest possible. Equation (7.14a) requires that $s_2 \geq s_1$. The $T$-$s$ diagram shows that the lowest possible value of $T_2$ occurs at the intersection of a constant entropy line through state (1) and the pressure $p_2$.

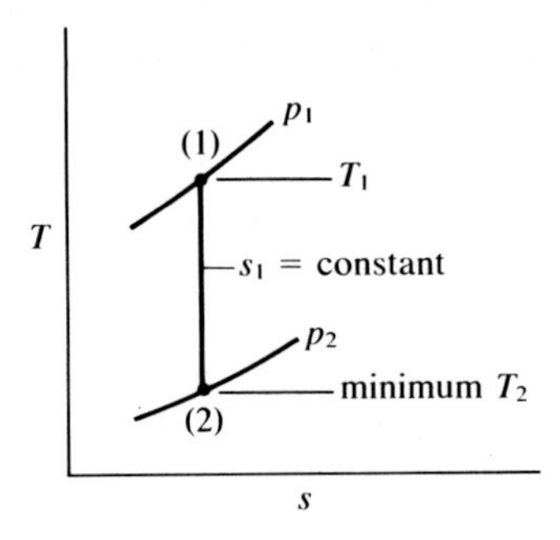


Therefore, the maximum exit velocity is associated with an isentropic expansion of the helium. Equation (7.10a) then applies to find the minimum value of $T_2$. Using helium properties from Table 5.5,

$$T_2 = 2000\left(\frac{150}{1400}\right)^{0.667/1.667} = 818 \text{ K}$$

Therefore

$$V_2 = \left(\frac{2 \mid 1 \text{ kg m} \mid 5.19 \times 10^3 \text{ (N m)} \mid (2000 - 818) \text{ K}}{\mid \text{N s}^2 \mid \text{kg K} \mid}\right)^{1/2}$$

$$= 3503 \text{ m/s}$$

An irreversible expansion would result in $s_2 > s_1$, a larger value of $T_2$, and a smaller exit velocity.

The entropy principle can also be stated for isolated systems. We recall that an isolated system is one that does not in any way interact with its surroundings, that is, neither the work interaction nor the heat interaction occurs between the system and its surroundings and no mass crosses the system boundary. Therefore, all isolated systems are closed and adiabatic. As a result, eq. (7.13a) applies and we can write

$$(dS \geq 0)_{\text{isolated system}} \tag{7.16a}$$

and

$$(\Delta S \geq 0)_{\text{isolated system}} \tag{7.16b}$$

While an isolated system cannot have work, heat, or mass transmitted across its boundary, such transfers between subsystems of the isolated system are permitted. Further, since the surroundings of a system usually include regions relatively close to the system, an isolated system in many cases can be formed by including a system (which may interact with its surroundings) and its surroundings within a single boundary across which no mass, heat, or work transfers occur. Application of eqs. (7.16) to such a system must include entropy changes for all subsystems of the isolated system.

An important fact is revealed by further examination of eqs. (7.13a) and (7.16a). If in a closed adiabatic system the entropy increases, the system undergoes a change in state. As we have noted, only changes of state for which $dS$ is positive are permitted. When $S$ ceases to increase, no further state change occurs and the process terminates. The final state is then the equilibrium state for the system. Thus we see that the equilibrium state for the system is attained when the entropy of the system reaches its maximum. Since time is required for any process to take place, we can speak of entropy increases in terms of increases with respect to time. An alternate way of writing the entropy principle is then

$$\left(\frac{dS}{d\tau} \geq 0\right)_{\text{adiabatic closed system}} \tag{7.17}$$

From the above discussion it is evident that the entropy principle restricts the direction in which processes in adiabatic closed systems can proceed. Various interpretations of the entropy principle are applied to problems in disciplines that include thermodynamics, chemistry, and economics.

## Example 7.1

Argon and carbon dioxide initially at different states are separated by an adiabatic partition in a rigid perfectly insulated container as shown in the diagram. Each gas is in an equilibrium state. Let the insulating effect of the partition be removed without removing the partition itself.

(a) Using the first law and your intuition and/or experience, determine the final temperature of the gases.
(b) Determine the final temperature of the gases by application of the principle of increase of entropy.

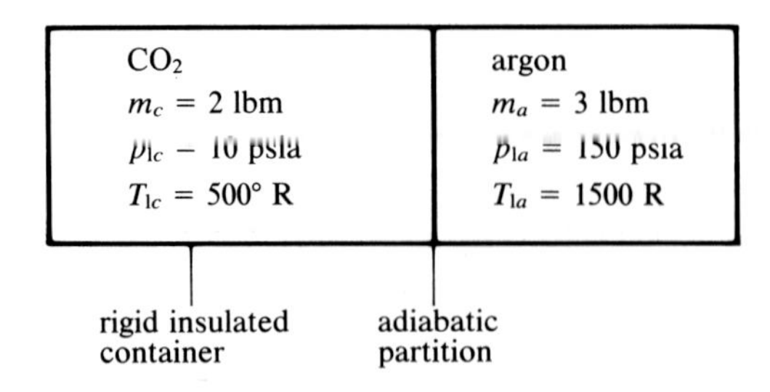


**Solution:**

(a) Taking the surface of the rigid container as the boundary of an isolated system and the argon and the carbon dioxide as subsystems, we write the first law for the isolated system as

$$Q - W = \Delta U = m_a u_{2a} + m_c u_{2c} - (m_a u_{1a} + m_c u_{1c}) = 0$$

where the notation in the diagram has been used and the subscript 2 indicates second states of the two gases at which the temperatures are not necessarily equal. The above equation can be written for the perfect gases in the forms

$$m_a(u_{2a} - u_{1a}) + m_c(u_{2c} - u_{1c}) = 0$$

$$m_a C_{va}(T_{2a} - T_{1a}) + m_c C_{vc}(T_{2c} - T_{1c}) = 0$$

By substitution of given values and use of $C_v$ values from Table 5.5, the above equation yields

$$T_{2c} = 1569.1 - 0.7127 T_{2a} \tag{a}$$

This expression is as far as the first law takes us. As long as it is satisfied, the first law is satisfied for the process, but it does not yield the solution sought. From experience we recognize that heat will be transferred from the argon to the colder carbon dioxide until both are at the same uniform temperature. Thus we set the temperatures in eq. (a) equal and obtain $T_{2a} = T_{2c} = 916$ R.

(b) Let us ignore the answer obtained in part (a) and solve the problem by application of the principle of increase of entropy. From this principle we recognize that a process in an adiabatic closed system (in this case, an isolated system) can proceed only as long as $dS > 0$ and that the process will terminate when $dS = 0$. We express the entropy change for the isolated system in terms of the two subsystems as

$$\begin{aligned} \Delta S = S - S_{\text{initial}} &= m_a s_{2a} + m_c s_{2c} - (m_a s_{1a} + m_c s_{c1}) \\ &= m_a(s_{2a} - s_{1a}) + m_c(s_{2c} - s_{1c}) \end{aligned}$$

where as before the subscript 2 denotes second states of the two gases at which the temperatures are not necessarily equal. $S$ is the corresponding value of entropy for the system. Noting that $v$ is constant for each gas, we use eq. (7.8c) to express the entropy change for each gas. Further, since the reference for entropy is unimportant when we are considering changes, we set $S_{initial}$ equal to zero. The expression for the entropy of the system at any equilibrium state thus becomes

$$S_{sys} = m_a C_{va} \ln \frac{T_{2a}}{T_{1a}} + m_c C_{vc} \ln \frac{T_{2c}}{T_{1c}}$$

By substituting known values, we obtain

$$S_{sys} = 0.2238 \ln \frac{T_{2a}}{1500} + 0.314 \ln \frac{T_{2c}}{500} \quad \text{(b)}$$

$T_{2a}$ and $T_{2c}$ are not independent in the above equation. They are related by the first law, eq. (a), which must be satisfied. Substitution of eq. (a) into eq. (b) gives the expression for $S_{sys}$ in terms of $T_{2a}$.

$$S_{sys} = 0.2238 \ln \frac{T_{2a}}{1500} + 0.314 \ln \left(\frac{1569.1 - T_{2a}}{500}\right) \quad \text{(c)}$$

Equation (c) is shown on the accompanying graph. The value of $T_{2a}$ at which $dS_{sys} = 0$, that is, the value at which $S_{sys}$ is at its maximum value, is the final temperature of the argon. As noted on the graph, this value is 916 R. This is, of course, the final temperature of the carbon dioxide also, and is the same answer as obtained in part (a). States that are associated with decreasing values of entropy ($dS_{sys} < 0$), as denoted by the broken segment of the curve, are not possible states for the isolated system. They satisfy the first law but not the second law. However, states on the segment of the curve for which $dS$ is positive (the solid curve) are states toward which the system progresses in a natural way. If the partition were restored to its original adiabatic nature to stop the heat-transfer process before the isolated system reached the final uniform temperature, the system would assume a state on the solid curve at $S_{sys}$ less than $S_{sys,max}$ and $T_{2a}$ would be greater than $T_{2c}$. Regardless of how the heat-transfer process proceeds, it can take place only in the direction shown by the arrows on the solid curve.

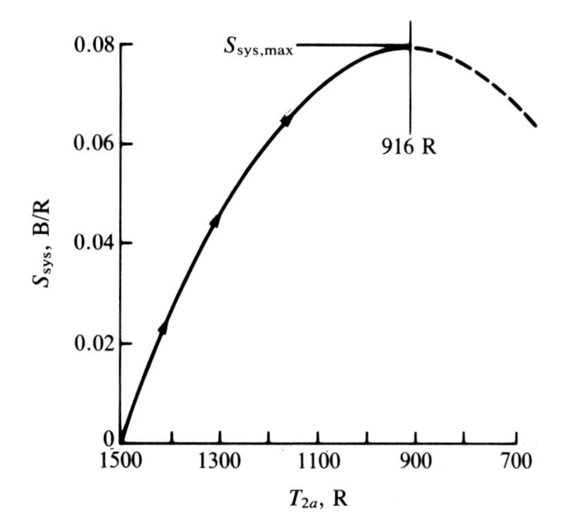

## 7–9 Available Energy

The laws governing energy and its use are central to the study of thermodynamics. As we have seen, the first law is the basis for the accounting of energy in its various interrelated forms. Limitations that exist in the conversion of energy from one form to another are imposed by the second law. In practice, the usefulness of energy depends not only on the amount of energy, but also on the form or condition in which it is available. For example, large amounts of energy are stored in ocean currents, but in this form are of limited usefulness. An important form of energy that is of fundamental interest in thermodynamics is heat energy. We can create heat sources by such means as the burning of a variety of fuels, the fission of nuclear materials, and the collection of solar energy. There are numerous practical and important uses of heat energy, but one of paramount interest is the conversion of heat to work. This is true because a large fraction of the shaft power used in an industrialized society can be traced to the conversion of heat to work in heat engines. In this section we shall focus attention on some additional aspects of heat as a form of energy.

Let us first establish the fact that not all heat energy is of equal usefulness in producing work or power. That this is true is evident from the second law. Consider a heat source (a heat reservoir) at $T_H = 2000$ R that delivers 100 B of heat energy to a heat engine that rejects heat to its surroundings at $T_0$, as shown in Figure 7.11a. We will take 40 F (500 R) as a typical temperature for the surroundings. The maximum fraction of the heat supplied that any heat engine could convert to work is $\eta_{max} = \eta_{rev}$, as given by eq. (6.4). For the case in Figure 7.11a, this fraction is 0.75. Therefore, the maximum work that could be produced from the heat supplied is 75 B. We will use $W_{max}$ as a measure of the usefulness of the heat supplied.

Consider now the case in which 100 B of heat energy is supplied from a source at $T_H = 1000$ R as shown in Figure 7.11b. The maximum fraction that could be converted to work is now 0.50 and the maximum work is 50 B. Thus we conclude that the 100 B supplied from the source at 2000 R is more useful in the production of work than is the same amount supplied from a source at 1000 R. To describe this fact we say that the heat supplied from the higher-temperature source has a higher grade or is of higher quality than is the energy supplied from the lower-temperature source. To quantify this observation, we introduce the term *available energy,* which is defined as that part of the heat supplied by a source that could be converted to work by a reversible heat engine operating between the source and the surroundings. The amount of heat not converted to work by the reversible heat engine is called the *unavailable energy.* In Figure 7.11a the available energy is 75 B, while the unavailable energy is 25 B. Similarly, in Figure 7.11b the available energy is 50 B and the unavailable energy is 50 B. Using the subscripts *a* and *u,* respectively, to designate the available and unavailable energy, we write

$$Q_H = Q_a + Q_u$$

or

$$Q_a = Q_H - Q_u \tag{7.18}$$

**Figure 7.11** Diagrams Related to the Concept of Available and Unavailable Energy. *a*. Heat from a source at 2000 R. *b*. Heat from a source at 1000 R. *c*. $Q_a$ and $Q_u$ as areas on the *T-S* diagram. *d*. Irreversible heat transfer.

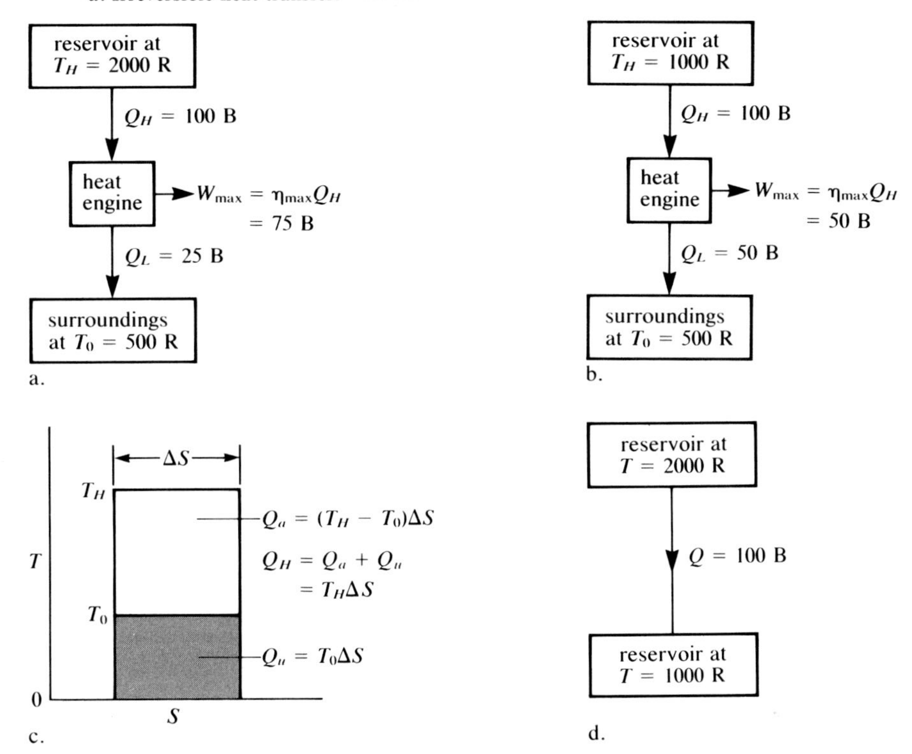


where $Q_H$ is the heat delivered by the source. The quantities in eq. (7.18) can be represented as areas on a temperature–entropy diagram, as shown in Figure 7.11c, where for convenience it has been assumed that the reversible engine is a Carnot engine.

Another point related to its usefulness is that heat energy is degraded as it is transferred irreversibly to a region at a lower temperature. This will be illustrated with the aid of Figure 7.11d, in which a heat transfer of 100 B occurs irreversibly from a reservoir at 2000 R to one at 1000 R.

Let us examine the available energy before and after the heat transfer. This quantity for the 100 B from the source at 2000 R is given in Figure 7.11a as 75 B, and for the 100 B from the source at 1000 R, it is given in Figure 7.11b as 50 B. From these values it is evident that the heat-transfer process illustrated in Figure 7.11d results in a degradation of the energy. The decrease in $Q_a$ is a measure of the degradation. It is also one measure of the irreversibility of a heat-transfer process. Thus we reach the conclusion that the usefulness of heat energy for producing work depends not only on the amount of heat supplied from a given source, but also on the temperature of the source.

From the above discussion, we see that when we desire energy in the form of work, efficient conversion of heat energy to work requires that we create heat sources with the highest possible temperature. In practice the maximum temperature of a heat source is limited more

**Figure 7.12** Available and Unavailable Energy for a Varying-Temperature Heat Source

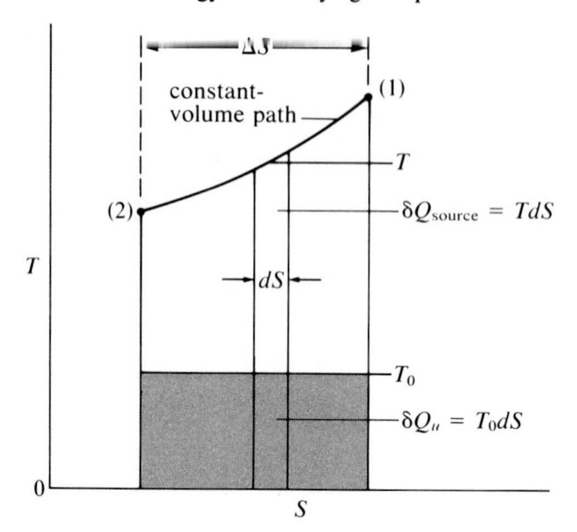


by the nature of materials of the components that contain and convey the working fluid of the heat engine than by our ability to create high temperatures. A limiting factor is typically the diminishing strengths that materials exhibit as temperature increases. Practical limits for continuous operation are, at the present time, in the range of 1500 to 2500 F, with the exact value depending on the particular type of energy conversion device considered.

The preceding discussion dealt exclusively with available energy in relation to constant-temperature heat sources, that is, heat reservoirs. It is of interest to examine heat sources for which the temperature varies since these are also encountered. The means for determining the available part of heat transfer from a source with a changing temperature can be illustrated by considering as a heat source a rigid tank containing a gas initially at $T_1$, a temperature greater than $T_0$. As heat transfer from the gas takes place, the gas temperature will decrease. Therefore, to determine the available and unavailable portions of the heat transferred from the tank, we can think of a Carnot engine (or a series of Carnot engines) that operates between the source at a varying temperature and the surroundings at the fixed temperature $T_0$. For our illustration the varying source temperature is described on *T-S* coordinates by a constant-volume path through state (1), as shown in Figure 7.12. When the source temperature is *T*, the Carnot engine receives heat $\delta Q = TdS$ and rejects heat $T_0dS$. These heat transfers are depicted as areas on the *T-S* diagram in Figure 7.12. The reversible work produced is the available energy. Therefore

$$\delta W_{rev} = \delta Q_a = (T - T_0)dS$$

or

$$Q_a = \int \delta Q_a = \int_2^1 (T - T_0)dS = \int_2^1 TdS - \int_2^1 T_0 dS$$

$$Q_a = Q_{source} - T_0 \Delta S \tag{7.19}$$

where $Q_{source}$ is the heat received by the Carnot engine from the varying-temperature source. As noted by the integral, this heat transfer is represented by the area below path (2) to (1) in Figure 7.12. We note also that $T_0 \Delta S$ is the heat rejected by the Carnot engine and therefore is the unavailable portion of $Q_{source}$. This is the same result that was obtained for the unavailable energy when the source was a heat reservoir.

The following examples illustrate some of the features of the concept of available and unavailable energy. Chapter 8 extends second-law concepts to additional topics on available energy, maximum work, and irreversibility.

## Example 7.m

Air initially at a pressure of 60 psia and 300 F is contained in an insulated rigid tank that has a volume of 100 ft³. The surroundings are at 40 F. What is the available energy when the tank containing the air is viewed as a heat source?

**Solution:**

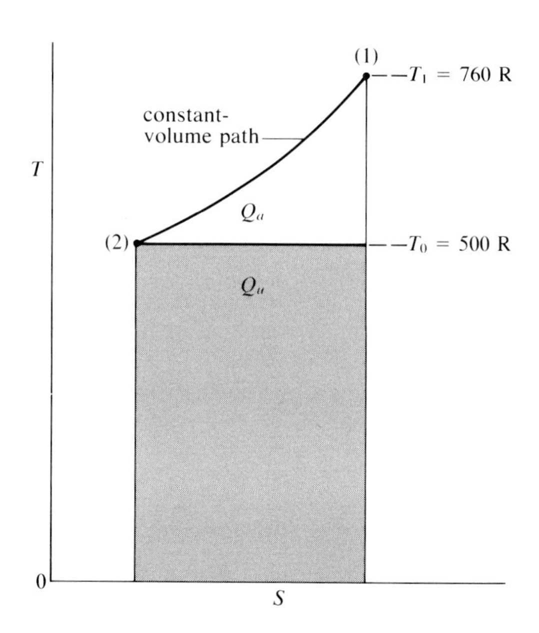


To utilize the tank as a heat source we must remove the insulation. To visualize the production of the available energy we imagine that the heat from the tank is supplied to a Carnot engine operating between the tank and the surroundings. As heat transfer occurs, the air in the tank will undergo a decrease in temperature. As long as the air temperature is greater than $T_0$ there is a potential to produce work. Thus to compute the total available energy we allow the tank temperature to decrease to the lowest possible value, namely, $T_0$. The *T*-*S* diagram shows $Q_a$ and $Q_u$ as areas. To compute $Q_a$ we apply eq. (7.19). We compute

$Q_{source}$ by application of the first law to the air in the tank. Only heat energy crosses the boundary of the air (the tank wall). Therefore

$$
\begin{aligned}
{}_1Q_2 &= U_2 - U_1 = m(u_2 - u_1) \\
&= \frac{p_1 V_1}{R\,T_1} C_v(T_2 - T_1) \\
&= \frac{60(144)100}{(1545/29)(760)}(0.172)(500 - 760) = -954 \text{ B}
\end{aligned}
$$

$Q_{source}$ is then 954 B. By eqs. (7.19) and (7.8c)

$$
\begin{aligned}
Q_a &= Q_{source} - mT_0|s_2 - s_1| \\
&= 954 - (500)(21.34)(0.172)\, ln\,\frac{760}{500} = 186 \text{ B}
\end{aligned}
$$

The unavailable energy $Q_u$ is 768 B.

---

## Example 7.n

Consider a process in which heat from a reservoir at 300 C is supplied to steam flowing in a tube in a steady-flow manner. The steam undergoes a constant-pressure process at 0.70 MPa from a temperature of 170 to 250 C. Changes in kinetic and potential energies are small. The surroundings are at 27 C.

(a) What is the available energy for the heat as it is supplied at the reservoir temperature?
(b) What is the available energy after the heat transfer to the steam has taken place?

**Solution:**

We must first determine the amount of heat transferred. The process that the steam undergoes is shown as process (1) to (2) on the *T-s* diagram. This process is treated as internally reversible. With work equal to zero, the heat transfer for the steady-flow process is, by the first law,

$$
\begin{aligned}
{}_1q_2 &= h_2 - h_1 \\
&= 2953.6 - 2775.6 = 178 \text{ kJ/kg}
\end{aligned}
$$

Properties are obtained from Appendix B.

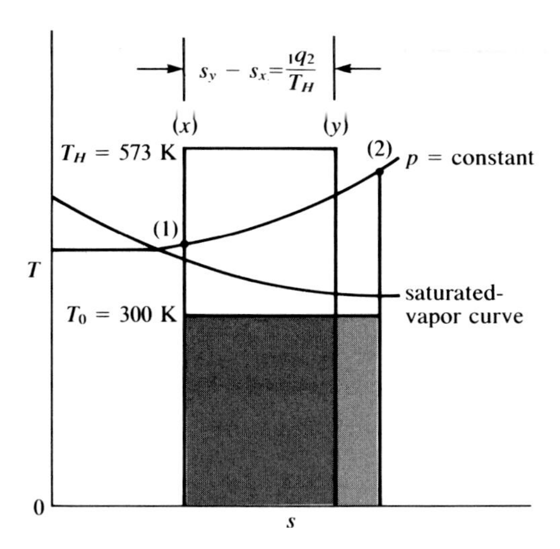


(a) The available energy at the reservoir temperature can be determined by use of the $T$-$s$ diagram. The diagram represents a Carnot engine receiving $q_H = {_xq_y} = {_1q_2}$ at $T_H$ and rejecting heat to the surroundings at $T_0$. Using eq. (7.19)

$$\begin{aligned} q_a &= q_H - T_0\Delta s = q_H - T_0\frac{q_H}{T_H} = q_H\left(1 - \frac{T_0}{T_H}\right) \\ &= {_1q_2}\left(1 - \frac{T_0}{T_H}\right) = 178\left(1 - \frac{273 + 27}{273 + 300}\right) \\ &= 84.8 \text{ kJ/kg} \end{aligned}$$

The unavailable energy is

$$q_u = 178 - 84.8 = 93.2 \text{ kJ/kg}$$

This is represented by the dark-shaded area on the $T$-$s$ diagram and is equal to the product $T_0\,(s_y - s_x)$.

(b) We note that the actual process involves the irreversible transfer of 178 kJ/kg of heat from the reservoir to the steam. The fact that this process is irreversible has no bearing on the calculation procedure to determine the available portion of the heat after the heat has been transferred to the steam. We now visualize the reversible transfer of 178 kJ/kg of heat from the steam to a series of Carnot engines that operates between the steam flowing in the tube and the surroundings. Application of eq. (7.19) to the constant-pressure process (1) to (2) yields

$$\begin{aligned} q_a &= {_1q_2} - T_0\,(s_2 - s_1) \\ &= 178 - (300)(7.1053 - 6.7354) \\ &= 67 \text{ kJ/kg} \end{aligned}$$

and

$$q_u = 178 - 67 = 111 \text{ kJ/kg}$$

The unavailable energy is equal to $T_0(s_2 - s_1)$ and is the sum of the dark- and light-shaded areas on the $T$-$s$ diagram. From the answers to parts (a) and (b) we see that the available energy has decreased during the process by 84.8 − 67 = 17.4 kJ/kg. The $T$-$s$ diagram is particularly useful in dealing with the available energy quantities. As has been noted, ${}_1q_2 = {}_xq_y$. Hence the areas for each are equal on the $T$-$s$ diagram. The area given by $T_0(s_2 - s_1)$ is greater than that given by $T_0(s_y - s_x)$. Therefore, by eq. (7.19) it is evident without calculations that there will be a decrease in available energy during the heat-transfer process. This is in agreement with previous discussion.

## Example 7.o

The diagram shows the arrangement for a steady-flow steam cycle. Although the components are the same as those for a simple steam-electric power plant cycle, the cycle described in this example is, for simplicity, somewhat different than such cycles, which are discussed in Chapter 9. The table gives information about the states at the various sections on the diagram. The processes within the cycle are assumed to be internally reversible and the processes through the turbine and the pump are adiabatic. Heat is supplied to the cycle from a reservoir at 700 F and rejected to the surroundings at 40 F by cooling water being passed from the surroundings through the condenser.

(a) Prepare a first-law accounting of energy for the cycle.
(b) Prepare an available-energy (second-law) accounting for the cycle.
(c) Compare the results for (a) and (b).

| State | $p$, psia | $t$, F | Condition |
|---|---|---|---|
| 1 | 98 | | sat. liq. |
| 2 | 98 | 660 | |
| 3 | 1 | | |
| 4 | 1 | | $s_4 = s_1$ |

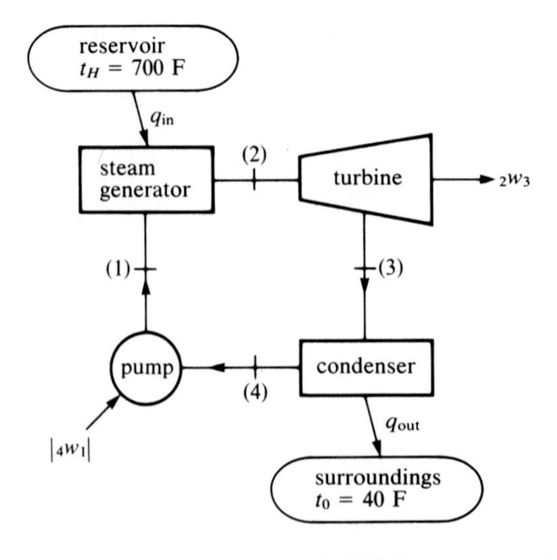

**Solution:**

The following table listing the required properties was completed using the *Steam Tables*. The $T$-$s$ diagram further describes the cycle and shows the reservoir and surroundings temperatures.

| State | $p$, psia | $t$, F | $s$, B/(lbm R) | $h$, B/lbm | Condition |
|---|---|---|---|---|---|
| 1 | 98 | 326.40 | 0.47247 | 297.09 | sat. liq. |
| 2 | 98 | 660 | 1.7880 | 1359.4 | superheat |
| 3 | 1 | 101.70 | 1.7880 | — | mixture |
| 4 | 1 | 101.70 | 0.47247 | — | mixture |

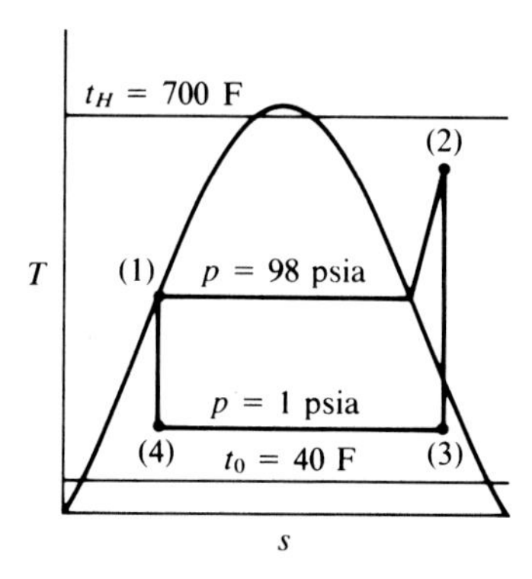


(a) Applying the steady-flow energy equation and assuming all changes in kinetic and potential energies are zero, we get

$$\begin{aligned} q_{in} &= {}_1q_2 = h_2 - h_1 \\ &= 1359.4 - 297.09 \\ &= 1062 \text{ B/lbm} \end{aligned}$$

Also

$$\begin{aligned} q_{out} &= |{}_3q_4| = T_3(s_3 - s_4) \\ &= (460 + 101.7)(1.788 - 0.47247) \\ &= 739 \text{ B/lbm} \end{aligned}$$

$$\begin{aligned} w_{net} &= q_{in} - q_{out} \\ &= 1062 - 739 = 323 \text{ B/lbm} \end{aligned}$$

The thermal efficiency for the cycle is then

$$\eta = \frac{w_{net}}{q_{in}} = \frac{323}{1062} = 0.304$$

(b) The available part of $q_{in}$ when delivered at the reservoir temperature is

$$(q_a)_{Res} = q_{in} - T_0\Delta s = q_{in} - T_0 \frac{q_{in}}{T_H}$$
$$= 1062\left(1 - \frac{500}{1160}\right) = 604 \text{ B/lbm}$$

The available part of $q_{in}$ after it has been transferred to the steam in process (1) to (2) is

$$(q_a)_{1,2} = q_{in} - T_0(s_2 - s_1)$$
$$= 1062 - (500)(1.788 - 0.47247) = 404 \text{ B/lbm}$$

The available part of the heat rejected in process (3) to (4) is

$$(q_a)_{3,4} = q_{out} - T_0(s_3 - s_4)$$
$$= 739 - (500)(1.788 - 0.47247) = 81 \text{ B/lbm}$$

This completes the calculations of the available energy quantities.

(c) The results for parts (a) and (b) are displayed graphically below. All quantities are in B/lbm units.

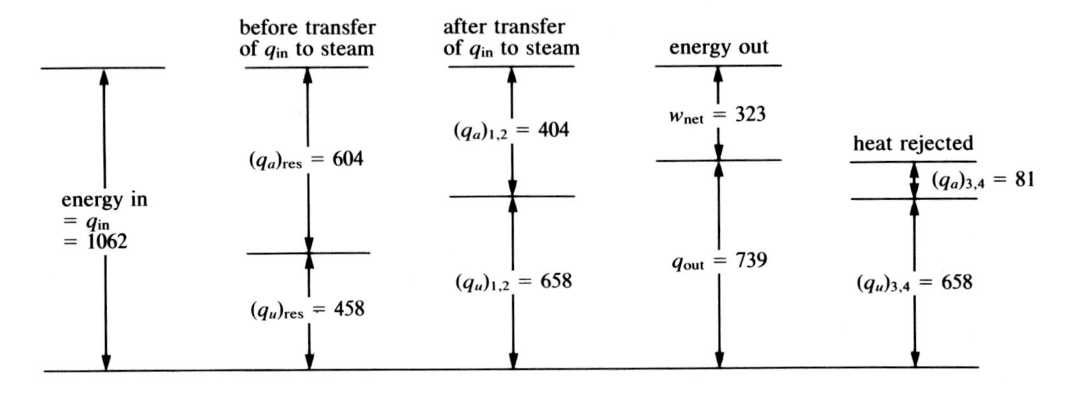


The following observations can now be made.

1. $(q_a)_{Res}/q_{in} = \eta_{rev} = 604/1062 = 0.569$
2. $w_{net}$ is 30.4 percent of $q_{in}$, while $w_{net}$ is 53.5 percent of $(q_a)_{Res}$, $(323/604 = 0.535)$. Comparison of $w_{net}$ with $(q_a)_{Res}$ is valid since $(q_a)_{Res}$ is the maximum net work obtainable from $q_{in}$ delivered at the reservoir temperature. The system produces about half of the maximum possible net work.
3. $w_{net}$ is 80 percent of $(q_a)_{1,2}$, $(323/404 = 0.80)$. This comparison is useful in assessing the extent to which the cycle itself produces work from the available energy transferred to the steam.
4. $q_{out}$ is 69.6 percent of $q_{in}$. However, only 11 percent of $q_{out}$ $(81/739 = 0.11)$ could be converted to work. Hence we see that the relatively large amount of heat rejected by the cycle is low-grade energy.

From the preceding example we see that the second-law analysis via the available–unavailable energy concept contributes significantly to the understanding of the conversion of heat to work in cyclic devices. The method yields information that the first law alone does not provide and allows us to compare the various energy quantities for the cycle in a number of different ways. While such comparisons are interesting and useful, thermal efficiency (net work divided by heat supplied) is widely used to describe heat-engine performance since, among other factors, economic considerations are nearly always important and the heat energy supplied is obtained from a fuel that is usually a costly item.

## 7–10 Summary

The basis for the definition of the property entropy is the equality of Clausius, which has its foundation in the second law. The definition of entropy in differential form is

$$dS = \left(\frac{\delta Q}{T}\right)_{\text{rev}}$$

For a fixed mass undergoing a change in state, the entropy change is

$$S_2 - S_1 = \int_1^2 \left(\frac{\delta Q}{T}\right)_{\text{rev}}$$

or on a unit mass basis

$$s_2 - s_1 = \int_1^2 \left(\frac{\delta q}{T}\right)_{\text{rev}}$$

The entropy change between any two states can be determined by integration of either of the above equations along a reversible path between the two states. However, since entropy is a property, any process, reversible or irreversible, between the two states will result in the same change in entropy.

The inequality of Clausius leads to the expression

$$dS \geq \frac{\delta Q}{T}$$

or

$$S_2 - S_1 \geq \int_1^2 \frac{\delta Q}{T}$$

and on a unit mass basis

$$s_2 - s_1 \geq \int_1^2 \frac{\delta q}{T}$$

From the definition of entropy it is evident that the equal sign applies to reversible processes. The inequality applies to processes that are internally irreversible. The heat transfer in the equations is then that for the actual internally irreversible process. From the expression for entropy it is seen that an adiabatic reversible process produces no change in entropy. Such processes are termed isentropic processes. The equations above that are applicable to reversible processes are also applicable to processes that are only internally reversible.

The *Tds* equations

$$Tds = du + pdv$$
$$Tds = dh - vdp$$

are useful relations between properties and apply to all pure substances. When applied to perfect gases with constant specific heats, the following expressions for entropy change are obtained.

$$s_2 - s_1 = C_v \ln \frac{T_2}{T_1} + R \ln \frac{v_2}{v_1}$$

$$s_2 - s_1 = C_p \ln \frac{T_2}{T_1} - R \ln \frac{p_2}{p_1}$$

The equation $pv^\gamma =$ constant describes a constant entropy path on $p$-$v$ coordinates for perfect gases with constant specific heats.

Entropy values based on an appropriate reference are tabulated with other properties in tables of properties for many pure substances encountered in engineering thermodynamics. Such tabulations are necessary because of the complex relations that exist between properties of substances that do not behave as perfect gases.

Heat transfer for a reversible process is given by

$$({}_1Q_2)_{\text{rev}} = \int_1^2 TdS$$

Therefore, heat transfer for a reversible process can be interpreted as an area on a temperature–entropy diagram.

For a fixed mass undergoing an adiabatic process in a closed system or in a single-stream steady-flow system

$$(dS \geq 0)_{\text{adiabatic}}$$

Since an isolated system is an adiabatic system, the above equation applies to isolated systems also. An isolated system may include subsystems that interact with each other. The preceding equation is a statement of the entropy principle. The equation restricts processes for adiabatic systems to those for which entropy either increases or remains constant.

A measure of the usefulness of heat energy to produce work is available energy. Available energy is defined as that part of the heat from a source that could be converted into work by a reversible engine (a Carnot engine, for example) operating between the heat source and the surroundings at temperature $T_0$. The unavailable energy is that part of the heat from the source that is rejected to the surroundings by the reversible engine. Therefore

$$Q_{\text{source}} = Q_a + Q_u$$

The unavailable energy for any process of a system that serves as a heat source is given by $T_0\,\Delta S$, where $\Delta S$ is the entropy change for the process. The concepts of available and unavailable energy are second-law concepts and, as such, provide understanding of the conversion of heat to work that is not evident from the first law.

## Selected References

Clausius, R. *The Mechanical Theory of Heat* (trans. by W. R. Browne). London: Macmillan, 1879.
Cork, J. M. *Heat*, 2nd ed. New York: Wiley, 1942.
Keenan, J. H. *Thermodynamics*. New York: Wiley, 1941.
Lee, J. F., and F. W. Sears. *Thermodynamics*. Reading, Mass.: Addison-Wesley, 1955.
Mott-Smith, M. *The Story of Energy*. New York: Appleton-Century, 1934.
Obert, E. F. *Thermodynamics*. New York: McGraw-Hill, 1948.
Obert, E. F., and R. A. Gaggioli. *Thermodynamics*, 2nd ed. New York: McGraw-Hill, 1963.
Potter, J. H. "On the Inequality of Clausius." *Combustion*, June 1964.
Roberts, J. K. *Heat and Thermodynamics*, 5th ed. (revised by A. R. Miller). New York: Interscience, 1960.
Stoever, H. J. *Engineering Thermodynamics*. New York: Wiley, 1951.
Wark, K. *Thermodynamics*, 3rd ed. New York: McGraw-Hill, 1977.
Zemanskey, M. W., and R. H. Dittman. *Heat and Thermodynamics*, 6th ed. New York: McGraw-Hill, 1981.

---

## Problems

7.1 A closed system of 1 lbm of air undergoes a cycle consisting of the following reversible processes.

$1 \rightarrow 2$ Adiabatic
$2 \rightarrow 3$ Constant volume
$3 \rightarrow 1$ Constant temperature

Initially the system is at 100 psia and 800 R. The volume at state (2) is twice that at state (1). Calculate the $\int(\delta Q/T)$ for each reversible process and the summation for the cycle. For this problem let $C_v = 0.172$ B/lbm R, $C_p = 0.2408$ B/lbm R, $\gamma = 1.400$, and $R = 53.53$ ft lbf/lbm R.

7.2 One kilogram of helium is at 500 K and 690 kPa. A second state of the helium is at 280 K and 140 kPa. For a closed system evaluate $\int(\delta Q/T)$ between the two states (a) along a constant-pressure and constant-volume reversible-path combination and (b) along an isothermal and constant-pressure reversible-path combination. Sketch the paths on *p-v* coordinates. Compare the results for (a) and (b).

7.3 A perfect gas has a molecular weight of 28.00 and its specific heat $C_v$ is

$$C_v = 0.200 + 0.3 \times 10^{-6}T^2$$

where $C_v$ is in B/(lbm R) and $T$ is Rankine temperature. A lbm of this gas undergoes a change in state from 100 psia and 1200 R to 20 psia and 560 R. Evaluate $\int(\delta q/T)$ for the following combinations of closed-system reversible paths and compare the results.

(a) Constant pressure, then constant volume
(b) Constant temperature, then constant pressure
(c) Adiabatic, then constant temperature

For each, sketch the paths on a single $p$-$v$ diagram.

7.4 A rigid vessel contains 2.3 lbm of air at an initial pressure and temperature of 15 psia and 500 R. Stirring (paddle-wheel) work in the amount of 79 B is done on the air during a process. Also, during the process 10 B of heat transfer takes place from the system to the surroundings. Calculate the entropy change of the air for the process.

7.5 A closed system of 0.46 kg of air initially at 275 kPa and having a specific volume of 0.312 m³/kg undergoes a process during which 95 kJ heat transfer to the system occurs and 43,680 J of work is done by the system. At the final state the pressure is 172 kPa. Determine the entropy change of the air.

7.6 Air expands through a turbine from 100 psia, 2000 R, to 15 psia, 1220 R. There is heat loss of 3.00 B/lbm from the air during the process. Calculate the work and the entropy change per lbm of air. Assume that kinetic and potential energy changes are negligible.

7.7 A closed system of 0.260 kg of nitrogen, $N_2$, undergoes the following reversible cycle:

1 → 2 Constant pressure
2 → 3 Constant volume
3 → 1 Adiabatic

If $T_1$ is 390 K and $T_3$ is 278 K, what is the temperature at state (2)?

7.8 During a reversible isothermal process the entropy of 0.157 lbm of hydrogen, $H_2$, increases by 0.0315 B/(lbm R). What is the ratio of the final to initial pressure of the closed system?

7.9 An unknown perfect gas has a molecular weight of 23.02. A closed system of the gas undergoes a reversible, constant-volume process during which the entropy increases 0.0950 B/lbm R and the internal energy increases 58.5 B/lbm. During the process the pressure increases from 84 psia to 113 psia. Determine the initial specific volume of the gas and the specific heat ratio, $\gamma$, for the gas.

7.10 A rigid, perfectly insulated container is divided into two equal volumes by a partition. One side is completely evacuated and the other contains air at 700 kPa and 24 C. The partition is punctured and the air comes to equilibrium. Determine (a) the pressure and temperature at the final equilibrium state and (b) the entropy change per kilogram of air.

7.11 One pound mass of an unknown perfect gas, initially at 15 psia and 70 F, undergoes a process to a final state where the pressure is 100 psia. For the gas $C_p$ is 1.210 B/lbm R and the specific heat ratio is 1.660. During the process the entropy of the gas increases by 0.2500 B/lbm R. What is the specific volume of the gas at the second state?

7.12 Complete the following table for $H_2O$:

| $p$, psia | $t$, F | $h$, B/lbm | $s$, B/(lbm R) | $v$, $ft^3$/lbm |
|---|---|---|---|---|
| 100.0 | | 1095.0 | | |
| 500 | 250 | | | |
| | 300 | | 1.5493 | |
| 1400 | | | | 0.5815 |
| | −5 | | 0.8767 | |
| 20 | | 68.04 | | |

7.13 A 6-$ft^3$ tank contains 2 lbm of steam at 500 F. If the steam is cooled to 300 F, by how much does its entropy change?

7.14 A closed system of 1 kg of $H_2O$ is initially at 100 kPa and 120 C. It undergoes a reversible, constant-temperature process during which 262 kJ heat transfer from the system takes place. Find the final pressure and calculate work for the process.

7.15 Solve the above problem if nitrogen instead of water is the working fluid.

7.16 Steam is supplied to a nozzle at 100 psia, 350 F, with negligible velocity. Expansion through the nozzle is reversible and adiabatic to a pressure of 14 psia at the outlet section. Determine the outlet area if the flow rate is 1.379 lbm/sec.

7.17 Solve the above problem if air instead of water is the working fluid.

7.18 A steam turbine operates reversibly and adiabatically. Steam enters the machine at the rate of 100,000 lbm/hr at 450 psia and 720 F. At the turbine exhaust the pressure is 1.00 psia and the velocity 500 ft/sec. If the inlet velocity is negligible, what is the power, in kilowatts, delivered by the turbine?

7.19 A closed system of $H_2O$, initially at 70 F, undergoes a reversible constant-pressure process at 60 psia. If heat transfer to the steam is 1200 B/lbm, what is the entropy change of the steam during the process?

7.20 Steam initially at 4.10 MPa and 600 C expands isentropically until it is saturated vapor ($x = 100$ percent). How much work will each kilogram of steam do (a) if it is a closed system and (b) if it is a steady-flow process with negligible change in kinetic energy? Show the process on a $T$-$s$ diagram.

7.21 Steam at 220 psia and 540 F is throttled to 160 psia and then enters a turbine. Expansion through the turbine is reversible and adiabatic to a pressure of 1.20 psia at the exhaust. The velocities entering and leaving the turbine are negligible. Determine (a) the entropy change during the throttling process and (b) the turbine work per pound of steam. Show the processes on $T$-$s$ and $h$-$s$ diagrams.

7.22 Solve the above problem if air instead of water is the working fluid.

7.23 For steam at 1.00 MPa ($t_{sat}$ = 179.91 C) the latent heat of vaporization, $h_{fg}$, is 2015.3 kJ/kg. Using eq. (7.11), calculate $s_{fg}$ for steam at 1.00 MPa. Compare the result with data from the *Steam Tables* in Appendix B.

7.24 At 10 F the entropy change $s_{fg}$ for refrigerant F-12 is 0.144026 B/lbm R. Using eq. (7.11), determine $h_{fg}$ for the substance at 10 F. Compare your result with tabulated data in Appendix E.

7.25 A closed system of 3 lbm of air, initially at 560 R and occupying a volume of 36.0 ft$^3$, undergoes a reversible polytropic process for which $n = 1.285$. The final temperature of the air is 910 R. Determine the heat transfer and entropy change of the air during the process. Show the process on $p$-$v$ and $T$-$s$ coordinates.

7.26 A closed system of 1 lbm of $H_2O$, initially saturated vapor at 196 psia, undergoes the following internally reversible processes:

$1 \rightarrow 2$ Constant temperature
$2 \rightarrow 3$ Constant pressure
$3 \rightarrow 1$ Constant entropy

During process $1 \rightarrow 2$, 200 B/lbm heat transfer to the system takes place. Calculate the thermal efficiency for the cycle. Calculate the thermal efficiency of a Carnot cycle having the same maximum and minimum temperatures as the given cycle. Explain the differences in the two efficiencies. Sketch a complete $T$-$s$ diagram.

7.27 One kilogram of air is initially at 105 kPa and 21 C. The closed system undergoes a reversible, polytropic process to a final pressure of 1.4 MPa. During the process the entropy of the air increases by 0.315 kJ/(kg K). Calculate work for the process.

7.28 Consider a Carnot heat engine using $H_2O$ as the working fluid. The state at the beginning of the heat-reception process is in the liquid–vapor mixture region at a quality of 20 percent and the state at the end of this process is in the superheat region. The heat-rejection process begins in the superheat region. Sketch the cycle on $p$-$v$ and $T$-$s$ coordinates.

7.29 Except as noted, at least one property remains constant during each of the processes for the following cycles. (Note that in the mixture region pressure and temperature are not independent.)

(a) Sketch on a $T$-$s$ diagram the reversible $H_2O$ cycle shown on the $p$-$v$ diagram. Show saturation lines.

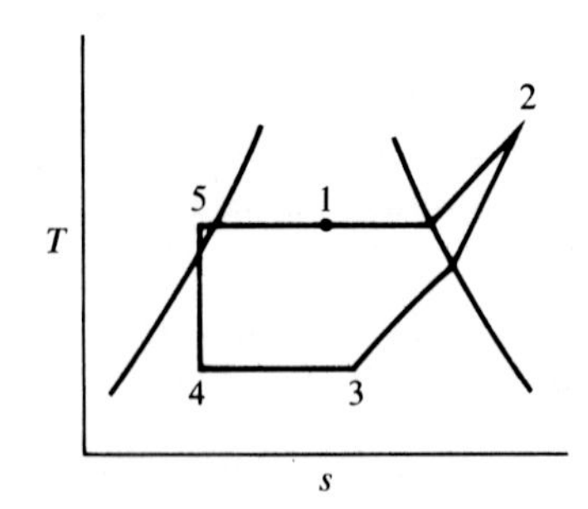

(b) Sketch on a $p$-$v$ diagram the reversible $H_2O$ cycle shown on the $T$-$s$ diagram. Show saturation lines.

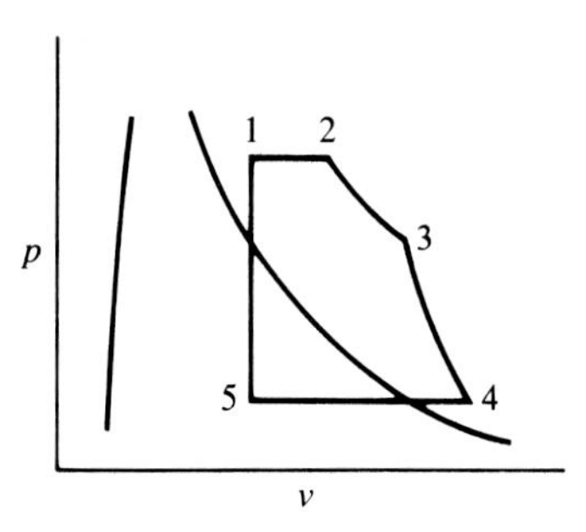


(c) Sketch on a $T$-$s$ diagram the reversible cycle shown on the $p$-$v$ diagram. The fluid is a perfect gas with constant specific heats.

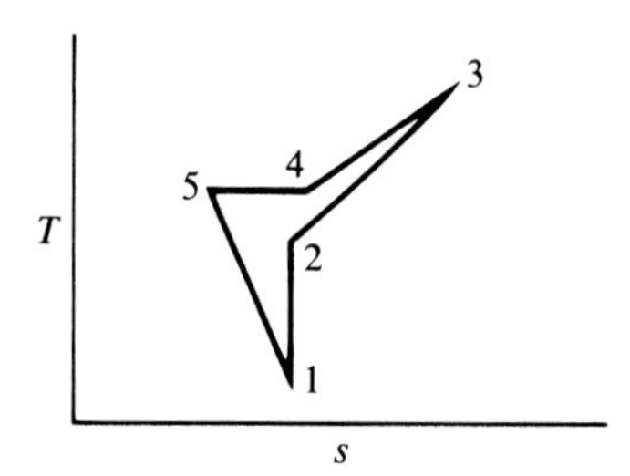


(d) Sketch on a $p$-$v$ diagram the reversible cycle shown on the $T$-$s$ diagram. The fluid is a perfect gas with constant specific heats. One property remains constant during each process, *except* between 5 and 1.

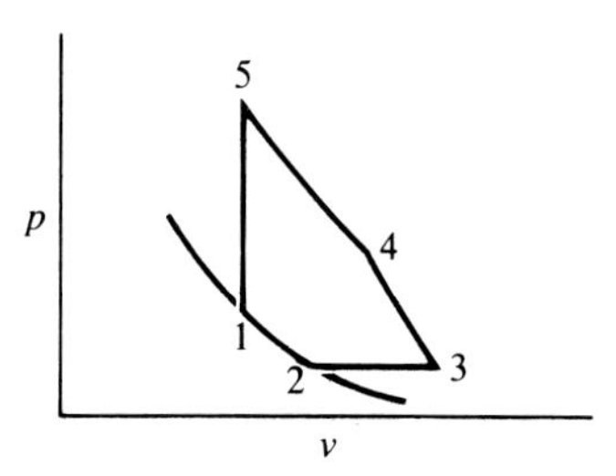


7.30 For a closed system of air with constant specific heats, sketch $p$-$v$ and $T$-$s$ diagrams showing:

(a) Several lines of constant temperature, indicating the highest and lowest temperatures
(b) Several reversible adiabatic paths
(c) Several lines of constant pressure, indicating highest and lowest
(d) Several lines of constant volume, indicating highest and lowest
(e) Through a given state (1) sketch the polytropic $pv^{1.3}$ = constant and $pv^{1.6}$ = constant

(f) Answer the following on the basis of the diagrams:
  (1) Does temperature increase or decrease during a reversible adiabatic expansion?
  (2) Is the internal energy change positive or negative during a reversible constant-pressure compression?
  (3) Does temperature increase during a compression along $pv^{1.3}$ = constant?
  (4) Is heat transfer to or from the system during an expansion along $pv^{1.6}$ = constant?

7.31 Using a *T-s* diagram, show for a Carnot cycle that

(a) The thermal efficiency is $(T_H - T_L)/T_H$.
(b) The coefficient of performance of a Carnot refrigerator is $T_L/(T_H - T_L)$.

7.32 A closed system of 1 lbm of air undergoes the following reversible cycle:

$1 \rightarrow 2$ Constant volume from 15 psia, 530 R, to 3000 R
$2 \rightarrow 3$ Constant temperature with $s_3 - s_2$ equal to 0.0515 B/lbm R
$3 \rightarrow 4$ Polytropic with $pv^{1.30}$ = constant
$4 \rightarrow 1$ Constant pressure

Sketch the cycle on *p-v* and *T-s* coordinates. Determine the pressure at state (3) and the entropy change during the polytropic process. Calculate $Q_{in}$ for the cycle.

7.33 One kilogram of air, initially at 1.0 MPa and 260 C, undergoes a polytropic process along the path $pv^{1.673}$ = constant to a final state at 200 kPa.

(a) Calculate the entropy change of the air during the process.
(b) Sketch the path on *p-v* and *T-s* coordinates.
(c) Is heat transfer positive, negative, or zero for the process?

7.34 Sketch the following processes for a perfect gas steady-flow cycle on *T-s* and *p-v* coordinates.

$1 \rightarrow 2$ Constant pressure with increasing entropy
$2 \rightarrow 3$ Throttling to $p_3$ less than $p_2$
$3 \rightarrow 4$ Isentropic, $p_4$ less than $p_3$
$4 \rightarrow 5$ Constant pressure
$5 \rightarrow 1$ Isentropic

All processes except the throttling process are internally reversible. What is the net change in entropy of the gas for the cycle, that is, what is $\oint ds$? Will the integral of $\delta q/T$ for the cycle be equal to zero? State the basis for each answer.

7.35 Consider an adiabatic closed system process or series of adiabatic processes occurring between states (1) and (2). (a) State any limitations regarding state (2). (b) Show that the net work for any adiabatic process between the two states is the same. (c) Let the adiabatic system be a piston–cylinder paddle-wheel combination such as that shown in Figure 2.2b. Describe several possible adiabatic processes between states (1) and (2) for the system. Describe the nature of the work involved for each.

7.36 Student A argues that, as a result of the entropy principle, entropy cannot decrease regardless of the process involved. Student B disagrees, citing examples of processes when entropy does decrease. Give several examples in support of the position taken by student B.

7.37 The following data are claimed to exist for the adiabatic flow of air through a nozzle. Inlet conditions: $p_1 = 35$ psia, $t_1 = 340$ F, $V_1 = 245$ ft/sec. Exit conditions: $p_2 = 20$ psia, $V_2 = 1320$ ft/sec. (a) On the basis of the given data, what is the temperature at (2)? (b) Is the flow described by the given data and the calculated temperature possible? If so, is the process reversible or irreversible? (c) Assuming that the given inlet conditions and the exit pressure are correct, what is the maximum possible exit velocity?

7.38 Air flows in a steady-flow process through a perfectly insulated pipe with a constant cross-sectional area. The inside pipe diameter is 4.13 inches. At a particular section the air is at 50 psia, 760 R and has a velocity of 350 ft/sec. Due to friction there will be a pressure decrease in the direction of flow. The pressure drop is accompanied by a decrease in temperature for the type of flow involved. Assume a series of states downstream of the given section, each at a progressively lower temperature. Calculate the pressure at each downstream state and then calculate the entropy change of the air between the given state and the downstream state. Graph the results on temperature-entropy coordinates beyond the point where the entropy change is a maximum. What is the maximum possible velocity of the air? At what pressure and temperature does this velocity occur?

7.39 A rigid container holding 10 pounds of air initially at 50 psia and 800 R serves as a heat source for a heat engine. The engine rejects heat to the surroundings at 60 F. Calculate the maximum net work the engine could deliver; that is, calculate the available energy.

7.40 Heat is transferred from a reservoir at 2000 F to water undergoing a phase change from saturated liquid to saturated vapor at a pressure of 200 psia in a steady-flow process. The temperature of the surroundings is 80 F. On the basis of a unit mass of water determine the available energy of (a) the heat transferred at the reservoir temperature, and (b) the energy received by the boiling water. Show all quantities as areas on a *T-s* diagram.

7.41 A quantity of 3.2 kg of oxygen in a closed rigid container, initially at atmospheric temperature 21 C, is to be cooled by a refrigerator to −73 C. Determine the minimum work required to accomplish the cooling process. Assume that the oxygen behaves as a perfect gas.

7.42 One kilogram of nitrogen in a piston–cylinder assembly undergoes an internally reversible cycle consisting of the following processes.

1–2: Isentropic compression, $p_1 = 100$ kPa, $t_1 = 8$ C, $v_2 = 0.17v_1$
2–3: Constant-volume heating to $t_3 = 560$ C
3–4: Isentropic expansion to $v_1$
4–1: Constant-volume cooling

The cycle receives heat from a reservoir at 750 C and the temperature of the surroundings is 5 C. Prepare a first-law energy analysis and a second-law analysis (available and unavailable energy) for the cycle. Show all quantities as areas on a *T-s* diagram.

# 8

# Second-Law Analysis

*Procedures for establishing whether or not a process is reversible were discussed in Chapter 6. Also, the expression for the maximum thermal efficiency of a heat engine was developed in Chapter 6, as were expressions for the maximum coefficients of performance for refrigerators and heat pumps. A study of the usefulness of energy in the form of heat for producing work in Section 7–9 led to the concept of available and unavailable energy. Each of these topics has its foundation in the second law. With increasing attention directed to efficient use of energy, the engineer is prompted to optimize the use of heat energy and to seek maximum work output for work-producing processes and minimum work input for work-consuming processes. For each case a goal is to approach as closely as possible (within certain economic limits) a reversible process, or, stated in another way, to minimize irreversibilities.*

*In this chapter we will expand on these topics and develop procedures to calculate the optimum work for a process as well as a procedure to evaluate irreversibility quantitatively. We will also further examine energy-conversion processes from the second-law point of view. Since these procedures follow from second-law concepts, they constitute what is commonly called second-law analysis. We shall limit this analysis to processes for pure substances since parallel methods for chemically reactive systems require background in absolute entropy, a consequence of the third law of thermodynamics, which is not addressed in this book.*

### 8–1 Some Conditions for Second-Law Analysis

Any system undergoing a process operates within a surrounding atmosphere or environment. The atmosphere (environment) is of such an extent that finite heat transfer to or from it will produce no change in the state of the atmosphere. Thus the atmosphere is a heat reservoir that can serve as a heat source or sink. The terms environment and atmosphere will be used interchangeably. (The system's surroundings *may* also include work-producing or work-receiving devices or heat reservoirs and sources not at the temperature of the atmosphere.) In all cases we consider, the environment will be at its most stable state. Under this condition the environment alone cannot be a source of work. In a practical sense this means that the environment is at uniform pressure at any level in a gravitational field, at uniform temperature, and is at rest with respect to the inertial reference for any system in the environment. More generally, there are no potentials in the environment that would permit the environment alone to serve as a source of work. Additionally, the environment remains distinct from any system within it. There is no mass transfer between a defined system and its environment. In developing the expressions for maximum work, irreversibility, and so on, *the only heat transfer is between a system and its environment,* regardless of the nature of any actual process.

Figure 8.1

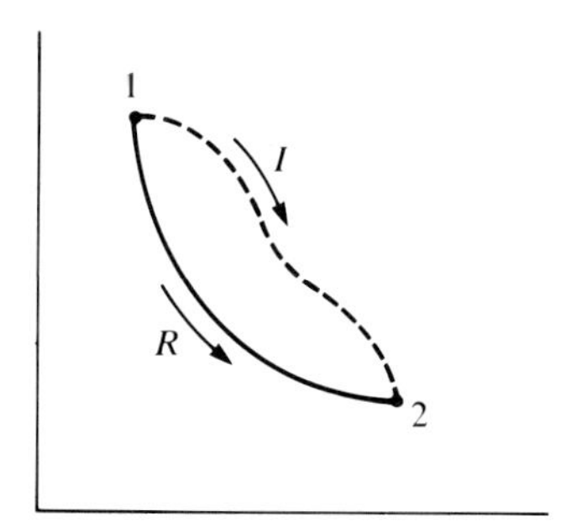


## 8–2 Maximum Work and Reversible Processes

Under the condition of heat exchange only with the atmosphere the maximum work for a system undergoing a given change of state will be realized during a reversible process. Consider any two equilibrium states such as (1) and (2) in Figure 8.1 with $R$ any reversible processes between the two states and $I$ any irreversible process between the same equilibrium states. The system can be any pure substance. *Assume* the work delivered during the irreversible process $I$ is greater than that for the reversible process $R$. With $R$ a reversible process, its direction can be reversed and a work input equal in magnitude to the work output during the original reversible process will be required. The result is a cycle consisting of the irreversible process $I$ from (1) to (2) and the reversed process $R$ from (2) to (1). With the original assumption that $W_I$ is greater than $W_R$, the cycle would have a net work output of magnitude $|W_I| - |W_R|$. For the cycle, heat exchange is with a single reservoir—the atmosphere—at a fixed temperature. The cycle, having a net work output while exchanging heat with a single reservoir at a fixed temperature, would result in a direct violation of the Kelvin-Planck statement of the second law. Thus we conclude the only assumption made is erroneous, and under the condition of heat exchange only with the atmosphere, no process for a given system can yield more work than that for a reversible process between two known equilibrium states.

By similar logic applied to processes requiring work done on a system, no process between two given equilibrium states can be accomplished with less work input (absolute value) than that of a reversible process when the only heat exchange is with the atmosphere.

Figure 8.2 shows any two reversible paths between any two equilibrium states for a given system. Heat exchange is with the atmosphere, only. Let us *assume* the work delivered during the process $R_1$ to be greater than that for the process $R_2$. By reversing the process $R_2$ we have a cycle consisting of the processes (1) to (2) along $R_1$ and (2) to (1) along $R_2$. A net work output $|W_{R_1}| - |W_{R_2}|$ would be realized during each cycle. Heat exchange is with a single reservoir, the atmosphere. The assumption that $W_{R_1}$ is greater than $W_{R_2}$ is erroneous because it produces a violation of the Kelvin-Planck statement of the second law. Next assume $W_{R_2}$ is greater than $W_{R_1}$. Following the same logic we find this assumption also would yield a violation of the second law. With (1) and (2) any two states of a pure substance and $R_1$ and $R_2$ any two reversible processes between those equilibrium states, we reach the only possible conclusion. When heat exchange is with the atmosphere only, the work is the same for all reversible processes between two given equilibrium states of a pure substance.

**Figure 8.2**

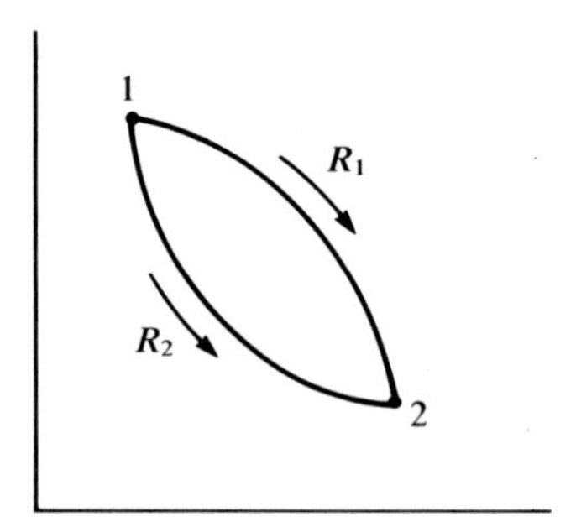


## 8–3 Evaluation of Maximum Work

To establish the maximum work for a system undergoing a process in a given atmosphere, we begin with a generalized open system, and, from the above discussion, consider a reversible change in state of the system. Maximum work for steady-flow and closed-system state changes then follows as a particular application of this general case. The most general case is illustrated in Figure 8.3, which shows an open system and its environment at the beginning of the process. In a manner identical to that used in developing the open-system energy equation, an inbound mass $dm_1$ crosses the open system boundary at section 1 and an outbound mass $dm_2$ leaves the open system, crossing section 2 during the time interval $d\tau$. During this time interval the open system will have work $\delta W_s$, which can be either positive or negative. The subscript $s$ refers to quantities associated with the open system.

For reversible heat transfer between the environment at $T_0$ and any part of the open system at local temperature $T$ ($T$ and $T_0$ generally differing by a finite amount), a reversible engine is utilized. Heat transfer $\delta Q_0$ from the environment to the working fluid of the reversible engine, work $\delta W_e$ for the reversible engine, and heat rejection $\delta Q_e$ at temperature $T$ from the working fluid of the reversible engine also take place during the time interval $d\tau$. Note that the heat transfer $\delta Q_e$ from the reversible engine and $\delta Q_s$, the heat transfer to the portion of the open system at $T$, are of equal magnitude but opposite sign; that is, $\delta Q_s = -\delta Q_e$. With $T$ varying over the open-system boundary, a multiplicity of such reversible engines would be required. We shall consider only one engine, and hence one location on the open-system boundary, to simplify the derivation. The summation for any number of reversible engines could be made by integration over the open-system boundary.

With heat exchange between the system and its environment only, the maximum work for the reversible process is

$$\delta W_{\text{max}} = \delta W_e + \delta W_s \quad \textbf{(a)}$$

The work of the reversible engine $\delta W_e$ can be expressed in terms of $\delta Q_s$ and the temperatures $T_0$ and $T$ using the definition of absolute thermodynamic temperature (eq. [6.8]) as

$$\frac{\delta Q_{\text{in}}}{\delta Q_{\text{out}}} = \frac{T_{\text{in}}}{T_{\text{out}}} = \frac{\delta W_e + \delta Q_s}{\delta Q_s} = \frac{T_0}{T} \quad \textbf{(b)}$$

**Figure 8.3** General System Exchanging Heat Reversibly with Its Environment

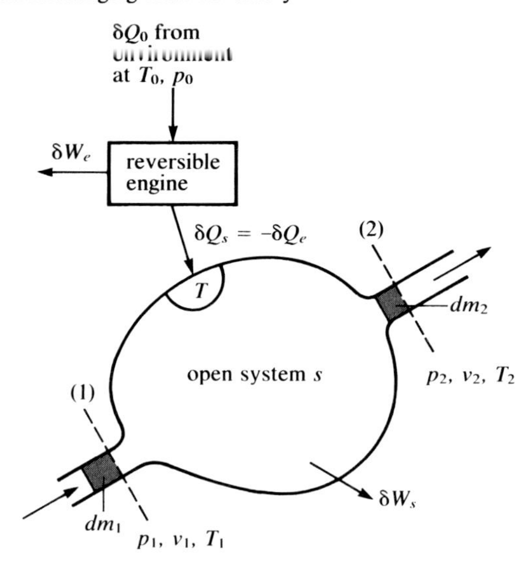


In Figure 8.3 the heat transfer is from the environment and to the open system, and thus $T_0$ is greater than $T$ for the illustrated direction of heat transfer. Rearranging the right-hand equality in (b) and solving for $\delta W_e$ we get

$$\delta W_e = \left(\frac{T_0}{T} - 1\right)\delta Q_s \tag{c}$$

Note that $\delta W_e$ always will be greater than zero if heat transfer to or from the system takes place and, further, $\delta W_e$ will never be less than zero if $T_0$ and $T$ are not equal. For the direction of heat transfer shown in Figure 8.3, $T_0$ is greater than $T$ and $\delta Q_s$ is positive. For $\delta Q_s$ to be from the system, and hence negative, $T$ must be greater than $T_0$. Equation (c), applied to these later conditions, yields a positive $\delta W_e$.

Combining eqs. (a) and (c), we get

$$\begin{aligned}\delta W_{\max} &= \delta W_s + \left(\frac{T_0}{T} - 1\right)\delta Q_s \\ &= \delta W_s - \delta Q_s + T_0\frac{\delta Q_s}{T}\end{aligned} \tag{d}$$

For the open system the general energy equation can be written as

$$\begin{aligned}\delta W_s - \delta Q_s = &-dE_s + (e_1 + p_1v_1)dm_1 \\ &- (e_2 + p_2v_2)dm_2\end{aligned} \tag{3.3}$$

Substituting from eq. (3.3) into eq. (d) yields

$$\delta W_{\max} = -dE_s + (e_1 + p_1 v_1)dm_1 - (e_2 + p_2 v_2)dm_2 + T_0 \frac{\delta Q_s}{T} \tag{e}$$

From Section 7–4, the differential entropy change for the open system is

$$dS_s = s_1 dm_1 - s_2 dm_2 + \left(\frac{\delta Q_s}{T}\right)_{\text{rev}}$$

which, when multiplied by $T_0$, becomes

$$T_0 dS_s = T_0 s_1 dm_1 - T_0 s_2 dm_2 + T_0 \left(\frac{\delta Q_s}{T}\right)_{\text{rev}}$$

or

$$T_0 \left(\frac{\delta Q_s}{T}\right)_{\text{rev}} = T_0 dS_s - T_0 s_1 dm_1 + T_0 s_2 dm_2 \tag{f}$$

Noting that $\delta Q_s$ in eq. (e) is for a reversible process, we combine eqs. (e) and (f) by solving each for $T_0\ (\delta Q_s/T)_{\text{rev}}$ to obtain

$$\delta W_{\max} = -\,dE_s + T_0 dS_s + (e_1 + p_1 v_1 - T_0 s_1)dm_1 - (e_2 + p_2 v_2 - T_0 s_2)dm_2 \tag{8.1}$$

Although eq. (8.1) has been developed for a single entering stream and a single leaving stream, it can be applied to multiple streams in a manner akin to that used for the general energy equation. For multiple streams

$$\delta W_{\max} = -dE_s + T_0 dS_s + \Sigma(e + pv - T_0 s)dm_{\text{in}} - \Sigma(e + pv - T_0 s)dm_{\text{out}} \tag{8.1a}$$

Keep in mind that eqs. (8.1) and (8.1a) are developed for a reversible change of state of a pure substance undergoing heat exchange only with the environment.

For a closed system eq. (8.1) reduces to

$$\delta W_{\max} = -\,d(E - T_0 S)$$

or, by integration,

$$W_{\max} = -\,(E_2 - E_1) + T_0(S_2 - S_1) \tag{8.2}$$

Expressed on a unit mass basis and noting that $e$ is equal to $u$ for a simple system, eq. (8.2) becomes

$$w_{\max} = -(u_2 - u_1) + T_0(s_2 - s_1) \tag{8.2a}$$

With $e$ equal to the sum of internal energy, kinetic energy, and potential energy, eq. (8.1), when applied to $m$ mass units undergoing a single-stream steady-flow process, integrates to

$$\begin{aligned} W_{\max} = & \left(mh_1 + \frac{mV_1^2}{2g_0} + m\frac{g}{g_0}z_1 - T_0S_1\right) \\ & - \left(mh_2 + \frac{mV_2^2}{2g_0} + m\frac{g}{g_0}z_1 - T_0S_2\right) \end{aligned} \tag{8.3}$$

since $dS_s$ and $dE_s$ are zero for steady-flow processes. This equation includes potential energy terms. For single-stream steady-flow processes with negligible potential energy changes, eq. (8.3) for a unit mass becomes

$$w_{\max} = \left(h_1 + \frac{V_1^2}{2g_0} - T_0s_1\right) - \left(h_2 + \frac{V_2^2}{2g_0} - T_0s_2\right) \tag{8.3a}$$

When more than one stream is involved in a steady-flow process, it is convenient to use a rate basis for calculations of the maximum work. Writing eq. (8.3) on a rate basis we get

$$\begin{aligned} \dot{W}_{\max} = & \Sigma\dot{m}\left(h + \frac{V^2}{2g_0} + \frac{g}{g_0}z - T_0s\right)_{\text{in}} \\ & - \Sigma\dot{m}\left(h + \frac{V^2}{2g_0} + \frac{g}{g_0}z - T_0s\right)_{\text{out}} \end{aligned} \tag{8.3b}$$

## Example 8.a

Suppose a simple, closed system consisting of a unit mass of a pure substance undergoes a reversible process from any state (1) to any other state (2) and, during the process, the only heat transfer is with the environment at $T_0$. By selecting a series of processes satisfying these conditions, show that the work is the same as $w_{\max}$ determined using eq. (8.2a).

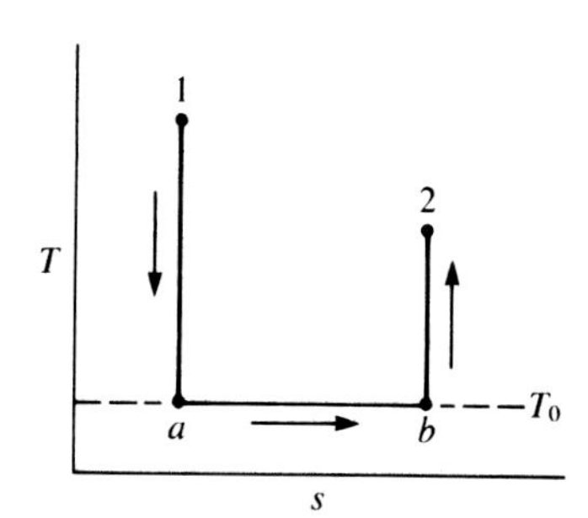

**Solution:**

A series of processes satisfying the stated conditions is the isentropic $1 \rightarrow a$, the isothermal $a \rightarrow b$ at $T_0$ during which reversible heat transfer with the environment takes place, and the isentropic path $b \rightarrow 2$. Applying the closed-system energy equation to each process we get

$$ {}_1q_a = u_a - u_1 + {}_1w_a = 0 $$

$$ {}_aq_b = u_b - u_a + {}_aw_b $$

$$ {}_bq_2 = u_2 - u_b + {}_bw_2 = 0 $$

Summing for the three processes yields

$$ {}_aq_b = u_2 - u_1 + {}_1w_a + {}_aw_b + {}_bw_2 $$

From the definition of change in entropy heat transfer during the reversible isothermal $a \rightarrow b$ is

$$ \int_a^b ds = \int_a^b \left(\frac{\delta q}{T}\right)_{\text{rev}} $$

or

$$ T_a(s_b - s_a) = {}_aq_b $$

Substituting for ${}_aq_b$ in the summation yields

$$ T_a(s_b - s_a) = u_2 - u_1 + {}_1w_a + {}_aw_b + {}_bw_2 $$

Noting that $T_a$ is equal to $T_0$, $s_b$ equal to $s_2$ and $s_a$ equal to $s_1$ the preceding equation becomes

$$ -(u_2 - u_1) + T_0(s_2 - s_1) = {}_1w_a + {}_aw_b + {}_bw_2 $$

By eq. (8.2a)

$$ w_{\max} = -(u_2 - u_1) + T_0(s_2 - s_1) $$

Thus the sum of the work quantities for the reversible processes between (1) and (2) when heat exchange is with the environment only is equal to $w_{\max}$. This specific example supports the more general conclusion derived in Section 8–2 that the maximum work is independent of the reversible path between the two states if heat exchange occurs only with the environment.

A heat reservoir presents another case warranting specific analysis. Such reservoirs are at a uniform temperature, say $T_r$, and have no mass entering or leaving. Under these conditions the reservoir can be treated as a closed system and eq. (8.1) becomes

$$\delta W_{\max,r} = -\,dE_r + T_0 dS_r \tag{g}$$

where the subscript $r$ refers to a heat reservoir. The entropy change of the reservoir during reversible heat transfer $\delta Q_r$ to or from the reservoir is

$$dS_r = \frac{\delta Q_r}{T_r}$$

which integrates to

$$(S_2 - S_1)_r = \frac{Q_r}{T_r}$$

since the reservoir temperature remains constant. Noting this entropy change, eq. (g) integrates to

$$W_{\max,r} = (E_1 - E_2)_r + \frac{T_0}{T_r} Q_r \tag{h}$$

With no work done on or by the heat reservoir, the closed-system energy equation is

$$Q_r = (E_2 - E_1)_r \tag{i}$$

Substituting for $(E_1 - E_2)_r$ in eq. (h) yields

$$\begin{aligned} W_{\max,r} &= -\,Q_r + \frac{T_0}{T_r} Q_r \\ &= Q_r\left(\frac{T_0}{T_r} - 1\right) \end{aligned} \tag{8.4}$$

When heat is transferred from a reservoir $Q_r$ will, by the sign convention we have used, be negative. $W_{\max,r}$ is identical to the work for a reversible engine having $|Q_r|$ heat transfer to the working fluid at temperature $T_r$ and rejecting heat to the environment at $T_0$. This is also, in the terminology of Section 7–9, the available part of the heat transfer from the reservoir.

## Example 8.b

A heat reservoir is at 1000 K and the environment is at 300 K. Determine $W_{max}$ when 1500 units of heat transfer from the reservoir take place. Compare the result with the work that would be delivered by a reversible engine having 1500 units of heat transfer to its working fluid at 1000 K while rejecting heat to a sink at 300 K.

**Solution:**

By eq. (8.4)

$$W_{max,r} = Q_r\left(\frac{T_0}{T_r} - 1\right)$$

Heat transfer is *from* the reservoir, thus

$$W_{max,r} = -1500\left(\frac{300}{1000} - 1\right)$$
$$= 1050 \text{ units}$$

For a reversible engine

$$\eta = \frac{W_{net}}{Q_{in}} = \frac{T_{in} - T_{out}}{T_{in}}$$

The work done by the reversible engine having heat transfer *to* its working fluid is

$$W_{net} = \frac{1000 - 300}{100}(1500)$$
$$= 1050 \text{ units}$$

---

It has been stressed in the development to this point that $W_{max}$ in the equations in this chapter is for the case in which heat transfer occurs between the system and the atmosphere only. We will now examine the situation in which a system exchanges heat with a reservoir as well as with the atmosphere. This case is readily handled with the equations previously developed. Consider a case in which a system receives heat from a reservoir at temperature $T_r$. Taking advantage of the fact that $W_{max}$ is the same for any reversible process that accomplishes the specified process while exchanging heat only with the atmosphere, we first compute $W_{max}$ for the reservoir–atmosphere combination. Next the maximum work is calculated for the process of delivering the specified amount of heat to the system from the atmosphere. The net maximum work is then the algebraic sum of the two $W_{max}$ values. This procedure is illustrated in the next example.

## Example 8.c

During a steady flow process $H_2O$ at 80 F and 15 psia is heated to 250 F and 15 psia by heat transfer from a reservoir at 2500 F. During the process there is no work and changes in potential and kinetic energies are negligible. Assuming the environment is at 77 F, calculate $W_{max}$ for the process per pound mass of $H_2O$.

**Solution:**

The first step is to determine the heat transfer to the $H_2O$ system. By the steady-flow equation for the given conditions

$$\begin{aligned} {}_1q_2 &= h_2 - h_1 \\ &= 1168.7 - 48.09 = 1120.6 \text{ B/lbm} \end{aligned}$$

This amount of heat transfer to the $H_2O$ is supplied from the 2500 F reservoir.

The maximum work for heat transfer of 1120.6 B/lbm from the 2500 F reservoir with the environment at 77 F is calculated using eq. (8.4) as

$$W_{max,r} = -1120.6\left(\frac{537}{2960} - 1\right) = 917.3 \text{ B/lbm}$$

For the $H_2O$ $W_{max}$ is, by eq. (8.3a),

$$\begin{aligned} w_{max} &= h_1 - h_2 + T_0(s_2 - s_1) \\ &= -1120.6 + 537(1.7809 - 0.09332) \\ &= -214.4 \text{ B/lbm} \end{aligned}$$

This quantity has added significance in that it represents the minimum work input required by a heat pump that would operate with the 77 F environment as a heat source and reject heat to the $H_2O$, resulting in the specified state change of the $H_2O$.

The net maximum work for the process is

$$\begin{aligned} W_{max} &= -214.4 + 917.3 \\ &= 702.9 \text{ B/lbm} \end{aligned}$$

## 8–4 Maximum Useful Work

In Section 2–9 we first noted that work is done on or by the atmosphere when the volume of a system changes during a process. For example, a closed system undergoing a reversible expansion will deliver work equal to the integral of $pdV$ for the process. However, during the expansion the moving boundary of the system pushes back the atmosphere. Thus a portion of the work is done on the atmosphere, with the remainder delivered to some other system or systems. That remainder we shall call the useful work for a process or, generally, the *useful work*, $W_u$, is equal to the work for the process less the work done on the atmosphere. This definition of useful work can be applied to both closed and open systems. Practically, however, few open systems experience a volume change during a process.

With the atmosphere (environment) at its most stable state, the pressure of the atmosphere, $p_0$, is constant throughout the atmosphere. Thus the *work done on the atmosphere* during any process is

$$W_{atm} = p_0(V_2 - V_1) \tag{8.5}$$

where $V_2$ and $V_1$ are, respectively, the final and initial volumes of the system being considered. During an expansion, the system volume increases, and thus $W_{atm}$ will be greater than zero. During a compression with the system volume decreasing, $V_2$ will be less than $V_1$ and $W_{atm}$ will be less than zero, indicating that the atmosphere is contributing to work done on the system. If the system volume does not change during a process ($V_2 = V_1$), $W_{atm}$ will be zero.

Equations to evaluate the maximum work for a process during which heat exchange is with the atmosphere only were developed in Section 8–3. For each of these the maximum useful work is

$$W_{max,u} = W_{max} - W_{atm} \tag{8.6}$$

For a closed system $W_{max,u}$ is established from eqs. (8.2) and (8.5) as

$$W_{max,u} = -(E_2 - E_1) + T_0(S_2 - S_1) - p_0(V_2 - V_1)$$

Combining terms for states (1) and (2) of the system exchanging heat only with the atmosphere we get, for a closed system,

$$W_{max,u} = (E_1 + p_0V_1 - T_0S_1) - (E_2 + p_0V_2 - T_0S_2) \tag{8.7}$$

Similar expressions to evaluate the maximum useful work for a general open system are obtained by subtracting $p_0\Delta V$ from eq. (8.1).

A perfect vacuum is a system that has zero mass. Can a perfect vacuum provide useful work? Consider the vacuum system shown in Figure 8.4. The system has no mass, and thus its stored energy and entropy at all states will be zero, that is, $E_1$, $E_2$, $S_1$, and $S_2$ will be zero because $m$ is zero. Further, there will be no heat exchange with the atmosphere since the

Figure 8.4 Perfect-Vacuum System

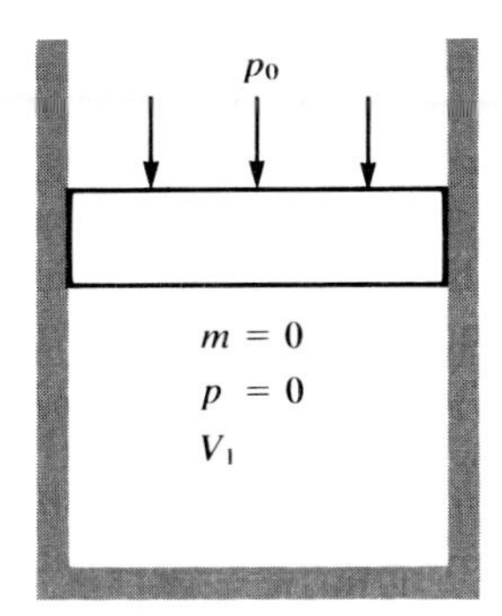


zero mass system can neither receive nor transfer heat. Under these conditions eq. (8.7) reduces to

$$W_{max,u} = p_0(V_1 - V_2)$$

If the process proceeds to a final condition where $V_2 = 0$, the maximum useful work for the process will be

$$W_{max,u} = p_0(V_1 - V_2) = p_0V_1 \quad \textbf{(8.8)}$$

Conversely, to produce a vacuum of volume $V_2$ starting with the piston face in contact with the cylinder head ($V_1 = 0$), the maximum useful work would be

$$W_{max,u} = -p_0(V_2 - V_1) = -p_0V_2 \quad \textbf{(8.8a)}$$

which is always negative. The product $p_0V_2$ is the minimum work required to produce the perfect vacuum. This illustration, as well as Example 8.c, clearly points out that in some applications $W_{max}$ is negative. It is appropriate to continue the notation $W_{max}$ for such applications since the equations yield $W_{max}$ values on an algebraic scale. Thus, in the above illustration for which $W_{max} = -p_0V_2$, the actual work must be algebraically less, say, $-1.1p_0V_2$. Accordingly, the actual work *input* would be larger, $1.1p_0V_2$, than the minimum work input, $p_0V_2$.

## Example 8.d

A simple closed system of 2 lbm of $O_2$ is initially at 60 psia and 10 F. The system undergoes a process to a final state at 15 psia and 150 F. Determine the maximum useful work for the process if the environment is at 77 F and 14 psia.

**Solution:**

On a unit mass basis the maximum useful work is by eq. (8.7)

$$w_{max,u} = u_1 - u_2 + T_0(s_2 - s_1) - p_0(v_2 - v_1)$$

The specific volumes are

$$v_1 = \frac{(48.3)(470)}{(60)(144)} = 2.627 \text{ ft}^3/\text{lbm}$$

and

$$v_2 = \frac{(48.3)(610)}{(15)(144)} = 13.64 \text{ ft}^3/\text{lbm}$$

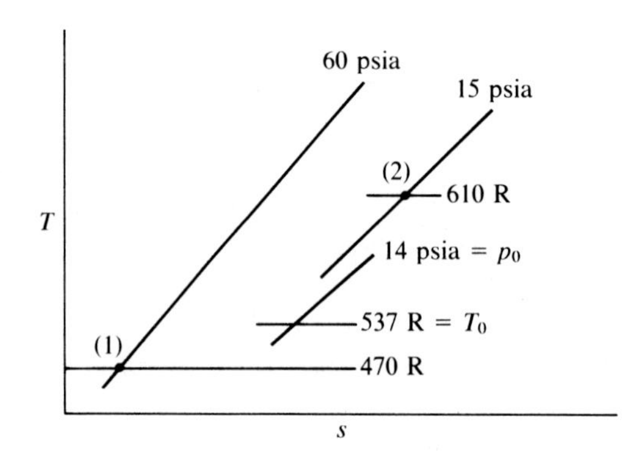


The $T$-$s$ diagram shows the two states and the state of the atmosphere. The entropy change of the oxygen is

$$s_2 - s_1 = 0.219\ ln\ \frac{610}{470} - \frac{48.3}{778}\ ln\ \frac{15}{60} = 0.1432 \text{ B/lbm R}$$

The maximum useful work is

$$\begin{aligned} W_{max,u} &= 0.157(470 - 610) + 537(0.1432) \\ &\quad - \frac{(14)(144)}{778}(13.64 - 2.627) \\ &= 26.38 \text{ B/lbm} \end{aligned}$$

For the 2 lbm of $O_2$

$$W_{max,u} = (2)(26.78) = 52.76 \text{ B}$$

For the given state change, the maximum useful work under the condition of heat exchange only with the environment is 52.76 B net output.

## Example 8.e

An axial flow compressor takes in air at 100 kPa and 25 C with negligible velocity. Discharge is at 1035 kPa with negligible velocity. The compressor operates adiabatically but not reversibly. Due to the irreversibilities the work input to the compressor is 1.15 times that required for isentropic compression from the given inlet state to the given discharge pressure. Determine the actual useful work and the maximum useful work per unit mass of air passing through the compressor if the environment is at 100 kPa psia and 25 C.

**Solution:**

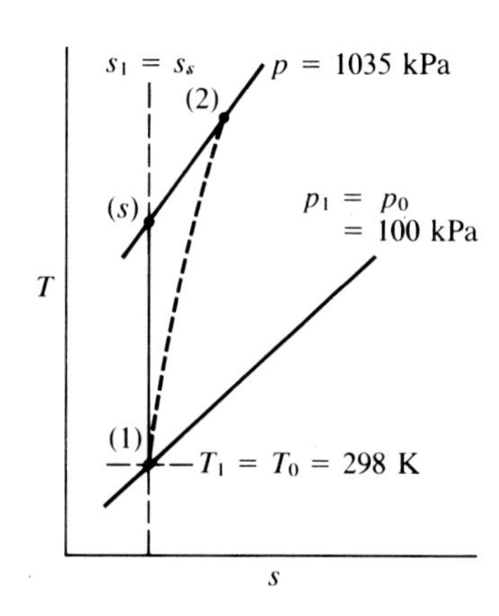


The compressor casing is a rigid body; thus there will be no change in the open-system volume for the actual process, the reversible compression, or for the determination of the maximum useful work. For reversible and adiabatic compression from the given inlet state to the discharge pressure of 1035 kPa, the discharge temperature would be

$$T_s = T_1\left(\frac{p_s}{p_1}\right)^{(\gamma-1)/\gamma}$$

$$= 298\left(\frac{1035}{100}\right)^{0.2857} = 581\text{ K}$$

and, from the steady-flow energy equation, the ideal work is

$${}_1w_s = h_1 - h_s = C_p(T_1 - T_s)$$

$$= 1.0(298 - 581) = -283\text{ kJ/kg}$$

Due to the irreversibilities the actual work is

$${}_1w_2 = (1.15)(-283) = -325.5\text{ kJ/kg}$$

The actual temperature at the compressor discharge is determined from the steady-flow energy equation for the actual process.

$${}_1w_2 = -325.5 = C_p(T_1 - T_2) = 1.0(298 - T_2)$$

which yields $T_2$ equal to 623.5 K.

With no work done on or by the atmosphere, the maximum work is the same as the maximum useful work and, by eq. (8.3a),

$$w_{max} = h_1 - h_2 + T_0 (s_2 - s_1)$$

Between the inlet state and the actual discharge state

$$s_2 - s_1 = 1.0 \ln \frac{623.5}{298} - 0.287 \ln \frac{1035}{100}$$
$$= 0.06754 \text{ kJ/(kg K)}$$

The maximum work for the actual state change is

$$w_{max} = h_1 - h_2 + T_0 (s_2 - s_1)$$
$$= -325.5 + (298)(0.6754) = -305.4 \text{ kJ/kg}$$

In summary, there are three work quantities to compare. The ideal (isentropic) process would be accomplished with a work input of 283 kJ/kg. For the actual process with no heat transfer, the work input is 325.5 kJ/kg. Had the actual state change taken place in a reversible manner with heat exchange between the system and the atmosphere only, the work input would have been 305.4 kJ/kg. We shall discuss the difference between the maximum work and the actual work in Section 8–7.

---

## 8–5 Availability for Closed Systems

When a system is in equilibrium with its environment (atmosphere), the system itself is in a state of stable equilibrium, and there are no unbalanced potentials that could be used to realize work under the condition of heat exchange only with the atmosphere. The temperature of the system will be equal to that of its environment and the pressure identical to that of the environment. There will be no relative velocity between the system and the environment. With the environment at rest, the system velocity will be zero. When these conditions exist, the system is at its *dead state.* Note that the dead state is a state of the system. When a system is at its dead state, work cannot be obtained from the system, the environment, or the system–environment combination.

The *availability* of a system existing in a given environment is the maximum useful work that could be obtained from the system–environment combination when heat exchange takes place only between the system and the environment. From this definition it is apparent that the availability of a system at a given state is dependent on the state of its environment also. Stated in another manner, the availability of a system in a given environment is the maximum useful work for a process during which a system proceeds from a known equilibrium state to its dead state while exchanging heat only with the atmosphere.

The maximum useful work that can be obtained during a state change of a closed system exchanging heat with the environment alone is given by eq. (8.7). The availability $A_1$ of a closed system at a state such as (1) on Figure 8.5 is, then, the maximum useful work when

**Figure 8.5**

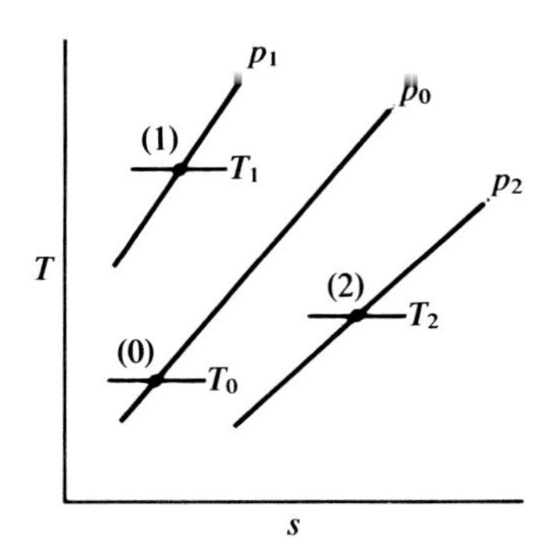


the system that exchanges heat only with the atmosphere undergoes the state change 1 → 0 in an environment at $T_0$ and $p_0$. The states (1) and (0) on Figure 8.5 are states of the system, with (0) being the dead state of the system. No path is shown in Figure 8.5 since, as discussed in Section 8–2, the maximum useful work for all reversible processes would be the same when heat exchange occurred only between the system and its environment. From the definition of availability and eq. (8.7) we get

$$\begin{aligned} A_1 &= (E_1 + p_0V_1 - T_0S_1) - (E_0 + p_0V_0 - T_0S_0) \\ &= {}_1W_{0,\max,u} \end{aligned} \tag{8.9}$$

with $E_0$, $V_0$, and $S_0$ properties of the closed system at its dead state. The maximum useful work for a state change of a closed system, as from (1) to (2) on Figure 8.5, will be the decrease in the availability, $A_1 - A_2$, of the closed system, which is

$$\begin{aligned} A_1 - A_2 &= (E_1 + p_0V_1 - T_0S_1) \\ &\quad - (E_2 + p_0V_2 - T_0S_2) \\ &= W_{\max,u} \end{aligned} \tag{8.9a}$$

The validity of this expression is evident when one recognizes that the quantity $E_0 + p_0V_0 - T_0S_0$ will be common for all states of any one system existing in a given environment. For processes involving more than one closed system, the maximum work is the decrease in availability for all closed systems associated with the process or

$$W_{\max,u} = \Sigma(A_1 - A_2) \tag{8.9b}$$

where $(A_1 - A_2)$ is the decrease in availability of one system included in the process. Suppose, for example, that a closed system and a heat reservoir were to interact during a process. The decrease in availability for the process would be

$$(A_1 - A_2)_{\text{process}} = \Sigma(A_1 - A_2) = (A_1 - A_2)_s + (A_1 - A_2)_r$$

where the subscripts $s$ and $r$ refer to the closed system and the heat reservoir respectively.

When a closed system exists at any state other than its dead state, it is capable of proceeding spontaneously to its dead state due to unbalanced potentials between the system and its environment. Such spontaneous state changes of a system require no work input from sources external to the system and its environment. Thus the availability $A$ of a closed system is, in all instances, equal to or greater than zero. The availability of a closed system, given in eq. (8.9), will be zero when the system is at its dead state.

## Example 8.f

A closed system of 1 lbm air is at $-10$ F and 10 psia. The environment is at 77 F and 14.7 psia. The system undergoes a process to a final state at 50 psia and 200 F. Determine the availability of the system at the initial state and at the final state, and the maximum useful work for the process.

**Solution:**

At the initial state the temperature and pressure of the simple closed system are each less than the corresponding properties of the atmosphere. The initial availability is

$$A_1 = (E_1 + p_0V_1 - T_0S_1) - (E_0 + p_0V_0 - T_0S_0)$$

by eq. (8.9). For the simple system this rearranges, for a unit mass, to

$$a_1 = (u_1 - u_0) + p_0\,(v_1 - v_0) + T_0\,(s_0 - s_1)$$

The specific volumes of the system are $v_1 = RT_1/p_1 = 16.66$ ft³/lbm and $v_0 = RT_0/p_0 = 13.52$ ft³/lbm. With pressures and temperatures known, the entropy change is determined by

$$\begin{aligned} s_0 - s_1 &= C_p\,ln\frac{T_0}{T_1} - R\,ln\frac{p_0}{p_1} \\ &= 0.24\,ln\frac{537}{450} - \frac{53.3}{778}\,ln\frac{14.7}{10} \\ &= 0.01603 \text{ B/lbm R} \end{aligned}$$

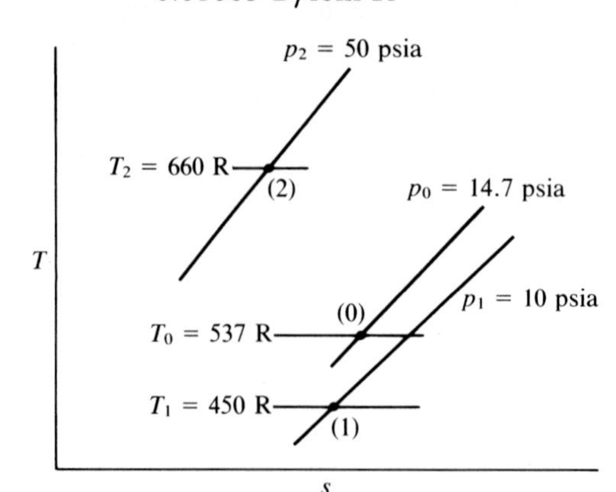

The availability $a_1$ is then

$$\begin{aligned} a_1 &= 0.172(450 - 537) + \frac{(14.7)(144)}{778} \\ &\quad \times (16.66 - 13.52) + 537(0.01603) \\ &= 2.456 \text{ B/lbm} \end{aligned}$$

Note that the availability $a_1$ is greater than zero even though the system pressure and temperature are less than those of the environment. At the second state

$$a_2 = (u_2 - u_0) + p_0 (v_2 - v_0) + T_0 (s_0 - s_2)$$

with

$$v_2 = \frac{RT_2}{p_2} = 4.886 \text{ ft}^3\text{/lbm}$$

The entropy difference $s_0 - s_2$ is

$$\begin{aligned} s_0 - s_2 &= 0.24 \ ln \frac{537}{660} - \frac{53.3}{778} \ ln \frac{14.7}{50} \\ &= 0.03435 \text{ B/lbm R} \end{aligned}$$

The availability of the closed system at the second state is

$$\begin{aligned} a_2 &= 0.172 (660 - 537) + \frac{(14.7)(144)}{778} \\ &\quad \times (4.886 - 13.52) + 537(0.03435) \\ &= 16.12 \text{ B/lbm} \end{aligned}$$

The maximum useful work for the state change is

$$\begin{aligned} a_1 - a_2 &= 2.456 - 16.12 \\ &= -13.66 \text{ B/lbm} \end{aligned}$$

Note that the availability at both states is greater than zero; however, the availability decrease—equal to the maximum useful work—is less than zero. Thus the *minimum* work input to achieve the given state change while exchanging heat only with the atmosphere would be work input of 13.66 B/lbm.

---

## Example 8.g

A rigid, insulated tank has a volume of 2 ft³ and contains helium at 300 psia and 120 F. The environment is at 77 F and 14.7 psia. Determine the availability of the helium in the tank.

**Solution:**

The availability is given by

$$A_1 = U_1 - U_0 + p_0 (V_1 - V_0) + T_0 (S_0 - S_1)$$

The mass of helium is

$$m = \frac{pV}{RT} = \frac{(300)(144)(2)}{(386)(580)} = 0.3859 \text{ lbm}$$

At the dead state the helium would be in equilibrium with its environment at 77 F and 14.7 psia. Thus, even though the helium is in a rigid vessel, it must expand to reach the dead state at which the volume is

$$V_0 = \frac{(0.3859)(386)(537)}{(14.7)(144)} = 37.79 \text{ ft}^3$$

The availability of the helium is then

$$A_1 = 0.3859[0.753(580 - 537)] + \left[\frac{(14.7)(144)}{778}(2 - 37.79)\right] + [(537)(0.3859)(s_0 - s_1)]$$

The specific entropy difference, $s_0 - s_1$, for the helium is

$$s_0 - s_1 = 1.25 \, ln \frac{537}{580} - \frac{386}{778} \, ln \frac{14.7}{300} = 1.3997 \text{ B/lbm R}$$

The availability of the helium is, then,

$$A_1 = 12.495 + (-97.358) + 290.06 = 205.2 \text{ B}$$

## 8-6 Availability for Open Systems

The definition of availability in Section 8–5 applies to either closed or open systems. Thus the availability of an open system is the maximum useful work for a reversible process in which the system proceeds to its dead state while exchanging heat with the atmosphere alone. Open systems can be defined for nozzles, compressors, pumps, turbines, heat exchangers, tanks, and many similar devices. In reflecting on the nature of such equipment, one notes that the open-system boundary does not change during processes involving such equipment. Thus work done on or by the atmosphere during open-system processes usually is zero and the maximum useful work is equal to the maximum work evaluated using eqs. (8.1) or (8.1a). Should one encounter an open system that experiences a boundary change during a process, the maximum useful work could be evaluated, of course, by subtracting $p_0\Delta V$ from the maximum work.

It then follows that the availability $B$ of an open system is equal to the integral of eq. (8.1) from any given state (1) to the dead state. This integration yields

$$\begin{aligned} B_1 = W_{\max,1\to 0} &= (E_1 - E_0)_s + T_0(S_0 - S_1)_s \\ &+ \Sigma[m(e + pv - T_0 s)]_{\text{in}} \\ &- \Sigma[m(e + pv - T_0 s)]_{\text{out}} \end{aligned} \tag{8.10}$$

where $E$ is the stored energy of the open system and $e$ is the specific stored energy of the working fluid entering or leaving the open system. For the systems we shall consider the stored energy $e$ is equal to the sum of the internal, kinetic, and potential energies at any state. Further, we shall consider only open systems having the stored energy $E_s$ equal to the internal energy of the open system.

The availability of open systems operating under steady-flow conditions follows, in like manner, from eq. (8.3a). For single stream, steady flow the availability per unit mass $b$ is obtained by applying eq. (8.3a) from any state (1) to the dead state (0) to yield

$$b_1 = w_{\max,1\to 0} = \left(h_1 + \frac{V_1^2}{2g_0} - T_0 s_1\right) - (h_0 - T_0 s_0) \tag{8.11}$$

where quantities carrying the subscript 0 are evaluated at the dead state of the steady-flow system. At the dead state the system velocity will be zero by definition of the dead state. Further, it should be noted that the potential energy at state (1) has been assumed to be negligible.

With the quantity $h_0 - T_0 s_0$ common for all states of a particular system in a given environment, the maximum work, equal to the maximum useful work, is equal to the decrease in $b$, the steady-flow availability or, on a unit mass basis,

$$\begin{aligned} w_{\max,1\to 2} &= b_1 - b_2 \\ &= \left(h_1 + \frac{V_1^2}{2g_0} - T_0 s_1\right) - \left(h_2 + \frac{V_2^2}{2g_0} - T_0 s_2\right) \end{aligned} \tag{8.11a}$$

For a single-stream steady-flow process during which the boundaries remain fixed, the atmospheric work is zero. In the case of multiple stream steady flow with fixed boundaries, the availability decrease—equal to the maximum useful work—also is equal to the maximum work. This maximum work, expressed on a rate basis, is determined using eq. (8.3b). The maximum useful work for a process in which multiple systems interact is determined by evaluating the decrease in availability of each system during the process. For processes involving a combination of open and closed systems, the decrease in availability is calculated for each system using the applicable equation. These individual decreases are then summed algebraically to determine the availability change for the process. A common example of this approach is encountered when a heat reservoir and a steady-flow system interact during a process. The availability change of the heat reservoir—treated as a closed system—is determined using eq. (8.4). The availability change of the steady-flow system is determined by application of eq. (8.11a). The summation must be on a consistent basis of a unit mass or a mass rate for the steady-flow system.

In summary, when heat exchange is with the atmosphere only the maximum useful work is equal to the decrease in availability of the system or systems involved in a process. For any process of a simple system—closed, steady flow open, or unsteady flow open—the maximum useful work can be determined from eq. (8.1) and the definition of atmospheric work as

$$\delta W_{max,u} = -dE_s + T_0 dS_s + (e_1 + p_1 v_1 - T_0 s_1)\, dm_1 - (e_2 + p_2 v_2 - T_0 s_2) dm_2 - p_0 dV_s \quad \textbf{(8.12)}$$

This is the most basic equation for determination of useful work, and hence for the change in availability of a system. For a closed system with $dV_s = 0$ eq. (8.12) reduces to eq. (8.2). Applied to a steady-flow open system with fixed boundaries ($dE_s$, $dS_s$, and $dV_s$ equal to zero), the equation simplifies to eq. (8.3a). When considering unsteady-flow open systems, eq. (8.12) must be applied. For such systems it is quite common to have one or more terms in this general equation equal to zero by the nature of the process and the defined open-system boundary.

## Example 8.h

During a steady-flow process $H_2O$ at 200 psia and 100 percent quality is throttled to 50 psia. The environment is at 77 F and 14.696 psia. Determine, per unit mass of $H_2O$, (a) the availability of the $H_2O$ before the throttling process, (b) the availability of the $H_2O$ after it has been throttled to 50 psia, and (c) the maximum useful work for the process.

**Solution:**

(a) At the inlet state the availability of the steam is, from eq. (8.11),

$$\begin{aligned} b_1 &= (h_1 - T_0 s_1) - (h_0 - T_0 s_0) \\ &= (h_1 - h_0) + T_0(s_0 - s_1) \\ &= (1199.3 - 45.09) + 537\ (0.08775 - 1.5464) \\ &= 370.91 \text{ B/lbm} \end{aligned}$$

With the dead state at 14.696 psia and 77 F the $H_2O$ will be subcooled at this state. Since the pressure is moderate, the enthalpy and entropy of the $H_2O$ at the dead state are taken as $h_f$ and $s_f$ at 77 F, neglecting the effect of pressure on these properties.

(b) Using the same procedures as those in (a)

$$\begin{aligned} b_2 &= (h_2 - T_0 s_2) - (h_0 - T_0 s_0) \\ &= (h_2 - h_0) + T_0(s_0 - s_2) \end{aligned}$$

For the throttling process the steady-flow energy equation reduces to $h_1 = h_2$. The entropy at (2) is fixed by the pressure of 50 psia and the enthalpy $h_2 = h_1 = 1199.3$ B/lbm. By linear interpolation in the superheat table for $H_2O$

$$s_2 = 1.6914 \text{ B/lbm R}$$

The availability after throttling is then

$$\begin{aligned} b_2 &= (1199.3 - 45.09) + 537\ (0.08775 - 1.6914) \\ &= 293.05 \text{ B/lbm} \end{aligned}$$

(c) The maximum useful work is equal to the decrease in the availability

$$\begin{aligned} w_{\text{max}} = w_{\text{max},u} &= b_1 - b_2 \\ &= 370.91 - 293.05 \\ &= 77.86 \text{ B/lbm} \end{aligned}$$

Note that the maximum useful work and the maximum work are equal for the throttling process since the system boundary does not change during the process.

An alternate procedure to determine the maximum useful work for this process is

$$w_{\text{max},u} = b_1 - b_2 = (h_1 - h_2) + T_0(s_2 - s_1)$$

With $h_1$ equal to $h_2$ in the throttling process

$$\begin{aligned} w_{\text{max},u} = b_1 - b_2 &= T_0(s_2 - s_1) \\ &= 537\ (1.6914 - 1.5464) \\ &= 77.86 \text{ B/lbm} \end{aligned}$$

## Example 8.i

Steam flows slowly from a line where the pressure is 500 psia and the temperature 700 F into a rigid tank that is initially evacuated ($p_1 = 0$). There is no heat transfer during the process in which the pressure in the tank increases to 500 psia. Determine (a) the final temperature of the steam in the tank and (b) $w_{max}$ for the process per pound of steam if the environment is at 50 F and 14.7 psia.

**Solution:**

(a) For the unsteady-flow process the energy equation, eq. (3.3), reduces to

$$0 = E_2 - E_1 - h_l(m_2 - m_1)$$

where 1 and 2 refer to the initial and final states of the open system and $h_l$ is the specific enthalpy of the steam in the line. For the open system $E_2 = U_2 = m_2u_2$, $m_1$ is zero and $E_1$ is zero. Thus

$$m_2u_2 = h_lm_2$$

or the final specific internal energy of the steam in the tank is equal to the specific enthalpy of steam in the line. At 500 psia and 700 F $h_l$ is 1356.7 B/lbm, and thus

$$u_2 = 1356.7 \text{ B/lbm}$$

The end state in the tank is fixed by this internal energy and the final pressure of 500 psia. Interpolating linearly in the superheat table, the end-state temperature in the tank is 984 F.

(b) During the process there is no outbound flow and velocities are negligible. Under these conditions eq. (8.1) integrates to

$$W_{max} = (E_1 - E_2)_s + T_0(S_2 - S_1) + (h_l - T_0s_l)(m_2 - m_1)$$

where 1 and 2 refer to the initial and final states of the open system, the subscript $s$ to the open system, and the subscript $l$ to steam in the line before it enters the open system. With no mass in the tank at the beginning of the process, $E_1$, $S_1$, and $m_1$ are zero and the above equation reduces to

$$W_{max} = -E_{2,s} + T_0S_{2,s} + m_2h_l - m_2T_0s_l$$

From part (a) of this example

$$E_{2,s} = m_2u_2 = m_2h_l$$

Substituting the quantity $-m_2h_l$ for $-E_{2,s}$ yields

$$\begin{aligned} W_{max} &= T_0S_{2,s} - m_2T_0s_l \\ &= T_0m_2s_2 - T_0m_2s_l \end{aligned}$$

Thus on a unit mass basis

$$\begin{aligned} w_{max} &= T_0s_2 - T_0s_l \\ &= T_0(s_2 - s_l) \end{aligned}$$

At the final state where the pressure is 500 psia and the internal energy 1356.7 B/lbm the specific entropy is 1.7312 B/lbm R. The specific entropy $s_l$ of the steam in the line is 1.6112 B/lbm R; thus

$$\begin{aligned} w_{max} &= 510(1.7312 - 1.6112) \\ &= 61.2 \text{ B/lbm} \end{aligned}$$

With no change in volume of the rigid tank during the process, there is no work done on or by the atmosphere; hence the maximum work is equal to the maximum useful work, which, in turn, is the decrease in availability for the process.

---

## Example 8.j

For the conditions given in Example 8.i determine the initial and final availability of each component in the process on the basis of 1 lbm of steam. Then determine the net decrease in availability. Compare the result with that obtained in Example 8.i.

**Solution:**

The initial availabilities are those for 1 lbm of steam in the line and a vacuum having the volume to contain 1 lbm of steam at the completion of the process. The availability of the steam flowing in the line is, by eq. (8.11),

$$b_1 = (h_l - T_0s_l) - (h_0 - T_0s_0)$$

For the line pressure and temperature ($p_1$ and $T_1$) of 500 psia and 700 F, respectively, and the dead state at 50 F and 14.7 psia

$$\begin{aligned} b_l &= h_l - h_0 + T_0\,(s_0 - s_l) \\ &= (1356.7 - 18.06) + 510(0.03607 - 1.6112) \\ &= 535.32 \text{ B/lbm} \end{aligned}$$

Note that the $H_2O$ is a subcooled liquid at the dead state.

The availability of a vacuum with a volume that will hold 1 lbm of steam at the end state where the pressure is 500 psia and the internal energy (from Example 8.i) is 1356.7 B/lbm is, by eq. (8.8),

$$b_{1,\text{vac}} = \frac{(14.7)}{778}(144)(1.6808) = 4.57 \text{ B/lbm}$$

At the end state the only availability will be that of the steam, considered a closed system, in the tank. Its availability $a_2$ is calculated using eq. (8.9) as

$$a_2 = (u_2 + p_0 v_2 - T_0 s_2) - (u_0 + p_0 v_0 - T_0 s_0)$$

with the dead state being subcooled liquid. Properties at state (2) were determined previously. Thus

$$\begin{aligned} a_2 &= (u_2 - u_0) + p_0(v_2 - v_0) + T_0(s_0 - s_2) \\ &= (1356.7 - 18.06) + \frac{(14.7)(144)}{778} \\ &\quad \times (1.6808 - 0.01602) + 510(0.03607 - 1.7312) \\ &= 478.65 \text{ B/lbm} \end{aligned}$$

The decrease in availability for the process is

$$\begin{aligned} \Sigma \text{ availability decrease} &= 535.32 + 4.57 - 478.65 \\ &= 61.2 \text{ B/lbm} \end{aligned}$$

which is equal to the result in Example 8.i. By these examples we see then the sum of the initial availabilities of all components in any process—closed or open—less the summation at the end state is equal to the maximum useful work for the process, which is, by definition, the decrease in availability for the process. While this statement may seem redundant, a review of the component availabilities shows that the equality of maximum useful work and availability decrease is not limited to a single closed or open system.

---

## 8–7 Irreversibility

The concept of irreversibility and the logic pattern to test for the irreversibility of a process were presented in Section 6–2. Now we shall extend those concepts and develop the necessary procedures to quantify the irreversibility of a process. These steps will be limited to systems that do not undergo chemical reaction for the reasons noted in the introduction to this chapter.

The irreversibility of a process is defined as the difference between the maximum work that could be obtained when heat exchange is with the atmosphere only and the actual work during a given change of state. The maximum work is determined using eqs. (8.1), (8.2),

(8.3), or (8.4), depending on the system involved, that is, unsteady flow open, closed, steady flow, or a heat reservoir respectively. The irreversibility of a process is, by definition,

$$I \equiv W_{max} - W \tag{8.13}$$

where $W$ is the actual work for the process. When more than one system is involved in a process, the irreversibility is

$$I = \Sigma(W_{max} - W) \tag{8.13a}$$

that is, the summation of the $W_{max}$ minus $W$ quantities for each system involved in a process. Perhaps the most common combination of systems is that of a heat reservoir (one system) plus an open or closed system.

The atmospheric work $p_0(V_2 - V_1)$ is fixed by the pressure of the environment and established end states of a process. Thus an alternate definition of irreversibility is

$$I = [W_{max} - p_0(V_2 - V_1)] - [W - p_0(V_2 - V_1)] \tag{8.13b}$$

The quantity $[W_{max} - p_0(V_2 - V_1)]$ is the maximum useful work between fixed end states when heat exchange is with the atmosphere only and is equal to the decrease in availability during a process. The second bracket is the actual useful work for the process. Thus the irreversibility of any one system undergoing a given state change can be found as

$$I = \text{decrease in system availability} - \text{actual useful work} \tag{8.14}$$

Again, if more than one system is involved in a process, a summation similar to eq. (8.13a) is used. Consider, for example, a process involving two closed systems $x$ and $y$. The irreversibility of the process would be

$$\begin{aligned} I = {} & [(A_1 - A_2)_x + (A_1 - A_2)_y] \\ & - [W_x - p_0(V_2 - V_1)_x + W_y - p_0(V_2 - V_1)_y] \end{aligned} \tag{8.14a}$$

Since the mass of system $x$ generally will not be equal to that of system $y$, care must be taken to have all terms in eq. (8.14a) evaluated on a consistent mass basis for each system. If it is necessary to evaluate availabilities or changes in availabilities, eq. (8.14) usually will yield the most direct solution for the irreversibility of a process. When irreversibility alone is desired, eq. (8.13) will provide a more direct solution.

---

## Example 8.k

Refer to Example 7.c. For the system and process described in that example, determine (a) the actual work, (b) $W_{max}$, and (c) the irreversibility of the process if the atmosphere is at 77 F and 14.70 psia.

**Solution:**

(a) For the adiabatic process of the closed system

$$_1Q_2 = U_2 - U_1 + {}_1W_2 = 0$$

and the actual work is

$$_1W_2 = m(u_1 - u_2) = mC_v(T_1 - T_2)$$

From Example 7.c the mass is 5.151 lbm and the final temperature is 884 R. With these values known the actual work is

$$\begin{aligned} W &= (5.151)(0.172)(520 - 884) \\ &= -322.5 \text{ B} \end{aligned}$$

(b) Had the state change taken place reversibly under the condition of heat exchange only with the atmosphere, the maximum work (minimum input for this case) would be, by eq. (8.2),

$$\begin{aligned} W_{max} &= -(1.151)(0.172)(884 - 520) + (5.151)(537)(0.09127) \\ &= -70.03 \text{ B} \end{aligned}$$

(c) The irreversibility of the process is, by eq. (8.13),

$$\begin{aligned} I &= -70.03 - (-322.5) \\ &= 252.5 \text{ B} \end{aligned}$$

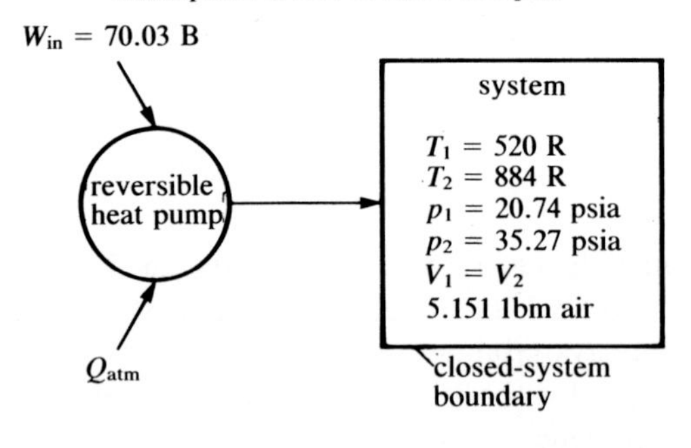


The results tell us that the state change of the system could have been accomplished with a work input of 70.03 B had the process taken place reversibly using the atmosphere as a heat source for a reversible heat pump as shown in the above illustration. Thus the excess work input over the minimum possible is 252.5 B, an excess that could have been used for other purposes. For example, suppose the paddle wheel had been driven by an electric motor and the reversible heat pump by an electric motor. The state change of the air could have been achieved and 252.5 B of electric energy still would have been "available" to perform work on another system or systems if each electric motor had an efficiency of 100 percent.

## Example 8.1

Steam is supplied to a stationary nozzle at 100 psia and 500 F with negligible velocity. The steam expands adiabatically through the nozzle to a pressure of 10 psia at the outlet section. Because of friction and other irreversibilities in the nozzle, the kinetic energy at the outlet is 95 percent of that which would have been achieved if the expansion had been reversible and adiabatic. Determine (a) the availability of the steam entering the nozzle, (b) the availability of the steam at the nozzle outlet section, and (c) the irreversibility of the process. The environment is at 70 F and 14.00 psia.

**Solution:**

(a) By eq. (8.11) the availability of the steam entering the nozzle is

$$\begin{aligned} b_1 &= (h_1 - T_0 s_1) - (h_0 - T_0 s_0) \\ &= (h_1 - h_0) + T_0 (s_0 - s_1) \end{aligned}$$

where the subscript 0 indicates properties of the steam at its dead state, a subcooled liquid. Using data from the *Steam Tables,* we get

$$\begin{aligned} b_1 &= (1279.1 - 38.09) + 530(0.07463 - 1.7085) \\ &= 375.1 \text{ B/lbm} \end{aligned}$$

(b) To establish the state at the nozzle outlet section we first determine the outlet enthalpy for an isentropic expansion to 10 psia. For the nozzle operating reversibly and adiabatically

$$h_1 - h_i = \frac{V_i^2}{2g_0} \tag{a}$$

while the actual nozzle will have

$$h_1 - h_2 = \frac{V_2^2}{2g_0} \tag{b}$$

With $V_2^2/2g_0$ equal to 95 percent of $V_i^2/2g_0$ it follows that

$$h_1 - h_2 = 0.95\,(h_1 - h_i) \tag{c}$$

The isentropic state $i$ is fixed by $s_i = s_1 = 1.7085$ B/lbm R and 10 psia. This is in the mixture region ($s_i < s_g$ at 10 psia) and the quality $x_i$ is determined from

$$s_1 = s_i = 1.7085 = 0.28358 + x_i\,(1.5041)$$

to yield $x_i = 0.9474$. The enthalpy $h_i$ (isentropic) is

$$h_i = 161.23 + (0.9474)(982.1) = 1091.6 \text{ B/lbm}$$

From eq. (c)

$$\begin{aligned} h_2 &= h_1 - 0.95\,(h_1 - h_i) \\ &= 1279.1 - 195\,(1279.1 - 1091.6) \\ &= 1101.0 \text{ B/lbm} \end{aligned}$$

The actual outlet state is fixed by the pressure 10 psia and the enthalpy 1101.0 B/lbm. At this state the quality is

$$\begin{aligned} 1101.0 &= 161.23 + (x_2)\,982.1 \\ x_2 &= 0.9569 \end{aligned}$$

and the entropy $s_2$ is

$$\begin{aligned} s_2 &= 0.28358 + (0.9569)(1.5041) \\ &= 1.7229 \text{ B/lbm R} \end{aligned}$$

The actual outlet velocity is, from eq. (b),

$$\begin{aligned} V_2 &= [(2)(32.174)(778)(1279.1 - 1101.0)]^{1/2} \\ &= 2986 \text{ ft/sec} \end{aligned}$$

The availability of the steam at the nozzle outlet is

$$\begin{aligned} b_2 &= \left(h_2 + \frac{V_2^2}{2g_0} - T_0 s_2\right) - (h_0 - T_0 s_0) \\ &= h_2 + \frac{V_2^2}{2g_0} - h_0 + T_0\,(s_0 - s_2) \\ &= 367.4 \text{ B/lbm} \end{aligned}$$

(c) With the availability at the nozzle inlet and outlet known, the irreversibility of the process is established by eq. (8.14) as

$$i = (b_1 - b_2) - w$$

Since the nozzle walls are fixed, the term $p_0 \Delta V$ for the system is zero. Also, by the nature of a nozzle there is no work ($W = 0$) during flow through the stationary nozzle. Under these conditions

$$\begin{aligned} i &= b_1 - b_2 \\ &= 375.1 - 367.4 \\ &= 7.7 \text{ B/lbm} \end{aligned}$$

Another approach to this example is through eq. (8.13), recognizing again that the actual work for the stationary nozzle is zero. By eq. (8.11a) the maximum work—equal to the irreversibility for processes during which no work is done on or by the system—is

$$w_{\max} = \left(h_1 + \frac{V_1^2}{2g_0} - T_0 s_1\right) - \left(h_2 + \frac{V_2^2}{2g_0} - T_0 s_2\right)$$

This rearranges to

$$w_{\max} = (h_1 - h_2) - \left(\frac{V_2^2 - V_1^2}{2g_0}\right) + T_0\,(s_2 - s_1)$$

For any stationary nozzle process the steady-flow energy equation reveals that the enthalpy decrease is equal to the kinetic energy increase when there is no heat transfer or work during the actual process. Under these conditions

$$w_{\max} = T_0\,(s_2 - s_1)$$

for the nozzle process. Using previously calculated values we get, for this example,

$$\begin{aligned} i = w_{\max} &= T_0\,(s_2 - s_1) \\ &= 530\,(1.7229 - 1.7085) \\ &= 7.7 \text{ B/lbm} \end{aligned}$$

---

The preceding example suggests that still another procedure for evaluating irreversibility is possible. We have defined the irreversibility of a process as

$$I \equiv W_{\max} - W$$

A general expression for $W_{\max}$, the maximum work that could be realized when heat exchange is solely with the atmosphere, is, in differential form,

$$\begin{aligned} \delta W_{\max} = &- dE_s + T_0 dS_s + (e_1 + p_1 v_1 - T_0 s_1) dm_1 \\ &- (e_2 + p_2 v_2 - T_0 s_2) dm_2 \end{aligned} \tag{8.1}$$

where the subscript $s$ refers to a generalized system, $dm_1$ to an inbound mass, and $dm_2$ to an outbound mass. The analogous general energy equation, also expressed in differential form for one inbound and one outbound stream, is

$$\delta W_s = \delta Q_s - dE_s + (e_1 + p_1 v_1) dm_1 - (e_2 + p_2 v_2) dm_2$$

Subtracting we get

$$\delta W_{max} - \delta W_s = -\delta Q_s + T_0 dS_s - T_0 s_1 dm_1 + T_0 s_2 dm_2 \qquad (8.15)$$
$$= dI$$

where $\delta Q_s$ is the *actual* heat transfer for the system. In eq. (8.15) the subscript $s$ indicates the particular system (control volume) to be considered and the subscripts 1 and 2 are, respectively, for inbound and outbound streams.

For a closed system $dm_1$ and $dm_2$ are zero (no mass crosses the boundary) and eq. (8.15) integrates to

$$I = -Q_s + T_0 (S_2 - S_1)_s \qquad \textbf{(8.15a)}$$

for any closed system. During a single-stream steady-flow process the mass flow rate is constant ($dm_1 = dm_2$) and on a unit mass basis eq. (8.15), for one system undergoing a steady-flow process, becomes

$$i = -{}_1q_2 + T_0 (s_2 - s_1) \qquad \textbf{(8.15b)}$$

since $dS_s$ will be zero for steady flow. In the case of multiple-stream steady flow irreversibility can be determined on a consistent time-rate basis or on the basis of a unit mass of any entering or leaving stream. Expressed on a rate basis for multiple-stream steady-flow processes eq. (8.15) integrates to

$$\dot{I} = -\dot{Q}_s + (\Sigma \dot{m} T_0 s)_{out} - (\Sigma \dot{m} T_0 s)_{in} \qquad \textbf{(8.15c)}$$

During any adiabatic process the heat transfer $Q_s$ will be zero. For closed systems and single-stream steady-flow open systems undergoing *adiabatic processes* the irreversibility on a unit mass basis is then

$$i = T_0 (s_2 - s_1) \qquad \textbf{(8.15d)}$$

The irreversibility for an unsteady-flow adiabatic process follows from eq. (8.15) and is

$$I = T_0 (S_2 - S_1)_s + \Sigma (m T_0 s)_{out} - \Sigma (m T_0 s)_{in} \qquad \textbf{(8.15e)}$$

---

## Example 8.m

A heat exchanger uses steam to raise the temperature of air. The heat exchanger operates adiabatically, that is, all heat transferred from the steam goes to the air. The terminal temperature difference for the unit (temperature of the leaving $H_2O$ minus the temperature of the leaving heated air) is 15 F. Steam enters the heat exchanger as saturated vapor at 90 psia and leaves as saturated liquid at the same pressure. Air enters the exchanger at 60 F and 15 psia at the rate of 100 ft$^3$/min and leaves at 13.50 psia. Calculate the irreversibility of the steady-flow process in B/min assuming the environment is at 77 F and 14.7 psia.

**Solution:**

The irreversibility of this process is determined using eq. (8.15) written as a rate equation. Thus for this example,

$$\dot{I} = (\Sigma T_0 \dot{m}s)_{out} - (\Sigma T_0 \dot{m}s)_{in}$$

or

$$\dot{I} = [(T_0 \dot{m}s)_{out,air} + (T_0 \dot{m}s)_{out,H_2O}] - [(T_0 \dot{m}s)_{in,air} + T_0 \dot{m}s_{in,H_2O}]$$

which rearranges to

$$\dot{I} = [T_0 \dot{m}(s_{out} - s_{in})]_{air} + [T_0 \dot{m}(s_{out} - s_{in})]_{H_2O}$$

for the adiabatic heat exchanger.

The mass rate of flow of the air is

$$\dot{m} = \frac{p\dot{V}}{RT} = \frac{(15)(144)(1000)}{(53.3)(520)} = 77.93 \text{ lbm/min}$$

The $H_2O$ flow rate is established by the steady-flow energy equation as

$$(77.93)(0.24)(305.31 - 60) = \dot{m}_{H_2O}\,(1185.9 - 290.76)$$

where the leaving air temperature of 305.31 F is fixed by the terminal temperature difference. Solving for $\dot{m}_{H_2O}$ we get

$$\dot{m}_{H_2O} = 5.126 \text{ lb/min}$$

The specific entropy change of the air is

$$\begin{aligned} s_{out} - s_{in} &= 0.24\, ln \frac{765}{520} - \frac{53.3}{778}\, ln \frac{13.50}{15.00} \\ &= 0.09987 \text{ B/lb R} \end{aligned}$$

and that of the steam is

$$s_{out} - s_{in} = -s_{fg} \text{ at 90 psia} = -1.1475 \text{ B/lbm R}$$

With the specific entropy changes and mass rates of flow determined, the rate of irreversibility for the process is

$$\begin{aligned} \dot{I} &= [(537)(77.93)(0.09987)]_{air} + [(537)(5.126)(-1.1475)]_{H_2O} \\ &= 1021 \text{ B/min} \end{aligned}$$

## Example 8.n

A rigid tank contains 5 ft$^3$ of nitrogen at 10.00 psia and 20 F. The nitrogen undergoes a process during which 3.66 B of paddle-wheel (stirring) work is done on the system while heat is transferred to the nitrogen from a reservoir at 260 F. At the end of the process the nitrogen is at 15.00 psia. The environment is at 77 F and 14.70 psia. Calculate (a) the decrease in availability of the heat reservoir, (b) the decrease in availability of the nitrogen, (c) the irreversibility of the process using eq. (8.14a), and (d) the irreversibility of the process using eq. (8.15a).

**Solution:**

(a) The mass of the nitrogen is

$$m = \frac{(10)(144)(5)}{(55.1)(480)} = 0.2722 \text{ lbm}$$

and its end-state temperature is

$$T_2 = \left(\frac{15.00}{10.00}\right) 480 = 720 \text{ R}$$

The actual heat transfer from the reservoir to the nitrogen is determined by applying the closed-system energy equation to the nitrogen as

$$\begin{aligned} {}_1Q_2 &= U_2 - U_1 + {}_1W_2 \\ &= mc_v\,(T_2 - T_1) + {}_1W_2 \\ &= (0.2722)(0.178)(720 - 480) + (-3.66) \\ &= 7.968 \text{ B} \end{aligned}$$

The heat transfer *from* the 260 F reservoir is, then, 7.968 B. The decrease in availability of the reservoir is, by eq. (8.4),

$$\begin{aligned} (A_1 - A_2)_r = Q_r\left(\frac{T_0}{T_r} - 1\right) &= W_{\max,r} \\ = -7.968\left(\frac{537}{720} - 1\right) & \\ = 2.025 \text{ B} & \end{aligned}$$

For heat reservoirs the maximum work and the maximum useful work (equal to the decrease in availability) are the same since there is no work done on or by the atmosphere.

(b) The decrease in availability of the nitrogen is determined using eq. (8.9a). For the given conditions the decrease in availability of the nitrogen is

$$(A_1 - A_2)_{N_2} = (U_1 - U_2) + p_0(V_1 - V_2) - T_0(S_1 - S_2)$$

With no change in volume of the nitrogen, the preceding equation becomes

$$(A_1 - A_2)_{N_2} = mC_v(T_1 - T_2) + T_0(S_2 - S_1)$$

For the constant-volume process of the nitrogen

$$S_2 - S_1 = mC_v \ ln \frac{T_2}{T_1}$$

and the decrease in availability of the nitrogen is

$$\begin{aligned}(A_1 - A_2)_{N_2} &= (0.2722)(0.178)(480 - 720) \\ &\quad + (537)(0.2722)(0.178) \ ln \frac{720}{480} \\ &= -1.079 \text{ B}\end{aligned}$$

(c) Using the decreases in availability, the irreversibility of the process is, by eq. (8.14a),

$$I = (A_1 - A_2)_r + (A_1 - A_2)_{N_2} - W_{N_2}$$

there being no atmospheric work associated with either the nitrogen or the heat reservoir and no actual work for the heat reservoir. Thus, for the given conditions,

$$I = 2.025 - 1.079 - (-3.66) = 4.606 \text{ B}$$

(d) Determination of the irreversibility of the process using eq. (8.15a) requires calculation of the irreversibility for each system involved in the process, and then a summation of these individual irreversibilities to determine the irreversibility of the process. Applying eq. (8.15a) to the reservoir we get

$$I_r = -Q_r + T_0(S_2 - S_1)_r$$

The entropy change of the constant-temperature reservoir is, by the definition of change in entropy,

$$(S_2 - S_1)_r = \frac{Q_r}{T_r} = \frac{-7.968}{720} = -0.0110667 \text{ B/R}$$

Thus

$$\begin{aligned}I_r &= -(-7.968) + 537(-0.110667) \\ &= 2.025 \text{ B}\end{aligned}$$

Note that this is equal to the decrease in availability of the reservoir for which there is no useful work during the process.

Application of eq. (8.15a) to the nitrogen yields

$$\begin{aligned} I_{N_2} &= -Q_{N_2} + T_0(S_2 - S_1)_{N_2} \\ &= -7.968 + (537)(0.2722)(0.178)\, ln \frac{720}{480} \\ &= 2.582 \text{ B} \end{aligned}$$

Summing the irreversibility for the nitrogen and the heat reservoir we get the irreversibility of the process

$$I - 2.025 + 2.582 = 4.606 \text{ B}$$

the same result as that calculated using eq. (8.14a). Note that the net irreversibility, the sum of that for the nitrogen and the heat reservoir, is

$$(A_1 - A_2)_r = -Q_r + T_0(S_2 - S_1)_r \tag{a}$$

plus

$$(A_1 - A_2)_{N_2} = -Q_{N_2} + T_0(S_2 - S_1)_{N_2} \tag{b}$$

The heat transfer from the reservoir is equal in magnitude but opposite in sign to that for the nitrogen. By adding (a) and (b) we get

$$(A_1 - A_2) = T_0(S_2 - S_1)_r + T_0(S_2 - S_1)_{N_2} = I$$

for the process. For the more general case it follows that the irreversibility $I$ for a process can be determined as

$$I = \Sigma[T_0(S_2 - S_1)]$$

when all components comprising a system can be enclosed within an adiabatic boundary of a closed system.

---

Let us now examine the significance of irreversibilities in relation to work. From eq. (8.13) we relate the actual work to irreversibility $I$ (which is zero for a reversible process, and otherwise a positive quantity) in the form

$$W_{actual} = W_{max} - I \tag{8.16}$$

When an energy-conversion process is carried out by use of a cyclic device, or if more than one system is involved with work production or consumption, eq. (8.16) applies provided each term is replaced by a summation for all of the subprocesses involved. It is clear from eq. (8.16) that irreversibilities of any kind will reduce the actual work for a process. If the process produces work, irreversibilities reduce the work output. If work input is required ($W_{act}$ is negative), irreversibilities serve to increase the work required. It is not possible in practice to eliminate all causes of irreversibility (heat transfer across a finite temperature difference, friction, etc.). Further, the reduction of irreversibilities may require large economic expenditures. For example, a decrease in the temperature difference between the cold and hot streams in a heat exchanger to reduce the irreversibility would require a larger, and thus more expensive, heat exchanger. However, the reduction of irreversibilities is, within prescribed economic limits, always a desirable goal since in processes involving work it promotes more effective use of energy.

### 8–8 Second-Law Effectiveness

The second-law analysis developed in this chapter permits us to view energy interactions and interchanges in an entirely different manner than is afforded by first-law analysis alone. In this section we shall examine a means for describing in terms of second-law analysis how effectively such interactions take place.

Numerous efficiency expressions defined in terms of energy ratios are in use. One familiar example is heat-engine thermal efficiency, the ratio of net work output to the heat supplied. Another example is furnace efficiency, which is defined as the ratio of the heat delivered to the energy content of the fuel burned. Expressions such as these are termed first-law efficiencies and for energy-conversion devices they are usually defined as the ratio of the desired output quantity to the required energy input. In an analogous manner, we can define ratios of energy quantities that are based on second-law analysis. There are many such definitions in the literature. We shall confine our discussion to energy-conversion devices such as heat engines and to heat-transfer processes. Illustrations of definitions and the calculations of effectiveness are given in the following example. Other definitions and applications of second-law effectiveness are given in the chapter references. Since there are numerous definitions and applications of first-law efficiencies and second-law effectivenesses, it is important that the definitions be fully understood before comparisons are made in a given application.

## Example 8.o

(a) Define and compute the second-law effectiveness for a Carnot heat engine.
(b) Compute the effectiveness for the heat engine described in Example 7.o.
(c) Define and compute the effectiveness for the heat-exchange process in Example 8.m.

**Solution:**

(a) We define effectiveness for a heat engine as the ratio of the net work output to the decrease in availability of the energy source. For a reservoir as an energy source,

$$\epsilon = \frac{W_{net}}{(A_1 - A_2)_r}$$

where for this case the denominator is the decrease in availability of the reservoir supplying the heat to the Carnot engine. From eqs. (8.4) and (8.9a),

$$(A_1 - A_2)_r = -Q_r\left(\frac{T_0}{T_r} - 1\right) = Q_{in}\left(1 - \frac{T_0}{T_r}\right)$$

The net work for a Carnot engine is

$$W_{net} = \eta_{max} Q_{in} = \left(1 - \frac{T_0}{T_r}\right)Q_{in}$$

Therefore, $\epsilon$ for a Carnot engine rejecting heat to the atmosphere at $T_0$ is unity.

(b) We refer to the graphical description of the energy quantities in Example 7.o for numerical values. For the complete energy-conversion system (the reservoir, the heat engine, and the surrounding atmosphere) we have, on a unit mass basis, for the decrease in availability of the reservoir

$$(a_{initial} - a_{final})_r = (q_a)_r = 604 \text{ B/lbm}$$

The net work is 323 B/lbm. Thus

$$\epsilon = \frac{w_{net}}{(q_a)_r} = \frac{323}{604} = 0.535$$

For the cycle alone, the decrease in availability for the process (1) to (2) (the maximum work) is the denominator in the effectiveness expression. Hence

$$\epsilon' = \frac{W_{net}}{(q_a)_{1,2}} = \frac{323}{404} = 0.80$$

The fact that $\epsilon'$ is large has significance only in that it indicates that the cycle is quite effective in converting the availability decrease to work. The large and undesirable decrease in availability (available energy in the terminology of Section 7–9) attributable to the irreversible transfer of heat from the reservoir to the steam accounts for the difference in $\epsilon$ and $\epsilon'$.

(c) The case in Example 8.m may be viewed as a process in which decrease in the availability of the condensing steam is the input quantity and the output is the availability increase of the heated air stream. Thus we define the effectiveness for the heat-transfer process as the ratio of the increase in availability of the air to the decrease in availability of the steam. Therefore

$$\epsilon = \frac{[\dot{m}\,(b_{out} - b_{in})]_{air}}{[\dot{m}\,(b_{in} - b_{out})]_{steam}}$$

Using eq. (8.11a), we obtain for the steam

$$\begin{aligned} b_{in} - b_{out} &= h_{in} - h_{out} + T_0(s_{out} - s_{in}) \\ &= 895.1 + (537)(-1.1475) = 278.9 \text{ B/lbm} \end{aligned}$$

and for the air

$$\begin{aligned} b_{out} - b_{in} &= h_{out} - h_{in} + T_0\,(s_{in} - s_{out}) \\ &= 0.24(305.3 - 60) + (537)(-0.09987) \\ &= 5.244 \text{ B/lbm} \end{aligned}$$

Substituting into the above expression for $\epsilon$

$$\epsilon = \frac{77.93(5.244)}{5.126(278.9)} = 0.286$$

A significant decrease in availability is indicated and is due to the inherently irreversible heat-transfer process.

---

## 8–9 Summary

Thermodynamic systems undergo processes in an atmosphere that is assumed to be at a uniform pressure $p_0$ and temperature $T_0$ and incapable of producing work by itself. In second-law analysis the atmosphere, also termed the environment, serves as the only heat source or sink regardless of the nature of the actual heat transfer for a process. With heat exchange between the system and the atmosphere only, the maximum work for a given state change of a system will be realized during any reversible process between the end states. The general expression for maximum work in differential form is

$$\begin{aligned} \delta W_{max} = &-dE_s + T_0 dS_s + \Sigma(e_1 + p_1 v_1 - T_0 s_1)dm_1 \\ &- \Sigma(e_2 + p_2 v_2 - T_0 s_2)dm_2 \end{aligned} \quad \textbf{(8.1a)}$$

where the subscript $s$ indicates the system, $dm_1$ is an inbound mass crossing the system boundary, and $dm_2$ is mass flowing out across the system boundary. For a simple closed system the maximum work per unit mass is

$$w_{max} = -(u_2 - u_1) + T_0(s_2 - s_1) \tag{8.2a}$$

and for heat transfer $Q_r$ from a heat reservoir at $T_r$

$$W_{max,r} = Q_r\left(\frac{T_0}{T_r} - 1\right) \tag{8.4}$$

Steady-flow processes with one inlet and one outlet stream have, on a unit mass basis,

$$w_{max} = \left(h_1 + \frac{V_1^2}{2g_0} - T_0 s_1\right) - \left(h_2 + \frac{V_2^2}{2g_0} - T_0 s_2\right) \tag{8.3a}$$

when potential energies are neglected.

Useful work is the work for a process minus the work done on the atmosphere during the process. The work done on the atmosphere is

$$W_{atm} = p_0(V_2 - V_1) \tag{8.5}$$

where $p_0$ is the pressure of the atmosphere and $V_1$ and $V_2$ are the initial and final volumes of the system respectively. Actual useful work and maximum useful work are found by subtracting $p_0(V_2 - V_1)$ from the actual work for a process and the maximum work to give

$$W_u = W - p_0(V_2 - V_1)$$

and

$$W_{max,u} = W_{max} - p_0(V_2 - V_1) \tag{8.6}$$

The availability of a system at a given state (1) is the maximum useful work that could be obtained if the system, exchanging heat only with the atmosphere, proceeds reversibly to its dead state, denoted by the subscript 0, where it is in equilibrium with the atmosphere. The availability, on a unit mass basis, for a simple closed system is then

$$a_1 = {}_1w_{0,\,max,u} = (u_1 + p_0 v_1 - T_0 s_1) - (u_0 + p_0 v_0 - T_0 s_0)$$

The availability of a fluid at any state during a steady-flow process is, per unit mass,

$$b_1 = {}_1w_{0,max,u} = \left(h_1 + \frac{V_1^2}{2g_0} - T_0 s_1\right) - (h_0 - T_0 s_0) \tag{8.11}$$

Note that there is no kinetic energy at the dead state where the fluid is in equilibrium with the atmosphere. Further, it is assumed that the atmospheric work is zero since steady-flow systems do not experience a change in control volume. With no change in the volume of a system operating in steady flow, the maximum work (equal to the maximum useful work) will be the decrease in $b$, the steady-flow availability.

The irreversibility of a process is, by definition,

$$I \equiv W_{max} - W \tag{8.13}$$

where $W$ is the actual work for the process. Alternatively the irreversibility of a process is

$$I = W_{max,u} - W_u$$

where the maximum useful work is equal to the decrease in availability during a process in which $W_u$ is the actual useful work. For a closed system the irreversibility of a process also can be determined as

$$I = -Q_s + T_0(S_2 - S_1)_s \tag{8.15a}$$

where the subscript $s$ indicates quantities for the system and $Q_s$ is the actual heat transfer for the system. The irreversibility of a single-stream steady-flow process is, per unit mass,

$$i = -{}_1q_2 + T_0(s_2 - s_1) \tag{8.15b}$$

From the two preceding equations the irreversibility of any *adiabatic* process of either a closed system or a steady-flow process is

$$i = T_0(s_2 - s_1) \tag{8.15d}$$

on a unit mass basis.

## Selected References

Gaggioli, R. A. "The Concept of Available Energy." *Chemical Engineering Science,* vol. 16, 1961.
Holman, J. P. *Thermodynamics,* 3rd ed. New York: McGraw-Hill, 1980.
Keenan, J. H., and G. N. Hatsopoulos. *Principles of General Thermodynamics.* New York: Wiley, 1965.
Moran, M. J. *Availability Analysis: A Guide to Efficient Energy Use.* Englewood Cliffs, N.J.: Prentice-Hall, 1982.
Reistad, G. M. "Available Energy Conversion and Utilization in the United States." American Society of Mechanical Engineers, Paper No. 74-WA/Pwr-1, November, 1974.
Sonntag, R. E., and G. J. Van Wylen. *Fundamentals of Classic Thermodynamics,* 2nd ed. New York: Wiley, 1976.
Wark, K. *Thermodynamics,* 3rd ed. New York: McGraw-Hill, 1977.

## Problems

For each of the problems below the atmosphere is at 77 F (25 C) and 14.70 psia (101.4 kPa) unless otherwise specified.

8.1 A closed system of 0.105 lbm of hydrogen undergoes a process from an initial state at 60 F and 12.5 psia to a final state where the pressure is 20 psia and the temperature is 200 F. During the process 5 B of heat transfer from a reservoir at 300 F to the system occurs. Calculate the maximum work.

8.2 A rigid, insulated container has air and a metal block as its contents. The metal block, initially at 370 C, has a mass of 1.14 kg and a specific heat of 0.0955 kJ/(kg·K). The air has a volume of 0.23 $m^3$ and is initially at −7 C and 97 kPa. Determine the maximum work for a process in which the air and the block come to thermal equilibrium.

8.3 During a steady-flow process $H_2O$ is throttled from 150 C, saturated liquid, to 85 kPa. What is the maximum work, per kilogram of $H_2O$, for the process?

8.4 A closed system of 1.5 lbm of nitrogen is compressed isothermally and reversibly from 14.7 psia to 147 psia at a temperature of 300 F. Calculate the actual work and the maximum work. Compare the results. Sketch a *T-s* diagram showing the given process and the paths that would be required for reversible processes between the end states if heat transfer were with the atmosphere only.

8.5 A rigid tank with a volume of 50 $ft^3$ initially contains air at 2.50 psia and 10 F. A valve is opened and air from the atmosphere flows slowly into the tank. As the air flows into the tank, heat is transferred from it to the atmosphere. Flow continues until the pressure and temperature of the air in the tank are equal to atmospheric pressure and temperature. Calculate the maximum work.

8.6 A closed system of 2.00 kg of carbon dioxide is at 20 C and 100 kPa. The carbon dioxide undergoes a process to a final state where the pressure is 250 kPa and the temperature is 150 C. During the process 15 kJ of heat transfer from a heat reservoir at 200 C to the system takes place. What is the maximum work for the process?

8.7 A rigid, insulated tank is completely evacuated. Atmospheric air flows slowly into the tank until the pressure is 101.4 kPa. Calculate (a) the final temperature of the air in the tank and (b) maximum work for the process.

8.8 Two streams of air mix adiabatically during a steady-flow process to form one leaving stream. One of the entering streams flows at the rate of 40 lbm/min and enters the mixing chamber at 15.0 psia and 300 F with a velocity of 100 ft/sec. The other entering stream is at 15.0 psia, 40 F, and flows at the rate of 1100 $ft^3$/min. The leaving stream is at 15.0 psia. Calculate the maximum work for the mixing process in B/min.

8.9 A rigid, insulated container is divided into two equal volumes by a partition. On one side of the partition is 1 kg of air at 0.350 MPa and 25 C. The other side is a perfect vacuum. The partition is ruptured and the system proceeds to equilibrium. Calculate the maximum work for this unrestrained expansion.

8.10 Fifteen kg/min of water at 5 C mix adiabatically with 5 kg/min of water at 70 C, with the pressure remaining constant at 101.4 kPa. Calculate the maximum work for the steady-flow mixing process.

8.11 In a steam generator steam enters the superheater section at 1500 psia with a quality of 99.0 percent and leaves at 1450 psia and 1000 F. Neglect velocities at the superheater inlet and outlet. Assume the heat transfer to the steam comes from products of combustion that serve as a constant-temperature heat reservoir at 2000 F. Calculate the maximum work on the basis of a unit mass of steam.

8.12 A rigid tank with a volume of 0.328 $m^3$ contains nitrogen at 483 kPa and 260 C. What is the availability of the nitrogen? If heat exchange took place between the nitrogen and the atmosphere until they were at the same temperature, what would be the availability of the nitrogen?

8.13 A steam turbine operates adiabatically in steady flow with negligible kinetic energies at the inlet and outlet. Steam enters the turbine at 2.75 MPa and 370 C. The exhaust pressure is 0.085 MPa. If the expansion through the turbine is reversible, what is the work per kilogram? How does this compare with the decrease in availability per kilogram? Now suppose the actual turbine work were 85 percent of that for the isentropic expansion. What would be the ratio of work to the decrease in availability? Note that the exhaust state will be at a different entropy and enthalpy than for the isentropic expansion.

8.14 A rigid, uninsulated tank has a volume of 4.736 $ft^3$. The tank contains carbon dioxide initially at 40 F and 12.0 psia. Paddle-wheel (stirring) work is done on the $CO_2$ in the amount of 27.34 B as it goes to a final state at which the temperature is 300 F. Determine the availability of the $CO_2$ at the initial state and at the final state.

8.15 Refer to problem 8.9. Calculate the availability of the system bounded by the container walls at the initial state and at the final state.

8.16 At the outlet of a steam nozzle the pressure is 0.150 MPa, the quality 90 percent, and the velocity 457 m/s. What is the availability per kilogram of steam?

8.17 In a vapor-compression refrigeration cycle Freon 12 enters the adiabatic compressor as saturated vapor at $-5$ F and leaves at 200 psia and 180 F. Assuming negligible velocities, determine the actual compressor work and the change in availability of the Freon 12, each on the basis of 1 pound mass of the refrigerant.

8.18 Steam enters a turbine at 600 psia, 900 F, and expands adiabatically to 1.50 psia at the turbine exhaust. Due to irreversibilities the actual turbine work is 85 percent of that which would be realized had the process been reversible. Assuming negligible kinetic energies, calculate the actual turbine work and the decrease in availability per pound mass of steam. If the turbine had operated reversibly and adiabatically, how would the turbine work compare with the decrease in availability of the steam?

8.19 A piston and cylinder contain 1 $ft^3$ of air at initial conditions of 50 psia and 20 F. During a process that takes place at constant pressure, 4.75 B of paddle-wheel work is done on the air along with 10 B of heat transfer to the air from a 1540 F reservoir. Calculate (a) the work done on the moving piston face by the air, (b) the decrease in availability of the heat reservoir, and (c) the change in availability of the air.

8.20 A rigid container is divided into two equal volumes by a partition. One part contains 1 kg of air at 210 kPa and 50 C. The other is a perfect vacuum. The partition is ruptured and 116 kJ of heat is transferred from a reservoir at 315 C to the container contents, which reach equilibrium after the heat transfer is completed. Determine (a) the initial availability of the system within the container, (b) the final availability of the system within the container, and (c) the maximum useful work for the process.

8.21 A rigid tank with a volume of 15 ft³ contains air at 2.00 psia and 40 F. A valve between the tank and the atmosphere is opened, allowing atmospheric air to flow into the tank until the air in the tank is at the same pressure and temperature as the atmosphere. Noting that the process is not adiabatic, determine the maximum useful work and the irreversibility for the process.

8.22 Determine the irreversibility of the process described in problem 8.9.

8.23 Determine the irreversibility of the throttling process described in problem 8.3.

8.24 An insulated heat exchanger has air flowing over the outside of a tube bank while steam flows through the tubes. Steam enters the tubes as dry and saturated vapor at 150 kPa and leaves as saturated liquid at the same pressure. Air enters the heat exchanger at 20 C and 110 kPa and leaves at 100 kPa. The terminal temperature difference (temperature of the leaving heating fluid minus the temperature of the leaving heated fluid) is 8 C. Air enters the heat exchanger at the rate of 15 m³/min. Neglecting kinetic energies, calculate (a) the ratio of the availability increase of the air to the availability decrease of the steam on a consistent basis and (b) the irreversibility in kJ/min.

8.25 Steam at 200 psia and 520 F flows at the rate of 20,000 lbm/hr into a desuperheater. Liquid $H_2O$ at 200 F is sprayed into this steam to produce saturated vapor at 190 psia leaving the unit. The desuperheater is perfectly insulated and the process is steady flow with negligible velocities. Determine the amount of liquid $H_2O$ introduced and the irreversibility of the process, each on an hourly basis.

8.26 Determine the irreversibility of the mixing process described in problem 8.10.

8.27 Consider the system shown below. The rigid container is perfectly insulated. The mass of the piston is such that the air pressure remains constant at 20 psia. For the process in which the system comes to thermal equilibrium (assume the piston and container walls have specific heat of zero), calculate (a) the equilibrium temperature of the air–aluminum composite system; (b) the decrease in availability of the vacuum, the air, the aluminum ball, and the piston; and (c) the irreversibility of the process.

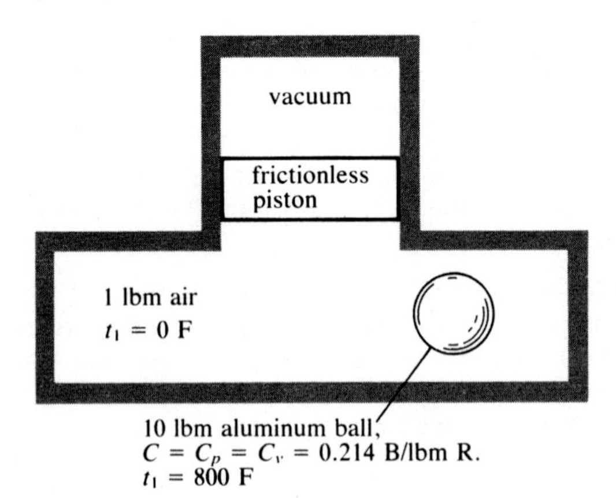


8.28 Steam at 400 psia and 600 F enters a turbine with negligible velocity. At the turbine exhaust the pressure is 50 psia, the quality 100 percent, and the velocity 400 ft/sec. Assuming the expansion through the turbine is adiabatic, determine (a) the ratio of turbine work to the decrease in availability of the steam and (b) the irreversibility of the process per pound mass of steam.

8.29 In Chapter 6 it was noted that direct heat transfer across a finite temperature difference is an irreversible process. Consider a process in which a metal block having a specific heat of 0.300 B/lbm-R and a mass of 5 pounds is initially at 1000 F. Heat is transferred from the metal block to 1.0 lbm of $H_2O$, which is initially saturated liquid at 68 psia. Consider the metal block and the $H_2O$ to be closed systems. For processes during which the pressure of the $H_2O$ remains constant, determine the irreversibility when the temperature of the metal block has decreased to 800 F, 600 F, 400 F, and the minimum possible value for the direct heat interaction.

8.30 An insulated steady-flow heat exchanger utilizes the exhaust from a gas turbine to generate low-pressure steam. The exhaust gas enters the heat-recovery steam generator at 425 C and leaves at 190 C. Feed water enters at 15 C and the steam leaves at 200 C. The exhaust gas pressure is constant at 110 kPa and the water side pressure is constant at 0.36 MPa. Assuming the exhaust gas has the same properties as dry air, calculate the irreversibility of the process per kilogram of $H_2O$.

8.31 One kilogram of steam at 6.90 MPa and 300 C is heated to 540 C during a steady-flow process. The process takes place at constant pressure with no work and negligible change in kinetic energy. A single constant-temperature heat reservoir at the lowest possible temperature is the heat source. Determine the irreversibility of the process.

8.32 A pressure vessel that has a volume of 10 $ft^3$ contains air at 500 psia and 77 F. A valve on the tank is opened and air flows slowly from the tank until the pressure of the air drops to 200 psia when the valve is closed. Assume there is no heat transfer during this process. Calculate (a) the availability of the air in the tank before outflow begins and (b) the availability of the air in the tank when its pressure is 200 psia. After some time the air remaining in the tank comes to thermal equilibrium with the atmosphere due to heat exchange with the atmosphere. Calculate the irreversibility of this heat-exchange process.

8.33 Steam flows slowly from a line where the pressure is 2.75 MPa and the temperature is 370 C into an insulated tank that is initially evacuated. Assuming there is no heat transfer from steam in the tank and the final pressure in the tank is 2.75 MPa, calculate (a) the final temperature of the steam in the tank and (b) the irreversibility of the process.

8.34 Based on your experience with availability, define an effectiveness for a turbine. Apply your definition to the steam turbine in problem 8.13.

8.35 Apply heat engine effectiveness as defined in Example 8.o to the energy conversion system in problem 7.42.

8.36 What is the value of the second-law effectiveness for the heat-transfer process in problem 8.30?

# 9

# Vapor Cycles for Power and Refrigeration

*In this chapter we study cycles for which the working fluid undergoes phase changes alternating between liquid and vapor. Such cycles are of great importance in the production of power and in providing refrigeration, as well as in the operation of heat pumps. Actual cycles for the generation of power in large quantities, such as those used in the production of electricity, utilize water as the working fluid. While, as we shall see, water is not an ideal fluid for such cycles, its availability and low cost adequately offset its undesirable characteristics. The relatively high triple-point temperature and low triple-point pressure of water make it an unsuitable working fluid for vapor refrigeration cycles. Several other suitable refrigerants exist. Heat sources for power cycles include the hot gases (products of combustion) produced by burning fossil fuels and the core of a boiling water (BWR) or pressurized water (PWR) nuclear reactor. Heat from the working fluid of a vapor power cycle is usually rejected to circulating water flowing through a condenser. Heat sources and sinks for refrigeration cycles are discussed in the sections pertaining to those cycles. As the several power, refrigeration, and heat pump cycles are encountered, note that each cycle and its associated end use result in the eventual dissipation of all energy to the atmosphere.*

*It is convenient and instructive to describe cycles on property diagrams, particularly the temperature–entropy diagram. In the discussion of the various cycles in this chapter, we will assume that the heat-transfer processes are at least internally reversible. This assumption is a reasonable one and it permits us to treat heat transfers as areas on* ***T-s*** *diagrams.*

## 9–1 The Carnot Vapor Power Cycle

The Carnot corollaries establish the upper bound of the thermal efficiency of heat engines. A Carnot power cycle is a reversible heat engine cycle theoretically capable of operating between a heat source at $T_H$ and a heat sink at $T_L$ at this maximum thermal efficiency. Figure 9.1 shows a Carnot power cycle operating within the saturation lines of a pure substance. The thermal efficiency of the cycle would not be changed if states (1) and (2) were in the superheat region or if states (3) and (4) were in the compressed-liquid region.

For all heat engines the thermal efficiency is defined as

$$\eta = \frac{w_{\text{net}}}{q_{\text{in}}} = \frac{q_{\text{in}} - q_{\text{out}}}{q_{\text{in}}} \tag{3.5a}$$

**Figure 9.1** A Carnot Vapor Power Cycle. *a.* $p$-$v$ diagram. *b.* $T$-$s$ diagram

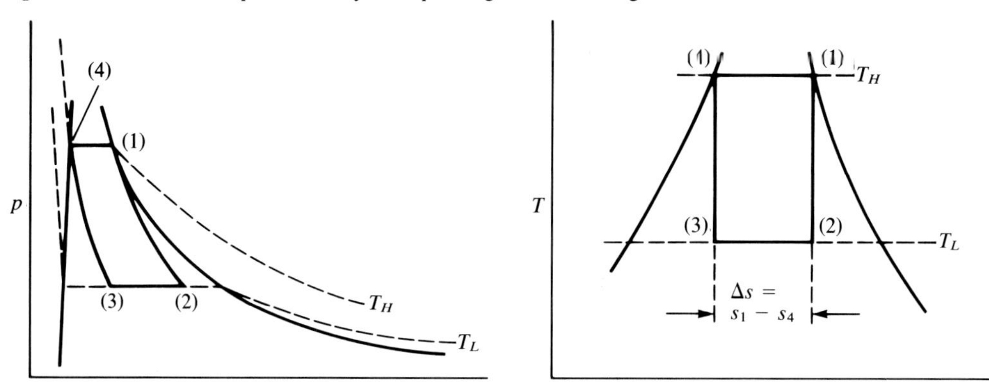


From the definition of entropy change

$$\delta q_{\text{rev}} = T ds$$

Thus the heat-transfer quantities for the Carnot cycle illustrated are

$$_4q_1 = q_{\text{in}} = T_H \, \Delta s$$

and

$$|_2q_3| = q_{\text{out}} = T_L \, \Delta s$$

The thermal efficiency of the Carnot cycle is then

$$\eta = \frac{T_H \, \Delta s - T_L \, \Delta s}{T_H \, \Delta s} = \frac{T_H - T_L}{T_H}$$

which is eq. (6.4).

## 9–2 The Rankine Cycle

There are at least two major difficulties encountered in attempting to design a vapor power plant that will operate on the Carnot power cycle. The first of these deals with the fact that heat transfer across the infinitesimal temperature difference required is completely impractical since, as discussed in Section 6–2, either an infinite time or an infinite heat-transfer area is required for a given amount of heat to be transferred. Second, the isentropic compression process (3) to (4) in Figure 9.1 is difficult to carry out in practice. Therefore, we will consider the simplest practical vapor power cycle, the Rankine cycle, the components of which are described in Figure 9.2. In this cycle vapor expands through the turbine, producing work. On passing through the condenser, it is condensed to liquid. The liquid is returned by

**Figure 9.2** The Rankine Vapor Power Cycle

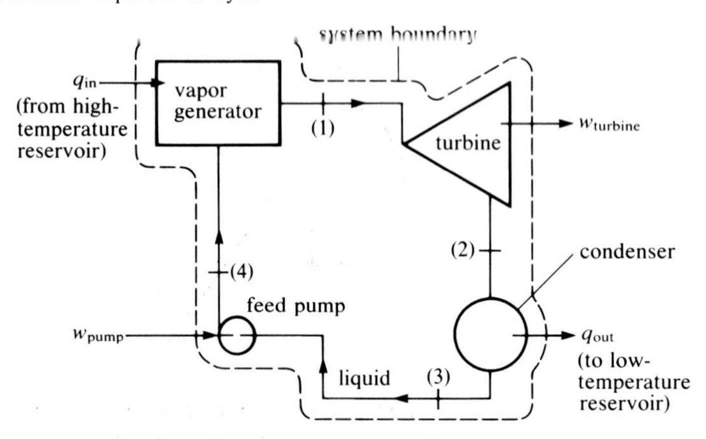


the feed pump to the vapor generator, where it is vaporized by heat addition. We shall assume for the ideal Rankine cycle that no state changes of the working fluid occur between the components and that the feed pump and the prime mover—usually a turbine—operate reversibly and adiabatically. Further, we assume that the heat-transfer processes in the vapor generator and the condenser are internally reversible. Therefore, there are no pressure changes as the working fluid passes through these components, and the areas below the paths for these processes on a *T-s* diagram represent the heat transfer. While each component in the cycle operates in a steady-flow manner, the composite system enclosed within the indicated boundary is a closed system. Thus the thermal efficiency of the cycle can be determined as

$$\eta = \frac{w_{net}}{q_{in}} = \frac{q_{net}}{q_{in}} = \frac{{}_4q_1 + {}_2q_3}{{}_4q_1} = \frac{{}_4q_1 - |{}_2q_3|}{{}_4q_1}$$

The heat or work quantities for any one device in the cycle are determined by application of the steady-flow energy equation. In this as well as other vapor cycles, changes in kinetic and potential energies are assumed to be small. Evaluation of the numerator of the thermal efficiency expression by determination of either the net work or the net heat is satisfactory, but the analysis for simple cycles is usually simplified by use of the latter approach.

An ideal Rankine cycle with $H_2O$ as the working fluid is shown on the *T-s* diagram in Figure 9.3. The indicated states are numbered to correspond to the sections of Figure 9.2. Note that state (3) is on the saturated-liquid line and that process (3) to (4) involves the pumping of liquid only from the condenser pressure to the pressure of the vapor generator. Thus state *(f)* on the *T-s* diagram is located inside the vapor generator. The temperatures $T_H$ and $T_L$ in the figure represent, respectively, the temperatures of the heat source and the heat sink (the atmosphere) for the cycle. For simplicity we assume that the heat source is a heat reservoir (a constant-temperature heat source). The transfer of heat across the finite temperature difference between the source at $T_H$ and the working fluid during the heat-reception process (4) to *(f)* to (1) is an inherently irreversible process, but the finite temperature difference is required to promote the heat transfer. A similar irreversible heat transfer exists between the system and $T_L$ for process (2) to (3).

**Figure 9.3** The Temperature–Entropy Diagram for an Ideal Rankine Cycle

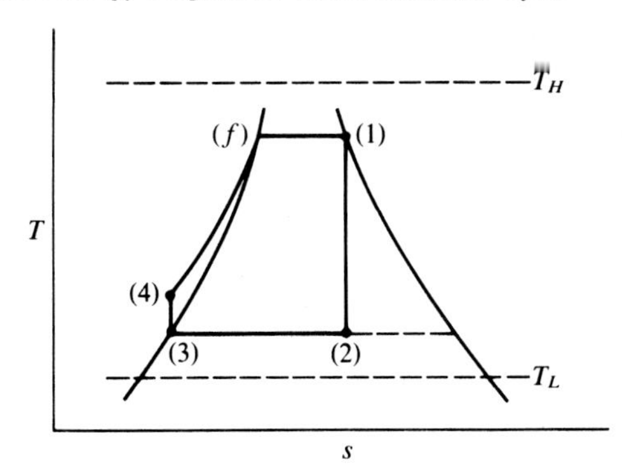


As noted, the Rankine cycle components operate in a steady-flow manner. As the various cycles are studied, it will become apparent that enthalpies must be determined at each section to perform a complete cycle analysis. For the process in which heat is transferred to the working fluid

$$q_{in} = {}_4q_1 = h_1 - h_4$$

The heat transferred from the working fluid is

$$q_{out} = |{}_2q_3| = |h_3 - h_2|$$

The enthalpy at (4) is determined using the second $Tds$ equation

$$Tds = dh - vdp$$

For the pump operating reversibly and adiabatically eq. (7.9) reduces to

$$dh = vdp$$

For moderate pressure increases through the pump the specific volume can be assumed constant with the result

$$\int_3^4 dh = v_3 \int_3^4 dp = v_3(p_4 - p_3) \qquad \textbf{(a)}$$

The steady-flow energy equation for the boiler feed pump operating adiabatically is

$$h_3 = h_4 + {}_3w_4 \qquad \textbf{(b)}$$

Combination of eqs. (a) and (b) for the pump yields

$$|{}_3w_4| = w_p = h_4 - h_3 \simeq v_3(p_4 - p_3) \qquad \textbf{(9.1)}$$

The same result can be obtained by application of eq. (2.14). The turbine expansion, assumed reversible and adiabatic, and thus isentropic, produces work

$$w_t = {}_1w_2 = h_1 - h_2 \tag{9.2}$$

Each process has been treated on the basis of a unit mass of the working fluid. Such a basis is recommended for vapor cycles whenever possible. Conversion to a rate basis from a unit mass basis is readily accomplished by multiplying work or heat-transfer quantities by the mass rate of flow.

## Example 9.a

An ideal Rankine cycle has saturated steam at 190 C entering the turbine and saturated liquid at 35 C leaving the condenser. The cycle receives heat from a reservoir at 370 C and rejects heat to the atmosphere at 26 C. Calculate the cycle thermal efficiency and compare the result with the efficiency of a Carnot power cycle operating between the two reservoirs.

**Solution:**

For the Rankine cycle, referring to Figures 9.2 and 9.3,

$$\eta = \frac{q_{in} - q_{out}}{q_{in}} = \frac{(h_1 - h_4) - (h_2 - h_3)}{h_1 - h_4}$$

From Table B–1, Appendix B, $h_1 = 2786.4$ kJ/kg and $h_3 = 146.68$ kJ/kg. For the turbine

$$\begin{aligned} s_1 = 6.5079 = s_2 &= s_{f_2} + (x_2)(s_{fg_2}) \\ &= 0.5053 + x_2\,(8.3531 - 0.5053) \end{aligned}$$

Solving for $x_2$ we have

$$x_2 = 0.7649$$

The enthalpy at the turbine exhaust is

$$\begin{aligned} h_2 &= 146.68 + 0.7649(2565.3 - 146.68) \\ &= 1996.7 \text{ kJ/kg} \end{aligned}$$

The boiler feed pump increases the pressure from 5.628 kPa (the saturation pressure at 35 C) to 1.2544 MPa (the saturation pressure for 190 C). Assuming the specific volume remains constant at $v = v_3$ during the pump process, the pump work input is, by eq. (9.1),

$$\begin{aligned} |{}_3w_4| = w_p &= 1.006 \times 10^{-3}(1254.4 - 5.628) \\ &= 1.256 \text{ kJ/kg} \end{aligned}$$

Because the pump work is small compared with the turbine work, $h_1 - h_2$, it can be neglected when the pressure rise is small. We shall neglect pump work only when such an omission is specifically stated. From eq. (9.1)

$$\begin{aligned} h_4 &= w_p + h_3 \\ &= 1.256 + 146.68 = 147.94 \text{ kJ/kg} \end{aligned}$$

The thermal efficiency of this Rankine cycle is

$$\begin{aligned} \eta &= \frac{(2786.4 - 147.94) - (1996.7 - 146.68)}{2786.4 - 147.94} \\ &= \frac{789.7}{2638.5} = 29.9 \text{ percent} \end{aligned}$$

Note that the numerator, the net heat transfer for the cycle, is equal to the net work of the cycle by the first law. Therefore, the turbine work is

$$\begin{aligned} w_t &= w_{net} + w_p \\ &= 789.7 + 1.256 \\ &= 791.0 \text{ kJ/kg} \end{aligned}$$

For a Carnot cycle operating between the two reservoirs the thermal efficiency is

$$\eta = \frac{T_H - T_L}{T_H} = \frac{370 - 26}{370 + 273} = 0.535$$

Thus, in the terminology of Section 7–9, 53.5 percent of the heat supplied at the reservoir temperature is available energy, that is, 53.5 percent could be converted to work by a reversible engine. However, because of the irreversibility of the heat-transfer processes, particularly that from the source to the working fluid during which a significant degradation of heat energy occurs, this Rankine cycle converts only about 30 percent of the heat supplied to work.

---

## 9–3 Some Effects of Pressure, Temperature, and Properties of the Working Fluid on the Rankine Cycle

In Figure 9.4 two temperature–entropy diagrams for Rankine cycles are shown. A segmented temperature–entropy diagram for a Rankine cycle with $H_2O$ as the working fluid is given in Figure 9.4a. Since each process is internally reversible, the area below the path is equal to the heat transfer for the process. Let us consider portions of the Rankine cycle shown on the diagram. First, inspect the segment described by the paths $f$–$x$–3–4–$f$. This constitutes a cycle that receives heat in the process (4) to ($f$) and rejects heat in the process ($x$) to (3). The enclosed area $A$ is the net heat transfer for the cycle and, by the first law for a

**Figure 9.4** Temperature–Entropy Diagrams for the Rankine Cycle. *a.* Rankine cycle with $H_2O$ as the working fluid. *b.* *T-s* diagram for a desirable vapor power cycle working fluid.

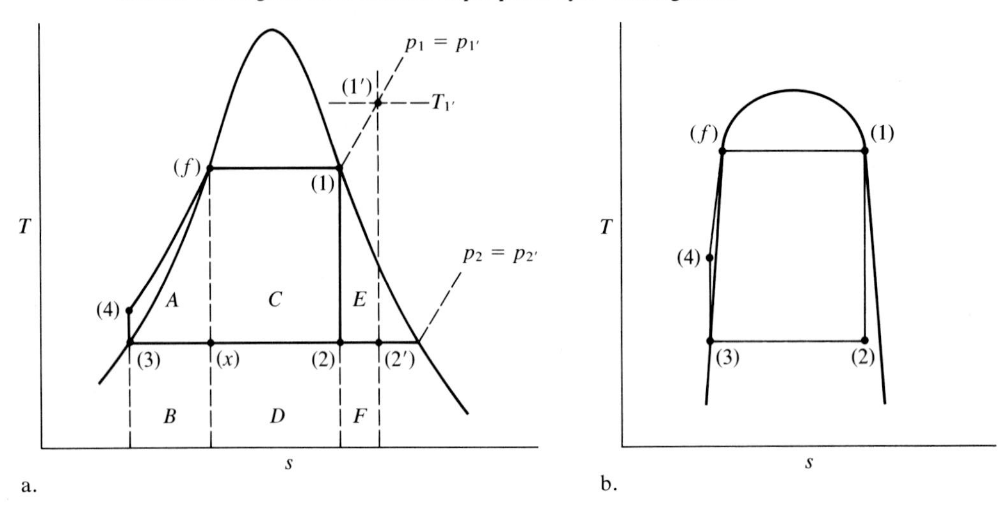


cycle, this area is also the net work. Therefore, this segment of the Rankine cycle has a thermal efficiency of

$$\eta_{AB} = \frac{\text{area } A}{\text{area } A + \text{area } B}$$

Next consider the segment described by 1–2–*x*–*f*–1 enclosing area *C*. By reasoning similar to the foregoing

$$\eta_{CD} = \frac{\text{area } C}{\text{area } C + \text{area } D}$$

It is evident from the *T-s* diagram that $\eta_{AB}$ is less than $\eta_{CD}$, and that the thermal efficiency of the Rankine cycle 1–2–3–4–*f*–1 lies between $\eta_{AB}$ and $\eta_{CD}$. Now let us consider the Rankine cycle with superheat, which is the cycle in Figure 9.4a defined by 1′–2′–3–4–*f*–1–1′. Considering the segment 1′–2′–2–1–1′ that encloses area *E* and the area below, area *F*, we obtain

$$\eta_{EF} = \frac{\text{area } E}{\text{area } E + \text{area } F}$$

Comparing the thermal efficiency of the segments of the Rankine cycle with superheat, we obtain

$$\eta_{AB} < \eta_{CD} < \eta_{EF}$$

From this we draw two important conclusions. First, we note from the *T-s* diagram that the Rankine cycle with superheat has a greater thermal efficiency than the Rankine cycle without superheat. Second, if we think of adding cycle segments *CD* and *EF* to the cycle composed of areas *A* and *B*, we note that each segment added increases the thermal efficiency of the resulting cycle. As the segments are added, the average temperature of the working fluid for the heat-reception process increases. From this we state a general rule applicable to all power cycles that any process that tends to increase the average temperature of the working fluid during the cycle heat-reception process will increase the thermal efficiency.

An increase in the steam-generator pressure $p_1$ will result in an increase in the cycle thermal efficiency for a fixed condenser pressure $p_2$. However, due to the diminishing strength with increasing temperature of the materials containing the working fluid in the steam generator, the temperature at the turbine inlet has an upper limit. Therefore, when operating at this limiting temperature, an increase in the steam-generator pressure will result in a decrease in the quality of the mixture at the turbine exhaust. This can be illustrated by considering increases in $p_1$ in Figure 9.4a with temperature $T_{1'}$ fixed. As $p_1$ increases, states (1′) and (2′) move to the left and the quality $x_{2'}$ decreases. This condition is significant in actual power plant cycles since high liquid content ($1 - x$ greater than about 10 percent), tends to reduce turbine performance and to increase erosion of the exhaust-end turbine blading. Reduction of condenser pressure for a given turbine inlet state also results in an increase in the thermal efficiency of the Rankine cycle. This effect, reflected by a decrease in areas *B, D,* and *F* of Figure 9.4a, is limited by the temperature of the circulating water to which heat is rejected in the condenser. Practically, colder circulating water is available during winter months. The result for a fixed turbine inlet state is a modest increase in cycle efficiency during winter months as contrasted with summer months.

The use of $H_2O$ as the working fluid in vapor power cycles is dictated by its availability, nontoxicity, and low cost. However, $H_2O$ has a number of characteristics that are undesirable. These include the extremely high saturation pressures for high temperatures and, at the condenser, the low saturation pressures at ambient temperatures. The first of these demands heavier walls for piping from the boiler feed pump to the steam-generator inlet, in the tubes of the steam generator, and the casing walls of the high-pressure section of the turbine. The second requires the exhaust end of the turbine and the condenser to operate at high vacuums if heat rejection is to take place at the desirable low temperatures.

The properties of the working fluid influence the shape of the *T-s* diagram for a Rankine cycle. The slope of a constant-pressure line in the liquid region on the *T-s* diagram can be determined from the relation $Tds = dh - vdp$ in which $dh = C_p dT$ for a liquid. This produces the slope as

$$\left(\frac{\partial T}{\partial s}\right)_p = \frac{T}{C_p}$$

Referring to Figure 9.4b, the desirability of a low-constant-pressure specific heat of the liquid is evident. For a liquid with a very small $C_p$, process 4–*f* approaches a vertical line, a condition that would allow an ideal Rankine cycle to approach Carnot cycle behavior.

Other characteristics of an optimum vapor power cycle fluid include a high critical temperature, high values for $h_{fg}$ at the steam-generator pressure, good heat-transfer characteristics at both high and low temperatures, and a nearly vertical saturated-vapor line on a *T-s* diagram. The last condition would reduce the percentage of liquid in the turbine exhaust. Unfortunately readily available substances that are otherwise suitable as working fluids for power cycles do not exhibit the characteristics displayed in Figure 9.4b.

Figure 9.5 Schematic Diagram of Components and $T$-$s$ Diagram for the Reheat Cycle

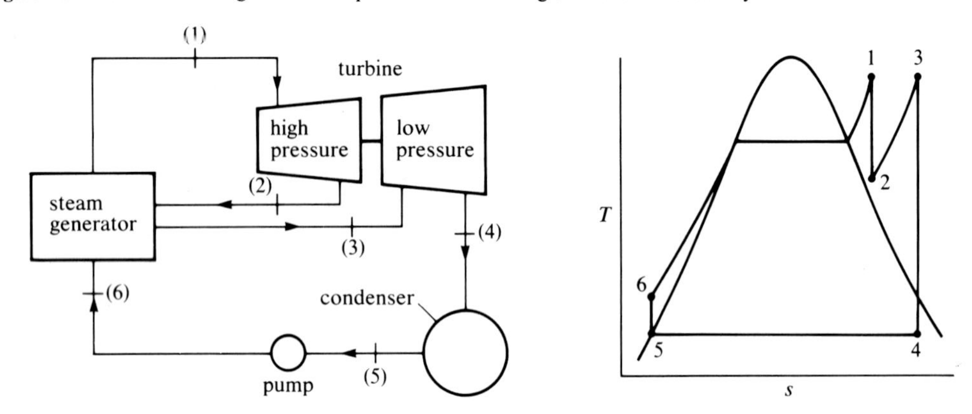


## 9–4 The Reheat Cycle

One Rankine cycle limitation discussed in the previous section is that as increasing steam-generator pressures are considered at the limiting working fluid temperature, the liquid content of the turbine exhaust steam increases to unacceptable values. A modification of the basic Rankine cycle that alleviates this problem and offers the possibility of increasing the cycle efficiency is the reheat cycle. Figure 9.5 shows a diagram of the components and the routing of the working fluid as well as the $T$-$s$ diagram for a reheat cycle with $H_2O$ as the working fluid. The steam from the steam generator expands through the high-pressure section of the turbine to an intermediate pressure and is returned to the steam generator to be reheated before it is expanded through the low-pressure section of the turbine to the condenser pressure. It is evident from the $T$-$s$ diagram that the quality of the mixture at the exhaust of the low-pressure turbine is greater than that which would result if reheat had not occurred.

The thermal efficiency of the reheat cycle is not necessarily greater than the simple Rankine cycle efficiency. Only if the average temperature of the working fluid for the reheat process (2) to (3) is greater than the average temperature for the heat-reception process (6) to (1) will there be an increase. Calculations taking into account the quality of the low-pressure turbine exhaust must be performed to determine the optimum reheat pressure for each case. In the computation of the thermal efficiency it is important to recognize for the reheat cycle that heat is transferred to the working fluid directly from the high-temperature heat source in process (2) to (3) as well as process (6) to (1), and that the work for both the high-pressure and low-pressure turbine sections must be included.

## Example 9.b

The following information is given for a reheat cycle such as that shown in Figure 9.5

| State | $p$, psia | $t$, F |
|---|---|---|
| 1 | 400 | 700 |
| 2 | 100 | |
| 3 | 100 | 700 |
| 4, 5 | 1 | |
| 6 | 400 | |

(a) Find the thermal efficiency of the reheat cycle and compare it with that of the similar cycle without reheat.
(b) For a turbine inlet mass flow rate of 5000 lbm/hr find the power produced by the turbine.

**Solution:**

(a) The thermal efficiency for the cycle is

$$\eta = \frac{{}_1w_2 + {}_3w_4 - w_p}{{}_6q_1 + {}_2q_3}$$

in which each term has B/lbm as its units. Using the *Steam Tables* and the given information, the following properties are obtained.

| State | $p$, psia | $t$, F | $h$, B/lbm | $s$, B/(lbm R) | Condition |
|---|---|---|---|---|---|
| 1 | 400 | 700 | 1362.5 | 1.6397 | Superheat |
| 2 | 100 | 380 | 1216.8 | 1.6397 | Superheat |
| 3 | 100 | 700 | 1379.2 | 1.8033 | Superheat |
| 4 | 1 | 101.7 | 1007.7 | 1.8033 | $x = 0.905$ |
| 5 | 1 | 101.7 | 69.74 | — | $x = 0$ |
| 6 | 400 | — | 70.93 | — | Subcooled |

The pump work was used in the evaluation of $h_6$ and is equal to 1.12 B/lbm. Writing the steady-flow energy equation for each process, we obtain

$${}_1w_2 = h_1 - h_2 = 1362.5 - 1216.8 = 145.7 \text{ B/lbm}$$

$${}_3w_4 = h_3 - h_4 = 1379.2 - 1007.7 = 371.5 \text{ B/lbm}$$

$${}_6q_1 = h_1 - h_6 = 1362.5 - 70.93 = 1291.6 \text{ B/lbm}$$

$${}_2q_3 = h_3 - h_2 = 1379.2 - 1216.8 = 162.4 \text{ B/lbm}$$

Substitution into the thermal efficiency expression yields

$$\eta = \frac{145.7 + 371.5 - 1.22}{1291.6 + 162.4} = 0.356$$

An isentropic expansion from state (1) to the condenser pressure of 1 psia yields a quality of 0.817 and an enthalpy of 915.8 B/lbm. On the basis of this value and other values from the foregoing table, the thermal efficiency of the Rankine cycle without reheat is 0.345. Thus the addition of the reheat process has resulted in a small increase in efficiency and a decrease in the liquid content of the exhaust steam to an acceptable value $(1 - 0.905 = 0.095)$.

(b) Since all streams entering and leaving both the low-pressure and high-pressure sections of the turbine have the same mass rate of flow, we can determine the power by multiplying the total turbine work by the mass flow rate at (1).

$$\begin{aligned} \dot{W}_t &= \dot{m}_1({}_1w_2 + {}_3w_4) = 5000(145.7 + 371.5) \\ &= 2.59 \times 10^6 \text{ B/hr} \\ &= 1016 \text{ HP} \end{aligned}$$

## 9–5 The Regenerative Vapor Power Cycle

A vapor power cycle with a thermal efficiency equal to that of a Carnot cycle is shown in Figure 9.6. With all heat transfer to the working fluid taking place at a constant temperature $T_H$ and all heat transfer from the working fluid at $T_L$, such a cycle, by the second Carnot corollary, would have the same thermal efficiency as that of a Carnot power cycle if each process were reversible. Saturated steam at $T_1 = T_H$ would enter the turbine at section (1) and expand reversibly but not adiabatically to the condenser pressure $p_2$. Saturated liquid at the condenser pressure would enter an axial-flow pump at (3). The feed-water flow is countercurrent to the steam expanding through the turbine. The steam-side temperature would never have to be greater than $dT$ above that of the liquid moving through the pump. The liquid would leave the pump at $T_4 = T_H = T_1$. The temperature–entropy diagram for such a cycle is illustrated in Figure 9.7. The turbine–pump combination operates reversibly and adiabatically and, if the turbine and pump are considered separate systems, the heat transfer from the turbine steam is equal to the heat transferred to the water moving through the axial-flow pump. The latter condition is described in Figure 9.7 with the area under (1)–(2) equal to the area under (3)–(4).

This perfect regenerative cycle is impractical for a number of reasons. They include the difficulty in bringing about, even in an irreversible manner, the necessary heat transfer from the steam flowing through the turbine to the liquid flowing through the pump, and the difficulties encountered with the relatively high liquid content of the steam in the turbine.

Practical approaches to the regeneration cycle utilize steam bled from the turbine and devices called feed-water heaters to increase the temperature of the feed water prior to its entering the steam generator. Two simple cycles employing feed-water heaters are shown in

**Figure 9.6** A Perfect Regenerative Vapor Power Cycle

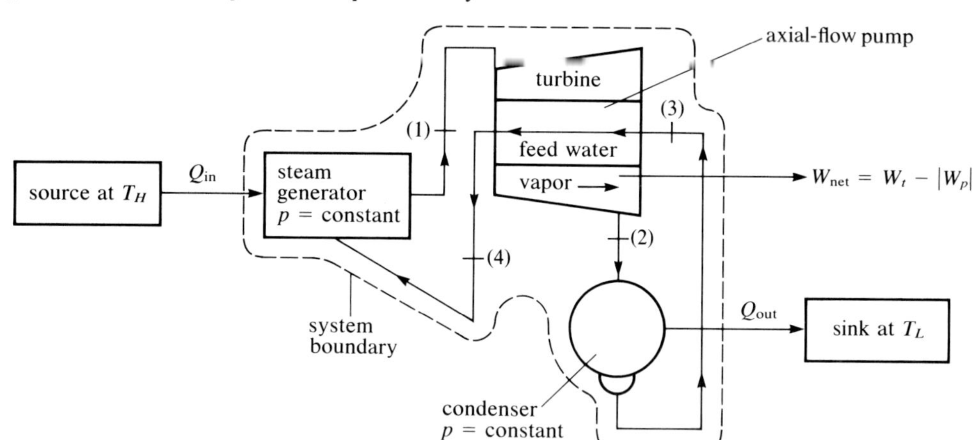


**Figure 9.7** Temperature–Entropy Diagram for a Perfect Regenerative Cycle

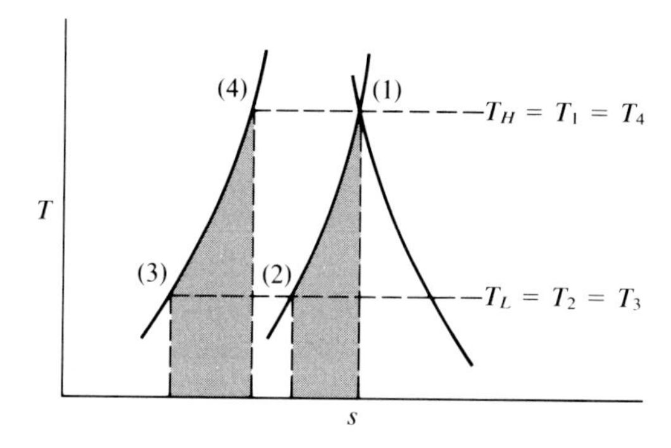


Figure 9.8. The accompanying *T-s* diagrams show the various states associated with the cycles. The ports in the turbine casing through which steam flows to supply the heaters are called bleed points. The pressure in a given feed-water heater is assumed to be equal to the bleed-point pressure. Due to the bleeding of the steam, the mass rates of flow at the various sections are not the same. Hence caution must be exercised to establish a consistent mass basis if areas on the *T-s* diagram are to be interpreted as heat transfer and compared.

Two types of feed-water heaters are illustrated in Figure 9.8. The open heater, Figure 9.8a, receives bleed steam at (2) and liquid at (5) from the condensate pump. These streams mix in the heater. The operating conditions are such that only saturated liquid at the heater pressure leaves at (6). The closed heater, Figure 9.8b, operates as a heat exchanger. The steam and the feed water do not mix. Heat is transferred to the feed water flowing through the heat exchanger tubes from the bleed steam as it condenses on the outside of the tubes. The feed-water pressure is higher than the pressure at which the steam condenses. Only saturated liquid leaves at (7). In practice the boundaries of the feed-water heaters are well insulated and therefore they are treated as adiabatic devices.

**Figure 9.8** Regenerative Power Cycles. *a.* Regenerative cycle with one open feed-water heater. *b.* Regenerative cycle with one closed feed-water heater.

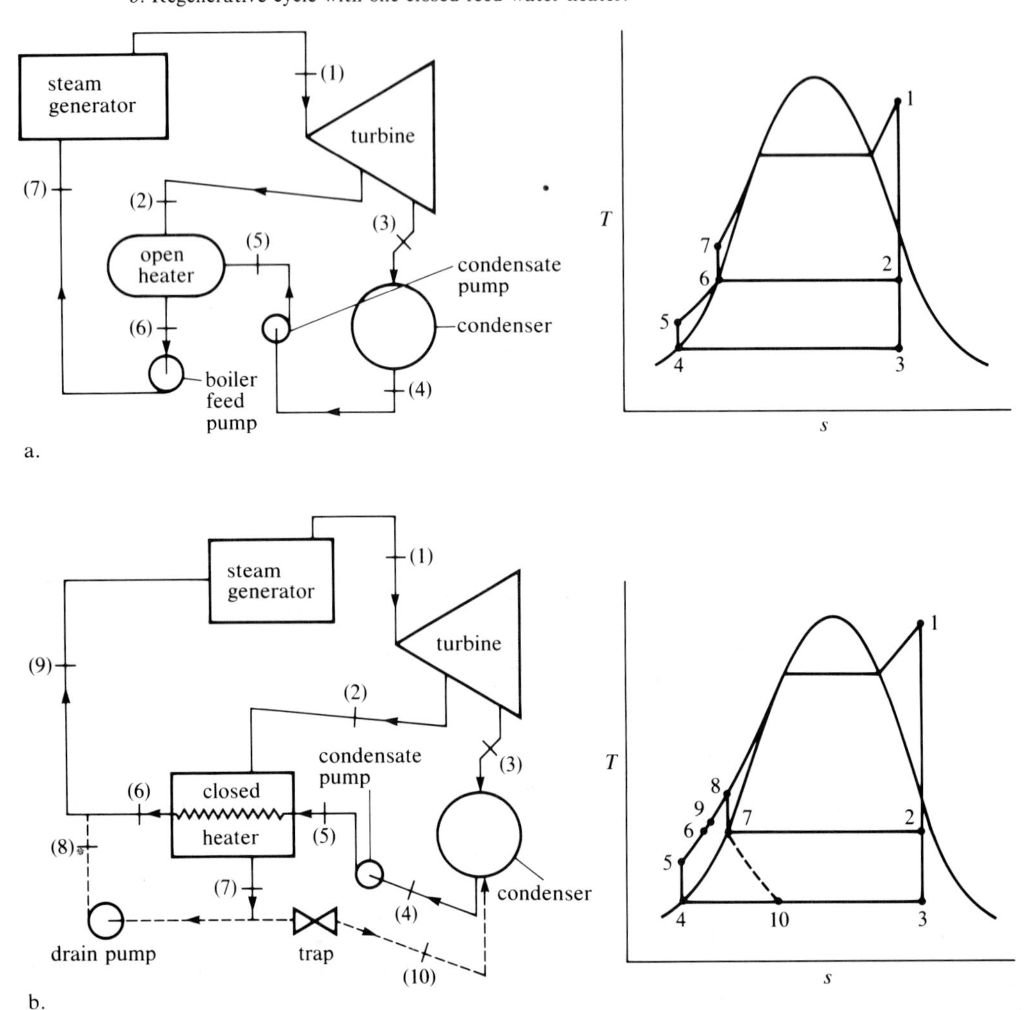


In addition to the assumptions made for the ideal Rankine cycle, we shall assume that the feed water leaves the heaters at the saturation temperature for the bleed pressure. This assumption fixes $t_6$ for the open heater as the saturation temperature for the bleed pressure. The same assumption establishes state (6) for the closed feed-water heater as subcooled liquid at the steam-generator pressure: $p_9 = p_1$ and $t_6 = t_7$. With these conditions the closed feed-water heater operates at its limiting condition having the *terminal temperature difference* $t_7 - t_6$ equal to zero.

Two methods of handling the drain (section [7]) from a closed feed-water heater are illustrated in Figure 9.8b. A drain pump, increasing the pressure from $p_2 = p_7$ to the steam-generator pressure $p_1 = p_9 = p_8 = p_6$, is slightly more efficient. The streams at (6) and (8)

**Figure 9.9** Optimum Number of Feed-Water Heaters for a Regenerative Cycle

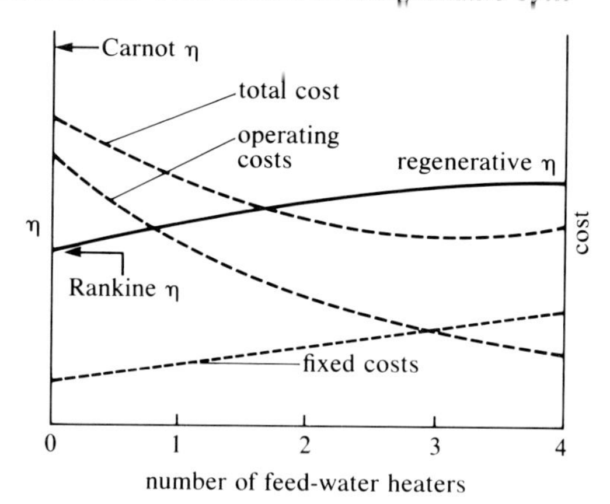


merge to form stream (9). Use of a trap (a device that acts as a throttling valve and permits only liquid to be throttled to a region of lower pressure), as indicated by the flow from section (7) to (10), is an irreversible process that will result in a slightly lower thermal efficiency for a given turbine inlet state and condenser pressure.

Figure 9.9 indicates qualitatively the optimization of the number of stages of feed-water heating for a regenerative power cycle. (See Fig. 9.10.) As the number of heaters is increased, the regenerative cycle efficiency increases from the base Rankine cycle efficiency. This produces a decrease in operating costs, a sizable fraction of which is the cost of fuel. However, due to the additional cost of equipment for the regenerative cycle, the fixed costs (investment costs) increase as the number of heaters increases. The total cost of producing power is the sum of the operating costs and the fixed costs. For the illustration in Figure 9.9, the minimum total cost of power production is found when three feed-water heaters are used. The optimum number depends, of course, on factors related to the steam-generator pressure and exit temperature, the specific cycle under consideration, and the components used in the cycle. Experience indicates that the optimum operation of a given number of feed-water heaters occurs when the temperature rise across each heater is approximately the same.

## Example 9.c

A regenerative steam cycle has a single feed-water heater. Steam enters the turbine at 400 psia, 700 F. Bleed steam to the feed-water heater is at 35 psia and the condenser pressure is 1.0 psia. Calculate the cycle thermal efficiency assuming:

(a) the feed-water heater is an open heater.
(b) the feed-water heater is a closed heater and the drain is trapped to the condenser. Also, determine for this part the power produced by the turbine if the mass rate of flow at (1) is 10,000 lbm/hr.

**Solution:**

(a) The flow diagram for the cycle for part (a) is given in Figure 9.8a. As indicated earlier, calculation of the cycle thermal efficiency first requires that enthalpy values for each section be established. Using the states indicated in Figure 9.8a, properties for the cycle are as follows:

| Section | Pressure, psia | Enthalpy, B/lbm | Entropy, B/(lbm R) |
|---|---|---|---|
| 1 | 400 | 1362.5 | 1.6397 |
| 2 | 35 | 1133.1 | 1.6397 |
| 3 | 1.00 | 915.83 | 1.6397 |
| 4 | 1.00 | 69.74 | $s_4 = s_f$ |
| 5 | 35 | 69.84 | $s_4 = s_5$ |
| 6 | 35 | 228.04 | $s_6 = s_f$ |
| 7 | 400 | 229.19 | $s_7 = s_6$ |

In addition to establishing enthalpy values, attention must be given to the fact that mass flow rates at the various sections in cycles such as those in Figure 9.8 are not all the same. Therefore, care must be taken to ensure that all quantities in the thermal efficiency expression are on a common basis. To proceed with this, we first apply mass conservation and first-law equations to the closed heater in Figure 9.8a. Using eqs. (2.13) and (3.4a) we obtain

$$\dot{m}_2 + \dot{m}_5 = \dot{m}_6$$

$$\dot{Q} + \dot{m}_2 h_2 + \dot{m}_5 h_5 = \dot{W} + \dot{m}_6 h_6$$

$\dot{Q}$ and $\dot{W}$ are zero since the heater boundary is assumed to be adiabatic and no work is involved at the heater. We note that none of the mass flow rates are specified. However, we can choose a suitable reference section, section (1), and proceed by dividing the above equations by $\dot{m}_1$ to yield

$$\frac{\dot{m}_2}{\dot{m}_1} + \frac{\dot{m}_5}{\dot{m}_1} = \frac{\dot{m}_6}{\dot{m}_1} = 1$$

$$\frac{\dot{m}_2}{\dot{m}_1} h_2 + \frac{\dot{m}_5}{\dot{m}_1} h_5 = \frac{\dot{m}_6}{\dot{m}_1} h_6 = h_6$$

We now introduce the notation $m'_{(\ )} = \dot{m}_{(\ )}/\dot{m}_1$, where, for example, $m'_2$ is the mass flowing at section (2) per unit mass flowing at section (1); $m'_2$ is less than unity. This establishes a unit mass at (1) as our basis. Using this notation and combining the above equations we obtain

$$m'_2 h_2 + (1 - m'_2) h_5 = h_6$$

Using the known enthalpy values, we solve for $m'_2$ and $m'_5$

$$m'_2 = 0.1488 \text{ lbm(2)/lbm(1)}$$

$$m'_5 = 0.8512 \text{ lbm(5)/lbm(1)}$$

In the thermal efficiency expression

$$\eta = \frac{q_{in} - q_{out}}{q_{in}}$$

each of the $q$ values must be written on the chosen basis, a unit mass at (1). We note that

$$q_{in} = {}_7q_1 = h_1 - h_7 = 1133.3 \text{ B/lbm(1)}$$

We determine $q_{out}$ on the basis of a unit mass at (1) as

$$\begin{aligned} q_{out} &= |{}_3q_4| = |(h_4 - h_3)m_3'| \\ &= \left[(915.8 - 69.74)\frac{\text{B}}{\text{lbm(3)}}\right]\left[\frac{0.8512 \text{ lbm(3)}}{\text{lbm(1)}}\right] \\ &= 720.2 \text{ B/lbm(1)} \end{aligned}$$

where it has been recognized that $m_3' = m_4' = m_5'$. The above expression for $q_{out}$ can also be obtained by writing eq. (3.4a) for the condenser and dividing by $\dot{m}_1$. Substitution of the above values into the thermal efficiency expression yields

$$\eta = \frac{1133.3 - 720.2}{1133.3} = 36.45 \text{ percent}$$

(b) Figure 9.8b applies. The trap and section (10) are included but the drain pump and section (8) do not apply. A property table similar to that of part (a) is as follows:

| Section | Pressure, psia | Enthalpy, B/lbm | Entropy, B/(lbm R) |
|---|---|---|---|
| 1 | 400 | 1362.5 | 1.6397 |
| 2 | 35 | 1133.1 | 1.6397 |
| 3 | 1.00 | 915.83 | 1.6397 |
| 4 | 1.00 | 69.74 | $s_4 = s_f$ |
| 5 | 400 | 70.93 | $s_5 = s_4$ |
| 6 | 400 | 228.04 | — |
| 7 | 35 | 228.04 | — |
| 10 | 1.00 | 228.04 | — |

Some of the enthalpies correspond to those of part (a). The feed water leaving the heater (section [6]) is compressed liquid at 400 psia and temperature $t_6 = t_7 = t_{sat}$ at 35 psia. Neglecting the effect of pressure on the enthalpy of the subcooled liquid, $h_6 = h_7$.

As was done in part (a), mass and energy balances for the feed-water heater are used to determine the bleed to the heater per unit mass at section (1). Following the procedure outlined in part (a) we obtain for the adiabatic heater

$$m_2'h_2 + (1.0)h_5 = m_2'h_7 + (1.0)h_6$$

and

$$m'_2 = \frac{h_6 - h_5}{h_2 - h_7} = \frac{228.04 - 70.93}{1133.1 - 228.04} = 0.1736 \text{ lbm(2)/lbm(1)}$$

On the basis of a unit mass at section (1)

$$q_{in} = {}_6q_1 = h_1 - h_6 = 1134.5 \text{ B/lbm(1)}$$

On the same basis, the condenser heat transfer, $q_{out}$, is determined by writing the steady-flow equation for the condenser

$$(1 - m'_2)h_3 + m'_2h_{10} + {}_3q_4 = (1.0)\ h_4$$

The enthalpy at section (10) is the same as that at section (7) since the process across the trap is a throttling process. Solving for $q_{out}$

$$q_{out} = |{}_3q_4| = 726.70 \text{ B/lbm(1)}$$

The cycle thermal efficiency is

$$\eta = \frac{1134.5 - 726.70}{1134.5} = 35.95 \text{ percent}$$

The decrease in thermal efficiency, as compared with that of part (a), can be attributed in part to the irreversibility of the throttling process across the trap.

The turbine power is determined by writing the time-rate energy equation for a boundary enclosing the turbine. Since the turbine is adiabatic, we obtain

$$\dot{m}_1h_1 = \dot{W}_t + \dot{m}_2h_2 + \dot{m}_3h_3$$

The mass rates of flow at (2) and (3) are

$$\dot{m}_2 = m'_2\dot{m}_1 = (0.1736)(10{,}000) = 1736 \text{ lbm/hr}$$

$$\dot{m}_3 = \dot{m}_1 - \dot{m}_2 = 10{,}000 - 1736 = 8264 \text{ lbm/hr}$$

Substitution of these values and enthalpy values into the above energy equation yields

$$\begin{aligned}\dot{W}_t &= (10{,}000)(1362.5) - (1736)(1133.1) - (8264)(915.8)\\ &= 4.09 \times 10^6 \text{ B/hr}\\ &= 1200 \text{ kW}\end{aligned}$$

It should also be noted that the thermal efficiencies of the cycles in this example are independent of the mass rates of flow. However, the physical sizes of the components and the power outputs differ as cycles with different mass rates of flow are considered.

**Figure 9.10** Reheat-Regenerative Steam Power Cycle

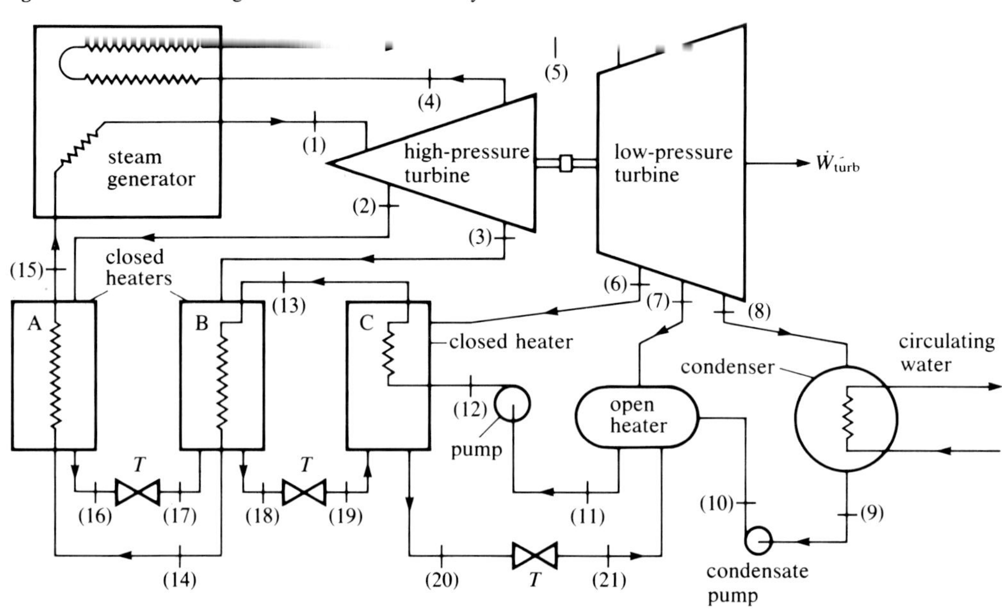


Modern high-pressure, high-temperature steam power cycles use several stages of feed-water heating and incorporate reheating to promote high thermal efficiency. The only open heater, usually called the deaerating heater, operates at a pressure slightly above atmospheric pressure. A flow diagram representative of such cycles is shown in Figure 9.10. Steam is supplied at (1) to the high-pressure turbine by the steam generator. The turbine has bleed points at (2) and (3) ($p_1 > p_2 > p_3 > p_4$) through which small fractions of the flow at (1) are routed to the closed feed-water heaters *A* and *B*, in which steam condenses and increases the feed-water temperature. Since $p_2$ is greater than $p_3$, the temperature in heater *A* is greater than that in heater *B*. The condensate from heater *A* at (16) is trapped to heater *B* and the liquid at (18) is trapped to heater *C*. Reheat occurs from (4) to (5), at which state the steam is supplied to the low-pressure turbine. Small fractions of the flow at (5) are routed past (6) and (7) ($p_5 > p_6 > p_7 > p_8$) to heater *C* and to the open heater. The flow rate at (11) is the same as that at (1), and at (12) the pressure is that at (15) and (1). The temperature at (11) is higher than that at (9) due to the mixing in the open heater of the entering streams with the incoming steam at (7). The temperature of the feed water progressively increases as it passes from (12) through heaters *C*, *B*, and *A* to (15). Analysis of a cycle such as this one follows the procedures outlined in Example 9.c. Mass and energy balances for the closed heaters starting with heater *A* establish the relative mass rates of flow at the various sections. Caution must be exercised to ensure that all calculations are on a consistent basis.

Figure 9.11 presents a detailed schematic diagram of a modern coal-fired steam-electric power-generating station. Included are the cycle components as well as the coal-preparation and coal-handling equipment and the electrostatic precipitator that removes ash particles

**Figure 9.11** Schematic Representation of a Steam–Electric Generating Station

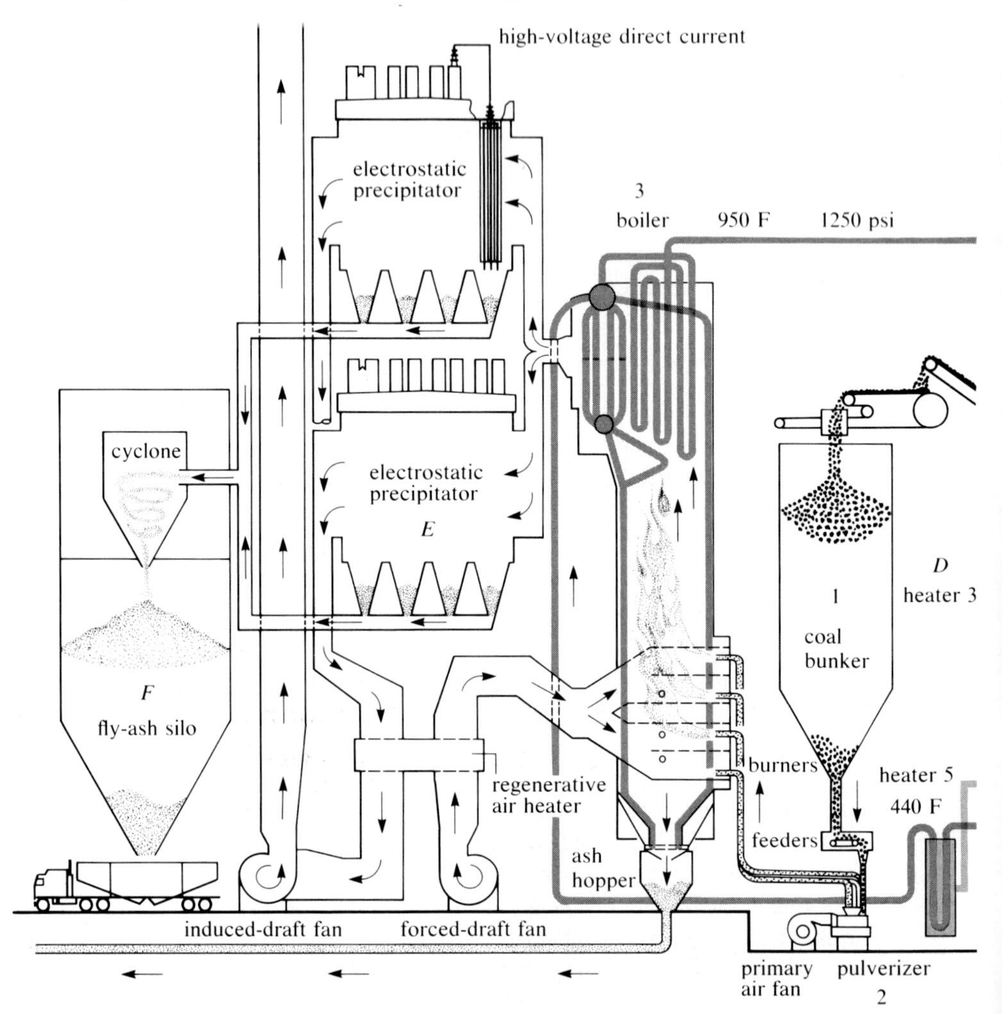


## UNIT #8, AMES, IOWA MUNICIPAL POWER PLANT

### Energy Flow and Transformations

1. Coal is delivered to the Ames Municipal Power Plant by rail and truck. Belt conveyors transfer the coal to the coal bunkers within the plant.
2. The coal flows by gravity from the bunkers into the pulverizers which grind the coal into a dust that is finer than face powder. The primary air fan blows this dust into the boiler where it mixes with air and burns in suspension like a liquid fuel.
3. The boiler consists of many steel tubes filled with water. The heat of combustion boils the water, and steam is collected in the steam drum at the top of the boiler. The steam is further heated in the superheater before it is piped to the turbine. Unit #8 burns refuse-derived fuel as a supplemental fuel. Bottom ash is collected in the ash hopper at the bottom of the boiler and is flushed to the ash pond.
4. The turbine consists primarily of blades attached to a shaft. Steam flows against the blades causing the shaft to turn at 3600 rev/min.
5. The generator and turbine are connected by a shaft.
6. The power transformer increases the generated voltage from 13,800 volts to 69,000 volts. From the power transformer the electrical energy goes to a substation for distribution to the consumer.

(Courtesy Ames Municipal Electric System, Ames, IA)

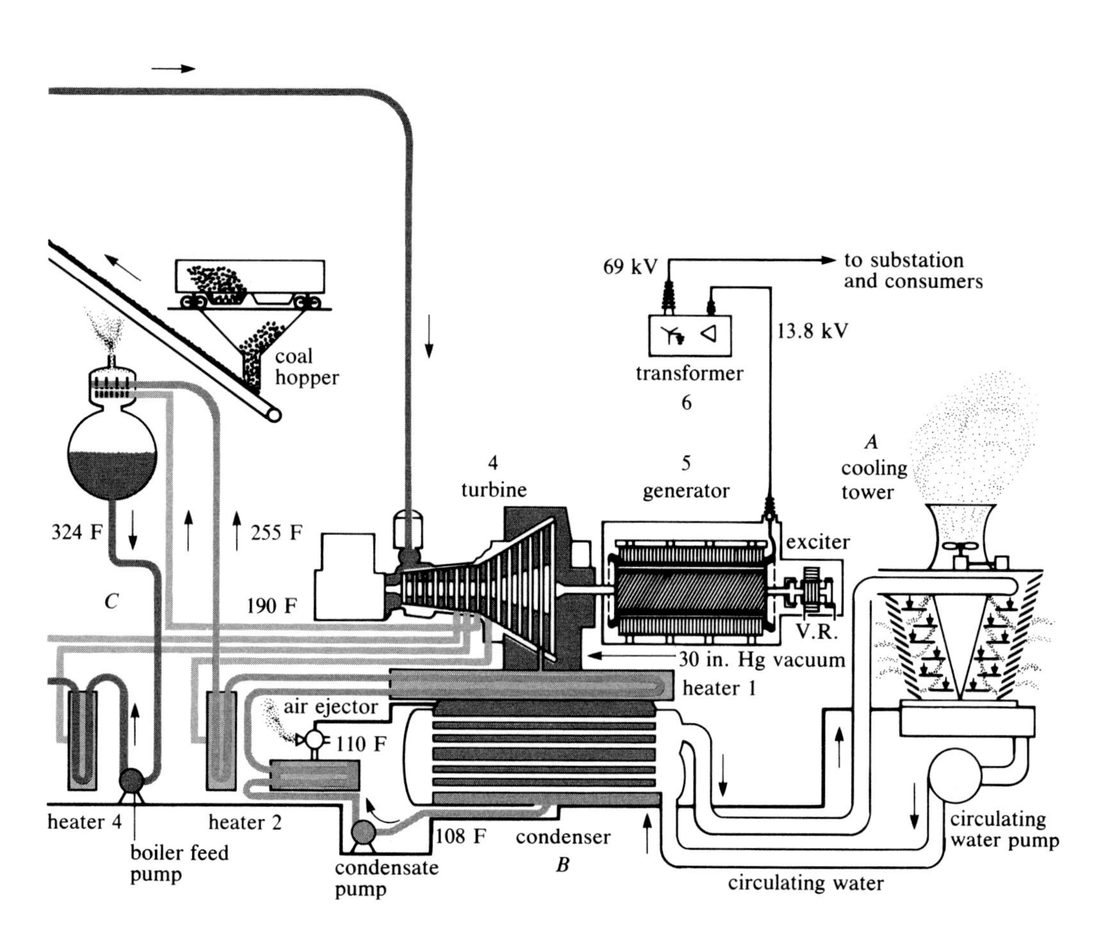


**Major Associated Equipment**

**A.** The cooling tower is used to provide a source of cool circulating water. This is done by moving air across the water that is falling through the cooling tower.

**B.** The exhaust steam from the turbine is admitted into the condenser where it is cooled and condensed by the circulating water. The condensate (water) is then pumped back towards the boiler.

**C.** Various heaters are used to heat the water as it returns to the boiler from the condenser. The small amount of steam required by the heaters is obtained from different points in the turbine.

**D.** One of the heaters is called the "deaerator" because in addition to heating the water it removes air and other dissolved gases.

**E.** The hotside electrostatic precipitator eliminates 99.7% of the fly ash from the boiler flue gases by means of a 20,000 Volts DC electric field so that ash will not be emitted from the chimney.

**F.** The fly ash that is collected in the electrostatic precipitator hoppers is pneumatically conveyed to the fly ash silo where it is temporarily stored.

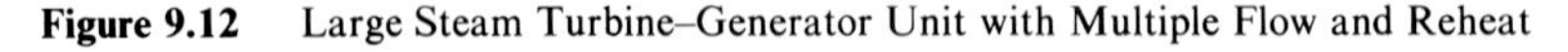

**Figure 9.12** Large Steam Turbine–Generator Unit with Multiple Flow and Reheat

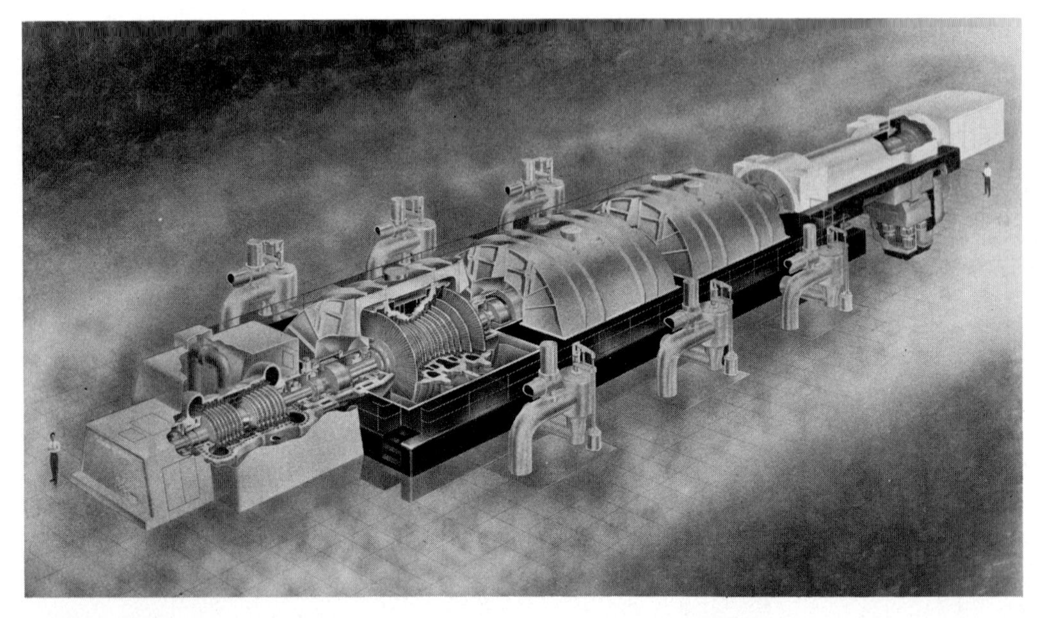

(Courtesy General Electric Co.)

from the stack gases. Figure 9.12 is a cutaway diagram of a large steam turbine-generator used in a reheat cycle. The condenser in Figure 9.13, one of several common designs, receives exhaust steam from a turbine through the large opening in its top. Steam condenses on the surface of the tubes shown, through which cooling water passes. The condensate collects in the bottom and is pumped back toward the steam generator by the condensate pump.

## 9–6 Cogeneration

In a broad sense *cogeneration* is the production of more than one useful form of energy from the same primary energy source. At this time the most common cogeneration systems involve the generation of steam for simultaneous use in producing electric energy (with a steam turbine driving an electric generator) and for industrial process heating or for space heating. Although the concept is not new (it was introduced early in this century), it is receiving increased attention as a result of escalating fuel prices. The early application of cogeneration was in district heating where turbine exhaust steam at pressures greater than atmospheric was piped to buildings for space heating. Such an application is known as a *topping cycle,* to indicate that the higher-temperature fluid is used for power generation and the heating is done with lower-temperature fluid. Cogeneration is found on many college and university campuses where the central "heating plant" supplying steam for building heating also serves

**Figure 9.13** One-Pass Rectangular Surface Condenser

(Courtesy Westinghouse Electric Corp.)

as an electric generating station. Cogeneration of steam for heating and electric power production is widely used in process industries. Examples include food-processing plants, chemical industries, and petroleum production and refining facilities. With increased use of steam injection for secondary recovery from oil wells, the cogeneration of electric power has become an economically desirable addition.

To assess the benefits of cogeneration let us first consider a conventional process steam plant without cogeneration or, alternatively, taking steam at the boiler outlet state of a steam power cycle and using it for the process requirement. This latter case is found in some systems where the process heating—perhaps district heating—was added after an electric generating station was in operation. As we shall see, this is not as desirable as a cogeneration system from both the thermodynamic and economic viewpoints. A schematic drawing for

**Figure 9.14** Process Steam Schematic

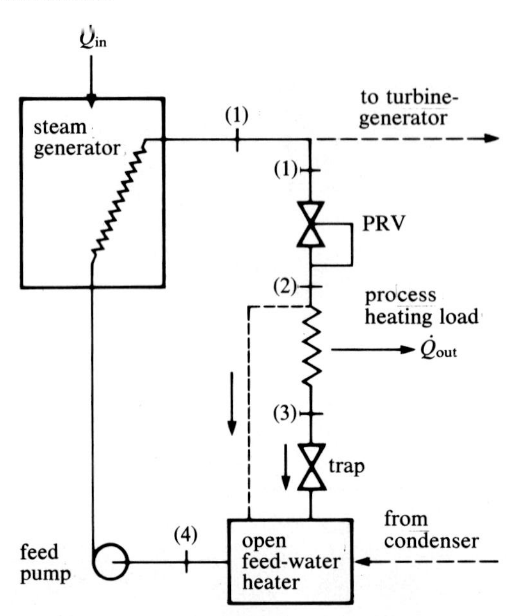


such a system is shown in Figure 9.14. A simple pressure-reducing valve (PRV), as shown, would reduce the steam pressure from that at the steam-generator outlet to that required for the process heating load, perhaps district heating. Recognizing the irreversibility of the throttling process across the PRV, one can reasonably expect this "no-work" process not to be the most desirable alternative. In practice the pressure reduction is usually combined with desuperheating the steam by injection of liquid $H_2O$. The desuperheating is desirable because the heat-transfer characteristics of condensing steam are superior to those of superheated vapor. The broken line in Figure 9.14 indicates steam supply to produce desired conditions at the feed-water heater outlet (section [4]).

Next let us consider two cogeneration systems that would satisfy the process steam requirements illustrated in Figure 9.14 and, additionally, generate electric energy. In assessing the merits of such systems, the conventional cycle thermal efficiency is not the significant parameter. Rather, one must consider the primary operating cost, which is that of the fuel required to produce the desired steam for both process and electric generation. In particular, the incremental increase in fuel required is the significant factor. Other considerations relate to fixed costs: the increase in steam-generator capacity and the cost of the turbine-generator and other auxiliary equipment. Thus a sound engineering economic evaluation is required in addition to the thermodynamic analysis. Studies of cogeneration systems have indicated that significantly lower costs for electric power can be achieved through cogeneration even though the turbine cost per kilowatt is typically greater than that of a large central station unit. The desirability of cogeneration is enhanced as fuel costs escalate since fuel chargeable

**Figure 9.15** Cogeneration with Back-Pressure Turbine

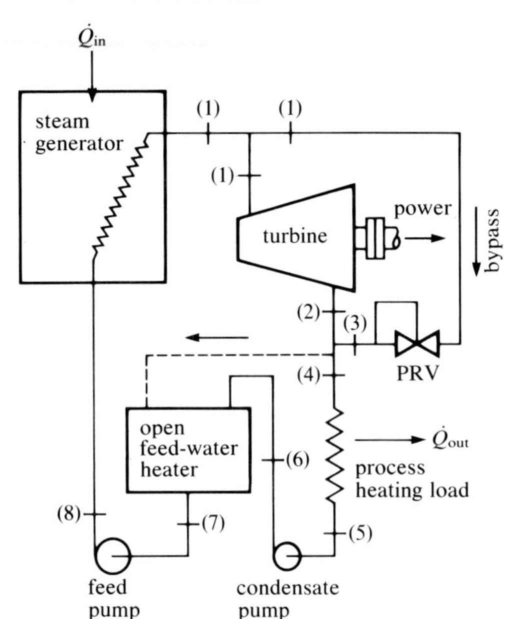


to power generation is an incremental increase over that already required to generate process steam. The incremental increase in fuel for power generation in a cogeneration system is such that the fuel cost is 50 percent or less of that for a conventional generating station.

One simple cogeneration system uses a back-pressure turbine, a turbine designed to operate at a high exhaust pressure. Rather than taking the required pressure drop across a pressure-reducing valve as shown in Figure 9.14, the pressure drop is through a turbine that delivers power and exhausts at the desired process steam pressure. A schematic of such a system is shown in Figure 9.15. In this figure no condenser is shown. Thus it is assumed that the process heating load condenses all steam passing through the turbine. The bypass line provides added flexibility for situations wherein the process heating load occasionally may exceed the turbine design flow rate. Depending on the desired feed-water state leaving the heater (section [7]), steam may be taken from the turbine exhaust to satisfy this requirement. The system illustrated is a definite improvement over that shown in Figure 9.14. However, it does not offer a high degree of flexibility in serving varying power and process requirements while minimizing fuel consumption. Its primary advantage is the simple system design for cases where little change in power or process steam demand is anticipated. As can be seen, a decrease in power demand will necessitate use of the bypass line to serve a fixed process steam demand. Use of the bypass line, with its pressure-reducing valve, moves the system in the direction of the steam system in Figure 9.14, with its inherent irreversibility.

**Figure 9.16** Single Extraction Turbine Schematic

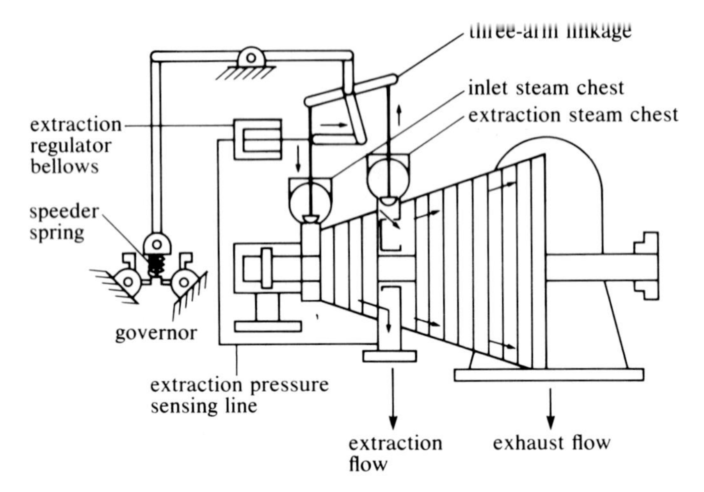


(Courtesy Elliot Company, a Subsidiary of United Technologies Corp.)

A more flexible system for cogeneration results when an extraction turbine is used. The bleed points for feed-water heating in power plant turbines are uncontrolled openings in the turbine casing, and thus bleed-point pressures vary with the power produced. An extraction turbine, in contrast, is one where the steam pressure at the extraction point is controlled even though the turbine load may vary. This is accomplished by feedback control from the extraction point to the turbine inlet valves. This feature results in a higher first cost for the turbine, which is offset by greater flexibility of the system. A simplified schematic of a turbine with a single extraction point is shown in Figure 9.16. This unit is for constant turbine speed—as would be required in driving a synchronous electric generator—and varying extraction steam-flow rates. Turbines with more than one extraction pressure are available for applications requiring process steam at more than one pressure (temperature). Obviously these units are more expensive and the control system more complex.

A cogeneration system with an extraction turbine is shown in Figure 9.17. In this schematic no bleed points for feed-water heaters are shown. These would be found in larger-capacity turbines. In addition a bypass line similar to the one shown in Figure 9.15 may be added for increased flexibility. Extraction steam, taken ahead of the process heating load, may be used to supply the feed-water heater and produce the desired feed-water state leaving the heater (section [8]).

It is useful to examine cogeneration from the second-law viewpoint, using the concept of available energy developed in Section 7–9. We recall that the available portion of a quantity of heat is the work that could be produced by a reversible engine operating between the source of the heat and the atmosphere. The unavailable portion is the heat rejected by the reversible engine. Heat that has a high fraction as available energy is termed high-grade energy. It is evident that high-grade heat energy is more valuable than low-grade energy in the production of power. Many industrial processing and domestic heating applications do not require high-grade energy. Further, it is unwise to use high-grade energy to accomplish

Figure 9.17 Cogeneration with an Extraction Turbine

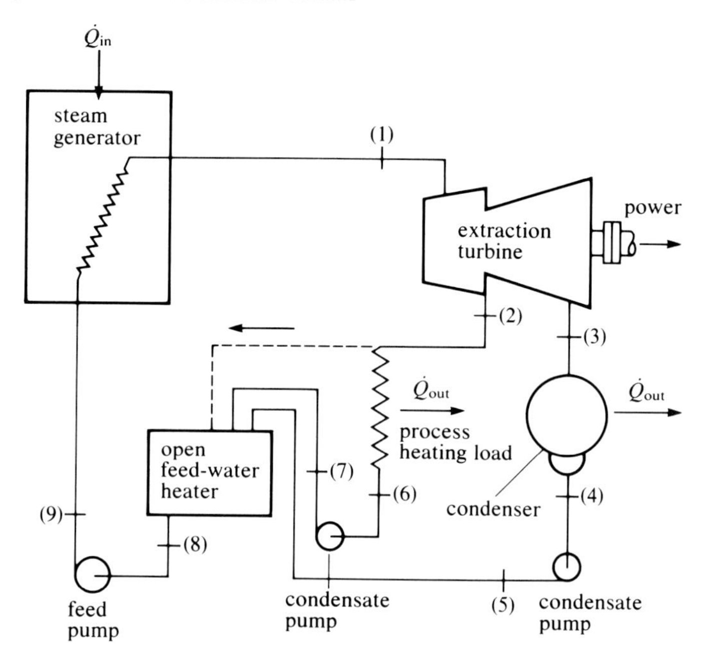


a task that could be performed with low-grade energy. A cogeneration steam system moves toward optimum utilization of available energy by using high-grade energy to produce power and then using the low-grade heat from the condensing turbine exhaust steam for a low-grade application, process or domestic heating. Although the attractiveness of a cogeneration system stems largely from economic considerations in certain cases, it also offers a means of using fuels in a manner that is attractive when viewed from the standpoint of second-law concepts. The following example illustrates some of the second-law considerations in relation to a cogeneration system.

## Example 9.d

The diagram shows two options for supplying a process heating requirement of $\dot{Q}_p = 10^8$ B/hr by utilizing steam supplied by the steam generator. Option I has the turbine in series with the process heat exchanger. The expansion through the turbine is isentropic and the pump work may be neglected. Option II uses a throttling valve in place of the turbine to reduce the pressure to that of the process heat exchanger, which is 17 psia in each option. The *T-s* diagram shows the various states and the processes and the following table lists the properties at the several states. The surrounding atmosphere is at 40 F.

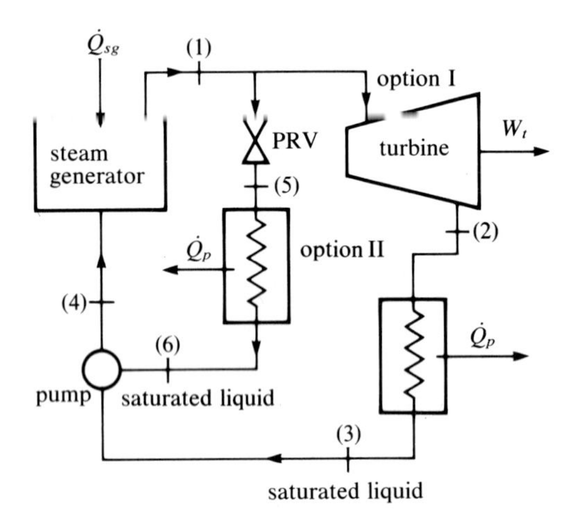


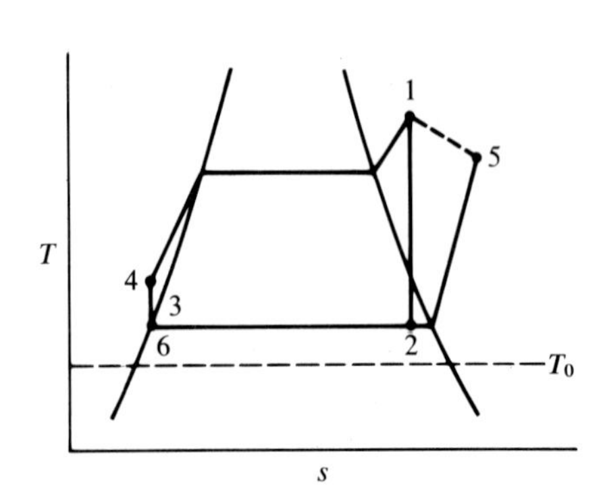


| State | $p$, psia | $t$, F | $h$, B/lbm | $s$, B/(lbm R) |
|---|---|---|---|---|
| 1 | 300 | 560 | 1292 | 1.604 |
| 2 | 17 | 219.43 | 1057 | 1.604 |
| 3 | 17 | 219.43 | 188 | 0.3232 |
| 4 | 300 | — | 188 | 0.3232 |
| 5 | 17 | — | 1292 | 1.915 |
| 6 | 17 | 219.43 | 188 | 0.3232 |

(a) Evaluate the two options in terms of available energy utilization.
(b) Determine the mass rate of flow required for each option.

**Solution:**

(a) For option I the available portion of the energy supplied to the steam generator, ${}_4q_1$, is determined using eqs. (7.18) and (7.19) as

$$\begin{aligned} {}_4(q_a)_1 &= {}_4q_1 - q_u = h_1 - h_4 - T_0(s_1 - s_4) \\ &= 1292 - 188 - 500(1.604 - 0.3232) = 464 \text{ B/lbm} \end{aligned}$$

Similarly, the available portion of the heat rejected in condensing the steam in process (2) to (3) is

$$\begin{aligned} {}_2(q_a)_3 &= |{}_2q_3| - T_0(s_2 - s_3) = |h_3 - h_2| - T_0(s_2 - s_3) \\ &= 1057 - 188 - 500(1.604 - 0.3232) = 229 \text{ B/lbm} \end{aligned}$$

When the pump work is neglected, the turbine work is the net cycle work. Thus

$$w_{net} = w_t = h_1 - h_2 = 1292 - 188 = 235 \text{ B/lbm}$$

In option I the available energy, 464 B/lbm, has been partially utilized to produce 235 B/lbm work. The process heat is obtained using relatively low-grade energy—energy with an availability of 229 B/lbm.

For option II there is no utilization of the available energy, that is, no work is produced. The availability is destroyed either in the throttling process or in the process of delivering the process heat. The available energy after the throttling process is

$$
\begin{aligned}
{}_5(q_a)_6 &= |{}_5q_6| - T_0(s_5 - s_6) = |h_6 - h_5| - T_0(s_5 - s_6) \\
&= 1292 - 188 - 500(1.915 - 0.3232) = 308 \text{ B/lbm}
\end{aligned}
$$

Thus the throttling process has reduced the available energy from 464 B/lbm to 308 B/lbm. Nonetheless, high-grade energy with an availability of 464 B/lbm is used in option II to supply the low-grade need of the process heating. It is obvious that option I is more desirable from the second-law point of view. (While this problem is illustrative of the usefulness of cogeneration, it is not meant to imply that steam at high pressure and temperature is generated for the purpose of throttling to a much lower pressure in process heating applications.)

(b) The different mass flow rates required for each option are determined by writing the time-rate steady-flow energy equation for the process heat exchanger. For option I

$$
\dot{m}_{\text{I}} = \frac{{}_2\dot{Q}_3}{(h_2 - h_3)} = \frac{10^8}{1057 - 188} = 1.15 \times 10^5 \text{ lbm/hr}
$$

Similarly

$$
\dot{m}_{\text{II}} = \frac{{}_5\dot{Q}_6}{(h_5 - h_6)} = \frac{10^8}{1292 - 188} = 9.06 \times 10^4 \text{ lbm/hr}
$$

Of course $\dot{Q}_{sg}$ is greater for option I; it exceeds $\dot{Q}_p$ by the turbine power $\dot{W}_t$. But note that $\dot{W}_t$ is produced in an amount equal to the additional rate of heat input at the steam generator. Therefore, power produced by a cogeneration plant where there is an established need for processing heat or other heating needs incurs a significantly lower fuel cost than does power produced by a conventional power plant.

---

Situations are encountered in which high-temperature (high-grade) energy is discharged from industrial processes that must take place at high temperature. Gases discharged from cement kilns and blast furnaces are two examples. Such sources can serve as the heat supply for power cycles operating between the source and the atmosphere. Cycles of this nature are called *bottoming cycles*. They typically operate on a Rankine cycle or a variation thereof. While the working fluid of the bottoming cycle in some cases may be steam, a number of other substances have been investigated for bottoming cycle applications since the temperature at which the heat is available may be below that at which steam serves as an acceptable working fluid. Organic compounds such as toluene and butane, as well as ammonia and other refrigerants, are included in this group. While such systems, with their lower-pressure and lower-temperature turbine inlet states, do not have the thermodynamic advantage of topping cycles, they do represent another opportunity for using fuels more efficiently.

**Figure 9.18** Simple Heat-Recovery Bottoming Cycle

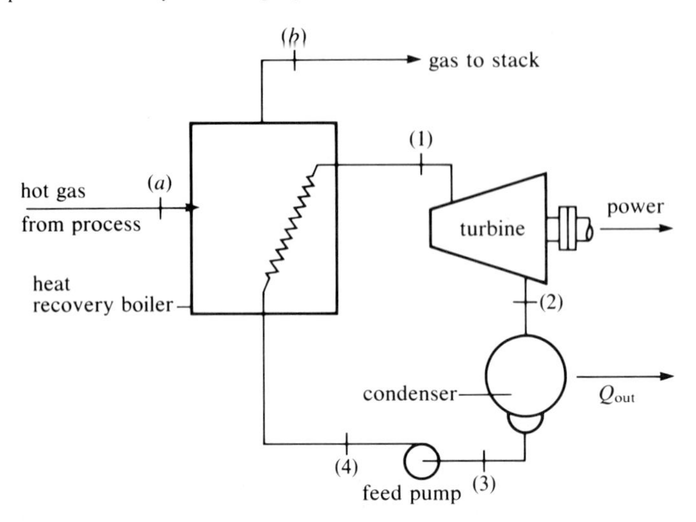


Figure 9.18 shows a bottoming cycle that utilizes heat from hot gases from a process (sometimes called waste heat) to vaporize the working fluid of a Rankine cycle. As a result, the temperature at (*b*) is less than that at (*a*) and lower-grade energy is discharged to the atmosphere from the gases and from the condenser of the bottoming cycle. Cycles such as this offer a means of utilizing some of the available energy that otherwise would be lost in discharging the hot process gases directly to the atmosphere.

A variation on the bottoming cycle is found in combined cycles. A typical combined cycle has a gas-turbine topping cycle exhausting into a heat-recovery boiler that generates steam for a bottoming steam cycle. The combined gas turbine–steam cycle has a higher thermal efficiency than can be achieved with a gas turbine or a steam cycle operating separately. There are numerous variations of combined cycles. Some of these are presented in Section 10–8.

## 9–7 The Carnot Refrigeration Cycle

A reversible refrigeration cycle is the ideal cycle for providing refrigeration since such a cycle would require the minimum work input for a given refrigerating effect. The coefficient of performance for a refrigerator operating on a reversible cycle is

$$\beta_{R,\max} = \frac{\text{refrigeration effect}}{\text{net work supplied}} = \frac{T_L}{T_H - T_L} \tag{6.5}$$

To begin the study of vapor refrigeration cycles, let us consider the reversed Carnot cycle operating between a refrigerated region at $T_L$ and the atmosphere at $T_H$, as described in the $T$-$s$ and $p$-$v$ diagrams of Figure 9.19. Let each of the processes be steady flow with negligible changes in kinetic and potential energies. Process (1) to (2) is the heat-reception process

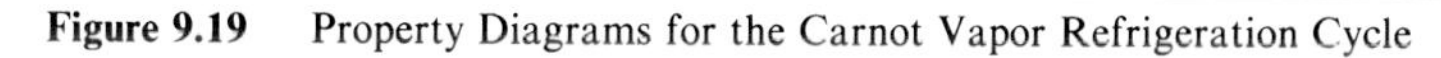

**Figure 9.19** Property Diagrams for the Carnot Vapor Refrigeration Cycle

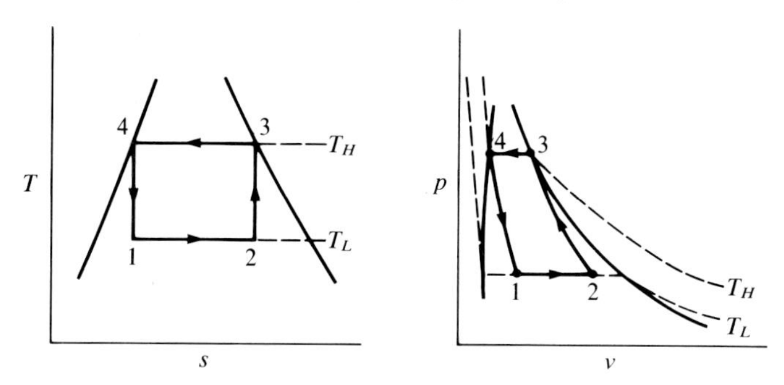


and takes place at constant pressure as well as constant temperature. Work is required to compress the mixture in process (2) to (3), after which heat rejection takes place to the atmosphere in process (3) to (4), also a constant-temperature, constant-pressure process. Work is produced in the expansion process (4) to (1). By the first law and the fact that all processes are reversible, the enclosed areas on both the *T-s* and *p-v* diagrams represent both the net heat and the net work for the cycle. Also, the work for processes (2) to (3) and (4) to (1) is given by eq. (2.14) as $-\int v dp$. Therefore, these quantities can be represented by areas on the *p-v* diagram.

Although the Carnot refrigeration cycle serves as a standard for comparison purposes, it is impractical for at least two reasons. First, the stipulation of heat transfer across an infinitesimal temperature difference requires either an infinite time or an infinite heat-transfer area to transfer a given amount of heat. Second, the work-producing process (4) to (1) is difficult to carry out since a mixture that is mostly liquid must be expanded. Further, the work produced is relatively small. Therefore, in practical cycles this process is replaced by a throttling process. The practical vapor refrigeration cycle is discussed in the next section.

## 9–8 The Simple Vapor-Compression Refrigeration Cycle

The refrigeration cycle most commonly used today is the simple vapor-compression cycle. Figure 9.20 shows a flow diagram for this cycle. The refrigerant enters the evaporator at section (1) as a low-quality liquid–vapor mixture, and as it passes through the evaporator at constant pressure, heat is transferred to it from the refrigerated region, which is at temperature $T_L$. This heat transfer causes evaporation of liquid refrigerant; hence the component is called the evaporator. The evaporator temperature increases or decreases as the evaporator pressure is increased or decreased. The refrigerant may leave the evaporator as saturated vapor, as a high-quality mixture, or as slightly superheated vapor. Process (2) to (3) through the compresser is isentropic to the condenser pressure in the ideal cycle. While the compression process may be specified as polytropic in some cases, for simplicity we shall consider only cycles with isentropic compression. Heat transfer from the refrigerant to condenser cooling water or directly to the surrounding atmospheric air temperature $T_H$ takes

**Figure 9.20** Flow Diagram for a Simple Vapor Compression Refrigeration Cycle

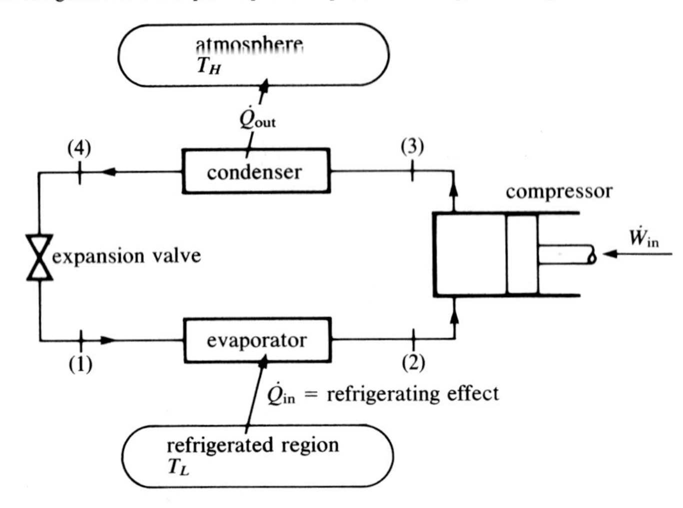


**Figure 9.21** The Temperature–Entropy Diagram for an Ideal Vapor Compression Refrigeration Cycle

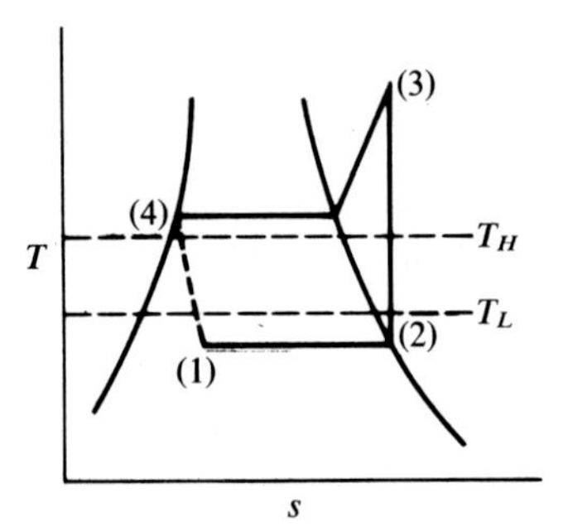


place at constant pressure in the condenser. On leaving the condenser the refrigerant is a liquid, saturated in the ideal cycle, and often subcooled in actual cycles. From (4) to (1) a throttling process takes place through an expansion valve or a capillary tube with part of the liquid undergoing a phase change (flashing) to vapor. The process taking place in each component is steady flow. As in all vapor cycles, kinetic and potential energy changes are assumed to be negligible. The temperature–entropy diagram for an ideal simple vapor-compression refrigeration cycle with saturated vapor leaving the evaporator is given in Figure 9.21. All processes are assumed to be internally reversible, except the irreversible throttling process (4) to (1). Areas below the paths (1) to (2) and (3) to (4) on the *T-s* diagram represent heat transfers. However, this is not true for the irreversible process (4) to (1). Therefore, the enclosed area is not the net heat transfer or the net work for the cycle. The finite temperature difference shown between the refrigerant in the evaporator and the refrigerated region at $T_L$ is necessary to promote heat transfer from the refrigerated region to the refrigerant. A similar and necessary temperature difference exists between the refrigerant in the condenser and the atmosphere at $T_H$.

The coefficient of performance for the ideal simple vapor-compression refrigeration cycle is given by

$$\beta_R = \frac{{}_1\dot{Q}_2}{|{}_2\dot{W}_3|} = \frac{\dot{m}\ {}_1q_2}{\dot{m}\,|{}_2w_3|} = \frac{h_2 - h_1}{h_3 - h_2} \tag{9.3}$$

The steady-flow energy equation has been used to evaluate ${}_1q_2$ and ${}_2w_3$. The compression process is called dry compression when the refrigerant entering the compressor is saturated or slightly superheated vapor, and wet compression if the compression process takes place within the mixture region. If the refrigerant leaves the evaporator as saturated vapor, the enthalpy at (2) will be that of saturated vapor, $h_g$, at the evaporator pressure or temperature. Should state (2) be in the mixture region, adequate information to establish the quality must be known. If state (2) is superheated vapor, sufficient information to fix state (2) must be available. The state leaving the compressor is fixed by knowledge of the condenser pressure and the entropy entering the isentropic compressor. For a throttling process through either an expansion valve or a capillary tube, $h_1$ is equal to $h_4$.

As noted in Section 3–9, a common measure of refrigerating capacity is the ton of refrigeration, which is defined as

$$\begin{aligned} 1 \text{ ton} &= 200 \text{ B/min} = 12{,}000 \text{ B/hr} \\ &= 210.9 \text{ kJ/min} = 12{,}658 \text{ kJ/hr} \end{aligned}$$

Parameters that are useful in comparing various refrigeration systems are the power required per ton of refrigeration and the mass rate of flow of the refrigerant per ton of refrigeration. From the definition of coefficient of performance, the power required is

$$\dot{W}_{\text{in}} = \frac{\dot{Q}_{\text{in}}}{\beta_R}$$

where $\dot{Q}_{\text{in}}$ is the refrigerating effect. Also, from the steady-flow energy equation written for the evaporator, the mass rate of flow of the refrigerant is

$$\dot{m} = \frac{\dot{Q}_{\text{in}}}{h_2 - h_1}$$

where $h_2 - h_1$ is the enthalpy increase of the refrigerant as it flows through the evaporator. By taking $\dot{Q}_{\text{in}}$ as 1 ton of refrigeration, we obtain the power and mass rate of flow per ton. In the English system of units

$$\dot{W}_{\text{in}} = \left[\frac{12{,}000 \text{ B}}{\text{ton hr}}\,\bigg|\,\frac{\text{HP hr}}{2545 \text{ B}}\right]\left[\frac{1}{\beta_R}\right] = \frac{4.715}{\beta_R}, \text{ HP/ton} \tag{9.4}$$

$$\dot{m} = \left[\frac{12{,}000 \text{ B}}{\text{ton hr}}\,\bigg|\,\frac{\text{lbm}}{(h_2 - h_1) \text{ B}}\right], \text{ lbm/ton hr} \tag{9.4a}$$

**Table 9.1** Comparison of Several Refrigerants*

| Refrigerant | Volume Flow Rate at Compressor Inlet, $ft^3$/ton min | $\beta_R$, Ideal | $\dot{W}$, HP/ton |
|---|---|---|---|
| Ammonia, $NH_3$ | 3.44 | 4.75 | 0.99 |
| Monochlorodifluoromethane (F-22) | 3.56 | 4.56 | 1.03 |
| Dichlorodifluoromethane (F-12) | 5.82 | 4.72 | 1.00 |
| Methyl chloride | 6.09 | 4.90 | 0.97 |
| Sulfur dioxide | 9.08 | 4.73 | 1.00 |
| Dichloromonofluoromethane (F-21) | 20.4 | 5.09 | 0.93 |
| Trichloromonofluoromethane (F-11, Carrene-2) | 36.3 | 5.14 | 0.92 |
| Methylene chloride (Carrene-1) | 74.5 | 4.90 | 0.97 |
| Trichlorotrifluoromethane (F-113) | 100.8 | 4.79 | 0.99 |

*Data are for an ideal, simple vapor-compression cycle with a 5 F evaporator temperature and 86 F condensing temperature.

Another useful quantity is the volume rate of flow into the compressor, which is

$$\dot{V}_2 = (VA)_2 = \dot{m}\, v_2 \tag{9.4b}$$

where the subscript 2 denotes the compressor inlet. Performance characteristics of several refrigerants, based on a simple vapor-compression cycle, a common evaporator temperature (5 F), and a common condensing temperature (86 F) are given in Table 9.1. Note that the horsepower per ton of refrigerating effect is approximately unity for each of the refrigerants. The tabulated values of the volume rate of flow at the compressor inlet per ton-minute are significant in that they dictate the type of compressor most likely to be used. Low-volume flow rates ($NH_3$, F-12, and F-22) are readily handled by positive displacement compressors. Refrigerants with high-volume flow rates per ton ordinarily are handled by centrifugal compressors.

In Figure 9.22 the influence of evaporator temperature on horsepower per ton of refrigerating effect is shown for one refrigerant, F-12. The figure is for a constant condensing temperature at 150 F. The inverse relation of power per ton and evaporator temperature is characteristic of all vapor refrigerants.

Abbreviated tables of properties for the refrigerants ammonia and Freon 12 (F-12) are given in Appendices D and E respectively.

Figure 9.22 The Effect of Evaporator Temperature on Power per Ton of Refrigeration for a Simple Vapor Compression Refrigeration Cycle. Refrigerant F-12 condensing at 150 F.

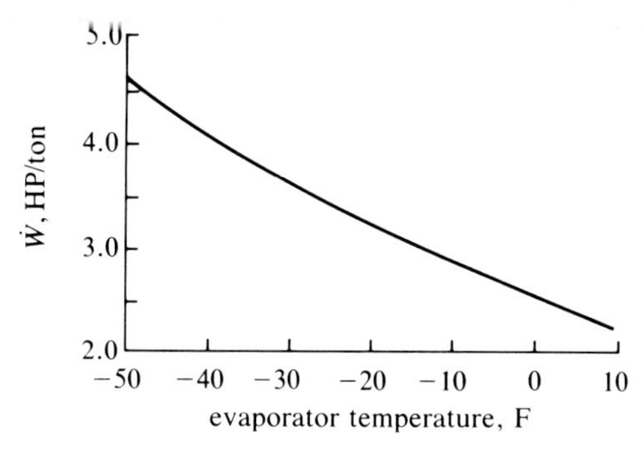


## Example 9.e

An ideal, simple vapor-compression cycle (Fig. 9.20) uses ammonia ($NH_3$) as the refrigerant. Saturated vapor at 0 F leaves the evaporator and saturated liquid at 90 F leaves the condenser. The refrigerating capacity is 20 tons. Calculate:

(a) The coefficient of performance for the cycle
(b) The mass rate of flow of the refrigerant, lbm/hr
(c) The horsepower required
(d) The volume rate of flow of the refrigerant entering the compressor, ft³/min

**Solution:**

The properties for ammonia are given in Appendix D. For the cycle and refrigerant the properties are

| Section | *t*, F | *p*, psia | *h*, B/lbm | *s*, B/(lbm·R) |
|---|---|---|---|---|
| 1 | 0 | 30.42 | 143.5 | — |
| 2 | 0 | 30.42 | 611.8 | 1.3352 |
| 3 | — | 180.6 | 724.71 | 1.3352 |
| 4 | 90 | 180.6 | 143.5 | — |

The enthalpy at section (4) is equal to that at section (1) due to the nature of the process (throttling) from (1) to (4). The enthalpy at (4) is $h_f$ at 90 F taken from Appendix D. At section (2), leaving the evaporator, the ammonia is saturated vapor at 0 F, and hence its enthalpy is $h_g$ at 0 F. At section (3), the compressor outlet, the state is fixed by $s_3 = s_2 = s_g$ at 0 F and $p_3 = p_4 = p_{sat}$ for 90 F. With the limited data in Appendix D a two-way linear interpolation has been used to determine the enthalpy at section (3). The coefficient of performance is

(a) $$\beta_R = \frac{h_2 - h_1}{h_3 - h_2} = 4.148$$

(b) The mass rate of flow is calculated from the 20-ton refrigeration capacity as

$$(\dot{m})(h_2 - h_1) = {}_1\dot{Q}_2 = (20)(12{,}000) \text{ B/hr}$$

or

$$\dot{m} = \frac{(20)(12{,}000)}{611.8 - 143.5} = 512.5 \text{ lbm/hr}$$

(c) While the power required could be calculated using eq. (9.4), application of the steady-flow energy equation is also appropriate. This equation yields

$$\begin{aligned}\dot{W}_{\text{in}} &= \dot{m}(h_3 - h_2)\\ &= \frac{(512.5)(724.71 - 611.8)}{2545} = 22.74 \text{ HP}\end{aligned}$$

(d) Refrigerant enters the compressor at (2), the saturated vapor state at 0 F. The specific volume is $v_2 = v_{g_2} = 9.116 \text{ ft}^3\text{/lbm}$. From eq. (9.4b) the volume rate of flow at the compressor inlet is

$$\dot{V}_2 = \frac{512.5 \text{ lbm}}{\text{hr}}\left|\frac{\text{hr}}{60 \text{ min}}\right|\frac{9.116 \text{ ft}^3}{\text{lbm}} = 77.87 \text{ ft}^3\text{/min}$$

---

## 9–9 A Vapor-Compression Cycle for Refrigeration at Two Temperatures Using a Single Compressor

Home refrigerator–freezer units are the major market for vapor-compression units producing refrigeration at two temperatures using a single compressor. In this section an ideal cycle with this capability is presented. Figure 9.23 shows the flow diagram for the cycle. A typical unit of this type operates with the refrigerator portion at approximately 40 F while the refrigerant passing through that evaporator is at a temperature of the order of 10 to 15 F. The refrigerant leaves the refrigerator evaporator as a mixture. It then is throttled through a second capillary tube or expansion valve to the minimum pressure in the cycle, that in the freezer evaporator. Refrigerant temperatures in the freezer evaporator are of the order of $-10$ to $-15$ F with the freezer compartment maintained at approximately 0 F. An ideal refrigerator–freezer cycle is described by the *T-s* diagram, Figure 9.24. The unit described by Figures 9.23 and 9.24 has been simplified in that a heat exchanger utilizing the cold refrigerant at $t_4$ to cool the fluid as it is throttled from (6) to (1) and an accumulator between the freezer evaporator and the compressor have been omitted. The thermodynamic analysis of this ideal cycle follows the same procedures used in studying the ideal simple vapor-compression cycle, with one exception. State (2), at a given saturation pressure or temperature, must be fixed by knowledge of the refrigerating effect, ${}_1q_2$, or by knowledge of the relative magnitudes of the refrigerating effects of the freezer and refrigerator if state (4) is fixed.

Figure 9.23 Refrigerator–Freezer with One Compressor

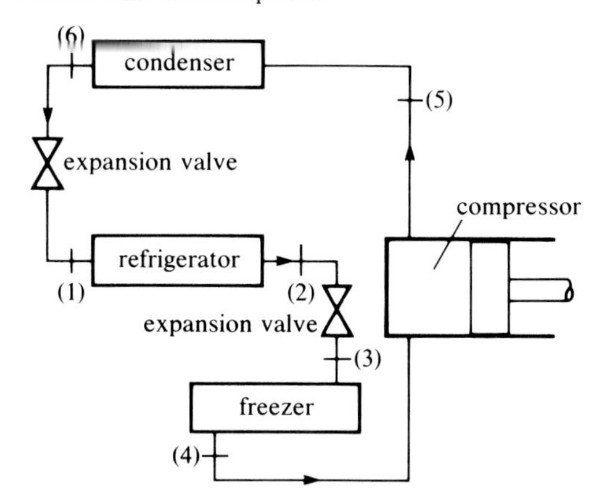


Figure 9.24 Temperature–Entropy Diagram for a Refrigerator–Freezer Combination

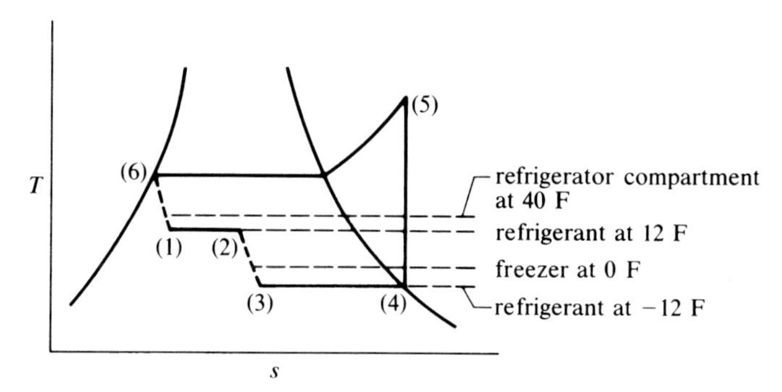


## 9–10 Heat Pumps

Heat pump cycles are, in a thermodynamic sense, identical to refrigeration cycles. The primary difference between a heat pump cycle and the equivalent refrigeration cycle is the desired effect. For refrigeration the desired effect is to maintain a region at a temperature less than that of the surroundings by heat transfer from the region to the refrigerant as it flows through the evaporator. The desired effect for heat pump cycles is heat transfer from the working fluid to the surroundings as that fluid passes through the condenser or an equivalent high-temperature heat exchanger. From this comparison it is apparent that the same unit, or cycle, can be used for either heating or cooling purposes. The desired effect for a refrigeration cycle is the refrigerating effect whereas that for a heat pump is the heating effect. This desirable combination has led to the design of units that serve as coolers for air conditioning during warm weather and provide heating during cold weather.

**Figure 9.25** Simple Vapor-Compression Cycle for a Heat Pump. *a*. Temperature–entropy diagram. *b*. Flow diagram.

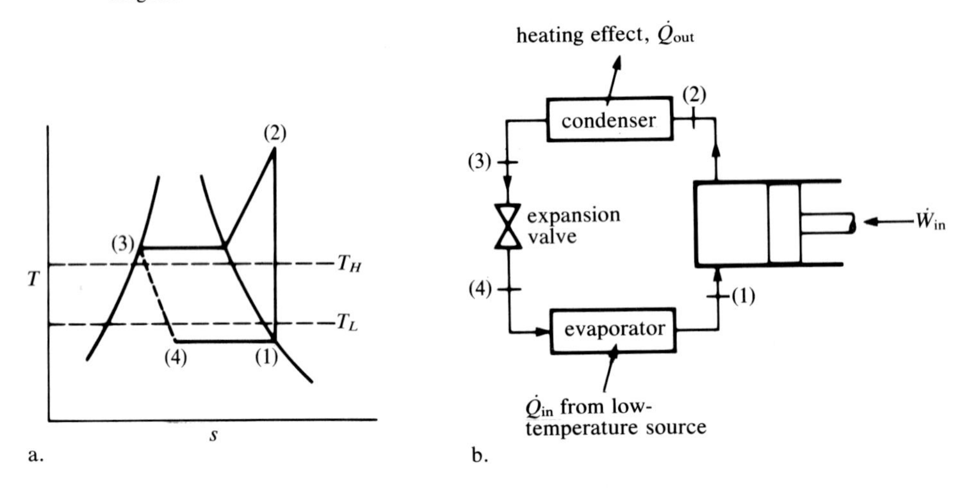


A simple vapor-compression cycle (Figure 9.20) can be used to provide cooling (air conditioning) as well as cold-weather heating. Figure 9.25 shows the vapor-compression cycle operating in the heat pump mode. When the heat pump is operated as a refrigerator to provide refrigeration for air conditioning, the heat pump condenser serves as the refrigerator evaporator and the heat pump evaporator becomes the refrigerator condenser. One method used to change the mode of operation of the cycle is to include a reversing valve to change the direction of the flow of the refrigerant through the condenser–throttling valve–evaporator combination. The use of the flow-reversing valve is necessary since the compressor can operate only when the flow through it is in one direction. This arrangement allows all components to remain in fixed locations regardless of the mode of operation.

In locations where the temperature of the atmosphere does not drop below about 20 F, a heat pump may be the only heating unit required. Hybrid heat pump/solar energy systems are being studied at this time as one possible aid in the reduction of fuel and electric energy for heating homes and larger buildings. As we shall see, heat pumps require less electric energy than do resistance heating installations. In view of the increasing costs of natural gas and other petroleum fuels, it is conceivable that combinations of solar, heat pump, and electric resistance heating may well become the common heating systems of the future. Environmental considerations would seem to enhance the desirability of electric energy generated at a large generating station where pollution control could be centralized and either coal or nuclear fuels would be employed.

A heat pump with the maximum possible coefficient of performance operates on a reversible heat pump cycle, the reversed Carnot cycle, for example. For such a heat pump the coefficient of performance is

$$\beta_{P,\max} = \frac{\text{heating effect}}{\text{net work supplied}} = \frac{T_H}{T_H - T_L} \tag{6.6}$$

**Table 9.2** Comparison of the Performance of a Vapor-Compression Heat Pump Cycle with the Performance of a Carnot Heat Pump for a Delivered Heating Effect of 100,000 B/hr. $T_H$ = 535 R (75 F). Vapor-compression cycle refrigerant: F-22.

| | | | **Carnot Heat Pump Cycle** | | **Vapor-Compression Heat Pump Cycle, Figure 9.25** | | | |
|---|---|---|---|---|---|---|---|---|
| $T_L$, R | $t_L$, F | $T_H - T_L$, R | $\beta_{P,\max}$ | Power Required, kW | $T_3$, R | $T_1$, R | $\beta_P$ | Power Required, kW |
| 520 | 60 | 15 | 35.7 | 0.821 | 560 | 500 | 7.98 | 3.67 |
| 500 | 40 | 35 | 15.3 | 1.92 | 560 | 480 | 5.79 | 5.06 |
| 480 | 20 | 55 | 9.37 | 3.01 | 560 | 460 | 4.44 | 6.60 |
| 460 | 0 | 75 | 7.13 | 4.11 | 560 | 440 | 3.67 | 7.98 |

where $T_H$ is the temperature at which heat is transferred from the working fluid to a region that is to be heated and $T_L$ is the temperature of the reservoir—the atmosphere, the earth, or an equivalent—from which heat is transferred to the working fluid of the Carnot heat pump.

It is instructive to compare the performance of a simple vapor-compression heat pump cycle with that of a reversible heat pump cycle. Let us select the heat pump cycle in Figure 9.25, using F-22 as the refrigerant for this illustration. A heat pump operating on this cycle is to provide a heating effect of 100,000 B/hr to a region at temperature $T_H$ = 535 R (75 F). The temperature of the refrigerant in the condenser must be greater than 75 F to promote heat transfer from the refrigerant to the heated space. For this illustration let $T_3 - T_H$ in Figure 9.25 be 25 R, that is, $T_3$ = 560 R (100 F). Also, the temperature of the refrigerant in the evaporator must be less than the source temperature $T_L$ to cause heat to be transferred from the source—the outside air, for example—to the refrigerant. Let $T_L - T_1$ be 20 R. A comparison of the heat pump coefficient of performance and the power required for this cycle and the Carnot heat pump cycle is presented in Table 9.2 for several values of $T_L$. The coefficients of performance shown for the Carnot heat pump were computed using eq. (6.6). Note that the Carnot heat pump would operate directly between the inside and outside temperatures. It is evident from inspection of Table 9.2 that the coefficients of performance for the vapor-compression cycle are significantly lower than the corresponding ones for the Carnot cycle. The power required is correspondingly higher. As might be expected, the power requirement goes up as the outside temperature goes down. The departure of the performance of the vapor-compression cycle from that of the Carnot cycle is attributable to the irreversible heat-transfer processes and to the irreversible throttling process. Although an actual vapor-compression heat pump cycle would depart somewhat from that described in this illustration and exhibit a lower coefficient of performance at a given source temperature, the comparison in Table 9.2 is indicative of the trends that are present as various practical considerations come into play for real cycles.

In summary, the heat pump is an economically attractive alternate to fossil-fuel-fired heating systems in many locations. The possible combinations of solar, heat pump, and electric resistance heating should be considered in evaluating systems for new buildings, keeping in mind the cooling effect that can be obtained from the refrigeration cycle for air-conditioning purposes.

## Example 9.f

It is necessary to supply heat to a certain building at the rate of 35,000 B/hr to maintain the inside temperature at 70 F. The outside air temperature that serves as a source for a heat pump is at 20 F. An ideal vapor-compression heat pump is available to provide heat at the specified rate. The following information is given: The refrigerant, F-12, leaves the evaporator as saturated vapor. The temperature in the evaporator is −5 F and the condenser pressure is 150 psia. Assume the cost of electricity to be 7 cents/(kWhr). Compute the cost of electricity per day to provide the heating if (a) the heat pump is used, (b) the heat could be supplied using a Carnot heat pump, and (c) electric resistance heating is used.

**Solution:**

(a) From Appendix E we obtain information for the following table. The states correspond to those in Figure 9.25.

| State | *p*, psia | *t*, F | *s*, B/(lbm R) | *h*, B/lbm |
|---|---|---|---|---|
| 1 | 21.422 | −5 | 0.16937 | 76.74 |
| 2 | 150 | 130 | 0.16937 | 91.62 |
| 3 | 150 | 109.35 | — | 33.53 |
| 4 | 21.422 | −5 | — | 33.53 |

From the steady-flow energy equation on a time-rate basis we obtain

$$\dot{m} = \frac{{}_2\dot{Q}_3}{h_2 - h_3} = \frac{35{,}000}{(91.62 - 33.53)} = 602.5 \text{ lbm/hr}$$

The power required is

$$\begin{aligned}\dot{W} &= \dot{m}\,|{}_1w_2| = \dot{m}(h_2 - h_1) = 602.5(91.62 - 76.74)\\ &= 8965 \text{ B/hr} = 2.63 \text{ kW}\end{aligned}$$

Therefore

$$\text{Cost} = 2.63(24)(0.07) = \$4.42 \text{ per day}$$

(b) From eq. (6.6)

$$\beta_{P,\max} = \frac{T_H}{T_H - T_L} = \frac{(70 + 460)}{(70 - 20)} = 10.6 = \frac{\dot{Q}_H}{\dot{W}_{\text{in}}}$$

Therefore, the power required is

$$\dot{W}_{\text{in}} = 35{,}000/10.6 = 3302 \text{ B/hr} = 0.967 \text{ kW}$$

and the cost is

$$\text{Cost} = 0.967(24)(0.07) = \$1.63 \text{ per day}$$

(c) An electric resistance heater converts all of the electric energy supplied to heat. The cost is computed as

$$\text{Cost} = \frac{35{,}000}{3413}(24)(0.07) = \$17.22 \text{ per day}$$

---

It is evident that the heat pump has an advantage over resistance heating when electricity costs alone are considered. The minimum possible cost of electricity for the specified heating application is $1.63 per day, part (b). A complete economic analysis would require that fixed costs and additional operating costs be included. In many climates the economics favor the heat pump over the resistance heating, especially if the need for air conditioning exists.

## 9–11 Summary

Vapor cycles are cycles in which the working fluid alternately undergoes phase changes between liquid and vapor. Such cycles are widely used to produce power and to provide refrigeration. While the Carnot power cycle, because of its maximum thermal efficiency, is the ideal power cycle, practical considerations require that actual cycles depart from the Carnot cycle. Therefore, such cycles have lower thermal efficiencies. Actual power cycles are modifications of the ideal Rankine cycle, which is composed of four internally reversible processes: a constant-pressure heat-addition process taking place at high pressure in which the working fluid undergoes a phase change from liquid to saturated or superheated vapor; an isentropic expansion through a turbine; a condensation process occurring at low pressure, resulting in saturated liquid; and an isentropic pumping process of the liquid to the high pressure. The thermal efficiency of the Rankine cycle is affected by the condenser and vapor generator pressures as well as by the temperature at which the vapor is delivered to the turbine. Any modification that increases the average temperature of the working fluid during the heat-reception process will increase the cycle thermal efficiency. To avoid high liquid content in the low-pressure sections of steam turbines, and possibly to improve thermal efficiency, reheat is used. In a reheat cycle steam expands to an intermediate pressure through the high-pressure section of the turbine, after which it is reheated at constant pressure and then expanded to the condenser pressure in the low-pressure section of the turbine. To increase the thermal efficiency of steam power cycles, regeneration is used. This consists of bleeding small amounts of steam from the turbine to preheat the feed water by means of feed-water heaters. Modern high-pressure high-temperature steam cycles are regenerative and employ reheat.

Cogeneration typically involves the generation of power by means of a power cycle and subsequent use of the heat rejected from the cycle for heating purposes. It offers maximum utilization of the energy from a primary source in that high-grade energy is used to produce power and low-grade energy is used for heating. Numerous cogeneration schemes exist and their adoption hinges primarily on economic considerations.

The reversed Carnot cycle is the ideal refrigeration cycle. However, it is impractical. The practical refrigeration cycle is the vapor-compression cycle in which vapor is compressed from a low pressure to a high pressure, condensed to a liquid, and throttled back to the low pressure. The liquid remaining after the throttling process undergoes a phase change, thereby absorbing heat from the refrigerated region. Numerous refrigerants that satisfy various applications are in use. The vapor-compression cycle can also serve as a heat pump cycle. The heat pump utilizes the heat rejected from the condenser to provide a heating effect. Many heat pumps can also operate as air conditioners.

## Selected References

*ASHRAE Guide and Data Book.* New York: American Society of Heating, Refrigerating and Air Conditioning Engineers.

*ASHRAE Handbook of Fundamentals.* New York: American Society of Heating, Refrigerating and Air Conditioning Engineers.

*Combustion Engineering.* New York: Combustion Engineering (published periodically).

Grant, I. *Thermodynamics and Heat Power.* Reston, Va.: Reston Publishing Co., 1974.

Potter, P. J. *Power Plant Theory and Design,* 2nd ed. New York: Ronald Press, 1959.

Solberg, H. L., O. C. Cromer, and A. R. Spalding. *Thermal Engineering.* New York: Wiley, 1960.

*STEAM. Its Generation and Use.* New York: Babcock & Wilcox Co. (published periodically).

Stoecker, W. F. *Refrigeration and Air Conditioning.* New York: McGraw-Hill, 1958.

Stoever, H. J. *Engineering Thermodynamics.* New York: Wiley, 1951.

*Thermodynamic Properties of Refrigerants.* New York: American Society of Heating, Refrigerating and Air Conditioning Engineers, 1969.

Threlkeld, J. L. *Thermal Environmental Engineering,* 2nd ed. Englewood Cliffs, N.J.: Prentice-Hall, 1970.

## Problems

9.1 A Carnot power cycle having $H_2O$ as the working fluid operates between a source at 400 F and a sink at 100 F. During heat transfer to the $H_2O$ the state change is from saturated liquid to saturated vapor. Sketch *p-v* and *T-s* diagrams for the cycle and calculate, per pound mass of $H_2O$, heat transfer to the system, heat transfer from the system, and the thermal efficiency of the cycle.

9.2 Suppose that a Carnot power cycle had the same conditions as those given in the preceding problem except that heat transfer to the $H_2O$ continued until the pressure reached 100 psia. Sketch *p-v* and *T-s* diagrams for the cycle, showing the saturation lines. Calculate the net work per pound mass of $H_2O$ and the cycle thermal efficiency. Compare the results with those from the preceding problem.

9.3 During a Carnot power cycle heat rejection takes place at 100 F. Determine the thermal efficiency of the cycle for heat addition at 400 F, 500 F, 600 F, 900 F, and 1300 F. Plot thermal efficiency versus the source temperature.

9.4 A Rankine cycle using $H_2O$ as the working fluid has steam entering the turbine at 400 F, saturated vapor. The turbine exhaust is at 100 F. Determine the heat transfer to the $H_2O$, the heat transfer from the $H_2O$, and the thermal efficiency of the cycle. Compare the results with those from problem 9.1.

9.5 Suppose that a Rankine cycle utilizing Freon-12 operates with F-12 entering the turbine at 200 F, saturated vapor. If the turbine exhaust is at 100 F, what is the thermal efficiency of the cycle? See Appendix E for properties.

9.6 Consider a Rankine cycle with $H_2O$, the working fluid, entering the turbine at 400 psia and 600 F. Calculate the turbine work and the thermal efficiency of the cycle for condenser pressure of 14.696 psia, 5.00 psia, 1.20 psia, and 0.80 psia. Graph the cycle thermal efficiency and turbine exhaust quality versus condenser pressure for the given turbine inlet state.

9.7 Compute the available energy (Section 7–9) for the heat-rejection process at the four condenser pressures in the previous problem for a surroundings temperature of 70 F. What do you conclude regarding the desirability of low condenser pressures in relation to available energy utilization by a power cycle?

9.8 A Rankine cycle has steam entering the condenser at 7.5 kPa. The turbine inlet pressure is 2.75 MPa. Determine the thermal efficiency of the cycle for turbine inlet temperatures of 229.12 C (saturated vapor), 320 C, and 420 C.

9.9 Suppose that the steam leaving the steam generator of a Rankine cycle plant is at 600 psia, 900 F, and is throttled to 400 psia before entering the turbine. Expansion through the turbine is isentropic to 1.50 psia at the exhaust. Calculate the turbine work per unit mass. What would the work be if no throttling were to take place, that is, if the turbine inlet were at 600 psia, 900 F?

9.10 Suppose steam is supplied to a turbine at 600 psia and the turbine exhausts at 1.20 psia. What turbine inlet temperature would be required if the minimum acceptable quality at the turbine exhaust were 90 percent? Assume the turbine operation to be reversible and adiabatic. How would the exhaust quality be affected if the turbine operated adiabatically but not reversibly?

9.11 Suppose that an imaginary fluid having the following properties were used in a Rankine cycle: (a) Its liquid phase is incompressible. (b) The value of $C_p$ for the liquid phase is constant and equal to 0.40 B/lbm R. (c) At 400 F $h_{fg}$ is 300 B/lbm. (d) The saturated-vapor line on $T$-$s$ coordinates has a negative slope. Neglecting pump work, determine the thermal efficiency of the cycle if saturated vapor at 400 F enters the turbine and the condenser temperature is 100 F.

9.12 Consider an ideal Rankine cycle with reheat in which the steam generator supplies steam at 700 psia and 900 F. The reheat pressure is 75 psia and the steam is reheated to 860 F. (a) Find the cycle thermal efficiency. (b) Compare your answer in part (a) with the thermal efficiency of the ideal Rankine cycle without the reheat process. The condenser pressure is 2.0 psia.

9.13 A Rankine cycle with reheat using steam as the working fluid has a high-pressure turbine inlet temperature of 800 F at a pressure of 700 psia. The exhaust of the high-pressure turbine has a quality of 100 percent. Reheat occurs to 800 F and the condenser pressure is 1.0 psia. Assume isentropic flow in the turbine and in the pump and find (a) the reheat pressure, (b) the cycle thermal efficiency, and (c) the net power produced if the mass rate of flow for the cycle is 15,000 lbm/hr.

9.14 A steam power plant operates on a regenerative cycle with one open feed-water heater. At the turbine inlet the pressure is 600 psia and the temperature is 720 F. Steam is bled to the feed-water heater at 50 psia. The condenser pressure is 1.20 psia. Determine the thermal efficiency of the cycle.

9.15 A steam power plant operates on a regenerative cycle with one closed feed-water heater. At the turbine inlet the pressure is 600 psia and the temperature is 720 F. Steam is bled to the feed-water heater at 50 psia and the drain from the heater is trapped back to the condenser, which operates at 1.20 psia. Refer to Figure 9.8b. Calculate the thermal efficiency of the cycle.

9.16 Repeat the preceding problem with the heater drain pumped ahead and into the feed water leaving the closed heater, as shown in Figure 9.8b.

9.17 A regenerative steam power cycle has one closed and one open feed-water heater. The turbine inlet is at 600 psia and 720 F. Steam is bled to the closed heater at 200 psia and to the open heater at 50 psia. The closed-heater drain is trapped back into the open heater. The condenser pressure is 1.20 psia. Sketch a flow diagram for the cycle and calculate the cycle thermal efficiency.

9.18 A reheat–regenerative steam cycle has a single open feed-water heater. Steam enters the high-pressure turbine at 600 psia and 720 F. Part of the steam exhausted from the high-pressure turbine at 50 psia goes to the feed-water heater and the remainder is reheated to 700 F. The reheated steam expands through the low-pressure turbine to the condenser pressure of 1.20 psia. Calculate the turbine work per pound of steam entering the high-pressure turbine and the cycle thermal efficiency.

9.19 Steam is supplied to the turbine in a regenerative cycle at 600 psia, 720 F. The turbine exhaust is at 1.20 psia. Two open feed-water heaters are used and the water leaves each as saturated liquid at the heater pressure. Ten percent of the steam entering the turbine is bled to the higher-pressure heater at 100 psia. Neglecting pump work, determine the pressure of the steam bled to the low-pressure heater.

9.20 Consider a steam power plant operating on a reheat–regenerative cycle with a single closed feed-water heater. Steam enters the turbine at 600 psia, 900 F, and expands to 100 psia. A portion of the 100-psia steam goes to the feed-water heater and the remainder is reheated to 800 F. Condensate from the heater is trapped back to the condenser as shown in Figure 9.8b. The reheated steam is returned to the turbine and expands to the condenser pressure of 1.00 psia. Sketch a flow diagram for the cycle. Neglecting the effect of pressure on the enthalpy of feed water leaving the feed-water heater, calculate the thermal efficiency of the cycle. If the turbine delivers 33,000 kW, what is the flow rate, lbm/hr, through the steam generator? Assuming heat transfer in the steam generator to be 9,125 B/lbm of coal fired, how many tons of coal per 24-hour day would be required? What is the power required to drive the feed-water pump? If the circulating water passing through the condenser tubes has a 15 F temperature rise, what is the circulating water flow rate in gallons per minute?

9.21 A reheat–regenerative cycle has steam entering the turbine at 2400 psia, 1000 F. Reheat is at 400 psia to 1000 F. Closed feed-water heater bleeds are at 1000 psia, 480 psia, 160 psia, and 8 psia. There is, also, one open feed-water heater at 35 psia. Drain from each closed heater is trapped back to the next lower pressure heater, except that the drain from the 8-psia heater is pumped ahead into the open heater. The condenser pressure is 1.20 psia. Sketch a flow diagram for the cycle.

9.22 An industrial power plant requires 100,000 lbm/hr of saturated steam ($x = 100$ percent) at 5.3 psi gage. The steam is to be supplied from the exhaust of a turbine that drives a 5500-kW electric generator. Assuming ideal operations of the turbine and generator, what would be the required pressure and temperature at the turbine inlet? Hint: Use Mollier diagram.

9.23 Suppose a process industry needs steam at 40 psia for a heating load of $40 \times 10^6$ B/hr. Further, suppose the plant has a steam generator delivering saturated vapor at 220 psia. Condensate is to leave the heating system at 40 psia and 200 F. Two plans are to be considered. In plan *A* the 40-psia steam is obtained by throttling, through a pressure-reducing valve, steam leaving the steam generator. Plan *B* utilizes a turbine to expand the saturated steam at 220 psia to 40 psia with the 40-psia exhaust steam then going to the process heating load. Determine the steam flow rate, lbm/hr, for each plan and the power produced by the turbine of plan *B*. Assume the turbine operates isentropically. Comment on the economic factors to be considered in comparing the two plans.

9.24 Suppose a system as shown in Figure 9.14 is used only to provide steam for process heating (no turbine or condenser is used). Steam leaves the steam generator at 1450 psia, 1000 F, and is supplied to the process heating load of 130 million B/hr at 50 psia. Condensate from the heating load is at 180 F. Feed water leaves the open heater as saturated liquid at 50 psia. Neglecting pump work, calculate (a) the steam generated, lbm/hr, and (b) $\dot{Q}_{in}$ for the steam generation, MB/hr. Do not neglect steam supplied to the feed-water heater.

9.25 Refer to Figure 9.15. Suppose steam is supplied to the isentropic turbine at 1450 psia and 1000 F. The turbine exhausts at 50 psia to the process heating load of 130 million B/hr. No bypass steam is used for process heating. Condensate from the heating load is at 180 F. Feed water leaves the open heater as saturated liquid at 50 psia. Neglecting pump work, calculate (a) the steam generated, lbm/hr; (b) $\dot{Q}_{in}$ for the steam generator, MB/hr; and (c) the turbine power in kilowatts. Do not neglect steam supplied to the feed-water heater. Compare the results with those of problem 9.24. Determine the incremental increase in $\dot{Q}_{in}$ and steam flow per kilowatt of turbine power when compared with problem 9.24.

9.26 A single extraction steam turbine, used for power generation and supply of process steam, operates on the cycle shown in Figure 9.17. The turbine operates reversibly and adiabatically with inlet at 1450 psia and 1000 F. Extraction is at 50 psia. The condenser pressure is 1.50 psia. Condensate from the 130 million B/hr process heating load is at 180 F. Feed water leaves the open heater as saturated liquid at 50 psia. Neglect pump work but not steam supply to the feed-water heater. Assuming the turbine power is 25 percent greater than that determined in problem 9.25, calculate (a) $\dot{Q}_{in}$ for the steam generator, MB/hr; (b) steam generated, lbm/hr; and (c) the turbine power in kilowatts. Determine the incremental increase in $\dot{Q}_{in}$ and steam flow, each per kilowatt of power, when compared with problem 9.24. Compare the incremental turbine steam rate, pounds per kilowatt, with that of problem 9.25.

9.27 Hot gases are available from an industrial process at 1000 F and 16 psia. Suppose these gases are used in a heat-recovery boiler and leave the boiler at 400 F. A bottoming vapor power cycle similar to that shown in Figure 9.18 uses toluene as the working fluid. Toluene enters the turbine as saturated vapor at 500 F and leaves the condenser as saturated liquid at 100 F. Assume $C_p$ for liquid toluene to be constant at 0.410 B/lbm R and the enthalpy of vaporization equal to 102.0 B/lbm at 500 F. Further assume the hot gas supplied to the boiler has the same specific heats as air. For a hot-gas inlet flow rate of 1 million ft$^3$/hr, determine the turbine power output, assuming that all components operate ideally and pump work is negligible.

9.28 A Carnot refrigeration cycle rejects heat at 100 F. Determine the coefficient performance for refrigerated region temperatures of 0 F, −100 F, −200 F, −300 F, −400 F, and −450 F. Plot the coefficient of performance versus the temperature of the refrigerated region. Comment on the work required to produce a given refrigerating effect at the progressively lower temperatures.

9.29 Suppose that a Carnot refrigeration cycle, having air as the working fluid, operates between 0 and 100 F. The pressure at the beginning of heat transfer to the air is 50 psia and the volume is doubled during that process. Assume a closed system with 0.0875 lbm air. If the cycle were repeated 150 times per minute, what refrigerating capacity, in tons, would be realized? What would be the power requirement, in kilowatts, to operate the unit? Calculate the power per ton of refrigerating effect and the coefficient of performance.

9.30 Consider the refrigeration cycle described in problem 5.48 which has air as the working fluid. Let each of the processes for the cycle be steady-flow processes. Sketch the flow diagram for the cycle and describe the components required. Show the cycle on *T-s* coordinates and find the power required per ton of refrigerating effect.

9.31 Would it be possible to have a refrigeration cycle with heat addition at 0 F, heat rejection at 77 F, and a coefficient of performance equal to 6.00? Justify your conclusion.

9.32 A refrigeration unit operates on the simple vapor-compression cycle using ammonia as the refrigerant. On leaving the evaporator the ammonia vapor is saturated vapor at 5 F. Saturated liquid at 140 psia leaves the condenser. Determine (a) the coefficient of performance, (b) the horsepower per ton of refrigerating effect, and (c) the volume rate of flow into the compressor per ton of refrigerating effect.

9.33 A refrigeration unit operates on the simple vapor-compression cycle using Freon-12 as the refrigerant. The refrigerant leaves the evaporator as saturated vapor at 5 F and leaves the condenser as saturated liquid at 75 F. Determine (a) the coefficient of performance, (b) the horsepower per ton of refrigerating effect, and (c) the volume rate of flow into the compressor per ton of refrigerating effect. Compare the results with those obtained in the preceding problem. Discuss other relevant factors that would be considered in comparing this cycle with that of the preceding problem.

9.34 A simple vapor-compression refrigeration cycle with Freon-12 as the refrigerant has a capacity of 3 tons. The refrigerant leaves the compressor as saturated vapor at 90 F. The evaporator temperature is 10 F. Calculate (a) the coefficient of performance, (b) the power, in kilowatts, to drive the compressor, and (c) the volume rate of flow going into the compressor.

9.35 A simple vapor-compression refrigeration cycle uses ammonia as the refrigerant. The intake to the compressor is at the rate of 25.00 $ft^3/min$ and ammonia leaves the condenser as saturated liquid at 200 psia. The ammonia entering the compressor is saturated vapor. Determine the refrigerating capacity and compressor power, in kilowatts, for evaporator temperatures of $-8$ F, 12 F, and 30 F. Plot the results versus evaporator temperature and determine if either capacity or power per ton is related linearly to the evaporator temperature.

9.36 The refrigeration unit in a small air conditioner has a capacity of 9000 B/hr. The refrigerant, Freon-12, leaves the evaporator of the simple vapor-compression cycle as saturated vapor at 40 F. Liquid leaving the air-cooled condenser is saturated at 150 psia. The compressor operates adiabatically, but not reversibly, and has an efficiency of 80 percent. The compressor efficiency is the ratio of isentropic to actual work based on the given inlet state and discharge pressure. The compressor is motor driven and electric energy cost is 8 cents/kWh. Determine the cost of electric energy to operate the unit for 24 hours.

9.37 A simple vapor-compression refrigeration cycle uses ammonia as the refrigerant. Individual components in the system do not operate ideally, however, no pressure drop occurs in the evaporator, condenser, or piping between components. The compressor discharge is at 220 F and 140 psia, the condenser outlet at 70 F, and the evaporator outlet at $-20$ F, saturated vapor. During compression 10 B of heat is transferred from each pound of ammonia. The power input to the compressor is 15 HP. Determine (a) the coefficient of performance, (b) the refrigeration capacity in tons, and (c) the compressor intake volume in cubic feet per minute. Sketch a *T-s* diagram for the cycle.

9.38 A refrigeration system operates on the cycle illustrated below. The heat exchanger between the evaporator outlet and the compressor inlet is similar to a closed feed-water heater in a steam power cycle; that is, heat transfer to the refrigerant between sections (4) and (5) is equal in magnitude to heat transfer from the refrigerant between sections (1) and (2). Suppose the refrigerant, Freon-12, leaves the condenser as saturated liquid at 110 F, the temperature drops to 80 F at section (2), and saturated vapor at −10 F enters the isentropic compression. Calculate the coefficient of performance and horsepower per ton of refrigerating effect. For a simple vapor-compression cycle having the same evaporator outlet and condenser outlet states, calculate the coefficient of performance and horsepower per ton of refrigerating effect. Sketch *T-s* diagrams for the cycles.

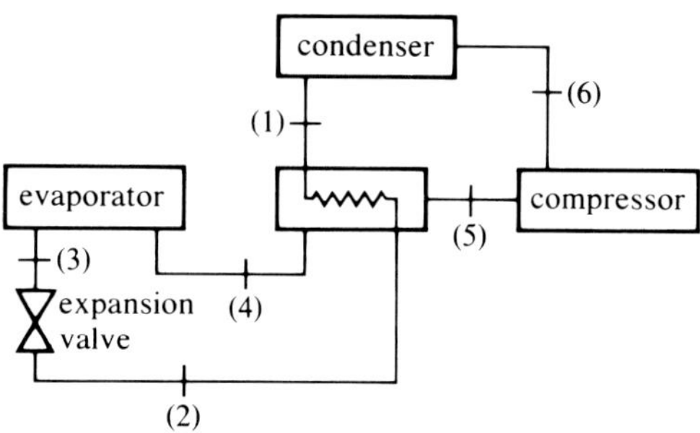


9.39 A refrigerator–freezer combination operates on the cycle shown in Figure 9.23 with Freon-12 as the refrigerant. At the compressor outlet the pressure is 150 psia and the temperature 200 F. The refrigerator evaporator operates at 15 F and the freezer evaporates at −15 F. Refrigerant leaving the refrigerator evaporator has a quality of 57.0 percent. Determine the refrigerating effect for each evaporator in B/lbm and the compressor work, B/lbm.

9.40 Another vapor cycle to produce refrigeration at two temperatures using a single compressor is illustrated in the following figure. Suppose the system uses ammonia as the refrigerant, each component operates ideally, the capacity of the first evaporator is 10 tons, and the capacity of the second evaporator is 5 tons. The pressure in the condenser is 170 psia, the temperature in the first evaporator is 0 F, and the pressure in the second evaporator is 15 psia. If the quality leaving each evaporator is 100 percent, what are the mass rate of flow through each evaporator and the power required to drive the compressor?

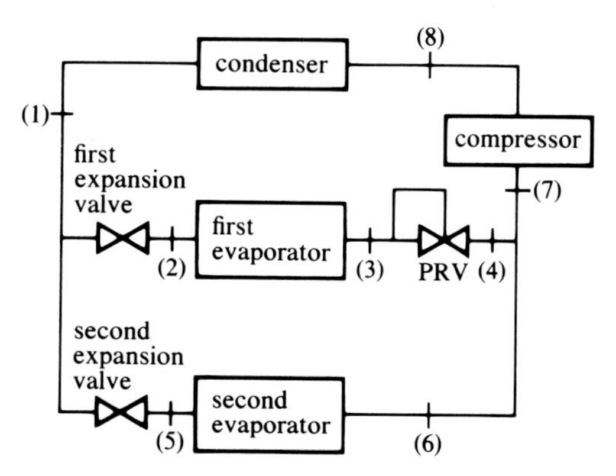

9.41 A simple vapor-compression cycle heat pump using Freon-12 operates with saturated vapor at 10 F leaving the evaporator and saturated liquid at 150 psia leaving the condenser. For a heating effect of 50,000 B/min, determine the compressor power required.

9.42 What is the numerical relation between the coefficient of performance for a heat pump and the coefficient of performance for a refrigeration cycle when each operates with identical states around the cycle?

9.43 A certain building requires heating at the rate of 80,000 B/hr. Three alternative methods of satisfying the requirement are to be considered. The first is a natural-gas-fueled furnace in which 75 percent of the heating value of the fuel, 1000 B/ft$^3$, is converted to energy for heating. The gas cost is $2.54 per thousand cubic feet. A second method being considered is electric resistance heating. Electric energy cost is 4.953 cents per kilowatt-hour. The third alternative is a heat pump using 57 F well water as an energy source. The heat pump is electric motor driven and operates on the simple cycle using Freon-12. Saturated liquid leaves the condenser at 110 F. The temperature difference between the well water and the refrigerant in the evaporator is 12 F. For each of the three alternatives, determine the cost of operation per 24-hour day. If the combined motor–compressor efficiency were 70 percent (isentropic work equal to 70 percent of actual work), what would be the cost of heat pump electric energy?

9.44 A heat pump operating on a simple vapor compression cycle uses Freon-12 as the working fluid and supplies heat to maintain a building at 70 F. The building heat loss is given by $10{,}000(70 - t_{atm})$ B/hr, where $t_{atm}$ is the temperature of the outside atmospheric air. Atmospheric air serves as the energy source for heat transfer to the working fluid as it passes through the evaporator. The evaporator temperature is 15 F less than $t_{atm}$ in order to provide the desired heat transfer. The state leaving the evaporator is saturated vapor and that leaving the condenser is saturated liquid at 125 psia. Assuming ideal operation of all components, determine the cost per hour to operate the heat pump when $t_{atm}$ is 60 F, 40 F, and 0 F. Assume that the cost of electrical energy is 8 cents per kilowatt-hour.

9.45 Consider a simple vapor-compression refrigeration cycle using ammonia as the refrigerant. The system has a capacity of 100 tons when operating with an evaporator temperature of 0 F and a compressor discharge at 170 psia, 240 F. Liquid leaving the condenser is subcooled to 80 F. The compressor is driven by a steam turbine that has throttle conditions (inlet state) of 200 psia and 500 F. The turbine exhaust is at 14.696 psia. Assuming ideal conditions, determine the mass rate of flow through the turbine in pounds per hour.

# 10

# Air-Standard Cycles

*Fossil-fuel-fired power plants that operate on vapor cycles, gas turbines, and internal combustion engines share a commonality in that they all are systems that convert to work a portion of the energy released by the combustion of a fuel. Actual gas turbines and internal combustion engines do not, in the thermodynamic sense, operate in cycles since the same working fluid does not undergo repetitive cycles as does the working fluid of a vapor power cycle. In this chapter we shall study several* ***air-standard*** *cycles for internal combustion engines and gas turbines. In each the combustion process is replaced by heat transfer from a high-temperature source and the exhaust process is replaced by heat transfer to a low-temperature reservoir. Air-standard cycles operate within closed boundaries and the working fluid is air rather than a fuel–air mixture that undergoes a chemical reaction and an accompanying change in chemical composition in a real gas turbine or internal combustion engine. We restrict our considerations to air-standard cycles to avoid dealing with combustion processes, a topic that is the subject of Chapter 12. The processes comprising the air-standard internal-combustion reciprocating-engine cycles are assumed to occur in piston–cylinder assemblies. Individual processes for the air-standard gas-turbine cycle are steady flow. While the differences between actual engine or turbine processes and the air-standard cycles might be substantial in some cases, the analysis of an air-standard cycle reveals some important parameters affecting the efficiency and performance of the device modeled by the cycle.*

*Air-standard cycles are based on perfect gas behavior. Each of the processes of the cycle is assumed to be internally reversible. The cycles can be studied on either a "cold" air or a "warm" air basis. Cold-air-standard cycles are those for which specific heats of the air are assumed to be constant. Warm-air-standard cycles take into account the changes in specific heats with temperature. While accurate equations expressing the variation of specific heats of air as a function of temperature are available, the use of such equations involves cumbersome iterations, particularly when isentropic processes are involved, as they are in most air-standard cycles. Therefore, tables of properties for air such as the Keenan and Kaye* ***Gas Tables,*** *which incorporate variations of specific heats with temperature, presently provide the most convenient approach to warm-air-standard cycle analysis. The results obtained using cold- and warm-air analyses do not differ greatly unless very high temperatures are involved. Therefore, we will consider only cold-air-standard cycles.*

## 10–1 Some Historical Aspects of Combustion Engine Development

The history of internal combustion engine development provides several of the more interesting insights into the thought processes of humankind. Huygens, in the 1680s, attempted to build an engine that used gunpowder as the fuel. A student who has accumulated some appreciation of the past and present efforts and problems of engine technology might view this attempt as ridiculous. What motivated Huygens to try it? Could it have been the result of his observations of musketry and cannons? This seems quite possible, if one reflects on the way technology, at any given time, is influenced by existing, established practice. The Newcomen condensing engines, developed in the late 17th century to pump water from English coal mines, are a case in point. These cumbersome, atmospheric steam engines were, in a most charitable sense, the pioneer reciprocating engines. The first of James Watt's steam engine patents, granted in 1769, followed the reciprocating engine theme—a theory that has dominated engine development for more than two centuries. Manufacture of the Boulton and Watt engines, beginning in 1775, initiated the Industrial Revolution, whereby the productivity of one worker was increased far beyond that of one craftsman.

Further evidence of an "accepted" technology logic is evidenced by the many years spent in attempting to develop a reciprocating (linear motion) electric motor, rather than the rotating machines that finally evolved. One of the earliest internal combustion engine patents was granted to Street, an Englishman, in 1794. Street proposed the use of turpentine as the engine fuel. Beau de Rochas was granted a patent in 1862 on an internal combustion engine cycle having processes essentially the same as those of the four-stroke, spark-ignition engine cycle in use today. Unfortunately he did not translate theory into practice. Lenoir had produced more than 300 engines by 1865. These engines, fueled with benzene, had a thermal efficiency of approximately 4 percent. Little did Lenoir know that 80 years later the "buzz bombs" falling on England would use the Lenoir cycle.

Internal combustion engine development moved rapidly during the 1860–1880 period. Otto and Langen built their first spark-ignition engine in 1867. Several thousand of these single-cylinder engines that delivered approximately 0.5 HP were sold. The four-stroke, spark-ignition engine developed by Otto and first marketed in 1876 was, by far, the most efficient engine at that time. Clerk, an Englishman, is credited with development of the two-stroke, spark-ignition engine in 1879. In 1892 Rudolf Diesel proposed a compression-ignition internal combustion engine that would use coal dust as the fuel. Although his coal-dust-fueled engines did not succeed, the cycle proved successful in 1897 when liquid fuels were used.

Rotating prime movers did not enter the picture until de Laval proved the feasibility of steam turbines in the 1880s. As late as the 1930s gas turbines could not produce power adequate to drive the associated compressor. But in less than a decade the increased efficiencies of both the compressor and turbine units resulted in units for aircraft and marine propulsion. One cannot help but speculate on where vehicle propulsion system technology would be today if rotating engine designs had been pursued for more than a century, as were reciprocating engines. Twenty–twenty hindsight, in this one aspect, teaches us the value of history and, conversely, the rigidity or lack of ingenuity that engineers must combat.

**Figure 10.1** Pressure–Volume Diagrams for Air-Standard Otto Cycles. *a.* Two-stroke cycle. *b.* Four-stroke cycle.

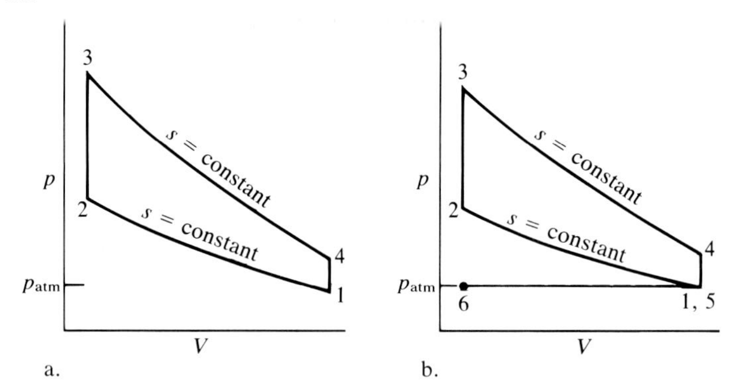


## 10–2 The Air-Standard Otto Cycle

The air-standard Otto cycle is comparable to the cycle used in most spark-ignition (SI) engines. These, and other internal combustion engine cycles, are either two- or four-stroke cycles. One stroke is represented by the piston motion from bottom dead center to top dead center. This motion is coincident with 180-degree rotation of the engine crankshaft. In Figure 10.1 pressure-volume diagrams for two- and four-stroke Otto cycles are shown for one cyclinder of an air-standard Otto cycle engine. The abscissa $V$ is the cylinder volume, that is, the volume contained within the cylinder head, piston face, and cylinder walls at any point during the cycle.

The four-stroke cycle taking place during two revolutions of the crankshaft is illustrated in Figure 10.1b. We will use that figure to describe the operation of a spark-ignition engine. The fuel–air mixture is drawn into the cylinder during the stroke (6) to (1) while the inlet valve remains open. This valve closes at approximately point (1). During the compression stroke (1) to (2) the fuel–air mixture is compressed and the charge is ignited (spark advance or retard) at approximately point (2). In an engine the combustion process, with a number of chemical reactions, continues beyond (3) until the reactions are frozen during the power stroke (3) to (4). The process (4) to (5) represents the pressure drop when the exhaust valve first opens. The final process in the engine, (5) to (6), is the exhaust stroke during which the exhaust valve remains open and the products of combustion are discharged to the atmosphere as the piston moves toward top dead center. Each of these processes in an engine differs from those of an air-standard cycle. For example, the intake process takes place at pressures less than $p_1 = p_6$ due to pressure drop across the valves. The exhaust pressures are not equal to $p_1 = p_6$, again due to pressure drops across the exhaust valve and through the exhaust system. In general the diagram for an operating engine, with time-dependent processes, would show roundings rather than the sharp intersections shown at the several points of Figure 10.1. Figure 10.2 is a cutaway diagram of a spark-ignition engine.

Now let us consider the air-standard Otto cycle, again referring to Figure 10.1. The net work *per cycle* will be the same for the two- and four-stroke cycles since process (5) to (6) cancels, exactly, process (6) to (1). This work is equal to the enclosed area on the diagram.

**Figure 10.2** A 1.6 Liter, 4 Cylinder Spark Ignition Engine

(Courtesy Ford Motor Co.)

Thus we shall treat each cycle as a closed-system cycle. The compression process between (1) and (2) is reversible and adiabatic in the air-standard Otto cycle. The constant-volume process from (2) to (3), replacing ignition and combustion in an SI engine, is accomplished by heat transfer from a reservoir at temperature $T_3$ or greater. In the air-standard cycle the process between (3) and (4) is reversible and adiabatic. The process (4) to (1), completing the cycle, is a constant-volume process during which heat is transferred from the air to a sink at a temperature less than $T_1$.

Let us now develop an expression for the thermal efficiency of the air-standard Otto cycle. Each process of the cycle takes place within the closed system defined by the inner surfaces of the piston–cylinder combination. It is convenient to choose as a basis a unit mass. Thus the applicable energy equation is

$$ {}_a q_b = u_b - u_a + {}_a w_b \tag{3.2e} $$

where $a$ and $b$ represent the initial and final states, respectively, for any process in the cycle.

With two reversible and adiabatic processes in the cycle, a convenient expression to use in determining the thermal efficiency is

$$\eta = \frac{q_{in} - q_{out}}{q_{in}}$$

For the cold-air-standard cycle

$${}_2q_3 = q_{in} = u_3 - u_2 = C_v(T_3 - T_2)$$

and

$$|{}_4q_1| = q_{out} = u_4 - u_1 = C_v(T_4 - T_1)$$

The thermal efficiency is then

$$\eta = \frac{C_v(T_3 - T_2) - C_v(T_4 - T_1)}{C_V(T_3 - T_2)} = 1 - \frac{T_4 - T_1}{T_3 - T_2} \qquad \textbf{(10.1)}$$

for cold-air-standard conditions. The compression ratio $r_c$ is defined as

$$r_c \equiv \frac{V_1}{V_2} \qquad \textbf{(10.1a)}$$

Equation (10.1) can be simplified when the cycle is considered on temperature–entropy coordinates, Figure 10.3. The entropy change during the constant-volume processes is equal, that is, $s_3 - s_2$ is equal to $s_4 - s_1$. From the first $Tds$ equation, eq. (7.8),

$$Tds = du$$

for constant-volume processes. With cold-air-standard conditions

$$ds = \frac{C_v dt}{T}$$

from which

$$s_3 - s_2 = C_v \ln \frac{T_3}{T_2}$$

and

$$s_4 - s_1 = C_v \ln \frac{T_4}{T_1}$$

Equating the entropy changes we have

$$C_v \ln \frac{T_3}{T_2} = C_v \ln \frac{T_4}{T_1}$$

**Figure 10.3** Temperature–Entropy Diagram for an Air-Standard Otto Cycle

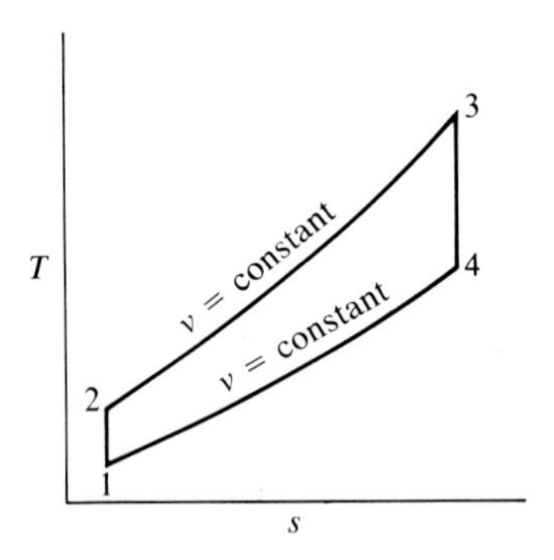


which reduces to

$$\frac{T_3}{T_2} = \frac{T_4}{T_1}$$

Subtracting 1 from each side of the equality gives

$$\frac{T_3}{T_2} - 1 = \frac{T_4}{T_1} - 1$$

or

$$\frac{T_3 - T_2}{T_2} = \frac{T_4 - T_1}{T_1}$$

By rearranging terms the preceding equality becomes

$$\frac{T_1}{T_2} = \frac{T_4 - T_1}{T_3 - T_2}$$

which, when substituted into eq. (10.1), yields

$$\eta = 1 - \frac{T_1}{T_2} \tag{10.1b}$$

However, with the process between (1) and (2) reversible and adiabatic, we have

$$\frac{T_2}{T_1} = \left(\frac{V_1}{V_2}\right)^{\gamma - 1} = r_c^{\gamma - 1}$$

which, substituted into eq. (10.1a), yields

$$\eta = 1 - \frac{1}{r_c^{\gamma - 1}} \tag{10.2}$$

for the thermal efficiency of the cold-air-standard Otto cycle.

**Figure 10.4** Comparison of Air-Standard and Actual Thermal Efficiencies for Otto Cycle Engines

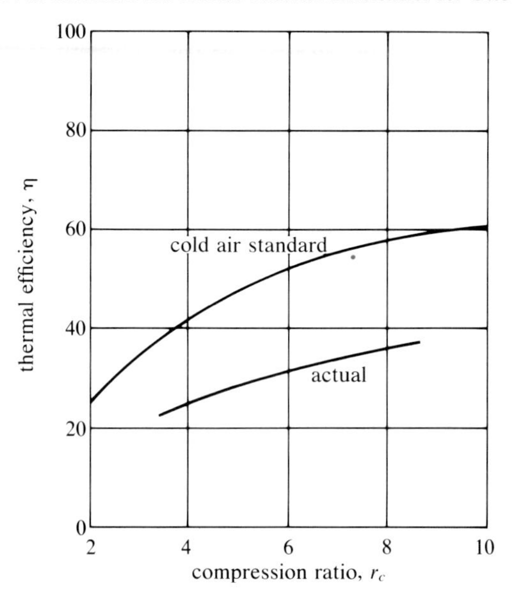


Equation (10.2) indicates the important effect of compression ratio on the thermal efficiency of air-standard Otto cycles. In Figure 10.4 the thermal efficiency of actual spark-ignition engines is shown to have a similar relation to the compression ratio. In view of the foregoing, the desirability of higher compression ratios would seem obvious. More work per cycle, which translates into more power at a given engine speed, in the past has been a desirable design criterion for spark-ignition engines, bounded only by the autoignition (detonation) temperature of the fuel–air mixture. Today environmental considerations override this engine characteristic, with power per unit volume of piston displacement and compression ratios of automotive engines being decreased as the industry strives to meet emission (exhaust gas) standards established by the Environmental Protection Agency. To offset decreased compression ratios, increased fuel economy is being realized through lighter-weight automotive vehicles.

## 10–3 The Air-Standard Diesel Cycle

Like the Otto cycle, the diesel cycle is a reciprocating engine cycle. Some of the significant differences between actual engines of the two types are as follows:

1. During the compression stroke of an Otto cycle engine the fuel–air mixture is compressed. The compression ratio is limited by the autoignition temperature of the mixture (detonation).
2. During the compression stroke of diesel engines only air is contained in the cylinder volume.
3. Combustion in the Otto-cycle engine is initiated by a spark discharge with peak pressure and temperature reached at approximately constant volume.

4. Combustion in a diesel engine is initiated when fuel is injected into the high-temperature air present in the cylinder volume as the compression stroke approaches completion. Because of this characteristic, diesel engines are frequently referred to as compression-ignition or CI engines. Injection continues to the cutoff point, with combustion occurring at approximately constant pressure.

From these comparisons of the two cycles it should be apparent that diesel engines have higher compression ratios than Otto-cycle engines—by about a factor of two. Diesel engines, particularly those operating at low speed, have less pollutants in their exhaust gases than do SI engines. Compression-ignition engines are commonly used as prime movers for standby (emergency) electric generators, truck and locomotive engines, and ship propulsion units. Diesel engines, particularly those for ship propulsion, are built to deliver power up to the 45,000-HP range. One of the largest units of this type was completed in 1964 by the FIAT Grandi Motori Works at Turin, Italy. The two-stroke, 12-cylinder in-line engine is rated for normal, continuous output of 25,200 HP at 122 rev/min with a peak of 32,500 HP. The cylinder bore is 35 1/2 inches and the stroke is 36 inches. Another example of a large diesel engine is that shown in Figure 10.5. Engines for standby or peaking power have power outputs ranging from a few to 5000 kW. The more powerful engines are supercharged and, of course, run at speeds synchronous with the electric system frequency. One example is a 5000-kW supercharged six-cylinder engine operating at approximately 167 rev/min. Cylinder dimensions of the engine are 29-inch bore and 54-inch stroke. Diesel engines operate with significantly higher actual thermal efficiencies than any other type of combustion engine.

Pressure–volume and temperature–entropy diagrams for a two-stroke, air-standard diesel cycle are shown in Figure 10.6. The cycle consists of four closed-system processes. Starting at (1) the air is compressed isentropically to (2). During the constant-pressure process from (2) to (3), heat transfer to the air replaces the fuel injection and combustion of the actual engine. The continued expansion from (3) to (4), which is isentropic, completes the power stroke. During the constant-volume process from (4) to (1), heat is rejected to a lower-temperature reservoir, completing the cycle.

The compression ratio $r_c$ is the volume ratio

$$r_c \equiv \frac{V_1}{V_2} \tag{10.1a}$$

The cutoff ratio is defined as

$$r_{co} \equiv \frac{V_3}{V_2} \tag{10.3}$$

As was true for the air-standard Otto cycle, each process is analyzed on the basis of the closed-system energy equation. With the cycle having two isentropic processes, the thermal efficiency is most conveniently calculated from

$$\eta = \frac{q_{\text{in}} - q_{\text{out}}}{q_{\text{in}}}$$

**Figure 10.5** Low Speed (43,000 HP) Marine Diesel Engine for Container Ship

(Courtesy Allis–Chalmers Corp.)

**Figure 10.6** *p*-*V* and *T*-*s* Diagrams for an Air-Standard Diesel Cycle

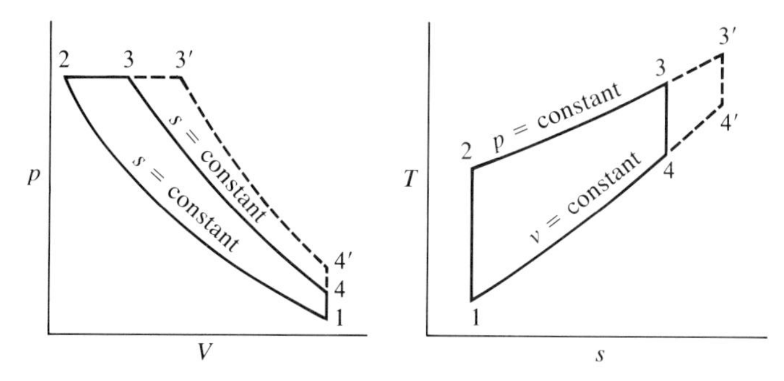

Heat transfer to the air during the constant-pressure process from (2) to (3) is

$$q_{in} = {}_2q_3 = u_3 - u_2 + \int_2^3 p\,dv = h_3 - h_2$$
$$= C_p(T_3 - T_2)$$

for cold-air-standard conditions. Heat rejected during the constant-volume process from (4) to (1) is

$$q_{out} = |{}_4q_1| = u_4 - u_1 = C_v(T_4 - T_1)$$

From these heat-transfer quantities the thermal efficiency of the cold-air-standard diesel cycle is

$$\eta = \frac{C_p(T_3 - T_2) - C_v(T_4 - T_1)}{C_p(T_3 - T_2)}$$
$$= 1 - \frac{1}{\gamma}\left(\frac{T_4 - T_1}{T_3 - T_2}\right) \qquad \textbf{(10.4)}$$

The relationships among the four temperatures in eq. (10.4) are conveniently established using the *T-s* diagram for the cycle, Figure 10.6. Once more we have a cycle bordered by two constant-entropy lines and

$$s_3 - s_2 = s_4 - s_1$$

By the first *Tds* equation, eq. (7.8),

$$s_4 - s_1 = C_v \, ln \frac{T_4}{T_1}$$

and by the second *Tds* equation, eq. (7.9),

$$s_3 - s_2 = C_p \, ln \frac{T_3}{T_2}$$

With $s_3 - s_2$ equal to $s_4 - s_1$, we have

$$C_v \, ln \frac{T_4}{T_1} = C_p \, ln \frac{T_3}{T_2}$$

or

$$\frac{T_4}{T_1} = \left(\frac{T_3}{T_2}\right)^{\gamma} \qquad \textbf{(10.5)}$$

While the analysis can be extended to express the thermal efficiency of the cycle in terms of the cutoff and compression ratios, the use of eqs. (10.4) and (10.5) provides an adequate approach.

**Figure 10.7** Air-Standard Otto and Diesel Cycles having the Same $r$, $q_{in}$, and Initial States

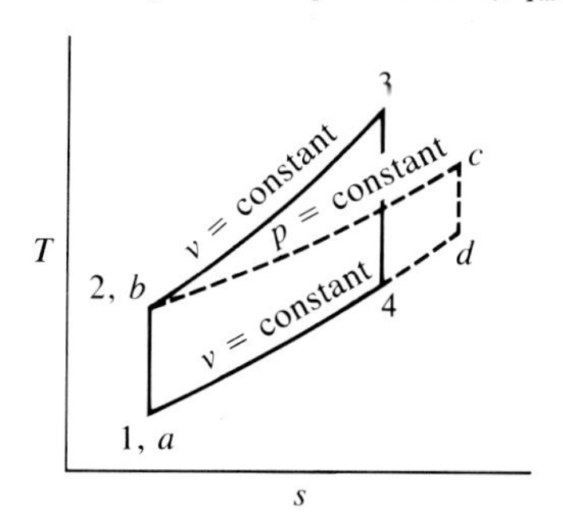


For a given engine piston displacement, the effect of more work per cycle is indicated in Figure 10.6 by points 3′ and 4′ and the broken-line paths to these states. For any internally reversible cycle

$$\oint dW = W_{net} = \oint p dV = m \oint dq = \oint dQ$$

The term $\oint p dV$ is equal to the enclosed area on the $p$-$V$ diagram and $m\oint dq$ is the enclosed area on the $T$-$s$ diagram multiplied by the mass of the system. An increase in the net work per cycle increases the cutoff ratio and the peak temperature of the cycle.

In an actual engine these effects are reflected in a shorter time interval for the expansion 3′ to 4′ and a higher exhaust gas temperature, $T_{4'}$. The consequent decrease in thermal efficiency is evident from the $T$-$s$ diagram of Figure 10.6 where the ratio of the area enclosed by 3–3′–4′–4–3 (net heat equal to net work) to the area under 3–3′ (net heat in) is less than that for the basic cycle 1–2–3–4–1.

## 10–4 Some Comparisons of Air-Standard Otto and Diesel Cycles

Temperature–entropy diagrams provide a convenient method for comparing the air-standard Otto and diesel cycles. Which cycle would be more efficient if each had the same compression ratio, the same state at the beginning of compression, and the same $q_{in}$ per cycle? For these comparisons we note again that the areas under internally reversible paths on the $T$-$s$ diagram are equal to the heat transfer per unit mass and that the enclosed area $q_{net}$ is equal to $w_{net}$ for the cycle. The first comparison is made in Figure 10.7. The numbers 1 through 4 are for the Otto cycle and *a, b, c,* and *d,* respectively, replace points 1 through 4 of the diesel cycle. For the same $q_{in}$ the constant-pressure line *b*–*c* for the diesel cycle must extend beyond line 3–4 to have the area under *b*–*c* equal that under 2–3. Recall that the lines of constant volume have steeper slopes than those of constant pressure. For the Otto cycle $q_{out}$ is the area under path 4–1 and is less than the area under *d*–*a*, $q_{out}$, for the diesel cycle. Thus we conclude that the Otto cycle is, for the stated conditions, more efficient than the diesel cycle.

Figure 10.8 Air-Standard Otto and Diesel Cycles having the Same Maximum Pressures and Temperatures

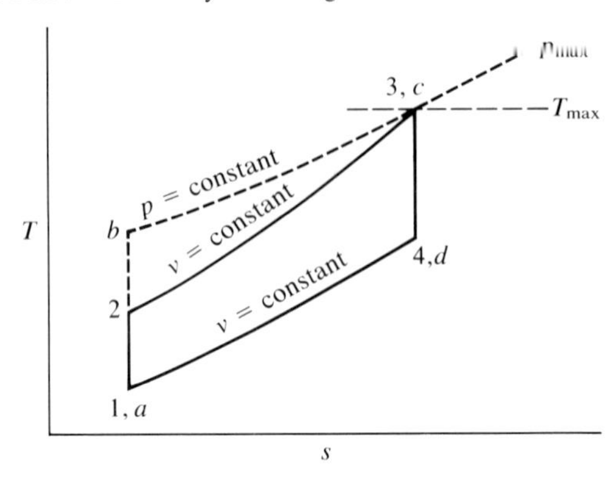


To illustrate this method of comparison further, consider an Otto cycle and a diesel cycle having the same maximum pressure and temperature and the same state at the beginning of the compression stroke. The cycles are shown in Figure 10.8 with $T_3$ equal to $T_c$ and $T_a$ equal to $T_1$. For the stated conditions $q_{out}$ is equal for each cycle since states (4) and ($d$) are coincident due to $s_3 = s_4 = s_c = s_d$ and $v_1 = v_a = v_4 = v_d$. For the diesel cycle, $q_{in}$ is greater than that for the Otto cycle. The work for the diesel cycle, $q_{in} - q_{out}$, is also greater than that of the Otto cycle since $q_{out}$ is equal for the two. To compare thermal efficiencies we shall choose the form

$$\eta = \frac{q_{in} - q_{out}}{q_{in}} = 1 - \frac{q_{out}}{q_{in}}$$

With $q_{out}$ quantities equal and $q_{in}$ greater for the diesel, we conclude that the diesel cycle is the more efficient of the two under the stated conditions.

As a final illustration of this method of comparison consider the two cycles operating with the same initial states, maximum pressures, and $q_{in}$. Under these conditions states (1) and ($a$) are coincident. The line of constant maximum pressure $p_{max}$ is next drawn on the temperature–entropy diagram, Figure 10.9. With isentropic compression for each cycle, states (2) and ($b$) must be on the constant-entropy line $s_1 = s_a$. For the Otto cycle the process from (2) to (3) takes place at constant volume while heat addition for the diesel, from ($b$) to ($c$), is at constant pressure $p_b = p_c = p_{max}$. Noting once more the steeper slope of constant-volume lines on the $T$-$s$ diagram, it is apparent that the Otto cycle must have a compression ratio less than that of the diesel if the two cycles are to satisfy the maximum-pressure criterion. On this basis state (2) is located for the Otto cycle at a pressure less than $p_{max}$ and a temperature less than $T_b$. The constant-volume line $v_2 = v_3$ is extended until it intersects $p_{max}$. To satisfy the condition of equal heat transfer to the air for each cycle, it is now apparent that point $c$ for the diesel cycle must lie to the left of point 3. The next line placed on the diagram is the constant-specific-volume line $v_a = v_1$ since heat transfer from the working fluid takes place at constant volume in each of the cycles.

**Figure 10.9** Air-Standard Otto and Diesel Cycles that have the Same Maximum Pressures, $q_{in}$, and Initial States

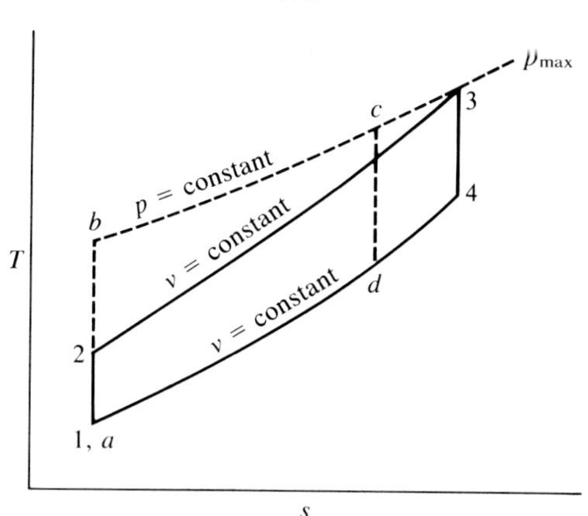


The final steps necessary for the comparison are the isentropic lines $s_3 = s_4$ and $s_c = s_d$, each extended until it intersects the constant-volume line $v_1 = v_a$. Inspection of the temperature–entropy diagram for each cycle reveals that the heat rejected by the diesel, $|_d q_a|$, is less than $|_4 q_1|$, the heat rejected during the Otto cycle. With $q_{in}$ the same for each, we conclude that the thermal efficiency of the diesel cycle is greater than that of an Otto cycle when each operates with the given constraints.

---

## Example 10.a

A cold-air-standard Otto cycle and a cold-air-standard diesel cycle operate under the following conditions.

Beginning of compression = 540 R, 14.50 psia

$q_{in}$ = 370 B/lbm

Maximum temperature = 3400 R

Calculate:

(a) The compression ratio for each cycle
(b) The thermal efficiency of each cycle
(c) The maximum pressure for each cycle
(d) The cutoff ratio for the diesel cycle
(e) The horsepower output of the Otto cycle if it operates as a four-stroke engine at 4000 rev/min and has a piston displacement of 300 in$^3$

Refer to Figures 10.1 and 10.6 for the state notation used in this example.

**Solution:**

From the maximum temperature and $q_{in}$ data the temperature at the end of the compression stroke is calculated for each cycle as

$$370 = 0.240(3400 - T_2)$$

$$T_2 = 1858 \text{ R (diesel)}$$

and

$$370 = 0.172(3400 - T_2)$$

$$T_2 = 1249 \text{ R (Otto)}$$

(a) The compression stroke is reversible and adiabatic for each of the air-standard cycles. For the diesel

$$r_c \equiv \frac{v_1}{v_2} = \left(\frac{T_2}{T_1}\right)^{1/(\gamma-1)} = \left(\frac{1858}{540}\right)^{1/0.4} = 21.96$$

and for the Otto cycle

$$r_c \equiv \frac{v_1}{v_2} = \left(\frac{T_2}{T_1}\right)^{1/(\gamma-1)} = \left(\frac{1249}{540}\right)^{1/0.4} = 8.136$$

(b) The thermal efficiency of the diesel cycle is, from eq. (10.4),

$$\eta = 1 - \frac{1}{\gamma}\left(\frac{T_4 - T_1}{T_3 - T_2}\right)$$

and from eq. (10.5)

$$\frac{T_4}{T_1} = \left(\frac{T_3}{T_2}\right)^{\gamma}$$

Solving eq. (10.5) for $T_4$ gives

$$T_4 = 540\left(\frac{3400}{1858}\right)^{1.40} = 1258 \text{ R}$$

The thermal efficiency of the diesel cycle is

$$\eta = 1.00 - \frac{1}{1.40}\left[\frac{1258 - 540}{3400 - 1858}\right]$$
$$= 66.74 \text{ percent}$$

For the Otto cycle the thermal efficiency according to eq. (10.1a) is

$$\begin{aligned}\eta &= 1 - \frac{T_1}{T_2} \\ &= 1 - \frac{540}{1249} \\ &= 56.77 \text{ percent}\end{aligned}$$

(c) The maximum pressure for the diesel cycle is that at (2), the end of the compression stroke, where

$$\begin{aligned}p_2 &= p_1\left(\frac{v_1}{v_2}\right)^{\gamma} = (p_1)(r_c)^{1.40} \\ &= (14.50)(21.96)^{1.40} \\ &= 1096 \text{ psia}\end{aligned}$$

In the Otto cycle the maximum pressure occurs at the end of the process in which heat is transferred to the system, state (3). For the stated conditions

$$\begin{aligned}p_2 &= p_1\left(\frac{v_1}{v_2}\right)^{\gamma} = p_1(r_c)^{1.40} \\ &= (14.50)(8.136)^{1.40} \\ &= 272.9 \text{ psia}\end{aligned}$$

The pressure at (3) is

$$\begin{aligned}p_3 &= p_2\left(\frac{T_3}{T_2}\right) \\ &= (272.9)\left(\frac{3400}{1249}\right) \\ &= 742.9 \text{ psia}\end{aligned}$$

(d) By definition the cutoff ratio is

$$r_{co} \equiv \frac{v_3}{v_2}$$

For the constant-pressure process from (2) to (3)

$$\frac{v_3}{v_2} = \frac{T_3}{T_2}$$

and

$$r_{co} = \frac{3400}{1858} = 1.830$$

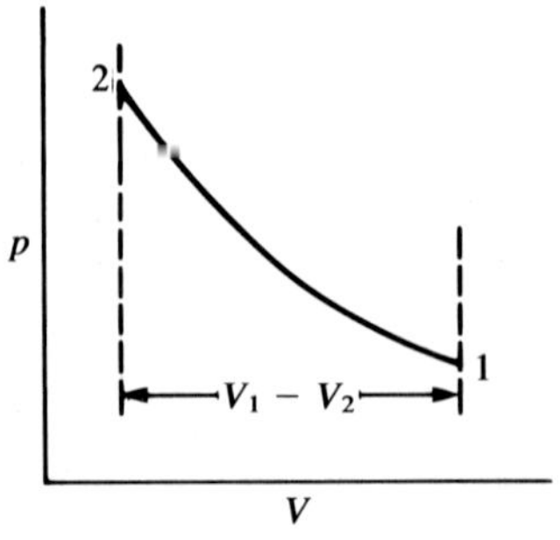

(e) Each of the preceding calculations was made on the basis of a unit mass of air. Solution of this part of the problem is clarified by first sketching a $p$-$V$ diagram for the given engine displacement. The volume ratio $V_1/V_2$ is the compression ratio, 8.136, calculated in part (a). The cylinder volume $V_1$ at bottom dead center is calculated from

$$\frac{V_1}{V_2} = 8.136 \quad \text{and} \quad V_1 - V_2 = 300 \text{ in}^3$$

with the result

$$V_1 = 342.0 \text{ in}^3 = 0.1979 \text{ ft}^3$$

The mass of air contained in the cylinders is

$$(14.50)(144)(0.1979) = (m)(53.3)(540)$$
$$m = 0.01436 \text{ lbm}$$

The heat rejected per pound of air is

$$|{}_4q_1| = q_{out} = 0.172(T_4 - T_1)$$

For a cold-air-standard Otto cycle

$$T_4 = T_1\left(\frac{T_3}{T_2}\right) = (540)\left(\frac{3400}{1249}\right) = 1470 \text{ R}$$

Thus

$$q_{out} = 0.172(1470 - 540)$$
$$= 160 \text{ B/lbm}$$

The net work per pound of air is

$$w_{net} = q_{net} = 370 - 160 = 210 \text{ B/lbm}$$

Noting from Figure 10.1b that two revolutions of the engine occur for each power cycle executed, the power is

$$\dot{W}_{net} = \frac{210\text{ B}}{\text{lbm}}\left|\frac{.01436\text{ lbm}}{}\right|\frac{1\text{ cycle}}{2\text{ rev}}\left|\frac{4000\text{ rev}}{\text{min}}\right|\frac{60\text{ min}}{\text{hr}}\left|\frac{\text{HP hr}}{2545\text{ B}}\right.$$

$$= 142.2\text{ HP}$$

---

**Figure 10.10** Flow Diagram for a Simple Gas Turbine

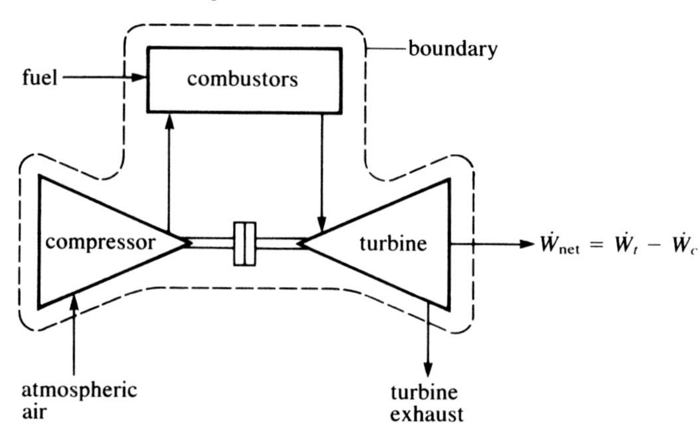


## 10–5 The Air-Standard Simple-Gas-Turbine Cycle

A simple gas turbine is shown schematically in Figure 10.10. Atmospheric air crosses the system boundary and enters the compressor. The compressor is connected to the turbine by a shaft and consumes part of the power produced by the turbine. The volume rate of flow that can be handled by reciprocating compressors is limited by the nature of these devices, specifically by piston displacement and the mass of reciprocating parts. These limitations have resulted in the development of centrifugal and axial-flow compressors for use in gas turbines. The air is compressed to a pressure approximately 10 times that of the atmosphere, and on leaving the compressor it enters several combustors. Fuel is sprayed into the combustors and steady-flow combustion takes place at approximately constant pressure. The high-temperature products of combustion leave the combustors, expand through the turbine producing power, and are exhausted to the atmosphere. The net power out is that produced by the turbine less the power required to drive the compressor. In this section we will denote the absolute turbine and compressor work and power terms by the subscripts $t$ and $c$. Thus the net power is the turbine power $\dot{W}_t$ minus the compressor power $\dot{W}_c$ as shown in Figure 10.10. Figure 10.11 is a cutaway diagram of a simple gas-turbine unit.

**Figure 10.11** Simple Single Shaft Gas Turbine

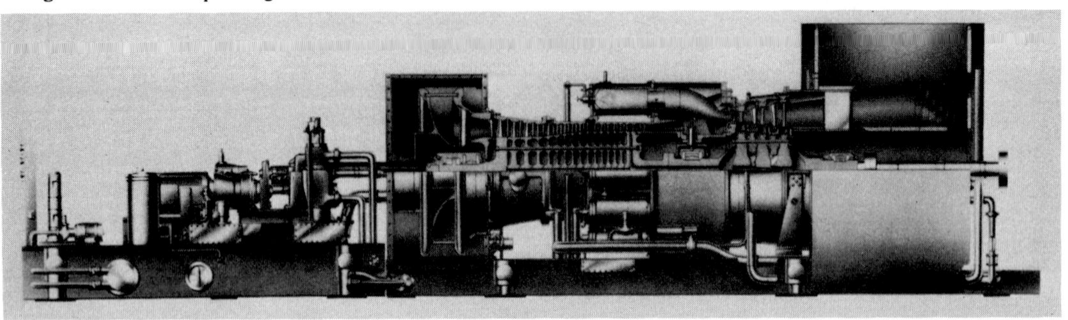

(Courtesy General Electric Co.)

The maximum allowable temperature of the gases entering the turbine is dictated by the temperature–strength characteristics of the turbine blading. The temperature is kept within limits by using high excess air (significantly more than is required for combustion of the fuel), utilizing compressor discharge air to cool combustors and mix with gases leaving the combustors, and, in some units, cooling hollow blades with bypass air from the compressor. By use of these cooling methods, it is at present possible to operate with temperatures as high as 2600 F at the turbine inlet. There is one distinct difference between maximum allowable temperatures for reciprocating engines and for gas turbines. The maximum temperature in reciprocating engines occurs intermittently—once per power cycle—whereas in a gas turbine the maximum temperature does not vary with time in steady operation. As a result, higher combustion temperatures can be tolerated in reciprocating engines.

Let us now consider the cold-air-standard gas-turbine cycle. Each of the processes in this cycle is a steady-flow process. In addition to assuming that all processes are internally reversible, we assume that changes in kinetic and potential energies are negligible. The combustion process is replaced by heat transfer to the air from a high-temperature reservoir and the exhaust is replaced by a cooler (heat exchanger) in which heat is transferred from the working fluid to a low-temperature sink. The diagram for the air-standard simple-gas-turbine cycle is given in Figure 10.12. Note that the cycle operates within a closed boundary. This air-standard cycle is commonly called the Brayton cycle. The processes comprising the Brayton cycle are shown on the *p-v* and *T-s* diagrams of Figure 10.13. The isentropic compression process (1) to (2), by eq. (2.14), requires work in the amount

$$w_c = \int_1^2 v\,dp$$

Graphically the integral is the area between path (1) to (2) and the ordinate. Heat transfer to the air takes place during the steady-flow constant-pressure process (2) to (3). The heat transfer to the working fluid is

$$q_{in} = {}_2q_3 = h_3 - h_2 = C_p(T_3 - T_2)$$

Figure 10.12 Flow Diagram for an Air-Standard, Simple Gas Turbine

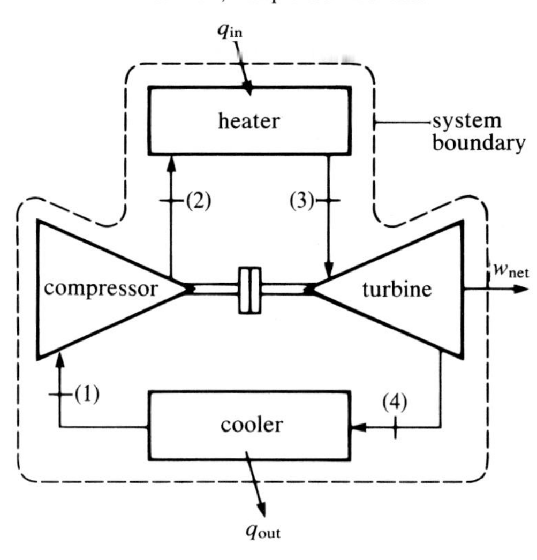


Figure 10.13 Pressure–Volume and Temperature–Entropy Diagrams for an Air-Standard, Simple Gas Turbine Cycle

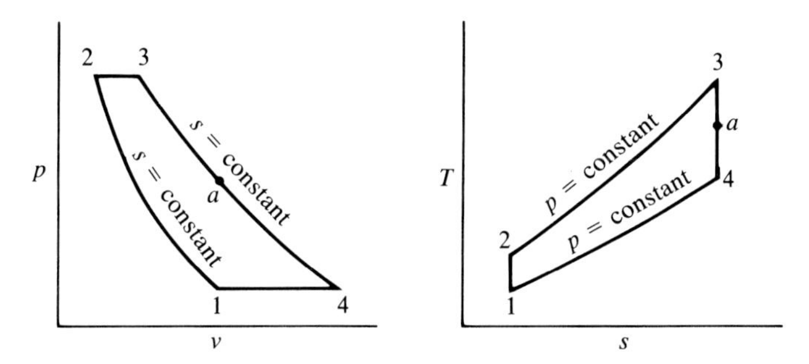


A substantial portion of the turbine power is used to drive the compressor. The part of the turbine work produced in the isentropic expansion (3) to (4) that is used to drive the compressor is

$$w_c = \int_3^a v dp$$

This work is the area between path (3) to ($a$) and the ordinate in the $p$-$v$ diagram, and must be equal to the area between path (1) to (2) and the ordinate. The air-standard cycle is completed by the steady-flow constant-pressure process from (4) to (1) for which

$$q_{out} = |{}_4q_1| = |h_1 - h_4| = C_p(T_4 - T_1)$$

Note that the air-standard gas-turbine cycle would not be able to deliver net work if constant-pressure lines on the $T$-$s$ diagram did not diverge as entropy increases. This is true because, to deliver net work, $w_t$ must be greater than $w_c$. Since $w_t = h_3 - h_4$ and $w_c = h_2 - h_1$, $(T_3 - T_4)$ must be greater than $(T_2 - T_1)$.

The thermal efficiency of the air-standard Brayton cycle is

$$\eta = \frac{w_{net}}{q_{in}} = 1 - \frac{q_{out}}{q_{in}} = 1 - \frac{h_4 - h_1}{h_3 - h_2}$$

which becomes

$$\eta = 1 - \frac{T_4 - T_1}{T_3 - T_2} \tag{10.6}$$

As was true for the cold-air-standard Otto and diesel cycles, the cold-air-standard simple-gas-turbine cycle is bordered by two isentropic processes. Thus

$$s_3 - s_2 = s_4 - s_1$$

or

$$C_p \ln \frac{T_3}{T_2} = C_p \ln \frac{T_4}{T_1}$$

Therefore

$$\frac{T_3}{T_2} = \frac{T_4}{T_1} \tag{10.6a}$$

In subtracting 1 from each side of this equality, we get

$$\frac{T_3 - T_2}{T_2} = \frac{T_4 - T_l}{T_1}$$

which rearranges to

$$\frac{T_1}{T_2} = \frac{T_4 - T_1}{T_3 - T_2}$$

Substitution into eq. (10.6) yields

$$\eta = 1 - \frac{T_1}{T_2} \tag{a}$$

for the cold-air-standard, simple gas turbine. However

$$\frac{T_2}{T_1} = \left(\frac{p_2}{p_1}\right)^{(\gamma-1)/\gamma}$$

**Figure 10.14** Air-Standard Gas Turbine for Electric Power Generation

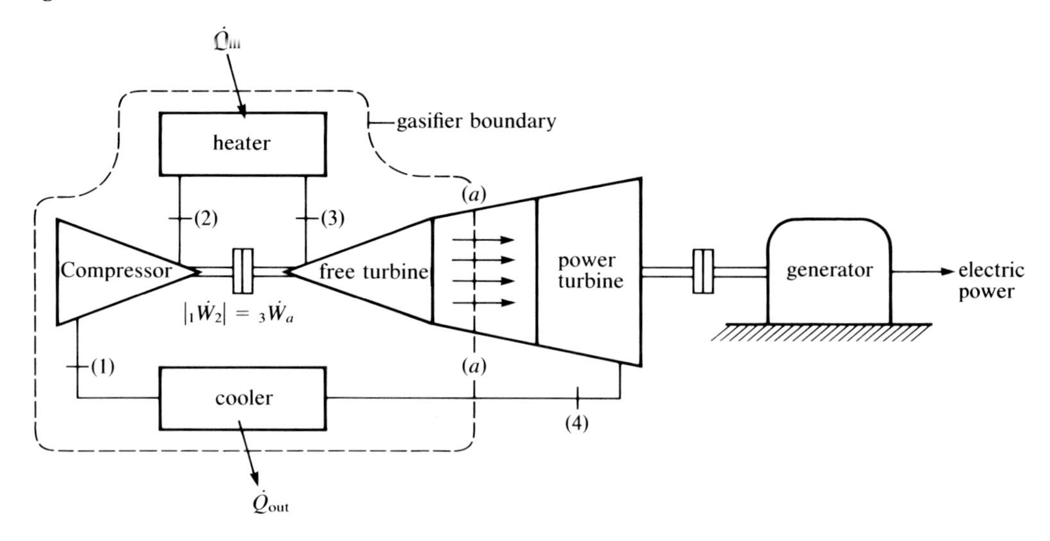


for the reversible adiabatic compressor process when specific heats are constant. The pressure ratio $r_p$ is defined as

$$r_p = \frac{p_2}{p_1}$$

for gas-turbine cycles. It follows then from eq. (a) that

$$\eta = 1 - \frac{1}{r_p^{(\gamma-1)/\gamma}} \tag{10.7}$$

for cold-air-standard simple-gas-turbine cycles. Thus the thermal efficiency of the cold-air-standard gas-turbine cycle varies with the compressor *pressure ratio*, $r_p$, similar to that of the cold-air-standard Otto cycle where the thermal efficiency was found to depend solely on the compression ratio $r_c$, a *volume ratio*.

One important use of stationary gas turbines is in the electric-power-generating industry. Gas-turbine-driven electric generators are used for peaking and standby (emergency) service. One modification of the simple gas turbine employed for this application is shown as an air-standard cycle in Figure 10.14. Two turbines that operate at different speeds are used. The free turbine of the gasifier section drives the compressor only and operates at speeds two to three times that of the power turbine. The power turbine drives only the electric generator, which operates at a synchronous speed, typically 3600 rev/min. This arrangement offers greater flexibility of operation than does the simple gas turbine. The free turbine exhaust, state (*a*) in Figures 10.13 and 10.14, is the state required to provide the power necessary to drive the compressor. The gases at state (*a*) expand through the power turbine to state (4).

## Example 10.b

An air-standard gas-turbine cycle, Figure 10.14, operates with the compressor intake at 88 F and 14.20 psia. The pressure ratio $r_p$ is 10 and the maximum temperature in the cycle is 2300 F. The unit, to be used for peaking power demands, is to deliver 10,000 kW. Assuming that the generator has an efficiency of 100 percent, calculate:

(a) The pressure at section ($a$), the power turbine inlet
(b) The cycle thermal efficiency
(c) The volume rate of flow entering the compressor in cubic feet per minute

**Solution:**

(a) The $T$-$s$ diagram in Figure 10.13 applies. Assuming kinetic and potential energy changes to be negligible, the compressor work is

$$w_c = |{}_1w_2| = C_p(T_2 - T_1)$$

The compressor discharge temperature $T_2$ is

$$T_2 = T_1 \left(\frac{p_2}{p_1}\right)^{(\gamma-1)/\gamma} = (548)(10)^{0.286} = 1058 \text{ R}$$

The free turbine work ${}_3w_a$ is equal to the work supplied to the compressor; thus

$${}_3w_a = |{}_1w_2|$$

or

$$C_p(T_3 - T_a) = C_p(T_2 - T_1)$$

This yields

$$T_a = 2250 \text{ R}$$

The exhaust pressure of the free turbine is

$$p_a = p_3 \left(\frac{T_a}{T_3}\right)^{\gamma/(\gamma-1)} = 142 \left(\frac{2250}{2760}\right)^{3.50} = 69.46 \text{ psia}$$

(b) The thermal efficiency can be calculated using either of the following forms

$$\eta = \frac{w_{\text{net}}}{q_{\text{in}}} = \frac{q_{\text{in}} - q_{\text{out}}}{q_{\text{in}}}$$

For either approach it is necessary to determine $T_4$, the temperature at the power turbine exhaust. By eq. (10.6a)

$$T_4 = T_1\left(\frac{T_3}{T_2}\right) = 548\left(\frac{2760}{1058}\right) = 1430 \text{ R}$$

and the cycle thermal efficiency is

$$\eta = \frac{w_{\text{net}}}{q_{\text{in}}} = \frac{C_p(T_3 - T_4) - C_p(T_2 - T_1)}{C_p(T_3 - T_2)} = 48.18 \text{ percent}$$

Alternative methods for calculation of the thermal efficiency are, from the $q_{\text{net}}$ form, eq. (10.6),

$$\eta = 1 - \frac{(T_4 - T_1)}{(T_3 - T_2)} = 48.18 \text{ percent}$$

or, by eq. (10.7),

$$\eta = 1 - \frac{1}{(10)^{0.286}} = 48.2 \text{ percent}$$

(c) The work for the power turbine is

$$_aw_4 = 0.24(2250 - 1430) = 196.8 \text{ B/lbm}$$

The power produced by the power turbine is $_a\dot{W}_4 = \dot{m}_a w_4$. Thus

$$\frac{10{,}000 \text{ kW}}{} \left| \frac{3413 \text{ B}}{\text{kW hr}} \right. = \frac{196.8 \text{ B}}{\text{lbm}} \left| \frac{\dot{m} \text{ lbm}}{\text{min}} \right| \frac{60 \text{ min}}{\text{hr}}$$

and

$$\dot{m} = 2890 \text{ lbm/min}$$

The volume rate of flow entering the compressor is

$$\dot{V}_1 = \frac{\dot{m}RT_1}{p_1} = \frac{(2890)(53.3)(548)}{(14.20)(144)}$$
$$= 41{,}280 \text{ ft}^3\text{/min}$$

**Figure 10.15** Air-Standard Gas Turbine with a Regenerator

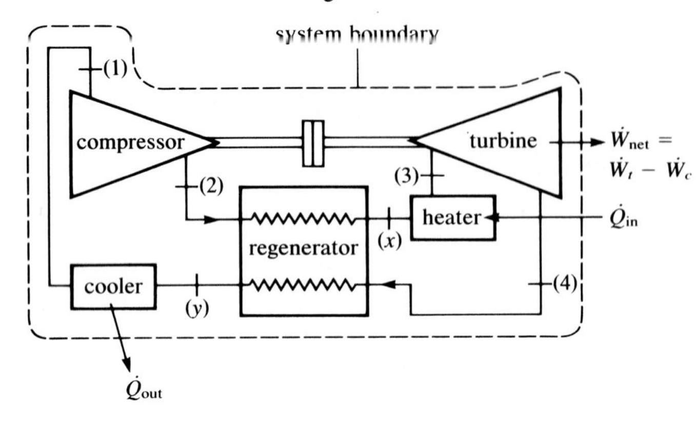


**Figure 10.16** *T-s* Diagram of Air-Standard Gas Turbine with a Regenerator

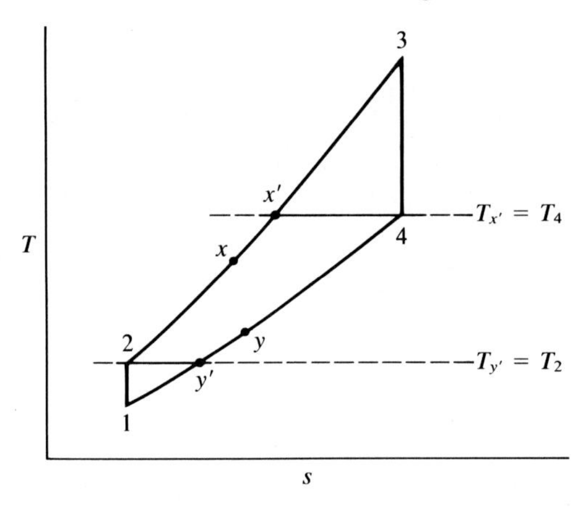


## 10–6 The Air-Standard Regenerative Gas-Turbine Cycle

The exhaust gases discharged from a simple gas turbine unit are typically at a temperature that is higher than the temperature at the compressor exit. This is evident for the air-standard simple-gas-turbine cycle from the *T-s* diagram of Figure 10.13 in which $T_4$ is greater than $T_2$. This condition permits regeneration to be employed. The addition of a regenerative heat exchanger (a ***regenerator***) significantly increases the thermal efficiency of a gas turbine. The arrangement of the components is shown for the air-standard cycle in Figure 10.15. Figure 10.16 is the corresponding *T-s* diagram. The turbine exhaust flowing through the regenerator decreases in temperature from $T_4$ to $T_y$ and heats the lower-temperature stream flowing from the air compressor, thereby increasing its temperature from $T_2$ to $T_x$. For a fixed state (3), the heat input to the heater is reduced. For a fixed compressor inlet state, pressure ratio, and turbine inlet temperature, the net work for a cycle with a regenerator is the same as that for the air-standard simple-gas-turbine cycle. Therefore, the thermal efficiency for the regenerative cycle is larger due to the decreased heat input. The regenerator is always positioned to heat the air stream leaving the compressor. Placing it ahead of the compressor would be self-defeating in that the compressor work requirement would increase due to the increased specific volume of the incoming air. See eq. (2.14).

The *regenerator effectiveness* $\epsilon$ is defined as

$$\epsilon = \frac{T_x - T_2}{T_4 - T_2} \tag{10.8}$$

The effectiveness of an actual regenerator is less than unity. The maximum possible effectiveness, $\epsilon = 1$, occurs when the heated-stream exit temperature $T_x$ is equal to $T_4$. The state for this case is shown in Figure 10.16 as state $x'$. Note that as $T_x$ increases, $T_y$ decreases, and when $T_x = T_{x'} = T_4$, $T_y = T_{y'} = T_2$. The regenerator is treated as an adiabatic heat exchanger similar to the closed feed-water heaters discussed in Chapter 9. For the regenerator we obtain from the steady-flow energy equation

$$h_4 + h_2 = h_y + h_x$$

which becomes, on the cold-air-standard basis,

$$C_p(T_4 - T_y) = C_p(T_x - T_2) \tag{10.9}$$

The heat transferred to the air from the high-temperature reservoir is then

$$q_{\text{in}} = C_p(T_3 - T_x)$$

and the heat transfer from the air to the low-temperature reservoir is

$$q_{\text{out}} = C_p(T_y - T_1)$$

Thus for the conditions noted it is evident that both $q_{\text{in}}$ and $q_{\text{out}}$ are reduced when a regenerator is used.

---

## Example 10.c

A cold-air-standard gas turbine with a regenerator operates with the compressor intake at 88 F and 14.20 psia. The pressure ratio $r_p$ is 10 and the maximum temperature during the cycle is 2300 F. For a regenerator effectiveness of 75 percent, calculate:

(a) $q_{\text{in}}$, the heat transfer to the air, B/lbm
(b) The net work, B/lbm
(c) $q_{\text{out}}$, the heat transferred from the air, B/lbm
(d) The cycle thermal efficiency

**Solution:**

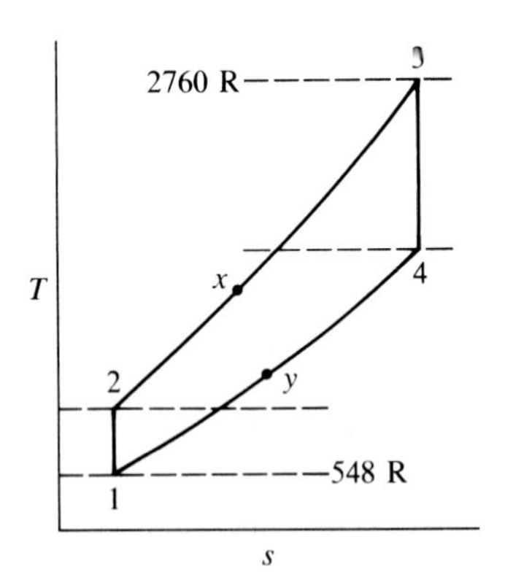


Note that the pressure ratio, compressor inlet state, and the turbine inlet state are the same as those in Example 10.b. With a regenerator effectiveness of less than 100 percent, the air entering the heater will be at a temperature less than that at the turbine exhaust. These factors are illustrated on the temperature–entropy diagram. The compressor discharge temperature $T_2$ is calculated by the same procedure used in Example 10.b, namely,

$$T_2 = 548(10)^{0.286} = 1058 \text{ R}$$

The temperature at the turbine exhaust $T_4$ is, from eq. (10.6a),

$$T_4 = T_1\left(\frac{T_3}{T_2}\right) = 1430 \text{ R}$$

By definition the regenerator effectiveness is

$$\epsilon \equiv \frac{T_x - T_2}{T_4 - T_2}$$

which, for this problem, becomes

$$0.75 = \frac{T_x - 1058}{1430 - 1058}$$

The solution for $T_x$, the temperature of the air entering the heater (combustors), yields

$$T_x = 1337 \text{ R}$$

Using the steady-flow energy equation and the above values, we obtain the following.

(a)

$$\begin{aligned} q_{in} &= h_3 - h_x = C_p(T_3 - T_x) \\ &= 0.24(2760 - 1337) = 341.5 \text{ B/lbm} \end{aligned}$$

(b) The net work is

$$w_{net} = {}_3w_4 - |{}_1w_2| = w_t - w_c$$
$$= 0.24[(2760 - 1430) - (1058 - 548)]$$
$$= 196.8 \text{ B/lbm}$$

which is the same as that for the power turbine in Example 10.b.

(c) With the regenerator, the temperature entering the cooler is decreased and the heat transferred from the working fluid is reduced to

$$q_{out} = |{}_yq_1| = C_p(T_y - T_1)$$

The temperature $T_y$ is determined from the steady-flow equation written for the adiabatic heat exchanger, eq. (10.9),

$$C_p(T_4 - T_y) = C_p(T_x - T_2)$$

Therefore

$$T_y = T_4 - T_x + T_2 = 1151 \text{ R}$$

and

$$q_{out} = |{}_yq_1| = 0.24(1151 - 548) = 144.7 \text{ B/lbm}$$

(d) The thermal efficiency of the cycle can be calculated as

$$\eta = \frac{w_{net}}{q_{in}} \quad \text{or} \quad \eta = 1 - \frac{q_{out}}{q_{in}}$$

By the first equation

$$\eta = \frac{196.8}{341.5} = 57.6 \text{ percent}$$

and from the second equation

$$\eta = 1 - \frac{144.7}{341.5} = 57.6 \text{ percent}$$

The thermal efficiency for the air-standard gas-turbine cycle without a regenerator in Example 10.b is 48.2 percent. This trend to thermal efficiency by use of a regenerator applies to real gas turbines as well as air-standard gas turbines. Regenerators are useful in land and sea gas-turbine propulsion units, but because of their added weight and configuration difficulties, they are not at present practical for aircraft applications.

**Figure 10.17** Flow Diagram for an Aircraft Turbojet Engine

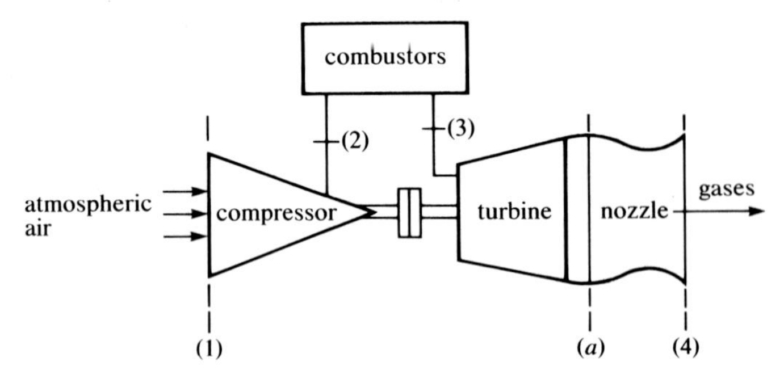


## 10–7 Gas Turbines for Aircraft Propulsion

The aircraft gas turbine for jet propulsion is similar in some respects to the gas-turbine unit illustrated in Figure 10.14, which is used for electric power generation. The aircraft unit employs a gasifier that is essentially the same as is used in the generating unit. The major difference between the two is in the use of the gases leaving the free turbine. For electric generation these gases expand through the power turbine that drives the generator. In a turbojet aircraft unit gases leaving the turbine accelerate as they expand through a nozzle to provide thrust. A schematic diagram of a turbojet aircraft engine is presented in Figure 10.17. It can be shown from Newton's second law of motion that the thrust developed by the turbojet under steady-flow conditions is

$$F = \frac{\dot{m}}{g_0}(V_4 - V_1) \tag{10.10}$$

where, as in Figure 10.17, $V_1$ is the inlet velocity and $V_4$ is the outlet velocity, both measured relative to the aircraft. Equation (10.10) is not applicable if the aircraft is accelerating.

In this section, only the cold-air-standard jet-aircraft propulsion system will be presented. The analysis will neglect any pressure rise that would occur in a diffuser ahead of the compressor and any pressure drop through the combustors, and it will be assumed that the velocities at all points except the nozzle outlet are negligible. Pressure-specific volume and temperature–entropy diagrams for an air-standard jet engine operating under these conditions are shown in Figure 10.18. The work required to drive the isentropic compressor is, by eq. (2.14), equal to the area on the *p-v* diagram between path (1) to (2) and the ordinate and must be equal to the area between path (3) to (*a*) and the ordinate, which is the work produced by the turbine. Specifically state (*a*) is fixed by

$$|{}_1w_2| = {}_3w_a = h_2 - h_1 = h_3 - h_a$$

For cold-air-standard conditions this becomes

$$C_p(T_2 - T_1) = C_p(T_3 - T_a) \tag{10.11}$$

**Figure 10.18** Property Diagrams for a Cold-Air-Standard Turbojet Engine

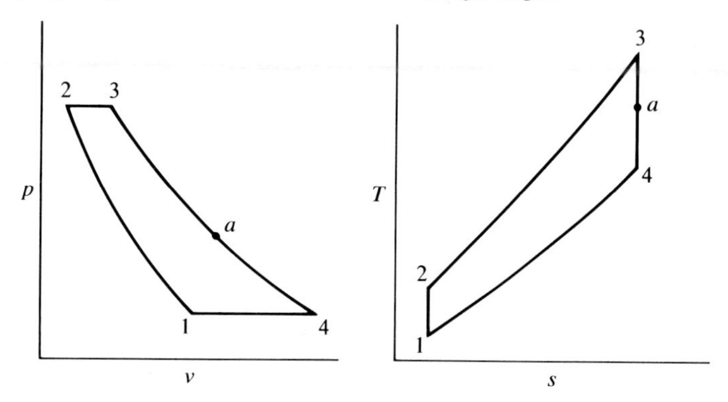


## Example 10.d

A cold-air-standard turbojet engine operates with the compressor intake at 88 F and 14.20 psia. The compressor pressure ratio is 10 and the maximum temperature is 2300 F. All components operate ideally. The velocity of the air entering the compressor is negligible. The engine is to develop a thrust of 10,000 lbf. Calculate:

(a) The pressure at the turbine exhaust
(b) The velocity at the nozzle exit
(c) The mass rate of flow necessary to develop the specified thrust

**Solution:**

(a) The turbine exhaust state is fixed by the compressor work and the isentropic turbine process. Using the notation of Figures 10.17 and 10.18, we observe that eq. (10.11) applies. Thus

$$C_p(T_2 - T_1) = C_p(T_3 - T_a)$$

for cold-air-standard conditions. From the compressor pressure ratio and inlet state

$$T_2 = 548(10)^{0.286} = 1058 \text{ R}$$

Substituting $T_2$ and the given values of $T_3$ and $T_1$ into eq. (10.11), we obtain

$$T_a = 2250 \text{ R}$$

For the isentropic turbine

$$p_a = p_3 \left(\frac{T_a}{T_3}\right)^{\gamma/(\gamma-1)}$$

The turbine inlet pressure $p_3$ is equal to the compressor discharge pressure in the ideal cycle; thus

$$p_3 = p_2 = (10)(14.20) = 142.0 \text{ psia}$$

With $p_3$ known, $p_a$ can be calculated as

$$p_a = (142.0)\left(\frac{2250}{2760}\right)^{3.500} = 69.46 \text{ psia}$$

(b) The nozzle exit velocity can be calculated by either of two procedures. Writing the steady-flow equation between the turbine inlet and the nozzle exit, we have

$$h_3 = h_4 + {}_3w_4 + \frac{V_4^2}{2g_0}$$

for isentropic processes through the turbine and the nozzle and negligible velocity entering the turbine. The alternative is to write the steady-flow equation for the isentropic nozzle only. Choosing this method, we have

$$h_a = h_4 + \frac{V_4^2}{2g_0}$$

with the turbine exhaust velocity negligible. For the nozzle process

$$T_4 = T_a\left(\frac{p_4}{p_a}\right)^{(\gamma-1)/\gamma}$$

with $p_4$ assumed equal to $p_1$. The nozzle exit temperature is then

$$T_4 = 2250\left(\frac{14.20}{69.46}\right)^{0.286} = 1430 \text{ R}$$

Note that the result is identical to that of Example 10.b only because the components operate isentropically in both cases. The velocity $V_4$ is

$$\begin{aligned} V_4 &= [2g_0\, C_p(T_a - T_4)]^{1/2} \\ &= 3139 \text{ ft/sec} \end{aligned}$$

(c) The required mass rate of flow is calculated using eq. (10.10)

$$F = \frac{\dot{m}}{g_0}(V_4 - V_1)$$

which, for the given conditions, becomes

$$F = \frac{\dot{m}}{g_0}(3139 - 0) = 10{,}000 \text{ lbf}$$

Solving for $\dot{m}$

$$\dot{m} = \frac{(10{,}000)(g_0)}{3139.0} = 102.5 \text{ lbm/sec} = 6149 \text{ lbm/min}$$

---

**Figure 10.19** Turbofan Aircraft Engine Cutaway

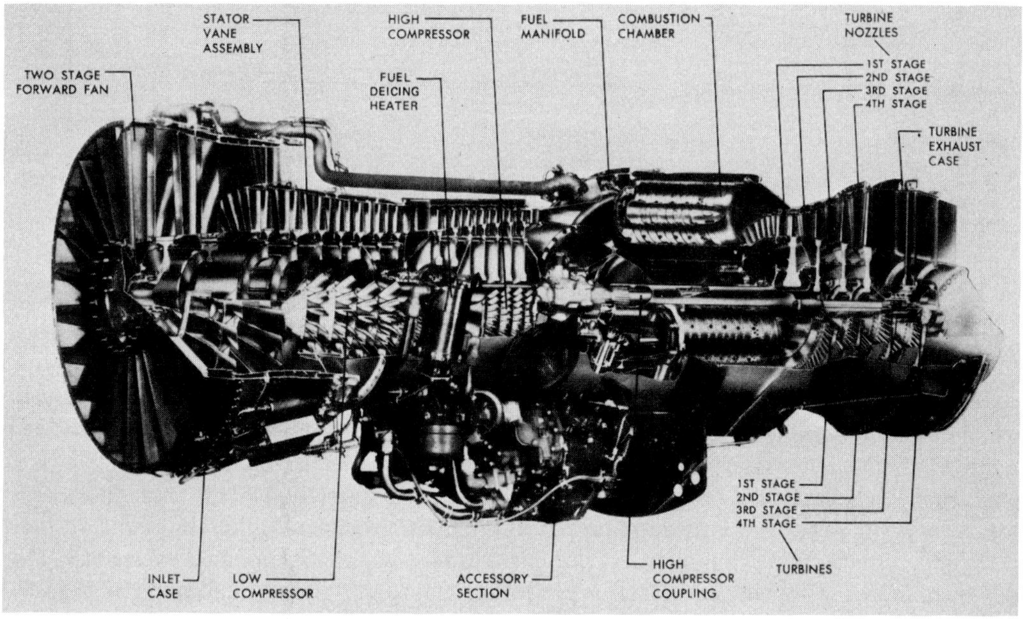

(Courtesy Pratt and Whitney Aircraft Group, Div. of United Technologies Corp.)

There are numerous variations of the aircraft gas-turbine cycle. Even the simple cycle presented is subject to significant changes when velocities are not neglected and the compressor, turbine, and nozzle are treated, more realistically, as adiabatic processes that are not reversible. The propjet engine utilizes approximately 90 percent of the net work, that is, turbine work minus compressor work, to drive propellers through gear-reduction units. The remaining expansion through an exhaust nozzle provides added propulsion force. Figure 10.19 is a diagram of a turbofan aircraft engine. This engine provides thrust by operating as a jet engine in discharging the turbine exhaust gases through a nozzle and by use of fans ahead of the compressors that operate in much the same way as conventional aircraft propellers. The turbine drives both compressors and the fans. Military aircraft in the fighter-intercepter category are equipped with afterburners for periodic high-thrust maneuvers. The afterburner utilizes injection of fuel into the oxygen-rich combustion gases leaving the turbine, thereby increasing the nozzle inlet temperature and, consequently, the thrust. Thermodynamic evaluations of these cycles follow the same general analysis techniques that have been applied to the turbojet engine.

**Figure 10.20** Schematic Diagram of Mercury–Steam Cycle Used in Schiller Station (1950)

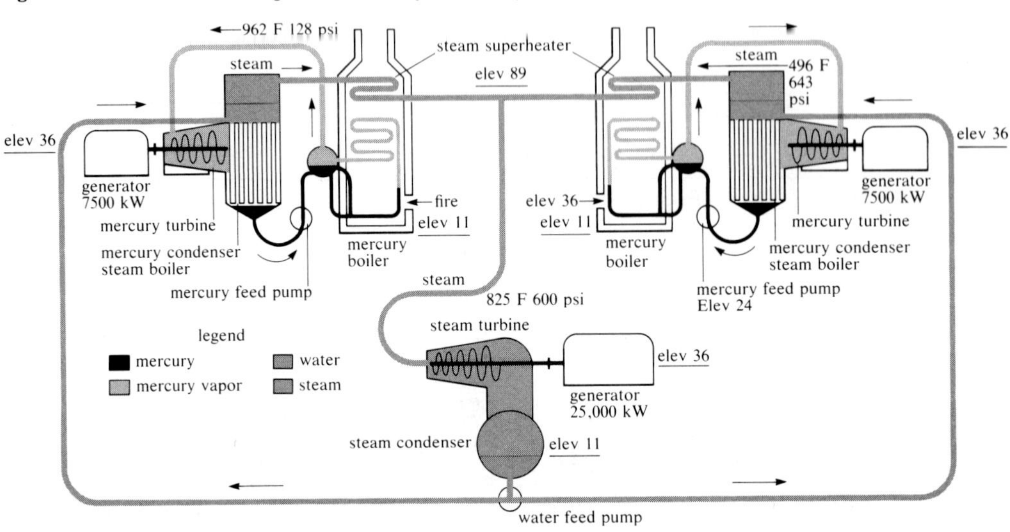


## 10–8 Combined Cycles

As the name implies, *combined cycles* are combinations of two or more systems (cycles) using different working fluids. Combined cycles, while not new, are receiving increased attention as they yield higher thermal efficiencies than can be achieved, for example, with a gas turbine or a vapor power cycle operated individually. These higher thermal efficiencies are reflected in lower fuel consumption per unit of power delivered, thereby reducing fuel costs and conserving fuel supplies. At present the most common combined cycles utilize a gas turbine as the topping cycle and a steam bottoming cycle with the hot exhaust gases from the combustion (gas) turbine serving as the energy source for steam generation.

The earliest combined cycles used in the United States were binary cycles employing mercury and steam as the working fluids. In 1928 a mercury cycle topping an existing steam cycle was placed in service by the Hartford Electric Light Company. Binary-cycle plants were built at several locations in the New England area where fuel costs typically were higher than in other parts of the United States. The selection of mercury for the topping cycle was based on several desirable thermodynamic characteristics of the substance. These include low saturation pressures for high saturation temperatures (e.g., 200 psia at 1017 F) and low specific heat of the liquid phase. Mercury exhibits many of the desirable characteristics for a vapor power cycle that were discussed in Section 9–3. However, the high toxicity of mercury vapor and its cost offset these advantages to some extent. Also, the low saturation pressures at relatively high temperatures mandate its use as a topping cycle fluid. For example, the saturation pressure is 0.40 psia (2.76 kPa) at approximately 402 F (206 C). Schiller Station, a unit of Public Service Company of New Hampshire, was placed in service in 1950. A schematic diagram of the cycle for that station is shown in Figure 10.20. The design plant net heat rate (higher heating value of fuel per net kilowatt-hour generated) was 9200 B/kWhr when burning Bunker C fuel oil and 9420 B/kWhr when burning pulverized coal.

**Figure 10.21** Combined Gas-Turbine Steam Power Cycle

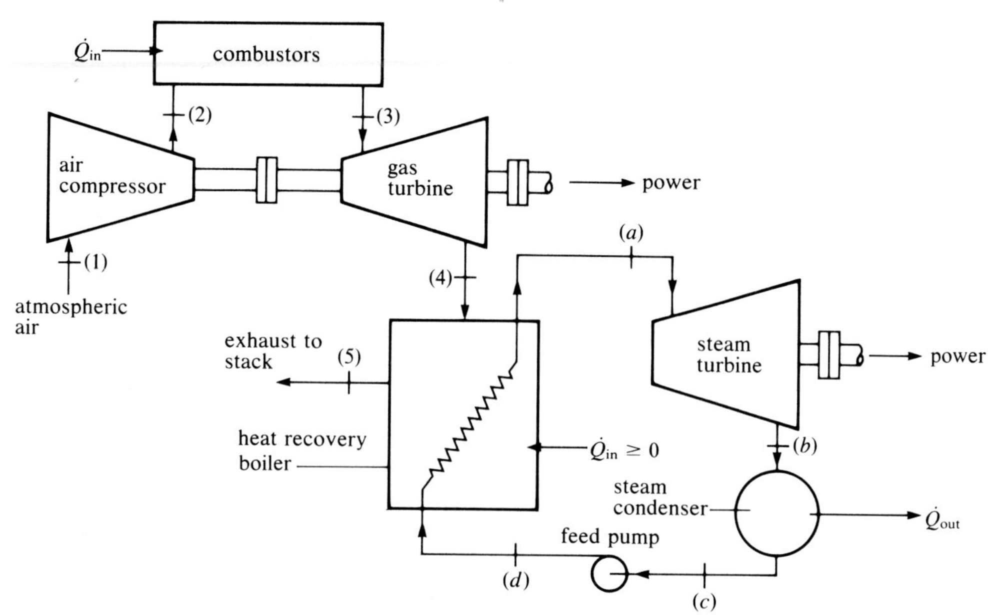


While the rated capacity of this 1950 plant was only 40,000 kW, its low heat rate surpassed that of the most modern large-capacity units in use at this time.

As can be seen in Figure 10.20, fuel was fired in the mercury boilers only. Exhaust from the mercury turbines passed into a heat exchanger, which served as the mercury condenser and the steam boiler where heat transfer from the condensing mercury was used to generate steam. The $Q_{out}$ for this combined cycle was the heat rejected from condensing steam to the steam condenser circulating water. The relative elevations shown on the figure are significant due to the high density of mercury. Note that this binary cycle is a combination of two Rankine cycles, with the mercury Rankine cycle topping a steam Rankine cycle.

As mentioned earlier, the combined cycle of current interest is one with a gas turbine topping a steam cycle. This combination utilizes the energy in the high-temperature gas-turbine exhaust to generate steam for the bottoming steam cycle, thereby recovering energy that otherwise would be rejected to the atmosphere by a gas turbine. While these systems may have more than one gas turbine unit exhausting to the heat-recovery steam boiler, we shall consider only a simple combination of one gas turbine and a Rankine steam cycle as shown in Figure 10.21. In practice the steam bottoming cycle may include regenerative feed-water heating and steam reheat. Since the gas-turbine exhaust contains a significant amount of free oxygen, additional fuel may be burned in the heat-recovery boiler. When this is done the cycle is termed a reheat combined cycle.

Combined cycles of the type described typically have higher thermal efficiencies (lower heat rates) than either a gas turbine or a steam cycle operating alone. A simple explanation of the higher efficiency is related to the higher temperatures permitted in the gas turbine. When cooled turbine blades are used, temperatures at the turbine inlet of the order of 2600 F are permitted. (The maximum working fluid temperature for steam at the turbine inlet in a steam cycle is about 1000 F.) The combined cycle rejects heat at the steam condenser at a temperature of about 100 F. Because of the wider range between the temperature at which heat is added to the combined cycle and the temperature at which heat is rejected, the increased thermal efficiency is expected. Since gas turbines typically use gas or liquid fuels, this combined cycle is particularly desirable for the use of synthetic gas produced from coal gasification. In addition to lower fuel cost per unit of energy generated, the combined gas-turbine/steam cycle plant has a lower capital cost than a steam-generating station of the same capacity.

It is instructive to examine the combined cycle in Figure 10.21 using the concept of available energy (Section 7–9). Because of the inherently high temperature at the turbine exhaust, the energy contained in the exhaust gases is relatively high-grade energy, that is, a sizable fraction is available energy. The addition of a steam bottoming cycle utilizes some of this high-grade energy to produce work. The following example illustrates several aspects of combined cycles.

## Example 10.e

Let us assume for the combined cycle in Figure 10.21 that the gas turbine operates as a simple air-standard gas turbine and that the steam cycle is an ideal Rankine cycle. The following information is given in the notation of Figure 10.21.

| | State | | | | | | | | |
|---|---|---|---|---|---|---|---|---|---|
| | **1** | **2** | **3** | **4** | **5** | ***a*** | ***b*** | ***c*** | ***d*** |
| $p$, kPa | 100 | 1150 | 1150 | 100 | 100 | 2750 | 7.5 | 7.5 | 2750 |
| $t$, C | 20 | | 1250 | | 200 | 420 | | | |

The net power produced by the air-standard gas turbine is 5000 kW. At the heat-recovery boiler, $\dot{Q}_{in}$ is zero. The pump work for the steam cycle is small and can be neglected.

(a) Compute the mass rate of flow of the air and of the steam.
(b) Compute the net power produced by the combined cycle.
(c) Compute the thermal efficiency for the combined cycle.
(d) What portion of the turbine exhaust energy is available energy? Compare this with the work produced by the steam cycle.

**Solution:**

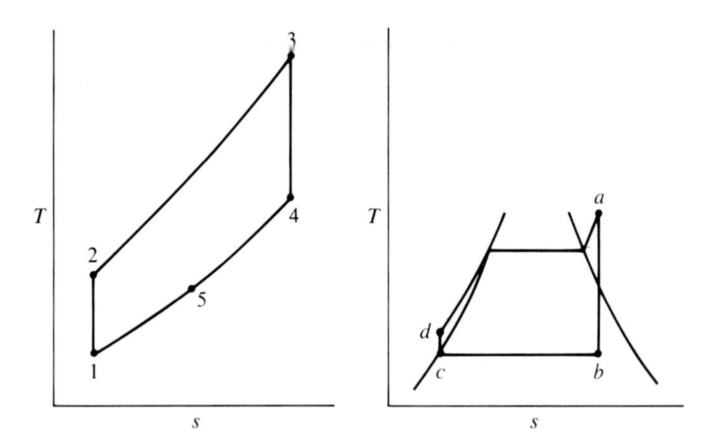


The states for the gas turbine and for the steam cycle are shown on the $T$-$s$ diagrams. We will use $A$ and $S$ to denote, respectively, the air and the steam. The following table lists the properties necessary for the solution of this problem. These values were determined by the methods illustrated in Examples 9.a and 10.b.

| | State | | | | | | | | |
|---|---|---|---|---|---|---|---|---|---|
| | **1** | **2** | **3** | **4** | **5** | ***a*** | ***b*** | ***c*** | ***d*** |
| $T$, K | 293 | 588 | 1523 | 758 | 473 | — | — | — | — |
| $h$, kJ/kg | — | — | — | — | — | 3280 | 2193 | 169 | 169 |

(a) The mass rate of flow of the air is given by

$$\dot{m}_A = \frac{\dot{W}_{A,\text{net}}}{w_{A,\text{net}}} = \frac{\dot{W}_{A,\text{net}}}{w_t - w_c}$$

where

$$\begin{aligned} w_t &= h_3 - h_4 = C_p(T_3 - T_4) = 1.00\ (1523 - 758) \\ &= 765 \text{ kJ/kg} \end{aligned}$$

and

$$\begin{aligned} w_c &= |{}_1w_2| = h_2 - h_1 = C_p\ (T_2 - T_1) = 1.00\ (588 - 293) \\ &= 295 \text{ kJ/kg} \end{aligned}$$

Therefore

$$\dot{m}_A = \frac{5000 \text{ kJ}}{\text{s}} \left| \frac{\text{kg}}{(765 - 295) \text{ kJ}} \right. = 10.64 \text{ kg/s}$$

The relationship between $\dot{m}_A$ and $\dot{m}_S$ is established by mass and energy balances for the heat-recovery boiler. Since $\dot{Q}$ and $\dot{W}$ are zero for the heat-recovery boiler, the steady-flow energy equation reduces to

$$\Sigma(\dot{m}h)_{\text{in}} = \Sigma(\dot{m}h)_{\text{out}}$$

$$\dot{m}_4 h_4 + \dot{m}_d h_d = \dot{m}_5 h_5 + \dot{m}_a h_a$$

The mass balance is satisfied by $\dot{m}_A = \dot{m}_4 = \dot{m}_5$ and $\dot{m}_S = \dot{m}_d = \dot{m}_a$. Therefore

$$\dot{m}_S = \dot{m}_A \frac{h_4 - h_5}{h_a - h_d} = \frac{(10.64)(1.00)(758 - 473)}{3280 - 169}$$
$$= 0.975 \text{ kg/s}$$

(b) The total power is the sum of the two power outputs. The net power for the steam cycle is the steam-turbine power since the power to drive the pump is small. The steam-turbine power is

$$\dot{W}_{S,\text{net}} = \dot{m}_S({}_a w_b) = \dot{m}_S(h_a - h_b)$$
$$= 0.975(3280 - 2193) = 1060 \text{ kW}$$

Thus

$$\dot{W}_{\text{total}} = \dot{W}_{A,\text{net}} + \dot{W}_{S,\text{net}}$$
$$= 5000 + 1060 = 6060 \text{ kW}$$

(c) For a combined cycle it is necessary to write the thermal efficiency in terms of time rates as

$$\eta_{\text{combined}} = \frac{\dot{W}_{\text{total}}}{\dot{Q}_{\text{in}}}$$

$\dot{Q}_{\text{in}}$ is supplied only at the combustor, for which the steady-flow energy equation yields

$$\dot{Q}_{\text{in}} = \dot{m}_A(h_3 - h_2) = 10.64(1.00)(1523 - 588)$$
$$= 9948 \text{ kJ/s}$$

Therefore

$$\eta_{\text{combined}} = \frac{6060}{9948} = 60.9 \text{ percent}$$

It is interesting to compare the combined cycle efficiency with that of the air-standard gas-turbine cycle operating alone. For the latter we obtain

$$\eta_{\text{gas turb}} = \frac{W_{A,\text{net}}}{\dot{Q}_{\text{in}}} = \frac{5000}{9948} = 50.3 \text{ percent}$$

Although the efficiency of an actual combined cycle would be less than that above, the significant increase in $\eta$ is indicative of the usefulness of combined cycles in the conversion of heat to work.

(d) To compute the available portion of the energy rejected by the gas turbine let us consider it to be a closed air-standard cycle with the processes 1–2–3–4–1 on the $T$-$s$ diagram. The heat rejected by this cycle is

$$\begin{aligned} q_{\text{out}} &= |{}_4q_1| = h_4 - h_1 = C_p(T_4 - T_1) = 1.00(758 - 293) \\ &= 465 \text{ kJ/kg } A \end{aligned}$$

By eq. (7.19) the available portion of $q_{\text{out}}$ is

$$\begin{aligned} q_a &= |{}_4q_1| - T_0(s_4 - s_1) = q_{\text{out}} - T_0\left(C_p \, ln \frac{T_4}{T_1} - R \, ln \frac{p_4}{p_1}\right) \\ &= 465 - 293(1.00) \, ln \frac{758}{293} = 187 \text{ kJ/kg } A \end{aligned}$$

In the above expressions $T_0$ has been taken as 293 K. $q_a$ is the maximum work that could be produced by using process (4) to (1) as a heat source and the atmosphere as a heat sink. The steam cycle receiving heat from the turbine exhaust and rejecting heat to the atmosphere produces work

$$w_S = h_a - h_b = 3280 - 2193 = 1086 \text{ kJ/kg } S$$

A comparison of the work produced by the steam cycle and the $q_{\text{avail}}$ must be made on a common basis. Selecting a kilogram of air as the basis, the steam cycle work per kilogram of air flowing through the gas turbine is

$$\begin{aligned} w_S &= (h_a - h_b) \frac{\dot{m}_S}{\dot{m}_A} = (3280 - 2193) \frac{0.975}{10.64} \\ &= 99.6 \text{ kJ/kg } A \end{aligned}$$

Therefore, the fraction of the available energy from the gas-turbine cycle that is converted to work by the steam cycle is $99.6/187 = 0.53$. Due to the requirements of finite temperature differences to promote heat transfer to and from the working fluids in operating the steam cycle, we would not expect this fraction to be unity. However, the steam cycle does utilize a sizable fraction of the high-grade energy that otherwise would be rejected to the atmosphere by the gas-turbine cycle.

## 10–9 Summary

Air-standard analysis is used to model gas turbines and reciprocating engines, both spark ignition and compression ignition (diesel). Air is the working fluid and the combustion process is replaced by a process in which the air is heated. While the engines modeled do not operate in thermodynamic cycles, air-standard cycles execute a cycle within a closed boundary. All processes are assumed to be internally reversible. For cold-air-standard analysis, $C_p$ and $C_v$ are assumed to be constant.

The air-standard Otto cycle, which is based on the spark-ignition engine, consists of four processes that occur in a closed piston–cylinder assembly: isentropic compression, constant-volume heating, isentropic expansion, and constant-volume heat rejection. The thermal efficiency of this cycle increases as the compression ratio increases. The compression ratio is the volume ratio for the isentropic compression process. The performance trends predicted from the air-standard Otto cycle are generally followed by actual spark-ignition engines.

The air-standard diesel cycle is similar to the air-standard Otto cycle except that the constant-volume heating process is replaced by a constant-pressure heating process. The thermal efficiency of this cycle depends on the compression ratio and the cutoff ratio, which is the ratio of the cylinder volume after heating to that before heating.

The air-standard simple-gas-turbine cycle is called the Brayton cycle. It consists of the following steady-flow processes: isentropic compression, constant-pressure heating, isentropic expansion through a turbine, and constant-pressure heat rejection. The thermal efficiency for this cycle increases as the cycle pressure ratio increases. A regenerator that recovers heat from the turbine exhaust also improves the thermal efficiency. A turbojet aircraft engine produces thrust by exhausting gases from the turbine through a nozzle. The turbine drives only the compressor.

Combined cycles typically consist of two cycles that have different working fluids. The topping cycle receives heat from a primary heat source and the heat it rejects serves as a heat source for the bottoming cycle. The thermal efficiency of a combined cycle is greater than either cycle would exhibit if operated alone. Combined cycles currently receiving attention include the combination of a gas turbine topping a steam cycle.

## Selected References

Dixon, J. R. *Thermodynamics I: An Introduction to Energy*. Englewood Cliffs, N.J.: Prentice-Hall, 1975.
Faires, V. M. *Thermodynamics of Heat Power*. New York: Macmillan, 1958.
Irey, R. K., A. Ansari, and H. H. Pohl. *Thermodyamics for Modular Instruction*, Vol. 1, *Unit IV B*. New York: Wiley, 1976.
Lichty, L. C. *Combustion Engine Processes*. New York: McGraw-Hill, 1967.
Obert, E. F. *Internal Combustion Engines*, 2nd ed. New York: International, 1950.
Rogowski, A. R. *Elements of Internal Combustion Engines*. New York: McGraw-Hill, 1953.
Stoever, H. J. *Essentials of Engineering Thermodynamics*. New York: Wiley, 1953.
Taylor, C. F. *The Internal Combustion Engine in Theory and Practice*. Cambridge, Mass.: M.I.T. Press, 1968.
Vincent, E. T. *The Theory and Design of Gas Turbines and Jet Engines*. New York: McGraw-Hill, 1950.
Zucrow, M. J. *Aircraft and Missile Propulsion*, Vol. 1. New York: Wiley, 1958.

## Problems

10.1 For an Otto cycle the temperature at the beginning of compression is 77 F. To avoid autoignition, the temperature at the completion of the compression stroke is not to exceed 700 F. Assuming cold-air-standard conditions, what is the maximum allowable compression ratio?

10.2 Suppose an air-standard Otto cycle has a temperature of 77 F at the beginning of compression, a compression ratio $r_c$ of 8, and heat rejection from the air equal to 200 B/lbm. What would be the maximum pressure during the cycle, the heat transfer to the air, and the cycle thermal efficiency if the pressure at the beginning of compression were 14 psia?

10.3 A spark-ignition engine has a compression ratio $r_c$ of 8.2. Ambient conditions are 14 psia and 77 F. Due to a change in operating conditions, it is necessary for the engine to deliver 20 percent more power when operating at the same speed. Using air-standard conditions, determine the approximate increase in fuel consumption of the engine as a result of the increased power requirement. Show the two cycles on $T$-$s$ and $p$-$V$ coordinates.

10.4 Consider a cold-air-standard Otto cycle having a temperature of 25 C at the beginning of the compression process and a maximum temperature of 1925 C during the cycle. To develop maximum net work per cycle, what compression ratio would be required? What would be the cycle thermal efficiency when the maximum net work per cycle was realized?

10.5 Suppose a cold-air-standard Otto cycle has a piston displacement of 250 in$^3$ and a compression ratio of 8. Assuming the maximum pressure during the cycle is 600 psia, the maximum temperature is 3200 F, and the temperature at the beginning of compression is 60 F, determine (a) the net power when the cycle is repeated 2000 times per minute, (b) the cycle thermal efficiency, and (c) the approximate fuel consumption per HP-hr if the engine fuel has a heating value of 20,000 B/lbm.

10.6 Suppose an IC engine has a clearance volume equal to 8 percent of the displacement volume. Calculate the compression ratio, assuming cold-air-standard conditions, for the cycle. What would be the thermal efficiency of a cold-air-standard Otto cycle having this clearance?

10.7 A cold-air-standard diesel cycle has a temperature of 60 F at the beginning of compression and a temperature of 1300 F at the end of compression. Heat transfer to the air is 550 B/lb. Determine (a) the compression ratio, (b) the cutoff ratio, and (c) the net work, B/lbm.

10.8 A diesel engine with a 29-inch bore and 54-inch stroke has 12 cylinders and a compression ratio of 20. The engine delivers 5000 HP when operating at 167 rev/min on a two-stroke cycle and taking in air at 15 psia and 90 F. Using cold-air-standard conditions, determine (a) heat transfer to the air in B/min, (b) maximum pressure and temperature during the cycle, (c) the cutoff ratio, and (d) the thermal efficiency of the air-standard cycle.

10.9 An air-standard cycle for a spark ignition engine with an exhaust turbine consists of the air-standard Otto cycle in which the constant volume heat rejection process is replaced by an isentropic expansion to the initial pressure and a constant pressure heat rejection process to complete the cycle. Consider the air-standard Otto cycle in Example 10.a. What would be the thermal efficiency if an exhaust turbine were added? Sketch the $p$-$v$ and $T$-$s$ diagrams for the cycle.

10.10 The expansion ratio $r_e$ for a diesel cycle is the ratio of the volume at the end of expansion to that at the end of the constant-pressure process, $(V_4/V_3)$ in Figure 10.6. The thermal efficiency of a cold-air-standard diesel cycle depends only on the compression ratio and the expansion ratio and is

$$\eta = 1 - \frac{1}{\gamma}\left[\frac{\left(\frac{1}{r_e}\right)^\gamma - \left(\frac{1}{r_c}\right)^\gamma}{\frac{1}{r_e} - \frac{1}{r_c}}\right]$$

For a cold-air-standard diesel cycle with fixed intake state and compression ratio show, on a $T$-$s$ diagram, the effect of power increase at constant engine speed on cycle thermal efficiency. Noting the change in expansion ratio for such a power increase, verify your conclusion using the equation given above.

10.11 Suppose an engine operates on a cold-air-standard cycle with the temperature at the beginning of compression 70 F and the pressure 14.7 psia. Assuming that heat transfer to the air is 300 B/lbm, and that the maximum temperature during the cycle is 2500 F, (a) calculate the thermal efficiency for an Otto cycle with these conditions, (b) calculate the thermal efficiency of a diesel cycle with these conditions, (c) calculate the compression ratio for each cycle, and (d) sketch a $T$-$s$ diagram showing the two cycles.

10.12 Consider a cold-air-standard diesel cycle with a compression ratio of 16. As shown in problem 10.10, the thermal efficiency of a cold-air-standard diesel cycle with a fixed compression ratio depends only on the expansion ratio. What are the limits of the expansion ratio for the cycle? What are the corresponding cutoff ratios? Sketch the $p$-$v$ and $T$-$s$ diagrams for diesel cycles operating at these limits. What is the maximum possible thermal efficiency of the cycle?

10.13 A cold-air-standard diesel cycle operates with a fixed state at the beginning of the compression process. For a given compression ratio, use a $T$-$s$ diagram to show the effect of increasing the maximum temperature on each of the following:

(a) Cutoff ratio
(b) Thermal efficiency
(c) $Q_{in}$
(d) Net work per cycle

10.14 A simple-cycle, cold-air-standard gas turbine has air entering the compressor at 14 psia and 520 R. At the turbine inlet the temperature is 2700 R and the pressure 168 psia. If all components operate ideally and velocities can be neglected, determine (a) the net work, B/lbm; (b) heat transfer to the air, B/lbm; (c) the turbine outlet temperature; and (d) the cycle thermal efficiency.

10.15 A simple-cycle, cold-air-standard gas-turbine cycle has air entering the compressor at 520 R. The turbine inlet temperature is 2700 R. The pressure ratio is that to give maximum net work per cycle. Determine (a) the pressure ratio, (b) the net work, B/lbm, (c) the turbine outlet temperature, and (d) the cycle thermal efficiency. Assume all components operate ideally and velocities are negligible. Using a $T$-$s$ diagram, compare this cycle with the one given in problem 10.14. Would this maximum net work cycle also yield the maximum thermal efficiency for the given compressor and turbine inlet temperatures?

10.16 Refer to problem 10.14. If a regenerator having 100 percent effectiveness were added to the unit, what would be the effect on compressor work, turbine work, and net work? With the regenerator, what would be (a) the heat transfer to the air and (b) the thermal efficiency of the cycle?

10.17 A gas-turbine-driven electric generator delivers 20 MW. The unit can be considered equivalent to that shown in Figure 10.14. The compressor inlet is at 14.0 psia and 537 R and the turbine inlet is at 150 psia and 2600 R. All components operate ideally and velocities can be neglected. Assuming cold-air-standard condition, calculate (a) the pressure at the free turbine exhaust, (b) the temperature at the power turbine exhaust, which is at 14 psia, and (c) the volume rate of flow, $ft^3/min$, entering the compressor. If fuel for the unit has a heating value of 18,900 B/lbm, estimate the fuel consumption in lbm/hr.

10.18 Suppose a simple-cycle, cold-air-standard gas turbine had the same compressor inlet state and pressure ratio and the same turbine inlet state as in problem 10.14. For a compressor efficiency of 86 percent and a turbine efficiency of 88 percent, determine (a) the net work, B/lbm; (b) heat transfer to the air, B/lbm; (c) the turbine outlet temperature; and (d) the cycle thermal efficiency. Compare the results with those from problem 10.14. Sketch a $T$-$s$ diagram showing the two cycles. *Note:* Turbine efficiency is the ratio of the actual turbine work to the work for an isentropic process from the actual turbine inlet state to the actual turbine exhaust pressure. Similarly, compressor efficiency is the ratio of the work required for an isentropic compression to the actual compressor work required, where the isentropic work is computed between the actual compressor inlet state and the actual discharge pressure.

10.19 Refer to problem 10.18. Suppose a regenerator having 80 percent effectiveness were added to the system. Determine (a) the heat transfer to the air, B/lbm; (b) the net work, B/lbm; (c) the thermal efficiency of the cycle; and (d) the temperature of the exhaust air (section $y$, Figure 10.15) leaving the regenerator. Using a $T$-$s$ diagram, compare this cycle with the one specified in problem 10.18.

10.20 A cold-air-standard turbojet operates with the compressor inlet at 7.00 psia and $-20$ F. The turbine inlet is at 2500 R and 79.0 psia. Neglecting velocities at all sections except the nozzle outlet, and assuming that each component operates ideally and the nozzle outlet is at 7.00 psia, calculate (a) the pressure at the nozzle inlet, (b) the velocity at the nozzle outlet, and (c) the mass rate of flow to produce thrust of 20,000 lbf.

10.21 Refer to the preceding problem. Suppose the engine is equipped with an afterburner to increase the temperature of the air entering the nozzle to that entering the turbine. If the mass rate of flow is not changed, by how much is the thrust increased? Using $Q_{in}$ as an approximation of the fuel consumption, what percent increase in fuel consumption will be required when using the afterburner?

10.22 Refer to the combined cycle shown in Figure 10.21. All gas-side processes are assumed to be cold-air standard. First consider the steam cycle with all components operating ideally. Steam enters the turbine at 400 psia, 700 F, and the condenser is at 1.00 psia. The steam turbine-generator delivers 10,000 kW. Calculate the heat transferred to the steam, B/hr, and the thermal efficiency of the steam cycle. Next consider the gas-turbine cycle with air entering the compressor at 14.0 psia and 60 F and the turbine inlet at 2600 R and 154 psia. Determine for the gas-turbine unit the heat transfer ${}_2q_3$, the net work, and the turbine exhaust temperature $T_4$. Suppose the stack gas temperature $T_5$ is 400 F. If no fuel were fired in the steam generator, that is, all heat transfer to the $H_2O$ were from the gas-turbine exhaust, what would be the required mass rate of flow through the gas-turbine unit? What would be the power developed by the gas turbine under these conditions? What would be the thermal efficiency of the combined cycle?

10.23 A gas-turbine/steam combined cycle has the components shown in Figure 10.21. Air enters the compressor at 14.5 psia and 60 F and is discharged at 145 psia. The gas-turbine inlet temperature is 2500 F and the exhaust pressure $p_4$ is 16 psia. The net gas-turbine system power is 5000 HP. Steam leaves the heat-recovery boiler at 1000 psia and 800 F and the steam turbine produces 3000 HP. The condenser pressure is 1.5 psia. The stack gas temperature at section (5) is 300 F and heat is added at the heat-recovery boiler. Assume air-standard conditions for the gas-turbine unit. On the premise that all components operate ideally, calculate (a) the rates at which heat is supplied at the air heater and at the heat recovery boiler and (b) the thermal efficiency of the combined cycle.

10.24 Repeat problem 10.23 using a compressor efficiency of 86 percent, gas-turbine efficiency of 88 percent, and steam-turbine efficiency of 80 percent. See problem 10.18 for the definitions of compressor and turbine efficiency.

10.25 Consider a two-fluid cycle using $H_2O$ and F-12 in which each fluid undergoes a Rankine cycle. Steam enters the steam turbine at 400 psia, quality 100 percent. The steam condenser operates at 20 psia and all heat transferred from the condensing steam serves as energy to heat and vaporize the F-12. The Freon-12 side of the cycle has vapor entering the turbine at 200 F, quality 100 percent. The F-12 is condensed at 90 F. Draw a flow diagram for the cycle. Calculate the combined cycle thermal efficiency. Describe a procedure that could be used to determine the optimum steam condenser pressure, that is, that yielding maximum cycle thermal efficiency, with the given steam-turbine inlet state and the given condensing temperature of the Freon-12.

# 11

# Perfect Gas Mixtures, Psychrometry and Air-Conditioning Processes

*In this chapter we first study the properties of constant-composition perfect gas mixtures. Many of the relations presented are valid for chemically reactive gas systems at a fixed state of such systems. These, as well as other characteristics of reactive systems, will be treated in Chapter 12. Psychrometry, the study of atmospheric air–water vapor mixtures, is an important application of the theory of gas mixtures. Several common processes for ambient pressure air–water vapor mixtures will be analyzed in the section on air-conditioning processes. These include heating and cooling, humidification and dehumidification, and combinations of these processes.*

## 11–1 Avogadro's Law

In Chapter 5 the equation of state for a perfect gas was presented on both the unit mass and the mole basis. The equation of state on the basis of a mole is

$$p\bar{v} = \bar{R}T \tag{5.1e}$$

or

$$pV = n\bar{R}T \tag{5.1f}$$

These equations apply to any single perfect gas. Equation (5.1f) can be used to illustrate Avogadro's law for perfect gases. This law states: "Equal volumes of gases at the same pressure and temperature contain the same number of molecules." If $p$, $V$, and $T$ in eq. (5.1f) are fixed and various gases are considered, we note that $n$, the number of moles, is also fixed, since $\bar{R}$, the universal gas constant, is the same for all gases. Avogadro's number or constant is $6.024 \times 10^{26}$ molecules/kmol ($2.732 \times 10^{26}$ molecules/lbmol). Hence we recognize that when the number of moles is fixed, the number of molecules is also fixed. This leads us to the useful fact that a mole of a perfect gas is simply a fixed number of molecules of that gas and, at a given pressure and temperature, the densities of perfect gases are proportional to their molecular weights.

The existence of the universal gas constant $\bar{R}$ makes the mole basis a most convenient one when dealing with mixtures of perfect gases.

### 11–2 The Gibbs-Dalton Law and the Equation of State for Perfect Gas Mixtures

The Gibbs-Dalton law provides an important foundation for our consideration of perfect gas mixtures. The law is: "Any perfect gas is as a vacuum to any other perfect gas mixed with it." In other words, in a mixture of perfect gases, each constituent behaves in all respects as though it existed alone at the temperature and volume of the mixture. The nature of pressure as discussed in Chapter 2 provides additional insight into this concept.

A corollary of the Gibbs-Dalton law is: "The pressure of a mixture of perfect gases is the sum of the pressures of the constituents when the pressure of each constituent is measured as it exists alone at the mixture temperature and volume." This corollary, which is readily subjected to experimental verification, introduces the concept of *partial pressures*. Suppose we have a mixture of several perfect gases. When any one of these gases occupies the mixture volume at the mixture temperature, the pressure it exerts on the boundaries is, by definition, the partial pressure of that constituent. For a mixture having $N$ different perfect gases

$$p_m = \sum_{i=1}^{N} p_i \tag{11.1}$$

where $p_m$ is the mixture pressure and $p_i$ the partial pressure of constituent $i$. For any constituent $i$ in a perfect gas mixture occupying the volume $V_m$ at temperature $T_m$ we have

$$p_i V_m = n_i \overline{R} T_m$$

For all constituents in a perfect gas mixture existing at $T_m$ and volume $V_m$

$$\sum_{i=1}^{N} p_i V_m = \sum_{i=1}^{N} n_i \overline{R} T_m$$

which, by eq. (11.1), reduces to

$$p_m V_m = \sum_{i=1}^{N} n_i \overline{R} T_m \tag{11.2}$$

Noting that $\sum_{i=1}^{N} n_i$ is equal to $n_m$, the total number of moles of all gases in the mixture, eq. (11.2) can be expressed as

$$p_m V_m = n_m \overline{R} T_m \tag{11.3}$$

This is the equation of state for a perfect gas mixture of $n_m$ moles of mixture existing at the mixture pressure $p_m$ and the mixture temperature $T_m$. For 1 mole of mixture

$$p_m \overline{v}_m = \overline{R} T_m \tag{11.3a}$$

## 11–3 The Mole Fraction and Partial-Pressure Relation

The *mole fraction, x,* is defined as the ratio of the number of moles of a given constituent to the total moles of all constituents in a perfect gas mixture. For constituent $i$ in a perfect gas mixture

$$x_i \equiv \frac{n_i}{n_m}$$

Also

$$n_i = \frac{m_i}{M_i} \quad \text{and} \quad n_m = \frac{m_m}{M_m}$$

where $n_m$ is the total number of moles of all constituents in the mixture and $m_m$ is the mass of the mixture.

For any one constituent in a perfect gas mixture we have

$$p_i V_m = n_i \overline{R} T_m$$

while for the mixture

$$p_m V_m = n_m \overline{R} T_m$$

Dividing the two equations of state

$$\frac{p_i}{p_m} = \frac{n_i}{n_m} \equiv x_i \tag{11.4}$$

Thus the partial pressure of any constituent in a perfect gas mixture is equal to the product of its mole fraction and the mixture pressure. This relation, valid at any state of perfect gas mixture, is applicable to constant-composition mixtures at any mixture pressure. For chemically reactive systems having continually changing composition and pressure, eq. (11.4) can be applied only at equilibrium states of the reactive system.

## 11–4 Volumetric Analysis of Perfect Gas Mixtures

When any single constituent in a perfect gas mixture existing at $T_m$ and $p_m$ is brought to the mixture pressure and temperature, that constituent will occupy a volume $V'$, which is defined as its partial volume. Figure 11.1a illustrates a mixture of gases exhibiting a volume $V_m$. Figure 11.1b shows a constituent $i$ existing alone at $p_i = p_m$ and $T_i = T_m$. For this condition the constituent exhibits its partial volume $V'_i$, which is less than $V_m$. A constituent gas at its partial volume has the equation of state

$$p_m V'_i = n_i \overline{R} T_m$$

**Figure 11.1** A Perfect Gas Mixture *(a)* and One Component at its Partial Volume *(b)*

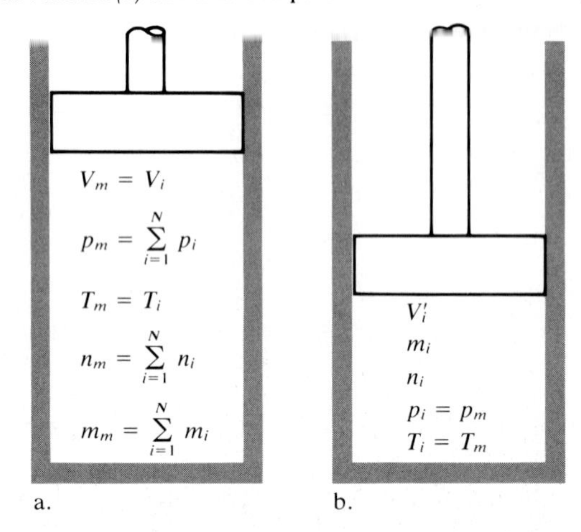


while

$$p_m V_m = n_m \overline{R} T_m$$

holds for the gas mixture. Dividing these equations, we obtain

$$\frac{V'_i}{V_m} = \frac{n_i}{n_m} \equiv x_i \qquad \textbf{(11.5)}$$

The left-hand side of the equality is defined as the *volume fraction* of constituent $i$. The right side of the equation has been defined as the mole fraction of constituent $i$ in the mixture. Thus, for a perfect gas mixture, the volume fraction and the mole fraction are identical.

The *apparent molecular weight* of a gas mixture is defined as the mass of the mixture per mole of mixture. One mole of mixture contains $x_i$ moles of constituent $i$. Therefore, the mass of constituent $i$ in 1 mole of mixture is

$$m_i = x_i M_i$$

The summation of the $x_i M_i$ products is the total mass of all constituents in 1 mole of mixture. Thus the apparent molecular weight of a mixture of $N$ constituents is

$$M_m = \sum_{i=1}^{N} x_i M_i \qquad \textbf{(11.6)}$$

## Example 11.a

A perfect gas mixture at 27.5 psia and 600 R has the composition

$$\begin{array}{rl} N_2 & 40.7 \text{ percent} \\ O_2 & 29.3 \text{ percent} \\ CO_2 & 30.0 \text{ percent} \end{array}$$

by volume. Calculate:

(a) The apparent molecular weight of the mixture, $M_m$
(b) The partial pressure of each constituent
(c) The volume occupied by 10.6 pounds of the mixture when it is at the given pressure and temperature

**Solution:**

Because of the convenient relation of the volume fraction and the mole fraction we select 1 mole of mixture as the basis for calculations. Use of the tabular method as illustrated is recommended. The $x_i$ values are, from eq. (11.5), the given volume fractions. The $M_i$ column is inserted for convenience. Equation (11.4) has been used to compute the values in the last column.

| Gas | $x_i$ | $M_i$ | $m_i$ | $p_i$ |
|---|---|---|---|---|
| $N_2$ | 0.407 | 28 | 11.396 | 11.193 |
| $O_2$ | 0.293 | 32 | 9.376 | 8.058 |
| $CO_2$ | 0.300 | 44 | 13.20 | 8.25 |
| | 1.000 | | 33.97 | 27.50 |

According to eq. (11.6), the summation of the $m_i$ column is the number of pounds of mixture per pound mole of mixture, which is the apparent molecular weight of the mixture. The answer to part (a) is then $M_m = 33.97$. The answers to part (b) are given in the $p_i$ column, the total of which must equal the mixture pressure. The answer to part (c) can be calculated in several ways. If we select the equation of state for the mixture

$$p_m V_m = n_m \overline{R} T_m$$

the value of $n_m$ is

$$n_m = \frac{10.6 \text{ lbm mix} \mid \text{lb mol mix}}{\mid 33.97 \text{ lbm mix}} = 0.312 \text{ lb mol mix}$$

The volume occupied by 10.6 pounds of mixture, equal to that occupied by 0.312 lb mol of mixture, is

$$V_m = \frac{n_m \overline{R} T_m}{p_m}$$

$$= \frac{(0.312)(1545)(600)}{(27.5)(144)} = 73.0 \text{ ft}^3$$

To solve part (c) in another way, we note that the volume occupied by any constituent existing alone at its partial pressure and the mixture temperature is the same as the mixture volume. We may select any gas in the mixture and write

$$V_m = \frac{n_i \overline{R} T}{p_i}$$

Thus for the oxygen in the mixture

$$n_{O_2} = \frac{0.293 \text{ lb mol } O_2}{\text{lb mol mix}} \left| \frac{\text{lb mol mix}}{33.97 \text{ lbm mix}} \right| \frac{10.6 \text{ lbm mix}}{}$$

$$= 0.0914 \text{ lb mol } O_2 \text{ in } 10.6 \text{ lbm of mixture}$$

Therefore

$$V_m = \frac{(0.0914)(1545)(600)}{(8.058)(144)}$$

$$= 73.0 \text{ ft}^3$$

---

## 11–5 Internal Energy and Enthalpy for Perfect Gas Mixtures

A second corollary of the Gibbs-Dalton Law is: "The internal energy of a mixture of perfect gases is equal to the sum of the internal energies of its constituents if the internal energy of each constituent is evaluated at the mixture volume and the mixture temperature." Since the internal energy of any perfect gas is independent of pressure and volume, that is, internal energy for a perfect gas is a function of temperature alone, the preceding statement reduces to evaluation of the internal energies at the mixture temperature. Under this condition

$$U_m = n_m \overline{u}_m = \sum_{i=1}^{N} n_i \overline{u}_i \qquad \textbf{(11.7)}$$

where $\overline{u}$ indicates the internal energy per mole. Dividing eq. (11.7) by $n_m$ yields, on the basis of a mole of mixture,

$$\overline{u}_m = \sum_{i=1}^{N} x_i \overline{u}_i \qquad \textbf{(11.7a)}$$

where $x_i$ is, by definition,

$$x_i = \frac{n_i}{n_m}$$

Enthalpy has been defined as

$$h \equiv u + pv$$

which becomes, on a mole basis for a perfect gas,

$$\overline{h} = \overline{u} + \overline{R}T \quad \textbf{(11.8)}$$

With $\overline{u}$ a pure temperature function, it is apparent that the enthalpy of any constituent in a mixture depends only on temperature. Therefore, the enthalpy of a mixture of perfect gases depends only on temperature. Thus

$$n_m\overline{h}_m = \sum_{i=1}^{N} n_i\overline{h}_i \quad \textbf{(11.9)}$$

or, on the basis of a mole of mixture,

$$\overline{h}_m = \sum_{i=1}^{N} x_i\overline{h}_i \quad \textbf{(11.9a)}$$

## 11–6 Specific Heats for Perfect Gas Mixtures

As noted in Chapter 5, molal specific heats of a single perfect gas are related to those on the basis of a unit mass by

$$\overline{C}_p = MC_p \quad \textbf{(5.14)}$$

Equation (5.15) relates $\overline{C}_p$, $\overline{C}_v$, and $\overline{R}$ as

$$\overline{C}_p - \overline{C}_v = \overline{R} \quad \textbf{(5.15)}$$

Table 5.5 lists on a unit mass basis approximate constant specific heats for several perfect gases. Because of the fixed difference between $\overline{C}_p$ and $\overline{C}_v$ as given by eq. (5.15), it is customary to present $\overline{C}_p$ data only. Molal specific heats corresponding to the $C_p$ values listed in Table 5.5 are given in Table 11.1, as are the values of $\overline{R}$ in both sets of units.

For a given gas, the specific heat ratio $\gamma$ has the same value on either the unit mass or the mole basis.

**Table 11.1** Approximate $\overline{C}_p$ Values for Several Perfect Gases: $\overline{R} = 8.314$ kJ/(kmol K); $\overline{R} = 1.985$ B/(lbmol R)

| Gas | M | B/(lbmol R) | kJ/(kmol K) |
|---|---|---|---|
| air, dry | 28.97 | 6.95 | 29.0 |
| argon | 39.95 | 4.95 | 20.8 |
| carbon dioxide | 44.01 | 8.89 | 37.2 |
| carbon monoxide | 28.01 | 6.97 | 29.1 |
| helium | 4.004 | 4.96 | 32.8 |
| hydrogen | 2.016 | 6.89 | 28.8 |
| nitrogen | 28.01 | 6.95 | 29.1 |
| oxygen | 32.00 | 7.01 | 29.4 |
| water vapor | 18.00 | 8.06 | 33.7 |

The molal specific heat for a perfect gas mixture of fixed composition is identified by differentiating eq. (11.9a) to

$$d\overline{h} = \sum_{i=1}^{N} x_i \, d\overline{h}_i$$

With each enthalpy differential equal to $\overline{C}_{p_i} dT$, we get

$$\overline{C}_{p_m} dT_m = \sum_{i=1}^{N} x_i \overline{C}_{p_i} dT_i$$

Since the temperature change of each constituent equals the mixture temperature change,

$$\overline{C}_{p_m} = \sum_{i=1}^{N} x_i \overline{C}_{p_i} \tag{11.10}$$

Also

$$\overline{C}_{v_m} = \overline{C}_{p_m} - \overline{R} \tag{11.11}$$

which is the same equation used to relate $\overline{C}_p$ and $\overline{C}_v$ for an individual perfect gas.

In those cases where approximate specific heats given in Table 11.1 are valid assumptions, the enthalpy change of 1 mole of a perfect gas mixture is calculated as

$$\overline{h}_2 - \overline{h}_1 = \overline{C}_{p_m} (T_2 - T_1) \tag{11.12}$$

The internal energy change, under the same conditions, is

$$\overline{u}_2 - \overline{u}_1 = \overline{h}_2 - \overline{h}_1 - \overline{R} (T_2 - T_1) \tag{11.13}$$

or

$$\overline{u}_2 - \overline{u}_1 = \overline{C}_{v_m} (T_2 - T_1) \tag{11.14}$$

where $\overline{C}_{v_m}$ is determined by eq. (11.11).

## Example 11.b

A perfect gas mixture consists of 4.54 kg of argon, 0.454 kg of hydrogen, and 7.71 kg of dry air. The mixture exists as a closed system initially at 87.1 kPa and 268 K. The mixture undergoes a constant-pressure reversible process to a final state where the temperature is 360 K. Calculate:

(a) The partial pressure of the hydrogen at the final state
(b) The internal energy change during the process
(c) The work for the process

**Solution:**

For the solution we shall assume a basis of the given masses of each gas. The moderate temperature change allows assumption of constant specific heats as given in Table 11.1. Using the tabular method we obtain:

| Gas | $m_i$, kg | $M_i$ | $n_i$, kmol | $x_i$ | $\overline{C}_{p_i}$ | $x_i\overline{C}_{p_i}$ |
|---|---|---|---|---|---|---|
| argon | 4.54 | 39.95 | 0.1136 | 0.1878 | 20.8 | 3.906 |
| hydrogen | 0.454 | 2.016 | 0.2252 | 0.3723 | 28.8 | 10.72 |
| dry air | 7.71 | 28.97 | 0.2661 | 0.4399 | 29.0 | 12.67 |
| | 12.70 | | 0.6049 | 1.0000 | | $27.38 = \overline{C}_{p_m}$ |

(a) For any mixture pressure the partial pressure of one constituent is

$$p_i = x_i p_m$$

With $p_{m_1} = p_{m_2}$ and $x_i$ not changing

$$p_{H_{2,1}} = p_{H_{2,2}} = (0.1878)(87.1) = 16.36 \text{ kPa}$$

(b) For the mixture

$$U_2 - U_1 = n_m(\overline{u}_2 - \overline{u}_1)_m = n_m\overline{C}_{v_m}(T_2 - T_1)$$

With $\overline{C}_{p_m}$ known, $\overline{C}_{v_m}$ can be computed as

$$\overline{C}_{v_m} = \overline{C}_{p_m} - \overline{R} = 27.38 - 8.314 = 19.07 \text{ kJ/(kmol K)}$$

Therefore, noting from the table that $n_m = 0.6049$

$$U_2 - U_1 = 0.6049(19.07)(360 - 268)$$
$$= 1061 \text{ kJ}$$

(c) The energy equation written for the process is

$$ {}_1W_2 = {}_1Q_2 - (U_2 - U_1) $$

For a reversible closed-system constant-pressure process, on the basis of a mole

$$ \begin{aligned} {}_1\overline{q}_2 &= \overline{u}_2 - \overline{u}_1 + \int_1^2 p d\overline{v} = \overline{u}_2 - \overline{u}_1 + p(\overline{v}_2 - \overline{v}_1) \\ &= \overline{h}_2 - \overline{h}_1 \end{aligned} $$

Therefore

$$ {}_1W_2 = n_m(\overline{h}_2 - \overline{h}_1)_m - n_m(\overline{u}_2 - \overline{u}_1)_m $$

where

$$ (\overline{h}_2 - \overline{h}_1)_m = \overline{C}_{p_m}(T_2 - T_1) $$

Substituting into the equation for work

$$ \begin{aligned} {}_1W_2 &= n_m(\overline{C}_{p_m} - \overline{C}_{v_m})(T_2 - T_1) = n_m \overline{R}\,(T_2 - T_1) \\ &= 0.6049(8.314)(360 - 268) \\ &= 463 \text{ kJ} \end{aligned} $$

---

## 11–7 Entropy of Perfect Gas Mixtures

A third corollary of the Gibbs-Dalton law treats the entropy of gas mixtures. This corollary may be stated as: "The entropy of a mixture of perfect gases is equal to the sum of the entropies of its constituents when each constituent is at the mixture volume and temperature." In Section 11–5 it was noted that these conditions for any constituent are satisfied when the constituent is at its partial pressure and the mixture temperature. Unlike internal energy or enthalpy, the entropy is dependent on pressure as well as temperature. Thus

$$ S_m = n_m\overline{s}_m = \sum_{i=1}^{N} n_i\overline{s}_i $$

where each $\overline{s}_i$ is dependent on its partial pressure as well as temperature.

For 1 mole of a perfect gas mixture undergoing a state change from (1) to (2)

$$ (\overline{s}_2 - \overline{s}_1)_m = \sum_{i=1}^{N} (x_i\overline{s}_i)_2 - \sum_{i=1}^{N} (x_i\overline{s}_i)_1 \qquad \textbf{(a)} $$

where each $x_i$ is constant for a constant-composition perfect-gas mixture. The entropy change for any single constituent can be determined using the second *Tds* equation, eq. (7.9).

$$ Tds = dh - vdp $$

This becomes

$$d\bar{s}_i = \frac{\overline{C}_{p_i} dT}{T} - \overline{R}\frac{dp_i}{p_i} \tag{b}$$

for any constituent in the mixture. In general $\overline{C}_p$ in the preceding equation depends on temperature. If precise answers are warranted, expressions for $\overline{C}_p$ as a function of temperature, such as those in Table 5.4, can be used. A reasonable approximation for most gases undergoing moderate changes in temperature is to assume that $\overline{C}_p$ is constant.

With this assumption eq. (b) integrates to

$$(\bar{s}_2 - \bar{s}_1)_i = \overline{C}_{p_i} \ln \frac{T_2}{T_1} - \overline{R} \ln \frac{p_{i,2}}{p_{i,1}} \tag{c}$$

for constituent $i$.

The partial-pressure ratio $p_{i,2}/p_{i,1}$ is related to the mixture pressure ratio, $(p_2/p_1)_m$, by eq. (11.4) as

$$\frac{p_{i,2}}{p_{i,1}} = \frac{(x_{i,2})(p_{m,2})}{(x_{i,1})(p_{m,1})} \tag{d}$$

However, the mole fraction does not change during a process of a constant-composition perfect-gas system. This allows us to express eq. (c) as

$$(\bar{s}_2 - \bar{s}_1)_i = \overline{C}_{p_i} \ln \left(\frac{T_2}{T_1}\right)_m - \overline{R} \ln \left(\frac{p_2}{p_1}\right)_m \tag{e}$$

For 1 mole of mixture we now can rewrite eq. (a), using eq. (e), in the form

$$(\bar{s}_2 - \bar{s}_1)_m = \sum_{i=1}^{N} x_i \overline{C}_{p_i} \ln \left(\frac{T_2}{T_1}\right)_m - \sum_{i=1}^{N} x_i \overline{R} \ln \left(\frac{p_2}{p_1}\right)_m \tag{f}$$

However, $\sum_{i=1}^{N} x_i \overline{C}_{p_i}$ is, by eq. (11.10), $\overline{C}_{p_m}$ and $\sum_{i=1}^{N} x_i = 1$. Substitution into eq. (f) yields

$$(\bar{s}_2 - \bar{s}_1)_m = \overline{C}_{p_m} \ln \left(\frac{T_2}{T_1}\right)_m - \overline{R} \ln \left(\frac{p_2}{p_1}\right)_m \tag{11.15}$$

Reversible adiabatic (isentropic) processes for perfect gas mixtures are encountered frequently. For such processes taking place with the conditions of constant composition and constant specific heats, eq. (11.15) becomes

$$\overline{C}_{p_m} \ln \left(\frac{T_2}{T_1}\right)_m = \overline{R} \ln \left(\frac{p_2}{p_1}\right)_m$$

The antilog of this equality is

$$\left(\frac{T_2}{T_1}\right)_m^{\overline{C}_{p_m}} = \left(\frac{p_2}{p_1}\right)_m^{\overline{R}} \tag{g}$$

Also

$$\overline{C}_p - \overline{C}_v = \overline{R}$$

Defining $\gamma_m$ as

$$\gamma_m \equiv \frac{\overline{C}_{p_m}}{\overline{C}_{v_m}} \tag{11.16}$$

eq. (g) becomes

$$\left(\frac{T_2}{T_1}\right)_m = \left(\frac{p_2}{p_1}\right)_m^{(\gamma_m - 1)/\gamma_m} \tag{11.17}$$

The result should not be surprising since the same relation was used for dry air, a perfect gas mixture, in numerous earlier examples and problems. Equation (11.17) in combination with the equation of state $p\overline{v} = \overline{R}T$ yields

$$\left(\frac{T_2}{T_1}\right)_m = \left(\frac{\overline{v}_1}{\overline{v}_2}\right)_m^{\gamma_m - 1} \tag{11.17a}$$

and

$$\left(\frac{p_2}{p_1}\right)_m = \left(\frac{\overline{v}_1}{\overline{v}_2}\right)_m^{\gamma_m} \tag{11.17b}$$

for constant-composition gas mixtures having constant specific heats. The specific heat ratio $\gamma_m$ is not a weighted value of $\gamma$ for the constituents in the mixture. It must be calculated using eq. (11.16).

---

## Example 11.c

Two hundred cubic feet per minute of a gas mixture consisting of 85 percent H and 15 percent $CO_2$ by volume enter a compressor at 14.7 psia and 520 R. The mixture is compressed reversibly and adiabatically to 75 psia. Velocities are negligible. Calculate:

(a) The horsepower required to drive the compressor
(b) The entropy change of the hydrogen per mole of mixture

**Solution:**

(a) With the volumetric analysis known, a basis of 1 mole of mixture is most convenient for the preliminary calculations.

| Gas | $x_i$ | $\overline{C}_{p_i}$ | $(x\,\overline{C}_p)_i$ |
|---|---|---|---|
| $H_2$ | 0.85 | 6.89 | 5.857 |
| $CO_2$ | 0.15 | 8.89 | 1.334 |
| | 1.00 | | $7.190 = \overline{C}_{p_m}$ |

From the steady-flow energy equation

$$_1\overline{w}_2 = \overline{h}_1 - \overline{h}_2 = \overline{C}_{p_m}(T_1 - T_2)$$

The outlet temperature $T_2$ is

$$T_2 = \left(\frac{p_2}{p_1}\right)_m^{(\gamma_m - 1)/\gamma_m} (T_1)$$

However $\gamma_m$ must be calculated as

$$\gamma_m = \frac{\overline{C}_{p_m}}{\overline{C}_{v_m}}$$

where $\overline{C}_{v_m}$ is

$$\begin{aligned}\overline{C}_{v_m} &= \overline{C}_{p_m} - \overline{R} \\ &= 7.190 - 1.985 = 5.205 \text{ B/(lbmol R)}\end{aligned}$$

For this mixture

$$\gamma_m = \frac{7.190}{5.205} = 1.381$$

and

$$T_2 = \left(\frac{75}{14.7}\right)^{0.2761} (520) = 815 \text{ R}$$

For 1 pound mole of mixture

$$\begin{aligned}_1\overline{w}_2 &= 7.190\,(520 - 815) \\ &= -2121 \text{ B/lbmol}\end{aligned}$$

The rate of flow is, from eq. (5.5),

$$\dot{n} = \frac{(14.7)(144)(200)}{(1545)(500)} = 0.5270 \text{ lbmol/min}$$

The power input to the compressor is

$$\dot{W}_{in} = \frac{0.5270 \mid 2121 \mid 60}{\mid \mid 2545}$$

$$= 26.4 \text{ HP}$$

(b) The entropy change of 1 pound mole of the hydrogen is

$$\bar{s}_2 - \bar{s}_1 = \bar{C}_p \, ln \frac{T_2}{T_1} - \bar{R} \, ln \frac{p_2}{p_1}$$
$$= 6.89 \, ln \frac{815}{520} - 1.985 \, ln \frac{75}{14.7}$$
$$= -\, 0.1388 \text{ B/(lbmol } H_2 \text{ R)}$$

On the basis of 1 pound mole of mixture

$$(\bar{s}_2 - \bar{s}_1)_{H_2} = x_{H_2} (\bar{s}_2 - \bar{s}_1)_{H_2}$$
$$= (0.85)(-0.1388) = -0.1179 \text{ B/(lbmol mix R)}$$

While the process is isentropic for the perfect gas mixture the entropy of each constituent changes. In this particular example the entropy change of the other gas, $CO_2$, would, on the basis of a mole of mixture, be equal in magnitude and opposite in sign to that of the hydrogen. The net result is no change in the entropy of the mixture during the reversible adiabatic process.

---

### 11–8 Perfect Gas Mixtures on a Unit Mass Basis

While the mole basis is preferable for the solution of perfect gas mixture problems, occasionally the unit mass basis is employed. On this basis the analysis of the mixture is by mass fraction (gravimetric). The mass fraction $m_f$ is defined as

$$m_{f_i} \equiv \frac{m_i}{m_m} \tag{11.18}$$

where $m_i$ is the mass of constituent $i$ in $m_m$ mass units of the gas mixture. Simple relations such as that of the mole fraction to the volume fraction or the partial-pressure to mixture-pressure ratio cannot be developed using the unit mass basis. Expressions for $u_m$, $h_m$, $C_{p_m}$, and $C_{v_m}$ follow from the procedure used for the mole basis. For a unit mass of a perfect gas mixture

$$u_m = \sum_{i=1}^{N} m_{f_i} u_i \tag{11.7b}$$

$$h_m = \sum_{i=1}^{N} m_{f_i} h_i \tag{11.9b}$$

$$C_{v_m} = \sum_{i=1}^{N} m_{f_i} C_{v_i}$$

and

$$C_{p_m} = \sum_{i=1}^{N} m_{f_i} C_{p_i} \tag{11.10a}$$

For each gas in the mixture

$$C_{p_i} - C_{v_i} = R_i$$

on a unit mass basis. For any one gas having the mass fraction $m_{f_i}$, the above equation becomes

$$m_{f_i} C_{p_i} - m_{f_i} C_{v_i} = m_{f_i} R_i$$

The summation for all constituents in a perfect gas mixture is

$$\sum_{i=1}^{N} m_{f_i} C_{p_i} - \sum_{i=1}^{N} m_{f_i} C_{v_i} = \sum_{i=1}^{N} m_{f_i} R_i$$

With the summations on the left side of the equality defined as $C_{p_m}$ and $C_{v_m}$ we get

$$C_{p_m} - C_{v_m} = \sum_{i=1}^{N} m_{f_i} R_i$$

The remaining summation is defined as the gas constant $R_m$ for the mixture or

$$R_m \equiv \sum_{i=1}^{N} m_{f_i} R_i \tag{11.19}$$

The apparent molecular weight of the mixture is then calculated as

$$M_m = \frac{\overline{R}}{R_m} \tag{11.20}$$

## Example 11.d

A perfect gas mixture has a mass fraction analysis of 19.7 percent oxygen, 49.8 percent helium, and 30.5 percent carbon monoxide. If the mixture is at 20.3 psia and 537 R, calculate:

(a) The volumetric analysis
(b) The gas constant $R_m$
(c) The apparent molecular weight of the mixture
(d) The specific heats $C_{p_m}$ and $C_{v_m}$ for the mixture

**Solution:**

A convenient basis for parts (a) and (b) is 100 pounds of mixture. Again the tabular form is recommended.

| Gas | $m_{f_i}$ | $m_i$ | $R_i$ | $(m_f R)_i$ | $M_i$ | $n_i$ | $x_i$ | $C_p$ | $(m_f C_p)_i$ |
|---|---|---|---|---|---|---|---|---|---|
| $O_2$ | 0.197 | 19.7 | 48.3 | 9.515 | 32.00 | 0.6156 | 0.04354 | 0.219 | 0.04314 |
| He | 0.498 | 49.8 | 386.0 | 192.2 | 4.003 | 12.44 | 0.8798 | 1.24 | 0.6175 |
| CO | 0.305 | 30.5 | 55.2 | 16.84 | 28.01 | 1.089 | 0.0770 | 0.249 | 0.07595 |
| | 1.000 | 100.0 | | 218.6 | | 14.14 | 1.000 | | 0.7366 |

(a) The mole fraction $x$ is equal to the volume fraction for perfect gas mixtures. From the table the volumetric analysis is 4.354 percent $O_2$, 87.98 percent He, and 7.70 percent CO.

(b) The gas constant for the mixture is $\sum_{i=1}^{N} (m_f R)_i$, which, from the table, is 218.6 ft lbf/(lbm R).

(c) The apparent molecular weight of the mixture is

$$M_m = \frac{\overline{R}}{R_m} = \frac{1545}{218.6} = 7.068$$

(d) From the table

$$C_{p_m} = 0.737 \text{ B/(lbm R)}$$

The constant volume specific heat of the mixture is

$$C_{v_m} = C_{p_m} - R_m$$
$$= 0.737 - \frac{218.6}{778} = 0.456 \text{ B/(lbm R)}$$

**Table 11.2** Volumetric Composition of Dry Air at Sea Level

| Constituent | Percent by Volume | Molecular Weight |
|---|---|---|
| $N_2$ | 78.03 | 28.01 |
| $O_2$ | 20.99 | 32.00 |
| A | 0.94 | 39.95 |
| $CO_2$ | 0.03 | 44.01 |
| $H_2$ | 0.01 | 2.024 |
| Ne | 0.0012 | 20.18 |
| He | 0.0004 | 4.003 |
| | Apparent molecular weight = 28.966 | |

## 11–9 Atmospheric Air

*Atmospheric air*—dry air plus water vapor—consists of several gases existing in constant proportions (dry air) and water vapor present in a wide range of proportions. The several constituents comprising dry air (see Table 11.2) represent a constant-composition perfect-gas mixture that can be treated as a single perfect gas.

The apparent molecular weight of dry air is frequently rounded to 29.0 and the composition to 21 percent $O_2$ by volume, 79 percent "atmospheric nitrogen" by volume for engineering calculations. The atmospheric nitrogen is actually a mixture of all gases in dry air except the oxygen. This approximation is used frequently in combustion calculations since the other gases (with the exception of the small amount of hydrogen) are inert. Under these conditions the apparent molecular weight of atmospheric nitrogen is approximately 28.2. The water vapor in atmospheric air is treated as a perfect gas since its partial pressure is very low. Another indication that the water vapor approximates perfect gas behavior is found in Table 3 of the *Steam Tables*. At typical ambient temperatures, say 70 F, the enthalpy is virtually constant for pressures representative of those of water vapor in atmospheric air. *Saturated air* is atmospheric air in which the water exists as saturated vapor.

The *dew point, $t_{dp}$*, of atmospheric air is the saturation temperature for the water-vapor partial pressure $p_w$. An alternative definition provides additional insight into the nature of the dew point. The alternate defines the dew point as that temperature to which atmospheric air must be cooled at constant pressure to cause condensation to occur. Condensation of water on window glass or cold water pipes is evidence the surface temperature of each is below the dew point of the atmospheric air coming into contact with these surfaces. The equivalence of the two definitions is established by

$$\frac{n_w}{n_w + n_{da}} \equiv x_w = \frac{p_w}{p}$$

where $p$ is the atmospheric pressure. From the equation it is apparent that the partial pressure of the water vapor will not change during constant-pressure cooling of the atmospheric air until the mole fraction, $x_w$, changes due to condensation of some water vapor.

**Figure 11.2** A Temperature–Entropy Diagram for the Water in Atmospheric Air

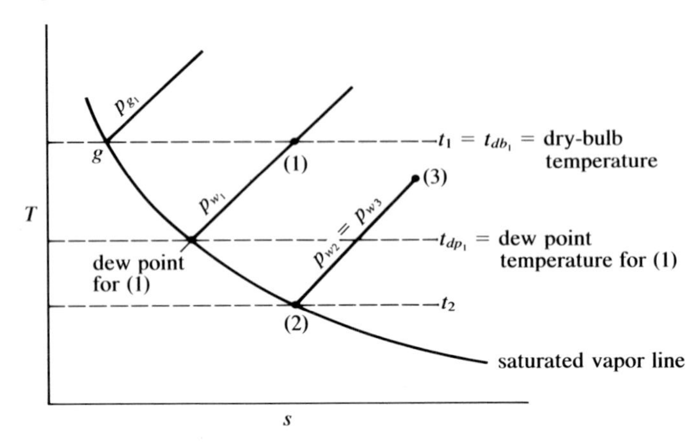


The *dry-bulb* temperature $t_{db}$ is the temperature of the air–water-vapor mixture determined by a conventional thermometer when the velocity of the atmospheric air relative to the thermometer is negligible. The terminology related to the water vapor in atmospheric air is illustrated in Figure 11.2. The figure is limited to states and processes of the water vapor in atmospheric air. At (1) the $H_2O$ is superheated vapor. In cooling the air to the dew point the path followed by the water vapor is $p_{w_1}$ = constant. The dew point lies on the saturated-vapor line. Cooling to temperatures less than $t_{d_{p_1}}$ results in the condensation of some water vapor. The water vapor remaining follows the saturated-vapor line as cooling continues to $t_2$. The state of the liquid formed on condensation is not shown on the diagram. If atmospheric air is heated at constant mixture pressure from $t_2$ to $t_3$, the partial pressure of the water vapor does not change since the mole fraction of the vapor remains constant.

## 11–10 Relative Humidity and Humidity Ratio

The *relative humidity* of atmospheric air is defined as

$$\phi \equiv \frac{p_w}{p_g} \tag{11.21}$$

where $p_w$ is the partial pressure of the water vapor and $p_g$ the saturation pressure of water at the dry-bulb temperature. See Figure 11.2 for an illustration of these pressures. The relative humidity is also the ratio of the water-vapor density to that of saturated water vapor at the same temperature.

$$\phi_1 = \frac{\rho_1}{\rho_{g_1}} \tag{11.21a}$$

The equivalence of the two definitions is established by the equation of state, assuming the water vapor follows perfect gas behavior. At state (1), Figure 11.2,

$$\rho_1 = \frac{p_{w_1}}{R_w T_1}$$

and

$$\rho_{g_1} = \frac{p_{g_1}}{R_w T_1}$$

Dividing

$$\phi_1 = \frac{\rho_1}{\rho_{g_1}} = \frac{p_{w_1}}{p_{g_1}} \tag{11.21b}$$

Saturated air, having $p_w = p_g$, has a relative humidity of 100 percent.

The specific humidity or humidity ratio $\omega$ of atmospheric air is defined as

$$\omega = \frac{m_w}{m_{da}} \tag{11.22}$$

Note that this is the ratio of the mass of water vapor to the mass of dry air and not the mass fraction of water vapor in atmospheric air. The humidity ratio is determined using eq. (11.22) and assuming that the water vapor and dry air are perfect gases. From the equation of state for each we have

$$m_w = \frac{p_w V}{R_w T}$$

and

$$m_{da} = \frac{p_{da} V}{R_{da} T}$$

Dividing the equalities

$$\frac{m_w}{m_{da}} = \frac{R_{da}}{R_w} \frac{p_w}{p_{da}}$$

For the water vapor or the dry air

$$R = \frac{\overline{R}}{M}$$

Thus

$$\omega = \frac{m_w}{m_{da}} = \frac{18}{28.96}\frac{p_w}{p_{da}} \tag{11.23}$$

$$\omega = 0.622\,\frac{p_w}{p_{da}}$$

For atmospheric air $p = p_w + p_{da}$ and

$$\omega = 0.622\,\frac{p_w}{p - p_w} \tag{11.23a}$$

## 11–11 The Adiabatic Saturation Process

For atmospheric air with a relative humidity $\phi$ less than 100 percent, the water-vapor pressure $p_w$ is less than the saturation pressure $p_g$ for the dry-bulb temperature. If the atmospheric air comes into contact with liquid $H_2O$, some of the water will evaporate and the humidity ratio $\omega$, as well as the relative humidity, will increase. If the evaporation occurs in an adiabatic manner, the energy required to evaporate the water (the latent heat of vaporization) will come from the surrounding air, thereby reducing the air temperature. An adiabatic steady-flow process with no work that results in sufficient evaporation of liquid $H_2O$ to produce saturated air ($\phi = 100$ percent) is termed an *adiabatic saturation process.* Such a process is illustrated in Figure 11.3. Atmospheric air enters the adiabatic steady-flow system at (1) and liquid $H_2O$ at (3) is sprayed into the air stream in such a manner that the air leaving at (2) is saturated.

One application of the adiabatic saturation process is in the evaporative cooling units used in areas where the relative humidity is typically low and the dry-bulb temperature is high. While we shall restrict our study to adiabatic processes that have the leaving stream at 100 percent relative humidity, the same analytic procedures could be applied to similar processes in which the leaving stream, section (2) in Figure 11.3, is not saturated. However, such evaporative cooling processes can be analyzed only when the state of the leaving stream is fixed by three independent properties, say, $p_2$, $t_{db,2}$, and $\phi_2$.

The analysis of an adiabatic saturation process follows from application of the steady-flow energy equation and the corresponding mass conservation equations. Neglecting kinetic and potential energy changes, the energy equation on a rate basis is

$$\dot{Q} + \Sigma\,(\dot{m}h)_{\text{in}} = \dot{W} + \Sigma\,(\dot{m}h)_{\text{out}} \tag{11.24}$$

for which $\dot{Q}$ and $\dot{W}$ are zero for the adiabatic saturation process. Therefore

$$\dot{m}_{da_1}h_{da_1} + \dot{m}_{w_1}h_{w_1} + \dot{m}_{w_3}h_{w_3} = \dot{m}_{da_2}h_{da_2} + \dot{m}_{w_2}h_{w_2} \tag{11.24a}$$

Mass conservation for the atmospheric air requires that expressions be written for both the dry air and the water. Thus

$$\dot{m}_{da_1} = \dot{m}_{da_2} = \dot{m}_{da} \tag{11.24b}$$

$$\dot{m}_{w_1} + \dot{m}_{w_3} = \dot{m}_{w_2}$$

**Figure 11.3** An Adiabatic Saturation Process

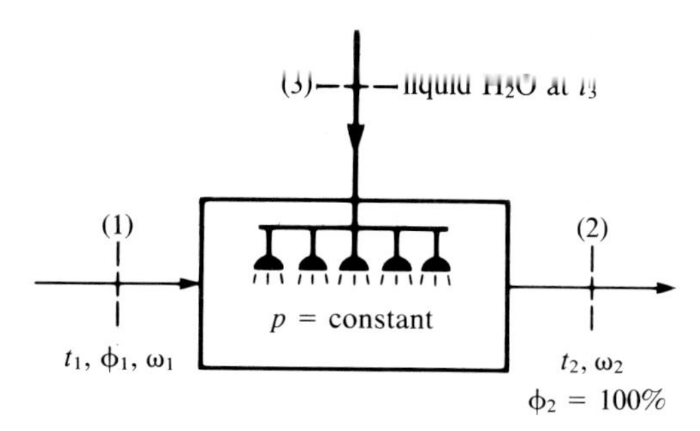


Dividing the above equations by $\dot{m}_{da}$ and recognizing $\dot{m}_w / \dot{m}_{da}$ as $\omega$, we obtain

$$h_{da_1} + \omega_1 h_{w_1} + \frac{\dot{m}_{w3}}{\dot{m}_{da}} h_{w_3} = h_{da_2} + \omega_2 h_{w_2}$$

$$\frac{\dot{m}_{w3}}{\dot{m}_{da}} = \omega_2 - \omega_1$$

Combining the above two equations yields

$$h_{da_1} + \omega_1 h_{w_1} + (\omega_2 - \omega_1) h_{w_3} = h_{da_2} + \omega_2 h_{w_2} \qquad \textbf{(11.24c)}$$

The specific enthalpies $h_{w_1}$ and $h_{w_2}$ can be taken as $h_g$ values at the dry-bulb temperatures $t_{db_1} = t_1$ and $t_{db_2} = t_2$ since the water vapor exhibits perfect gas behavior. Further, $h_{w_3}$ is equal to $h_f$ at $t_3$. Hence we can rewrite eq. (11.24c) as

$$h_{da_1} + \omega_1 h_{g_1} + (\omega_2 - \omega_1) h_{f_3} = h_{da_2} + \omega_2 h_{g_2} \qquad \textbf{(11.24d)}$$

It should be noted that each of the terms in eq. (11.24d) is on the basis of a unit mass of dry air. In other words, this equation constitutes an energy equation for the adiabatic saturation process written with a unit mass of *dry air* as the basis. The enthalpy of the water vapor in the above equations, $h_g$, exhibits a simple relationship with temperature for the range of temperatures encountered in dealing with atmospheric air. In the English system of units

$$h_g = [1061.2 + 0.43(t_F)], \text{ B/lbm} \quad (0 \leq t_F \leq 150) \qquad \textbf{(11.25)}$$

and in SI units

$$h_g = [2501 + 1.82(t_C)], \text{ kJ/kg} \quad (-20 \leq t_C \leq 70) \qquad \textbf{(11.25a)}$$

These expressions are useful in applications that involve an unknown temperature and also may be used in lieu of the *Steam Tables* to determine values of $h_g$.

### Example 11.e

Air at 90 F dry bulb having a relative humidity of 15 percent undergoes an adiabatic saturation process at a constant pressure of 14.12 psia. Atmospheric air is supplied to the unit at the rate of 750 ft³/min. Liquid $H_2O$ at 57 F is sprayed into the unit. Assuming the saturated air leaving the unit is at 70 F, calculate the mass of spray water required per hour and the volume rate of flow of the leaving air.

**Solution:**

Using the notation of Figure 11.3,

$$\omega_1 = 0.622\left(\frac{p_{w_1}}{p - p_{w_1}}\right) \quad \text{and} \quad p_{w_1} = \phi_1\, p_{g_1}$$

Therefore,

$$p_{w_1} = (0.15)(0.6988) = 0.1048 \text{ psia}$$

thus

$$\omega_1 = 0.622\left(\frac{0.1048}{14.12 - 0.1048}\right) = 0.004652\,\frac{\text{lbm H}_2\text{O}}{\text{lbm da}}$$

For the leaving stream, saturated at 70 F, $\phi_2 = 100$ percent and

$$\omega_2 = 0.622\left(\frac{p_{g2}}{p - p_{g2}}\right) = 0.622\left(\frac{0.3632}{14.12 - 0.3632}\right)$$
$$= 0.01642\,\frac{\text{lbm H}_2\text{O}}{\text{lbm da}}$$

The spray water per pound of dry air is

$$\omega_2 - \omega_1 = 0.01642 - 0.004652 = 0.01177\,\frac{\text{lbm H}_2\text{O}}{\text{lbm da}}$$

The mass rate of flow of the dry air, a constant, is by eq. (5.4)

$$\dot{m}_{da} = \frac{p_{da}\dot{V}_1}{R_{da}T_1} = \frac{(14.015)(144)(750)}{(53.3)(550)} = 51.63 \text{ lbm da/min}$$

The spray water per hour is

$$\dot{m}_3 = (51.63)(60)(0.01177) = 36.46 \text{ lbm/hr}$$

The volume rate of flow of the leaving stream $\dot{V}_2$ is

$$\dot{V}_2 = \frac{\dot{m}_{da}R_{da}T_2}{p_{da_2}} = \frac{(51.63)(53.3)(530)}{(14.12 - 0.3632)144} = 726.2 \text{ ft}^3/\text{min}$$

In this example the volume rate of flow of the dry air has been taken as the volume rate of flow of the atmospheric air. This is, of course, correct since each constituent in a mixture of gases occupies the mixture volume. Note, however, that the appropriate partial pressure has been used.

---

## 11–12 The Adiabatic Saturation and Wet-Bulb Temperatures

The state of atmospheric air, a binary mixture, is fixed by three independent properties. The mixture pressure is determined easily by using a barometer. The dry-bulb temperature can be established with a mercury-in-glass thermometer or another type of thermometer. One other independent property that could be used to fix the state would be the dew point. A polished metal plate that could be cooled and whose surface temperature could be measured would serve as a "dew-point meter." Using this device, the dew point of the surrounding atmosphere would be the plate surface temperature when the first condensation of the water became visible on the polished surface. Accurate measurement of the dew-point temperature by this means is difficult, however.

In this section we develop the equation for another method of establishing a third independent property. Figure 11.3 is a general schematic for adiabatic saturation processes. Consider a unique application of the device in which atmospheric air enters at section (1) where the dry-bulb temperature and the pressure have been measured. Further, let the spray water at section (3) be at the same temperature as the saturated air leaving the unit at section (2). Under these conditions the temperature $t_2$ is called the *adiabatic saturation temperature.*

The steady-flow energy equation for this particular adiabatic saturation process is, from eq. (11.24d), on the basis of a unit mass of dry air

$$h_{da_1} + \omega_1 h_{g_1} + (\omega_2 - \omega_1)h_{f_2} = h_{da_2} + \omega_2 h_{g_2} \qquad \textbf{(11.24e)}$$

where $\omega_1$, the humidity ratio of the atmospheric air is unknown. With the temperature at sections (2) and (3) identical, eq. (11.24e) can be written as

$$h_{da_1} + \omega_1(h_{g_1} - h_{f_2}) = h_{da_2} + \omega_2(h_{g_2} - h_{f_2}) \qquad \textbf{(11.24f)}$$

where (1) is the atmospheric air entering the adiabatic saturation unit and (2) indicates properties at the adiabatic saturation temperature.

Solving eq. (11.24f) for $\omega_1$, the humidity ratio of the atmospheric air supplied to the unit, we have

$$\omega_1 = \frac{\omega_2 h_{fg_2} - C_{P_{da}}(t_1 - t_2)}{h_{g_1} - h_{f_2}} \qquad \textbf{(11.26)}$$

For atmospheric air, the adiabatic saturation temperature has been found to be equal to the temperature that a thermometer attains as atmospheric air is passed over the thermometer, the bulb of which is covered with a wick soaked in water. The device used to measure the wet-bulb temperature is called a psychrometer.

The dry- and wet-bulb temperatures plus the barometric pressure are the three properties commonly used to establish the state of atmospheric air. The procedures followed in calculating psychrometric properties from these properties are illustrated in the following example.

## Example 11.f

The U.S. Weather Service reports a barometric pressure of 14.20 psia, 88 F dry bulb and 68 F wet bulb. For these conditions calculate:

(a) The humidity ratio
(b) The dew point
(c) The relative humidity
(d) The specific volume of the atmospheric air

**Solution:**

In the solution of this problem it is useful to refer to the adiabatic saturation process as described in Figure 11.3. As noted in this section, under the condition $t_3 = t_2$, the temperature $t_2$ is the wet-bulb temperature. Subscript 1 refers to the air entering the adiabatic saturation device at (1). Thus we associate the subscript 2 with the wet-bulb condition ($t_2 = t_{wb}$ and $\phi_2 = 100$ percent) and the subscript 1 with the atmospheric air for which we are to determine the humidity ratio, the dew point, the relative humidity, and the specific volume.

(a) We will apply eq. (11.26) to determine the humidity ratio. Using eq. (11.23a) we first determine $\omega$ at the wet-bulb temperature

$$\omega_2 = 0.622\left(\frac{p_{g2}}{p - p_{g2}}\right) = 0.622\left(\frac{0.3391}{14.20 - 0.3391}\right)$$
$$= 0.01521 \frac{\text{lbm H}_2\text{O}}{\text{lbm da}}$$

The result is the humidity ratio of saturated air at the wet-bulb temperature, 68 F. The humidity ratio of the atmosphere $\omega_1$ from eq. (11.26) is

$$\omega_1 = \frac{(0.01521)(1055.1) - 0.24(88 - 68)}{1099.9 - 36.09}$$
$$= 0.01058 \frac{\text{lbm H}_2\text{O}}{\text{lbm da}}$$

(b) The partial pressure of the water vapor in the air must be calculated to establish the dew point. With the humidity ratio and barometric pressure known, eq. (11.23a) applied at section (1) is

$$0.01058 = 0.622\left(\frac{p_{w_1}}{p - p_{w_1}}\right) = 0.622\left(\frac{p_{w_1}}{14.20 - p_{w_1}}\right)$$

which yields

$$p_{w_1} = 0.2375 \text{ psia}$$

The dew point is the saturation temperature for this pressure. Using Table 1 from the *Steam Tables*

$$t_{dp} = 58 \text{ F (closest value)}$$

Note that the dew-point temperature is less than the wet-bulb temperature, which, in turn, is less than the dry-bulb temperature. These three temperatures are equal only for saturated air, that is, atmospheric air having a relative humidity of 100 percent. For all values of relative humidity less than 100 percent, the relative magnitudes of the three temperatures will be the same as exhibited in this example.

(c) The relative humidity is

$$\phi_1 = \frac{p_{w_1}}{p_{g_1}} = \frac{0.2375}{0.6562} = 26.19 \text{ percent}$$

(d) With the proportions of dry air and water vapor established, the gas constant $R_m$ could be determined for the given mixture state. However, the gas constant changes as the humidity ratio changes. For this reason another approach is taken. For a volume $V_m$ of atmospheric air, the specific volume $v_m$ at a given state is

$$v_m = \frac{V_m}{(m_{da} + m_w)}$$

Multiplying and dividing by $m_{da}$ we obtain

$$v_m = \frac{V_m}{m_{da}\left[1 + \dfrac{m_w}{m_{da}}\right]} = \frac{(V_{da}/m_{da})}{(1 + \omega)} = \frac{v_{da}}{(1 + \omega)}$$

where

$$v_{da} = \frac{R_{da}T}{p_{da}} = \frac{(53.3)(548)}{(14.20 - 0.2375)(144)} = 14.53 \frac{\text{ft}^3}{\text{lb dry air}}$$

Therefore

$$v_m = \frac{14.527}{1 + 0.01058} = 14.38 \text{ ft}^3/\text{lbm atmospheric air}$$

Figure 11.4 A Psychrometric Chart for Standard Atmospheric Pressure

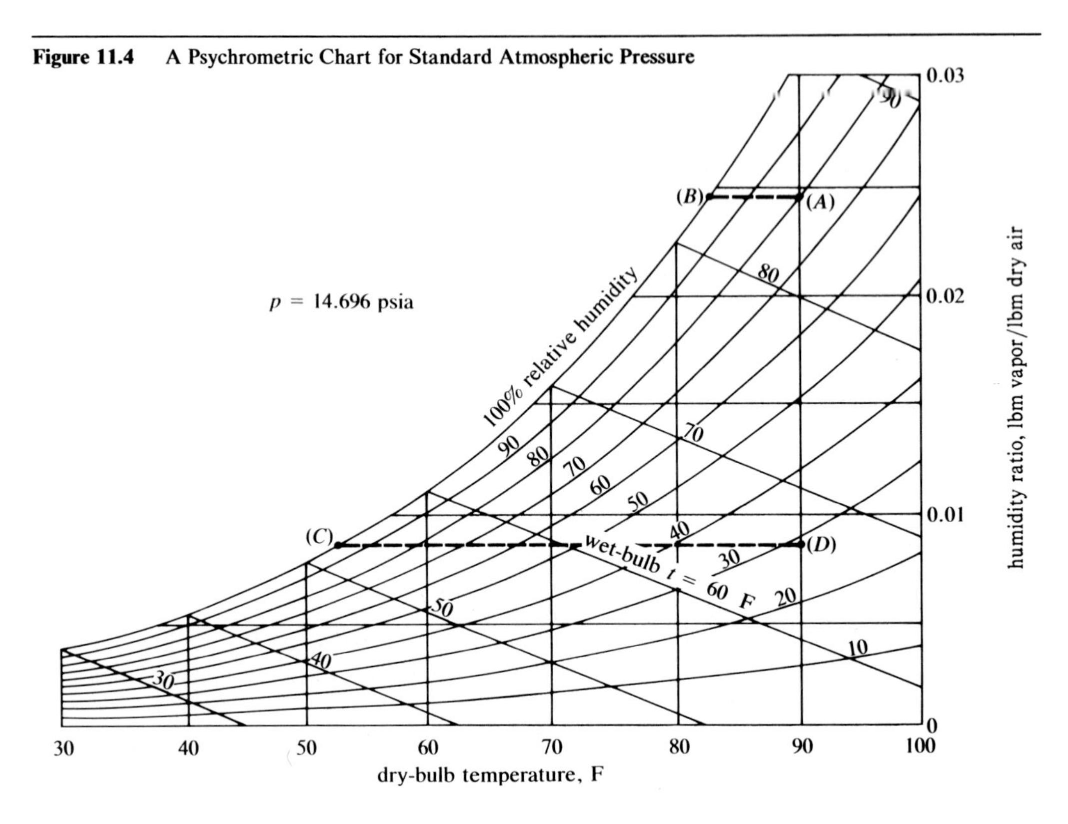


## 11–13 Psychrometric Charts

The relations of dry-bulb temperature, wet-bulb temperature, humidity ratio (specific humidity), relative humidity, and other properties of atmospheric air are shown graphically on diagrams called psychrometric charts. A chart of this type will be exact at only one barometric pressure because eqs. (11.23a) and (11.26), basic equations used to prepare such charts, depend on the atmospheric pressure. Figure 11.4 is a skeleton psychrometric chart for a pressure of 1 standard atmosphere. At elevations significantly above sea level, the changes in atmospheric pressure are of a magnitude large enough to introduce significant errors if a standard atmosphere psychrometric chart is used. As previously noted, three psychrometric properties are required to fix a state of atmospheric air. With the atmospheric pressure fixed, as in Figure 11.4, specification of any two of the following five psychrometric quantities will fix a state: dry bulb, wet bulb, and dew point temperatures, relative humidity, and humidity ratio. When a state is established, the remaining three quantities can be read from the psychrometric chart. Complete sea-level psychrometric charts are included in Appendix F in both English and SI units.

In addition to its utility in establishing the properties of atmospheric air, a psychrometric chart is useful in understanding processes in which atmospheric air is the working fluid. As an example, consider a process during which air at 90 F dry bulb, 80 percent relative humidity, and 1 standard atmosphere pressure is cooled, dehumidified, and reheated to its original dry-bulb temperature. Starting at the initial state (*A*) in Figure 11.4, cooling at constant-mixture pressure would proceed to (*B*) along the constant humidity ratio line, $\omega_A$ equal to

$\omega_B$, to the saturation line, $\phi = 100$ percent. Further cooling would result in condensation of a portion of the water vapor as the process proceeds to ($C$). Heating at constant atmospheric pressure would follow the constant humidity ratio line $\omega_C$ equal to $\omega_D$ to the point where the dry-bulb temperature at ($D$) would equal that at ($A$). The dehumidification from ($B$) to ($C$) would result in condensation of $\omega_B - \omega_C$ pounds of water per pound of dry air. The dew point of the original system (state [$A$]) is the temperature at ($B$). The dew point of the mixture at the final state ($D$) is the temperature at ($C$). Thus the constant humidity ratio lines are, additionally, lines of constant dew point and constant water-vapor partial pressure. The psychrometric chart ordinate $\omega$ is frequently given in the units of grains of water per pound of dry air. The conversion factor is 7000 grains equal to 1 pound.

## 11–14 Some Air-Conditioning Processes

Most problems that arise in air conditioning are of two kinds. In one a stream of atmospheric air is changed from one dry-bulb temperature and humidity ratio to some other dry-bulb temperature and humidity ratio. Problems of this type typically involve determination of the rates at which water is supplied or removed and heat is transferred to or from the air stream. Most problems of this kind are solved conveniently by first obtaining results on a basis of a unit mass of dry air and then converting the results to a rate basis after the dry-air-mass rate of flow has been determined. In the second kind, the air within an enclosure, a room, or a building, is to be maintained at a given psychrometric state while air flows into and out of the enclosure. The specified state within the enclosure is maintained by heat transfer to or from the system and by supplying or removing water. For problems of this nature, determination of the temperature and humidity ratio of the air entering the enclosure is commonly required. Another type of problem, not amenable to the unit-mass dry-air basis of solution, is the mixing of two or more entering streams to produce a leaving stream having an unknown psychrometric state.

The air-conditioning processes we shall consider proceed in a steady-flow manner. In general the velocities are sufficiently small to permit us to neglect the kinetic energy terms. Under these conditions the steady-flow equation on a time-rate basis, eq. (11.24), is applicable. In most applications the power $\dot{W}$ is zero. Several common air-conditioning processes are considered in more detail in the following.

**Heating or Cooling at Constant Humidity Ratio** Perhaps the simplest air-conditioning process is one in which atmospheric air flows without change in composition over a heating or cooling unit. Such a process occurs, for example, when atmospheric air contacts the surfaces of a steam radiator (convector) or the surfaces of tubes through which chilled water is circulated. If the humidity ratio remains constant during such a process and the work is negligible or zero, the heat transfer can be determined by application of eq. (11.24).

$$ {}_1\dot{Q}_2 + \dot{m}_{da}h_{da_1} + \dot{m}_w h_{w_1} = \dot{m}_{da}h_{da_2} + \dot{m}_w h_{w_2} $$

Dividing by $\dot{m}_{da}$ we obtain

$$ {}_1q_2 = h_{da_2} - h_{da_1} + \omega(h_{g_2} - h_{g_1}) $$

The above equation is on the basis of a unit mass of dry air.

**Heating and Humidifying** Another common air-conditioning process is one in which a stream of atmospheric air is both heated and humidified. Such processes are commonly associated with air flowing through a furnace equipped with a humidifier (an open pan of water, porous ceramic sections, or atomizing sprayers). The work quantity for such units is not zero, because of the power supplied to the fan; however, it is small and frequently assumed to have a negligible effect on the entering and leaving air streams.

An example of heating and humidification is that for a building during winter months when the outside temperature is significantly less than that maintained within the building. Inside air, at a higher dry-bulb temperature and humidity than that of the surrounding atmosphere, leaks or is exhausted to the atmosphere. This is typically replaced with colder, drier air from the atmosphere. Additionally, the inside air is affected by heat transfer through the building, by people within the space, and by mechanical and electric devices from which heat may be transferred to the air within the space. Humidification is necessary because the replacement air from the atmosphere has a lower humidity ratio than that to be maintained within the building. The required humidification is determined using mass balances for the dry air and the water, and the heating requirement is calculated using the steady-flow energy equation. Problems of this nature may have temperature and water-vapor enthalpies as unknowns in the energy equation. In such cases iterative solutions can be avoided by the use of eq. (11.25) or (11.25a).

**Cooling and Dehumidifying** Many air-conditioning processes involve cooling and dehumidification. This process is quite common during summer months. The cooling and dehumidification are accomplished by causing the air to flow over the evaporator of a refrigeration cycle or a chilled-water heat exchanger. Typically the air is cooled to a temperature below its dew point, thereby condensing water vapor from the air stream. The condensate formed on the chilled tubes is drained from the system, and the air stream leaving the cooling section usually is saturated. The temperature of the condensate has only a minor effect on the required refrigeration. For this reason it is common to assume the condensate temperature to be equal to that of the air leaving the cooling section. The air leaving the cooling section ordinarily is too cold to be discharged directly to the region being air-conditioned. For this reason it is common to have the cold air pass through a heating section. The heating frequently is accomplished by passing the air over the condenser tubes of the refrigeration unit. If additional reheating is required, a supplementary unit, such as a steam coil, can be used. The desired end effect of these processes is air at a lower relative humidity and a comfortable dry-bulb temperature.

Other methods of achieving the desired effect include mixing of the cold, dehumidified air with a stream of air from the conditioned enclosure. Also, energy economy is often achieved with heat-recovery systems where warm air, exhausted from the enclosed space, passes countercurrent through a heat exchanger with the chilled air of low humidity being warmed by the warmer air.

In this section we have addressed only the thermal aspects of some air-conditioning processes. Filtering, odor control, toxicity, and acoustics are some of the added considerations in a more complete air-conditioning design. Specific equations for the processes discussed have not been presented. The solution of each problem should start with the steady-flow energy equation and the several equations for psychrometric properties derived in Sections 11–9 to 11–12.

## Example 11.g

The air in a building is to be maintained during the summer at 75 F and a relative humidity of 50 percent. It is estimated that heat will be transferred to the air at the rate of 15,000 B/hr (conduction through the walls, people in the building, electric lighting, etc.) and that water vapor will be supplied at the rate of 6.00 lbm/hr (from people in the building). It is proposed that the desired conditions be maintained by withdrawing air at the rate of 1000 ft³/min and replacing it with air at a lower temperature and humidity ratio. What must be the temperature and relative humidity of the air supplied to the building? Assume the barometric pressure is 14.70 psia and the water vapor (from the people) is at 98 F.

**Solution:**

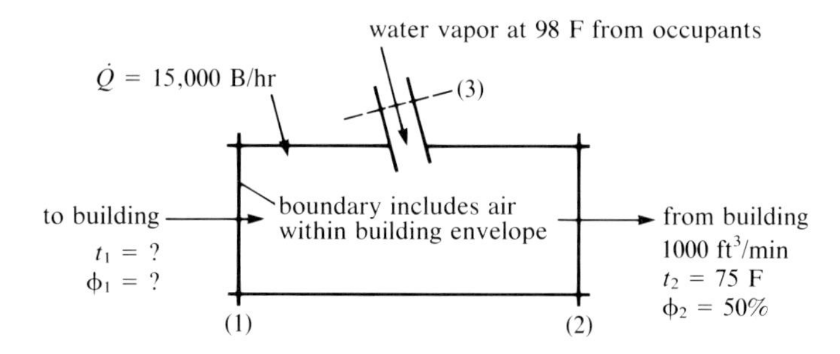


The process is shown schematically in the diagram.

The solution of this problem requires mass balances for dry air and $H_2O$ and an energy balance. First consider the air leaving the building. From the relative humidity $\phi_2$

$$p_{w_2} = (0.50)(0.4300) = 0.215 \text{ psia}$$

Therefore, the amount of water leaving the building is obtained using eq. (5.4)

$$p_{w_2}\dot{V}_2 = \dot{m}_{w_2}RT_2$$

or

$$\dot{m}_{w_2} = \frac{(0.215)(144)(1000)}{(85.8)(535)} = 0.6745 \text{ lbm/min}$$

For the dry air leaving

$$p_{da_2}\dot{V}_2 = \dot{m}_{da_2}R_{da}T_2$$

and

$$\dot{m}_{da_2} = \frac{(14.70 - 0.215)(144)(1000)}{(53.3)(535)} = 73.15 \text{ lbm/min}$$

For the water vapor

$$\dot{m}_{w_1} = \dot{m}_{w_2} - \dot{m}_{w_3} = 0.5745 - (6.00/60)$$
$$= 0.5745 \text{ lb vapor/min}$$

Therefore, the air entering the building at (1) must supply dry air at the rate of 73.15 lb da/min and water vapor at the rate of 0.5745 lb vapor/min to maintain the given conditions within the building. The energy equation written for the building air is

$$(\dot{m}h)_{w_3} + (\dot{m}h)_{da_1} + (\dot{m}h)_{w_1} + \dot{Q} = (\dot{m}h)_{da_2} + (\dot{m}h)_{w_2}$$

Rearranging and substituting, we obtain

$$\dot{Q} = \dot{m}_{da}(h_{da_2} - h_{da_1}) + (\dot{m}h_g)_{w_2} - (\dot{m}h_g)_{w_1} - (\dot{m}h_g)_{w_3}$$

$$\frac{15{,}000}{60} = (73.15)(0.24)(75 - t_1) + (0.6745)(1094.2)$$
$$- 0.5745\, h_{g_1} - (0.10)(1104.2)$$

Inspection of the above equation shows $t_1$ and $h_{g_1}$ to be unknowns. However, eq. (11.25) gives

$$h_{g_1} = 1061.2 + 0.43t_1$$

which can now be substituted into the steady-flow energy equation to yield

$$250.00 = 1316.7 - 17.554t_1 + 738.01 - 609.63$$
$$- 0.24702t_1 - 110.42$$

from which we get $t_1 = 61$ F. The relative humidity of the air supplied to the building is

$$\phi_1 = \frac{p_{w_1}}{p_{g_1}}$$

Hence $p_{w_1}$ must first be determined by

$$\omega_1 = \frac{\dot{m}_{w_1}}{\dot{m}_{da_1}} = \frac{0.5745}{73.15} = 0.622\left(\frac{p_{w_1}}{14.70 - p_{w_1}}\right)$$

from which

$$p_{w_1} = 0.1832 \text{ psia}$$

The relative humidity of the air supply at 61 F is then

$$\phi_1 = \frac{0.1832}{0.2655} = 69.04 \text{ percent}$$

## Example 11.h

To maintain the building air at 75 F and 50 percent relative humidity, as stated in Example 11.g, the air leaving could be cooled and dehumidified to 61 F and a relative humidity of 69.04 percent and returned to the building. The following illustration shows schematically the equipment that might be used to accomplish the desired result. Note that sections (1) and (3) in this diagram correspond respectively to sections (2) and (1) in Example 11.g. Calculate the required cooling load ${}_1\dot{Q}_2$ and the reheat load ${}_2\dot{Q}_3$ for the system, each in B/min. Assume the air at section (2) is saturated.

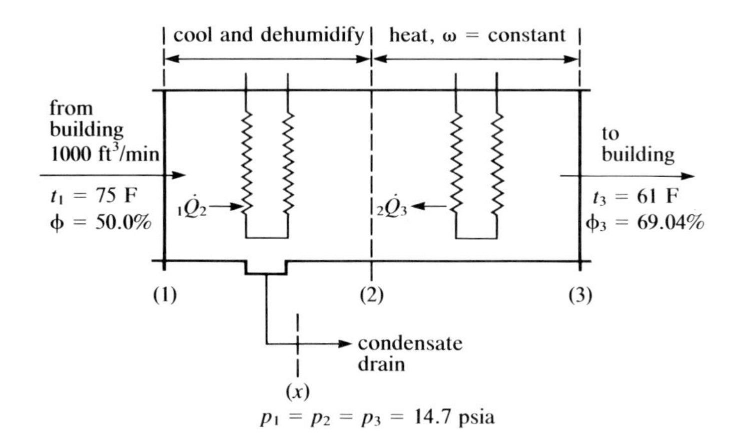


**Solution:**

A basis of 1 pound of dry air is selected. The humidity ratio does not change between sections (2) and (3). Thus partial pressures of the water vapor at sections (2) and (3) are equal. From the preceding example $p_{w_3}$ is equal to 0.1832 psia (section [1] of Example 11.g). The temperature of the saturated air at section (2) is the dew-point temperature for state (3), 51 F (closest value). The heating load, per pound of dry air, is given by

$$h_{da_2} + \omega_2 h_{g_2} + {}_2q_3 = h_{da_3} + \omega_3 h_{g_3}$$

with

$$\omega_3 = 0.622\left(\frac{p_{w_3}}{p - p_{w_3}}\right) = 0.622\left(\frac{0.1832}{14.517}\right)$$
$$= 0.007853 \text{ lbm H}_2\text{O/lbm da}$$

The heat transfer is

$$\begin{aligned} {}_2q_3 &= 0.24(61 - 51) + 0.007853\,(1088.1 - 1083.7) \\ &= 2.400 + 0.0346 = 2.4346 \text{ B/lbm da} \end{aligned}$$

Note the relative magnitude of the heat transfer to the dry air and the water vapor. Noting that ${}_2\dot{Q}_3 = \dot{m}_{da}({}_2q_3)$

$${}_2\dot{Q}_3 = (73.15)(2.4346) = 178.1 \text{ B/min}$$

For the cooling section

$$h_{da_1} + \omega_1 h_{g_1} + {}_1q_2 = h_{da_2} + \omega_2 h_{g_2} + (\omega_1 - \omega_2)h_{f_x}$$

Using information from Example 11.g for the air coming from the building

$$\omega_1 = \frac{0.6745}{73.15} = 0.009220 \text{ lbm H}_2\text{O/lbm da}$$

and per pound of dry air

$${}_1q_2 = (h_{da_2} - h_{da_1}) + \omega_2 h_{g_2} - \omega_1 h_{g_1} + (\omega_1 - \omega_2)h_{f_x}$$

The condensate temperature $t_x$ is assumed to equal $t_2$ unless another temperature is given. From the steady-flow equation

$$\begin{aligned} {}_1q_2 &= 0.24(51 - 75) + (0.007853)(1083.7) \\ &\quad - (0.009220)(1094.2) + (0.001367)(19.06) \\ &= -5.76 + 8.5106 - 10.091 + 0.02606 \\ &= -7.314 \text{ B/lbm da} \end{aligned}$$

The magnitude of the condensate term $(\omega_1 - \omega_2)h_{f_x}$, very small compared with the other terms, indicates that the solution is not sensitive to $t_x$. The cooling load is ${}_1\dot{Q}_2 = \dot{m}_{da}({}_1q_2)$ and

$${}_1\dot{Q}_2 = (73.15)(-7.314) = -535.0 \text{ B/min}$$

The refrigeration required is

$$\dot{Q}_{\text{refrig}} = \frac{535 \text{ B}}{\text{min}} \left| \frac{\text{ton min}}{200 \text{ B}} \right. = 2.68 \text{ tons}$$

## 11–15 Cooling Processes for Condenser Circulating Water

In Chapter 9 some vapor cycles for power and refrigeration were presented. For each of the cycles heat transfer from the working fluid takes place during flow through the condenser. In units of large capacity this heat transfer typically is from the working fluid to circulating water passing through the condenser.

Flow rates for condenser circulating water, particularly for vapor power cycles, are very large. For purposes of illustration consider a 300-MW unit with a cycle thermal efficiency of 38 percent and a 15 F temperature rise for the circulating water as it passes through the condenser. From the cycle thermal efficiency

$$\eta = \frac{W_{\text{net}}}{Q_{\text{in}}} = \frac{Q_{\text{in}} - Q_{\text{out}}}{Q_{\text{in}}} = 0.38$$

and the rate of heat transfer to the working fluid is

$$\dot{Q}_{\text{in}} = \frac{300{,}000 \text{ kW}}{} \left| \frac{3413 \text{ B}}{\text{kWhr}} \right| \frac{}{0.38}$$

$$= 2.694 \times 10^{9} \text{ B/hr}$$

By the first law

$$Q_{\text{out}} = (2.694)(10^{9})(0.62)$$
$$= 1.67 \times 10^{9} \text{ B/hr}$$

For the 15 F rise the circulating water temperature, the enthalpy increase would be approximately 15 B/lbm. Therefore

$$\dot{V} = \frac{1.67 \times 10^{9} \text{ B}}{\text{hr}} \left| \frac{\text{lbm}}{15 \text{ B}} \right| \frac{\text{ft}^3}{62.4 \text{ lbm}} \left| \frac{\text{hr}}{60 \text{ min}} \right.$$

$$= 29{,}740 \text{ ft}^3/\text{min} = 222{,}500 \text{ gal/min}$$

In the past a generating station of the given capacity would typically be located near a large body of water or a river and "once-through" cooling would have been used. Such cooling involves taking water from, say, a river, passing it through the condenser tubes, and discharging the heated water back into the river. With increased power generation and stringent ecologic and environmental standards imposed by state and federal agencies, once-through cooling is seldom permitted in new, high-capacity power stations.

Large ponds or canals where the water is cooled by evaporation and convection are one alternative to once-through cooling. Spray ponds, in which the return water is sprayed as droplets into the atmosphere to produce evaporative cooling, require significantly less area, particularly in regions where the atmosphere is quite arid. A third method to cool circulating water is the use of cooling towers, including evaporative induced- and natural-draft towers and, recently, dry towers that are closed water-to-air heat exchangers. An induced-draft

**Figure 11.5** Schematic Diagrams of Three Types of Cooling Towers. *a.* An induced draft evaporative cooling tower. *b.* A natural draft evaporative cooling tower. *c.* A dry cooling tower.

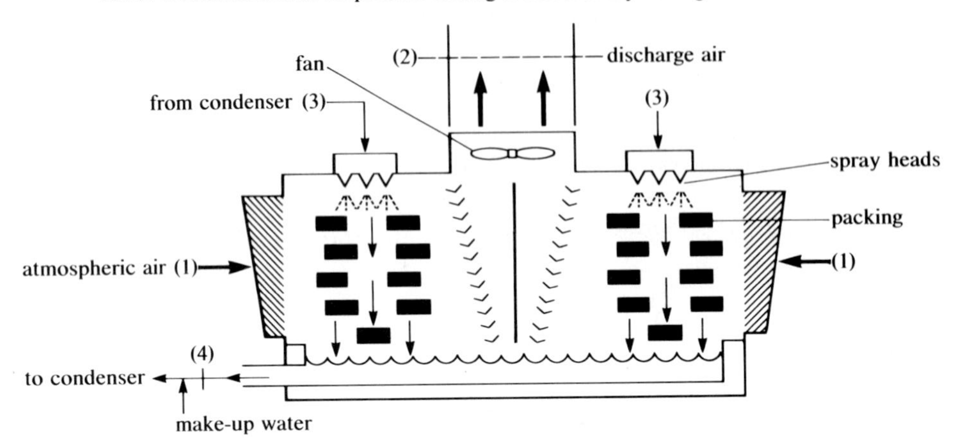


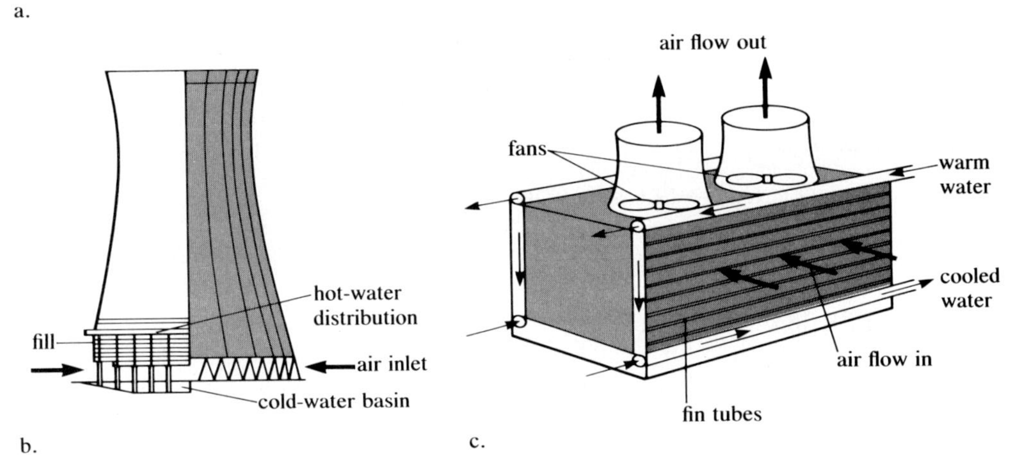


evaporative cooling tower is shown schematically in Figure 11.5a. Ambient atmospheric air enters at (1). Warm moist air ($\phi \cong 100$ percent) is discharged at (2). Warm water from the condenser of the power plant enters at (3) and the cooled water at (4) is mixed with makeup water and returned to the condenser. The makeup water flow rate is equal to the rate at which water evaporates in the tower. Towers of this type are constructed in cells, with the number of cells approximately proportional to the condenser circulating-water flow rate.

Figure 11.5b describes a natural-draft evaporative cooling tower. Natural-draft towers depend on buoyant effects to move the air through the tower. Fans are not used. Air warmed by the hot water in the lower part of the tower rises to create a natural draft through the tower. A typical height for such a tower is 300 feet. An inherent disadvantage of evaporative towers, both induced and natural draft, is that the discharge air is normally saturated and the temperature of the surrounding atmosphere may be well below the dew point temperature of the discharge air. These conditions can cause condensation of water vapor in the discharge air, with the condensate drifting and settling on land and buildings close to the towers. However, evaporative cooling towers are widely used not only in power plant applications but in other applications where a continuous supply of chilled water is required. Natural-draft towers are used in conjuction with large power plants, while induced-draft evaporative towers are employed over a wide range of cooling requirements.

Thermodynamic analysis of evaporative cooling towers is based on the steady-flow energy equation and mass balances for $H_2O$ and dry air. It is reasonable to assume that the tower boundary is adiabatic and that kinetic energies and work are negligible.

Figure 11.5c illustrates a dry cooling tower. Water is cooled inside finned tubes across which atmospheric air flows. Since the tower operates as a heat exchanger, water evaporation is not involved. Dry towers require more land area than other towers and, for steam power cycle applications, typically produce higher condenser pressures that result in lower cycle thermal efficiencies. Such towers are used in areas where adequate makeup water for evaporative cooling towers is not available.

## 11–16 Summary

The pressure exhibited by a mixture of perfect gases at an equilibrium state is the sum of the partial pressures of the constituents. For a mixture of $N$ constituents

$$p_m = \sum_{i=1}^{N} p_i$$

The partial pressure of a constituent is the pressure that constituent would exhibit if it existed alone at the volume and temperature of the mixture. The partial pressure of constituent $i$ is related to the mixture pressure by

$$p_i = x_i p_m$$

where $x_i = n_i/n_m$ is defined as the mole fraction of constituent $i$.

The volume $V'$ constituent $i$ exhibits when it exists alone at the mixture pressure and temperature is its partial volume. The ratio of $V'$ to the mixture volume is defined as the volume fraction and

$$\frac{V'_i}{V_m} = x_i$$

The equation of state of a mixture of perfect gases on the basis on 1 mole of mixture is

$$p_n \bar{v}_m = \overline{R} T_m$$

where $\overline{R}$ is the universal gas constant. The following expressions apply to a mole of mixture of $N$ perfect gases with constant specific heats.

$$M_m = \sum_{i=1}^{N} x_i M_i, \qquad \overline{C}_{p_m} = \sum_{i=1}^{N} x_i \overline{C}_{p_i}, \qquad \overline{C}_{p_m} - \overline{C}_{v_m} = \overline{R}$$

$$(\overline{u}_2 - \overline{u}_1)_m = \overline{C}_{v_m}(T_2 - T_1)$$

$$(\overline{h}_2 - \overline{h}_1)_m = \overline{C}_{p_m}(T_2 - T_1),$$

$$(\overline{s}_2 - \overline{s}_1)_m = \overline{C}_{p_m} \ln\left(\frac{T_2}{T_1}\right)_m - \overline{R} \ln\left(\frac{p_2}{p_1}\right)_m$$

The equation for entropy is applicable to a single constituent $i$ when the subscript $m$ is replaced by the subscript $i$. Isentropic processes for a gas mixture with constant specific heats is described by $pv^{\gamma}$ = constant, provided $\gamma = \overline{C}_{p_m}/\overline{C}_{v_m}$.

When a unit mass of a mixture of perfect gases with constant specific heats is considered, the following relationships apply. $m_f$ is the mass fraction.

$$C_{p_m} = \sum_{i=1}^{N} m_{f_i} C_{p_i}, \quad C_{p_m} - C_{v_m} = R_m = \sum_{i=1}^{N} m_{f_i} R_i, \quad M_m = \frac{\overline{R}}{R_m}$$

In atmospheric air (dry air plus water vapor) the water vapor as well as the dry air can be treated as a perfect gas. The relative humidity $\phi$ for atmospheric air is defined as $p_w/p_g$, where $p_w$ is the partial pressure of the water vapor and $p_g$ is the vapor saturation pressure corresponding to the air temperature. The humidity ratio $\omega$ is defined as the ratio of the mass of the water vapor to the mass of dry air and is given in terms of pressures as

$$\omega = 0.622 \frac{p_w}{p - p_w}$$

Three temperatures are used to describe atmospheric air. The dry-bulb temperature is that measured by a conventional thermometer. The dew-point temperature is the saturation pressure corresponding to the partial pressure of the water vapor. The wet-bulb temperature is that measured when atmospheric air is passed over a thermometer, the bulb of which is covered with a wick soaked with water. A psychrometric chart is a convenient means of displaying the various properties and processes for atmospheric air.

## Selected References

Holman, J. P. *Thermodynamics,* 2nd ed. New York: McGraw-Hill, 1974.
Reynolds, W. C., and H. C. Perkins. *Engineering Thermodynamics.* New York: McGraw-Hill, 1970.
Stoecker, W. F. *Refrigeration and Air Conditioning.* New York: McGraw-Hill, 1958.
Stoever, H. J. *Engineering Thermodynamics.* New York: Wiley, 1951.
Threlkeld, J. L. *Thermal Environmental Engineering,* 2nd ed. Englewood Cliffs, N.J.: Prentice-Hall, 1970.
Wark, K. *Thermodynamics,* 3rd ed. New York: McGraw-Hill, 1977.

## Problems

11.1 A mixture of several perfect gases is at 93 C. The partial pressure of gas $A$ in the mixture is 40 kPa. Calculate the volume of 1 kmol of the mixture if the mole fraction of gas $A$ is 0.254.

11.2 An ideal gas mixture consisting of 2.81 lbm of $N_2$, 0.537 lbm of $H_2$, and 4.72 lbm of $O_2$ is at 537 R and 20.7 psia. Calculate (a) the pound moles of each gas, (b) the partial pressure of each gas, and (c) the volume occupied by the mixture.

11.3 A mixture of 1.25 kmol of $O_2$ and 0.342 kmol of $H_2$ is at 0 C and 101.3 kPa. Determine the density of the mixture, kg/m$^3$.

11.4 A stream of $N_2$ at 15 psia, 120 F, flows at the rate of 300 ft$^3$/min. A stream of $O_2$ at 20 psia and 200 F flows at the rate of 50 lbm/min. The two streams mix, yielding a single stream that leaves the mixing chamber at 17 psia and 130 F. Determine (a) the mass rate of flow, lbm/min., of the leaving stream, (b) the mole fraction of each gas in the leaving stream, and (c) the volume rate of flow of the leaving stream.

11.5 A gas mixture consists of 0.703 kg of argon and 1.238 kg of nitrogen. The mixture, initially 101.3 kPa and 0 C, is heated at constant volume in a closed system until its pressure is 345 kPa. Determine (a) the mixture volume at the second state and (b) the partial pressure of each constituent at the second state. Compare the ratio of the partial pressure at state (2) with that at state (1) for each constituent.

11.6 A mixture of 1.50 lbmol of $O_2$ and 3.75 lbmol of $N_2$ is at 20 psia and 100 F. Determine the density, lbm/ft$^3$, of each gas and the density of the mixture. Is the mixture density equal to the sum of the constituent densities?

11.7 A constant volume vessel is to contain a mixture of argon and helium at 100 psia and 70 F. The vessel has a volume of 4.967 ft$^3$. Helium is first introduced into the evacuated vessel, after which the pressure is 40 psia and the temperature is 90 F. Determine the mass of argon that must be added.

11.8 A mixture at 105 kPa and 23 C consists of 25.66 percent $N_2$, 53.22 percent $H_2$, and 20.42 percent $CO_2$ by volume. Calculate the partial pressure of each gas, the density of the mixture, and the apparent molecular weight of the mixture.

11.9 Twenty pounds of $N_2$, 12 pounds of $O_2$, and 6 pounds of He make up a gas mixture that is at 530 R. The partial pressure of the helium is 6.25 psia. Calculate (a) the volumetric analysis of the mixture, (b) the apparent molecular weight of the mixture, and (c) the volume occupied by the mixture.

11.10 The exhaust from a gas turbine is at 1400 F and 15 psia and has the volumetric analysis: 3.27 percent $CO_2$, 3.27 $H_2O$, 77.71 percent $N_2$, and 15.75 percent $O_2$. Calculate the density of the exhaust gas mixture and the partial pressure of the $H_2O$ in the mixture. Is the assumption of ideal gas behavior for the $H_2O$ reasonable?

11.11 A gas mixture, consisting of $CO_2$, $N_2$, and He is at 600 kPa and 100 C. The partial pressure of the $CO_2$ is 65 kPa. The part by volume of $N_2$ is 0.270. Calculate (a) the volumetric analysis, (b) the molal specific heats of the mixture, and (c) the volume of mixture to contain 3.2 kg of $N_2$.

11.12 Ten pounds of a gas mixture initially at 20 psia and 80 F undergo a constant-volume closed system process to a final temperature of 250 F. The volumetric analysis of the mixture is 25 percent $N_2$, 55 percent $H_2$, and 20 percent $CO_2$. Calculate the heat transfer for the process, the volume of the mixture, and the partial pressure of the $CO_2$ at the final state.

11.13 Suppose a mixture containing 25.66 percent $N_2$, 53.92 percent $H_2$, and 20.42 percent carbon dioxide by volume is at 110 kPa and 24 C. A closed system of the mixture undergoes a reversible, constant-pressure process during which the volume is doubled. Calculate the work and heat transfer per kilomole and per kilogram of the mixture.

11.14 Suppose the stack gases from a furnace are used to heat air being supplied to the furnace. The stack gas mixture enters the steady-flow heat exchanger at 470 F and 14.0 psia at the rate of 30,000 $ft^3/min$ and leaves at 315 F. Air enters the heat exchanger at 60 F and 15 psia at the rate of 17,000 $ft^3/min$. The stack gas analysis, by volume, is 79 percent $N_2$, 10.3 percent $CO_2$, 6.1 percent $O_2$, and 4.6 percent $H_2O$. Assume all heat transferred from the stack gases goes to the air, kinetic energies are negligible, and the $H_2O$ behaves as a perfect gas. Calculate the temperature of the air leaving the unit.

11.15 A rigid, insulated container is divided into two volumes by a partition. In one part 0.907 kg of $O_2$ is at 104 kPa and 66 C. The other side, with a volume of 0.650 $m^3$, contains $N_2$ at 275 kPa and 22 C. The partition is ruptured, and two gases mix, eventually coming to equilibrium. Determine the volumetric analysis, the pressure, and the temperature of the equilibrium mixture.

11.16 A stream of $O_2$ and a stream of $N_2$ mix adiabatically during a steady-flow process. The entering oxygen flows at the rate of 300 $ft^3/min$ at 125 F and 15 psia with a velocity of 150 ft/sec. The nitrogen enters at the rate of 700 $ft^3/min$ at 80 F and 15 psia through a tube 0.375 ft in diameter. The leaving mixture is at 15 psia, velocity 100 ft/sec. Determine (a) the temperature of the leaving mixture, (b) the volume rate of flow of the mixture, and (c) the required area for the leaving stream.

11.17 Consider a rigid, perfectly insulated container divided into $N$ sections by partitions. Suppose each section contains a perfect gas at the same pressure $p_1$ and temperature $T_1$. The partitions are broken and the gases mix, coming to a final equilibrium state at $p_2$ and $T_2$. Is the final temperature dependent on the amount of each gas? What would be the internal energy and enthalpy changes for the mixing process? Would the answers to the preceding questions be different if variable, rather than constant, specific heats were used?

11.18 A perfect gas mixture of 10 percent helium and 90 percent nitrogen by mass undergoes a process from 500 R and 15 psia to 2000 R and 75 psia. Calculate (a) the enthalpy change per unit mass of mixture, (b) the entropy change per unit mass of mixture, and (c) the entropy change of each constituent per unit mass of mixture.

11.19 Suppose a mixture containing 30 percent $N_2$ and 70 percent $O_2$ by volume is compressed reversibly and adiabatically from 15 psia, 77 F, to 90 psia. For a closed system of 1 pound mass of the mixture, calculate the entropy change of each gas per unit mass of mixture and work for the process.

11.20 Two hundred cubic feet per minute of a gas mixture consisting of 85 percent $H_2$ and 15 percent $CO_2$ by volume enter a compressor at 14.7 psia and 60 F. The mixture is compressed isentropically and leaves the steady-flow compressor at 75 psia. Neglecting kinetic energies, calculate (a) the horsepower to drive the compressor and (b) the entropy change of each gas, per pound mass of mixture and per pound mole of mixture.

11.21 Suppose a closed system of a gas mixture is compressed polytropically ($n = 1.327$) from 105 kPa and 16 C to 620 kPa. The mixture consists of 50 percent helium, 20 percent carbon dioxide, and 30 percent nitrogen by volume. For 1 kg of the mixture, determine (a) the heat transfer and (b) the entropy change during the process.

11.22 A gas turbine is supplied with a mixture of 77 percent $N_2$, 12 percent $O_2$, and 11 percent $H_2O$ by volume. The mixture enters the turbine at 2000 R and 140 psia at the rate of 10,000 $ft^3$/min and velocity 500 ft/sec. The mixture expands isentropically through the turbine, exhausting at 37 psia at a velocity of 800 ft/sec. Calculate the power in kilowatts delivered by the turbine.

11.23 An isentropic nozzle is supplied with a gas mixture of 60 percent $N_2$ and 40 percent $O_2$ by volume. At the nozzle inlet the pressure is 690 kPa, the temperature 260 C, and the velocity negligible. If the mixture leaves the nozzle at 105 kPa, what is the mass rate of flow per unit area at the nozzle outlet?

11.24 Atmospheric air at 80 F and 14.23 psia has a relative humidity of 53 percent. Determine the humidity ratio of the air and the minimum temperature to which the air could be cooled without condensing out any water.

11.25 1000 $ft^3$ of atmospheric air at 75 F, 14.4 psia, with a dew point of 60 F, state (1), is cooled at constant pressure in a closed system to 50 F, state (2). Determine (a) the mass of the water condensed out, (b) the volume of the water condensed out, (c) the partial pressure of the water vapor in the air at (2), and (d) the volume of the atmospheric air at (2).

11.26 Air at 78 F and 14.696 psia has a humidity ratio of 0.01228 lbm $H_2O$ per lbm dry air. Calculate (a) the dew point, (b) the relative humidity, and (c) the density of the atmospheric air.

11.27 Occasionally reference is made to the atmosphere being "heavy" when the relative humidity is high. For 88 F dry-bulb temperature and 14.696 psia barometric pressure, calculate the density of atmospheric air having (a) 90 percent relative humidity and (b) 40 percent relative humidity.

11.28 The air in a room is at 14.32 psia and 80 F dry bulb. An uninsulated cold-water line running through the room has an outside surface temperature of 57 F. What maximum relative humidity would be allowable if condensation on the pipe surface were to be avoided?

11.29 In some locales windows will "frost" during winter months even though the room temperature is of the order of 70 F. How can the frost formation be explained?

11.30 Explain the natural disappearance of frost from a surface outdoors in the winter when the temperature is below freezing.

11.31 A small pressure cooker is equipped with a safety valve that opens when the pressure reaches 15 psi gage. Suppose the cooker initially contains liquid $H_2O$ at 70 F and saturated air at 70 F and 14.50 psia. Assuming the volume of liquid $H_2O$ remains constant, what will be the temperature in the cooker when the safety valve first opens?

11.32 Suppose atmospheric air is at $-10$ F, 14.696 psia, and has a relative humidity of 50 percent. If the air is heated to 68 F, what will be its relative humidity?

11.33 Air at 30 C and relative humidity of 65 percent is cooled at constant pressure of 101.3 kPa to 10 C. How much water is condensed per kilogram of dry air?

11.34 The following processes involving atmospheric air initially at 70 F and 75 percent relative humidity take place in a steady-flow system at 14.7 psia. Sketch each process on a psychrometric chart. (a) Simple heating to 100 F, (b) cooling to 55 F and reheating to 65 F, and (c) adiabatic saturation.

11.35 A stream of atmospheric air at 90 F dry bulb undergoes a steady-flow, adiabatic process during which no work is done and velocities are negligible. Liquid $H_2O$ at 55 F is sprayed into the air, producing a leaving stream that is at 70 F and 80 percent relative humidity. During the process the pressure remains constant at 14.70 psia. Calculate (a) the dew point of the leaving air stream, (b) the spray water evaporated per pound of dry air, and (c) the relative humidity of the entering air.

11.36 In hot, arid locations air can be cooled by evaporating water. Suppose air at 95 F and 20 percent relative humidity flows through a chamber in which liquid $H_2O$ at 58 F is sprayed into the air. Assuming the steady-flow process takes place adiabatically and kinetic energies are negligible, what minimum temperature of the leaving air stream could be achieved?

11.37 On a certain day the U.S. Weather Service reports a barometer reading of 14.20 psia, a dry bulb of 88 F, and a wet bulb of 68 F. On the basis of these data *calculate* (do not use a psychrometric chart) the following:

(a) Humidity ratio (derive necessary equation)
(b) Dew point
(c) Relative humidity
(d) Specific volume of the atmospheric air
(e) Pounds of dry air per 1000 $ft^3$ of mixture

11.38 At an altitude of 7500 feet above sea level it is observed that the dry bulb is 88 F, the wet bulb 68 F, and the atmospheric pressure is 11.13 psia. For this psychrometric state, calculate (a) the humidity ratio, (b) the relative humidity, (c) the dew point, and (d) the specific volume of the atmospheric air. Compare the results with those obtained for the preceding problem and with those read from the psychrometric chart in Appendix F. Note the errors introduced when a standard barometric pressure psychrometric chart is used at pressures other than 14.696 psia.

11.39 Air at 10 F, relative humidity 47 percent, is heated and humidified to 78 F and 40 percent relative humidity during a steady-flow process that takes place at a constant pressure of 14.696 psia. The flow rate of the inlet air is 700 $ft^3$/min. Liquid $H_2O$ for humidification is supplied at 60 F. Calculate (a) the liquid $H_2O$ evaporated per hour and (b) the heat transfer, B/hr.

11.40 One thousand cubic feet per minute of air at 85 F, 70 percent relative humidity, undergoes a steady-flow cooling and dehumidification process at 14.696 psia. The air leaves the unit at 50 F. Calculate (a) the heat transfer, B/hr, (b) the percent of the heat transfer required to cool the dry air, and (c) the volume rate of flow leaving the unit.

11.41 Three streams enter a chamber where they mix and heat is transferred to the system. Stream $A$ is liquid $H_2O$ at 55 F; stream $B$ is air at 60 F, 43 F dew point flowing at the rate of 1000 $ft^3$/min; and stream $C$ is dry air at 20 F flowing at the rate of 50 lbm/min. One stream at 120 F and 85 percent relative humidity leaves the unit. All streams are at 14.00 psia. Calculate (a) the flow rate of stream $A$, lbm/hr, and (b) the heat transfer, B/hr.

11.42 A stream of air at 14.50 psia and 65 F is flowing at the rate of 500 $ft^3$/min. A second stream at 14.50 psia and 125 F is flowing at the rate of 300 $ft^3$/min. The two streams are mixed adiabatically to form a third stream at 14.50 psia. If the relative humidity of the first stream is 40 percent and that of the second is 15 percent, what are the temperature and relative humidity of the third stream?

11.43 The relative humidity of a stream of air at 85 F is to be changed from 60 to 30 percent by first cooling the stream to condense out part of the water vapor and then heating the stream back to 85 F. If the stream enters the dehumidifying apparatus at the rate of 1000 $ft^3$/min, and if it leaves the cooler saturated, (a) at what rate must heat be transferred from the stream in the cooler and (b) at what rate must heat be transferred to the stream in the heater? Assume that the processes take place at a pressure of 14.70 psia.

11.44 During the winter the air within a building is to be maintained at a temperature of 70 F and a relative humidity of 40 percent. Because the building is not absolutely airtight, air will be lost continuously and will be replaced by air from the outside. If air is lost at the rate of 300 $ft^3$/min, and if the outdoor air is at a temperature of 0 F and has a relative humidity of 50 percent, at what rates must heat and water vapor be supplied to the building to compensate for the heat and humidity lost because of the infiltration of the outdoor air? Assume that the supply of water vapor is obtained by supplying liquid $H_2O$ initially at 60 F, and assume that the barometric pressure is 14.70 psia.

11.45 Air is cooled and dehumidified in the apparatus shown. The cooling is produced by a simple vapor-compression cycle using ammonia as the refrigerant. The refrigeration cycle operates with saturated vapor at 0 F entering the compressor and saturated liquid at 86 F leaving the condenser. There is no heat transfer across the walls of the dehumidification chamber, that is, all heat transferred from the air goes to the refrigerant. Assuming the refrigeration cycle to operate ideally, calculate (a) the condensate, lbm/hr; (b) the heat transferred from the air, B/min; (c) the tons of refrigerating capacity required; (d) the coefficient of performance of the refrigeration cycle; and (e) the compressor power, HP.

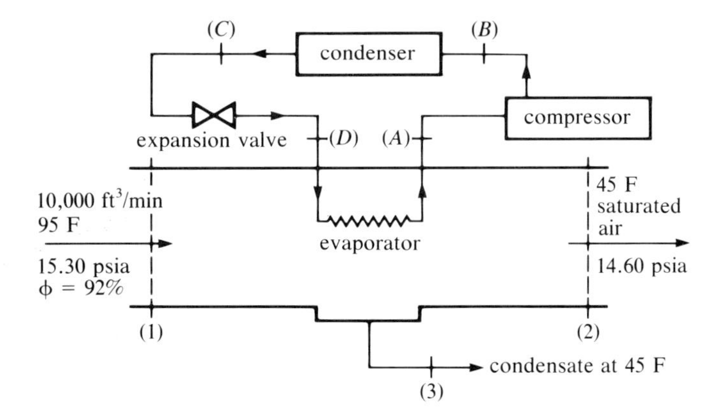

11.46 A simple-cycle, vapor-compression heat pump uses F-12 as the refrigerant and the atmosphere as the heat source. Refrigerant leaves the evaporator as saturated vapor at 10 F and enters the expansion valve as saturated liquid at 120 F. Air at 65 F, relative humidity 40 percent, is blown across the condenser and all heat transferred from the F-12 goes to this air, producing a 30 F rise in the dry-bulb temperature. The 3-HP compressor operates isentropically. Sketch a schematic diagram for the system and determine the volume rate of flow of the 65 F air approaching the condenser. Assume the pressure of the air remains constant at 14.23 psia.

11.47 Steam enters the condenser in a power plant at 1.50 psia, 90 percent quality, at the rate of 800,000 lbm/hr. Condensate leaves the condenser at 112 F. The circulating water leaves the condenser at 105 F and enters a cooling tower at the same temperature. Atmospheric air at 75 F, 46 percent relative humidity, enters the tower. The air discharged from the tower is saturated at 100 F. The pressure is assumed to be constant and equal to 14.13 psia. Assuming the circulating water leaving the tower is at 85 F, calculate (a) the mass rate of flow of the circulating water passing through the condenser, (b) the circulating water makeup, lbm/hr, and (c) the volume rate of flow of the atmospheric air entering the tower.

11.48 A simple vapor-compression refrigeration cycle with ammonia as the refrigerant has a capacity of 30 tons. Circulating water for the condenser is supplied from a cooling tower. The refrigeration cycle has saturated vapor at 5 F leaving the evaporator, liquid at 90 F and 200 psia leaving the condenser, and the compressor operating polytropically with $n = 1.34$. Atmospheric air at 65 F, relative humidity 40 percent, enters the tower and the air leaving the tower is saturated at 85 F. Assume a constant pressure of 14.70 psia for the tower. If the circulating water enters the condenser at 77 F and undergoes a 10 F rise as it passes through the condenser, what are (a) the circulating water rate of flow through the condenser and (b) the volume rate of flow of the saturated air leaving the cooling tower?

# 12

# Combustion Process Fundamentals

*In a broad sense combustion is a rapid oxidation process during which a fuel and an oxidizer undergo chemical reactions. Although fossil hydrocarbons (petroleum products, natural gas, and coal) are the most common fuels, the basic concepts presented in this chapter are applicable to fuels ranging from bagasse (sugar cane by-products) and black liquor (paper process by-product) to solid waste fired in a steam generator of an electric generating station, distillates used for vehicle fuels, and liquid hydrogen employed in rocket propulsion systems. Atmospheric air is by far the most common oxidant. However, other oxidizers, such as liquid oxygen and nitric acid, are used in specialized applications.*

*Perhaps the most common word associated with combustion is flame, the incandescent product of a combustion reaction. Combustion, like heat transfer, is a rate process, and hence the product composition is time dependent. One common example of this time dependency is encountered in the spark-ignition internal combustion engine where the time available for combustion is of the order of milliseconds. Expansion during the power stroke results in a rapid decrease in temperature of the combustion gases. As temperatures decrease, reaction rates decrease. Consequently the composition of the combustion products is "frozen" at temperatures of the order of 3000 R. The result of this particular time–temperature "squeeze" is evidenced in the carbon monoxide and unburned hydrocarbons present in the engine exhaust gases. Similar effects occur in gas turbines and jet engines, however, the formation of undesirable products in such units is related to the high excess air and the formation of oxides of nitrogen ($NO_x$). Large fossil-fuel-fired generating stations have not escaped the stigma of undesirable effluents. Particulates, sulfur oxides, and oxides of nitrogen, $NO_x$, are problems for such stations. The sulfur oxides produced during the burning of sulfur-bearing coal present one of the major challenges to current power station technology.*

## 12–1 Complete Combustion

Complete combustion takes place when each combustible element in the fuel is completely oxidized. For fuels consisting of carbon and hydrogen only, complete combustion is established when all carbon burns to carbon dioxide and all hydrogen burns to form water. These reactions are written as

$$C + O_2 \rightarrow CO_2, \quad H_2 + \frac{1}{2}O_2 \rightarrow H_2O \tag{12.1}$$

The achievement of complete combustion is, in practice, dependent on satisfying four conditions:

1. The fuel and oxidizer must be adequately mixed ($M$).
2. Sufficient oxidizer, usually air, must be present to permit each fuel element to burn to its most stable state ($A$).
3. The fuel and oxidizer must have adequate reaction time to reach the complete combustion state ($T$).
4. The temperatures during the combustion process must be such that complete combustion is achieved ($T$).

The *MATT* conditions described represent the basic requisites for complete combustion. The equilibrium composition of the products of combustion for given reactants can be predicted by application of second-law concepts.* The composition will depend on the temperature and pressure of the products. For example, a high temperature and a low pressure would result in an equilibrium product composition for a hydrocarbon fuel that would include species other than $CO_2$ and $H_2O$. However, these species are usually present in relatively small amounts and will not be considered in this introductory treatment of combustion.

In Chapter 11 the composition of dry air and the concept of "atmospheric nitrogen" were introduced. To the accuracy necessary for most engineering calculations, dry air was divided into two constituents—oxygen (21 percent by volume) and atmospheric nitrogen (79 percent by volume). For this breakdown the atmospheric nitrogen was found to have an apparent molecular weight of 28.2 while that for dry air was rounded to 29. We will use these approximations as we consider the air required for complete combustion. *Theoretical air* (stoichiometric air) is the exact amount of air required for complete combustion of a fuel, and is determined from the analysis of the fuel. The chemical analysis of a solid or liquid fuel is typically given as a gravimetric analysis, that is, by mass fraction. Analyses of gaseous fuels are reported on a volumetric basis (mole fraction). As an illustration let us compute the theoretical amount of air required for complete combustion of 1 mole of carbon monoxide, CO. We will first determine the amount of oxygen required. Complete combustion of CO with $O_2$ is written as

$$CO + \frac{1}{2}O_2 \rightarrow CO_2$$

Thus 1 mole of CO reacting with 0.5 mole of $O_2$ produces 1 mole of $CO_2$. To determine the dry air required, we will use the fact that there is 0.21 mole of oxygen per mole of dry air. Thus

$$TA = \frac{0.5 \text{ mole } O_2}{\text{mole CO}} \left| \frac{\text{mole da}}{0.21 \text{ mole } O_2} \right. = 2.38 \frac{\text{moles da}}{\text{mole CO}}$$

*Such analysis involves the entropy principle and requires knowledge of the third law of thermodynamics and its consequence, absolute entropy. This topic is not addressed in this book.

For complete combustion of 1 mole of CO with dry air we write

$$CO + \frac{1}{2}O_2 + 1.88N_2 \rightarrow CO_2 + 1.88N_2$$

where the number of moles of atmospheric nitrogen has been computed as $0.79(TA)$. The nitrogen is assumed to be inert and therefore does not enter into the chemical reaction.

In many combustion processes the actual air supplied exceeds the theoretical requirement. For such cases the excess air is defined as

$$\text{Excess air} = \frac{S - TA}{TA} \tag{12.2}$$

where $S$ represents the actual dry air supplied.

The *material balance* for a combustion process is a tabulation of the total mass of all reactants (input) and the total mass of all products (output). Mass must, of course, be conserved in any combustion process. This is satisfied by requiring that all atomic species be conserved. For example, all carbon atoms in the fuel must appear in the products. Before a material balance can be prepared, a basis must be chosen. The basis is usually dictated by the information given. For cases involving complete combustion of a known fuel, a basis of a mole or a unit mass of fuel is usually chosen. A convenient aid in preparing a material balance for complex combustion processes is the use of a tabular form in which the chemical species are broken down into their elements. The "bookkeeping" nature of a material balance is illustrated by the following example.

## Example 12.a

A fuel-gas mixture at 20 psia and 300 F has the volumetric analysis

| | |
|---|---|
| $CH_4$ | 20.0% |
| $C_2H_6$ | 40.0% |
| $N_2$ | 30.0% |
| $H_2O$ | 10.0% |
| | 100.0% |

Determine:

(a) The material balance when the fuel is burned completely with 15 percent excess air

(b) The volume of dry air at 15 psia and 100 F supplied per cubic foot of fuel

(c) The dew point of the products of combustion if the products are at 15 psia and 300 F

**Solution:**

(a) We assume that each constituent in the fuel exhibits perfect gas behavior. The basis chosen for the solution is 1 lb mol of fuel. The table lists the number of pound moles of each chemical element in the fuel per pound mole of fuel. For a material balance, the substances may be listed in any convenient form. For example, the table shows the number of pound moles of $H_2$ in each fuel constituent, rather than the number of pound moles of atomic hydrogen. A typical value in the table is the number of pound moles of $H_2$ in the ethane ($C_2H_6$). This value was determined as follows:

$$n_{H_2} \text{ in } C_2H_6 = \frac{3 \text{ lb mol } H_2}{\text{lb mol } C_2H_6} \left| \frac{0.400 \text{ lb mol } C_2H_6}{\text{lb mol fuel}} \right.$$

$$= 1.200 \frac{\text{lb mol } H_2}{\text{lb mol fuel}}$$

The summation of the $n_{H_2}$ column, 1.700, is the total number of pound moles of $H_2$ per pound mole of the fuel-gas mixture. The numbers for the other elements are interpreted in a similar manner. The $m_i$ values were determined from the product of $x_iM_i$, and according to eq. (11.6), the sum of the $m_i$ values is the apparent molecular weight of the fuel.

| Gas | $x_i$ | $M_i$ | $m_i$, lbm | $n_C$, lb mol | $n_{H_2}$, lb mol | $n_{O_2}$, lb mol | $n_{N_2}$, lb mol |
|---|---|---|---|---|---|---|---|
| $CH_4$ | 0.200 | 16 | 3.20 | 0.200 | 0.400 | — | — |
| $C_2H_6$ | 0.400 | 30 | 12.00 | 0.800 | 1.200 | — | — |
| $N_2$ | 0.300 | 28 | 8.40 | — | — | — | 0.300 |
| $H_2O$ | 0.100 | 18 | 1.80 | — | 0.100 | 0.050 | — |
| | 1.000 | | 25.40 | 1.000 | 1.700 | 0.050 | 0.300 |

The theoretical air required is

$$TA = \frac{n_{O_2} \text{ required}}{\text{lb mol fuel}} \left| \frac{\text{lb mol da}}{0.21 \text{ lb mol } O_2} \right.$$

The number of pound moles of $O_2$ required to burn the hydrogen and the carbon is determined by use of eq. (12.1). We must subtract the $O_2$ in the fuel. The oxygen requirement is

1.00 lb mol $O_2$ for complete combustion of 1.00 lb mol C

$\frac{1.700}{2}$ lb mol $O_2$ for complete combustion of 1.700 lb mol $H_2$

$-0.050$ lb mol $O_2$ present in each lb mol of fuel

1.80 lb mol $O_2$ required

Therefore

$$TA = \frac{1.80}{0.21} = 8.571 \text{ lb mol da/lb mol fuel}$$

With 15 percent excess air, the dry air supplied is, by eq. (12.2),

$$\begin{aligned} S &= (1.15)(8.571) = 9.857 \text{ lb mol da/lb mol fuel} \\ &= 9.857(29) = 285.86 \text{ lbm da/lb mol fuel} \end{aligned}$$

The atmospheric nitrogen is determined from the actual air supplied as

$$m_{N_2,\,\text{atmos}} = (0.79)(9.857)(28.2) = 219.7 \text{ lbm/lb mol fuel}$$

The excess oxygen, which appears in the products as $O_2$, can be determined from the air supplied as

$$\begin{aligned} m_{O_2,\,\text{excess}} &= (0.21)(1.15 - 1.0)(8.571)(32) \\ &= 8.64 \text{ lbm } O_2\text{/lb mol fuel} \end{aligned}$$

The material balance per lb mol of fuel is:

| **Input, lbm** | | **Output, lbm** | | |
|---|---|---|---|---|
| fuel | = 25.40 | carbon dioxide ($CO_2$) | = (1.00)(44) | = 44.0 |
| dry air | = 285.86 | water ($H_2O$) | = (1.70)(18) | = 30.6 |
| Total | = 311.3 lbm | atmospheric nitrogen | | = 219.7 |
| | | nitrogen ($N_2$) in fuel | = (0.03)(28) | = 8.4 |
| | | oxygen ($O_2$) | | = 8.64 |
| | | Total | | = 311.3 lbm |

(b) The volume of dry air supplied per cubic foot of fuel at the given pressures and temperatures is

$$\frac{V_{\text{da}}}{V_{\text{fuel}}} = \frac{n_{\text{da}}}{n_{\text{fuel}}} \cdot \frac{\overline{R}T_{\text{da}}}{\overline{R}T_{\text{fuel}}} \cdot \frac{p_{\text{fuel}}}{p_{\text{da}}}$$

$$= \frac{9.857}{1.000} \cdot \frac{560}{760} \cdot \frac{20}{15} = 9.684 \text{ ft}^3 \text{ da/ft}^3 \text{ fuel}$$

(c) The dew point of the products of combustion is the saturation temperature corresponding to the partial pressure of the water vapor in the products. The partial pressure of the water vapor is given by eq. (11.4) as the mole fraction of the water vapor multiplied by the pressure of the products. Thus

$$\begin{aligned} p_w &= x_w p_{\text{pr}} = \left(n_w \Big/ \sum_{i=1}^{N} n_i\right) p_{\text{pr}} \\ &= \frac{1.7}{1 + 1.7 + (219.7/28.2) + 0.03 + (8.64/32)}(15) \\ &= 2.364 \text{ psia} \end{aligned}$$

The corresponding saturation temperature from the *Steam Tables* is 132 F. If the products are cooled to a temperature lower than the dew-point temperature, 132 F, condensation of water vapor will occur.

## 12–2 Incomplete Combustion of Hydrocarbon Fuels

Many combustion processes do not reach an end state where combustion is complete. For some of these reactions the material balance can be established if the fuel analysis and the products of combustion analysis are determined by experiment. These analyses, subject to experimental error, are carried out using a variety of experimental techniques. Fuel composition is determined via methods suitable to the nature of the fuel. A fuel-gas analysis, for example, may be determined using a gas chromatograph. Other fuels may be analyzed using wet chemistry methods. The composition of gaseous combustion products may be reported on either a "wet" or a "dry" basis. A dry analysis does not include water vapor, whereas a wet analysis does. A common instrument used to determine the composition of gaseous products of combustion is the Orsat analyzer. A standard Orsat analysis will detect and measure the volume fraction of carbon dioxide, oxygen, and carbon monoxide on a dry basis. The nitrogen present in the products is determined by difference. Modified Orsat units have the capability of measuring total unburned hydrocarbons; however, they do not determine specific hydrocarbon species. A complete analysis of the products formed during a combustion reaction may require the use of Orsat analyzers, gas chromatography, and wet chemistry quantitative analysis procedures. The choice of equipment employed is dictated by the species that must be identified and the necessary precision of the analysis.

Many liquid hydrocarbon fuels are approximated by the formula $(CH_2)_n$. Such fuels include gasoline, kerosene, and diesel and other distillate fuels used in reciprocating internal combustion engines, gas turbines, and jet aircraft engines, as well as stationary applications for heating and steam generation. These fuels are mixtures of several hydrocarbons produced by refinery processing. Gasolines, one example of such mixtures, can be approximated as $(CH_2)_8$ for material balances. Jet-engine fuels and kerosenes are heavier cuts, with the approximate formula $(CH_2)_{11.6}$.

Currently internal combustion engines operate using less than the theoretical (stoichiometric) air supply. Rather than excess air, the term air–fuel ratio, or its reciprocal, fuel–air ratio, is used to describe the air supply to an internal combustion engine. The air–fuel ratio is defined as the ratio of the mass of air supplied to the mass of the fuel supplied. When it can be assumed that the fuel is composed entirely of carbon and hydrogen, with all of the carbon appearing in the dry product gases, it is possible to determine the air–fuel ratio and the carbon–hydrogen ratio of the fuel from the dry exhaust gas analysis. Also, when such fuels are burned with air, as in a steam generator, the excess air can be calculated when the dry product gas analysis is known. The following example illustrates the methods applicable to such determinations.

## Example 12.b

A fuel containing only carbon and hydrogen undergoes a combustion process with dry air as the oxidizer source. The *dry product gases* are found to have the volumetric analysis:

| | |
|---|---|
| $CO_2$ | 8.70% |
| $O_2$ | 0.30% |
| CO | 8.90% |
| $H_2$ | 3.70% |
| $N_2$ | 78.40% |
| | 100.00% |

Calculate:

(a) The carbon–hydrogen ratio of the fuel, kg C/kg $H_2$
(b) The air–fuel ratio
(c) The excess (or deficiency) of the air supply
(d) The material balance

**Solution:**

For the given conditions a basis of 1 kmol of dry product gas is the most convenient. The following table is on that basis.

| Dry Product Gas | $x_i$ | kmol C | kmol $H_2$ | kmol $O_2$ | kmol $N_2$ |
|---|---|---|---|---|---|
| $CO_2$ | 0.0870 | 0.0870 | — | 0.0870 | — |
| CO | 0.0890 | 0.0890 | — | 0.0445 | — |
| $O_2$ | 0.0030 | — | — | 0.0030 | — |
| $H_2$ | 0.0370 | — | 0.0370 | — | — |
| $N_2$ | 0.7840 | — | — | — | 0.7840 |
| | 1.0000 | 0.1760 | 0.0370 | 0.1345 | 0.7840 |

The air supply per kilogram mole of dry product gas is determined by a nitrogen balance. Let $S$ be the number of kilogram moles of dry air supplied per kilogram mole of dry products. Since the fuel contains only carbon and hydrogen, the nitrogen in the products entered as a constituent of the dry air. Thus

$$n_{N_2} \text{ supplied in dry air} = \frac{S \text{ kmol da}}{\text{kmol dry prod}} \left| \frac{0.79 \text{ kmol } N_2}{\text{kmol da}} \right.$$

From the table the nitrogen in the products per kilogram mole of dry products is 0.7840 kmol. Thus

$$n_{N_2} \text{ supplied in dry air} = n_{N_2} \text{ in the products}$$

Solving for $S$,

$$S = \frac{0.7840 \text{ kmol N}_2}{\text{kmol dry prod}} \left| \frac{\text{kmol da}}{0.79 \text{ kmol N}_2} \right.$$
$$= 0.9924 \text{ kmol da/kmol dry products}$$

The oxygen supplied in the dry air is (0.21)(0.9924) = 0.2024 kmol. In the dry product gases 0.1345 kmol of $O_2$ is accounted for. The remaining $O_2$ is assumed to have been used in forming $H_2O$ by combustion of the $H_2$ in the fuel. The oxygen balance is

$$n_{O_2} \text{ to } H_2O = 0.2024 - 0.1345$$
$$= 0.0739 \text{ kmol O}_2\text{/kmol dry products}$$

Complete combustion of 1 kmol of $H_2$ requires 0.5 kmol $O_2$. (See eq. [12.1].) Therefore

$$n_{H_2} \text{ to } H_2O = (0.0739)(2) = 0.1478 \text{ kmol H}_2\text{/kmol dry products}$$

The dry product gases contain, on the same basis, 0.0370 kmol $H_2$. The total $H_2$ in the fuel is then

$$\text{Total } n_{H_2} = 0.0370 + 0.1478$$
$$= 0.1848 \text{ kmol H}_2\text{/kmol dry products}$$

All carbon in the fuel is accounted for in the dry product gases. This assumes no carbon-bearing solid products. The total carbon, per kilogram mole of dry product gases, is

$$0.1760 \frac{\text{kmol C}}{\text{kmol dry prod}}$$

(a) The carbon–hydrogen ratio of the fuel is

$$\frac{m_C}{m_{H_2}} = \frac{0.1760 \text{ kmol C}}{\text{kmol dry prod}} \left| \frac{12 \text{ kg C}}{\text{kmol C}} \right| \frac{\text{kmol dry prod}}{0.1848 \text{ kmol H}_2} \left| \frac{\text{kmol H}_2}{2 \text{ kg H}_2} \right.$$
$$= 0.5714 \text{ kg C/kg H}_2$$

(b) The air–fuel ratio is the mass of the dry air supplied per unit mass of fuel. The mass of the fuel, which consists of 0.176 kmol C and 0.1848 kmol $H_2$, is

$$m_{\text{fuel}} = (0.176)(12) + (0.1848)(2)$$
$$= 2.482 \text{ kg fuel/kmol dry products}$$

Therefore

$$\text{Air–fuel ratio} = \frac{m_{da}}{m_{fuel}}$$

$$= \frac{0.9924 \text{ kmol da}}{\text{kmol dry prod}} \left| \frac{29 \text{ kg da}}{\text{kmol da}} \right| \frac{\text{kmol dry prod}}{2.482 \text{ kg fuel}}$$

$$= 11.59 \text{ kg da/kg fuel}$$

(c) The oxygen required for complete combustion of the fuel is

$$n_{O_2} \text{ required} = 0.176 + (0.5)(0.1848)$$
$$= 0.2684 \text{ kmol } O_2\text{/kmol dry products}$$

from which

$$\text{Theoretical dry air} = \frac{0.2684}{0.21} = 1.278 \frac{\text{kmol da}}{\text{kmol dry prod}}$$

The excess air is, by eq. (12.2),

$$\frac{S - TA}{TA} = \frac{0.9924 - 1.278}{1.278} = -22.4 \text{ percent}$$

Thus there is a deficiency of 22.4 percent in the air supply.

(d) The material balance below is based on 1 kmol of dry product gas, the apparent molecular weight of which has been determined from eq. (11.6) to be 28.60.

| **Input, kg** | | **Output, kg** | |
|---|---|---|---|
| fuel | = 2.482 | dry product gases | = 28.60 |
| dry air | = 28.78 | $H_2O$ = (0.1478)(18) | = 2.66 |
| Total | = 31.26 kg | Total | = 31.26 kg |

---

The procedures introduced in this and the preceding section involved dry air, not atmospheric air. It is usually assumed that the water vapor in the air supply does not enter into the combustion reaction when atmospheric air is supplied to a combustion process. In such cases the material balance includes the water in the atmospheric air in both the input and the output. Energy balances for combustion processes, as we shall see, are affected to a small degree by atmospheric water vapor in any process for which the initial and final temperatures are not the same.

## Example 12.c

Prepare a material balance for Example 12.b assuming atmospheric air with a wet-bulb temperature of 16 C and a dry-bulb temperature of 27 C at 101.3 kPa is supplied to the combustion process.

**Solution:**

The material balance can be obtained by altering the material balance of Example 12.b to include the water vapor in the air. From the psychrometric chart, Appendix F, we obtain $\omega = 0.0069$ kg vapor/kg dry air. Thus the water vapor entering per kilogram mole of dry products is

$$m_{H_2O} = \frac{0.0069 \text{ kg } H_2O}{\text{kg da}} \left| \frac{28.78 \text{ kg da}}{\text{kmol dry prod}} \right.$$

$$= 0.199 \frac{\text{kg } H_2O}{\text{kmol dry prod}}$$

The material balance is

**Input, kg**

| | | |
|---|---|---|
| fuel | = | 2.482 |
| atmos. air | | |
| dry air | = | 28.78 |
| $H_2O$ vapor | = | 0.199 |
| Total | = | 31.46 kg |

**Output, kg**

| | | |
|---|---|---|
| dry product gases | = | 28.60 |
| $H_2O$ from fuel | = | 2.66 |
| $H_2O$ from atmos. air | = | 0.199 |
| Total | = | 31.46 kg |

### 12–3 The First Law and Standard State Reactions for Combustion Processes

To illustrate the use of the first law in combustion processes let us apply it to determine the heat transfer for a simple combustion process in which carbon monoxide undergoes complete combustion with the theoretical amount of oxygen. Let the process take place at a pressure of 1 standard atmosphere in a steady-flow system with work and changes in kinetic and potential energies equal to zero. The combustion reaction is described by the following diagram.

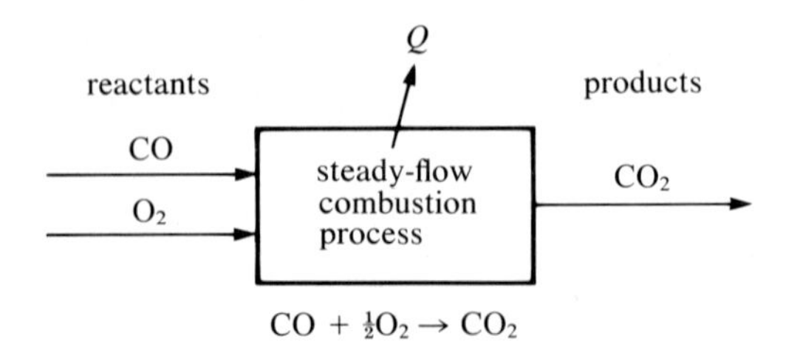

Figure 12.1 *H T* Diagram for the Combustion of One Mole of Carbon Monoxide.

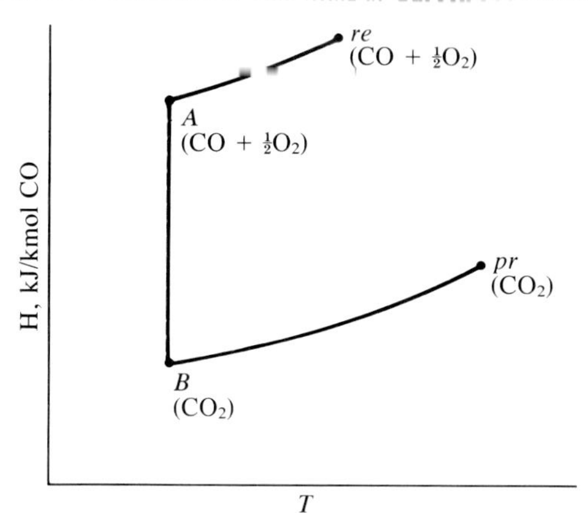


At the outset let us note that the first law as discussed in Chapter 3 and subsequent chapters is applicable to non-pure substances as well as pure substances, although to this point we have applied it only to processes involving pure substances. Recall that pure substances are those that are homogeneous and invariant in chemical composition. It is evident that in combustion processes we are not dealing with pure substances, but in view of the general nature of the first law, we can write the familiar steady-flow energy equation for the combustion process under consideration as

$$Q = H_{pr} - H_{re} \tag{a}$$

where the subscripts "pr" and "re" denote, respectively, the products and the reactants. (In this chapter we will generally use the symbol $H$ to represent enthalpy per unit mole or unit mass in relation to combustion processes. In each use the exact basis must be specified. Similarly $Q$ will be used to denote the corresponding heat transfer.) To determine $Q$ we must evaluate the enthalpy change in eq. (a). This is complicated by the fact that during the combustion process the composition changed and energy was released due to the chemical reaction. As a result the methods used for evaluating enthalpy changes for pure substances are not applicable. Another method must be employed. The *H-T* diagram is useful in describing the method we shall use to evaluate enthalpy changes for combustion processes. Figure 12.1 is the *H-T* diagram for the combustion process under consideration. The diagram illustrates the case in which the temperature of the products is greater than the temperature of the reactants. It is applicable to the pressure of 1 standard atmosphere at which the process occurs. Full description of a state on the diagram requires that the composition at that state be specified. The compositions of the reactants and the products are shown in parentheses. The basis selected is 1 kmol of CO and the enthalpy coordinate has the units kJ/kmol CO. It is important to note that the enthalpy of both the reactants and the products can be determined on the basis of a mole of CO. This is illustrated as follows: The enthalpy of the reactants, each of which behaves as a pure substance prior to the combustion process, at temperature $T_{re}$ is given by

$$H'_{re} = H'_{CO} + H'_{O_2} = n_{CO}\bar{h}_{CO} + n_{O_2}\bar{h}_{O_2}$$

where $H'$ is enthalpy in energy units (kJ) and $\overline{h}$ is the enthalpy per mole of the corresponding pure substance. On the basis of 1 mole of CO

$$H_{re} = H'_{re}/n_{CO} = \overline{h}_{CO} + (n_{O_2}/n_{CO})\overline{h}_{O_2} \tag{b}$$

Similarly, on the basis for 1 mole of CO we have for the products, 1 mole of $CO_2$, a pure substance, at $T_{pr}$

$$H_{pr} = H'/n_{CO} = (n_{CO_2}/n_{CO})\overline{h}_{CO_2} \tag{c}$$

In the above expressions a suitable reference for the $\overline{h}$ values must be chosen. Equations (b) and (c) show that the enthalpy of the reactants and the enthalpy of the products both can be represented as points on a single diagram as shown in Figure 12.1.

To evaluate the enthalpy change for the combustion process, we will make use of what is known as a standard state reaction. Standard state reactions are those in which each reactant and each product exists at a standard state. The most common standard state is 25 C (77 F) and a pressure of 1 standard atmosphere. Since enthalpy is a property, the enthalpy change $H_{pr} - H_{re}$ does not depend on the path for the process from state (re) to state (pr). Therefore, we can include a standard state reaction in the evaluation of the enthalpy change. This is the process $A$ to $B$ in Figure 12.1. From eq. (a) and the path shown in Figure 12.1, we obtain

$$\begin{aligned} Q &= H_{pr} - H_{re} \\ &= (H_{pr} - H_B) + (H_B - H_A) + (H_A - H_{re}) \\ &= (H_A - H_{re}) + (H_B - H_A) + (H_{pr} - H_B) \end{aligned} \tag{d}$$

The terms in eq. (d) can be interpreted as follows: $H_A - H_{re}$ represents the enthalpy change of the reactants (each of which behaves as a pure substance) between $T_{re}$ and $T_A$, which is the standard state temperature. This enthalpy change can be determined by standard methods applicable to perfect gases. The term $H_B - H_A$ represents the enthalpy change for 1 mole of CO and 0.5 mole of $O_2$ at $A$ reacting to produce 1 mole of $CO_2$ at $B$. This process can be viewed as a combustion process for which the reactants are supplied at the standard state temperature and the products are cooled to the standard state temperature. The enthalpy change $H_B - H_A$ is then the heat transfer necessary to return the products to the standard state temperature after the combustion process has occurred and would be a negative quantity. (Later we will introduce a particular name for the standard state reaction described.) Standard state data are available for a large number of reactions. The term $H_{pr} - H_B$ is the enthalpy change for the products, 1 mole of $CO_2$, a pure substance, between the standard state temperature $T_B$ and $T_{pr}$, and would be evaluated on the basis of perfect gas behavior. It is instructive to note that for the conditions shown in Figure 12.1, $Q$ is negative ($H_{pr} < H_{re}$) even though the temperature of the products is greater than the temperature of the reactants. This condition frequently occurs in combustion processes. This example illustrates the usefulness of standard state reactions in the application of the first law to combustion processes. Before proceeding further, enthalpy changes for three standard state reactions are discussed and illustrated.

*Standard State Enthalpy of Formation* The standard state enthalpy of formation, $\Delta H_f^0$, is the enthalpy change that takes place when a compound is formed from its chemical elements with each of the elements and the compound at the standard state. (The superscript $^0$ denotes quantities at the standard state.) Again, the enthalpy–temperature diagram is useful in describing the states and enthalpy changes. To illustrate, consider the enthalpy of formation of gaseous butane, $C_4H_{10}(g)$. From the formula of the compound we can write the formation process as

$$4C + 5H_2 \rightarrow C_4H_{10}$$

This formation process is illustrated in Figure 12.2a in which the state of the elements, the reactants, is state (1) and the state of the compound, the products, is state (2). The standard state enthalpy of formation is the enthalpy of the products minus the enthalpy of the reactants, or

$$\Delta H_{f,C_4H_{10}}^0 = \bar{h}_{C_4H_{10}}^0 - [(5)\bar{h}_{H_2}^0 + (4)\bar{h}_C^0] \tag{e}$$

As has been noted earlier, the enthalpy of a pure substance must be given with respect to a reference state. We shall fix the reference state as the standard state, setting the enthalpy of each element equal to zero at that state. In addition, the enthalpy of diatomic gases will be set equal to zero at the standard state. For this condition, as an example

$$\bar{h}_{H_2}^0 = 0$$

but

$$\bar{h}_H^0 \neq 0$$

With enthalpy reference states established, eq. (e) becomes

$$\Delta H_{f,C_4H_{10}}^0 = \bar{h}_{C_4H_{10}}^0$$

or, generally,

$$\Delta H_{f,\text{compound}}^0 = \bar{h}_{\text{compound}}^0 \tag{12.3}$$

Values of enthalpy of formation for many substances are negative, although some substances exhibit positive values. Table 12.1 lists enthalpy of formation values for several substances. While most of the values for $\Delta H_f^0$ are for the compound in the gaseous phase, the value listed for water is for the liquid phase. Values of $\Delta H_f^0$ between the liquid and vapor phases differ by the value of $\bar{h}_{fg}$ listed. To obtain $\Delta H_f^0$ for a substance in the liquid phase, subtract $\bar{h}_{fg}$ from $\Delta H_f^0$ listed for the gaseous phase; $\bar{h}_{fg}$ is added to $\Delta H_f^0$ for the liquid phase to obtain $\Delta H_f^0$ for the vapor phase of a given substance.

**Figure 12.2** Illustrations for Enthalpy of Formation and Enthalpy of Combustion for Gaseous Butane, $C_4H_{10}$. *a.* Enthalpy of formation. *b.* Enthalpy of combustion. *c.* States related to enthalpy of formation and enthalpy of combustion. *d.* States related to enthalpy of combustion for product $H_2O(l)$ and $H_2O(v)$. Enthalpy values are on the basis of a mole of butane.

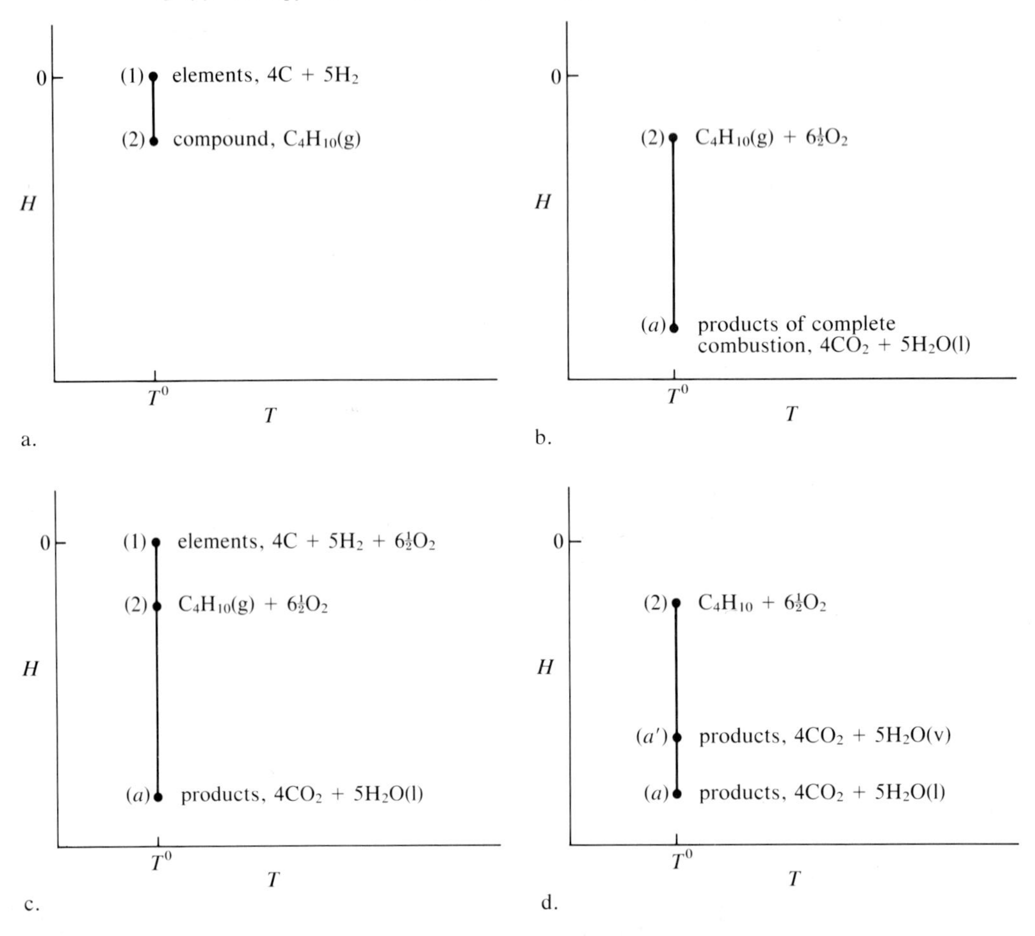


*Standard State Enthalpy of Combustion* The standard state enthalpy of combustion of a fuel, $\Delta H^0_C$, is the enthalpy change for a process in which a fuel is burned to completion with the reactants and the products each at the standard state. Again, using butane as an example, the complete combustion reaction is

$$C_4H_{10} + 6\frac{1}{2}O_2 \rightarrow 4CO_2 + 5H_2O$$

with each reactant and each product at the standard state. $\Delta H^0_C$ is given by the enthalpy of the products minus that of the reactants. For cases in which $H_2O$ is present in the products, the phase of the water must be specified since $\Delta H^0_C$ values are found in the literature for the product water in the liquid phase $H_2O(l)$, as well as in the vapor phase, $H_2O(v)$. Figure

**Table 12.1*** Some Thermochemical Properties of Combustion Components at 25 C (537 R) and One Standard Atmosphere (Values Rounded to Five Significant Digits)

| Substance | Chemical Formula | Formula Weight | Phase | $\Delta H_f^0$ | | $\Delta H_C^0$, $H_2O(l)$† | | $\bar{h}_{fg}^0$ | |
|---|---|---|---|---|---|---|---|---|---|
| | | | | B/lb mol | kJ/kmol | B/lb mol | kJ/kmol | B/lb mol | kJ/kmol |
| hydrogen | $H_2$ | 2.0159 | gas | 0 | 0 | −122,970 | −286,030 | — | — |
| oxygen | $O_2$ | 31.998 | gas | 0 | 0 | 0 | 0 | — | — |
| nitrogen | $N_2$ | 28.013 | gas | 0 | 0 | 0 | 0 | — | — |
| carbon | C | 12.011 | solid | 0 | 0 | −169,300 | −393,790 | — | — |
| carbon monoxide | CO | 28.011 | gas | −47,551 | −110,600 | −121,750 | −283,190 | — | — |
| carbon dioxide | $CO_2$ | 48.010 | gas | −169,300 | −393,790 | 0 | 0 | — | — |
| water | $H_2O$ | 18.015 | liquid | −122,970 | −286,030 | 0 | 0 | 18,916 | 43,999 |
| methane | $CH_4$ | 16.043 | gas | −32,211 | −74,923 | −383,020 | −890,900 | — | — |
| ethane | $C_2H_6$ | 32.070 | gas | −36,432 | −84,741 | −671,100 | −1,561,000 | — | — |
| propane | $C_3H_8$ | 44.098 | gas | −44,676 | −103,920 | −955,080 | −2,221,500 | 6,489 | 15,093 |
| n-butane | $C_4H_{10}$ | 58.125 | gas | −54,270 | −126,230 | −1,237,800 | −2,879,100 | 9,063 | 21,081 |
| n-heptane | $C_7H_{16}$ | 100.21 | gas | −80,802 | −187,950 | −2,088,000 | −4,856,700 | 15,723 | 36,572 |
| n-octane | $C_8H_{18}$ | 114.23 | gas | −89,676 | −208,590 | −2,371,400 | −5,515,900 | 17,847 | 41,512 |
| methanol | $CH_3OH$ | 32.042 | gas | −83,328 | −193,820 | −328,900 | −765,020 | 16,344 | 38,016 |
| ethanol | $C_2H_5OH$ | 46.070 | gas | −101,140 | −235,250 | −606,350 | −1,410,400 | 18,324 | 42,622 |
| acetylene | $C_2H_2$ | 26.038 | gas | 97,542 | 226,880 | −559,100 | −1,300,500 | — | — |
| benzene | $C_6H_6$ | 78.115 | gas | 35,676 | 82,982 | −1,420,400 | −3,303,900 | 14,562 | 33,871 |
| sulfur | S | 32.064 | solid | 0 | 0 | −127,700 | −297,030 | — | — |
| sulfur dioxide | $SO_2$ | 64.063 | gas | −127,700 | −297,030 | 0 | 0 | — | — |

*Data from *JANAF Thermochemical Tables*, 2nd ed., NSRDS-NBS 37, 1971; *Selected Values of Chemical Thermodynamic Properties*, Tech. Notes 270–3 and 270–4, NBS, 1968, 1969; and *Selected Values of Physical and Thermodynamic Properties of Hydrocarbons and Related Compounds*, API Research Project 44, 1953.

†Liquid $H_2O$ in products.

12.2b illustrates the standard state enthalpy of combustion for butane with $H_2O(l)$ in the products. $\Delta H_C^0$ is given by $H_a - H_2$ and is written as

$$\Delta H^0_{C,C_4H_{10}} = [4\bar{h}^0_{CO_2} + 5\bar{h}_{H_2O(l)}] - \left[\bar{h}^0_{C_4H_{10}} + \left(6\frac{1}{2}\right)\bar{h}^0_{O_2}\right] \quad \text{(f)}$$

Table 12.1 lists the value of $\Delta H_C^0$ with the product $H_2O(l)$ for butane as well as several other combustible substances. $\Delta H_C^0$ values are either negative or zero. Natural gas, blast furnace gas, and a number of other fuel gases are mixtures of several different gases, each of which exhibits essentially perfect gas behavior. For these gas mixtures the standard state enthalpy of combustion per unit mole of fuel gas mixture is given by

$$\Delta H^0_{C,m} = \sum_{i=1}^{N} (x\Delta H^0_C)_i \quad \textbf{(12.4)}$$

where $N$ is the number of constituents in the mixture and $x$ is the mole fraction of constituent $i$.

Enthalpy of combustion values can be obtained from enthalpy of formation data. To illustrate this let us again consider the butane example. In view of eq. (12.3), we can write eq. (f) as

$$\Delta H^0_{C,C_4H_{10}} = [4\,\Delta H^0_{f,CO_2} + 5\,\Delta H^0_{f,H_2O}] - [\Delta H^0_{f,C_4H_{10}}] \quad \text{(g)}$$

Figure 12.2c provides an illustration of eq. (g). The first bracketed term on the right side of eq. (g) is $H_a - H_1$ in Figure 12.2c. The second bracketed term in this equation is $H_2 - H_1$ in the figure. Therefore

$$\Delta H^0_{C,C_4H_{10}} = (H_a - H_1) - (H_2 - H_1) = H_a - H_2$$

Equation (g), extended to all fuels, becomes

$$\Delta H^0_{C,\ compound} = \sum_{i=1}^{N} (n_i \Delta H^0_{f,i})_{pr} - \Delta H^0_{f,\ compound} \tag{12.5}$$

For both eq. (g) and eq. (12.5), use of $\Delta H^0_f$ for $H_2O$ in the liquid phase will yield $\Delta H^0_C$ with the product $H_2O(l)$.

The standard state enthalpy of formation can be determined from standard state enthalpy of combustion data. To illustrate, for the butane example we can write from Figure 12.2c

$$\Delta H^0_{f,C_4H_{10}} = H_2 - H_1 = (H_a - H_1) - (H_a - H_2) \tag{h}$$

The enthalpy change $H_a - H_1$ is the enthalpy of combustion of the chemical elements forming butane, namely,

$$H_a - H_1 = [4\ \Delta H^0_{C,C} + 5\ \Delta H^0_{C,H_2}] - H_1 \tag{i}$$

However, $H_1$ is zero by the selected reference state for enthalpies of the elements. The other term in eq. (h) is

$$H_a - H_2 = \Delta H^0_{C,C_4H_{10}} \tag{j}$$

Combining eqs. (i) and (j)

$$\Delta H^0_{f,C_4H_{10}} = [4\ \Delta H^0_{C,C} + 5\ \Delta H^0_{C,H_2}] - \Delta H^0_{C,C_4H_{10}} \tag{k}$$

Generalizing the preceding, we have

$$\begin{aligned} H_2 - H_1 &\equiv \Delta H^0_{f,\ compound} \\ &= \sum_{i=1}^{N} (n_i \Delta H^0_C)_{elements} - \Delta H^0_{C,\ compound} \end{aligned} \tag{12.6}$$

Let us now consider the standard state enthalpy of combustion for the case in which the $H_2O$ in the products is in the vapor phase. For any combustible substance containing hydrogen there will be water in the products of complete combustion. $\Delta H^0_C$ with $H_2O(v)$ in the products will be larger (less negative) than $\Delta H^0_C$ with $H_2O(l)$ in the products by an amount

**Table 12.2** Approximate Data for Liquid Petroleum Fuels

| Substance | Average Formula | Formula Weight Range | Phase | Higher Heating Value | |
|---|---|---|---|---|---|
| | | | | B/lbm | kJ/kg |
| automotive gasoline | $C_8H_{17}$ | 113–126 | liquid | 20,260–20,460 | 47,130–47,600 |
| kerosene and jet-engine fuel | $C_{11.6}H_{23.2}$ | 154–160 | liquid | 19,750 | 45,940 |
| diesel, 1–D | $C_{12}H_{26}$ | 170 | liquid | 19,240 | 44,760 |
| diesel, 2–D | | 180 | liquid | 19,110 | 44,460 |

equal to the enthalpy of vaporization of the water in the products. This is illustrated for butane in Figure 12.2d, in which ($a'$) is the state of the products of complete combustion with $H_2O(v)$. From the figure, $\Delta H_C^0 = H_{a'} - H_2$, and $H_{a'} - H_a = 5\bar{h}_{fg,H_2O}$. It should be noted that the actual phase of the water in the products at the standard state depends on the amount of water present. However, by convention and for reference purposes, $\Delta H_C^0$ values are quoted for all $H_2O(l)$ or all $H_2O(v)$ in the products.

Two commonly used terms that are related to the enthalpy of combustion are the *lower heating value* (LHV) and the *higher heating value* (HHV). These quantities are always positive and are usually reported on the basis of a unit mass of fuel rather than on a mole basis. They are given on a unit mass basis by

$$\text{LHV} = -[\Delta H^0_{C,f},\ H_2O(v)]/M_f \tag{12.7}$$

$$\text{HHV} = -[\Delta H^0_{C,f},\ H_2O(l)]/M_f$$

where the subscript $f$ denotes the fuel and $M_f$ is the molecular weight of the fuel. From the above equations it is evident that HHV and LHV are related by the enthalpy of vaporization of the water formed per unit mass of fuel during the combustion process. In equation form the relation is

$$\text{LHV} = \text{HHV} - (mh_{fg})_{H_2O} \tag{12.8}$$

The enthalpy of combustion of gaseous fuels can be determined from an analysis of the composition of the fuel and application of eq. (12.4) or eq. (12.5) using data from Table 12.1. There are also standard combustion tests for determining the enthalpy of combustion of fuel gases. Liquid and solid fuels such as gasolines, kerosenes, and coals are complex combinations of several chemical compounds. For these fuels the higher heating values can be determined by use of an instrument known as an oxygen bomb calorimeter. The higher heating values and the lower heating values are sometimes referred to, respectively, as the gross and the net heating values. Some typical data for liquid fuels are given in Table 12.2. Table 1.4 also lists higher heating values for selected solid, liquid, and gaseous fuels.

## Example 12.d

Using enthalpy of combustion data from Table 12.1, calculate the enthalpy of formation of propane. Start with an appropriately labeled *H-T* diagram and compare the result with enthalpy of formation data from Table 12.1.

**Solution:**

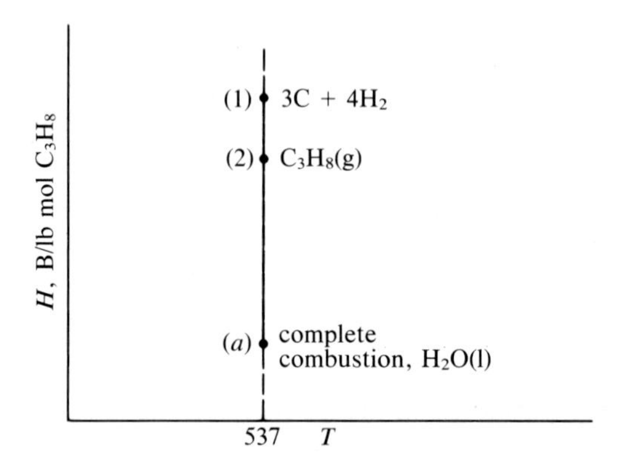


The formation equation is

$$3C + 4H_2 \rightarrow C_3H_8$$

By definition

$$\Delta H^0_{f,C_3H_8} = H_2 - H_1$$

and

$$H_2 - H_1 = (H_a - H_1) - (H_a - H_2)$$

For this example

$$H_a - H_1 = 3\ \Delta H^0_{C,C} + 4\ \Delta H^0_{C,H_2}$$

and

$$H_a - H_2 = \Delta H^0_{C,C_3H_8}$$

Combining terms

$$H_2 - H_1 = (3\ \Delta H^0_{C,C} + 4\ \Delta H^0_{C,H_2}) - \Delta H^0_{C,C_3H_8}$$

Substituting data from Table 12.1

$$\Delta H^0_{f,C_3H_8} = H_2 - H_1$$
$$= [3(-169{,}300) + 4(-122{,}970)] - (-955{,}080)$$
$$= -44{,}700 \text{ B/lb mol } C_3H_8$$

which is in close agreement with the value listed in Table 12.1 for the enthalpy of formation of propane (−44,676 B/lb mol).

---

## Example 12.e

Compute the higher heating value and the lower heating value in Btu per pound mass units for the fuel gas in Example 12.a using enthalpy of combustion data.

**Solution:**

Equation (12.4) applies. Values from Example 12.a and Table 12.1 are listed in tabular form below to determine the enthalpy of combustion for the mixture.

| Gas | $x_i$ | $\Delta H^0_{C,i}$ | $x_i \Delta H^0_{C,i}$ |
|---|---|---|---|
| $CH_4$ | 0.200 | −383,020 | −76,600 |
| $C_2H_6$ | 0.400 | −671,100 | −268,440 |
| $N_2$ | 0.300 | 0 | 0 |
| $H_2O$ | 0.100 | 0 | 0 |
| | 1.000 | | −345,040 B/lb mol fuel |

Since the enthalpy of combustion values listed in Table 12.1 are given per mole and are for the product $H_2O(l)$, we have, from eq. (12.7), for the fuel gas

$$\text{HHV} = -(-345{,}040)/25.40 = 13{,}584 \text{ B/lbm fuel}$$

where the molecular weight of the fuel is from Example 12.a. The lower heating value is determined from eq. (12.8) as

$$\text{LHV} = \text{HHV} - (mh_{fg})_{H_2O} = 13{,}584 - \frac{30.6}{25.40}(1050.6)$$
$$= 12{,}318 \text{ B/lbm fuel}$$

The ratio 30.6/25.4 is, from the material balance of Example 12.a, the mass of the water in the products per unit mass of fuel.

---

*Standard State Enthalpy of Reaction* The standard state enthalpy of reaction, $\Delta H_R^0$, is the enthalpy change for any combustion process in which the reactants and the products are each at the standard state. Enthalpies of formation and combustion therefore are special cases of the enthalpy of reaction. The enthalpy change for an incomplete combustion process at the standard state is an example of an enthalpy of reaction. This is illustrated in the following example.

## Example 12.f

One kilogram mole of ethane, $C_2H_6$, is burned with the theoretical amount of air in such a manner that 80 percent of the carbon burns to $CO_2$ and the remainder to CO. Determine the standard state enthalpy of reaction.

**Solution:**

The combustion reaction equation is

$$C_2H_6 + 3.5O_2 + \left(\frac{79}{21}\right)3.5N_2$$
$$\rightarrow 1.6CO_2 + 0.4CO + 3H_2O + 0.2O_2 + \left(\frac{79}{21}\right)3.5N_2$$

At the standard state the enthalpy change of the nitrogen will be zero, and hence it will be deleted from the analysis. An *H-T* diagram for the process is shown below.

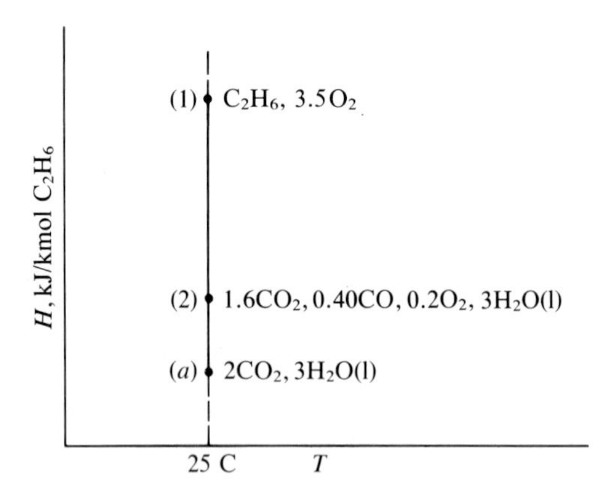


The standard state enthalpy of reaction is

$$H_2 - H_1 = (H_a - H_1) - (H_a - H_2)$$

or

$$\Delta H^0_R = \Delta H^0_{C,C_2H_6} - 0.40\ \Delta H^0_{C,CO}$$
$$= -1{,}561{,}000 - (0.40)(-283{,}190)$$
$$= -1{,}447{,}700 \text{ kJ/kmol } C_2H_6$$

This illustration shows a general characteristic of incomplete combustion reactions. Suppose the reaction were to take place in a steady-flow process during which velocities were negligible and no work interaction occurred. By the steady-flow energy equation

$$Q = H_2 - H_1$$

Referring to the *H-T* diagram, it is apparent the maximum energy release, $H_a - H_1$, has not been achieved. The maximum can be realized only when the process reaches state (*a*). Due to incomplete combustion the energy release is decreased by

$$|(0.4)(\Delta H^0_{C,CO})| = 113{,}280 \text{ kJ/kmol } C_2H_6$$

---

The general relation between enthalpies of combustion and enthalpies of reaction at the standard state is

$$\Delta H^0_R = \sum_{i=1}^{N} (n_i \Delta H^0_{C,i})_{re} - \sum_{i=1}^{N} (n_i \Delta H^0_{C,i})_{pr} \qquad \textbf{(12.9)}$$

An alternative expression for standard state enthalpies of reaction is obtained by application of eq. (12.3). This alternate is

$$\Delta H^0_R = \sum_{i=1}^{N} (n_i \Delta H^0_{f,i})_{pr} - \sum_{i=1}^{N} (n_i \Delta H^0_{f,i})_{re} \qquad \textbf{(12.9a)}$$

## 12–4 Internal Energy Changes During Combustion Processes

When a system undergoes a chemical reaction at constant volume the first law, expressed for a closed system, becomes

$$Q = U_2 - U_1 = U_2 - U_B + U_B - U_A + U_A - U_1$$

where $Q$ is the heat transfer and $U_1$ and $U_2$ are the summations of the internal energies of the reactants and the products respectively. States $A$ and $B$ are at the standard state pressure and temperature and $U_B - U_A$ is the standard state internal energy of reaction $\Delta U^0_R$. It is typical to find only enthalpy data (for example, $\Delta H^0_f$, or $\Delta H^0_C$, and $\overline{C}_p$ data) for combustible substances. For this reason internal energy changes are usually calculated from enthalpy changes using the definition of enthalpy. For a standard state reaction

$$U_B - U_A = H_B - H_A - p^0(V_B - V_A)$$

where $B$ represents the product state and $A$ the reactant state.

With each product and reactant at the standard state the preceding equation can be written as

$$U_B - U_A = H_B - H_A - (n_B\overline{R}T_B - n_A\overline{R}T_A)_{\text{gases}} - (p_BV_B - p_AV_A)_{\text{liquid and solid}} \quad \textbf{(12.10)}$$

For standard state reactions the product temperature $T_B$ is equal to the reactant temperature $T_A$. The difference in the $pV$ terms for liquids and solids is, in most cases, negligible when compared with the enthalpy change. With these conditions eq. (12.10) reduces to

$$\Delta U_R^0 = \Delta H_R^0 - \overline{R}T^0(n_B - n_A)_{\text{gases}} \quad \textbf{(12.10a)}$$

where $T^0$ is the standard state temperature and $n_B$ and $n_A$ are the moles of gas products and reactants respectively.

## 12–5 A Review of Variable Specific Heats for Perfect Gases

In cases where the reactants and the products for a combustion process are near the standard state temperature, the use of constant specific heats taken at the standard state temperature to evaluate enthalpy changes may yield results that are sufficiently accurate. However, for the temperature changes that occur during many combustion processes, an assumption of constant specific heats will introduce significant errors. The following is a presentation of a means of including variable specific heats for perfect gases in combustion analysis.

In Chapter 5 variations of specific heats with temperature were discussed for perfect gases. For a given perfect gas we observed that

$$\overline{C}_p = \frac{d\overline{h}}{dT}$$

where $\overline{h}$ is a function of temperature only. It then follows that

$$d\overline{h} = \overline{C}_p dT = f(T)dT$$

Therefore, the enthalpy change between states (1) and (2) is given by

$$\overline{h}_2 - \overline{h}_1 = \int_1^2 \overline{C}_p dT = \overline{C}_{p_a}(T_2 - T_1)$$

where

$$\overline{C}_{p_a} = \frac{\int_1^2 \overline{C}_p dT}{T_2 - T_1} \quad \textbf{(5.11)}$$

**Figure 12.3** Perfect Gas Molal Average Specific Heats $\overline{C}_{p_a}$ Between 537 R (298 K) and the Abscissa Temperature. Specific heat values in parentheses are in SI units.

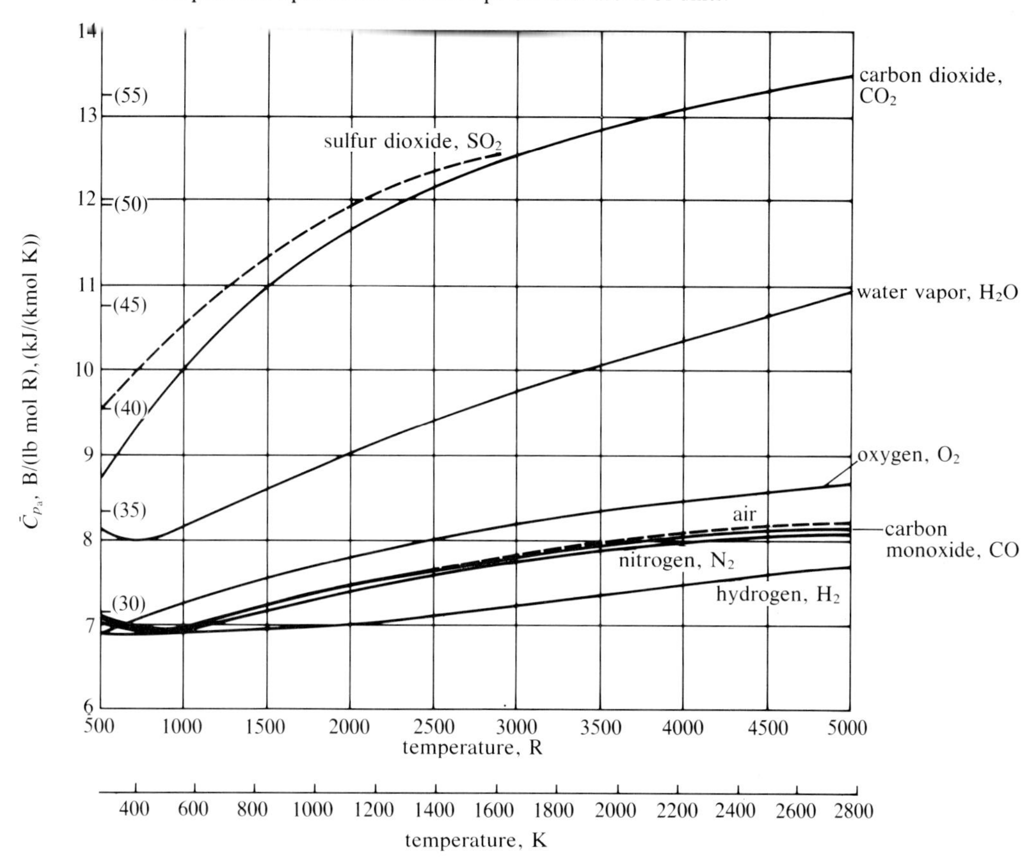


Figure 12.3 shows average molal specific heats for several gases between 537 R (298 K) and the abscissa temperature. The several curves were developed using the instantaneous specific heat equations listed in Table 12.3 in the manner

$$\underset{537\rightarrow T}{\overline{C}_{p_a}} = \frac{\int_{537}^{T} f(T)dT}{(T - 537)} \tag{12.11}$$

The same procedure was followed in preparing Figure 12.4, which shows average molal specific heats for several light hydrocarbon fuel gases between 537 R (298 K) and $T$. Appropriate conversions were made to obtain the SI scales in Figures 12.3 and 12.4. Such information or a table of properties for each gas is necessary for accurate energy analyses of combustion processes where the reactants and the product gases are at different temperatures. The average molal specific heats presented in Figures 12.3 and 12.4 are based on

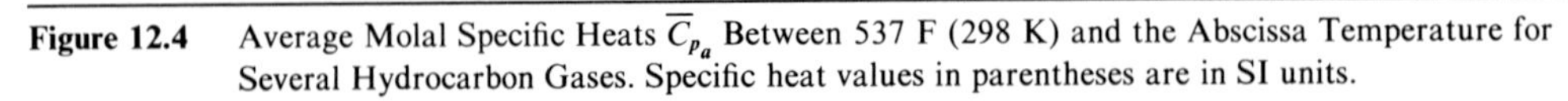

**Figure 12.4** Average Molal Specific Heats $\overline{C}_{p_a}$ Between 537 F (298 K) and the Abscissa Temperature for Several Hydrocarbon Gases. Specific heat values in parentheses are in SI units.

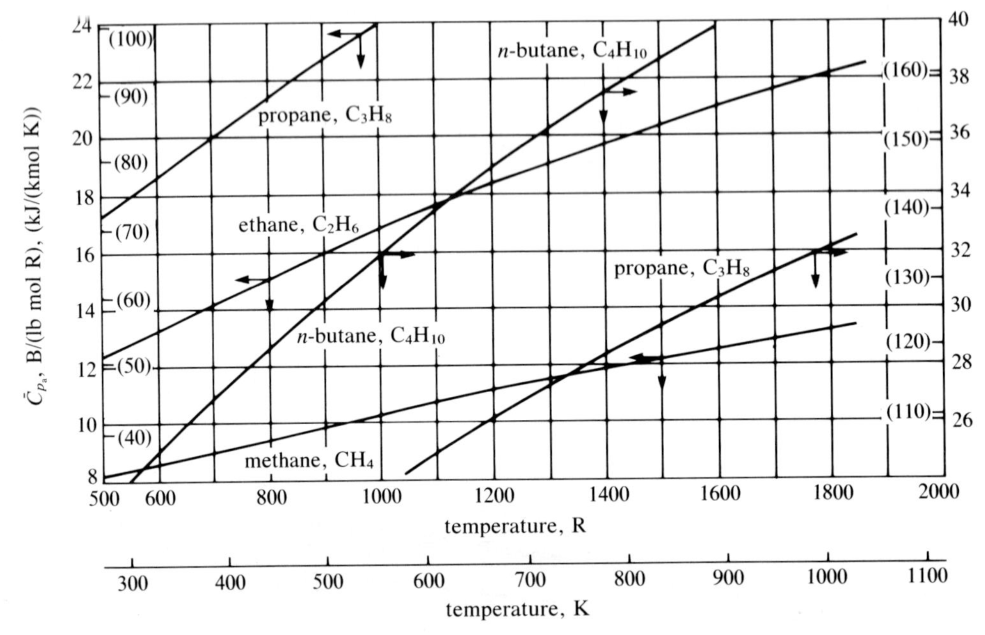


ideal gas behavior of each species; hence water vapor and the hydrocarbons must be at low pressures. As pressure increases and these gases approach their saturated vapor states, the accuracy of the graphed values of $\overline{C}_{p_a}$ decreases.

As an alternative to Figures 12.3 and 12.4, the average specific heats for the gases in Table 12.3 are listed as a function of temperature in Appendix G. Abridged listings from Appendix G of average molal specific heats for six gases are given in Table 12.4.

Using data from Figures 12.3, 12.4, Table 12.4, or Appendix G, the molal enthalpy change of a given gas between $T^0$ and any temperature $T$ is

$$\overline{h} - \overline{h}^0 = \underset{T^0 \to T}{\overline{C}_{p_a}} (T - T^0) \tag{12.12}$$

where $T^0$ is either 537 R or 298 K.

For processes having neither $T_1$ nor $T_2$ equal to $T^0$, the molal enthalpy change is calculated by

$$\overline{h}_2 - \overline{h}_1 = \underset{T^0 \to T_2}{\overline{C}_{p_a}} (T_2 - T^0) - \underset{T^0 \to T_1}{\overline{C}_{p_a}} (T_1 - T^0) \tag{12.13}$$

**Table 12.3** Equations for Instantaneous Specific Heats $\overline{C}_p$ for 12 Gases*

| Item | Substance | Formula | Equation $\overline{C}_p$ in Btu/(lb mol R), $T$ in Rankine | Temperature Range, R |
|---|---|---|---|---|
| 1 | oxygen | $O_2$ | $\overline{C}_p = 11.515 - \frac{172}{\sqrt{T}} + \frac{1530}{T}$ | 540–5000 |
| | | | $= 11.515 - \frac{172}{\sqrt{T}} + \frac{1530}{T} + \frac{0.05}{1000}(T - 5000)$ | 5000–9000 |
| 2 | nitrogen, air | $N_2$ | $\overline{C}_p = 9.47 - \frac{3.47 \times 10^3}{T} + \frac{1.16 \times 10^6}{T^2}$ | 540–9000 |
| 3 | hydrogen | $H_2$ | $\overline{C}_p = 5.76 + \frac{0.578}{1000}(T) + \frac{20}{\sqrt{T}}$ | 540–4000 |
| | | | $= 5.76 + \frac{0.578}{1000}(T) + \frac{20}{\sqrt{T}} - \frac{0.33}{1000}(T - 4000)$ | 4000–9000 |
| 4 | water vapor | $H_2O$ | $\overline{C}_p = 19.86 - \frac{597}{\sqrt{T}} + \frac{7500}{T}$ | 540–5400 |
| 5 | carbon monoxide | CO | $\overline{C}_p = 9.46 - \frac{3.29 \times 10^3}{T} + \frac{1.07 \times 10^6}{T^2}$ | 540–9000 |
| 6 | carbon dioxide | $CO_2$ | $\overline{C}_p = 16.2 - \frac{6.53 \times 10^3}{T} + \frac{1.41 \times 10^6}{T^2}$ | 540–6300 |
| 7 | sulfur dioxide | $SO_2$ | $\overline{C}_p = 6.945 + 5.56 \times 10^{-3}T - 1.17 \times 10^{-6}T^2$ | 540–2700 |
| 8 | methane | $CH_4$ | $\overline{C}_p = 3.42 + 9.91 \times 10^{-3}T - 1.28 \times 10^{-6}T^2$ | 510–1860 |
| 9 | ethane | $C_2H_6$ | $\overline{C}_p = 1.38 + 23.25 \times 10^{-3}T - 4.27 \times 10^{-6}T^2$ | 510–1860 |
| 10 | propane | $C_3H_8$ | $\overline{C}_p = 0.41 + 35.95 \times 10^{-3}T - 6.97 \times 10^{-6}T^2$ | 510–1860 |
| 11 | n-butane | $C_4H_{10}$ | $\overline{C}_p = 2.25 + 45.40 \times 10^{-3}T - 8.83 \times 10^{-6}T^2$ | 510–1860 |
| 12 | pentane | $C_5H_{12}$ | $\overline{C}_p = 3.14 + 55.85 \times 10^{-3}T - 10.98 \times 10^{-6}T^2$ | 510–1860 |

*Items 1–6 from Sweigert, R. L., and M. W. Beardsley. Empirical Specific Heat Equations Based on Spectroscopic Data. Georgia State Engineering Exp. Sta. Bulletin no. 2, 1938. Item 7 from Bryant, W. M. D. Empirical Molecular Heat Equations from Spectroscopic Data. *Industrial and Engineering Chemistry*, vol. 25, p. 820, 1933. Items 8–12 from Fallon, J. F., and K. M. Watson. *National Petroleum News*, Tech. Section, June 7, 1944.

For mixtures of the gases the average molal constant-pressure specific heat between $T^0$ and $T$ is calculated as

$$\overline{C}_{p_{a,m}} = \sum_{i=1}^{N} (x\overline{C}_{p_a})_i \qquad \textbf{(12.14)}$$

and the enthalpy change from $T^0$ to $T$ is

$$(\overline{h} - \overline{h}^0)_m = \overline{C}_{p_{a,m}}(T - T^0)$$

For those cases where neither $T_1$ nor $T_2$ is $T^0$, the molal enthalpy change for a constant-composition gas mixture is calculated using eq. (12.13) with each $\overline{C}_{p_{a,m}}$ calculated using eq. (12.14).

**Table 12.4** Selected Values of Average Molal Specific Heats $\overline{C}_{P_a}$ Between 537 R (298 K) and the Listed Temperatures for Six Gases (See Appendix G for Expanded Listings.)

| | | Oxygen, $O_2$ $\overline{C}_{P_a}$ | | Nitrogen, $N_2$ $\overline{C}_{P_a}$ | | Hydrogen, $H_2$ $\overline{C}_{P_a}$ | |
|---|---|---|---|---|---|---|---|
| *T*, R | *T*, K | $\frac{B}{lb\ mol\ R}$ | $\frac{kJ}{kmol\ K}$ | $\frac{B}{lb\ mol\ R}$ | $\frac{kJ}{kmol\ K}$ | $\frac{B}{lb\ mol\ R}$ | $\frac{kJ}{kmol\ K}$ |
| 537 | 298 | 6.941 | 29.08 | 7.034 | 29.47 | 6.933 | 29.03 |
| 900 | 500 | 7.222 | 30.23 | 6.934 | 29.03 | 6.928 | 29.00 |
| 1800 | 1000 | 7.736 | 32.39 | 7.347 | 30.76 | 7.045 | 29.49 |
| 3600 | 2000 | 8.329 | 34.87 | 7.915 | 33.13 | 7.437 | 31.13 |
| 4500 | 2500 | 8.524 | 35.69 | 8.089 | 33.86 | 7.648 | 32.02 |

| | | Water, $H_2O(v)$ $\overline{C}_{P_a}$ | | Carbon Dioxide, $CO_2$ $\overline{C}_{P_a}$ | | Carbon Monoxide, CO $\overline{C}_{P_a}$ | |
|---|---|---|---|---|---|---|---|
| *T*, R | *T*, K | $\frac{B}{lb\ mol\ R}$ | $\frac{kJ}{kmol\ K}$ | $\frac{B}{lb\ mol\ R}$ | $\frac{kJ}{kmol\ K}$ | $\frac{B}{lb\ mol\ R}$ | $\frac{kJ}{kmol\ K}$ |
| 537 | 298 | 8.067 | 33.80 | 8.928 | 37.38 | 7.046 | 29.52 |
| 900 | 500 | 8.074 | 33.80 | 9.827 | 41.14 | 6.994 | 29.28 |
| 1800 | 1000 | 8.841 | 37.01 | 11.40 | 47.75 | 7.416 | 31.05 |
| 3600 | 2000 | 10.16 | 42.55 | 12.87 | 53.89 | 7.970 | 33.36 |
| 4500 | 2500 | 10.65 | 44.60 | 13.28 | 55.60 | 8.138 | 34.07 |

## Example 12.g

A gas mixture has the volumetric composition

| | |
|---|---|
| $CO_2$ | 14.73% |
| CO | 3.64% |
| $N_2$ | 79.36% |
| $CH_4$ | 2.27% |
| | 100.00% |

Calculate the enthalpy and internal energy changes when 1 pound mole of the mixture is heated from 750 to 1500 R with no chemical reaction.

**Solution:**

The solution, assuming perfect gas behavior, requires determination of the molal average specific heat for the mixture between 537 and 750 R and that between 537 and 1500 R using eq. (12.14). Tabular calculation of these quantities is convenient.

| Gas | Mole Fraction, $x_i$ | $\overline{C}_{p_a}$ $537 \rightarrow 750$ | $x_i\overline{C}_{p_a}$ $537 \rightarrow 750$ | $\overline{C}_{p_a}$ $537 \rightarrow 1500$ | $x_i\overline{C}_{p_a}$ $537 \rightarrow 1500$ |
|---|---|---|---|---|---|
| carbon dioxide, $CO_2$ | 0.1473 | 9.459 | 1.393 | 10.98 | 1.617 |
| carbon monoxide, CO | 0.0364 | 6.957 | 0.253 | 7.279 | 0.265 |
| nitrogen, $N_2$ | 0.7936 | 6.908 | 5.482 | 7.209 | 5.721 |
| methane, $CH_4$ | 0.0227 | 9.262 | 0.210 | 12.08 | 0.274 |
| | 1.0000 | | 7.338 | | 7.877 |

The $\overline{C}_{p_a}$ values are taken from Appendix G. The enthalpy change is

$$\begin{aligned}\overline{h}_{1500} - \overline{h}_{750} &= (7.877)(1500 - 537) - (7.338)(750 - 537) \\ &= 6023 \text{ B/lb mol mixture}\end{aligned}$$

By definition of enthalpy the internal energy change per pound mole of mixture is

$$\overline{u}_2 - \overline{u}_1 = \overline{h}_2 - \overline{h}_1 - (p_2\overline{v}_2 - p_1\overline{v}_1)$$

and for perfect gases

$$\overline{u}_2 - \overline{u}_1 = \overline{h}_2 - \overline{h}_1 - [(n\overline{R}T)_2 - (n\overline{R}T)_1]$$

For 1 lb mol of the constant-composition mixture the preceding equation becomes

$$\overline{u}_2 - \overline{u}_1 = \overline{h}_2 - \overline{h}_1 - [n\overline{R}\,(T_2 - T_1)]$$

or

$$\begin{aligned}\overline{u}_2 - \overline{u}_1 &= 6023 - \frac{(1.00)(1545)}{778}[1500 - 750] \\ &= 4534 \text{ B/lb mol mixture}\end{aligned}$$

## 12–6 The Generalized Combustion Process

Once the enthalpies of combustion and other thermodynamic properties are known for the reactants and products of a combustion process, it is possible to proceed with the analysis of energy transformations and interactions for such systems. The general steady-flow energy equation for systems involving combustion and no elevation changes can be written on a molar basis (a mole of fuel, for example) as

$$Q + H_{\mathrm{re}} + \sum_{i=1}^{N}\left(\frac{M_iV_i^2}{2g_0}\right)_{\mathrm{re}} = W + H_{\mathrm{pr}} + \sum_{i=1}^{N}\left(\frac{M_iV_i^2}{2g_0}\right)_{\mathrm{pr}} \qquad \textbf{(12.15)}$$

where $M_i V_i^2/2g_0$ is the kinetic energy on the basis chosen and $H_{re}$ and $H_{pr}$ are the summations of the enthalpies for the reactants and the products respectively. For a steady-flow process with no work and negligible changes in kinetic energy, the above equation becomes

$$Q = H_{pr} - H_{re}$$

Common examples of steady-flow combustion processes involving heat transfer include residential, steam generator, and industrial furnaces. The combustor in a gas turbine or a jet-aircraft engine is another example of a device in which a steady-flow combustion process takes place. (See Figures 10.10 and 10.17.) To a good approximation such combustors may be treated as adiabatic with no kinetic energy change. Before proceeding to a generalized combustion process, two examples of simple combustion processes are presented.

## Example 12.h

Let the following conditions exist for the combustion process described in Figure 12.1. Reactants: $T_{re} = T_{CO} = T_{O_2} = 400$ K. Volume rate of flow of CO = 0.040 m³/s. Products: $T_{pr} = 900$ K. Process: Steady flow at a pressure of 1 atmosphere. Work and change in kinetic energy are zero. Determine the rate of heat transfer for the process.

**Solution:**

Starting with a basis of 1 mole of CO, the steady-flow energy equation yields $Q = H_{pr} - H_{re}$. From the path in Figure 12.1 we can write

$$\begin{aligned} Q = H_{pr} - H_{re} &= H_{pr} - H_B + (H_B - H_A) + H_A - H_{re} \\ &= (n\overline{C}_{p_a})_{CO_2}(T_{pr} - T^0) + \Delta H_R^0 \\ &\quad + (n\overline{C}_{p_a})_{CO}(T^0 - T_{CO}) + (n\overline{C}_{p_a})_{O_2}(T^0 - T_{O_2}) \end{aligned}$$

where in this case $\Delta H_R^0 = \Delta H_C^0$. Substitution of $\overline{C}_{p_a}$ values from Appendix G and mole values from Figure 12.1 yields

$$\begin{aligned} Q &= (1)(46.73)(900 - 298) + (-283{,}190) \\ &\quad + [1(29.12) + 0.5(29.67)](298 - 400) \\ &= -259{,}500 \text{ kJ/kmol CO} \end{aligned}$$

The heat-transfer rate is obtained by

$$\dot{Q} = \dot{n}_{CO}(H_{pr} - H_{re})$$

where from eq. (5.5)

$$\begin{aligned} \dot{n}_{CO} &= \frac{VA}{\overline{v}} = \frac{VAp}{\overline{R}T} = \frac{0.040(101.3)}{8.314(400)} \\ &= 0.001218 \text{ kmol/s} \end{aligned}$$

Therefore

$$\dot{Q} = (0.001218)(-259{,}500) \\ = -316 \text{ kJ/s}$$

---

## Example 12.i

A liquid fuel containing only carbon and hydrogen is burned with dry air as the oxygen source in a steady-flow process involving only heat transfer. The fuel and air are supplied at 298 K and the products are at 500 K. The fuel has a lower heating value of 43,000 kJ/kg. The volumetric analysis of the combustion products is as follows:

| | |
|---|---|
| $CO_2$ | 11.41% |
| $H_2O$ | 12.36% |
| $N_2$ | 74.12% |
| $O_2$ | 2.11% |
| | 100.00% |

Determine the heat transfer per kilogram of fuel burned.

**Solution:**

The initial basis for solution is a kilogram mole of products. Since the products contain hydrogen and carbon only in the form of $H_2O$ and $CO_2$, we conclude that the combustion was complete. Further, since $O_2$ is present in the products, we conclude that excess dry air was supplied. Using the *H-T* diagram, we obtain the following expression for the heat transfer.

$$Q = H_{\text{pr}} - H_{\text{re}} = H_{\text{pr}} - H_a + H_a - H_{\text{re}}$$

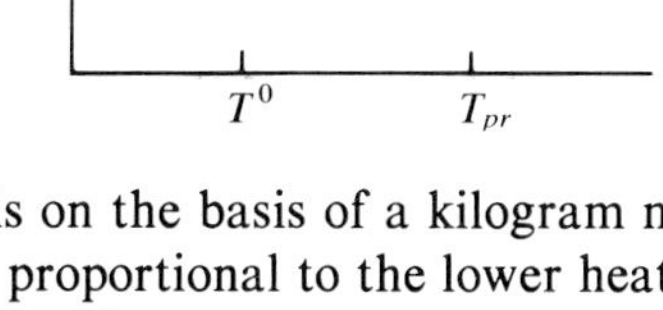

In this expression each term is on the basis of a kilogram mole of products, and $H_a - H_{\text{re}}$ is $\Delta H^0_{\text{C}}$, $H_2O$(v). This term is proportional to the lower heating value of the fuel. To obtain $H_a - H_{\text{re}}$ on the basis of a mole of products, we note that the mass of the fuel supplied per mole of products is

$$m_{\text{fuel}} = (0.1141)(12) + (0.1236)(2) = 1.616 \text{ kg fuel/kmol pr}$$

The expression for the heat transfer then becomes

$$Q = [(n\overline{C}_{p_a})_{CO_2} + (n\overline{C}_{p_a})_{H_2O} + (n\overline{C}_{p_a})_{N_2} + (n\overline{C}_{p_a})_{O_2}](T_{pr} - T^0) - m_{fuel}(LHV)$$

Using average specific heat values from Table 12.4, the above equation yields

$$\begin{aligned} Q &= [(0.1141)(41.14) + (0.1236)(33.80) + (0.7412)(29.03) \\ &\quad + (0.0211)(30.23)] \times (500 - 298) - 1.616(43{,}000) \\ &= -63{,}210 \text{ kJ/kmol products} \end{aligned}$$

To obtain the heat transfer per kilogram of fuel, we divide the above value by the mass of the fuel per kilogram mole of products. Thus

$$\begin{aligned} Q &= (-63{,}210 \text{ kJ/kmol pr})/(1.616 \text{ kg fuel/kmol pr}) \\ &= -39{,}100 \text{ kJ/kg fuel} \end{aligned}$$

---

To illustrate a more complex combustion process, consider a constant-pressure steady-flow process during which a gaseous hydrocarbon fuel is burned with excess atmospheric air to an end state where combustion is incomplete. The atmospheric air and the fuel are supplied at different temperatures. Let the work and kinetic energy change be zero. The steady-flow energy equation then reduces to

$$Q = H_{pr} - H_{re}$$

where the basis is a mole of fuel. The first step is to describe the process on an $H$-$T$ diagram, Figure 12.5. Each of the states and processes shown in the figure is identified in the following. The water vapor supplied with the atmospheric air will be treated as a separate item.

| State | | Process |
|---|---|---|
| 1 | Dry air at $T_{da}$ | |
| $a$ | Dry air at $T^0$ | $H_a - H_1 = [n\overline{C}_{p_a}]_{da}(T_a - T_1)$ |
| $b$ | Fuel at $T_f$ | |
| $c$ | Fuel at $T^0$ | $H_c - H_b = [n\overline{C}_{p_a}]_f(T_c - T_b)$ |
| $d$ | Products of complete combustion with all $H_2O$ formed in the liquid phase | $H_d - H_c = \Delta H^0_{c,f}$, $H_2O(l)$ |
| $e$ | Actual products, $H_2O$ liquid | $H_e - H_d = -\sum_{i=1}^{N}(n_i\Delta H^0_{c_i})_{pr}$ |
| $f$ | Actual products, $H_2O$ in vapor phase | $H_f - H_e = m_{w,e}h^0_{fg}$ |
| 2 | Actual products at $T_2$ | $H_2 - H_f = \sum_{i=1}^{N}(n\overline{C}_{p_a})_i (T_2 - T^0)$ |

**Figure 12.5** A General Enthalpy–Temperature Diagram for a Combustion Process.

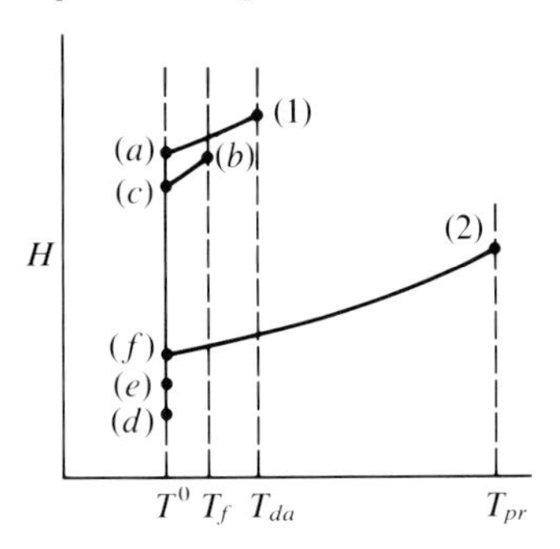


The enthalpy of the dry air at ($a$) is zero. Therefore, the enthalpy of the fuel and the dry air (the reactants) at $T^0$ is the enthalpy of the fuel $H_c$. The final term is the enthalpy change of the water vapor in the atmospheric air supply. For $n_{\text{da}}$ moles of dry air this enthalpy change is

$$(H_2 - H_1)_w = (n_{\text{da}}) \left(\frac{29}{18}\right) \omega \overline{C}_{p_{a,w}} (T_2 - T_1)$$

Another procedure that can be used to determine the enthalpy change of the water vapor in the air supply is to establish states (1) and (2) for the water vapor by its temperature and partial pressure at each state and then use the *Steam Tables* to evaluate $h_{w_1}$ and $h_{w_2}$. The above expression then becomes

$$(H_2 - H_1)_w = (n_{da})(29)(\omega)(h_2 - h_1)_w$$

With the enthalpy change established for each step, the summation of the changes is

$$\begin{aligned} Q = H_{\text{pr}} - H_{\text{re}} &= (H_2 - H_f) + (H_f - H_e) + (H_e - H_d) \\ &+ (H_d - H_c) + (H_c - H_b) + (H_a - H_1) + (H_2 - H_1)_w \end{aligned}$$

where each enthalpy change is on the basis of a mole of fuel.

The above procedures may be incorporated into the analysis of closed system combustion processes. The closed-system energy equation can be written in terms of enthalpies as

$$\begin{aligned} Q &= U_{\text{pr}} - U_{\text{re}} + W \\ &= H_{\text{pr}} - H_{\text{re}} - \left[\sum_{i=1}^{N} (pn_i\overline{v}_i)_{\text{pr}} - \sum_{i=1}^{N} (pn_i\overline{v}_i)_{\text{re}}\right] + W \end{aligned} \qquad \textbf{(12.16)}$$

Neglecting the $p\overline{v}$ terms for liquid and solids, eq. (12.16) applied to constant-volume combustion (work equal to zero) becomes

$$\begin{aligned} Q &= U_{\text{pr}} - U_{\text{re}} \\ &= H_{\text{pr}} - H_{\text{re}} - \overline{R}\left[\sum_{i=1}^{N} (n_iT)_{\text{pr}} - \sum_{i=1}^{N} (n_iT_i)_{\text{re}}\right]_{\text{gases}} \end{aligned} \qquad \textbf{(12.16a)}$$

In general $\Sigma n_{\text{pr}}$ and $\Sigma n_{\text{re}}$ are not equal and $T_{\text{pr}}$ is not equal to $T_{\text{re}}$.

## Example 12.j

Methane gas and atmospheric air are supplied to a burner at 57 F and 250 F respectively. The air supply has a humidity ratio of 0.00978 lbm w/lbm da. Ninety percent of the carbon burns to $CO_2$, and the remainder to CO; 95 percent of the hydrogen burns to $H_2O$, and the remainder appears as $H_2$ in the products. Fifteen percent excess air is supplied. Calculate the heat transferred during the constant-pressure steady-flow process, per pound mole of $CH_4$, assuming the products of combustion are at 1000 R. Work and kinetic energy changes are zero and $p_{\text{atmos}} = 14.7$ psia.

**Solution:**

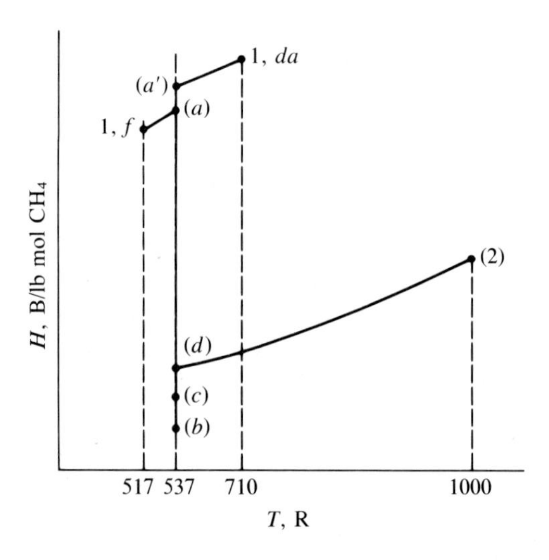


The basis selected for the solution is 1 lb mol of methane. The accompanying *H-T* diagram shows the various states related to the combustion process. The dry-air supply is $m_{da} = 1.15$ *TA*. Therefore

$$m_{da} = (1.15)\left[1.0 + \left(\frac{2.0}{2}\right)\right]\left(\frac{1}{0.21}\right)(29)$$
$$= 317.6 \text{ lbm da/lb mol } CH_4$$

The products of complete combustion, state (*b*), are

| | |
|---|---|
| $CO_2$ | 1.000 lb mol |
| $H_2O(l)$ | 2.000 lb mol |
| $O_2$ | 0.300 lb mol |
| $N_2$ | 8.652 lb mol |

At state ($c$), due to incomplete combustion, the composition is

| | | | |
|---|---|---|---|
| $CO_2$ | 0.900 lb mol | $H_2$ | 0.100 lb mol |
| CO | 0.100 lb mol | $O_2$ | 0.400 lb mol |
| $H_2O(l)$ | 1.900 lb mol | $N_2$ | 8.652 lb mol |

The only change between ($c$) and ($d$) is the vaporization of 1.900 lb mol of $H_2O$. The composition at state (2) is identical to that at ($d$). The enthalpy changes are, per pound mole of $CH_4$,

$$(H_a - H_1)_{\text{fuel}} = 8.29(537 - 517) = 165.8 \text{ B}$$

$$(H_{a'} - H_1)_{\text{da}} = 6.97(537 - 710) = -1205.8 \text{ B}$$

$$(H_b - H_a) = \Delta H^0_{C,CH_4} = -383{,}020 \text{ B}$$

$$\begin{aligned}(H_c - H_b) &= -\ [0.10\ (-122{,}970)] - [0.10\ (-121{,}750)] \\ &= 24{,}472 \text{ B}\end{aligned}$$

$$(H_d - H_c) = (1.900)(18{,}916) = 35{,}940 \text{ B}$$

Values used in determining the above enthalpy changes were obtained from Appendix G, Figure 12.3, and Table 12.1. The table below applies to process ($d$) to (2). The $\overline{C}_{p_a}$ values were obtained from Appendix G.

| **Gas** | ***n*, lb mol** | $\overline{C}_{p_a}$ | $n\overline{C}_{p_a}$ |
|---|---|---|---|
| $CO_2$ | 0.900 | 10.1 | 9.09 |
| CO | 0.100 | 7.03 | 0.703 |
| $H_2O$ | 1.900 | 8.14 | 15.47 |
| $H_2$ | 0.100 | 6.93 | 0.693 |
| $O_2$ | 0.400 | 7.29 | 2.916 |
| $N_2$ | 8.652 | 6.97 | 60.30 |
| | 12.05 | | 89.17 |

$$H_2 - H_d = 89.17(1000 - 537) = 41{,}290 \text{ B}$$

The final item is the enthalpy change of the water vapor in the air supply

$$\begin{aligned}m_w = (0.00978)(317.6) &= 3.106 \text{ lbm } H_2O \\ &= 0.1726 \text{ lb mol } H_2O\end{aligned}$$

Assuming the products of combustion are at a pressure of 1 atmosphere, the partial pressure of *all* $H_2O$ in the products is given by

$$\frac{p_w}{p_m} = x_w \equiv \frac{n_w}{n_m} = \frac{1.900 + 0.1726}{12.05 + 0.1726} = 0.170$$

and

$$p_w = 0.170(14.70) = 2.49 \text{ psia}$$

a pressure that makes perfect gas behavior a reasonable assumption. Using Appendix G and eq. (12.13), the enthalpy change of the water vapor from the atmospheric air supply is

$$\begin{aligned}(H_2 - H_1)_w &= 0.1726\,\{[(8.14)(1000 - 537)] \\ &\quad - [(8.00)(710 - 537)]\} \\ &= 411 \text{ B}\end{aligned}$$

Summing the several enthalpy changes, we get

$$\begin{aligned}Q &= (H_2 - H_d) + (H_d - H_c) + (H_c - H_b) + \Delta H_C^0 \\ &\quad + (H_{a'} - H_1)_{\text{da}} + (H_a - H_1)_{\text{fuel}} + (H_2 - H_1)_w \\ &= 41{,}290 + 35{,}914 + 24{,}472 - 383{,}020 - 1206 \\ &\quad + 166 + 411 \\ &= -282{,}000 \text{ B/lb mol } CH_4\end{aligned}$$

Several observations can be made on the basis of this example. These include:

1. Preheated air supply increases the heat output for a given $T_2$, $(H_{a'} - H_1)$ having the same sign as the enthalpy of combustion.
2. Incomplete combustion decreases the heat output.
3. Nitrogen dominates the enthalpy change for process $(d)$ to (2). Thus considerable energy goes into increasing the temperature of the nitrogen in the products. Greater heat output could be achieved in the example problem if pure oxygen or oxygen-enriched air were used in place of atmospheric air.
4. From the relative magnitudes of the enthalpy changes, it is apparent that a low fuel temperature and the water vapor in the air supply have little effect on the result.
5. In addition to promoting complete combustion to increase heat transfer for a combustion process, heat output could be increased not only by cooling the products to a low temperature, but also by condensing the water vapor in the products. To condense the water vapor the temperature of the products would have to be decreased to temperatures below the dew point of the products.

---

It should be noted that, for illustrative purposes, hydrogen was included in the products of incomplete combustion in the above example and in Example 12.b. However, in actual combustion processes it is unlikely that molecular hydrogen, $H_2$, will appear in large amounts in the products of combustion since hydrogen has a high affinity for oxygen and readily combines with oxygen to form water.

**Table 12.5** Ideal Adiabatic Flame Temperatures for Some Hydrocarbon Fuels

| Substance | Formula | Flame Temperature, R Gas/Liquid* | Flame Temperature, K Gas/Liquid* |
|---|---|---|---|
| methane | $CH_4$ | 4194 | 2330 |
| ethane | $C_2H_6$ | 4291 | 2384 |
| propane | $C_3H_8$ | 4315/4288 | 2394/2382 |
| n-butane | $C_4H_{10}$ | 4326/4296 | 2403/2387 |
| n-pentane | $C_5H_{12}$ | 4331/4304 | 2406/2391 |
| n-hexane | $C_6H_{14}$ | 4336/4308 | 2409/2393 |
| n-heptane | $C_7H_{16}$ | 4340/4312 | 2411/2396 |
| n-octane | $C_8H_{18}$ | 4343/4315 | 2413/2397 |
| acetylene | $C_2H_2$ | 5251 | 2917 |
| ethene | $C_2H_4$ | 4626 | 2570 |
| propene | $C_3H_6$ | 4518 | 2510 |
| benzene | $C_6H_6$ | 4558 | 2532 |

*Hydrocarbon phase. Combustion with stoichiometric dry air.

## 12–7 Ideal Adiabatic Flame Temperature

The maximum temperature that can be produced in the steady-flow combustion of a fuel with a given amount and source of oxygen is the product temperature when the combustion is complete and the process is adiabatic with no work or kinetic energy change. This temperature is called the *ideal adiabatic flame temperature.* The governing equation is obtained from the steady-flow energy equation, eq. (12.15), which reduces to

$$H_{pr} = H_{re}$$

For a closed-system constant-volume combustion process the equation for determining the ideal adiabatic flame temperature is, from eq. (12.16),

$$U_{pr} = U_{re}$$

As has been illustrated previously, a consistent basis must be chosen for the solution of such equations. Ideal adiabatic flame temperatures for steady-flow combustion of several hydrocarbons with the theoretical amount (stoichiometric) of dry air are given in Table 12.5. Note that these temperatures are in the approximate range of 4200 to 4600 R, except for $C_2H_2$. The hydrocarbon and dry-air supply are at 537 R for the tabulated flame temperatures.

The effect of some variables on flame temperature are illustrated by Figure 12.6. The process is one in which a fuel is burned during a steady-flow adiabatic process. To simplify the discussion it has been assumed that the fuel and oxidizer are supplied at the standard state temperature. For complete combustion the enthalpy change, $H_a - H_1$, equals $\Delta H^0_C$ of the fuel, and is independent of any excess oxidizer that may have been supplied. The enthalpy change $H_b - H_a$ is dependent on the amount of hydrogen in the fuel that burns to $H_2O$.

**Figure 12.6** *H-T* Diagram Illustrating Adiabatic Flame Temperature.

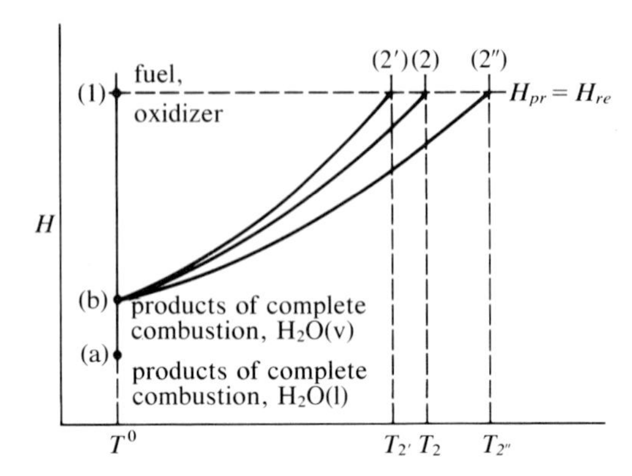


During the state change from ($b$) to (2), the gaseous products of combustion do not change composition. For this process the enthalpy change is

$$H_2 - H_b = \sum_{i=1}^{N} (n\overline{C}_{p_a})_i (T_2 - T_b) \tag{a}$$

Since the process is adiabatic, the several enthalpy changes must add to zero. Therefore

$$(H_2 - H_b) + (H_b - H_a) + (H_a - H_1) = 0$$

The temperature $T_2$ that satisfies this equation is the ideal adiabatic flame temperature. The effect of the oxidizer supplied on the adiabatic flame temperature for a given fuel can be approximated by differentiating eq. (a) and assuming that the effect of changes in $T_2$ are significantly more dependent on the number of moles of gas products than on specific heats. The differential is

$$dH = \sum_{i=1}^{N} (n\overline{C}_{p_a})_i \, dT$$

The slope of paths emanating from ($b$) is

$$\frac{dH}{dT} = \sum_{i=1}^{N} (n\overline{C}_{p_a})_i \tag{b}$$

Between ($b$) and (2) the number of moles is constant for a given fuel and oxidizer supply. Let the temperature $T_2$ be the adiabatic flame temperature for a given fuel when it is burned with the theoretical air supply. When excess air is supplied, the number of moles in eq. (b) is increased with a consequent increase in the slope $dH/dT$ of the path. This effect is shown in Figure 12.6 by path ($b$) to (2′). The result is a lower temperature $T_{2'}$. The effect of excess air on the ideal adiabatic flame temperature is further illustrated by the data in Table 12.6. The table is for the combustion of methane, $CH_4$, with dry air in steady-flow combustion. The methane and dry air are supplied at 537 R (298 K).

**Table 12.6** Effect of Excess Air on the Ideal Adiabatic Flame Temperature of $CH_4$

| Percent Theoretical Dry Air | Percent Excess Dry Air | Flame Temperature | |
|---|---|---|---|
| | | R | K |
| 100 | 0 | 4194 | 2330 |
| 110 | 10 | 3941 | 2189 |
| 120 | 20 | 3705 | 2058 |
| 130 | 30 | 3502 | 1946 |
| 140 | 40 | 3335 | 1853 |
| 150 | 50 | 3188 | 1771 |
| 160 | 60 | 3056 | 1698 |
| 170 | 70 | 2941 | 1634 |
| 180 | 80 | 2838 | 1577 |
| 190 | 90 | 2744 | 1524 |
| 200 | 100 | 2658 | 1477 |

Combustion with pure oxygen or oxygen-enriched air reduces the number of moles of product gases to a figure less than that for combustion with the theoretical air supply. The result of this change is a smaller slope and a path the nature of (*b*) to (2″), with $T_{2''}$ greater than $T_2$. With complete combustion assumed for each case, the effect of using oxygen, rather than air, in acetylene torches for welding and cutting is evident. A similar assessment also is apparent for the hydrogen–oxygen reactant system used in rocket propulsion.

Determination of ideal adiabatic flame temperatures for constant-volume combustion, a first approximation for spark-ignition internal combustion engines, is most conveniently approached using $\Delta H_C^0$ and $\overline{C}_{p_a}$ data. For this application eq. (12.16a) becomes

$$U_{pr} - U_{re} = 0 = (H_{pr} - H_{re}) - \overline{R}(n_{pr}T_{pr} - n_{re}T_{re}) \qquad \textbf{(12.16b)}$$

where $n_{pr}$ is the total number of moles of gas-phase products and $n_{re}$ is the total number of moles of gas-phase reactants. The ideal adiabatic flame temperature for constant-volume combustion is $T_{pr}$. When not all of the reactants are at the same temperature, the term $n_{re}T_{re}$ becomes $\sum_{i=1}^{N} (n_iT_i)_{re}$.

## Example 12.k

A liquid fuel that is 85 percent carbon and 15 percent hydrogen by mass undergoes complete combustion in an adiabatic process with excess dry air. The higher heating value for the fuel is 46,600 kJ/kg. The air and the fuel are supplied at 298 K and the products are at 1000 K. Find the percent excess air assuming:

(a) The process is steady flow with no work or kinetic energy changes.
(b) The process occurs in a closed system with no work.

**Solution:**

For both parts (a) and (b) we may interpret the temperature of the products, 1000 K, to be the adiabatic flame temperature. We choose as a basis 100 kg of fuel. On this basis the fuel consists of 85/12 = 7.083 kmol of carbon and 15/2 = 7.5 kmol $H_2$. The combustion reaction in equation form is

$$\begin{aligned} 7.083\,C + 7.5\,H_2 + 10.83\,(1 + x)\,O_2 + 40.75\,(1 + x)\,N_2 \\ \rightarrow 7.083\,CO_2 + 7.5\,H_2O + (10.83x)\,O_2 \\ + 40.75\,(1 + x)\,N_2 \end{aligned}$$

where the theoretical oxygen required is 7.083 + (7.5/2) = 10.83 kmol/100 kg fuel and $x$ is the fraction of excess air.

(a) The steady-flow energy equation reduces to $H_{pr} - H_{re} = 0$, which, from the $H$-$T$ diagram, can be written as

$$H_{pr} - H_{re} = (H_{pr} - H_b) + (H_b - H_a) + (H_a - H_{re}) = 0$$

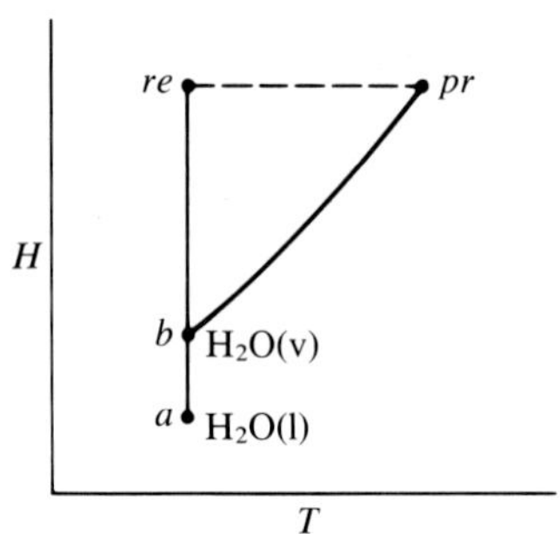


The first term is evaluated from $\overline{C}_{P_{a,m}}\,(T_{pr} - T_b)$ where the specific heat term is determined as follows:

| Product | $n$ | $\overline{C}_{P_a}$ | $n\overline{C}_{P_a}$ |
|---|---|---|---|
| $CO_2$ | 7.083 | 47.75 | 343.6 |
| $H_2O$ | 7.50 | 37.01 | 277.8 |
| $O_2$ | $10.83x$ | 32.39 | $350.8x$ |
| $N_2$ | $40.75(1 + x)$ | 30.76 | $1253.5(1 + x)$ |
| | $n_{pr}$ | | $\overline{C}_{P_{a,m}} = 1875 + 1604x$ |

The three enthalpy terms are

$$\begin{aligned} H_{pr} - H_b &= \overline{C}_{P_{a,m}}(T_{pr} - T_b) = (1875 + 1604x)(1000 - 298) \\ &= 1{,}316{,}300 + 1{,}126{,}000x \text{ kJ} \end{aligned}$$

$$H_b - H_a = (nMh^0_{fg})_{H_2O} = 7.5(18)(2442) = 329{,}670 \text{ kJ}$$

$$H_a - H_{re} = -m_f(\text{HHV}) = -100(46{,}660) = -4{,}666{,}000 \text{ kJ}$$

The sum of the above terms is zero.

$$1{,}316{,}300 + 1{,}126{,}000x + 329{,}670 - 4{,}666{,}000 = 0$$

Solving for $x$ we obtain

$$x = 2.68$$

Thus the excess air required to limit the products temperature to 1000 K is 268 percent.

(b) The governing equation is

$$U_{pr} - U_{re} = H_{pr} - H_{re} - \overline{R}(n_{pr}T_{pr} - n_{re}T_{re}) = 0$$

From the table in part (a), $n_{pr} = 55.34 + 51.59x$. From the reaction equation $n_{re} = 10.83(1 + x) + 40.75(1 + x) = 51.59(1 + x)$, and does not include the liquid fuel. Using the expression for the enthalpy change from part (a), substitution into the above equation yields

$$\begin{aligned} 1{,}316{,}300 + 1{,}126{,}000x + 329{,}670 - 4{,}666{,}000 \\ - 8.314\,[(55.34 + 51.59x)(1000) \\ - (51.59)(1 + x)(298)] = 0 \end{aligned}$$

from which $x = 4.06$. Thus for constant-volume combustion 406 percent excess air is required to limit the products temperature to 1000 K. It should be noted that the enthalpy change for part (b) is not zero. Therefore, state (pr) as shown in the $H$-$T$ diagram for part (a) does not apply to part (b).

---

## Example 12.1

A fuel mixture consists of 85.00 percent methane ($CH_4$) and 15.00 percent ethane ($C_2H_6$) by volume. The fuel is burned with 25 percent excess dry air during an adiabatic, steady-flow process. Work and kinetic energy changes are negligible. The fuel and dry air are supplied at 298 K each at a pressure of 1 standard atmosphere. The pressure remains constant during the process. Calculate the ideal adiabatic flame temperature.

**Solution:**

With the volumetric analysis of the fuel-gas mixture known, a convenient basis for calculations is 1 kmol of the fuel. The first step in the solution is the calculation of the amount of each reactant and product and the enthalpy of combustion of the fuel. The fuel table is as follows:

| Gas | $x_i$ | $\Delta H^0_{C,i}$ | $x_i \Delta H^0_{C,i}$ | kmol C | kmol $H_2$ |
|---|---|---|---|---|---|
| $CH_4$ | 0.8500 | − 890,900 | −757,270 | 0.8500 | 1.700 |
| $C_2H_6$ | 0.1500 | −1,561,000 | −234,150 | 0.3000 | 0.450 |
| | 1.0000 | | −991,420 | 1.1500 | 2.150 |

The dry-air supply is

$$n_{da} = \left(1.15 + \frac{2.15}{2.0}\right)\left(\frac{1}{.21}\right)(1.25) = 13.24 \text{ kmol da}$$

The products of complete combustion are

| Substance | $n_i$, kmol |
|---|---|
| $CO_2$ | 1.150 |
| $H_2O$ | 2.150 |
| $O_2$ | 0.556 |
| $N_2$ | 10.46 |

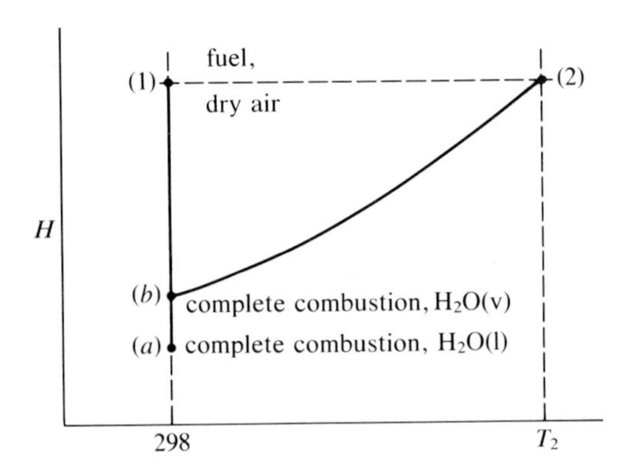


The adiabatic process is described on the $H$-$T$ diagram from which

$$H_2 - H_1 = 0 = (H_2 - H_b) + (H_b - H_a) + (H_a - H_1)$$

The enthalpy change $H_a - H_1$, $\Delta H^0_{C_m}$ of the fuel-gas mixture is

$$H_a - H_1 = \Delta H^0_{C_m} = -991{,}420 \text{ kJ}$$

From ($b$) to ($c$) the $H_2O$ formed by combustion of hydrogen in the fuel is vaporized and

$$H_b - H_a = (2.15)(18)(2442) = 94{,}580 \text{ kJ}$$

Since the sum of the enthalpy changes is zero

$$H_2 - H_b = -(H_a - H_1) - (H_b - H_a)$$
$$= 991{,}420 - 94{,}580 = 896{,}840 \text{ kJ}$$

The specific heats required to evaluate the enthalpy change ($b$) to (2) depend on the unknown temperature $T_2$. Therefore, an iterative solution for $T_2$ is required. Ideal adiabatic flame temperatures for hydrocarbon fuels burned with theoretical air are in the range of 2300 to 2600 K. To evaluate the average specific heats, let us assume $T_2$ to be 2400 K. The specific heats listed in the following table were obtained from Figure 12.3.

| Product | $n_i$ | $\overline{C}_{p_{a,i}}$ | $n_i \overline{C}_{p_{a,i}}$ |
|---|---|---|---|
| $CO_2$ | 1.150 | 55.6 | 64.0 |
| $H_2O$ | 2.150 | 44.0 | 94.7 |
| $O_2$ | 0.556 | 35.5 | 19.8 |
| $N_2$ | 10.46 | 33.7 | 352.8 |
| | | | $531.3 = \Sigma(n\overline{C}_{p_a})_i$ |

From the above equation for enthalpy change

$$H_2 - H_b = \Sigma(n\overline{C}_{p_a})_i(T_2 - T_b) = 896{,}840 = 531.3(T_2 - 298)$$

we obtain $T_2 = 1986$ K. This value is significantly less than that estimated. Therefore, the second estimate of $T_2$ used to obtain $\overline{C}_{p_a}$ values should be greater than 1986 K because the $n_i\overline{C}_{p_{a,i}}$ product decreases with temperature. Let us assume 2050 K for the iteration. Now the product table is as follows:

| Product | $n_i$ | $\overline{C}_{p_{a,i}}$ | $n_i \overline{C}_{p_{a,i}}$ |
|---|---|---|---|
| $CO_2$ | 1.150 | 54.1 | 62.1 |
| $H_2O$ | 2.150 | 42.8 | 92.0 |
| $O_2$ | 0.556 | 35.0 | 19.5 |
| $N_2$ | 10.46 | 33.2 | 347.3 |
| | | | $520.9 = \Sigma(n\overline{C}_{p_a})_i$ |

Solving for $T_2$ from

$$896{,}840 = 520.9(T_2 - 298)$$

yields $T_2 = 2020$ K. Additional iteration would not change the answer significantly.

---

Actual flame temperatures are less than the ideal adiabatic flame temperatures due to dissociation. At elevated temperatures, say 3000 R (1667 K) and above, the combustion process does not reach the complete combustion state. In terms of energy release, dissociation can be likened to incomplete combustion. The amount, or degree, of dissociation increases as temperature increases at a given pressure. Conversely, increased pressure, at a given temperature, moves the process closer to complete combustion. At a given pressure and temperature the dissociation of $CO_2$ to $CO$ and $O_2$ is significantly greater than the dissociation of $H_2O$ to $H_2$ and $O_2$ or the more complex dissociation to $H_2$, $O_2$, H, O, and OH.

## 12–8 An Alternative Approach to Combustion Process Energy Analysis

In Sections 12–6 and 12–7 enthalpy of combustion data were used in the analysis of some combustion processes. That choice permits direct identification of energy losses due to incomplete combustion and similar effects. Further, as illustrated in Examples 12.i and 12.k, the enthalpy of combustion method is applicable to liquid and solid fuels for which the higher heating value or the lower heating value is known. (In some cases this may be the only standard state reaction data available for a fuel.) While the enthalpy of combustion is adequate for many applications, an alternative approach using enthalpy of formation data is available for use when enthalpy of formation data are known for the substances involved. This approach is the subject of this section.

When the enthalpy of chemical elements in their usual composition, for example, $N_2$, $O_2$, $H_2$, was set equal to zero at a standard state of 537 R (298 K) and 1 atmosphere, it was noted in eq. (12.3) that

$$\bar{h}^0_{\text{compound}} = \Delta H^0_{f,\,\text{compound}}$$

or the standard state enthalpy is equal to the standard state enthalpy of formation.

If the enthalpy of a compound is independent of pressures and depends only on temperature, the molal enthalpy of that compound at any temperature $T$ is

$$\bar{h}(T) = \Delta H^0_f + \int_{T^0}^{T} \bar{C}_p dT = \Delta H^0_f + \underset{T^0 \to T}{\bar{C}_{p_a}} (T - T^0) \tag{12.17}$$

Equation (12.17) can be applied to any substance following perfect gas behavior, including $H_2O$, at low partial pressures. It also is applicable to subcooled liquids and solids at moderate pressures. Thus the general steady-flow energy equation can be written from eq. (12.15) as

$$\sum_{i=1}^{N} n_i \left(\bar{h}_i + \frac{M_i V_i^2}{2g_0}\right)_{\text{re}} + Q = \sum_{i=1}^{N} n_i \left(\bar{h}_i + \frac{M_i V_i^2}{2g_0}\right)_{\text{pr}} + W \tag{12.18}$$

The molal enthalpy of any reactant or product is evaluated using eq. (12.17). The energy equation for closed systems is

$$Q = \sum_{i=1}^{N} (n_i \bar{u}_i)_{\text{pr}} - \sum_{n=1}^{N} (n_i \bar{u}_i)_{\text{re}} + W \tag{12.19}$$

When using enthalpies of formation for closed-system processes, it is suggested that the enthalpy change for the process be determined first and the internal energy change be determined using the definition of enthalpy. Following this procedure

$$\begin{aligned} &\Sigma(n_i \bar{u}_i)_{\text{pr}} - \Sigma(n_i \bar{u}_i)_{\text{re}} \\ &= [\Sigma(n_i \bar{h}_i)_{\text{pr}} - \Sigma(n_i \bar{h}_i)_{\text{re}}] - [\Sigma(n_i p_i \bar{v}_i)_{\text{pr}} - \Sigma(n_i p_i \bar{v}_i)_{\text{re}}] \end{aligned}$$

where the summation is that of the number of moles of a constituent multiplied by the molal property of that constituent. Typically the $p_i \bar{v}_i$ product is negligible for solids and liquids. If the gas-phase constituents can be assumed to follow perfect gas behavior, the above equation reduces to

$$\begin{aligned}\Sigma(n_i u_i)_{pr} - \Sigma(n_i u_i)_{re} \\ = [\Sigma(n_i\bar{h}_i)_{pr} - \Sigma(n_i\bar{h}_i)_{re}] - \bar{R}[\Sigma(n_i T)_{pr} - \Sigma(n_i T)_{re}]\end{aligned}$$

where the right-hand bracket is for gas-phase reactants and products only.

---

## Example 12.m

Methane ($CH_4$) and 10 percent excess dry air are supplied to a steady-flow process at a pressure of 1 atmosphere. The methane is supplied at 537 R and the dry air at 710 R, each with negligible velocity. The products of complete combustion leave the system at 1000 R at a velocity of 600 ft/sec. There is no work during the process. On the basis of 1 lb mol of methane, determine the heat transfer for the process using enthalpy of formation data.

**Solution:**

The first step in the solution is to establish the amount of each reactant and each product. The basis chosen is a lb mol of $CH_4$. The air supplied is

$$S = 1.10\left(1.0 + \frac{2.0}{2}\right)\left(\frac{1}{0.21}\right) = 10.48 \text{ lb mol da/lb mol } CH_4$$

The complete chemical-reaction equation for the combustion process is

$$CH_4 + 2.2O_2 + 8.28N_2 \rightarrow CO_2 + 2H_2O + 0.2O_2 + 8.28N_2$$

For this process eq. (12.18) reduces to

$$\sum_{i=1}^{N} (n\bar{h}_1)_i + Q = \sum_{i=1}^{N} n_i\left(\bar{h}_2 + \frac{MV_2^2}{2g_0}\right)_i$$

The enthalpy of the reactants per pound mole of $CH_4$ is

$$\begin{aligned}\bar{h}_{CH_4} &= \Delta\bar{H}_{f,CH_4} + \bar{C}_{pa}\,(T_1 - 537)_{CH_4} = \Delta H^0_{f,CH_4} \\ &= -32{,}211 \text{ B/lb mol } CH_4\end{aligned}$$

$$\begin{aligned}\bar{h}_{da} &= (n\Delta H^0_f)_{da} + (nC_{p_a})_{da}(T_1 - 537) \\ &= 0 + (10.48)(6.97)(710 - 537) = 12{,}640 \text{ B/lb mol } CH_4\end{aligned}$$

The enthalpy of the reactants per pound mole of $CH_4$ is then

$$\begin{aligned}\Sigma n\bar{h}_{re} &= -32{,}211 + 12{,}640 \\ &= -19{,}570 \text{ B}\end{aligned}$$

Next the enthalpy and apparent molecular weight of the combustion products must be determined. A tabular approach is convenient. On the basis of 1 lb mol of $CH_4$ we have

| Gas | $n_i$, lb mol | $\Delta H^0_{f_i}$ | $n_i \Delta H^0_{f_i}$ | $\overline{C}_{pa_i}$ | $n_i \overline{C}_{pa_i} \Delta T$* | $n_i \overline{h}_{ipr}$ | $M_i$ | $n_i M_i$ |
|---|---|---|---|---|---|---|---|---|
| $CO_2$ | 1.000 | −169,300 | −169,300 | 10.050 | 4653. | −164,500 | 44 | 44 |
| $H_2O(v)$ | 2.000 | −104,054 | −208,108 | 8.142 | 7539. | −200,569 | 18 | 36 |
| $O_2$ | 0.200 | 0 | 0 | 7.292 | 675 | 675 | 32 | 6.4 |
| $N_2$ | 8.28 | 0 | 0 | 6.970 | 26,720 | 26,720 | 28 | 232 |
| | 11.5 | | | | $\Sigma nh_{pr}$ = | −337,700 | | 318 |

$^*\Delta T = T_{pr} - 537 = 1000 - 537$

The $\Delta H^0_f$ data are from Table 12.1 except for the vapor phase of $H_2O$, which has

$$\begin{aligned} \Delta H^0_f,\ H_2O(v) &= \Delta H^0_f,\ H_2O(l) + (n\overline{h}^0_{fg})_{H_2O} \\ &= -122{,}970 + (1)(18)(1050) \\ &= -104{,}054 \text{ B/lb mol } H_2O(v) \end{aligned}$$

The apparent molecular weight of the products is

$$M_{pr} = \frac{318}{11.5} = 38.4$$

Substitution into the energy equation above yields

$$\begin{aligned} Q &= \left[-337{,}700 + \frac{(11.5)(38.4)\ (600^2)}{(2)(32.17)(778)}\right] - [-19{,}570] \\ &= -315{,}000 \text{ B/lb mol } CH_4 \end{aligned}$$

## 12–9 Summary

Combustion is a rapid oxidation process during which a fuel and oxidizer undergo chemical reactions. Air is a common oxidizer. Complete combustion of a hydrocarbon fuel occurs when all of the carbon burns to form $CO_2$ and all of the hydrogen burns to form $H_2O$. The theoretical (stoichiometric) amount of air $TA$ is that required for complete combustion. When the air supplied $S$ exceeds that required, excess air is supplied.

$$\text{Excess air} = \frac{S - TA}{TA}$$

The air–fuel ratio is the ratio of the mass of air supplied to that of the fuel supplied.

A material balance for a combustion process is a tabulation of the total mass of the reactants (input) and the total mass of the products (output). By the mass conservation principle, the mass input must be equal to the mass output, although the chemical composition changes. Incomplete combustion may occur due to insufficient mixing of the reactants, or insufficient oxidizer, time, or temperature. An analysis of the products of combustion provides information that is useful in establishing the material balance and characterizing the combustion process. In the analysis of a combustion process, the selection of a basis is an important first step.

Application of the first law to combustion processes requires that an additional property, the composition, be specified at each state of the system. Previously developed energy relations are applicable provided internal energy and enthalpy terms include the effects of the chemical reaction and the composition change. These effects are included by means of standard state reactions. Standard state reactions are those in which the reactants and the products exist at the standard state (1 standard atmosphere pressure and 77 F [25 C]). The standard state enthalpy of reaction $\Delta H^0_R$ is the enthalpy of the products minus the enthalpy of the reactants. The enthalpy change for a combustion process is

$$H_{pr} - H_{re} = H_{pr} - H_B + \Delta H^0_R + H_A - H_{re}$$

where $B$ denotes the products at the standard state and $A$ denotes the reactants at the standard state. The internal energy change for a combustion process is determined from the definition of enthalpy. The standard state enthalpy of combustion is the enthalpy of reaction for a process in which the fuel is burned to completion. Values for the enthalpy of combustion are quoted for the product $H_2O$ in both the liquid and vapor phases. The standard state enthalpy of formation is the enthalpy of reaction when a compound is formed from its chemical elements. Standard state enthalpies of combustion and formation are related.

The ideal adiabatic flame temperature is the temperature of the products of combustion when complete combustion takes place with heat transfer, work, and changes in kinetic energy equal to zero. Its value depends on the fuel, the oxidizer, and the system (closed or steady flow) in which the process takes place.

## Selected References

Combustion of Gas. Reprint from *Gas Engineers Handbook*. American Gas Association. New York: Industrial Press, 1967.

Griswold, J. *Fuels, Combustion and Furnaces*. New York: McGraw-Hill, 1946.

Hottel, H. C., G. C. Williams, and C. N. Satterfield. *Thermodynamic Charts for Combustion Processes*. New York: Wiley, 1949.

Johnson, A. J., and G. H. Auth. *Fuels and Combustion Handbook*. New York: McGraw-Hill, 1951.

Jost, W. *Explosions and Combustion Processes in Gases,* translated by H. O. Croft. New York: McGraw-Hill, 1946.

Lewis, B., and G. von Elbe. *Combustion, Flames and Explosions of Gases*. New York: Academic Press, 1951.

Maxwell, J. B. *Data Book on Hydrocarbons*. Florence, Ky.: D. Van Nostrand, 1950.

McBride, B. J., S. Heimel, J. C. Ehlers, and S. Gordon. *Thermodynamic Properties to 6000 K for 210 Substances Involving the First 18 Elements*. NASA SP–3001, 1963.

Pearson, J. D., and R. C. Fellinger. *Thermodynamic Properties of Combustion Gases*. Ames, Iowa: Iowa State University Press, 1965.

Penner, S. S. *Thermodynamics*. Reading, Mass.: Addison-Wesley, 1968.

Razjevic, K. *Handbook of Thermodynamic Tables and Charts*. New York: McGraw-Hill, 1976.

*Selected Values of Chemical Thermodynamic Properties. NBS Tech. Note 270–3* and *270–4,* 1968, 1969.

Smith, M. L., and K. W. Stinson. *Fuels and Combustion*. New York: McGraw-Hill, 1952.

Stull, D. R., and H. Prophet. *JANAF Thermochemical Tables,* 2nd ed. NSRDS–NBS37. Washington, D.C.: U.S. Government Printing Office, 1971.

## Problems

12.1 A gas fuel at 60 F and 14.70 psia is burned with the theoretical amount of dry air at 77 F and 14.70 psia. Calculate the volume of air required per cubic foot of fuel and the air-fuel ratio if the fuel is (a) methane, (b) propane, and (c) a mixture of 90 percent methane and 10 percent ethane by volume.

12.2 Assuming a gasoline to have an average molecular formula of $C_8H_{16}$, determine the theoretical air requirement, in kilograms of dry air per kilogram of fuel, for complete combustion.

12.3 A fuel-gas mixture contains 60 percent $CH_4$, 30 percent $C_3H_8$, 5 percent CO, and 5 percent $N_2$ by volume. The fuel is burned completely with 10 percent excess air. The air supply is at 80 F dry bulb, 72 F wet bulb, and standard barometer. Calculate the material balance on the basis of 1 pound mole of fuel?

12.4 An Oklahoma natural gas has the volumetric analysis of 89.1 percent $CH_4$, 4.2 percent $C_2H_6$, 3.9 percent $N_2$, and 2.8 percent $CO_2$. The gas, at 60 F and 14.70 psia, is burned with 15 percent excess air having a dry-bulb temperature of 80 F, relative humidity of 57 percent, and pressure of 14.70 psia. Combustion is complete with the products at 400 F and 14.70 psia. Assuming the fuel is burned at the rate of 1000 $ft^3$/min and the combustion products are at 14.70 psia and 400 F, calculate (a) the rate of atmosphere air supply, cubic feet per minute; (b) the volume rate of flow of the products, cubic feet per minute; and (c) the dew point of the products of combustion.

12.5 Liquid octane, $C_8H_{18}$, is burned with the theoretical amount of dry air for complete combustion. Combustion, however, is incomplete with 90 percent of the carbon burning to $CO_2$ and 95 percent of the hydrogen burning to $H_2O$. Calculate the volumetric analysis of the combustion products assuming $H_2O$ in the products follows perfect gas behavior and the unburned hydrogen in the products is all in the form of $H_2$.

12.6 Forty-eight kilograms of carbon are burned to $CO_2$ and CO during combustion with 500 kg of dry air. Determine (a) the ratio of air supplied to air required for complete combustion and (b) the maximum mass of $CO_2$ that could be formed under the given conditions.

12.7 A fuel gas has the following volumetric analysis: $C_2H_6$, 30 percent; $C_3H_8$, 50 percent; CO, 15 percent; and $N_2$, 5 percent. When the fuel is burned with air of 90 F dry bulb, 78 F wet bulb, and standard barometer, the Orsat (dry gas) analysis of the combustion products is: $CO_2$, 9.79 percent; CO, 1.50 percent; $O_2$, 4.57 percent; and $N_2$, 84.14 percent. Using a nitrogen balance to determine the air supply, calculate (a) the percent excess air supplied, (b) the pounds of $H_2O$ in the products per pound mole of fuel, and (c) the overall material balance per pound mole of the fuel.

12.8 The dry exhaust gas analysis from an internal combustion engine shows 8.70 percent $CO_2$, 0.3 percent $O_2$, 8.9 percent CO, 3.7 percent $H_2$, and 78.4 percent $N_2$. Assuming the engine fuel contains only carbon and hydrogen, calculate the kilograms of dry air supplied per kilogram of fuel.

12.9 Suppose an internal combustion engine uses a fuel containing only carbon and hydrogen and the air supply to the engine is at 77 F and 14.20 psia with a relative humidity of 48 percent. The volumetric analysis of the dry exhaust gases from the engine is 10.8 percent $CO_2$, 5.7 percent CO, 1.1 percent $CH_4$, 2.3 percent $H_2$, 0.5 percent $O_2$, and 79.6 percent $N_2$. Calculate the pounds mass of atmospheric air supplied per 100 pounds mass of fuel. At a certain section in the engine exhaust system the pressure is 15 psia and the temperature 1400 R. If fuel is supplied to the engine at the rate of 1.14 lbm/min and if the exhaust gas velocity at the given section is 100 ft/sec, what is the area of the exhaust passage at the given section?

12.10 Using enthalpy-of-formation data, calculate the standard state enthalpy of combustion $CH_4$, $C_6H_6$, and CO.

12.11 Calculate the standard state enthalpy of formation of CO and $H_2O(l)$ from appropriate enthalpy-of-combustion data.

12.12 Carbon monoxide and dry air, each at the standard state, undergo a process in which all carbon monoxide goes to $CO_2$. Assuming all products are at the standard state, calculate the enthalpy change when (a) the reaction takes place with the theoretical air supply and (b) the reaction takes place with 100 percent excess air. Would the result be changed if theoretical oxygen or 100 percent excess oxygen were the oxidizer rather than air?

12.13 One alternative for the production of synthetic gas is to pass steam over a hot bed of coke (carbon). Assuming the reaction is

$$H_2O(g) + C(s) \rightarrow CO(g) + H_2(g)$$

calculate $\Delta H_R^0$ for the reaction at the standard state.

12.14 Suppose three fuels are to be considered for a steam power plant. The first is a coal having a higher heating value of 11,000 B/lbm and a cost of $37.00 per ton; the second, a number 6 (Bunker C) fuel oil with a higher heating value of 18,750 B/lbm, specific gravity 0.967 at $20.00 per barrel (42 gallons); and, third, a natural gas with the analysis, by volume, of 80.5 percent $CH_4$, 18.2 percent $C_2H_6$, and 1.3 percent $N_2$ available at a cost of $2.52 per 1000 cubic feet (60 F, 1 atmosphere). On the basis of cost per unit of heating value, which fuel would be the most desirable?

12.15 What proportions, by mass, will a mixture of liquid ethanol and gasoline have if the higher heating value of the mixture is 19,500 B/lbm? The gasoline mixed has the formula $C_8H_{17}$, and a higher heating value of 20,260 B/lbm.

12.16 Higher heating values for fuels are determined using an oxygen bomb calorimeter with combustion taking place in a constant-volume vessel in the presence of high excess oxygen. Data obtained from the constant-volume closed-system experiments are applied frequently to steady-flow constant-pressure combustion processes. Using $\Delta H_c^0$ (higher heating values), compare constant-volume ($\Delta U_c^0$) and constant-pressure ($\Delta H_c^0$) values for (a) carbon, (b) $H_2$ ($H_2O$ liquid in products), and (c) jet-engine fuel from Table 12.2.

12.17 For the standard state reaction

$$C_3H_8 + \text{oxygen} \rightarrow 2\ CO_2 + 0.7\ CO + 0.3\ CH_4 + 1.5\ H_2 + 1.9\ H_2O(l)$$

determine $\Delta H_R^0$.

12.18 Using $\overline{C}_{p_a}$ data from Appendix G, calculate the enthalpy change of 7.53 lbm of $N_2$ between 800 and 3000 R. Repeat the calculation assuming constant $C_p$ as given in Table 5.5. Compare the results.

12.19 Suppose a closed system of 0.73 kmol of dry air undergoes an adiabatic process at 105 kPa, from 300 K to a final state of 1100 K. Calculate the work for the process (a) assuming constant specific heat from Table 5.5, and (b) using average specific heat data.

12.20 Using average specific heat data from Appendix G, determine the enthalpy and internal energy change per pound mole of $H_2O$ between 760 R and 2660 R. Using the *Steam Tables* determine the enthalpy and internal energy changes between 300 F and 2200 F at pressures of 1.0 psia, 10 psia, and 50 psia. Compare the results on the basis of a pound mole.

12.21 Liquid methyl alcohol, $CH_3OH$, undergoes complete combustion with the theoretical amount of dry air in a rigid vessel at low pressure. The temperatures of the reactants and the products are respectively 537 R and 1000 R. Find the heat transfer per pound mole of fuel for the process.

12.22 One kilogram mole of a fuel gas consisting of 80 percent $CH_4$ and 20 percent $C_2H_6$, by volume, is burned completely with the theoretical air supply. The fuel and dry air are each at 25 C and 101.3 kPa. The steady-flow process takes place in such a manner that no work is done and kinetic energies can be neglected. The products of combustion are at 900 K. Calculate the heat transfer for the process.

12.23 Calculate the enthalpy of combustion of hydrogen ($H_2$) and methane ($CH_4$) at 800 R. Assume $H_2O$ in the products to be in the vapor phase. Compare the results to $\Delta H_c^0$ for the substances also with $H_2O$ in the vapor phase. Note the effect of temperature on $\Delta H_c$.

12.24 Using an *H-T* diagram explain why higher temperatures are obtained with an oxygen-acetylene cutting torch than could be obtained if the acetylene were burned with air.

12:25 Liquid octane, $C_8H_{18}$, is burned with 90 percent of the theoretical dry-air supply. Carbon monoxide is the only combustible in the products and there is no free $O_2$ in the products. If the process takes place in steady flow, if kinetic energies are negligible, and if there is no heat transfer during the process, what is the work per kilogram of octane if the products of combustion are at 800 K and the reactants are at 298 K?

12.26 During a constant-volume closed-system process, $CH_4$, initially at 77 F, and 20 percent excess dry air at 800 F burn completely. The products of combustion are at 1040 F. Determine the heat transfer per pound mole of $CH_4$.

12.27 During a steady-flow process gas fuel is burned in a furnace with dry air supplied from an air heater as shown in the figure. Fuel is supplied to the burner at 77 F and air enters the heater at $t_a$ equal to 77 F. The air leaving the heater is at $t_b$ equal to 240 F. Seventy-five percent of the heating value of the fuel is transferred from the combustion products in the furnace. The fuel analysis, by volume, is 83.2 percent $CH_4$, 10.6 percent $C_2H_6$, and 6.2 percent $N_2$. The volumetric analysis of the dry gases leaving the furnace (section [x]) is 11.0 percent $CO_2$, 3.76 percent $O_2$, and 85.24 percent $N_2$. Calculate the temperature of the combustion products entering the heater, $T_x$, and of the products leaving the heater, $T_y$.

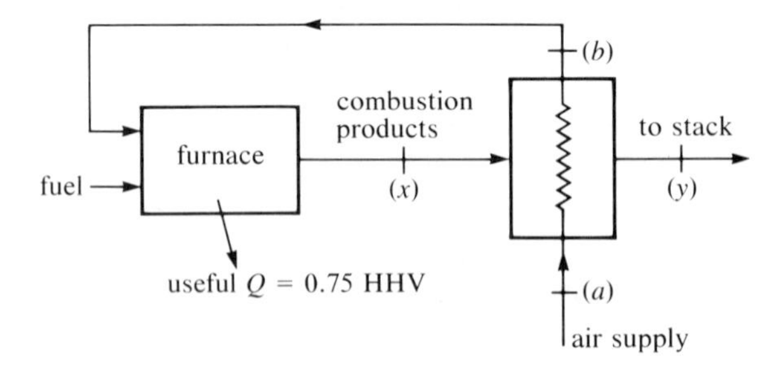


12.28 An unknown amount of $C_2H_6(g)$ is burned completely with dry air in an adiabatic, steady-flow process during which no work is done and kinetic energies can be neglected. The fuel and air are supplied at 25 C and the combustion products are at 1400 K. Using a basis of 100 kmol of air supplied, determine the mass of $C_2H_6$ supplied.

12.29 Suppose that an internal combustion engine operates under steady-flow conditions and delivers 150 HP. Further, suppose that the engine fuel is liquid octane, $C_8H_{18}$, and that the theoretical amount of dry air is supplied to the engine. If the air and fuel are supplied at 77 F, if the products of combustion are at 1400 R, and if fuel is supplied at the rate of 78 lbm/hr, what is the heat transfer rate from the engine, B/hr? Assume complete combustion.

12.30 Distillate fuel with an average molecular formula of $C_{11.6}H_{23.2}$ and a higher heating value of 46,300 kJ/kg is burned in a gas turbine. The fuel enters the burners at 25 C along with a dry air supply at 25 C. Products leaving the combustors must be at a temperature no greater than 1390 K due to the limitations imposed by the turbine blade materials. Assuming negligible kinetic energies, no heat transfer, and complete combustion, calculate the minimum excess air that must be supplied.

12.31 An internal combustion engine uses a fuel containing carbon and hydrogen only. The fuel and dry air are supplied to the engine at 77 F. Heat losses from the engine to cooling water and by radiation are 25 percent of the higher heating value of the fuel. The engine exhaust gases are at 1290 F, the engine delivers 390 HP, and the higher heating value of the fuel is 19,600 B/lbm. The volumetric analysis of the dry exhaust gases is 8.0 percent $CO_2$, 3.4 percent CO, 1.5 percent $O_2$, 2.3 percent $H_2$, and 84.8 percent $N_2$. Assuming steady-flow conditions and negligible velocities, determine the pounds mass of fuel supplied per hour. *Suggestion:* Make preliminary calculations on the basis of 1 lb mol of dry exhaust gas, including the net work and mass of fuel. Then proceed to a rate basis.

12.32 During a steady-flow process with negligible velocities and no work CO is burned completely with dry air. Determine the ideal adiabatic flame temperature (a) if the theoretical amount of air is supplied and (b) if 50 percent excess air is supplied.

12.33 Refer to Figure 12.6. Why do the constant composition lines, *b* to 2, for example, have an increasing slope as temperature increases?

12.34 Suppose a fuel, say $CH_4$, is supplied to a burner at 77 F. Using an *H-T* diagram, show the effect of the following variables on the adiabatic flame temperature: (1) excess air supplied; (2) combustion with pure oxygen rather than air; (3) preheating theoretical air supplied to the burner; (4) incomplete combustion. Use complete combustion with theoretical air as the base for the comparisons.

12.35 Acetylene is burned completely during a steady-flow process with negligible velocities and no work. Calculate the ideal adiabatic flame temperature if combustion is with theoretical air.

12.36 A natural gas contains, by volume, 76.8 percent $CH_4$ and 23.2 percent $C_2H_6$. The gas is burned with 15 percent excess air during a steady-flow process in which velocities are negligible and no work is done. The fuel is supplied to the burner at 25 C and the dry air at 127 C. Calculate the ideal adiabatic flame temperature.

12.37 A system of $C_3H_8$ liquid and the theoretical amount of dry air is at 77 F. Calculate the ideal adiabatic flame temperature for closed-system, constant-volume combustion.

12.38 Liquid octane and the theoretical amount of dry air undergo a constant-volume, adiabatic combustion process. Initially the octane is at 77 F, the dry air is at 740 F, and each is at 100 psia. The system is ignited and combustion goes to completion. Determine the final temperature and pressure. Indicate any assumptions made that introduce errors into the calculated results.

12.39 For the combustion process described in problem 12.3, let the fuel be supplied at 77 F and the products be at 800 F. Using enthalpy-of-formation data from Table 12.1, determine the heat transfer for the process if it occurs in a steady-flow system with no work or kinetic energy change.

# 13

# Some Aspects and Applications of One-Dimensional Compressible Flow

*In the description of steady flows in previous chapters the assumption that the flow was one dimensional—that is, that flow properties vary only in the direction of flow—was either stated or implied. This assumption requires that the flow be uniform in planes normal to the direction of flow. While most flows are not one dimensional, one-dimensional treatment is a good approximation for many flows. In this chapter we will study compressible flows that occur in passages for which the area of cross section varies in the direction of flow. These flows will be treated as one dimensional. Two of the more common devices that exhibit area variation in the flow direction are nozzles and diffusers. A nozzle is a varying-area flow passage that, when operated at design steady-flow conditions, increases the stream kinetic energy as pressure and enthalpy decrease. Diffusers also are varying-area passages. However, their purpose is to increase pressure at the expense of kinetic energy during steady flow. Other devices subject to one-dimensional flow analysis range from flow meters to reaction propulsion systems.*

## 13–1 Static and Stagnation States

To this point we have considered only static properties of the working fluid. Most simply, the *static properties,* for example, static pressure and static temperature, are those that would be measured by instruments whose velocity relative to the moving stream is zero. Figure 13.1 illustrates the nature of pressure measurements for one-dimensional flow. The static pressure can be measured by means of a small hole in the channel wall called a pressure tap. The pressure tap, as at section *A,* is normal to the flow, and thus the velocity at that section does not adversely affect the pressure reading. Frequently taps are placed at several locations around the periphery and a collector ring connected to each tap leads to the measuring instrument.

Stagnation states are often referred to as "total" states and are dependent on the local stream velocity. The stagnation pressure probe, an upstream-facing tube terminated by a pressure-measuring instrument, as shown in Figure 13.1, measures the pressure that results when the moving stream is decelerated to zero velocity with a consequent increase in pressure.

**Figure 13.1** Static and Stagnation Pressures

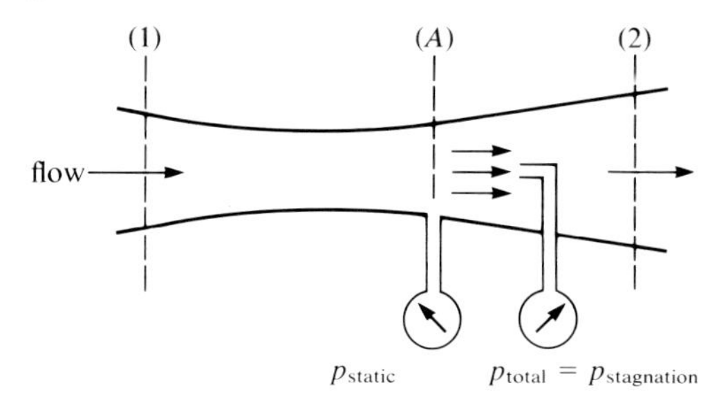


We shall define the isentropic stagnation state (hereafter denoted simply as the stagnation state) by

$$h_t = h + \frac{V^2}{2g_0} \tag{13.1}$$

and

$$s_t = s \tag{13.2}$$

where the subscript $t$ denotes properties at the stagnation state, while properties not carrying the subscript are static properties. In a physical sense this definition is based on isentropic deceleration of a stream having velocity $V$, static enthalpy $h$, and static entropy $s$, to a state where the velocity is zero. Figure 13.2a is an enthalpy–entropy diagram illustrating the static and stagnation states. (See Section 7–7 for a discussion of $h$-$s$ diagrams.)

Suppose flow from section (1) to section (2) through the varying-area passage in Figure 13.1 undergoes an expansion process that takes place reversibly and adiabatically. If the work is zero, the steady-flow energy equation becomes

$$h_1 + \frac{V_1^2}{2g_0} = h_2 + \frac{V_2^2}{2g_0}$$

or, by eq. (13.1),

$$h_{t_1} = h_{t_2} = h_t = \text{constant} \tag{13.1a}$$

Note that only the adiabatic no-work condition has been applied to this point. For isentropic flow

$$s_1 = s_2$$

or, from eq. (13.2),

$$s_{t_1} = s_1 = s_2 = s_{t_2} \tag{13.2a}$$

**Figure 13.2** Enthalpy–Entropy Diagrams Illustrating Static and Stagnation States. The diagrams apply to general pure substances. *a.* A static state and the corresponding stagnation state. *b.* *h-s* diagram for an isentropic expansion process between states (1) and (2), Figure 13.1. *c.* *h-s* diagram for an adiabatic irreversible expansion process between states (1) and (2), Figure 13.1.

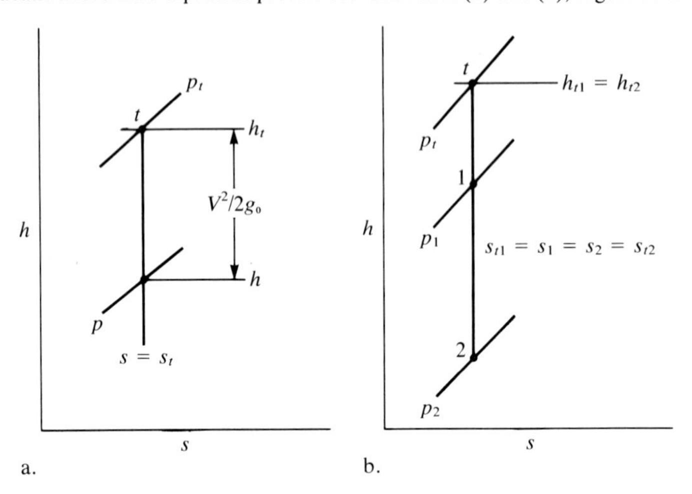


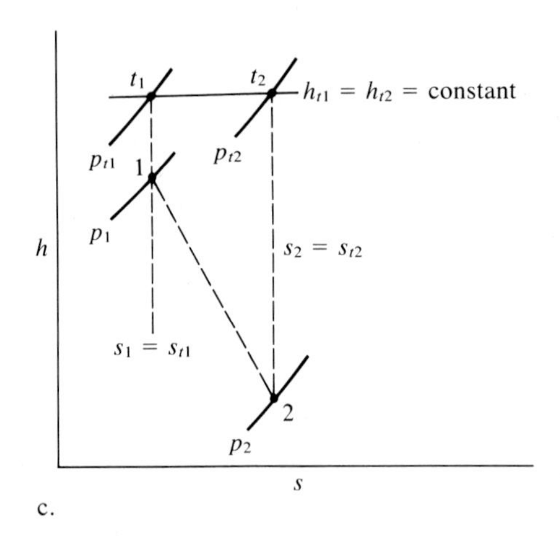


The *h-s* diagram for this process is shown in Figure 13.2b. From the above, two important characteristics of such flows are apparent. By the first law the stagnation enthalpy remains constant for all steady-flow processes, reversible or irreversible, during which no work or heat interactions with the surroundings take place. (The mechanical potential energy terms have been dropped because of negligible changes in elevation and the low density of the fluids undergoing such processes.) Second, for reversible and adiabatic flow, the static entropy does not change. Therefore

$$ds = ds_t = 0$$

It is evident from Figure 13.2b that the stagnation state is a useful reference state for isentropic flows.

It is useful to apply a previously developed property relation to isentropic processes that do not involve work. Earlier, the second $Tds$ equation $Tds = dh - vdp$ was demonstrated to be true for pure substances. Applied to stagnation states the equation becomes

$$T_t ds_t = dh_t - v_t dp_t \qquad \textbf{(a)}$$

From the preceding paragraph $dh_t = 0$ and $ds_t = 0$. It then follows that

$$dp_t = 0 \qquad \textbf{(13.3)}$$

for such processes involving pure substances. In summary, $h_t$, $s_t$, and $p_t$ remain constant for all steady-flow isentropic processes if no work interaction takes place.

Had the expansion from section (1) to section (2) of Figure 13.1 taken place with no work and adiabatically but not reversibly, the entropy would, by the entropy principle, increase. For such conditions eq. (a) becomes

$$T_t ds_t = -v_t dp_t > 0$$

From the preceding equation it is evident that the stagnation pressure must decrease. The $h$-$s$ diagram in Figure 13.2c shows the static states (1) and (2) and the corresponding stagnation states for the irreversible adiabatic flow. This flow would occur if the area of cross section of the passage were very small compared with the length. The irreversibility would be caused by fluid friction effects. This flow is in constrast to that illustrated in Figure 13.2b. Isentropic flow could be approached, for example, in the passage if the area of cross section were large compared with the passage length.

## Example 13.a

Air flows reversibly and adiabatically through a varying area passage. At the passage inlet the static temperature is 1500 R and the velocity is 500 ft/sec. The air leaves the passage with a velocity of 2000 ft/sec and a static pressure of 12.25 psia. No shaft work is done during the steady-flow process. Calculate:

(a) The stagnation temperature at the inlet
(b) The stagnation temperature at the outlet
(c) The static temperature at the outlet
(d) The stagnation pressure at the inlet
(e) The ratio of the inlet area to the outlet area

**Solution:**

The $h$-$s$ diagram in Figure 13.2b applies since the flow is isentropic. For perfect gases the $T$-$s$ and the $h$-$s$ diagrams are similar since for such gases enthalpy is a function of temperature only.

(a) From the definition of stagnation enthalpy

$$h_{t_1} = h_1 + \frac{V_1^2}{2g_0}$$

which, for a perfect gas with constant specific heats, becomes

$$C_p(T_{t_1} - T_1) = \frac{V_1^2}{2g_0}$$

Solving for $T_{t_1}$ we have

$$T_{t_1} = \frac{V_1^2}{2g_0\, C_p} + T_1 = 1521 \text{ R}$$

(b) With no work or heat interactions

$$T_{t_2} = T_{t_1} = 1521 \text{ R}$$

(c) The static temperature at the outlet can be determined as

$$T_2 = T_{t_2} - \frac{V_2^2}{2g_0\, C_p}$$
$$= 1188 \text{ R}$$

(d) For isentropic no-work flow

$$p_{t_1} = p_{t_2}$$

From the relation between pressure and temperature for isentropic flow for a perfect gas with constant specific heats (eq. [7.10a]) we can write

$$p_{t_2} = p_2 \left(\frac{T_{t_2}}{T_2}\right)^{(\gamma-1)/\gamma} = 12.25\left(\frac{1521}{1188}\right)^{3.500}$$
$$= 29.08 \text{ psia} = p_{t_1}$$

(e) To determine the area ratio for the steady-flow process, the mass conservation expression yields

$$\frac{A_1}{A_2} = \left(\frac{V_2}{V_1}\right)\left(\frac{v_1}{v_2}\right) = \frac{V_2}{V_1} \left|\frac{RT_1}{p_1}\right| \frac{p_2}{RT_2}$$

The static pressure at the inlet can be calculated as

$$p_1 = p_{t_1}\left(\frac{T_1}{T_{t_1}}\right)^{\gamma/(\gamma-1)} = 29.08\left(\frac{1500}{1521}\right)^{3.50} = 27.71 \text{ psia}$$

The area ratio is then

$$\frac{A_1}{A_2} = \frac{2000}{500}\,\Bigg|\,\frac{1500}{1188}\,\Bigg|\,\frac{12.25}{27.71} = 2.232$$

---

## Example 13.b

At a particular section in a steam nozzle the stagnation pressure is 120 psia, the stagnation temperature is 380 F, and the static pressure is 50 psia. Determine:

(a) The static temperature
(b) The velocity
(c) The mass rate of flow per unit area $\dot{m}/A$

**Solution:**

(a) From the *Steam Tables*

$$h_t = 1213.3 \text{ B/lbm and } s_t = 1.6157 \text{ B/lbm R}$$

With $s_t = s$ we have, at 50 psia,

$$s_t = s = 1.6157 < s_g \text{ at 50 psia}$$

Therefore the static temperature is the saturation temperature for 50 psia, 281.03 F. The states are shown on the *h-s* diagram below.

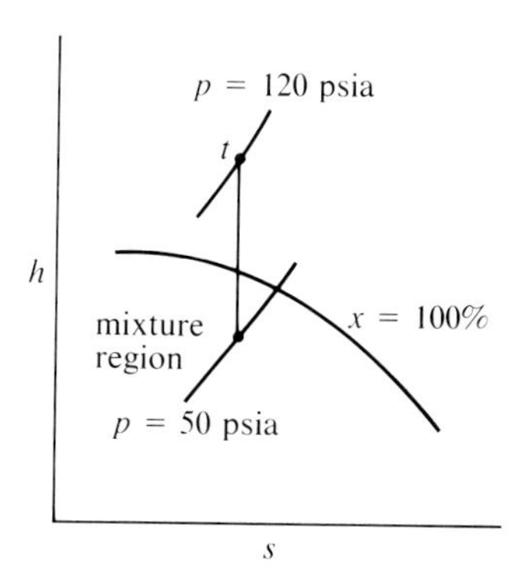


(b) The velocity, by eq. (13.1), is

$$V = [2g_0\,(h_t - h)]^{1/2}$$

The static enthalpy $h$ is fixed by the static pressure of 50 psia and $s = s_t = 1.6157$. From Table 2 of the *Steam Tables* at 50 psia

$$s_t = s = 1.6157 = 0.41129 + (x)(1.2476)$$

and the quality $x$ at the static state is 0.9654 which yields the static enthalpy

$$\begin{aligned} h &= 250.24 + (0.9654)(924.2) \\ &= 1142.2 \text{ B/lbm} \end{aligned}$$

Finally the velocity at the given section is

$$\begin{aligned} V &= [2(778)(32.17)(1213.3 - 1142.2)]^{1/2} \\ &= 1886 \text{ ft/sec} \end{aligned}$$

(c) From the continuity equation

$$\frac{\dot{m}}{A} = \frac{V}{v} = \frac{1866}{8.223} = 227.0 \text{ lbm/sec ft}^2$$

where $v$, the specific volume at the static state, was calculated as

$$v = (0.9654)(8.518) = 8.223 \text{ ft}^3\text{/lbm}$$

---

## 13–2 The Isentropic Speed of Sound and the Mach Number

An equation based on Newton's second law of motion relates the net force acting on a steady-flow system (control volume) to the change in velocity of the fluid passing through the control volume. For one-dimensional flow this equation is

$$F = \frac{\dot{m}}{g_0}(V_2 - V_1) \tag{13.4}$$

The velocities are taken as positive in the direction of flow.

One application of eq. (13.4) is the development of an equation for wave speeds in compressible fluids. A wave is a disturbance that propagates through a medium. Figure 13.3a shows a one-dimensional wave moving in a channel at a constant speed $a$ into a stationary fluid at state (1), leaving behind it the fluid at static state (2), which has velocity $u$. The reference frame is the fixed channel wall. In Figure 13.3b the reference frame is changed to the wave that is now stationary. The conditions in this figure were obtained by subtracting the wave speed $a$ from the velocities in Figure 13.3a. Static states are unchanged by this transformation. The channel wall now moves with velocity $a$ to the left and the flow is steady with respect to the stationary wave. The flow at both states (1) and (2) is to the left, with $V_1 = a$ and $V_2 = a - u$. We shall confine our discussion to compression waves. For such

**Figure 13.3** Moving and Standing Waves for a Compressible Fluid. *a.* A moving one dimensional wave. A state change is caused by the passing of the wave. *b.* A stationary wave with a state change occurring as the fluid passes through the wave.

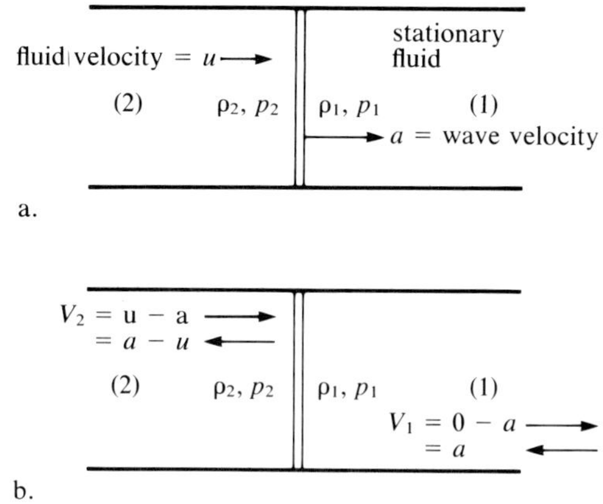

waves the wave thickness is very small, a fact that has been substantiated by experiments. We now apply eq. (13.4) to a control volume enclosing the stationary wave in Figure 13.3b. Since the wave is thin, the fluid shear at the wall will be neglected. The net force acting on the control volume then becomes the sum of the $pA$ forces. Noting that the sign convention is positive in the direction of flow, we see that eq. (13.4) yields

$$p_1A_1 - p_2A_2 = \frac{\dot{m}}{g_0}(V_2 - V_1) \tag{13.4a}$$

Because of the thinness of the wave, areas $A_1$ and $A_2$ can be taken as equal, even if the channel cross section varies in the direction of flow. For these conditions eq. (13.4a) becomes

$$(p_1 - p_2) = \frac{\dot{m}}{A}\frac{1}{g_0}(V_2 - V_1) \tag{a}$$

By the continuity equation, with $\dot{m}/A$ equal for each face of the wave

$$\frac{\dot{m}}{A} = \rho_1V_1 = \rho_2V_2$$

and

$$V_2 = \frac{\rho_1}{\rho_2}V_1 \tag{b}$$

Substitution of eq. (b) into eq. (a) yields

$$p_1 - p_2 = \frac{\dot{m}}{A}\frac{1}{g_0}\left[\frac{\rho_1}{\rho_2}V_1 - V_1\right]$$

$V_1$ from Figure 13.3b is the wave speed $a$. Therefore

$$p_1 - p_2 = \frac{\dot{m}}{A} \frac{a}{g_0} \left[ \frac{\rho_1}{\rho_2} - 1 \right] \tag{c}$$

However

$$\frac{\dot{m}}{A} = \rho_1 a$$

and eq. (c) becomes

$$p_1 - p_2 = \frac{\rho_1 a^2}{g_0} \left( \frac{\rho_1 - \rho_2}{\rho_2} \right) \tag{d}$$

Solving eq. (d) for $a$, the wave velocity, yields

$$a^2 = \frac{\rho_2}{\rho_1} g_0 \left( \frac{p_2 - p_1}{\rho_2 - \rho_1} \right) \tag{13.5}$$

Equation (13.5) is valid for waves producing pressure changes ranging from magnitude $dp$ to strong waves where $p_2 - p_1$ is a finite difference. An example of a strong compression wave is a blast wave caused by an explosion.

For weak waves, such as sound waves, the density $\rho_2$ approaches $\rho_1$, the static pressure $p_2$ differs only infinitesimally from $p_1$, and the process across the wave approaches an isentropic process as a limit. Waves of this nature propagate at the isentropic speed of sound $c$, which is given by

$$c = \left[ g_0 \left( \frac{\partial p}{\partial \rho} \right)_s \right]^{1/2} \tag{13.6}$$

For any pure substance in the gaseous phase, the expression $pv^k$ = constant can be used in the vicinity of a given state to describe the isentropic path through that state. $k$ is called the isentropic exponent. Writing this expression in the form $p/\rho^k$ = constant and differentiating it according to eq. (13.6) produces

$$c = [k g_0 p v]^{1/2} \tag{13.6a}$$

For perfect gases $k$ and $\gamma$ are identical at a given temperature, and thus

$$c = [\gamma g_0 R T]^{1/2} \tag{13.6b}$$

expresses the isentropic speed of sound in a perfect gas.

The Mach number $N_m$ is defined as

$$N_m = \frac{V}{c} \tag{13.7}$$

For internal flow the isentropic speed of sound must be calculated from static properties of the fluid at the section in a flow passage where the velocity is $V$. As we shall see, both $V$ and $c$ may change as a fluid moves from section to section through a flow passage. When aircraft flight speeds are expressed in terms of the Mach number it is customary to evaluate the sonic velocity $c$ from static properties of the atmosphere at the aircraft location and the velocity $V$ as the air speed, that is, the speed of the aircraft relative to its local atmosphere.

When conditions are such that the velocity is less than the speed of sound, that is, $N_m < 1$, the flow or air speed is said to be subsonic. For velocities greater than the local speed of sound the flow or air speed is termed supersonic. Hypersonic flows are associated with Mach numbers of approximately 5 or greater.

## Example 13.c

During a steady-flow process it is found that air has a velocity of 220 m/s at a location where the static pressure is 85 kPa and the static temperature is $-40$ C. Calculate the isentropic speed of sound and the Mach number at the given state.

**Solution:**

By eq. (13.6b) the isentropic speed of sound is

$$c = [(1.40)(8314/29)(273 - 40)]^{1/2} = 305.8 \text{ m/s}$$

The Mach number is

$$N_m \frac{V}{c} = \frac{220}{305.8} = 0.719$$

## Example 13.d

At a given section in a steady-flow system it is found that steam has a velocity of 1200 ft/sec where the static temperature is 350 F and the static pressure is 100 psia. Calculate and compare the Mach numbers:

(a) Assuming steam is a perfect gas (eq. [13.6b])
(b) Using eq. (13.6a)

**Solution:**

(a) For the perfect gas assumption

$$c = (\gamma g_0 RT)^{1/2}$$

At the given state $C_p$ can be approximated as

$$C_p = \left(\frac{\Delta h}{\Delta T}\right)_{p=100}$$

From the *Steam Tables* at 100 psia $h = 1194.8$ B/lbm at 340 F and $h = 1205.9$ B/lbm at 360 F. The approximate $C_p$ is

$$C_p = \frac{1205.9 - 1194.8}{360 - 340} = 0.555 \text{ B/lbm R}$$

For a perfect gas $\gamma = C_p/C_v$ and $C_v = C_p - R$. Therefore

$$\gamma = \frac{0.555}{0.555 - \dfrac{(1545)}{(18)(778)}}$$
$$= 1.248$$

By eq. (13.6b)

$$c = [(1.248)(32.17)(85.83)(810)]^{1/2}$$
$$= 1671 \text{ ft/sec}$$

(b) From Figure 6, isentropic expansion exponents, in the *Steam Tables,*

$$k = 1.304$$

and from Table 3 of the *Steam Tables*

$$v = 4.592 \text{ ft}^3/\text{lbm}$$

From eq. (13.6a)

$$c = [(1.304)(32.17)(144)(100)4.592]^{1/2}$$
$$= 1666 \text{ ft/sec}$$

While the results from parts (a) and (b) of the example do not show a large difference, similar calculations at higher pressures would show increasing errors when assuming that superheated steam behaves as a perfect gas. The substitution of $RT$ for $pv$ and $\gamma$ for $k$ should be limited to substances that follow perfect gas behavior.

## 13–3 Isentropic Flow of Perfect Gases in Varying-Area Passages

Some characteristics of isentropic steady flows with work equal to zero were established in Section 13–1. These characteristics—namely, constant stagnation entropy and constant stagnation pressure—were stated to hold true for all pure substances. In this section we shall examine such flow for perfect gases and, in so doing, determine the passage geometry required to produce desired exit conditions for nozzles and diffusers operating with a fixed stagnation state.

The mass rate of flow per unit area from the continuity equation is

$$\frac{\dot{m}}{A} = \frac{V}{v}$$

at any section of a flow passage. The stagnation enthalpy has been defined as

$$h_t = h + \frac{V^2}{2g_0}$$

which, when solved for the velocity, becomes

$$V = [2g_0\,(h_t - h)]^{1/2} \tag{a}$$

For a perfect gas having constant specific heats the enthalpy difference is

$$h_t - h = C_p(T_t - T) \tag{b}$$

where $h$ is the static enthalpy and $T$ the static temperature at a particular section of the passage. Substituting (*b*) into (*a*) we find

$$V = [2g_0\,C_p(T_t - T)]^{1/2} \tag{c}$$

The stagnation state is fixed by any two independent properties and, for a perfect gas,

$$p_t v_t = RT_t$$

is the $p$-$v$-$T$ relation at the stagnation state. The static and stagnation states, being at the same entropy, have properties related by

$$\left(\frac{p_t}{p}\right)^{(\gamma-1)/\gamma} = \frac{T_t}{T} = \left(\frac{v}{v_t}\right)^{\gamma-1} \tag{d}$$

for perfect gases. The specific volume is then

$$v = \left(\frac{p_t}{p}\right)^{1/\gamma} v_t \tag{e}$$

From eqs. (c) and (e) and the continuity equation, we obtain

$$\frac{\dot{m}}{A} = \frac{V}{v} = \frac{[2g_0\, C_p(T_t - T)]^{1/2}}{\left(\frac{p_t}{p}\right)^{1/\gamma} v_t} \tag{13.8}$$

The static temperature and the static pressure are related by eq. (d). Consequently $\dot{m}/A$ can be expressed as a function of either $T$ or $p$ for a fixed stagnation state. To illustrate, suppose we choose the static pressure as the independent variable. As an intermediate step eq. (13.8) can be written in the form

$$\frac{\dot{m}}{A} = \frac{\left[2g_0\, C_p\, T_t\left(1 - \frac{T}{T_t}\right)\right]^{1/2}}{\left(\frac{p_t}{p}\right)^{1/\gamma} v_t}$$

However, by eq. (d)

$$\frac{T}{T_t} = \left(\frac{p}{p_t}\right)^{(\gamma-1)/\gamma}$$

Making this substitution, we obtain

$$\frac{\dot{m}}{A} = \frac{\left\{2g_0\, C_p\left[1 - \left(\frac{p}{p_t}\right)^{(\gamma-1)/\gamma}\right]\right\}^{1/2}}{\left(\frac{p_t}{p}\right)^{1/\gamma} v_t} = f(p) \tag{13.8a}$$

for isentropic no-work flow of a perfect gas with constant specific heats. An analogous expression for $\dot{m}/A$ as a function of the local static temperature $T$ could be formed also.

For the several conditions established, it is apparent that a graph of $\dot{m}/A$ versus $p$ can be plotted for a given perfect gas. An alternate plot of $\dot{m}/A$ versus $p/p_t$ likewise could be developed from eq. (13.8a). Before proceeding with additional operations on eq. (13.8a) let us examine the limiting values of $p$, the static pressure. First consider the equation for the special case where $p$ and $p_t$ are equal. The denominator of eq. (13.8a) will be finite and equal to the stagnation specific volume for this condition. The numerator will be zero; thus $\dot{m}/A$ is zero when $p$ is equal to $p_t$. The lower bound of the static pressure is zero. Inspection of eq. (13.8a), with $p$ equal to zero, shows the denominator going to infinity while the numerator remains finite. This too yields $\dot{m}/A$ equal to zero for a fixed stagnation state of a perfect gas. It then follows that eq. (13.8a) will yield positive values of $\dot{m}/A$ for static pressures greater than zero up to the limit where $p$ and $p_t$ are equal. The graph of eq. (13.8a) is shown in Figure 13.4.

**Figure 13.4** Isentropic Flow of a Perfect Gas with Constant Specific Heats in a Varying-Area Flow Passage

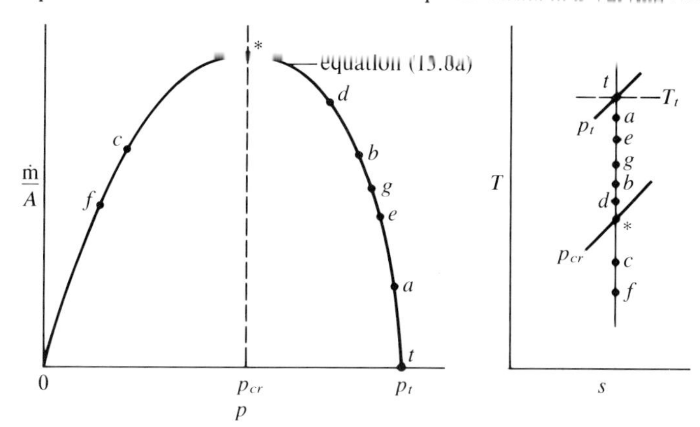


The pressure ratio at maximum $\dot{m}/A$ can be established by setting the derivative of $\dot{m}/A$ with respect to $p$ equal to zero. Such a procedure reveals that the maximum $\dot{m}/A$ occurs at

$$\frac{p}{p_t} = \left(\frac{2}{\gamma + 1}\right)^{\gamma/(\gamma-1)}$$

We shall define this ratio of $p$ to $p_t$ as the *critical pressure ratio.* Since the static and stagnation states are at the same entropy, the isentropic $p$-$T$ relation applies and

$$\left(\frac{p}{p_t}\right)_{cr} = \left(\frac{2}{\gamma + 1}\right)^{\gamma/(\gamma-1)} = \left(\frac{T}{T_t}\right)_{cr}^{\gamma/(\gamma-1)} \tag{13.9}$$

for constant specific-heat perfect gases flowing isentropically through varying area passages.

Figure 13.4 shows several processes applicable to isentropic flow in nozzles and diffusers with fixed stagnation states. The states on the curve based on eq. (13.8a) correspond to those in the $T$-$s$ diagram. As each of the processes is discussed, it is important to note that both the stagnation state and the mass rate of flow are assumed to be fixed. Thus only $A$ varies in the ratio $\dot{m}/A$, and as $A$ decreases $\dot{m}/A$ increases. The process ($a$) to ($b$) involves a pressure drop and will occur in a converging passage. The static temperature, which is related to the static pressure, will decrease from $T_a$ to $T_b$ as shown in the $T$-$s$ diagram in Figure 13.4. The passage configuration for the process ($a$) to ($b$) is shown in Figure 13.5a. Flow from ($a$) to ($c$) would take place in a passage that converges until the critical pressure is reached, after which the passage diverges as the pressure further decreases from the critical pressure to $p_c$. This configuration is shown in Figure 13.5b. The flow is said to be "choked" when the downstream pressure is less than or equal to the critical pressure. (This will be discussed in more detail later.) The maximum value of $\dot{m}/A$ occurs where the passage area is a minimum. This section is called the throat of the passage. The process from ($d$) to ($e$) would occur in a diverging passage and produce an increase in static pressure from $p_d$ to $p_e$. This passage is shown in Figure 13.5c. For the pressure to increase from $p_f$ to $p_g$, it is evident that a converging–diverging configuration is required, as shown in Figure 13.5d.

**Figure 13.5** Passage Geometries for Isentropic No-Work Flow. *a.* Converging passage (nozzle). *b.* Converging–diverging passage (nozzle). *c.* Diverging passage (diffuser). *d.* Converging–diverging passage (diffuser).

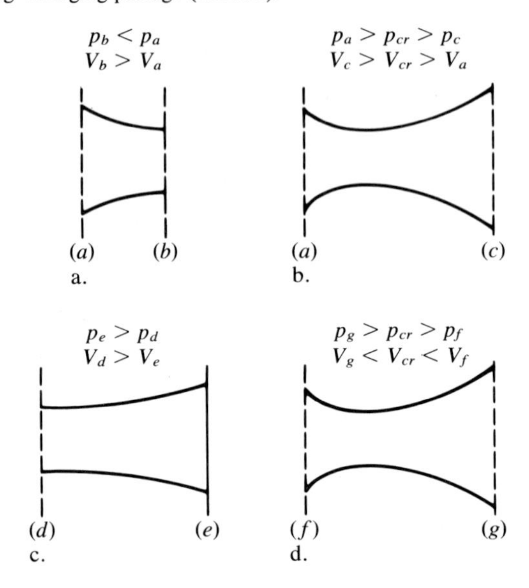


The *T-s* diagram is useful in establishing the directions of change of the velocities in the processes described above. The velocity at any state on the constant entropy path of the *T-s* diagram in Figure 13.4 in terms of $T_t$ and $T$, the static temperature, is

$$V = [2g_0 C_p (T_t - T)]^{1/2}$$

Since $T_t$ is fixed, decreases in $T$ yield increases in $V$. The relationships of the velocities are shown in each part of Figure 13.5. The configurations that increase velocity are identified as nozzles and those that increase the pressure are identified as diffusers.

We shall now investigate the flow at the critical pressure ratio in more detail by solving for the velocity $V_{cr}$. The critical temperature ratio from eq. (13.9) is

$$\left(\frac{T}{T_t}\right)_{cr} = \frac{2}{\gamma + 1}$$

From the definition of stagnation enthalpy

$$C_p(T_t - T_{cr}) = \frac{V_{cr}^2}{2g_0}$$

Substitution for $T_t$ from eq. (13.9) yields

$$C_p\left[\left(\frac{\gamma + 1}{2}\right) T_{cr} - T_{cr}\right] = \frac{V_{cr}^2}{2g_0} \tag{a}$$

However

$$C_p = \frac{\gamma R}{(\gamma - 1)}$$

for a perfect gas, and thus eq. (a) can be written as

$$\frac{\gamma R T_{cr}}{(\gamma - 1)}\left[\frac{\gamma + 1}{2} - 1\right] = \frac{V_{cr}^2}{2g_0} \qquad \textbf{(b)}$$

Solving for $V_{cr}$ we find

$$V_{cr} = [\gamma g_0 R T_{cr}]^{1/2}$$

The right side of the equality is, by eq. (13.6b), the isentropic speed of sound for a perfect gas at $T_{cr}$. From the above it is apparent that a perfect gas reaches a Mach number of 1 at the critical (choking) state. In subsequent discussion we shall denote the isentropic critical state as the star (*) state. Therefore, the state where the Mach number is unity is described by

$$\frac{T^*}{T_t} = \frac{2}{\gamma + 1} = \left(\frac{p^*}{p_t}\right)^{(\gamma - 1)/\gamma} \qquad \textbf{(13.10)}$$

for perfect gases having constant specific heats.

We can now identify regions of subsonic and supersonic flow for the isentropic flows illustrated in Figures 13.4 and 13.5 by examining the $T$-$s$ diagram in Figure 13.4. As previously noted, decreasing $T$ along the isentropic path in the figure corresponds to increasing velocity. The velocity at the stagnation state is zero and, from the above discussion, the Mach number where the critical pressure exists (the star state) is unity. Therefore, states ($a$), ($e$), ($g$), ($b$), and ($d$) are subsonic states. Any state with a temperature less than $T^*$ has a velocity greater than $V^*$ and a sonic velocity less than $c^*$, and therefore is a state at which supersonic flow exists. States ($c$) and ($f$) are such states. The nozzle in Figure 13.5b produces a supersonic velocity at the exit and the diffuser in Figure 13.5d receives supersonic flow and discharges subsonic flow.

## Example 13.e

Air at 100 psia, 400 F, enters a nozzle with negligible velocity. Expansion through the nozzle is isentropic. At a section where the static pressure is 80 psia calculate, for constant specific heats:

(a) The velocity
(b) The isentropic speed of sound
(c) The Mach number
(d) The mass rate of flow per unit area $\dot{m}/A$

Repeat the calculations for isentropic expansion to 70, 60, 52.8, 50, 40, and 30 psia. Plot the results (ordinate) versus the local static pressure.

**Solution:**

The intermediate values and the results of the calculations related to the solution are tabulated below. The *T-s* diagram shows selected states related to the solution.

| $p$, psia | $T$, R | $V$, ft/sec | $c$, ft/sec | $N_m$ | $v$, ft$^3$/lbm | $\dot{m}/A$, lbm/sec ft$^2$ |
|---|---|---|---|---|---|---|
| 100.0 | 860.0 | 0 | 1436.8 | 0 | 3.183 | 0 |
| 80.0 | 806.9 | 798.8 | 1391.8 | 0.5739 | 3.733 | 214.0 |
| 70.0 | 776.7 | 1000.5 | 1365.6 | 0.7327 | 4.107 | 243.6 |
| 60.0 | 743.2 | 1184.7 | 1335.7 | 0.8870 | 4.585 | 258.4 |
| 52.8 | 716.6 | 1312.7 | 1311.6 | 1.00 | 5.024 | 261.3 |
| 40.0 | 661.9 | 1542.9 | 1260.5 | 1.224 | 6.125 | 251.9 |
| 30.0 | 609.7 | 1734.3 | 1209.8 | 1.434 | 7.522 | 230.6 |

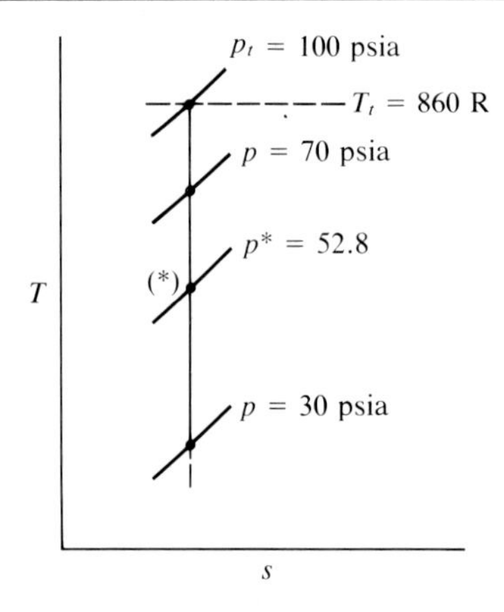


The static temperature at each pressure is calculated by

$$T = \left(\frac{p}{p_t}\right)^{(\gamma-1)/\gamma} T_t$$

with $p_t$ equal to 100 psia and $T_t$ equal to 860 R. Velocities are calculated from

$$V = [2g_0\, C_p(T_t - T)]^{1/2}$$

Isentropic speeds of sound are determined by

$$c = [\gamma g_0\, RT]^{1/2}$$

The Mach number at each section is

$$N_m = \frac{V}{c}$$

Specific volumes are calculated from the equation of state using static properties. Each value of $\dot{m}/A$ is calculated from

$$\frac{\dot{m}}{A} = \frac{V}{v}$$

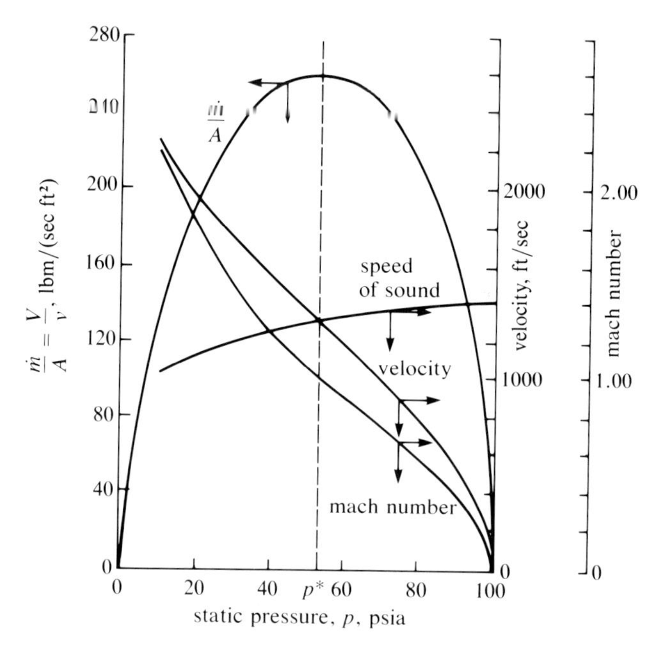


From the accompanying graph of the results it is apparent from the $\dot{m}/A$ curve that the nozzle must have a converging–diverging configuration. Subsonic flow exists in the converging section and supersonic flow exists in the diverging section. A Mach number of unity occurs at $p = 52.8$, which, according to eq. (13.9), is $p^*$.

---

## 13–4 Flow Through Converging Nozzles

The flow through a converging nozzle further illustrates the phenomenon of choking. Consider the isentropic flow of a perfect gas in the converging nozzle shown in Figure 13.6a. The stagnation state ($t$) at the nozzle entrance is assumed to be fixed and the discharge region pressure $p_d$ can be varied. When $p_d$ is equal to the stagnation pressure $p_t$ there is no flow. When $p_d$ is decreased by a small amount to $p_1$ the flow velocity will increase, the pressure at the nozzle exit plane $p_e$ will be equal to $p_d$, and $\dot{m}/A_e$ will be less than the maximum possible. (Note that in this illustration $A_e$ is fixed and $\dot{m}$ varies.) The conditions for $p_d = p_1$ are shown in parts (b) and (c) of Figure 13.6. As $p_d$ is decreased further, the velocity $V_e$ and $\dot{m}/A_e$ will increase. When $p_d$ reaches $p^*$, the star state exists at the nozzle exit plane ($e$), ($N_{m,e} = 1$), and $\dot{m}/A_e$ reaches its maximum as shown in Figure 13.6c. Decreasing $p_d$ to a pressure $p_2$ less than $p^*$ will have no effect on the flow upstream of the nozzle exit plane. The mass rate of flow does not change and the expansion of the gas from $p^*$ to $p_2 = p_d$ occurs in the nozzle discharge region in an irreversible manner as illustrated in the $T$-$s$ diagram in Figure 13.6b. When $p_d$ is less than or equal to $p^*$, the nozzle is said to be choked. Stated in another way, a discharge-region pressure equal to or less than $p^*$ chokes the nozzle. This phenomenon can be explained by considering the time-dependent events that take place immediately after $p_d$ is lowered. When $p_d$ is decreased, an expansion wave that increases the flow velocity is transmitted upstream at the speed of sound *relative to the fluid*. When the flow velocity at ($e$) is subsonic, the wave can propagate upstream of ($e$) and increase the

**Figure 13.6** Flow of a Perfect Gas through a Converging Nozzle. *a.* A converging nozzle. *b.* *T-s* diagram for converging nozzle flow. *c.* $\dot{m}/A_e$ versus discharge-region pressure.

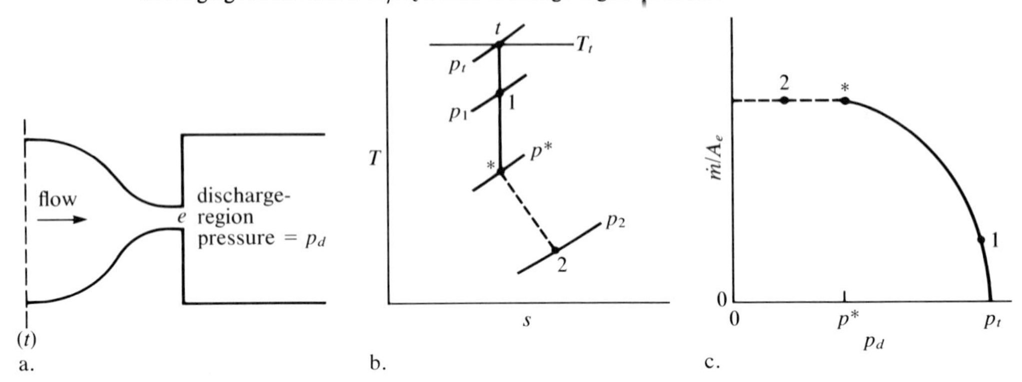


velocity. However, when $V_e$ is the speed of sound ($N_{m,e} = 1$), the wave is not transmitted upstream, and no change upstream of (*e*) occurs. Thus we see that when $p_d > p^*$, $p_e = p_d$ and the flow is isentropic from the stagnation state to $p_d$. However, when $p_d \leq p^*$, the nozzle is choked, $p_e = p^*$, and the flow is isentropic only to $p^*$. This condition is illustrated in the *T-s* diagram in Figure 13.6b. For perfect gases $p^*$ is given by eq. (13.9). Isentropic flow *within* the nozzle is a good approximation of real flows for both the choked and unchoked cases and will be assumed unless otherwise stated.

The phenomenon of choking also occurs in converging–diverging nozzles such as that shown in Figure 13.5b. If the nozzle is choked, that is, $N_m = 1$ at the throat, changes downstream of the throat will not affect the mass flow rate through the nozzle or the flow upstream of the throat. Although choking has been discussed in relation to perfect gases, it can also occur in the flow of the vapor phase of all pure substances.

## 13–5 Off-Design Performance of Supersonic Nozzles

In this section we consider the performance of supersonic nozzles when the discharge-region pressure differs from that for which the nozzle was designed. It is assumed that the nozzles were designed on the basis of isentropic flow. When the nozzle exit pressure is other than the design value, overexpansion or underexpansion occurs. An underexpanded nozzle flow is one in which the exit-plane static pressure is greater than the discharge-region pressure. An overexpanded flow is one for which the exit-plane static pressure is greater than the design value. Figure 13.7 shows a nozzle designed for supersonic flow discharging into a region where the pressure is $p_d$. The accompanying pressure-distribution diagram shows the pressure distribution in the direction of flow for several values of $p_d$. The inlet stagnation state is fixed, the inlet velocity $V_1$ is subsonic, the throat is at section (2), and section (3) is the nozzle exit plane. Two limiting solutions for choked flow (star state at [2]) occur at exit-plane static pressures $p_a$ and $p_b$. For each, the nozzle flow is isentropic and the discharge-region pressure is equal to the static pressure at the nozzle exit plane. The mass flow rate for either pressure will be the same. For $p_3 = p_a = p_d$ subsonic flow exists in the diverging section of the nozzle. For a static pressure $p_v$ greater than $p_a$, the throat pressure will be greater than $p^*$ and the passage will not be choked. The regime of flow for discharge pressures greater than $p_a$ is the *venturi* regime. Venturi are frequently used as flow meters. By measuring the pressures $p_2$ and $p_1$ for a given venturi, the mass flow rate can be determined.

**Figure 13.7** Some Effects of the Discharge Region Pressure on Flow of a Perfect Gas through a Converging–Diverging Passage

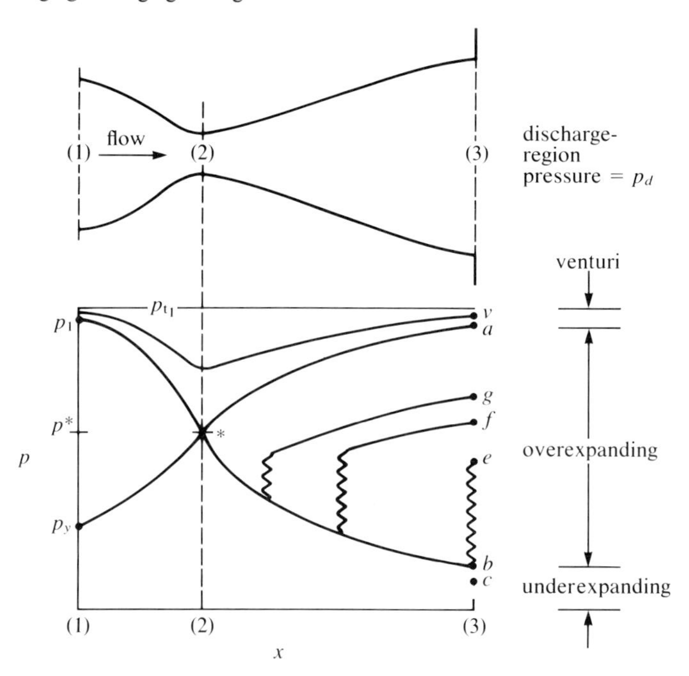


The isentropic expansion to $p_3 = p_b = p_d$ proceeds with continuing velocity and Mach number increase and static pressure decrease through the entire nozzle. The Mach number in the diverging section is supersonic. This represents ideal supersonic nozzle flow. If the discharge region is decreased from $p_b$ to $p_c$, the flow within the nozzle is not affected. At the nozzle exit plane (3) the static pressure will be $p_b$ and the gas will expand to the discharge-region pressure downstream of (3). If the discharge-region pressure $p_d$ is greater than $p_b$, shock waves form in the flow. Shock waves in one-dimensional flow occur in planes normal to the flow direction and are characterized by an irreversible step increase in static pressure. In steady flows shock waves can occur only when the flow is supersonic, and the flow downstream of a normal shock is always subsonic. When $p_d$ is equal to $p_e$ in Figure 13.7, a normal shock occurs at the nozzle exit plane (3). The flow for this case is supersonic between the throat and the shock wave and subsonic downstream of the shock wave. As $p_d$ is increased the shock wave assumes positions further upstream in the nozzle and subsonic flow exists in part of the diverging section of the nozzle. This part of the diverging section acts as a subsonic diffuser and the pressure increases and the velocity decreases in the flow direction. These conditions exist when $p_d$ assumes the values $p_f$ and $p_g$. When $p_d$ reaches $p_a$ the shock wave disappears and subsonic flow exists throughout the diverging section of the nozzle. Overexpanded flows, those that produce shock waves as described above, are highly undesirable and should be avoided when off-design operation is anticipated. Underexpansion is much less penalizing than overexpansion and is preferable when off-design flows are likely to occur.

The converging–diverging passage in Figure 13.7 can operate as a supersonic diffuser as well as a supersonic nozzle. When operating as a supersonic diffuser the inlet pressure will be less than $p^*$. The pressure $p_y$ in the figure is a possible value. For isentropic flow the diffuser would increase the static pressure to $p^*$ at the throat and to $p_a$ at the outlet. Supersonic flow would exist in the converging section and subsonic flow would exist in the diverging section.

### 13–6 Nozzle Flow for Steam and Other Vapors

As was noted earlier, many qualitative characteristics of nozzle flow are similar for fluids other than perfect gases. One of the more significant differences is the procedure for determination of the critical pressure ratio. For perfect gases having constant specific heats, the critical pressure and temperature ratios can be calculated readily. The complex equation of state for steam—saturated or superheated—along with the changes in the isentropic exponent $k$, a state function, prevent such a simple approach. Actually the critical pressure ratio for steam varies depending on the nozzle inlet state. The critical pressure ratio for steam, at a given stagnation state, can be calculated from

$$\frac{\dot{m}}{A} = \frac{V}{v} = \frac{[2g_0 (h_t - h)]^{1/2}}{v} \tag{13.11}$$

with the enthalpy and specific volume, at an arbitrary series of pressures, established by $s$ equal to $s_t$. The results plotted as $\dot{m}/A$ versus $p$ peak in a manner similar to that shown in Figure 13.4. The critical pressure ratio is $p/p_t$ for the maximum value of $\dot{m}/A$. Critical pressure ratios for steam fall in the range 0.55 (superheated steam) to 0.58 (saturated steam) when calculated in this manner.

An alternate procedure is to determine the critical pressure ratio by experiment. Results from this approach are dependent not only on the inlet stagnation state, but on the geometry and length of the converging nozzle as well. Steam at a known stagnation state is supplied to the nozzle. Discharge-region pressures are reduced in steps until the converging nozzle chokes. The critical pressure ratio is taken as the ratio of the static pressure at which choking was first observed to the stagnation pressure. Recognizing that the actual nozzle behavior is not isentropic, the procedure nevertheless is used in practice for geometrically similar nozzles. As steam or other vapors expand through nozzles, supersaturation often occurs, particularly in nozzles with small axial lengths. Condensation necessary to achieve stable equilibrium at each section in the passage is time dependent and requires a nucleus for condensed $H_2O$ to form. As a result it is not uncommon to observe steam at temperatures below the saturation temperature for an observed static pressure. This condition, a metastable state, disappears with adequate time; however, such time intervals frequently are not available in nozzle flow.

In Section 13–1 several characteristics of reversible adiabatic no-work flow were discussed. For such processes involving vapor flows, as well as perfect gas flows, stagnation enthalpy, stagnation pressure, and static and stagnation entropies all remain constant. For irreversible adiabatic no-work flows, entropy increases and stagnation pressure decreases.

The design and analysis of ideal steam and other vapor nozzles follow the same pattern as for perfect gas nozzle flow, except for the property relations and the critical pressure ratio. Unless otherwise stated, equilibrium flow with critical pressure ratios of 0.55 for initially superheated steam and 0.58 for saturated steam will be assumed.

## Example 13.f

Steam at 100 psia, 400 F, is supplied to a nozzle with negligible velocity. Assuming the critical pressure ratio to be 0.55, calculate for isentropic expansion to 15 psia:

(a) The throat velocity
(b) The sonic velocity for the vapor at the throat
(c) The velocity at the nozzle discharge
(d) The stagnation pressure at the nozzle discharge

Use the section notation in Figure 13.7.

**Solution:**

(a) For the given stagnation conditions

$$h_t = \text{constant} = 1227.5 \quad \text{and} \quad s_t = \text{constant} = 1.6517$$

The throat pressure $p_2$ is

$$(0.55)(100) = 55.0 \text{ psia}$$

At $s = 1.6517$ and $p = 55.0$ psia the steam is saturated vapor. From Table 2 of the *Steam Tables,*

$$h_2 = 1176.3 \text{ B/lbm}$$

From the definition of stagnation enthalpy

$$V_2 = [2(778)(32.17)(1227.5 - 1176.3)]^{1/2}$$
$$= 1601 \text{ ft/sec}$$

(b) From the *Steam Tables* the isentropic exponent $k$ is 1.31 at the throat conditions; thus from eq. (13.6a)

$$c_2 = [(1.31)(32.17)(55)(144)(7.789)]^{1/2} = 1612 \text{ ft/sec}$$

The small difference between $V_2$ and $c_2$ indicates that the critical pressure ratio is not exactly 0.55.

(c) The discharge section at 15 psia has $s_t = s_3 = 1.6517$, less than $s_g$ at 15 psia; therefore, the discharge state is in the mixture region and

$$1.6517 = 0.31367 + (x_3)1.4414$$

$$x_3 = 0.9033$$

The enthalpy at the section is

$$h_3 = 181.19 + (0.9033)(969.7) = 1057.1 \text{ B/lbm}$$

With constant stagnation enthalpy the velocity at the nozzle outlet is

$$\begin{aligned} V_3 &= [2(32.17)(778)(1227.5 - 1057.1)]^{1/2} \\ &= 2921 \text{ ft/sec} \end{aligned}$$

(d) For isentropic nozzle flow the stagnation pressure does not change, and thus

$$p_{t_1} = 100 \text{ psia} = p_{t_2} = p_{t_3}$$

---

### 13–7 Irreversible Adiabatic Flow in Varying-Area Passages

Actual flow through varying-area passages is subject to irreversibilities associated with fluid shear or friction in the boundary layer (viscous region) adjacent to the walls of the passage and irreversibilities created by turbulence and velocity differences between elements of the flow within the passage. From the steady-flow equation, stagnation enthalpy is constant for adiabatic no-work steady-flow processes. This condition will hold true for all pure substances whether the flow is reversible or irreversible. However, irreversible adiabatic flow in these passages results in increased static entropy, and consequently stagnation entropy. As shown in Figure 13.2c, the stagnation entropy increase causes a decrease in stagnation pressure at each downstream section in the flow passage.

Actual nozzle flow is related to ideal—isentropic—flow by the nozzle efficiency. The *nozzle efficiency, $\eta_N$,* is defined as the ratio of the actual kinetic energy of the stream at the nozzle outlet to the kinetic energy that would be realized in isentropic nozzle flow from the same inlet stagnation state to the actual discharge static pressure. These conditions and the associated states are illustrated in Figure 13.8 for perfect gases having constant or variable specific heats as well as vapors. From the definition of nozzle efficiency

$$\eta_N \equiv \frac{V_2^2}{2g_0}\frac{2g_0}{V_s^2} = \frac{V_2^2}{V_s^2} = \frac{h_{t_1} - h_2}{h_{t_1} - h_s} \tag{13.12}$$

for each fluid described above. For the special case of a perfect gas with constant specific heats the nozzle efficiency is

$$\eta_N = \frac{h_{t_1} - h_2}{h_{t_1} - h_s} = \frac{C_p(T_{t_1} - T_2)}{C_p(T_{t_1} - T_s)} = \frac{T_{t_1} - T_2}{T_{t_1} - T_s} \tag{13.12a}$$

Figure 13.8 Adiabatic Irreversible Nozzle Flow

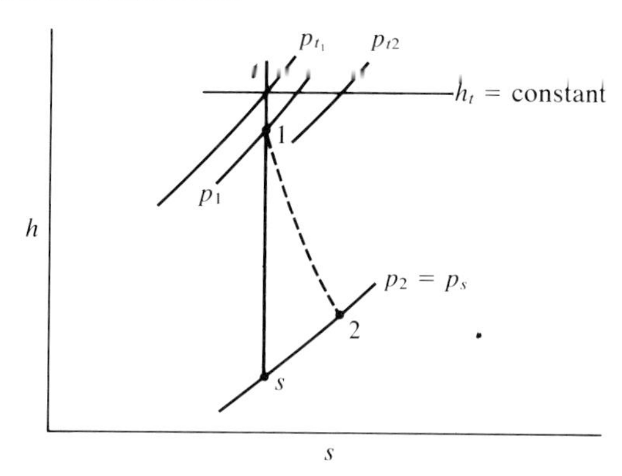


Figure 13.9 Apparatus for Determining Steam Nozzle Efficiencies

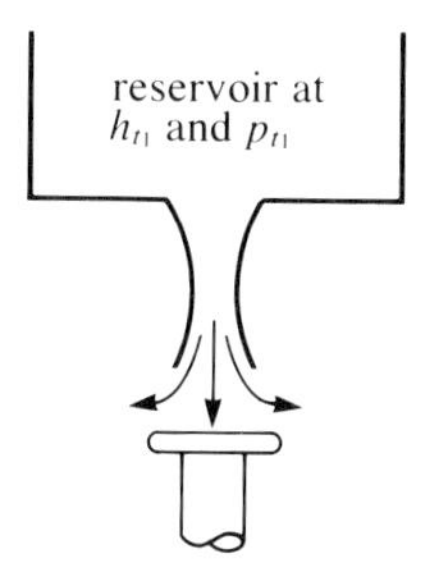


Well-designed converging nozzles or converging sections of supersonic nozzles in turbines have efficiencies of 95 to 98 percent. For such nozzles the efficiency is, in some cases, assumed to be 100 percent and the throat condition for choked flow is taken as the star state. Typically the diverging section of a nozzle is much longer than the converging section and the losses associated with wall friction are assumed to take place in the diverging section only.

Steam nozzle efficiencies are often determined using an experimental apparatus such as that shown schematically in Figure 13.9. The mass rate of flow through the nozzle is determined by condensing the steam discharged from the nozzle. The force exerted on the flat plate normal to the nozzle axis is measured, as are the reservoir (stagnation) state and the nozzle outlet static pressure. From these data the actual velocity at the nozzle discharge is calculated from

$$F = \frac{\dot{m}}{g_0} V_2$$

with $V_2$, the actual nozzle outlet velocity, the one unknown. The isentropic nozzle outlet velocity is calculated from the inlet stagnation properties and the static pressure at the outlet.

## Example 13.g

A converging–diverging nozzle is supplied with steam at 60 psia, 320 F, the stagnation state of the flow. For a nozzle efficiency of 90.0 percent, an outlet static pressure of 15.0 psia and a mass flow rate of 1.000 lbm/sec calculate the outlet area.

**Solution:**

The actual outlet conditions must be determined using the nozzle efficiency. Using the notation of Figure 13.8,

$$\eta_N = \frac{h_{t_1} - h_2}{h_{t_1} - h_s}$$

From the *Steam Tables,*

$$h_{t_1} = h_{t_2} = 1192.6 \text{ B/lbm}$$

$$s_{t_1} = s_s = 1.6634 \text{ B/lbm R}$$

To calculate $h_s$ we know $p_s = p_2 = 15.0$ psia and $s_s = 1.6634$, thus

$$1.6634 = 0.31367 + (x_s)(1.4414)$$

and $x_s = 0.9364$. The enthalpy $h_s$ at 15 psia is

$$\begin{aligned} h_s &= 181.19 + (0.9364)(969.7) \\ &= 1089.2 \text{ B/lbm} \end{aligned}$$

From the nozzle efficiency

$$(0.900)(1192.6 - 1089.2) = 1192.6 - h_2$$

from which $h_3 = 1099.5$ B/lb. The actual outlet velocity is

$$\begin{aligned} V_2 &= [2(32.17)(778)(1192.6 - 1099.5)]^{1/2} \\ &= 2159 \text{ ft/sec} \end{aligned}$$

The *actual* outlet state is fixed by $p_2 = 15.0$ psia and $h_2 = 1099.5$ B/lbm, a state at which the quality has yet to be calculated. From the known properties

$$h_2 = 1099.5 = 181.19 + (x_2)(969.7)$$

and $x_3 = 0.9470$. For nozzle efficiencies less than 100 percent the actual quality and static enthalpy of mixtures are always greater than the values for the isentropic outlet state *s*. With $x_2$ known

$$v_3 = (0.9470)(26.29) = 24.89 \text{ ft}^3\text{/lbm}$$

The continuity equation, written for the actual outlet conditions, is

$$A_2 = \frac{\dot{m}v_2}{V_2} = \frac{(1.00)(24.90)(144)}{2159} = 1.661 \text{ in}^2$$

---

## Example 13.h

A converging–diverging nozzle discharges air at 14.70 psia, 560 R (static conditions), and a Mach number of 2.00. Expansion through the nozzle is isentropic to the throat, however, losses in the diverging part of the nozzle result in a nozzle efficiency of 93.0 percent. The area of the outlet cross section is 5.42 in². Calculate:

(a) The stagnation temperature
(b) The inlet stagnation pressure
(c) The outlet stagnation pressure
(d) The throat area
(e) The entropy change between the throat and outlet

**Solution:**

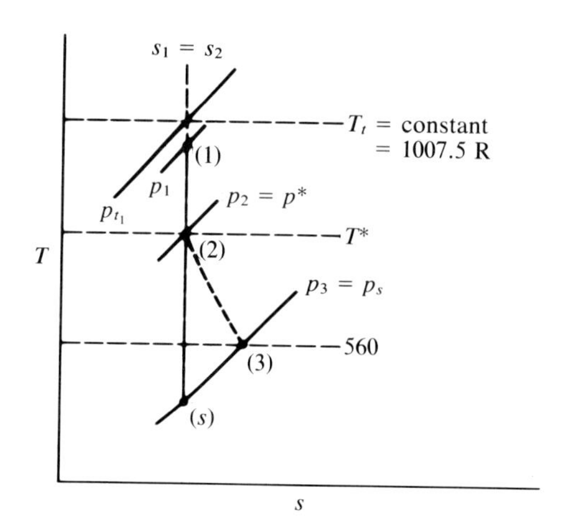


(a) The static and stagnation states for this example are shown on the temperature-entropy diagram. Since the nozzle has no heat or work interactions with the surroundings, the stagnation enthalpy is the same at all sections of the passage. This condition establishes constant stagnation temperature for a perfect gas, thus

$$C_p(T_t - T_3) = \frac{V_3^2}{2g_0} = [N_{m,3}c_3]^2/2g_0$$

The sonic velocity is

$$\begin{aligned} c_3 &= (\gamma g_0 R T_3)^{1/2} \\ &= [(1.40)(32.17)(53.3)(560)]^{1/2} \\ &= 1159.4 \text{ ft/sec} \end{aligned}$$

The outlet velocity $V_3$ is then

$$\begin{aligned} V_3 &= (2.00)(1159.4) \\ &= 2318.8 \text{ ft/sec} \end{aligned}$$

From the stagnation enthalpy equation

$$0.24(T_t - 560) = \frac{(2318.8)^2}{2(32.17)(778)}$$

which yields $T_t = 1007.5$ R.

(b) The temperature at ($s$) is determined from the nozzle efficiency (eq. [13.12a]).

$$\eta_N \equiv \frac{V_3^2}{V_s^2} = \frac{C_p(T_t - T_3)}{C_p(T_t - T_s)} = 0.930 = \frac{1007.5 - 560}{1007.5 - T_s}$$

from which $T_s = 526.3$ R. With the inlet stagnation state having the same entropy as state ($s$) the inlet stagnation pressure can be calculated as

$$\begin{aligned} p_{t_1} &= p_s \left(\frac{T_{t_1}}{T_s}\right)^{\gamma/(\gamma-1)} = (14.7)\left(\frac{1007.5}{526.3}\right)^{3.50} \\ &= 142.68 \text{ psia} \end{aligned}$$

(c) The stagnation pressure decreases during flow through a nozzle having an efficiency less than 100 percent. However, at any section in the nozzle, the static entropy and the corresponding stagnation entropy are equal (eq. [13.2]). The outlet stagnation pressure is then

$$\begin{aligned} p_{t_3} &= p_3 \left(\frac{T_{t_3}}{T_3}\right)^{\gamma/(\gamma-1)} = 14.7\left(\frac{1007.5}{560}\right)^{3.5} \\ &= 114.82 \text{ psia} \end{aligned}$$

(d) With isentropic flow from the inlet to the throat, the throat state will be the star state for the inlet stagnation state, and thus

$$\frac{p_2}{p_{t_1}} = \frac{p^*}{p_{t_1}} = \left(\frac{2}{\gamma + 1}\right)^{\gamma/(\gamma-1)}$$

and $p_2 = p^* = 75.37$ psia.

The static temperature at the throat can be calculated using the isentropic $p$-$T$ relation for the inlet stagnation and throat static state or by using the critical temperature ratio for a perfect gas having constant specific heats. Choosing the latter

$$T_2 = T^* = \left(\frac{2}{\gamma + 1}\right)T_{t_1} = 839.6 \text{ R}$$

The throat velocity can be determined as $V_2$ equal to $c_2$.

$$\begin{aligned} V_2 = c_2 &= (\gamma g_0 R T_2)^{1/2} \\ &= 1420 \text{ ft/sec} \end{aligned}$$

The throat area is calculated using the continuity equation at the throat and outlet sections to yield

$$\begin{aligned} A_2 &= \frac{v_2}{v_3}\,\frac{V_3}{V_2}\,A_3 \\ &= \frac{4.123}{14.10}\,\frac{2318.8}{1420.0}\,5.42 \\ &= 2.588 \text{ in}^2 \end{aligned}$$

where $v_2$ and $v_3$ have been calculated using the perfect gas equation of state.

(e) With reference to the temperature–entropy diagram, it is evident that

$$s_3 - s_2 = s_{3_t} - s_{1_t}$$

The equation for entropy change written for the stagnation states is

$$s_{t_3} - s_{t_1} = C_p \ln \frac{T_{t_3}}{T_{t_1}} - R \ln \frac{p_{t_3}}{p_{t_1}}$$

With the stagnation enthalpy constant, $T_{t_3}$ equals $T_{t_1}$ for a perfect gas and

$$s_{t_3} - s_{t_1} = -R \ln \frac{p_{t_3}}{p_{t_1}}$$

or

$$\begin{aligned} s_3 - s_2 &= -\frac{53.3}{778} \ln \frac{114.82}{142.68} \\ &= 0.0149 \text{ B/lbm R} \end{aligned}$$

## 13–8 Summary

One-dimensional steady flow provides a reasonably accurate description of the flow in passages such as nozzles and diffusers. Static properties are those properties that would be measured by instruments whose velocity relative to the moving fluid is zero. The stagnation state corresponding to a static state (1) is defined by $h_t = h_1 + (V_1^2/2g_0)$ and $s_t = s_1$, where the subscript $t$ denotes the stagnation state. The stagnation state would be achieved if the moving stream were decelerated to zero velocity isentropically with no work. For adiabatic flows with no work, $h_t$ is constant. For isentropic no-work flows, both $h_t$ and $p_t$ are constant. For adiabatic irreversible flows that do not involve work, both the static and the stagnation entropy increase and $p_t$ decreases. The isentropic speed of sound for all pure substances in the vapor phase is given by

$$c = \left[g_0\left(\frac{\partial p}{\partial \rho}\right)_s\right]^{1/2} = (g_0 kpv)^{1/2}$$

where $k$ is the isentropic exponent. For perfect gases, $k = \gamma$ and

$$c = (g_0 \gamma RT)^{1/2}$$

The Mach number is defined as

$$N_m = \frac{V}{c}$$

where both $V$ and $c$ are evaluated at the same point on the flow. Subsonic flows are those for which the Mach number is less than unity and supersonic flows have a Mach number greater than unity.

The behavior of the isentropic flow of a perfect gas in a diffuser or a nozzle is related to the Mach number. In subsonic flows a converging passage will increase the velocity and decrease the pressure, while a diverging passage increases the pressure and decreases the velocity. In a converging passage in supersonic flow, the pressure increases and the velocity decreases, and in a diverging passage the velocity increases and the pressure decreases. When the Mach number at the throat of a converging–diverging passage with a subsonic inlet velocity is unity, the passage is choked. When such a passage is choked, changes downstream of the throat will not affect the flow upstream of the throat. The pressure at the point where the Mach number is unity (the star state) is

$$\frac{p^*}{p_t} = \left(\frac{2}{\gamma + 1}\right)^{(\gamma - 1)/\gamma}$$

for perfect gases with constant specific heats. A converging nozzle will choke when the discharge region pressure $p_d$ is less than or equal to the star state pressure. For choked flow in a converging nozzle, the flow is isentropic only within the nozzle. If $p_d$ is greater than $p^*$, the flow is not choked and the flow can be treated as isentropic from the inlet to the discharge-region pressure.

When a supersonic nozzle is operated at a discharge pressure that differs from the design value, changes will occur at the nozzle outlet. If $p_d$ is greater than the design value, shock waves will form in the diverging section and the flow downstream of the shock will be subsonic. Flow with a shock wave cannot be treated as isentropic. If the discharge pressure is lower than the design value, supersonic flow in the diverging section is maintained and an expansion occurs in the downstream discharge region.

When significant viscous effects are present in nozzle flows, the assumption of isentropic flow does not yield accurate results. For such flows nozzle efficiency is used to describe the nozzle performance. Nozzle efficiency is given by

$$\eta_N = \frac{V_e^2}{V_s^2}$$

where $V_e$ is the actual nozzle exit velocity and $V_s$ is the velocity that would be realized in isentropic nozzle flow from the same inlet stagnation state to the actual discharge static pressure. Nozzle efficiencies are typically in the range 0.9 to 0.98.

### Selected References

Ellenwood, F. O., and C. O. MacKey. *Thermodynamic Charts,* 2nd ed. New York: Wiley, 1944.
Hall, N. A. *Thermodynamics of Fluid Flow.* Englewood Cliffs, N.J.: Prentice-Hall, 1956.
Holman, J. P. *Thermodynamics,* 2nd ed. New York: McGraw-Hill, 1974.
Owczarek, J. A. *Fundamentals of Gas Dynamics.* New York: International Textbook, 1964.
Reynolds, W. C., and H. C. Perkins. *Engineering Thermodynamics.* New York: McGraw-Hill, 1970.
Shapiro, A. H. *The Dynamics and Thermodynamics of Compressible Fluid Flow,* Vol. 1. New York: Ronald Press, 1953.

## Problems

13.1 Air enters an adiabatic nozzle at 100 psia, 700 R (static), with a velocity of 695 ft/sec. At the exit section of the nozzle the static pressure is 14 psia and the velocity 1850 ft/sec. Determine (a) the inlet stagnation pressure and temperature, (b) the stagnation pressure and temperature at the exit section, and (c) the entropy change between the inlet and exit sections of the nozzle. Sketch an *h-s* diagram showing static and stagnation states.

13.2 A stream of carbon dioxide at 345 kPa and 335 K has a velocity of 275 m/sec. Calculate the stagnation pressure and temperature of the stream.

13.3 Steam at 100 psia, 400 F, has a velocity of 1400 ft/sec. What are the stagnation enthalpy, stagnation pressure, and stagnation temperature of the steam?

13.4 A pure substance having a molecular weight of 31.07 expands adiabatically through a nozzle. At the nozzle inlet the stagnation pressure is reported to be 390 kPa and that at the outlet 425 kPa. Is this possible?

13.5 A stream of air undergoes a steady-flow process during which heat transfer to the air is 90 B/lbm. At the entering section the air is at 28 psia, 700 R, and has a velocity of 700 ft/sec. At the exit section the stagnation pressure is the same as that at the inlet section and the velocity is 1100 ft/sec. Determine the static pressure and temperature at the exit section. Sketch a *T-s* diagram showing static and stagnation states for the two sections.

13.6 Air expands isentropically at the rate of 50 lbm/sec through a turbine. At the turbine inlet the stagnation pressure is 140 psia and the stagnation temperature is 2000 R. At the turbine exhaust the stagnation pressure is 16 psia. Calculate (a) the stagnation temperature at the turbine exhaust and (b) the horsepower delivered by the turbine.

13.7 Air flows through a section of a straight tube with a 6-inch inside diameter at the rate of 5 lbm/sec. The entrance pressure and temperature are 80 psia and 100 F, and the exit pressure and temperature are 18 psia and 20 F.

(a) Using eq. (13.4) find the magnitude and direction of the force that the air exerts on the tube.
(b) Find the magnitude and direction of the force that must be applied to the tube to keep it stationary.

13.8 Air enters a horizontal diffuser at a pressure and temperature of 10 psia and 60 F with a velocity of 900 ft/sec. The inlet cross-sectional area is 1.4 ft². The flow is steady and isentropic between the entrance and the exit. The exit pressure is 12 psia. The pressure surrounding the diffuser walls is 14 psia. Determine (a) the magnitude and direction of the force of the flow on the diffuser, and (b) the magnitude and direction of the axial force necessary to anchor the diffuser.

13.9 Air enters a horizontal, diverging passage at 10 psia and 500 R with a velocity of 1050 ft/sec. The cross-sectional area of the inlet is 1.25 ft². Flow through the passage is isentropic and there is no work interaction. The velocity of the air at the outlet section is 200 ft/sec. Determine (a) the stagnation temperature and pressure, (b) the cross-sectional area of the outlet, and (c) the axial force acting on the inner wall of the passage.

13.10 A stream of air at 105 kPa is moving at 365 m/sec. Determine the stagnation temperature and stagnation pressure if the Mach number is (a) 0.50, (b) 1.00.

13.11 Steam has a velocity of 1500 ft/sec. Determine the Mach number if the steam is at (a) 100 psia saturated vapor, (b) 100 psia and 600 F, and (c) 600 psia and 600 F.

13.12 A perfect gas is supplied to a nozzle at a fixed stagnation state. The static pressure decreases at successive sections from the inlet to the outlet of the nozzle. What two factors show that the Mach number increases as the expansion through the nozzle takes place?

13.13 For a perfect gas with constant specific heats show that

$$\frac{T}{T_t} = \left(1 + \frac{\gamma - 1}{2} N_m^2\right)^{-1}$$

and

$$\frac{p}{p_t} = \left(1 + \frac{\gamma - 1}{2} N_m^2\right)^{\gamma/(1-\gamma)}$$

13.14 From a steady-flow equation and the second *Tds* equation (eq. [7.9]) show that

$$dp = -\frac{VdV}{vg_0}$$

for any pure substance undergoing a steady-flow, reversible adiabatic process during which no work is done.

13.15 For reversible, adiabatic steady flow with no work the relation

$$\frac{dA}{A} = -\frac{dV}{V}\left(1 - N_m^2\right)$$

is true for any pure substance. Using this relation and the equation from the preceding problem, show (a) for a substance undergoing an isentropic process from a subsonic state to a supersonic state, that the flow passage must have a converging-diverging configuration, (b) that for the pressure to increase in an isentropic supersonic flow, the passage must have a converging configuration, and (c) that for isentropic flow, the Mach number can be unity only at the section where the passage has a minimum cross-sectional area.

13.16 Air expands isentropically through a nozzle from 100 psia, 700 R, velocity 695 ft/sec at the inlet to 14 psia at the exit section. Calculate (a) the stagnation pressure and temperature at the inlet and (b) the Mach number at the exit section.

13.17 Carbon dioxide flows isentropically through a nozzle. At one section in the passage the static pressure is 27.0 psia, the static temperature 763 R, and the velocity 593 ft/sec. For these conditions calculate (a) the Mach number at the given section and (b) the velocity when the flow reaches a Mach number of 1.

13.18 Air at 1000 kPa, 390 K, is supplied to a nozzle at a negligible velocity. Expansion through the nozzle is isentropic to a static pressure of 100 kPa. For a flow rate of 35 kg/min, determine (a) the throat area and (b) the outlet area.

13.19 Air is supplied to a nozzle at a velocity of 600 ft/sec. Expansion through the nozzle is isentropic. At the nozzle outlet the pressure is 10 psia, the temperature 500 R, the Mach number 2.15, and the area 4.15 in$^2$. Determine the inlet and throat areas of the nozzle.

13.20 An ideal, supersonic wind tunnel consists of a converging–diverging nozzle, a constant-area test section, and a supersonic diffuser. Flow in the test section is to be at a Mach number of 2.50 with 10 psia and 400 R, static conditions. The test section is 1 foot by 1 foot. Air is to leave the diffuser at 500 ft/sec. Determine (a) the throat area of the nozzle, (b) the outlet area of the diffuser, and (c) the volume rate of flow in cubic feet per minute measured at 14.696 psia and 60 F. Assume reversible and adiabatic flow through the components.

13.21 A converging–diverging passage has an inlet area of 1.00 ft$^2$, a throat area of 0.500 ft$^2$, and an outlet area of 0.950 ft$^2$. Air enters the passage at 15 psia, 570 R, with a velocity of 150 ft/sec. The throat pressure is 12.27 psia. Determine the static pressure and the velocity at the outlet if the flow is isentropic.

13.22 At a given section in a converging–diverging nozzle nitrogen has a stagnation pressure of 690 kPa, a stagnation temperature of 445 K, and Mach number 1.407. Flow through the nozzle is isentropic. Determine (a) the static temperature at the given section, (b) the static pressure at the given section, (c) the velocity at the throat of the nozzle, and (d) the ratio of the area at the given section to that at the throat of the nozzle.

13.23 Air expands isentropically through a nozzle from 100 psia, 800 R, and negligible velocity to a static pressure of 15 psia. At what section or in what portion of the nozzle will the following occur?

(a) Maximum velocity
(b) Minimum specific volume
(c) Velocity that increases more rapidly than specific volume
(d) Minimum static temperature
(e) Lowest sonic velocity
(f) $T = [2/(\gamma + 1)]T_t$
(g) Mach number of 1
(h) Mach number greater than 1
(i) Minimum static enthalpy
(j) Maximum mass rate of flow per unit area

13.24 A converging nozzle is supplied with air at 80 psia, 700 R, and velocity 500 ft/sec. The nozzle discharges into a region at 39 psia. If the outlet area of the nozzle is 1.44 $in^2$, what is the mass rate of flow through the nozzle?

13.25 An ideal gas expands isentropically within a converging nozzle. The stagnation state is at 100 psia and 540 R and the gas is flowing subsonically at the nozzle inlet. The cross-sectional area of the nozzle outlet is 1.00 $in^2$. The pressure in the discharge region is 50 psia. Determine the mass rate of flow through the nozzle if the gas is (a) carbon dioxide and (b) helium. Also determine the Mach number at the outlet section for each gas.

13.26 Air is supplied to a converging nozzle at 700 kPa and 500 K (static conditions) with a velocity of 180 m/sec. The outlet area of the nozzle is $9 \times 10^{-4}$ $m^2$. Assuming the pressure in the discharge region is 380 kPa, calculate the mass rate of flow through the nozzle.

13.27 A converging nozzle has an outlet area of 1.00 $in^2$. Carbon dioxide enters the nozzle subsonically at a stagnation state of 25 psia and 600 R. The nozzle discharges into the atmosphere, which is at 14.7 psia. Under these conditions the mass rate of flow through the nozzle is 0.650 lbm/sec. Suppose the stagnation temperature is held constant at 600 R but the mass rate is increased to 2.00 lbm/sec. What inlet stagnation pressure would be required to achieve this higher flow rate?

13.28 Hydrogen is supplied to a converging nozzle at 400 psia and 1100 R with negligible velocity. Assuming the discharge region is at 100 psia and the mass rate of flow is 2.74 lbm/sec, calculate (a) the velocity of the hydrogen at the nozzle exit section and (b) the outlet area of the nozzle.

13.29 Air flows through a converging nozzle. At the nozzle outlet section the static pressure and temperature are 105 kPa and 250 K respectively. Assuming the outlet Mach number is 0.953 and the inlet Mach number is 0.157, determine the ratio of the inlet area to outlet area.

13.30 Steam with a stagnation state at 400 psia and 800 F is expanded isentropically in a nozzle to an exit pressure of 98 psia. Determine the exit velocity, Mach number, and mass rate of flow per unit area.

13.31 Saturated steam enters a converging nozzle at 400 psia with a velocity of 600 ft/sec. The discharge region is at 100 psia and the nozzle outlet area is 1.00 $in^2$. Determine (a) the inlet stagnation pressure and (b) the mass rate of flow.

13.32 A nozzle is to expand steam from 60 psia, 320 F, isentropically to 15 psia. Assuming the inlet velocity is negligible, determine the throat and exit section areas for a flow rate of 2000 lbm/hr.

13.33 An isentropic nozzle is to be designed to discharge steam at 30 psia and 400 F with a Mach number of 2.15 and at the rate of 10,000 lbm/hr. Determine the throat area of the nozzle.

13.34 Refer to problem 13.32. Suppose flow from the inlet to the throat of the nozzle is assumed to be isentropic but losses in the diverging section of the nozzle produce a nozzle efficiency of 88 percent. What is the required cross-sectional area at the nozzle outlet?

13.35 Air is supplied to a nozzle with a stagnation pressure of 100 psia and a stagnation temperature of 1200 R. At the nozzle outlet the velocity is 2280 ft/sec and the static pressure is 15 psia. Determine (a) the outlet stagnation pressure, (b) the nozzle efficiency, and (c) the mass rate of flow per unit area at the outlet.

13.36 Air expands adiabatically, but not reversibly, through a nozzle. At the nozzle inlet the stagnation pressure is 700 kPa and the stagnation temperature is 500 K. Due to the irreversibility of the process the stagnation pressure decreases by 70 kPa between the nozzle inlet and outlet sections. Assuming the static pressure at the nozzle outlet is 140 kPa, calculate the nozzle efficiency. Sketch a $T$-$s$ diagram showing the inlet and outlet static and stagnation states.

13.37 Refer to the nozzle test apparatus illustrated in Figure 13.9. Steam, saturated vapor, is supplied to the nozzle at 200 psia with a low velocity, and the static pressure at the nozzle outlet is 5 psia. The force due to the stream impacting on to the flat plate is found to be 96.74 lbf. The steam is condensed and the flow rate is determined to be 3212 lbm/hr. From these data determine the nozzle efficiency.

13.38 Nitrogen enters a nozzle at 800 R and 90 psia at a velocity of 500 ft/sec. At the nozzle exit section the static pressure is 15 psia. For a nozzle efficiency of 90 percent and an outlet area of 1.637 $in^2$ determine the volume rate of flow at the nozzle outlet.

13.39 For each of the following specify the geometry (converging-diverging, converging, or either) if the flow is isentropic:

(a) A perfect gas having a specific heat ratio of 1.667 is to be expanded from 1000 kPa, 500 K, and negligible velocity to 500 kPa.
(b) Steam at 400 psia and 600 F, with a low velocity, is to be expanded until the velocity is 2500 ft/sec.
(c) Air at 100 psia and 700 R with a Mach number of 0.100 is to be expanded until the static temperature is 400 R.
(d) A diffuser is to reduce the Mach number from 2.50 to 0.41.
(e) Air at 140 kPa and 220 K with Mach number 2.10 is to be brought to a static pressure of 350 kPa.

# 14

# Compressors, Fans, Pumps, and Turbines

*In this chapter some thermodynamic considerations for reciprocating and rotating machines are presented. Particular emphasis will be placed on reciprocating and rotating compressors, fans, pumps, and turbines. The selection of a compressor for a specific application depends on the pressure ratio and the volume of working fluid to be handled. In some applications a reciprocating compressor may handle, most effectively, a relatively small intake volume for a desired high-pressure ratio. The other bound is the use of axial or centrifugal compressors for the large intake volumes per unit time encountered in stationary and aircraft gas-turbine applications. Fans are used for high-volume, low-pressure rise applications involving air or other gases. Pumps are typically used to increase the pressure of liquids.*

*Turbines deliver power in a wide variety of applications, usually where high-power demands are involved. Thermodynamic analysis of selected representative machines are included in this chapter.*

## 14–1 Energy Relations for Compression Processes

For steady-flow compression processes, the steady-flow energy equation

$$h_1 + \frac{V_1^2}{2g_0} + \frac{g}{g_0} z_1 + {}_1q_2 = h_2 + \frac{V_2^2}{2g_0} + \frac{g}{g_0} z_2 + {}_1w_2 \qquad \textbf{(3.4b)}$$

must be satisfied regardless of the type or nature (reversible or irreversible) of the process. Using the definition of stagnation enthalpy

$$h_t = h + \frac{V^2}{2g_0} \qquad \textbf{(13.1)}$$

eq. (3.4b) can be written for applications where kinetic energy change is not small and potential energy change is neglected in the form

$${}_1q_2 + h_{t1} = {}_1w_2 + h_{t2} \qquad \textbf{(14.1)}$$

**Figure 14.1** Comparison of Compression Paths

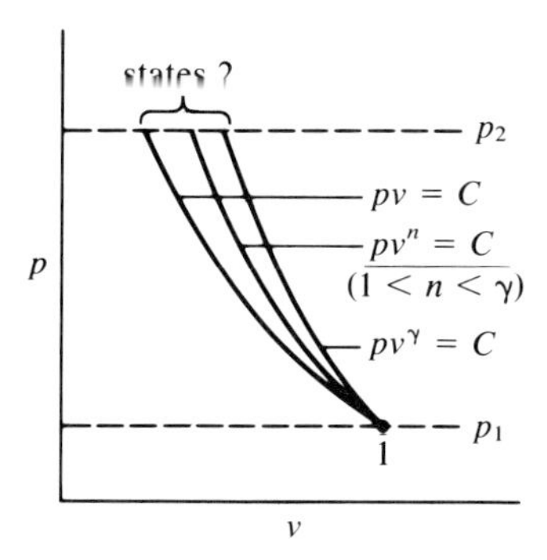


Potential energy change may be included if the terms are significant, as might be the case in the pumping of liquids.

Although irreversibilities occur in all real compression processes, a useful description of such processes is obtained by neglecting the irreversibilities. (Later in this chapter we will account for any irreversibilities that occur by introducing compressor efficiencies.) The work for a steady-flow, internally reversible process as given by

$$ {}_1w_2 = -\int_1^2 v\,dp - \frac{V_2^2 - V_1^2}{2g_0} - \frac{g}{g_0}(z_2 - z_1) \tag{2.14} $$

provides the basis for this approach, and is applicable to all substances. However, we at this point shall limit the discussion to the compression of perfect gases, as this topic has wide engineering application. It is evident from the equation that the manner in which the specific volume varies with pressure will affect the work. To demonstrate this we consider three processes: isothermal compression ($pv$ = constant), polytropic compression ($pv^n$ = constant, $1 < n < \gamma$), and isentropic compression ($pv^\gamma$ = constant). The paths for these processes are shown in Figure 14.1. We will neglect kinetic energy in this discussion. By comparing areas ($\int v\,dp$), it is evident that the path requiring the least work is the isothermal path, $pv$ = constant. The work per unit mass for this path is, from eq. (2.14),

$$ {}_1w_2 = -\int_1^2 v\,dp = -RT\int_1^2 \frac{dp}{p} = -RT \ln \frac{p_2}{p_1} \tag{14.2} $$

Application of eq. (3.4b) to this path yields ${}_1w_2 = {}_1q_2$ since enthalpy is a function of temperature only for perfect gases. The expression obtained from eq. (2.14) for the polytropic process $pv^n$ = constant is given by eq. (5.21) as

$$ {}_1w_2 = \frac{n}{n-1}(p_1v_1 - p_2v_2) \tag{5.21} $$

The above equation can be written as

$$_1w_2 = \frac{n}{n-1} R(T_1 - T_2) = \frac{n}{n-1} RT_1\left(1 - \frac{T_2}{T_1}\right)$$

$$_1w_2 = \frac{n}{n-1} RT_1\left[1 - \left(\frac{p_2}{p_1}\right)^{(n-1)/n}\right] \tag{14.3}$$

where the temperature–pressure relation for a polytropic process for perfect gases, eq. (5.19), has been used.

$$\frac{T_2}{T_1} = \left(\frac{p_2}{p_1}\right)^{(n-1)/n}$$

The change in enthalpy for the process is not zero but it may be expressed in terms of the end-point temperatures. The heat transfer for the process can be determined from application of eq. (3.4b).

The isentropic compression process $pv^\gamma$ = constant is a special case of the polytropic process. Therefore, eq. (14.3) with $n$ replaced by $\gamma$ yields an expression for the work for the isentropic process. Since the process is adiabatic, eq. (3.4b) also yields a work expression

$$_1w_2 = h_1 - h_2 = C_p(T_1 - T_2)$$

Note that the expression $pv^\gamma$ = constant applies only to perfect gases with constant specific heats.

Heat transfer for the isothermal and polytropic processes is negative. This is immediately evident for the isothermal process from the previously noted fact that $_1w_2 = {_1q_2}$. Since the work is negative, the heat transfer is negative. The cooling has the effect of reducing the specific volume during the process and of reducing the discharge temperature.

Expressions for compressor power can be obtained by multiplying the work equations by $\dot{m}$. For the polytropic case

$$_1\dot{W}_2 = \frac{n}{n-1} \dot{m}RT_1\left[1 - \left(\frac{p_2}{p_1}\right)^{(n-1)/n}\right] \tag{14.4}$$

Note that when $RT_1$ is replaced by $p_1v_1$, the product $\dot{m}v_1p_1$ appears in the expression. By the continuity equation, $mv_1$ is the inlet volume rate of flow $V_1A_1 = \dot{V}_1$. Therefore, the power for the polytropic compression process can be written in terms of the inlet volume rate of flow and the pressures $p_1$ and $p_2$. A similar expression can be written for the isothermal compression process.

## Example 14.a

An ideal air compressor takes in 500 ft³/min at 14 psia and discharges the air at 120 psia. Calculate the power required to drive the compressor if the compression process is (a) isothermal, (b) polytropic with $n = 1.31$, and (c) isentropic.

**Solution:**

(a) For isothermal compression the expression for the power is obtained using eq. (14.2) and $\dot{W} = \dot{m}w$ and takes the form

$$\dot{W}_{in} = \dot{m}RT_1 \, ln \frac{p_2}{p_1} = \dot{m}v_1 p_1 \, ln \frac{p_2}{p_1} = \dot{V}_1 p_1 \, ln \frac{p_2}{p_1}$$

$$= \frac{500 \text{ ft}^3}{\text{min}} \left| \frac{(14)(144) \text{ lbf}}{\text{ft}^2} \right| \frac{\text{HP sec}}{550 \text{ ft lbf}} \left| \frac{\text{min}}{60 \text{ sec}} \right. ln \frac{120}{14}$$

$$= 65.6 \text{ HP}$$

(b) For the polytropic compression eq. (14.4) gives the power required as

$$\dot{W}_{in} = \frac{n}{n-1} p_1 \dot{V}_1 \left[ \left( \frac{p_2}{p_1} \right)^{n-1/n} - 1 \right]$$

$$= \frac{1.31}{0.31} \left| \frac{(14)(144)(500)}{(550)(60)} \right| \left( \frac{120}{14} \right)^{0.2366} - 1$$

$$= 85.5 \text{ HP}$$

where $\dot{V}_1 p_1$ has been substituted for $\dot{m}RT_1$.

(c) The above equation with $n$ replaced by $\gamma$ is applicable to the isentropic case. The power required for the isentropic compression process is

$$\dot{W}_{in} = \frac{\gamma}{\gamma - 1} p_1 \dot{V}_1 \left[ \left( \frac{p_2}{p_1} \right)^{(\gamma-1)/\gamma} - 1 \right]$$

$$= \frac{1.4}{1.4 - 1} \left| \frac{(14)(144)(500)}{(550)(60)} \right| \left( \frac{120}{14} \right)^{0.286} - 1$$

$$= 90.7 \text{ HP}$$

**Figure 14.2** Illustration of Multistage Compression with Intercooling. *a.* Two-stage compressor with intercooler. *b.* *p-v* diagram for two-stage compressor.

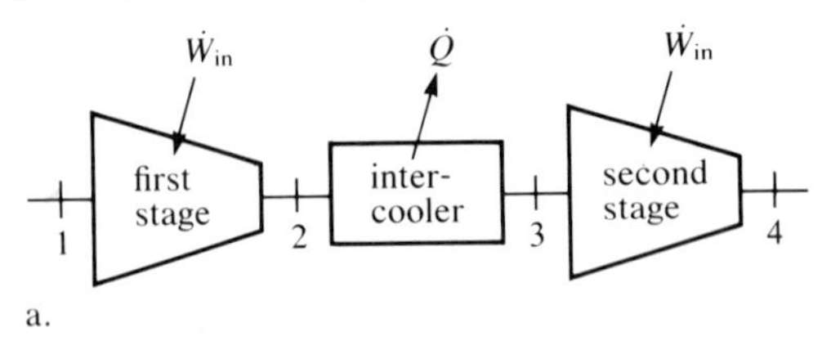


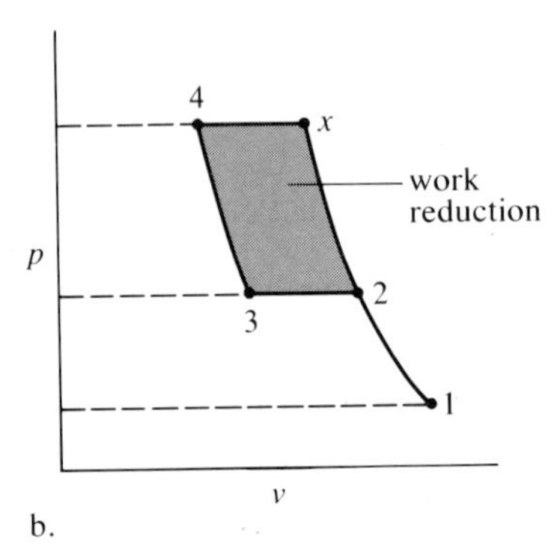


## 14–2 Ideal Multistage Intercooled Compressors

Although the reduction in the work requirement that results from cooling during the compression process is desirable, it is difficult or impractical to attain large heat-transfer rates during compression. For compressors that have discharge-to-inlet-pressure ratios greater than about five, it becomes economically desirable to consider using two or more stages of compression, with an intercooler between each stage. This arrangement is called multistaging with intercooling. Figure 14.2a shows the components for a two-stage intercooled compressor. Compression to the intercooler pressure $p_2$ occurs in the first stage, after which the gas is cooled at constant pressure and then passes to the second stage to be compressed to the discharge pressure $p_4$. The compression process for each stage is assumed to be polytropic, with the polytropic exponent $n$ the same for each stage. The effect of intercooling on the work is illustrated in Figure 14.2b. Since the operation of each stage is assumed to be internally reversible, eq. (2.14) applies and, with no change in kinetic energy, the area between the path and the ordinate on the *p-v* diagram represent the work. From the diagram it is seen that the constant-pressure cooling process (2) to (3) reduces the work required for the second-stage compression process (3) to (4). Had a single-stage compressor been used, the discharge state would be state ($x$) on the diagram. Hence the work reduction realized by utilizing two-stage compression with intercooling is the shaded area on the diagram. For air compressors the lowest temperature to which it is practical to attempt to cool the air is the temperature of the air entering the compressor. Complete intercooling is said to exist when $T_3 = T_1$.

The work required to drive the two-stage compressor with intercooling is the sum of the works required for each stage. Applying eq. (14.3) to each stage, we obtain

$$w_{in} = \frac{n}{n-1} R\, T_1 \left[ \left(\frac{p_2}{p_1}\right)^{(n-1)/n} - 1 \right] + \frac{n}{n-1} R\, T_3 \left[ \left(\frac{p_4}{p_3}\right)^{(n-1)/n} - 1 \right] \tag{14.5}$$

where complete intercooling has not been assumed. For complete intercooling the above equation reduces to

$$w_{in} = \frac{n}{n-1} R\, T_1 \left[ \left(\frac{p_2}{p_1}\right)^{(n-1)/n} + \left(\frac{p_4}{p_3}\right)^{(n-1)/n} - 2 \right] \tag{14.5a}$$

For a given inlet state and discharge pressure $p_4$, the work expressed by the above equation is a function of only the intercooler pressure $p_2 = p_3$. By differentiating this equation with respect to the intercooler pressure and setting the result equal to zero, the intercooler pressure corresponding to the minimum work requirement is obtained. The result is

$$p_2 = p_3 = \sqrt{p_1 p_4} \quad \text{or} \quad \frac{p_2}{p_1} = \frac{p_4}{p_3} \tag{14.6}$$

Equation (14.6) yields the optimum intercooler pressure and reflects the fact that the work requirement is a minimum when the pressure ratio across each state is the same. For ideal compressors with two or more stages, it can be shown that the optimum stage pressure ratio *PR* for complete intercooling is

$$PR = \left(\frac{p_f}{p_1}\right)^{1/N} \tag{14.6a}$$

where $p_f$ is the discharge pressure from the last stage, $p_1$ is the inlet pressure to the first stage, and $N$ is the number of compressor stages. From eq. (14.3) we can see that when the inlet temperature to each stage is the same, and when the pressure ratio across each state is the same, the work per unit mass will be identical for all stages. For steady flow the mass rate of flow $\dot{m}$ is the same for all stages. Thus we conclude that the power required to drive each stage is the same. The power for one stage is given by eq. (14.4). The power required to drive an $N$-stage, completely intercooled compressor with the optimum stage pressure ratio is then

$$\dot{W}_{in} = N \frac{n}{n-1} p_1\, \dot{V}_1 \left[ \left(\frac{p_f}{p_1}\right)^{(n-1)/nN} - 1 \right] \tag{14.7}$$

While the mass rate of flow through each stage is the same, the volume rate of flow, of course, is not. The product $\dot{m}v$ is the volume rate of flow, and since $v$ progressively decreases, the volume rate of flow decreases correspondingly.

## Example 14.b

A two-stage compressor with an intercooler takes in air at 14 psia and 70 F. The air leaves the intercooler at 70 F and 37 psia. The second stage delivers air at 120 psia. The compression for both stages is polytropic with $n = 1.31$, and the first-stage volume intake is 500 ft³/min.

(a) Calculate the power required to drive the compressor.
(b) Determine the temperature at the outlet of each stage and the heat transfer per unit mass for the first stage and for the intercooler.

**Solution:**

(a) First, let us check to determine whether or not the intercooler pressure is the optimum, since complete intercooling is indicated. From eq. (14.6) the optimum intercooler pressure would be

$$p_2 = [(14)(120)]^{1/2} = 41.0 \text{ psia}$$

Since the intercooler pressure is not the optimum, the power is not the same for each stage even though complete intercooling exists. Reflection on the development of the compressor equations leads to the conclusion that eq. (14.5a) gives the work per unit mass for both stages when complete intercooling exists, regardless of the intercooler pressure. Multiplication of this equation by $\dot{m}$ to obtain power yields

$$\begin{aligned}\dot{W}_{\text{in}} &= \frac{n}{n-1} p_1 \dot{V}_1 \left[\left(\frac{p_2}{p_1}\right)^{(n-1)/n} + \left(\frac{p_4}{p_3}\right)^{(n-1)/n} - 2\right] \\ &= \frac{1.31}{0.31}(14)(144)(500)\left[\left(\frac{37}{14}\right)^{0.2366} + \left(\frac{120}{37}\right)^{0.2366} - 2\right] \\ &\quad \times \left(\frac{1}{33{,}000}\right) \\ &= 74.8 \text{ HP}\end{aligned}$$

It is noteworthy that the single-stage compressor in part (b) of Example 14.a, which operates under similar intake conditions, has $n = 1.31$, and discharges to the same pressure, requires 85.8 HP.

(b) The stage outlet temperatures are found using eq. (5.19)

$$T_2 = 530\left(\frac{37}{14}\right)^{0.2366} = 667 \text{ R}$$

$$T_4 = 530\left(\frac{120}{14}\right)^{0.2366} = 700 \text{ R}$$

The steady-flow equation applied to the first stage is

$${}_1q_2 = {}_1w_2 + h_2 - h_1$$

By eq. (14.3) the work is

$$
\begin{aligned}
{}_1w_2 &= \frac{1.31\ (53.3)}{0.31\ (778)}(530)\left[1 - \left(\frac{37}{14}\right)^{0.2366}\right] \\
&= -39.7\ \text{B/lbm}
\end{aligned}
$$

Therefore, the heat transfer is

$$
{}_1q_2 = -39.7 + 0.24(667 - 530) = -6.82\ \text{B/lbm}
$$

The heat transfer for the intercooler is, from the steady-flow equation,

$$
{}_2q_3 = h_3 - h_2 = 0.24(530 - 667) = -32.9\ \text{B/lbm}
$$

The accompanying *T-s* diagram shows the paths for the two-stage compression process. Since each process is assumed to be internally reversible, the areas below the paths represent heat transfer. State ($x$) is that state that would exist at $p = 120$ psia if the compression process had occurred in a single-stage polytropic compressor with $n = 1.31$. Note that $T_4$ is less than $T_x$. This is an advantage in some applications.

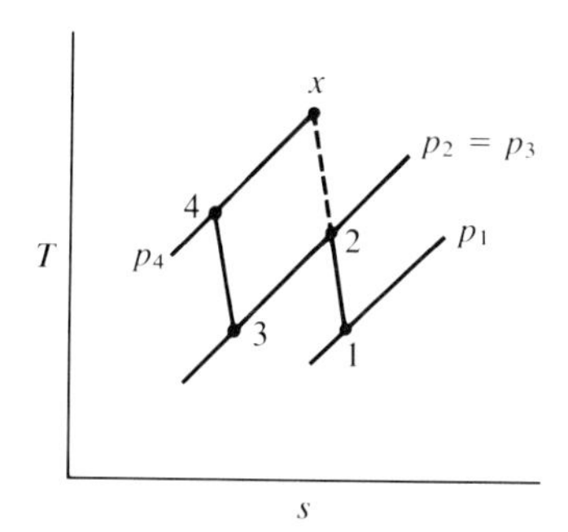


## 14–3 Ideal Gas Compressors

In this section we discuss two types of gas compressors that are representative of the general compressor classes. The first is the axial-flow compressor. This type of compressor consists of a series of moving blades that do work on the gas to increase the pressure and velocity, alternating with a series of fixed blades that form diffusing passages to decelerate the flow and increase the pressure. A section of a simplified axial-flow compressor is shown in Figure 14.3a. The rotors to which the moving blades are attached are mounted on a common shaft. The blades move in a plane perpendicular to the axis of rotation. The gas-flow passage has an annular configuration and the flow is generally in the direction of the axis of rotation of the shaft. A stage consists of a rotor (one set of moving blades) and a stator (one set of fixed blades). For the typical stage shown in the figure, the gas enters the rotor at (1) and is compressed to state ($a$), after which it enters the stator and is diffused to a higher pressure

**Figure 14.3** An Ideal Axial Flow Compressor. *a.* Section of an axial-flow compressor. *b.* *h-s* diagram for one stage. *c.* *h-s* diagram for complete compressor.

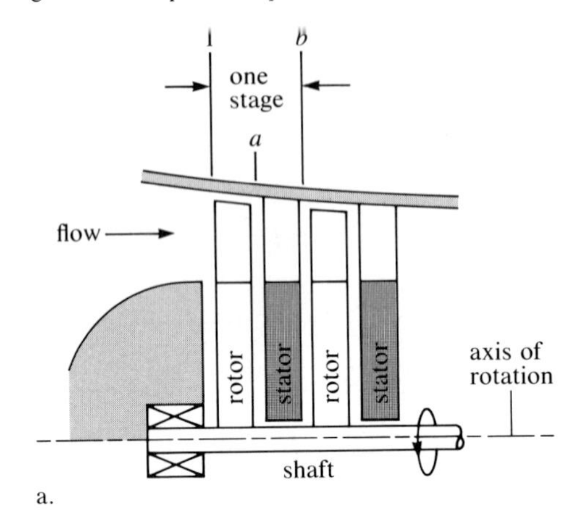


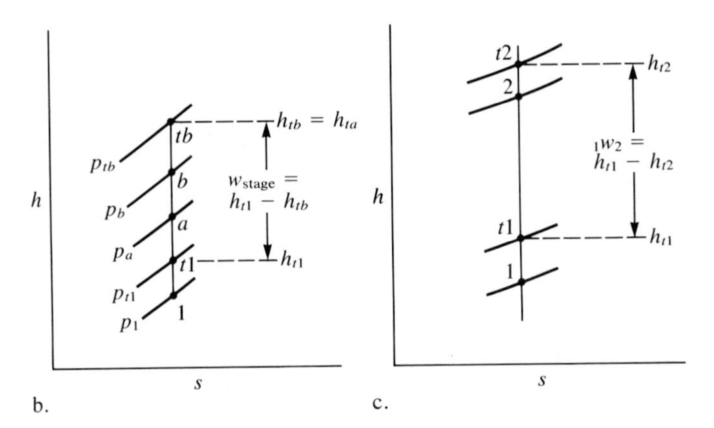


at state ($b$). Ideally the flow is isentropic. Figure 14.3b shows the *h-s* diagram for one stage of compression. Kinetic energies are usually significant. Therefore, the work for the stage by eq. (14.1) is $h_{t1} - h_{tb}$. Since the pressure increase across a single stage is small, several stages are required to produce a high compressor exit pressure. Figure 14.3c shows the *h-s* diagram for an ideal multistage axial flow compressor with an inlet state (1) and an exit state (2). States associated with the individual stages are not shown. As noted in the diagram, the work required is the change in stagnation enthalpy across the compressor.

A centrifugal compressor is similar to the axial flow compressor shown in Figure 14.3a in that the flow enters the compressor in the axial direction and a bladed rotor does work in compressing the fluid. However, instead of leaving the compressor axially, the flow leaves radially in a plane normal to the shaft axis. On leaving the rotor the flow enters a diffusing passage to further increase the pressure.

**Figure 14.4** A Zero-Clearance Volume Compressor

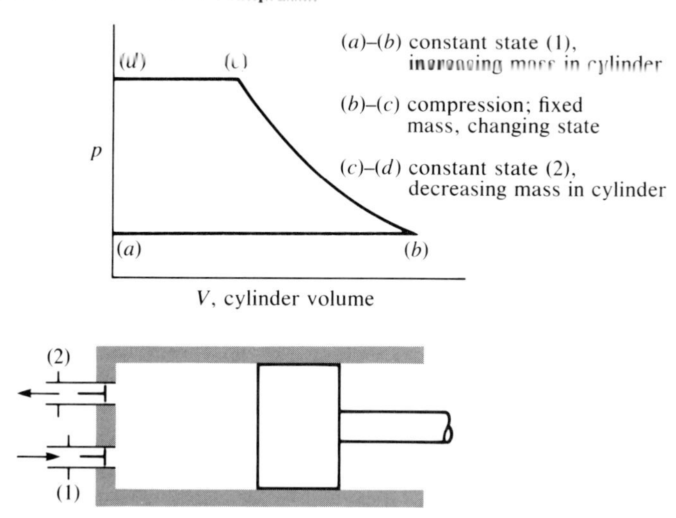


The next type of compressor is the reciprocating compressor, which consists of an arrangement of a cylinder, a piston, and valves that operate to take in low-pressure gas and discharge high-pressure gas. Kinetic energy changes are usually small and will be neglected here. Figure 14.4 shows a diagram of a zero-clearance compressor, one that has zero cylinder volume when the piston is in the leftmost position in the diagram. This compressor, while impractical, is useful in describing several aspects of reciprocating compressors. Let us now describe the events that occur in the cylinder as the system undergoes one cycle of operation. The pressure cylinder-volume diagram in the figure is useful for this description. Ideal behavior is assumed, that is, there is no pressure drop across the valves and all processes are internally reversible. With the piston initially at point (*a*), zero cylinder volume, rightward movement of the piston with the inlet valve open draws in gas at state (1). At the end of the stroke the cylinder volume is $V_b$ and the inlet valve closes. A closed-system compression process takes place in which the cylinder volume decreases to $V_c$. At this point the discharge valve opens and the piston moving to the left discharges high-pressure gas at state (2). At point (*d*) the discharge valve closes and the inlet valve opens, after which the system begins another cycle of operation. Since the system bounded by the cylinder walls and the piston face operates intermittently as an open and closed system, it is evident that the inlet and outlet flows will be intermittent in nature and that the flow is not truly steady. The matter of how to evaluate the work of compression is also important. To address this matter let us sum the work terms for each process that occurs in one cycle of operation. The total work for the gas is given by

$$W_{\text{total}} = {}_aW_b + {}_bW_c + {}_cW_d + {}_dW_a \qquad \textbf{(a)}$$

The last term is clearly zero. Each of the remaining processes involves piston face work, which is given by $\int pdV$. Noting that $V_a$ and $V_d$ are zero, eq. (a) becomes

$$W_{\text{total}} = \int_a^b pdV + \int_b^c pdV + \int_c^d pdV = p_b V_b + \int_b^c pdV - p_c V_c \qquad \text{(b)}$$

This expression involves only the pressures and cylinder volumes at points ($b$) and ($c$) on the diagram, which are, respectively, at thermodynamic states (1) and (2). Also, the mass in the cylinder is the same at these points. Therefore, we can divide eq. (b) by the mass $m$ to obtain the work per unit mass as

$$w_{\text{total}} = p_1 v_1 - p_2 v_2 + \int_1^2 pdv \qquad \text{(c)}$$

Consider now the differential of the product $pv$

$$d(pv) = pdv + vdp \quad \text{or} \quad -vdp = -d\,(pv) + pdv \qquad \text{(d)}$$

Integration between thermodynamic states (1) and (2) yields

$$-\int_1^2 vdp = p_1 v_1 - p_2 v_2 + \int_1^2 pdv \qquad \text{(e)}$$

By comparing eq. (e) with eq. (c) we see that the total work for compressing the gas between states (1) and (2) is

$${}_1w_2 = -\int_1^2 vdp \qquad \text{(f)}$$

which is the same as eq. (2.14), the equation for the work for an internally reversible steady-flow process, when kinetic and potential energy changes are zero. Thus, at least for an ideal zero-clearance volume reciprocating compressor, we can ignore the intermittent nature of the flow and treat the compressor as a steady-flow device when computing the work.

We now consider an ideal reciprocating compressor that has a finite clearance volume. Figure 14.5 illustrates such a compressor. The compression from ($b$) to ($c$) parallels that of the zero-clearance compressor. Discharge of gas at state (2) takes place from ($c$) to ($d$). The volume between the cylinder head and piston face at ($d$) is called the *clearance volume.* The volume change $V_b - V_d$, the volume swept through by the piston during one stroke, is the *displacement volume.* At ($d$) the discharge valve closes and the gas remaining in the cylinder expands from ($d$) to ($a$). This expansion commonly is assumed to take place with the same path equation as that for the compression process ($b$) to ($c$). The inlet valve opens at ($a$) and the intake volume is $V_b - V_a$.

For ideal compressors the clearance volume has no effect on the work required for compression. This can be shown by an approach similar to that taken in determining the work for the zero clearance. Thus eq. (2.14) in the form ${}_1w_2 = -\int_1^2 vdp$ is applicable to ideal reciprocating compressors with clearance volumes. An important effect of clearance

Figure 14.5 Reciprocating Compressor with Clearance Volume

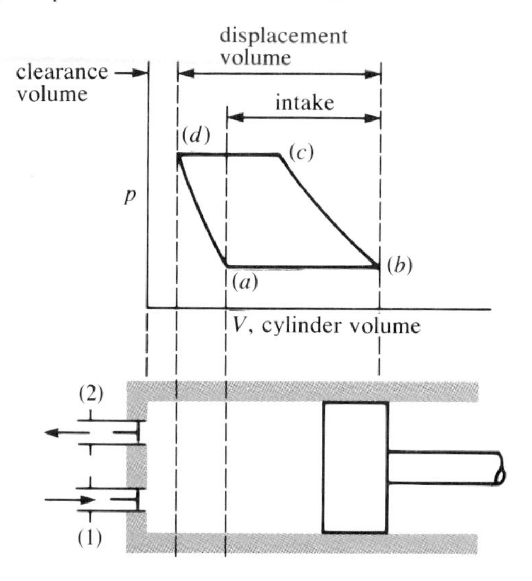


volume on compressor performance is related to the capacity in terms of volume rate of flow. Referring to Figure 14.5, it has been noted that the volume of gas drawn in per stroke is $V_b - V_a$, which is less than the piston displacement $V_b - V_d$. As the discharge pressure is increased for a compressor with a fixed geometry, point $(a)$ moves to the right. If the clearance volume is decreased at a fixed discharge pressure, point $(a)$ moves to the left. For these reasons the clearance volume is made as small as possible.

*Volumetric efficiency* is related to clearance volume effects. It is defined as

$$\eta_v = \frac{\text{actual mass intake}}{\text{mass to fill displacement volume}} \tag{14.8}$$

and is applicable to reciprocating engines as well as reciprocating compressors. The density for evaluating the denominator is taken at the gas inlet state. For ideal compressors as described in Figure 14.5, eq. (14.8) becomes

$$\eta_{v,\text{ideal}} = \frac{\rho_1(V_b - V_a)}{\rho_1(V_b - V_d)} = \frac{V_b - V_a}{V_b - V_d} \tag{14.8a}$$

A two-stage air-to-air intercooled reciprocating compressor is shown in Figure 14.6. Electric-motor-driven units of this design serve many needs for compressed air, ranging from gasoline-engine-driven units for tire dealer service trucks to the supply of compressed air for instrumented control systems. Discharge pressures of 200 psig, with power input of 50 HP, to handle displacements up to 250 ft$^3$/min are typical. As another example, heavy-duty units with four-stage compression are used for discharge pressures in the range of 3500 psig.

**Figure 14.6** A Two-Stage Intercooled Reciprocating Air Compressor

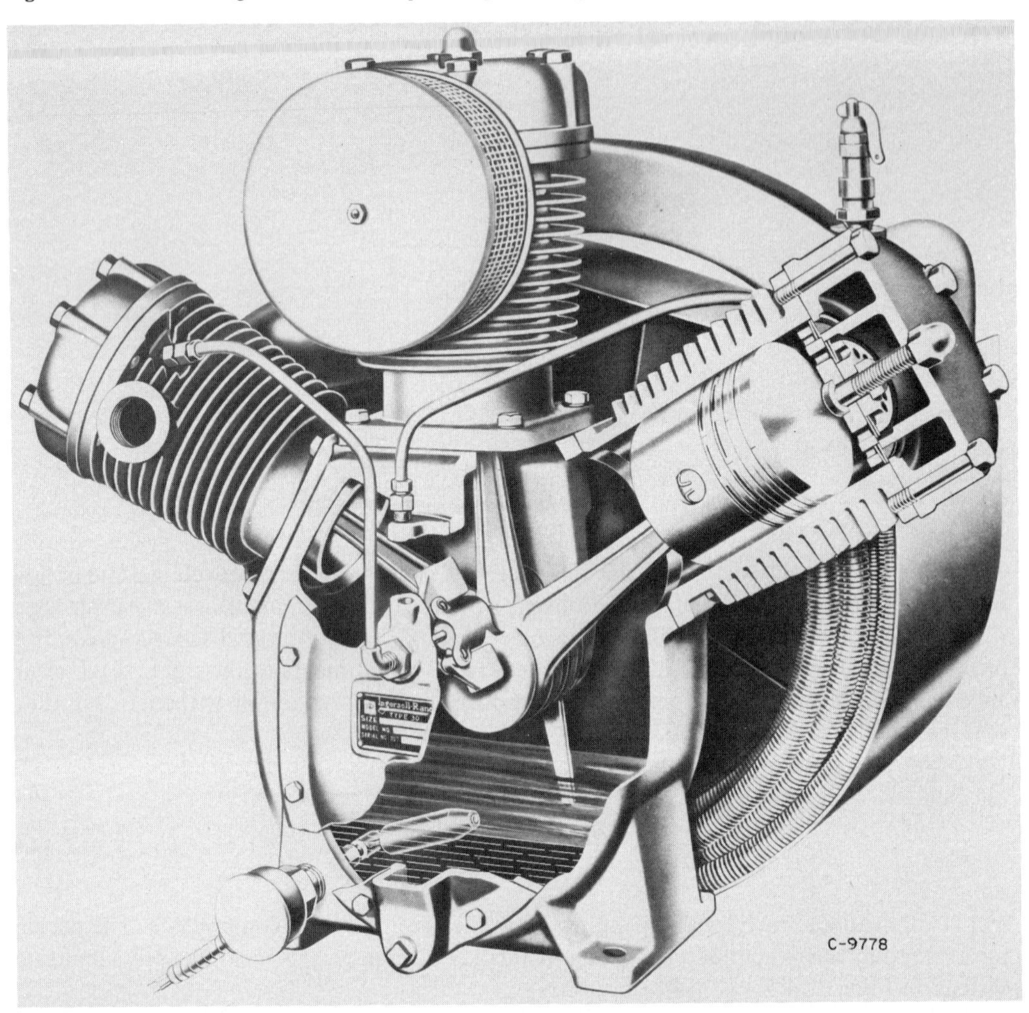


(Courtesy Ingersoll–Rand Co.)

## Example 14.c

An ideal reciprocating single-stage air compressor has a clearance volume equal to 3.15 percent of the displacement volume. If the compressor takes in air at 14.17 psia, discharges it at 79.36 psia, and operates polytropically with $n = 1.300$, what is the volumetric efficiency? If the piston displacement is 1.00 $ft^3$ and the machine operates at 1000 cycles per minute, what is the inlet volume rate of flow?

**Solution:**

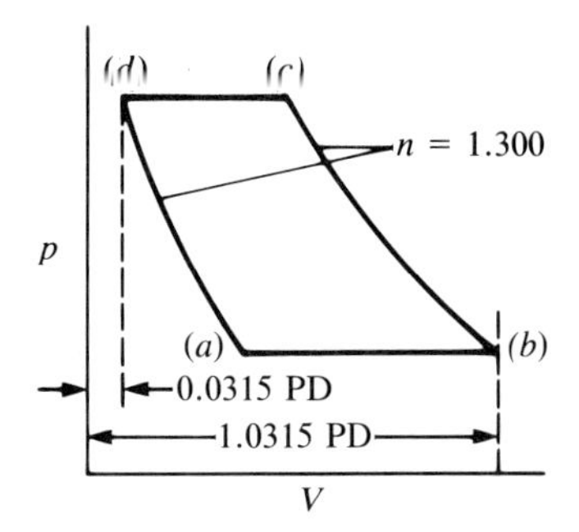


The pressure cylinder–volume diagram shows the volumes related to the problem. The volume at $(a)$ is expressed in terms of the piston displacement $PD$ by

$$V_a = \left(\frac{p_d}{p_a}\right)^{1/n} V_d = (0.0315\ PD)\left(\frac{79.36}{14.17}\right)^{0.769}$$
$$= 0.1185\ PD$$

The ideal volumetric efficiency, defined by eq. (14.8a), is

$$\eta_{v,\text{ideal}} = \frac{1.0315\ PD - 0.1185\ PD}{PD} = 91.3 \text{ percent}$$

The volume rate of flow at (1) for $\dot{N}$ cycles per minute is

$$\dot{V}_1 = \eta_v\ PD\ \dot{N} = 0.913(1.00)\ (1000) = 913\ \text{ft}^3/\text{min}$$

## 14–4 Compressor Efficiencies

Compressor efficiencies are used to relate actual (test) compressor performance to ideal performance. A clear definition of each efficiency should be obtained from an appropriate test code or other source when actual and ideal performances of a compressor are being compared since there are numerous definitions of compressor efficiency in the technical literature.

One efficiency for axial-flow and centrifugal compressors is defined as the ratio of isentropic work to actual work. Kinetic energies are accounted for by using stagnation enthalpies. For determining the actual or the isentropic work, eq. (14.1) is applicable, with the result that the work is the change in stagnation enthalpy, since the above-mentioned compressors operate adiabatically. Hence we define the adiabatic compressor efficiency as

$$\eta_c = \frac{w_s}{w_a} = \frac{(h_{ts} - h_{t1})_s}{(h_{t2} - h_{t1})_a} \tag{14.9}$$

**Figure 14.7** *h-s* Diagram Illustrating Terms for Adiabatic Compressor Efficiency

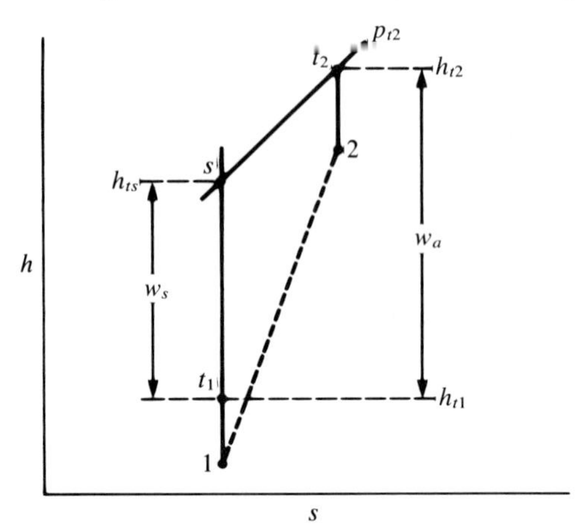


where the subscript $a$ denotes actual stagnation enthalpies and the subscript $s$ denotes the isentropic enthalpy change between the actual inlet stagnation state and a state established by the intersection of a line of constant entropy through state (1) and the outlet stagnation pressure $p_{t2}$. State $s$ and the enthalpy differences related to the adiabatic compressor efficiency are shown in Figure 14.7. If constant specific heats are assumed, the expression for adiabatic compressor efficiency is obtained by replacing the stagnation enthalpies in eq. (14.9) by stagnation temperatures. For high temperatures a method that incorporates the variation of specific heats is recommended.

Efficiencies of reciprocating compressors frequently are expressed in terms of the "indicated power," $\dot{W}_I$, for the compressor. The indicated power is determined from the operating speed of the compressor and an indicator diagram, which is the measured variation of cylinder pressure with cylinder volume. A typical indicator diagram for one cylinder of a reciprocating compressor is shown in Figure 14.8. The mean pressure of the gas in the cylinder during the intake stroke (expansion of the clearance gas and the intake of a fresh gas charge) is $p'$. During the intake stroke work equal to $p'AL$ is done on the piston face, where $A$ denotes the cross-sectional area of the piston and $L$ is the length of the stroke. The product $AL$ is the displacement volume. The mean pressure during compression and discharge is $p''$ and the corresponding work done on the gas by the piston is $p''AL$. The net work per piston cycle is $(p'' - p')AL$. The pressure difference $p'' - p'$ is called the *indicated mean effective pressure* and will be denoted by the symbol $p_m$. The net work per piston cycle, the product of the indicated mean effective pressure and the displacement volume, is called the *indicated work*. It is equal to the rectangular area in Figure 14.8, which is also equal to the enclosed area of the indicator diagram. Thus $p_m$ can be determined by measuring the area of the indicator diagram and dividing by the displacement volume. The indicated power $\dot{W}_I$ is given by the product $W_I \dot{N}$, where $\dot{N}$ is the number of piston cycles (the number of revolutions of the crank) per unit time. Thus, for one cylinder of a reciprocating compressor, the indicated power is

$$\dot{W}_I = p_m L A \dot{N} \tag{14.10}$$

**Figure 14.8** Compressor Indicator Diagram

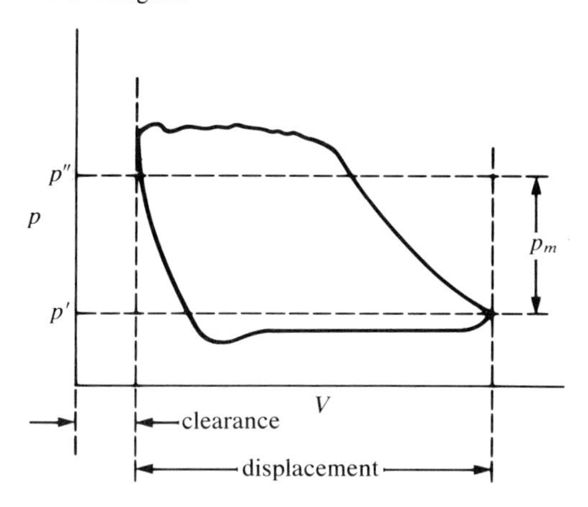


An alternative definition of mean effective pressure, applicable to reciprocating engines as well as compressors, is "that pressure which, acting unopposed on the piston face for one stroke, will produce work equal to the net work per cycle of the compressor (engine)."

The *compression efficiency* of a single-stage reciprocating compressor, or of any stage of a multistage compressor, is defined as the ratio of the ideal power required for adiabatic compression to the indicated power

$$\text{Compression efficiency} = \frac{\dot{W}\text{, ideal, adiabatic}}{\dot{W}_{\text{I}}} \tag{14.11}$$

The equivalent compression efficiency of a multistage reciprocating compressor is defined as the ideal power required to operate the compressor (adiabatic compression in each stage with complete intercooling and optimum stage pressure ratio) divided by the sum of the indicated powers for each cylinder, that is,

$$\eta_{c,\text{equiv}} = \frac{\dot{W}\text{, ideal, adiabatic}}{\Sigma \dot{W}_{\text{I}}} \tag{14.12}$$

The numerator of eq. (14.12) is calculated using eq. (14.7) with $n = \gamma$.

For motor-driven, reciprocating compressors the *mechanical efficiency* is defined as the sum of the indicated powers for the compressor cylinders divided by the motor power delivered to the compressor shaft

$$\eta_{c,m} = \frac{\Sigma \dot{W}_{\text{I}} \text{ of compressor cylinders}}{\dot{W} \text{ at motor-compressor coupling}} \tag{14.13}$$

## Example 14.d

An axial-flow compressor takes in air at 9.05 psia and 467 R, static conditions, with a velocity of 400 ft/sec. The compressor stagnation pressure ratio is 11. For an adiabatic compressor efficiency of 88 percent, as defined by eq. (14.9), determine the power required to drive the compressor if the mass rate of flow is 79.0 lbm/sec. Assume constant specific heats.

**Solution:**

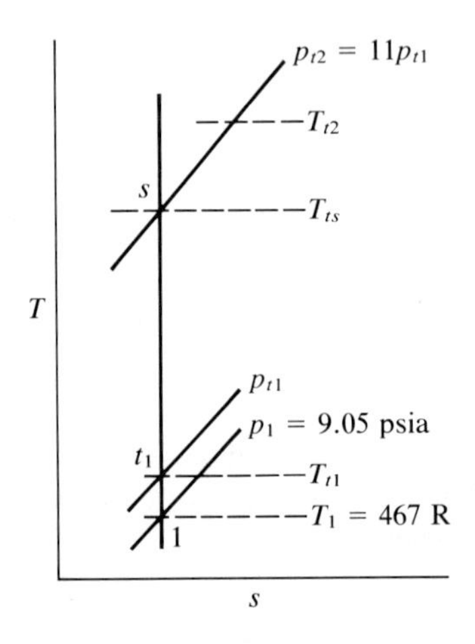


The power will be calculated after the actual work per unit mass has been determined. The static and stagnation states, along with known data, are shown on the *T*-*s* diagram. Note that the static state at the compressor discharge cannot be established since the discharge velocity is unknown.

The inlet stagnation temperature is calculated using the definition of stagnation enthalpy, from which we obtain

$$T_{t1} = \frac{V_1^2}{2g_0\,C_p} + T_1 = \frac{400^2}{2g_0(0.24)(778)} + 467 = 480.3 \text{ R}$$

The inlet stagnation pressure is

$$p_{t1} = p_1\left(\frac{T_{t1}}{T_1}\right)^{\gamma/(\gamma-1)} = 9.05\left(\frac{480.3}{467}\right)^{3.5}$$
$$= 9.986 \text{ psia}$$

For the compressor stagnation pressure ratio of 11 the discharge stagnation pressure is

$$p_{t2} = (11)(9.986)$$
$$= 109.85 \text{ psia}$$

The stagnation temperature at state $s$ is

$$T_{ts} = T_{t1}\left(\frac{p_{t2}}{p_{t1}}\right)^{(\gamma-1)/\gamma} = 480.3(11)^{0.286} = 952.9 \text{ R}$$

Replacing the stagnation enthalpies in eq. (14.9) with stagnation temperatures, we can solve for the actual stagnation temperature change as

$$T_{t2} - T_{t1} = \frac{T_{ts} - T_{t1}}{\eta_c} = \frac{952.9 - 480.3}{0.88} = 537.1 \text{ R}$$

By eq. (14.1), the actual work for the adiabatic process is

$$w_{\text{in},a} = h_{t2} - h_{t1} = C_p(T_{t2} - T_{t1}) = 0.24\ (537.1)$$
$$= 128.9 \text{ B/lbm}$$

Finally, the actual power required for the given mass rate of flow is

$$\dot{W}_{\text{in},a} = \dot{m}w_{\text{in},a} = 128.9(79)(3600)/(2545) = 14{,}400 \text{ HP}$$

---

## 14–5 Constant-Density Pump Processes

Pumps are used to increase the pressure of liquids during steady-flow processes. Some examples of pump applications were mentioned in the vapor power cycles discussed in Chapter 9. For a pump operating adiabatically, the steady-flow equation yields

$$-{}_1w_2 = w_{\text{in}} = h_2 - h_1 + \frac{V_2^2 - V_1^2}{2g_0} + \frac{g}{g_0}(z_2 - z_1)$$

on a unit mass basis. In many applications there are negligible changes in head, $z_2 - z_1$, and kinetic energy between the suction and discharge planes of a pump. With these added restrictions the pump work reduces to

$$w_{\text{in}} = h_2 - h_1$$

For the same restrictions the work input to an ideal (reversible) pump is given by eq. (2.14) as

$$w_{p,\text{in}} = \int_1^2 v\,dp$$

**Table 14.1** Pump Work for Several Discharge Pressures and a Fixed Inlet at 100 F, Saturated Liquid ($H_2O$)

| Inlet Specific Volume, ft³/lbm | Discharge Pressure, psia | $v_1(p_2 - p_1)$, B/lbm | $\left(\frac{v_1 + v_2}{2}\right)(p_2 - p_1)$, B/lbm |
|---|---|---|---|
| 0.01613 | 500 | 1.449 | 1.447 |
| 0.01613 | 1000 | 2.941 | 2.937 |
| 0.01613 | 2000 | 5.926 | 5.908 |
| 0.01613 | 5000 | 14.88 | 14.77 |

In general a functional relation of specific volume and pressure would be necessary to integrate this equation. However, for many applications, a reasonable approximation is to assume the specific volume of the liquid remains constant, and thus

$$w_{p,\text{in}} = v(p_2 - p_1) = h_2 - h_1 \tag{14.14}$$

The error introduced in using the preceding equation increases as the ratio of pump discharge to suction pressure increases. Table 14.1, prepared using data from the *Steam Tables,* indicates the error trend when constant specific volume, equal to the intake specific volume, is assumed for a pump handling liquid $H_2O$. The added assumption of equal suction and discharge temperature has been used in preparing the last column in the table. The tabulated values indicate that utilization of the specific volume at the pump inlet yields quite accurate results for the pressure range indicated.

Pump efficiency is defined as the ratio of ideal work input to actual work input. For applications where kinetic and potential energies are important, the expression for pump efficiency is

$$\eta_p = \frac{\text{ideal work input}}{\text{actual work input}} = \frac{v_1(p_2 - p_1) + (V_2^2 - V_1^2)/2g_0 + (g/g_0)(z_2 - z_1)}{\text{actual work input}} \tag{14.15}$$

The denominator in this expression can be determined by measurements of the pump power input and the mass rate of flow. For a given pump geometry, the kinetic energy term in the numerator can be computed if the mass flow rate is known since the specific volume is essentially constant. An alternate expression for pump efficiency is obtained by multiplying the numerator and the denominator of eq. (14.15) by the mass rate of flow $\dot{m}$. This yields an equation for pump efficiency in terms of ideal and actual power inputs.

Figure 14.9 is a cutaway diagram of a large boiler feed pump.

## 14–6 Fans

Fans or blowers normally are used in high-volume-rate, low-pressure-rise applications involving air and other gases. Typical of these applications are the forced- and induced-draft fans in fossil-fuel-fired steam-generating stations and those used for circulating air in the

**Figure 14.9** Large Boiler Feed Pump

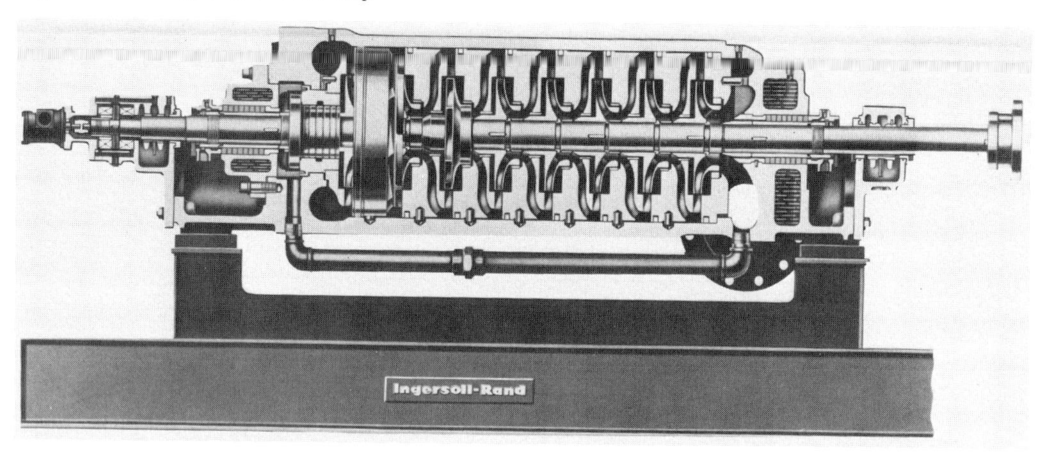


(Courtesy Ingersoll–Rand Co.)

**Table 14.2** Comparison of $v_1(p_2 - p_1)$ and $(h_2 - h_1)_s$ for Air Initially at 14.696 psia and 537 R ($v_1 = 13.530$ ft³/lbm)

| $p_2 - p_1$, lbf/ft² | Equivalent Head, inches $H_2O$ | $v_1(p_2 - p_1)$, ft lbf/lbm | $T_2 = T_1\left(\frac{p_2}{p_1}\right)^{(\gamma-1)/\gamma}$, R | $(h_2 - h_1)_s$, ft lbf/lbm |
|---|---|---|---|---|
| 5.19 | 1.00 | 70.2 | 537.3 | 70.2 |
| 15.56 | 3.00 | 210.1 | 538.1 | 210.2 |
| 31.12 | 6.00 | 421.1 | 539.2 | 419.2 |
| 46.68 | 9.00 | 631.6 | 540.3 | 627.1 |
| 62.24 | 12.00 | 842.2 | 541.4 | 834.0 |
| 124.50 | 24.00 | 1684.5 | 545.8 | 1651.3 |

heating or cooling mode of an air-conditioning system. For an ideal (reversible) fan operating under steady-flow conditions, the work input is given by eq. (2.14). For static pressure increases equivalent to several inches of water the assumption of constant specific volume is a reasonable approximation. With this added assumption the work supplied to an ideal fan is, from eq. (2.14),

$$-{}_1w_2 = w_{in} = v_1(p_2 - p_1) + \frac{V_2^2 - V_1^2}{2g_0}$$

Kinetic energy terms have been retained since they usually are significant. The error introduced by assuming incompressible flow through a fan is indicated in Table 14.2 where $v_1(p_2 - p_1)$ values are compared with the enthalpy change, $h_2 - h_1$, for a reversible and adiabatic process between the same inlet static state and discharge static pressure. In the comparison the kinetic and potential energy changes are not included, and hence they are assumed to be the same for the two calculation procedures. From the data of Table 14.2 it

**Figure 14.10** Pressure–Volume Diagrams for Two Constant Density Steady Flow Processes. *a.* Pump. *b.* Fan.

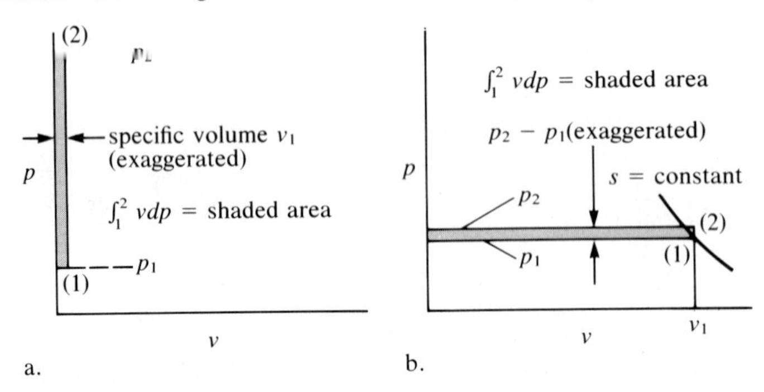


is apparent the assumption of constant-density (incompressible) flow results in errors of 1 percent or less for static pressure rises of up to 12 inches $H_2O$ when compared with isentropic fan performance.

The power required to drive an ideal fan, assuming constant specific volume through the unit, is

$$\dot{W}_{\text{in, ideal}} = \dot{m}\left[v_1(p_2 - p_1) + \frac{V_2^2 - V_1^2}{2g_0}\right] \tag{14.16}$$

Fan efficiency is defined as

$$\eta_f = \frac{\text{ideal power}}{\text{actual power}} \tag{14.17}$$

where the ideal power is calculated using eq. (14.16) and the actual power is determined by test for the fan.

In the preceding section an approximation for pump work was discussed. The approximation was for large pressure increases in a fluid experiencing small changes in specific volume. In this section an analogous situation was discussed for high-volume, low-pressure rise devices, or fans. Figure 14.10 compares these two examples of constant-density steady-flow processes.

## Example 14.e

Four thousand cubic feet of air per minute at 14.70 psia and 60 F flow through a fan. Assuming the velocities of the air at the fan inlet and discharge ducts are 75 and 150 ft/sec, respectively, and the static pressure at the fan discharge is 4.50 inches higher than that in the inlet duct, calculate:

(a) The ideal horsepower to drive the fan
(b) For a fan efficiency of 60 percent, the actual power required to drive the fan

**Solution:**

(a) A convenient basis for calculation is a unit mass of air at the inlet state where

$$v_1 = \frac{RT_1}{p_1} = 13.09 \text{ ft}^3/\text{lbm}$$

The ideal work for the given static pressure rise is

$$\int_1^2 v\,dp = v_1(p_2 - p_1)$$

The relation between the static pressure rise and the water column height is given by eq. (2.4b) as the product of the column height and the specific weight of the water. Specific weight is related to density by eq. (2.3)—specific weight $= \rho g/g_0 = g/(vg_0)$. With the specific volume of water taken from the *Steam Tables* at 60 F, the pressure rise that occurs at standard gravity is

$$p_2 - p_1 = \frac{4.5 \text{ in}}{} \left| \frac{\text{ft}}{12 \text{ in}} \right| \frac{\text{lbf}}{0.016035 \text{ ft}^3} = 31.16 \text{ lbf/ft}^2$$

The mass rate of flow through the fan is

$$\dot{m} = \frac{\dot{V}}{v_1} = \frac{4000 \text{ ft}^3}{\text{min}} \left| \frac{\text{min}}{60 \text{ sec}} \right| \frac{\text{lbm}}{13.09 \text{ ft}^3} = 5.092 \text{ lbm/sec}$$

The ideal power calculated using eq. (14.16) is

$$\begin{aligned} \dot{W}_{\text{in,ideal}} &= \left[(13.09)(31.16) + \frac{150^2 - 75^2}{2g_0}\right]\left[\frac{5.092}{550}\right] \\ &= (407.96 + 262.28)\left[\frac{5.092}{550}\right] \\ &= 6.20 \text{ HP} \end{aligned}$$

Note that a significant error would be introduced if the kinetic energy terms were neglected.

(b) For a fan efficiency of 60 percent, the actual power required is, by eq. (14.17),

$$\dot{W}_{\text{in},a} = \frac{6.20}{0.60} = 10.3 \text{ HP}$$

**Figure 14.11** *h-s* Diagram Illustrating Adiabatic Turbine Efficiencies. *a. h-s* diagram displaying terms in eq. (14.18). *b. h-s* diagram displaying terms in eq. (14.19).

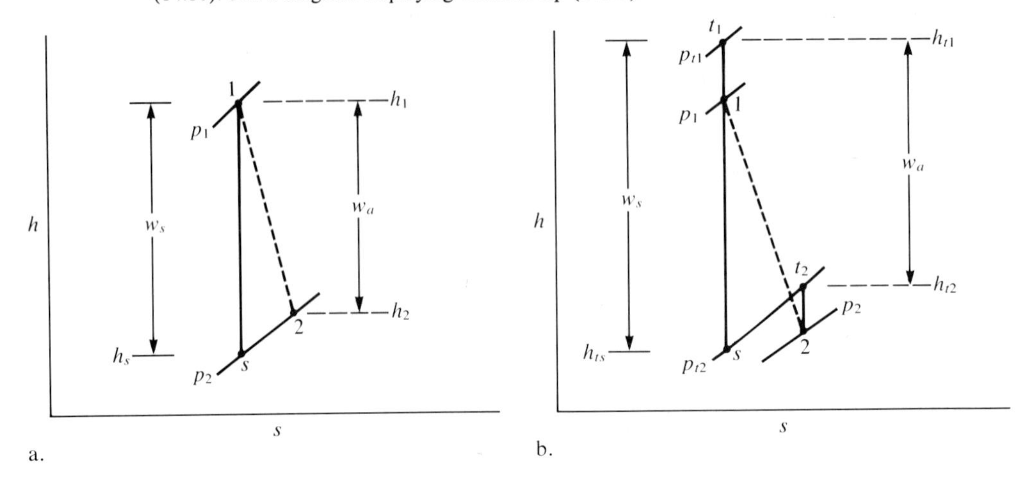


## 14–7 Turbines

In this section we first consider some general aspects of turbine operation and then examine, for selected cases, the internal flow characteristics of turbines.

Turbines with compressible working fluids, steam and gas turbines, usually operate adiabatically. Therefore, the ideal turbine process is an isentropic expansion. Adiabatic turbine efficiency is defined as the ratio of the actual work to the isentropic work. For turbines in which changes in kinetic energy are negligible, the turbine efficiency is expressed as

$$\eta_t = \frac{w_a}{w_s} = \frac{h_1 - h_2}{h_1 - h_s} \tag{14.18}$$

where the subscripts 1 and 2 denote the inlet and outlet states and state $s$ is the state that would exist if the expansion took place isentropically from the inlet state to the actual outlet pressure. Figure 14.11a illustrates the terms related to eq. (14.18).

For cases in which the kinetic energy terms are not negligible, adiabatic turbine efficiency is defined as

$$\eta_t = \frac{w_a}{w_s} = \frac{(h_{t1} - h_{t2})_a}{(h_{t1} - h_{ts})_s} \tag{14.19}$$

where state $s$ is established by the intersection of a line of constant entropy through state (1) and the outlet stagnation pressure. The terms in eq. (14.19) are illustrated in Figure 14.11b. The irreversibilities that produce the entropy increases shown in Figure 14.11 result in turbine efficiencies in the range of 80 to 90 percent. Equations (14.18) and (14.19) apply to either gas or steam turbines.

Figure 14.12 Cross-Sectional Drawing of an Impulse Steam Turbine with One Curtis Stage and Two Rateau Stages

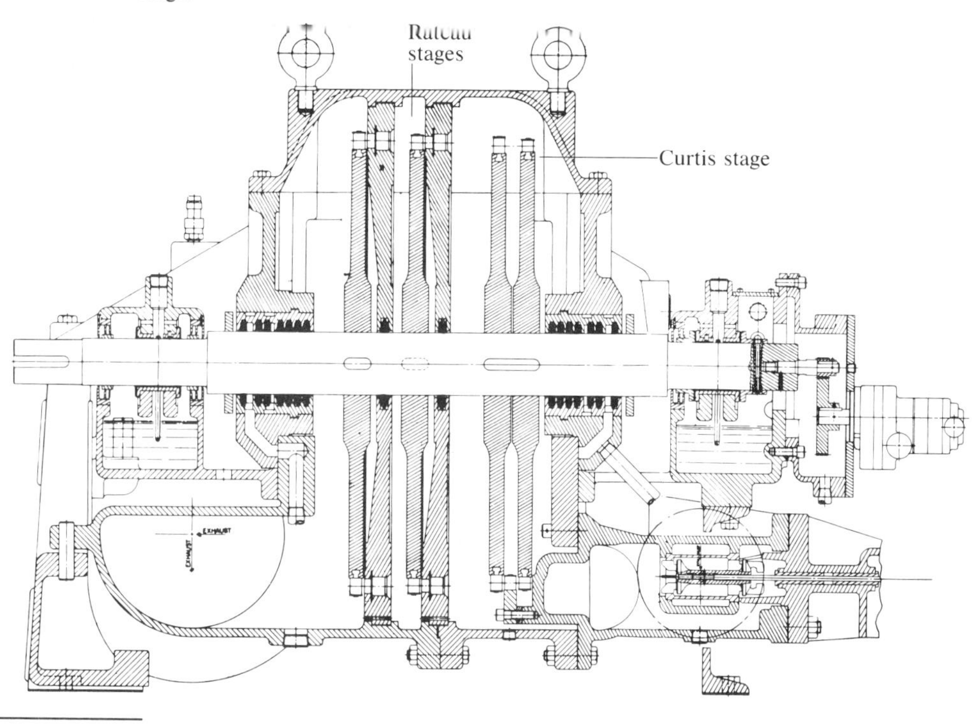


(Reprinted with permission of the Trane Co.)

Turbines are composed of a series of stages, each of which consists of a set of stationary nozzles that inject the working fluid into a moving blade row. The dynamics of the interaction of the working fluid with the blades produces a force on the blades that, in turn, transmits a torque to the rotating turbine shaft to produce power. In the remainder of this section we examine the flow through steam-turbine stages. The principles discussed apply to gas turbines also.

**Impulse Turbine Staging** Impulse turbines are characterized by the use of nozzles to provide a high-velocity stream, which, in turn, enters the rotating *blades* or *buckets* fixed to the wheels or disks of the turbine. The turbine blades, disks, and shaft comprise the turbine rotor. The tangential force exerted on each blade, or blade row, is dependent on the blade speed and the change in the tangential velocity components of the flow entering and leaving the blading. The tangential components of force and velocity are those components that are in the planes of rotation of the disks and that are perpendicular to radii from the shaft centerline to the blades. Figure 14.12 shows a cross-sectional drawing of a three-stage, impulse steam turbine. The first stage, consisting of a nozzle block, a rotating row fixed to the right hand disk, a stationary bladed segment fixed to the turbine casing, and a second rotating row fixed to the next disk, is called a Curtis stage. The other two stages in the unit illustrated have one rotating row for each nozzle block or diaphragm. These are called Rateau stages.

**Figure 14.13** Sectional View through One Nozzle and Several Adjacent Blades of a Rateau Stage

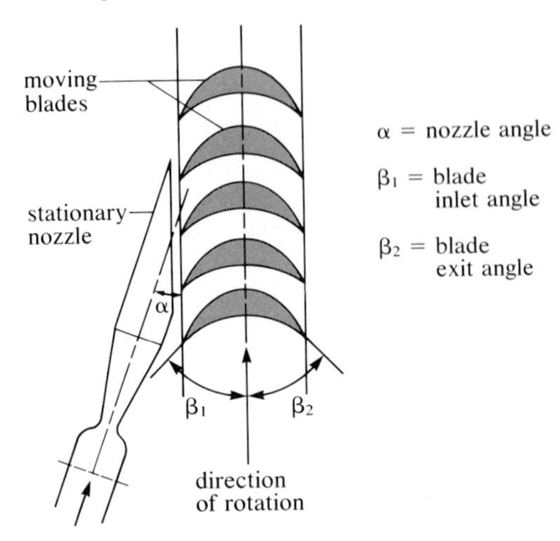


The Curtis stage is used to take a large pressure drop through its nozzles, thereby decreasing the number of stages required for a turbine operating with fixed inlet and exhaust pressures. The use of a Curtis stage reduces the length of the turbine, and hence its cost. However, the higher velocities through the nozzles and blading tend to reduce the efficiency since higher velocities typically reduce nozzle efficiencies and increase losses through the rotating and stationary blading.

A Rateau stage with one blade row for each nozzle block or diaphragm is shown in more detail in Figure 14.13. The velocity at the nozzle outlet is dependent on the nozzle inlet conditions, pressure ratio, and nozzle efficiency. Procedures for calculating the absolute velocity leaving the nozzle were presented in Chapter 13. The static pressure and absolute velocity changes through an ideal Curtis stage are shown in Figure 14.14. A Curtis stage with two or more rows of impulse blading for each nozzle block, as shown in the figure, is called a *velocity-compounded* stage. An impulse turbine that has several Rateau stages in series, that is, has lower static pressure at the inlet to each set of nozzles as the working fluid moves from the inlet to the exit, is termed *pressure compounded.* Thus the units shown in Figures 14.12 and 14.15 are impulse turbines incorporating a combination of velocity and pressure compounding.

We shall focus, principally, on the analysis of impulse turbine stages having frictionless flow through the blades, that is, flow for which no change of any static property occurs. With this limitation the power to the blading in one impulse stage can be calculated by either of two methods. For steady flow through the blades the work per unit mass of fluid passing through the blade row is

$$w = \frac{V_1^2 - V_2^2}{2g_0} \tag{14.20}$$

**Figure 14.14** A Curtis Impulse Turbine Stage

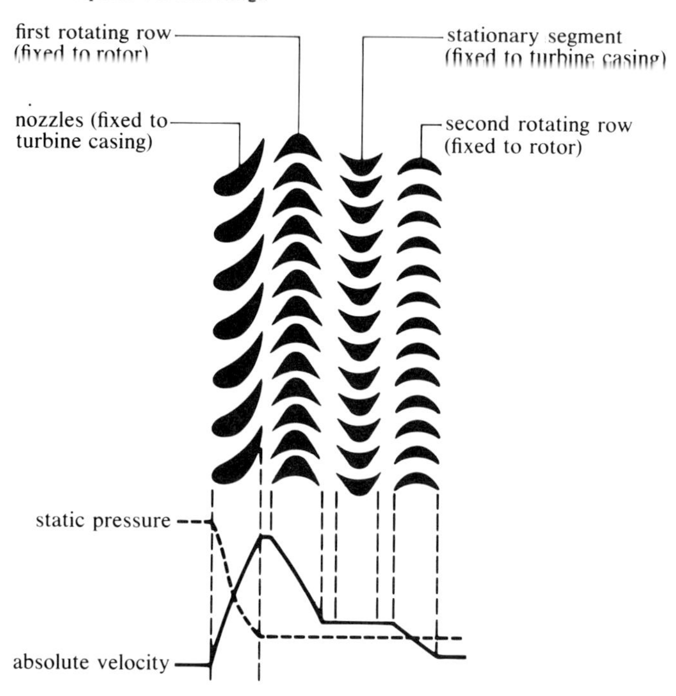


**Figure 14.15** An Impulse Steam Turbine with One Curtis Stage and Six Rateau Stages

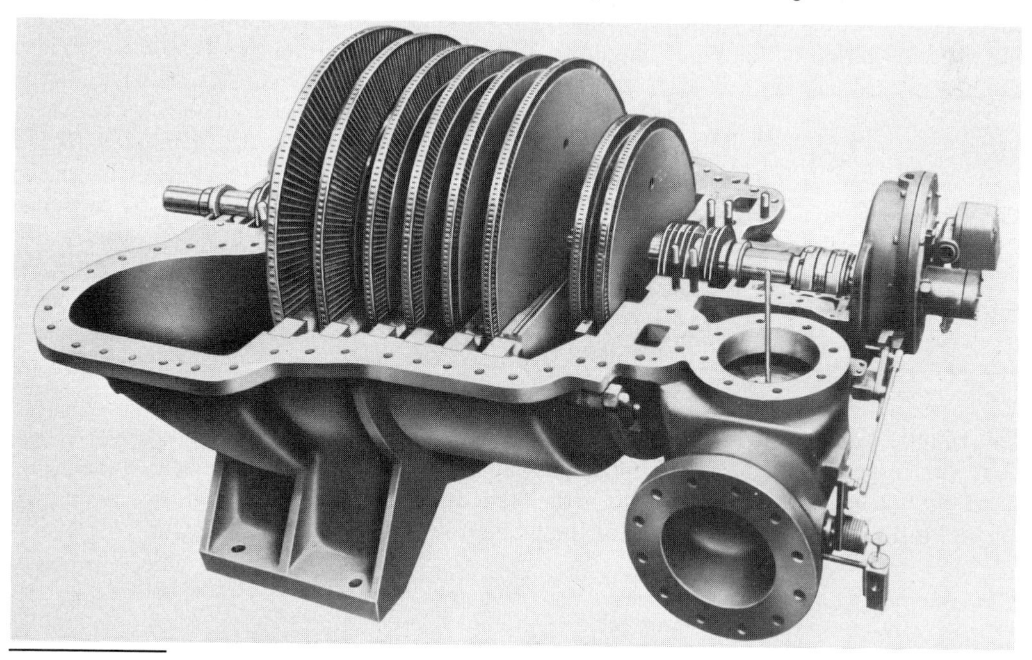

with the intertial reference for the velocities being the turbine casing. For a mass flow $\dot{m}$ the power delivered to the blades is

$$\dot{W} = \dot{m}\left[\frac{V_1^2 - V_2^2}{2g_0}\right] \tag{14.20a}$$

Equations (14.20) and (14.20a) are valid only for adiabatic flow and no change in static enthalpy.

Alternative procedures for evaluating the power delivered to an impulse blade row follow from eq. (13.4). This equation, written across the blades, can be expressed in terms of the *tangential* (whirl) *velocity components,* either absolute or relative to the moving blade. The blade speed $V_b$ is dependent on the turbine speed and the radial distance from the center line of the shaft to a specific location on the blade between its root and tip. For short blades the blade speed frequently is calculated using the pitch diameter (midpoint to midpoint distance between blades 180 degrees apart) and the turbine speed. Large turbines, with exhaust end blades up to 44 inches in length, require warped (twisted) blading to take into account the increase in blade speed with radial position. We shall limit this discussion to cases in which the blade speed can be assumed as single valued for a given stage.

The tangential force acting on the fluid at the radius where the blade speed is $V_b$ is given by eq. (13.4) as

$$F = \frac{\dot{m}}{g_0}(V_2' - V_1')$$

where the prime notation indicates tangential components of the velocities. The subscripts 1 and 2 denote the entrance and the exit of the blade row. The rate of energy transfer to the blade is the product of the blade force and the blade velocity. The force on the blade is equal and opposite to that on the fluid. Therefore, the power developed on one Rateau stage is

$$\dot{W} = \frac{\dot{m}}{g_0} V_b (V_1' - V_2') \tag{14.21}$$

The sign convention for this equation is positive in the direction of blade motion; $V_b$ and $V_1'$ are always positive, but $V_2'$ can have either sign.

The velocity diagram for a Rateau stage is shown in Figure 14.16. This diagram is based on the vector equation

$$\mathbf{V}_f = \mathbf{V}_r + \mathbf{V}_b$$

This equation states that the absolute velocity of the fluid $\mathbf{V}_f$ is equal to the velocity of the fluid with respect to the blade ($\mathbf{V}_r$) plus the blade velocity $\mathbf{V}_b$. Absolute velocities are measured relative to the turbine casing, as is the blade velocity. The vector diagrams shown are for the entrance (1) and the exit (2) of the blade row. Only one blade is shown.

**Figure 14.16** Velocity Diagram for a Rateau Stage. Flow is in the plane of the page.

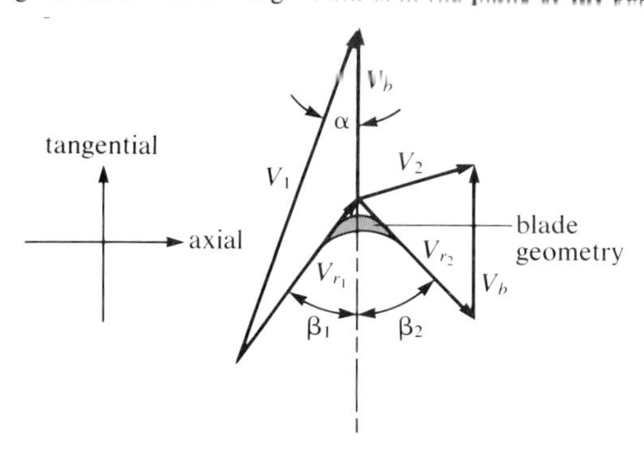


The absolute velocity leaving the nozzle, $V_1$, is equal to the absolute velocity entering the blade row. For frictionless flow the magnitudes of $V_{r_1}$ and $V_{r_2}$, the velocities relative to the blade, are equal. A symmetric blade is one for which the blade entrance angle $\beta_1$ and the blade exit angle $\beta_2$ are equal. The optimum relation of blade speed and the nozzle outlet velocity, $V_1$, that is, the one producing maximum power for a fixed nozzle outlet velocity, can be established by any one of several procedures. Inspection of eq. (14.20) reveals that the work or power for a fixed mass rate of flow increases as the absolute velocity leaving the blade, $V_2$, decreases. The upper limit would be achieved if $V_2$ became zero. This, of course, is an impossibility since the flow would also become zero when $V_2$ was zero.

For a given nozzle angle $\alpha$ (Figure 14.16), the optimum value of $V_b$ for a given $V_1$ can be established from eq. (14.21). At the blade entrance the equation for the tangential (whirl) component is

$$V_1' = V_{r_1}' + V_b \tag{a}$$

while that at the exit of the blade is

$$V_2' = V_{r_2}' + V_b \tag{b}$$

For a *frictionless* flow the magnitude of $V_{r_2}$ is equal to that of $V_{r_1}$. For blading that is symmetric (equal entrance and exit angles)

$$V_{r_2}' = -V_{r_1}' \tag{c}$$

Let us first substitute eq. (b) for $V_2'$ in eq. (14.21) with the result

$$\dot{W} = \frac{\dot{m}}{g_0} V_b \left[ V_1' - V_{r_2}' - V_b \right] \tag{d}$$

To express the preceding equation in terms of $V_1'$ and $V_b$ for frictionless flow and symmetric blading, eqs. (a), (b), and (c) are utilized in the following steps:

$$V_{r_1}' = V_1' - V_b$$

$$V_{r_2}' = -V_{r_1}'$$

$$V_{r_2}' = V_b - V_1'$$

Substituting from the last equation for $V_{r_2}'$ in eq. (d) yields

$$\begin{aligned} \dot{W} &= \frac{\dot{m}}{g_0} V_b \left[ V_1' - V_b + V_1' - V_b \right] \\ &= 2\frac{\dot{m}}{g_0} V_b \left[ V_1' - V_b \right] = 2\frac{\dot{m}}{g_0} \left[ V_1' V_b - V_b^2 \right] \end{aligned} \qquad \textbf{(e)}$$

For a fixed $V_1'$ eq. (e), differentiated with respect to $V_b$, becomes

$$\frac{d\dot{W}}{dV_b} = \frac{2\dot{m}}{g_0}\left[ V_1' - 2V_b \right] \qquad \textbf{(f)}$$

The optimum relation of $V_b$ and $V_1$ is obtained by setting $d\dot{W}/dV_b$ equal to zero. The result is

$$V_b = \frac{V_1'}{2} = \frac{V_1 \cos \alpha}{2} \qquad \textbf{(14.22)}$$

for *frictionless flow and symmetric blading*. For a given value of $V_1$, the absolute velocity entering the blading of an impulse stage, the optimum value of $V_b$ and the magnitude of $V_1'$ increase as the nozzle angle $\alpha$ approaches its limiting value of zero.

For a zero-nozzle angle impulse turbine stage—a physical impossibility—the blade speed would be one half the absolute velocity entering the blade and $V_1' = V_1$. For this limiting condition eq. (e) becomes

$$\begin{aligned} \dot{W} &= \frac{2\dot{m}}{g_0} \frac{V_1}{2} \left[ V_1' - \frac{V_1'}{2} \right] \\ &= \frac{\dot{m}}{g_0} \frac{V_1^2}{2} \end{aligned} \qquad \textbf{(14.23)}$$

which is identical to eq. (14.20a) when $V_2$ is zero.

## Example 14.f

Steam leaves the nozzle of a Rateau turbine stage with a velocity of 1400 ft/sec. The nozzle angle is 20 degrees. If the blading is frictionless and symmetric, calculate (a) the blade entrance angle and (b) the maximum power delivered to the blades for a mass rate of flow of 5.49 lb/sec. Draw the velocity diagram for the blade row.

**Solution:**

(a) For the stated conditions the maximum power will be developed when

$$V_b = \frac{1400 \cos 20}{2} = \frac{1315.8}{2} = 657.9 \text{ ft/sec}$$

The velocity diagram for the stage is shown in the figure. The tangential component of the entering relative velocity $V'_{r_1}$ is

$$V'_{r_1} = V'_1 - V_b = 1315.8 - 657.9$$
$$= 657.9 \text{ ft/sec}$$

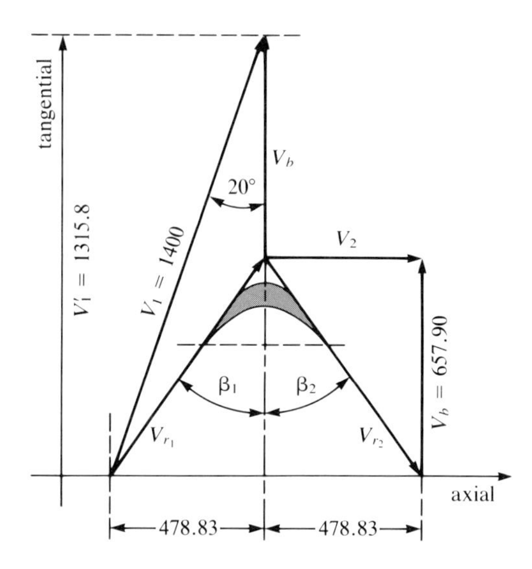


The axial component of $V_1$, equal to the axial component of $V_{r_1}$, is

$$V_1 \sin 20 = 478.83 \text{ ft/sec}$$

The blade entrance angle necessary to have the steam enter parallel to the blade entrance surface is

$$\tan \beta_1 = \frac{478.83}{657.90} \quad \text{and} \quad \beta_1 = 36.05 \text{ degrees}$$

With symmetric blading the exit angle $\beta_2$ will be equal to the entrance angle.

(b) The power delivered to the blade row can be calculated by two methods. From eq. (14.21)

$$\dot{W} = \frac{5.49}{g_0}(657.90)\ [1315.8]\left[\frac{1}{550}\right]$$
$$= 268.6 \text{ HP}$$

The second method that can be used to calculate the power utilizes eq. (14.20a). The absolute velocity $V_2$ leaving the blade is, in this example, 478.83 ft/sec since the blade is frictionless and symmetric and the blade speed is the optimum. The power is

$$\dot{W} = \frac{5.49}{2g_0}[1400^2 - 478.83^2]\left[\frac{1}{550}\right]$$
$$= 269 \text{ HP}$$

---

**Reaction Turbine Staging** In the preceding paragraphs examples of impulse turbine stages were presented. Common to each was the increase in absolute velocity only through the stationary turbine nozzle block. Additionally, the magnitude of the fluid velocity relative to the moving blades did not change if the blading was frictionless. The high absolute velocity of the working fluid as it leaves the nozzles and enters the rotating blade row or rows is characteristic of impulse turbine stages. With losses due to fluid friction increasing as the velocity of a moving fluid relative to its confining surfaces increases, it is apparent that lower velocities are desirable.

A pure reaction stage has stationary vanes that direct the flow into the moving blades. A pure reaction turbine stage would, ideally, have no state or velocity change through the stationary vanes. The decreases in static pressure and enthalpy, with an accompanying increase in velocity, would occur only through the rotating blade rows. With these conditions the rotating row of a reaction turbine stage would act as a series of moving nozzles. The tangential force resulting from the state change through the moving blades, analogous to the thrust forces developed in reaction propulsion systems such as jet-propelled aircraft and rockets, would be due solely to reaction effects. Thus a pure reaction stage is termed 100 percent reaction. The typical reaction stage, in contrast, has the enthalpy drop divided approximately equally through the stationary vanes and the rotating blades. Such a stage is called a 50 percent reaction stage. Significant consequences of such staging are the reduced absolute velocities through the stationary passages and the decreased relative velocity entering the rotating blades. In contrast to the frictionless impulse blade there is an increase in the magnitude of the relative velocity as the fluid expands through the rotating blades of a reaction stage. A more significant difference is the change in area between the inlet and outlet sections of the reaction blading. This geometry is necessary to produce the state change and increase in relative velocity magnitude during flow through reaction blading. Thus while ideal impulse blading is symmetric, the ideal reaction blading is nonsymmetric. The static pressure and absolute velocity changes through two reaction stages in series are shown in Figure 14.17.

The determination of the absolute velocity leaving the nozzles or stationary vanes of a reaction turbine stage follows the procedures discussed in Chapter 13. These were applied to impulse turbine nozzles in the preceding material. Understanding of some desirable characteristics of reaction blading follows from consideration of the steady-flow energy equation and impulse–momentum relations. For adiabatic reaction staging the steady-flow equation reduces to

$$\frac{V_1^2}{2g_0} + h_1 = \frac{V_2^2}{2g_0} + h_2 + w \tag{a}$$

**Figure 14.17** Absolute Velocity and Static Pressure Distribution through Two Adjacent Reaction Stages

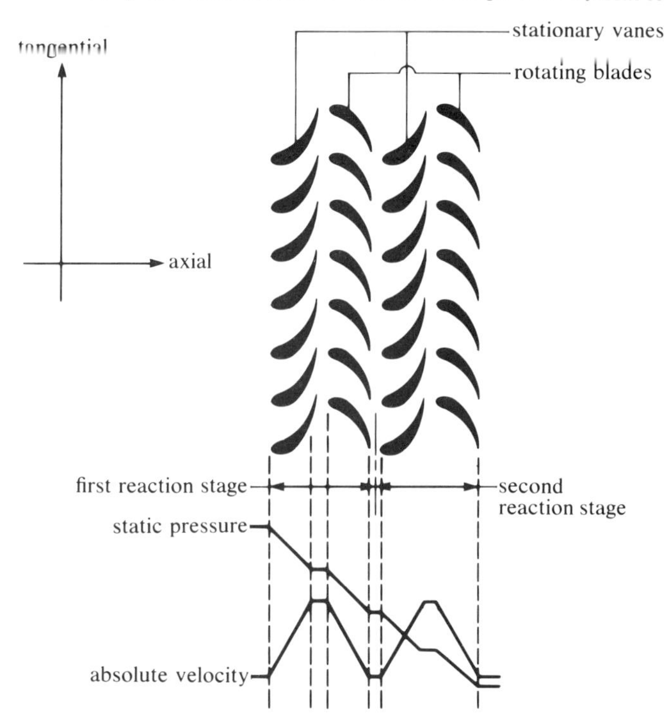


where the subscripts 1 and 2 refer to the blade entrance and exit sections respectively. Inspection of this equation reveals, for a given enthalpy drop and inlet velocity $V_1$, that work approaches a maximum as the absolute velocity leaving the blade $V_2$ approaches zero. This condition is, of course, a physical impossibility since the mass rate of flow would also become zero when $V_2$ was zero. Additionally, the accomplishment of zero absolute velocity leaving the rotating row could be achieved only with a zero-angle stage, that is, the nozzle angle, blade entrance angle, and the blade exit angle would each be zero. The condition of equal blade entrance and exit angles is not possible for reaction blading with its requisite change in passage area through the rotating row. An alternate form of the steady-flow equation for the blade row is obtained when the moving blades are chosen as the inertial reference. For this case the energy equation is

$$h_1 + \frac{V_{r1}^2}{2g_0} = h_2 + \frac{V_{r2}^2}{2g_0} \quad \textbf{(b)}$$

where each $V_r$ is a velocity relative to the moving blades.

Equation (14.21), earlier applied to impulse turbine blading, is equally valid for reaction blading.

$$\dot{W} = \frac{\dot{m}}{g_0} V_b \left[ V_1' - V_2' \right] \quad \textbf{(14.21)}$$

Considered in conjunction with eq. (a), it is evident that a minimum value of $V_2'$ is desirable. As was noted earlier, the absolute velocity leaving the blade, $V_2$, cannot be zero. However, the tangential component of $V_2$, $V_2'$ in eq. (14.21), will be zero when the fluid leaves the rotating reaction blade in an axial direction, that is, parallel to the turbine shaft.

The optimum relation of the absolute velocity entering the rotating row, $V_1$, and the blade speed, $V_b$, is established by the same procedure applied to impulse blading. The result, for frictionless, adiabatic flow yields maximum power for a given nozzle angle $\alpha$ when

$$V_b = V_1' = V_1 \cos \alpha \tag{14.24}$$

To illustrate the analysis of reaction blading let us first consider an example where the working fluid enters each a frictionless, rotating row parallel to the center line of the turbine shaft. Further, assume that the criterion of optimum blade speed and entering absolute velocity is satisfied, namely, $V_b$ will equal $V_1 \cos \alpha$. The blade exit angle, $\beta_2$, is dependent on the enthalpy drop through the rotating row (eq. [b]) and, in this example, is that necessary to produce an absolute velocity, $V_2$, having no tangential component. The velocity diagram for reaction blading satisfying these conditions is illustrated in Figure 14.18.

To identify more specifically the conditions associated with the velocity diagram of Figure 14.18, let us consider the several relations following from the specific blade geometry. From the figure,

$$V_1^2 = V_{r_1}^2 + V_b^2 \tag{c}$$

and

$$V_2^2 = V_{r_2}^2 - V_b^2 \tag{d}$$

Subtracting eq. (c) from eq. (d) we have

$$V_2^2 - V_1^2 = V_{r_2}^2 - V_{r_1}^2 - 2V_b^2$$

Substituting from eqs. (a) and (b)

$$2g_0\,(h_1 - h_2) - 2g_0 w = 2g_0\,(h_1 - h_2) - 2V_b^2$$

The preceding equation, solved for the work, yields

$$w = \frac{V_b^2}{g_0} = \frac{(V_1 \cos \alpha)^2}{g_0} = \frac{2\,(V_1 \cos \alpha)^2}{2g_0}$$

for frictionless, reaction blading of the specified geometry.

Figure 14.18 Velocity Diagram and Blade Geometry for an Axial Entry Reaction Turbine Blade Row

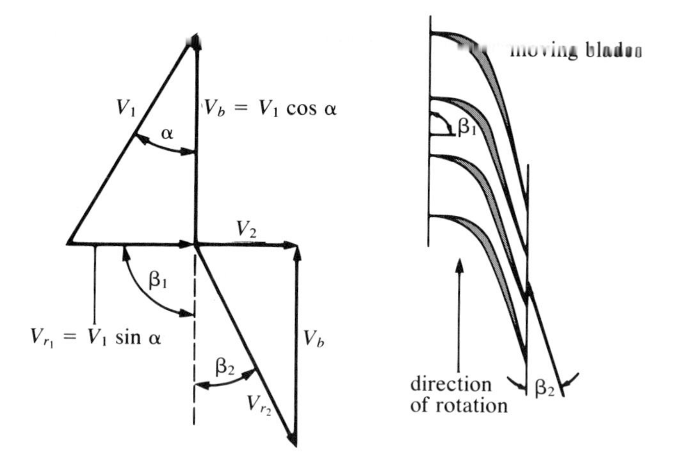


## Example 14.g

A reaction steam-turbine blade row is to have the geometry and other characteristics indicated in Figure 14.18. The blade speed is 800 ft/sec, the steam enters the blade at 50 psia, 100 percent quality, and the angle $\alpha$ is 27 degrees. The flow through the rotating row is isentropic. Steam leaves the blade at 30 psia. Calculate:

(a) The absolute velocity leaving the stationary vanes
(b) The relative velocity entering and leaving the rotating blade row
(c) The blade exit angle, $\beta_2$
(d) The ratio of exit to inlet area for the rotating blade row
(e) The power delivered for a flow of 1 lbm/sec

**Solution:**

(a) For maximum power the absolute velocity entering the rotating row is

$$V_1 = \frac{V_b}{\cos\alpha} = \frac{800}{0.8910} = 897.9 \text{ ft/sec}$$

(b) The magnitude of the inlet relative velocity is

$$V_{r_1} = V_1 \sin\alpha = 407.6 \text{ ft/sec}$$

at the inlet angle $\beta_1$ equal to 90 degrees. At state (1) $h_1 = 1174.4$. The state leaving the rotating row is established by the pressure, 30 psia, and the entropy, $s_2 = s_1$. From these properties

$$s_1 = 1.6589 = s_2 = 0.36821 + (x_2)1.3314$$

which yields $x_2 = 0.9694$. The enthalpy leaving the rotating row is then

$$\begin{aligned} h_2 &= 218.93 + (0.9694)(943.4) \\ &= 1135.4 \text{ B/lbm} \end{aligned}$$

The magnitude of the relative velocity leaving the blade is

$$\begin{aligned} V_{r_2} &= [2g_0(778)(1174.4 - 1135.4) + (407.6)^2]^{1/2} \\ &= 1455 \text{ ft/sec} \end{aligned}$$

(c) Referring to Figure 14.18, the blade exit angle $\beta_2$ is

$$\begin{aligned} \beta_2 &= \text{arc cos } \frac{V_b}{V_{r_2}} \\ &= 56.6 \text{ degrees} \end{aligned}$$

(d) The area change through the reaction stage is established using the moving blades as the reference. For steady flow the continuity equations for the inlet and exit sections are, respectively,

$$\dot{m}v_1 = A_1 V_{r_1}$$

and

$$\dot{m}v_2 = A_2 V_{r_2}$$

with $\dot{m}$ constant for steady flow. The area ratio is then

$$\frac{A_2}{A_1} = \frac{V_{r_1}}{V_{r_2}} \left| \frac{v_2}{v_1} \right.$$

The areas in this ratio are in planes normal to the relative velocities. The specific volume entering the rotating blades is 8.518 ft³/lbm ($v_g$ at 50 psia) and that leaving the blades is

$$v_2 = x_2 v_{g_2} = (0.9694)(13.748) = 13.328 \text{ ft}^3/\text{lbm}$$

The ratio of outlet to inlet area is

$$\frac{A_2}{A_1} = \frac{407.6}{1455.0} \left| \frac{13.328}{8.518} \right. = 0.438$$

(e) The power delivered to the blades can be calculated using the steady-flow energy equation. The absolute velocity leaving the blade is

$$\begin{aligned} V_2 &= [V_{r_2}^2 - V_b^2]^{1/2} \\ &= [1455^2 - 800^2]^{1/2} = 1215 \text{ ft/sec} \end{aligned}$$

The work is

$$w = \frac{V_1^2 - V_2^2}{2g_0} + h_1 - h_2$$
$$= \frac{(897.9)^2 - (1215)^2}{2g_0(778)} + 1174.4 - 1135.4$$
$$= 25.6 \text{ B/lbm}$$

For the flow of 1 lbm/sec the power is

$$\dot{W} = \frac{25.6 \;\big|\; 3600}{\;\big|\; 2545} = 36.2 \text{ HP}$$

---

## 14-8 Summary

This chapter has dealt with some thermodynamic considerations related to compressors, pumps, fans, and turbines. The flow through such machines is treated as steady. Hence the steady-flow energy equation is applicable. When the machine has one inbound and one outbound stream and the flow is treated as internally reversible as well as steady, the expression

$$_1w_2 = -\int_1^2 v\,dp - \frac{V_2^2 - V_1^2}{2g_0} - \frac{g}{g_0}(z_2 - z_1)$$

is applicable. From consideration of perfect gases and the internally reversible process $pv^n = C$, with $1 \leq n \leq \gamma$, it was observed that the work required for compression between an initial state and a final pressure is smallest when the compression process is isothermal ($n = 1$). Cooling of the gas during compression is required. Although the reduction in the work required that results from such cooling is desirable, it is difficult or impractical to attain large heat transfer rates during compression. In many applications it becomes economically desirable as a substitute to employ multistaging with intercooling, that is, two or more compression stages with an intercooler between each stage. For ideal multistage compression, the optimum stage pressure ratio $PR$ is given by

$$PR = \left[\frac{p_f}{p_1}\right]^{1/N}$$

where $p_f$ is the discharge pressure of the last stage, $p_1$ is the inlet pressure to the first stage, and $N$ is the number of compression stages. This expression requires that the gas temperatures at the inlet of each stage be the same.

Two types of compressors are representative of compressor classes. An axial-flow compressor consists of several stages, with each stage composed of a rotor (a rotating disk with attached blades) that does work on the gas, increasing its pressure, and a stator, a series of fixed blades that form diffusing passages that further increase the pressure. The reciprocating compressor consists of an arrangement of cylinders, pistons, and valves that operates to intake low-pressure gas and discharge high-pressure gas. Because of clearance volume effects, the volume of gas drawn in each intake stroke is less than the volume displaced by the pistons. Volumetric efficiency for a reciprocating compressor is defined as the ratio of the actual mass taken in to the mass of gas (density taken at the inlet state) required to fill the displacement volume. For ideal reciprocating compressors the volumetric efficiency is the ratio of the actual volume drawn in per intake stroke to the displacement volume.

Compressor efficiency relates ideal (reversible) performance to actual performance. Adiabatic compressor efficiency for axial-flow compressors is defined as the ratio of isentropic work to actual work. The efficiency of a reciprocating compressor is defined as the ratio of the power required for isentropic compression to the indicated power $\dot{W}_I$, which is equal to $p_m L A \dot{N}$; $p_m$ is the mean effective pressure, $L$ is the stroke, $A$ is the piston face area, and $\dot{N}$ is the number of piston cycles per unit time.

Pumps are used to increase the pressure of liquids during steady-flow processes. Fans and blowers are used in high-volume-rate low-pressure-rise applications involving air and other gases. For most engineering applications the specific volume of the flow through fans, blowers, and pumps can be assumed constant. Efficiencies for fans, blowers, and pumps are defined as the ratio of ideal power input (computed for reversible flow at constant specific volume) to the actual power required.

The ideal process through a gas or a steam turbine is an isentropic expansion. Turbine efficiency is defined as the ratio of the work for the actual expansion to that for isentropic expansion. A turbine is composed of a series of stages, each of which consists of a set of stationary nozzles that inject the working fluid into a moving blade row. Analysis of the flow through ideal impulse and reaction stages by means of velocity diagrams permits the power output to be computed.

## Selected References

Church, E. F. *Steam Turbines,* 3rd ed. New York: McGraw-Hill, 1950.

*Compressed Air and Gas Data.* Gibbs, C. W. (Ed.) Ingersoll-Rand, 1971.

*Compressed Air Handbook.* Compressed Air and Gas Institute, 1947.

Skrotzki, B. G. A., and W. A. Vopat. *Steam and Gas Turbines.* New York: McGraw-Hill, 1950.

Solberg, H. L., O. C. Cromer, and A. R. Spalding. *Thermal Engineering.* New York: Wiley, 1960.

Steam Turbines. Fans. Special reprints from *Power* magazine.

Stoever, H. J. *Engineering Thermodynamics.* New York: Wiley, 1951.

Van Wylen, G. A., and R. E. Sonntag. *Fundamentals of Classical Thermodynamics,* 2nd ed., SI version. New York: Wiley, 1976.

## Problems

14.1 An ideal reciprocating compressor takes in air at 14.20 psia and 87 F at the rate of 500 ft$^3$/min. The compressor operates isothermally and discharges at 105 psia. Neglecting kinetic energies, determine the power required to drive the compressor. If a heat exchanger, installed ahead of the compressor, cooled the air entering the compressor to 60 F at 14.20 psia, and if the volume rate at the compressor inlet and the compressor pressure ratio were not changed, what reduction in power would be realized?

14.2 At the inlet to a single-stage compressor air is at 14.7 psia and 77 F. The compressor discharge is at 100 psia. Neglecting kinetic energies, calculate the work per unit mass of air delivered by the compressor for (a) isothermal compression, (b) isentropic compression, and (c) polytropic compression with $n$ equal to 1.313.

14.3 Suppose a single-stage compressor delivers 150 ft³/min of air at 117 psia and 270 F. The compressor operates polytropically with $n$ equal to 1.30 and takes in air at 14.0 psia. Neglecting kinetic energies, calculate (a) the power to drive the compressor and (b) the rate at which heat is transferred from the air as it passes through the compressor.

14.4 A water-cooled gas compressor takes in helium at 100 F and 13 psia at the rate of 15,000 ft³/min with negligible velocity. On leaving the compressor the helium is at 250 F and has a velocity of 500 ft/sec. The power required is 800 HP. The cooling water undergoes a temperature rise of 23 F as it passes through the unit. Determine the pounds of cooling water required per minute.

14.5 The axial-flow compressor of a gas-turbine unit takes in air at 9.00 psia and 10 F. The velocity of the air entering the compressor is 500 ft/sec and of that leaving the compressor is 300 ft/sec. The compressor operates insentropically with a stagnation pressure ratio of 11. Calculate the power required to drive the compressor if the intake volume rate is 100,000 ft³/min.

14.6 An ideal, two-stage compressor takes in air at 14.5 psia and 77 F and discharges it at 200 psia. Intercooling is complete, and the intercooler pressure is that for minimum power requirement. Determine the power per 1000 ft³/min entering the compressor if the compression for each stage is (a) isentropic and (b) polytropic with $n$ equal to 1.300. Calculate the intercooler heat transfer rate for the two compression processes.

14.7 A gas mixture of 50 percent oxygen and 50 percent helium, by volume, is compressed isentropically in a two-stage compressor. The mixture enters the first stage at 12 psia and 70 F at the rate of 900 ft³/min. The intercooling is complete and the intercooler pressure is that for minimum power. Assuming the second-stage discharge pressure is 275 psia, calculate (a) the power required to drive the compressor and (b) the heat transferred from the mixture, B/min, as it passes through the intercooler.

14.8 A two-stage air compressor delivers 80 lbm of air per minute at 150 psia. Air enters the first stage at 14.7 psia and 70 F. The intercooler pressure is 52 psia and air leaves the intercooler at 120 F. Assuming each stage operates polytropically with $n$ equal to 1.305, calculate (a) the power to drive the compressor, (b) the heat-transfer rate for each stage, and (c) the volume rate of flow into the second stage.

14.9 Consider a two-stage compressor taking in a perfect gas at the rate of $\dot{V}_1$ ft³/min at $p_1$ and $T_1$ and discharging the gas at $p_2$ and $T_2$. If the compressor operates polytropically with the same $n$ for each stage, and if the intercooling is complete and the intercooler pressure is that for minimum work requirement, how does the heat transfer rate for the first stage compare with that for the second stage?

14.10 Suppose a two-stage, isentropic compressor takes in air at 14.5 psia and 60 F and discharges the air at 200 psia. If the intercooling is not complete, air will be taken into the second stage at a temperature greater than 60 F. Consider a situation where air enters the second stage at 120 F. What intercooler pressure would yield the minimum work requirement for the compressor? Compare the intercooler pressure with that for the conditions given in problem 14.6.

14.11 A compressor takes in air at 20 F and 10 psia at the rate of 12,000 ft³/min with a velocity of 150 ft/sec. The compressor operates adiabatically with a stagnation pressure ratio of 8. Assuming the adiabatic compressor efficiency, based on stagnation conditions, is 85 percent, calculate the power required to drive the compressor.

14.12 An ideal single-stage reciprocating compressor takes in air at 14 psia and 90 F while operating at 200 cycles per minute. Compression is isentropic to 98 psia. For a clearance volume of 5 percent of the piston displacement, calculate the piston displacement if the entering volume rate of flow is 100 ft³/min.

14.13 A two-stage compressor operates at 150 rev/min. Air enters the first stage at 15 psia, 80 F. Air leaves the intercooler at 80 F and is compressed to 160 psia in the second stage. The optimum intercooler pressure exists and for each stage the clearance volume is 10 percent of the piston displacement. The first-stage piston displacement is 1.000 $ft^3$ and each stage operates isentropically. Calculate (a) the ideal volumetric efficiency for each stage, (b) the power required to drive the compressor, and (c) the second-stage piston displacement.

14.14 A single-stage compressor operates polytropically with $n$ equal to 1.300. The piston displacement is 1.50 $ft^3$ and the clearance is 4 percent. Air enters the compressor at 14 psia, 60 F, and is discharged at 75 psia. Assuming the compressor operates at 180 rev/min, determine (a) the ideal volumetric efficiency, (b) the power to drive the compressor, and (c) the mean effective pressure.

14.15 A simple vapor-compression refrigeration cycle has ammonia leaving the evaporator as saturated vapor at 5 F. The compressor has an adiabatic efficiency of 80 percent and the condenser pressure is 170 psia. Assuming ammonia leaves the condenser at 80 F, calculate (a) the volume rate of flow entering the compressor and (b) the compressor power per ton of refrigerating effect.

14.16 A reciprocating compressor with a clearance volume of 5 percent of its piston displacement and a piston displacement of 0.70 $ft^3$ serves as the compressor for a simple vapor compression refrigeration cycle. Saturated vapor leaves the evaporator and the compressor operates polytropically with $n = 1.05$ at a speed of 500 piston cycles per minute. Freon 12 is the refrigerant, the evaporator temperature is $-15$ F, and the condenser pressure is 125 psia.

(a) What is the power required to drive the compressor?
(b) What is the refrigerating capacity in tons for the cycle?
(c) At what rate is heat rejected at the condenser?

14.17 Water enters a condensate pump at the rate of 22.5 $ft^3$/sec at a vacuum of 13.5 psi. The pump discharge, 9.53 feet above the inlet, is at 27.43 psi gage. The suction and discharge pipes are of equal diameter. Assuming constant density, calculate the power required to drive the pump. Would the power requirement change if the inlet temperature changed?

14.18 A fire pump delivers 1050 gal/min during an acceptance test. The suction, 4 feet below the center line of the pump, is at 3 psi vacuum and 60 F. The discharge, 3 feet above the center line of the pump, is at 86 psi gage. If the power input to the pump is 97 HP, and if kinetic energies are negligible, what is the pump efficiency?

14.19 Solve problem 14.17 if liquid mercury is pumped instead of water. The specific gravity of mercury is 13.6.

14.20 A pump is to handle a liquid that has a specific gravity of 0.673. The liquid enters the pump at an absolute pressure equivalent to a head of 7.5 feet of the liquid. The pump discharge is at an absolute pressure equivalent to 125 feet of the liquid. The kinetic energy change across the pump is negligible. Calculate the ideal pump power required per cubic foot per minute of the liquid.

14.21 The forced-draft fan in a power plant takes in air at 14.2 psia and 80 F with negligible velocity. The static pressure rise through the fan is 15 inches of $H_2O$ and air leaves the fan at a velocity of 30 ft/sec. Calculate the ideal power required per 1000 $ft^3$/min entering the fan. Assume dry air flows through the fan.

14.22 Refer to the preceding problem. Suppose the air entering the fan is at 14.2 psia and 80 F but has a relative humidity of 60 percent. Calculate (a) the ideal power per 1000 $ft^3$/min entering the fan and (b) the ideal power requirement for the same dry-air-mass flow rate as that in the preceding problem.

14.23 A ventilating fan with an efficiency of 74 percent draws air from a building at 14.5 psia and 78 F at the rate of 2500 $ft^3/min$. The static pressure rise through the fan is 2.75 in $H_2O$ and the fan discharges into a duct with a 1.00-$ft^2$ cross-sectional area. Determine the ideal power required to drive the fan.

14.24 A turbojet operates on a cold-air-standard cycle with a Mach number of 1.35 in an atmosphere at 6.08 psia and 440 R. The compressor stagnation pressure ratio is 10.5 and the compressor efficiency is 100 percent. The stagnation temperature at the turbine inlet is 2650 R and the turbine efficiency is 85 percent. The static pressure at the exit section of the 95 percent efficient nozzle is 6.08 psia. The stagnation pressure remains constant during heat addition and the compressor and turbine efficiencies are based on stagnation conditions. Calculate (a) the stagnation temperature and pressure at the compressor inlet, (b) the stagnation pressure at the turbine outlet, and (c) the mass rate of flow required to develop 40,000 lbf thrust.

14.25 Steam enters the nozzles of a zero-angle, impulse turbine stage at 50 psia and 300 F with a negligible velocity. The nozzle is isentropic and discharges at 15 psia static pressure. The steam is turned through 180 degrees as it passes through the frictionless blade row. For a flow rate of 10,000 lbm/hr, determine (a) the blade speed $V_b$ for maximum power and (b) the maximum power that could be delivered. Sketch the velocity diagram for the blading.

14.26 In a Rateau stage of a turbine, steam enters the nozzles at 60 psia, 460 F, with a velocity of 500 ft/sec and leaves the nozzle at 40 psia. The nozzle efficiency is 89 percent. The blades are frictionless and symmetric. The nozzle angle is 20 degrees. Determine the optimum blade speed and the kinetic energy change across the blade row. For a flow rate of 10,000 lbm/hr, calculate the power delivered to the blades from (a) the steady-flow energy equation, (b) eq. (14.21), and (c) eq. (14.21) using relative velocity tangential components. Finally, determine the entrance and exit angles for the blades.

14.27 Consider a Rateau stage of a steam turbine where the velocity leaving the nozzles is 1270 ft/sec. The nozzle angle is 22 degrees. Steam is to leave the blading parallel to the turbine shaft center line. Assuming the blading is frictionless and symmetric, determine (a) the required blade speed and (b) the blade entrance angle. Sketch the velocity diagram for the blading.

14.28 Suppose, in a specific Rateau stage of an impulse turbine, the pitch diameter of the blades is 28 inches and the turbine shaft rotates at 9000 rev/min. If steam is supplied to the isentropic nozzles at 600 psia, saturated vapor, with negligible velocity, if the blading is frictionless and symmetric, and if the nozzle angle is $\alpha$ degrees, determine the blade entrance angle as a function of $\alpha$ to yield the maximum power.

14.29 A Rateau stage of a turbine has a nozzle angle of 15 degrees, a blade speed of 600 ft/sec, and an absolute stream velocity of 1400 ft/sec entering the frictionless, symmetric blading. Determine the blade entrance angle and the axial force (parallel to the turbine shaft center line) for a flow rate of 1 lbm/sec.

14.30 A small steam turbine has four Rateau stages. Steam is supplied to the unit at 60 psia, 400 F, with negligible velocity and the exhaust is at 1.00 psia. The static enthalpy drop through the turbine is divided equally among four stages. The nozzle angle for each stage is 20 degrees, the blading is frictionless and symmetric, and the nozzles operate isentropically. For a flow rate of 10,000 lbm/hr and a turbine speed of 3600 rev/min, determine (a) the blade speeds for maximum power, (b) the kinetic energy of the steam leaving the last stage, and (c) the power delivered by the ideal turbine.

# Appendix A

## Physical Constants, Prefixes, and Conversion Factors

### Physical Constants*

| | | |
|---|---|---|
| Avagadro's number | $N_A$ | $= 6.024 \times 10^{26}$ kmol$^{-1}$ |
| universal gas constant | $\overline{R}$ | = 8314 N m/(kmol K) |
| | | = 1545 lbf ft/(lb mol R) |
| standard gravity | $g$ | = 9.807 m/s$^2$ |
| | | = 32.17 ft/s$^2$ |

*Values rounded to four significant figures.

### Prefixes

The Names of Multiples and Submultiples of SI Units May Be Formed by Application of the Prefixes Given Below.

| Factor by Which Unit is Multiplied | Prefix | Symbol | Example |
|---|---|---|---|
| $10^{12}$ | tera | T | terahertz (THz) |
| $10^{9}$ | giga | G | gigawatt (GW) |
| $10^{6}$ | mega | M | megajoule (MJ) |
| $10^{3}$ | kilo | k | kilometer (km) |
| $10^{-2}$ | centi | c | centimeter (cm) |
| $10^{-3}$ | milli | m | milliwatt (mW) |
| $10^{-6}$ | micro | $\mu$ | microsecond ($\mu$s) |
| $10^{-9}$ | nano | n | nanosecond (ns) |

**Conversion Factors***

| Quantity | To Convert From | To | Multiply by† |
|---|---|---|---|
| acceleration | $m/s^2$ | $ft/sec^2$ | 3.281 |
| area | $m^2$ | $ft^2$ | 10.76 |
| | $in^2$ | $m^2$ | $6.452 \times 10^{-4}$ |
| angle | degree | radian | 0.01745 |
| angular speed | rev/min | radian/sec | 0.1047 |
| density | $lbm/ft^3$ | $kg/m^3$ | 16.02 |
| energy, work, heat | N m | J | 1 |
| | ft lbf | J | 1.356 |
| | B | ft lbf | 778.2 |
| | B | kJ | 1.055 |
| force | lbf | N | 4.448 |
| gas constant | ft lbf/(lbm R) | J/(kg K) | 5.380 |
| length | in | cm | 2.54 |
| | ft | m | 0.3048 |
| mass | kg | lbm | 2.205 |
| | slug | lbm | 32.17 |
| | kmol | lb mol | 2.205 |
| mass rate of flow | kg/s | lbm/sec | 2.205 |
| power, heat transfer rates | kJ/s | kW | 1 |
| | kW | B/hr | 3413 |
| | HP | B/hr | 2545 |
| | HP | ft lbf/sec | 550 |
| | kW | HP | 1.341 |
| pressure | $N/m^2$ | Pa | 1 |
| | atm (Std) | kPa | 101.3 |
| | atm (Std) | $lbf/in^2$ | 14.70 |
| | psi | kPa | 6.893 |
| | in hg | psi | 0.491 |
| | in $H_2O$ | psi | 0.0361 |
| | cm $H_2O$ | kPa | 1.333 |
| specific energy | B/lbm | kJ/kg | 2.326 |
| | B/lb mol | kJ/kmol | 2.326 |
| specific entropy | B/(lbm R) | kJ/(kg K) | 4.187 |
| specific heat | B/(lbm R) | kJ/(kg K) | 4.187 |
| | B/(lb mol R) | kJ/(kmol K) | 4.187 |
| specific volume | $m^3/kg$ | $ft^3/lbm$ | 16.02 |
| temperature‡ | K | R | 1.8 |
| torque | lbf ft | N m | 1.356 |
| velocity | ft/sec | m/s | 0.3048 |
| volume | US gal | $in^3$ | 231.0 |
| | $m^3$ | $ft^3$ | 35.31 |
| volume rate of flow | $m^3/s$ | $ft^3/sec$ | 35.31 |
| | $ft^3/min$ | $m^3/s$ | $4.719 \times 10^{-4}$ |

*Values rounded to four significant figures.

†Numerical values listed have units equal to the ratio of the units in the "To" column to those in the "To Convert from" column. For example, the first two table entries are 3.281 $(ft/sec^2)/(m/s^2)$; 10.76 $ft^2/m^2$.

‡Temperature-conversion equations:
$t_C = (t_F - 32)/1.8, \quad t_F = 1.8t_C + 32, \quad T_K = t_C + 273.15, \quad T_R = t_F + 459.67$

# Appendix B

## Steam Tables (SI Units)

### Symbols and Units

$h$ specific enthalpy, kilojoules per kilogram, kJ/kg
$p$ pressure, megapascals, MPa (except Table B–3)
$s$ specific entropy, kilojoules per kilogram kelvin, kJ/(kg K)
$t$ temperature, C
$T$ temperature, K
$u$ specific internal energy, kilojoules per kilogram, kJ/kg
$v$ specific volume, cubic meters per kilogram, $m^3$/kg

**Note 1: Tabulated values of specific volumes are expressed as $10^3 v$. Thus the actual specific volume, $v$, is**

$$(10^3 v)(10^{-3}) = v, \text{ m}^3/\text{kg}$$

**Note 2: Tabulated values of saturation pressures in Table B–3, only, have the units kilopascals, kPa. (kPa)($10^3$) = MPa.**

### Subscripts

$f$ refers to a property of liquid in equilibrium with vapor
$g$ refers to a property of vapor in equilibrium with liquid (above the triple point) or in equilibrium with solid (below the triple point)
$i$ refers to a property of solid in equilibrium with vapor

---

**Table B–1** $H_2O$—Saturated Liquid and Saturated Vapor (SI Units)
Units: $v$ in $m^3/kg$; $u$ in kJ/kg; $h$ in kJ/kg; $s$ in kJ/kg K

| | | Specific Volume | | Internal Energy | | Enthalpy | | Entropy | |
|---|---|---|---|---|---|---|---|---|---|
| Temp. C $t$ | Pressure MPa $p$ | Sat. Liq. $10^3 v_f$ | Sat. Vapor $10^3 v_g$ | Sat. Liq. $u_f$ | Sat. Vapor $u_g$ | Sat. Liq. $h_f$ | Sat. Vapor $h_g$ | Sat. Liq. $s_f$ | Sat. Vapor $s_g$ |
| **0.01** | **0.0006113** | 1.0002 | 206,136 | 0.00 | 2375.3 | 0.01 | 2501.4 | 0.0000 | 9.1562 |
| **1.0** | 0.0006567 | 1.0002 | 192,577 | 4.15 | 2376.7 | 4.16 | 2503.2 | 0.0152 | 9.1229 |
| **2.0** | 0.0007056 | 1.0001 | 179,889 | 8.36 | 2378.1 | 8.37 | 2505.0 | 0.0305 | 9.1035 |
| **3.0** | 0.0007577 | 1.0001 | 168,132 | 12.56 | 2379.5 | 12.57 | 2506.9 | 0.0457 | 9.0773 |
| **5.0** | 0.0008721 | 1.0001 | 147,120 | 20.97 | 2382.3 | 20.98 | 2510.6 | 0.0761 | 9.0257 |
| 6.98 | **0.0010** | 1.0002 | 129,208 | 29.30 | 2385.0 | 29.30 | 2514.2 | 0.1059 | 8.9756 |
| 10.0 | 0.0012276 | 1.0004 | 106,379 | 42.0 | 2389.2 | 42.01 | 2519.8 | 0.1510 | 8.9008 |
| 13.03 | **0.0015** | 1.0007 | 97,980 | 54.71 | 2393.3 | 54.71 | 2525.3 | 0.1957 | 8.8279 |
| 15.0 | 0.0017051 | 1.0009 | 77,926 | 62.99 | 2396.1 | 62.99 | 2528.9 | 0.2245 | 8.7814 |
| 17.5 | **0.0020** | 1.0013 | 67,004 | 73.48 | 2399.5 | 73.48 | 2533.5 | 0.2607 | 8.7237 |
| **20.0** | 0.002339 | 1.0018 | 57,791 | 83.95 | 2402.9 | 83.96 | 2538.1 | 0.2966 | 8.6672 |
| 22.94 | **0.0028** | 1.0024 | 48,742 | 96.26 | 2407.0 | 96.27 | 2543.4 | 0.3384 | 8.6024 |
| 25.01 | 0.003169 | 1.0029 | 43,360 | 104.88 | 2409.8 | 104.89 | 2547.2 | 0.3674 | 8.5580 |
| **30.0** | 0.004246 | 1.0043 | 32,894 | 125.78 | 2416.6 | 125.79 | 2556.3 | 0.4369 | 8.4533 |
| 32.88 | **0.0050** | 1.0053 | 23,192 | 137.81 | 2420.5 | 137.82 | 2561.5 | 0.4764 | 8.3951 |
| **35.0** | 0.005628 | 1.0060 | 25,216 | 146.67 | 2423.4 | 146.68 | 2565.3 | 0.5053 | 8.3531 |
| 40.29 | **0.0075** | 1.0079 | 19,238 | 168.78 | 2430.5 | 168.79 | 2574.8 | 0.5764 | 8.2515 |
| **45.0** | 0.009593 | 1.0099 | 15,258 | 188.44 | 2436.8 | 188.45 | 2583.2 | 0.6387 | 8.1648 |
| 53.97 | **0.015** | 1.0141 | 10,022 | 225.92 | 2448.7 | 225.94 | 2599.1 | 0.7549 | 8.0085 |
| 60.06 | **0.020** | 1.0172 | 7649 | 251.38 | 2456.7 | 251.40 | 2609.7 | 0.8320 | 7.9085 |
| 64.97 | **0.025** | 1.0199 | 6204 | 271.90 | 2463.1 | 271.93 | 2618.2 | 0.8931 | 7.8314 |
| **70.0** | 0.03119 | 1.0228 | 5042 | 292.95 | 2469.6 | 292.98 | 2626.8 | 0.9549 | 7.7553 |
| 75.87 | **0.040** | 1.0265 | 3993 | 317.53 | 2477.0 | 317.58 | 2636.8 | 1.0259 | 7.6700 |
| **80.0** | 0.04739 | 1.0291 | 3407 | 334.86 | 2482.2 | 334.91 | 2643.7 | 1.0753 | 7.6122 |
| 85.94 | **0.060** | 1.0331 | 2732 | 359.79 | 2489.6 | 359.86 | 2653.5 | 1.1453 | 7.5320 |
| **90.0** | 0.07014 | 1.0360 | 2361 | 386.85 | 2494.5 | 376.92 | 2660.1 | 1.1925 | 7.4791 |
| 95.14 | **0.085** | 1.0398 | 1972 | 398.48 | 2500.7 | 398.57 | 2668.4 | 1.2517 | 7.4141 |
| **100** | 0.10135 | 1.0435 | 1672.9 | 418.94 | 2506.5 | 419.04 | 2676.1 | 1.3069 | 7.3549 |
| 111.37 | **0.150** | 1.0528 | 1159.3 | 466.94 | 2519.7 | 467.11 | 2693.6 | 1.4336 | 7.2233 |
| **120** | 0.19853 | 1.0603 | 891.9 | 503.50 | 2529.3 | 503.71 | 2706.3 | 1.5276 | 7.1296 |
| 127.44 | **0.250** | 1.0672 | 718.7 | 535.10 | 2537.2 | 535.37 | 2716.9 | 1.6072 | 7.0527 |
| **135** | 0.3130 | 1.0746 | 582.2 | 567.35 | 2545.0 | 567.69 | 2727.3 | 1.6870 | 6.9777 |
| 143.63 | **0.400** | 1.0836 | 462.5 | 604.31 | 2553.6 | 604.74 | 2738.6 | 1.7766 | 6.8959 |
| **150** | 0.4758 | 1.0905 | 392.8 | 631.68 | 2559.5 | 632.20 | 2746.5 | 1.8414 | 6.8379 |
| 158.85 | **0.600** | 1.1006 | 315.7 | 669.90 | 2567.4 | 670.56 | 2756.8 | 1.9312 | 6.7600 |
| **170** | 0.7917 | 1.1143 | 242.8 | 718.33 | 2576.5 | 719.21 | 2768.7 | 2.0419 | 6.6663 |
| 179.91 | **1.00** | 1.1273 | 194.44 | 761.68 | 2583.6 | 762.81 | 2778.1 | 2.1387 | 6.5865 |
| **190** | 1.2544 | 1.1414 | 156.54 | 806.19 | 2590.0 | 807.62 | 2786.4 | 2.2359 | 6.5079 |
| 201.41 | **1.60** | 1.1587 | 123.80 | 856.94 | 2596.0 | 858.79 | 2794.0 | 2.3442 | 6.4218 |
| **220** | 2.318 | 1.1900 | 86.19 | 940.87 | 2602.4 | 943.62 | 2802.1 | 2.5178 | 6.2861 |
| 240.94 | **3.40** | 1.2311 | 58.77 | 1037.61 | 2603.9 | 1041.82 | 2803.7 | 2.7101 | 6.1370 |
| **260** | 4.688 | 1.2755 | 42.21 | 1128.39 | 2599.0 | 1134.37 | 2796.9 | 2.8838 | 6.0019 |
| 279.88 | **6.40** | 1.3317 | 30.23 | 1226.85 | 2586.2 | 1235.37 | 2779.7 | 3.0657 | 5.8580 |
| **300** | 8.581 | 1.4036 | 21.67 | 1332.0 | 2563.0 | 1344.0 | 2749.0 | 3.2534 | 5.7045 |
| **315** | 10.547 | 1.4720 | 16.867 | 1415.5 | 2536.6 | 1431.0 | 2714.5 | 3.3982 | 5.5804 |
| 333.30 | **13.4** | 1.5841 | 12.242 | 1526.2 | 2488.6 | 1547.4 | 2652.6 | 3.5858 | 5.4082 |
| **350** | 16.513 | 1.7403 | 8.813 | 1641.9 | 2418.4 | 1670.6 | 2563.9 | 3.7777 | 5.2112 |
| 368.28 | **20.6** | 2.126 | 5.320 | 1817.3 | 2258.8 | 1861.1 | 2348.4 | 4.0663 | 4.8571 |
| **374.136*** | **22.09** | 3.155 | 3.155 | 2029.6 | 2029.6 | 2099.3 | 2099.3 | 4.4298 | 4.4298 |

*Critical point

**Table B–2** $H_2O$—Superheated Vapor (SI Units)
Units: $v$ in $m^3/kg$; $u$ in kJ/kg; $h$ in kJ/kg; $s$ in kJ/kg K

| | Pressure, MPa (saturation temperature in parentheses) | | | | | | | | | | | |
|---|---|---|---|---|---|---|---|---|---|---|---|---|
| | 0.002 (17.50 C) | | | | 0.004 (28.96 C) | | | | 0.008 (41.51 C) | | | |
| Temp. C $t$ | $10^3v$ | $u$ | $h$ | $s$ | $10^3v$ | $u$ | $h$ | $s$ | $10^3v$ | $u$ | $h$ | $s$ |
| 20 | 67,582 | 2403.0 | 2538.2 | 8.7396 | — | — | — | — | — | — | — | — |
| 30 | 69,897 | 2417.1 | 2556.8 | 8.8023 | 34920 | 2416.6 | 2556.3 | 8.4810 | — | — | — | — |
| 40 | 72,211 | 2431.1 | 2575.6 | 8.8630 | 36080 | 2430.8 | 2575.1 | 8.5419 | — | — | — | — |
| 60 | 76,836 | 2459.4 | 2613.1 | 8.9791 | 38398 | 2459.1 | 2612.7 | 8.6583 | 19179 | 2458.5 | 2611.9 | 8.3366 |
| 80 | 81,460 | 2487.8 | 2650.7 | 9.0888 | 40714 | 2487.5 | 2650.4 | 8.7682 | 20341 | 2487.0 | 2649.8 | 8.4469 |
| 100 | 86,081 | 2516.3 | 2688.4 | 9.1928 | 43028 | 2516.1 | 2688.2 | 8.8724 | 21501 | 2515.7 | 2687.7 | 8.5514 |
| 150 | 97,628 | 2588.3 | 2783.6 | 9.4320 | 48806 | 2588.2 | 2783.4 | 9.1118 | 24395 | 2588.0 | 2783.1 | 8.7914 |
| 200 | 109,170 | 2661.6 | 2879.9 | 9.6471 | 54580 | 2661.5 | 2879.8 | 9.3271 | 24284 | 2661.4 | 2879.6 | 9.0069 |

| | Pressure, MPa (saturation temperature in parentheses) | | | | | | | | | | | |
|---|---|---|---|---|---|---|---|---|---|---|---|---|
| | 0.10 (99.63 C) | | | | 0.16 (113.32 C) | | | | 0.36 (139.87 C) | | | |
| Temp. C $t$ | $10^3v$ | $u$ | $h$ | $s$ | $10^3v$ | $u$ | $h$ | $s$ | $10^3v$ | $u$ | $h$ | $s$ |
| 100 | 1695.8 | 2506.7 | 2676.2 | 7.3614 | — | — | — | — | — | — | — | — |
| 120 | 1792.9 | 2537.3 | 2716.6 | 7.4668 | 1112.2 | 2532.4 | 2710.4 | 7.2374 | — | — | — | — |
| 140 | 1888.7 | 2567.7 | 2756.5 | 7.5659 | 1173.5 | 2563.8 | 2751.5 | 7.3395 | 510.8 | 2550.1 | 2734.0 | 6.9318 |
| 160 | 1983.8 | 2597.8 | 2796.2 | 7.6597 | 1234.0 | 2594.7 | 2792.1 | 7.4354 | 539.4 | 2583.7 | 2777.9 | 7.0355 |
| 180 | 2078.2 | 2627.9 | 2835.8 | 7.7489 | 1293.9 | 2625.3 | 2832.4 | 7.5263 | 567.4 | 2616.3 | 2820.6 | 7.1320 |
| 200 | 2172. | 2658.1 | 2875.3 | 7.8343 | 1353.3 | 2655.9 | 2872.4 | 7.6127 | 594.9 | 2648.4 | 2862.5 | 7.2225 |
| 250 | 2406. | 2733.7 | 2974.3 | 8.0333 | 1500.6 | 2732.2 | 2972.3 | 7.8135 | 662.2 | 2727.2 | 2965.6 | 7.4295 |
| 300 | 2639. | 2810.4 | 3074.3 | 8.2158 | 1646.8 | 2809.3 | 3072.8 | 7.9969 | 728.3 | 2805.6 | 3067.8 | 7.6161 |

| | Pressure, MPa (saturation temperature in parentheses) | | | | | | | | | | | |
|---|---|---|---|---|---|---|---|---|---|---|---|---|
| | .70 (164.97 C) | | | | 2.05 (213.67 C) | | | | 2.75 (229.12 C) | | | |
| Temp. C $t$ | $10^3v$ | $u$ | $h$ | $s$ | $10^3v$ | $u$ | $h$ | $s$ | $10^3v$ | $u$ | $h$ | $s$ |
| 170 | 276.9 | 2581.8 | 2775.6 | 6.7354 | — | — | — | — | — | — | — | — |
| 190 | 292.4 | 2617.5 | 2822.2 | 6.8382 | — | — | — | — | — | — | — | — |
| 210 | 307.3 | 2651.9 | 2867.0 | 6.9331 | — | — | — | — | — | — | — | — |
| 230 | 322.0 | 2685.3 | 2910.7 | 7.0217 | 102.48 | 2636.9 | 2847.0 | 6.4267 | 72.98 | 2606.1 | 2806.8 | 6.2266 |
| 250 | 336.3 | 2718.2 | 2953.6 | 7.1053 | 108.47 | 2678.0 | 2900.3 | 6.5306 | 78.07 | 2653.5 | 2868.2 | 6.3464 |
| 280 | 357.4 | 2766.9 | 3017.1 | 7.2233 | 116.88 | 2735.2 | 2974.8 | 6.6691 | 84.95 | 2716.8 | 2950.0 | 6.4993 |
| 320 | 385.2 | 2831.3 | 3100.9 | 7.3697 | 127.44 | 2806.9 | 3068.2 | 6.8322 | 93.33 | 2793.4 | 3050.1 | 6.6733 |
| 370 | 419.4 | 2912.1 | 3205.6 | 7.5391 | 140.04 | 2893.4 | 3180.5 | 7.0140 | 103.13 | 2883.3 | 3166.9 | 6.8625 |
| 420 | 453.3 | 2993.7 | 3310.9 | 7.6968 | 152.25 | 2978.7 | 3290.9 | 7.1793 | 112.50 | 2970.8 | 3280.2 | 7.0321 |
| 500 | 507.0 | 3126.8 | 3481.7 | 7.9299 | 171.32 | 3115.8 | 3467.0 | 7.4198 | 127.00 | 3110.0 | 3459.3 | 7.2767 |

*Continued on next page*

**Table B-2—*Continued***

| Temp. C $t$ | Pressure, MPa (saturation temperature in parentheses) | | | | | | | | | | | |
|---|---|---|---|---|---|---|---|---|---|---|---|---|
| | 4.1 (251.87 C) | | | | 6.9 (284.91 C) | | | | 9.7 (308.83 C) | | | |
| | $10^3v$ | $u$ | $h$ | $s$ | $10^3v$ | $u$ | $h$ | $s$ | $10^3v$ | $u$ | $h$ | $s$ |
| 255 | 49.17 | 2611.1 | 2812.7 | 6.0822 | — | — | — | — | — | — | — | — |
| 275 | 52.98 | 2664.5 | 2881.7 | 6.2105 | — | — | — | — | — | — | — | — |
| 300 | 57.20 | 2722.7 | 2957.2 | 6.3453 | 30.06 | 2635.9 | 2843.3 | 5.9442 | — | — | — | — |
| 330 | 61.79 | 2785.6 | 3038.9 | 6.4843 | 33.73 | 2722.4 | 2955.1 | 6.1345 | 21.38 | 2640.8 | 2848.2 | 5.8324 |
| 360 | 66.08 | 2844.1 | 3115.0 | 6.6075 | 36.85 | 2794.5 | 3048.7 | 6.2861 | 24.25 | 2736.0 | 2971.3 | 6.0317 |
| 400 | 71.51 | 2918.6 | 3211.8 | 6.7557 | 40.58 | 2880.0 | 3160.0 | 6.4567 | 27.39 | 2837.3 | 3103.0 | 6.2336 |
| 450 | 77.99 | 3009.1 | 3328.9 | 6.9235 | 44.86 | 2979.1 | 3288.6 | 6.6409 | 30.79 | 2946.9 | 3245.6 | 6.4381 |
| 500 | 84.25 | 3098.7 | 3444.1 | 7.0776 | 48.88 | 3074.3 | 3411.5 | 6.8053 | 33.90 | 3048.6 | 3377.4 | 6.6144 |
| 540 | 89.16 | 3170.3 | 3535.9 | 7.1933 | 51.98 | 3149.3 | 3507.9 | 6.9269 | 36.24 | 3127.4 | 3478.9 | 6.7424 |
| 600 | 96.39 | 3278.5 | 3673.7 | 7.3568 | 56.49 | 3261.3 | 3651.1 | 7.0968 | 39.61 | 3243.6 | 3627.9 | 6.9192 |

| Temp. C $t$ | Pressure, MPa (saturation temperature in parentheses) | | | | | | | | | | | |
|---|---|---|---|---|---|---|---|---|---|---|---|---|
| | 17.0 (352.37 C) | | | | 25 (> critical pressure) | | | | 38 (> critical pressure) | | | |
| | $10^3v$ | $u$ | $h$ | $s$ | $10^3v$ | $u$ | $h$ | $s$ | $10^3v$ | $u$ | $h$ | $s$ |
| 370 | 10.706 | 2558.5 | 2740.5 | 5.4830 | 1.8470 | 1742.8 | 1789.0 | 3.9407 | 1.6226 | 1654.3 | 1716.0 | 3.7926 |
| 390 | 12.346 | 2657.5 | 2867.4 | 5.6775 | 4.6126 | 2275.7 | 2391.0 | 4.8585 | 1.8163 | 1792.7 | 1861.7 | 4.0156 |
| 420 | 14.212 | 2764.9 | 3006.5 | 5.8828 | 7.577 | 2581.7 | 2771.1 | 5.4218 | 2.5517 | 2085.8 | 2182.8 | 4.4882 |
| 440 | 15.264 | 2824.0 | 3083.5 | 5.9925 | 8.687 | 2680.1 | 2897.3 | 5.6013 | 3.611 | 2324.9 | 2462.1 | 4.8856 |
| 470 | 16.679 | 2903.2 | 3186.8 | 6.1343 | 10.011 | 2792.1 | 3042.4 | 5.8008 | 5.038 | 2559.2 | 2750.7 | 5.2826 |
| 500 | 17.967 | 2975.6 | 3281.1 | 6.2587 | 11.123 | 2884.3 | 3162.4 | 5.9592 | 6.101 | 2708.3 | 2940.1 | 5.5328 |
| 550 | 19.930 | 3088.1 | 3426.9 | 6.4416 | 12.724 | 3017.5 | 3335.6 | 6.1765 | 7.484 | 2890.4 | 3174.8 | 5.8273 |
| 600 | 21.74 | 3195.0 | 3564.6 | 6.6040 | 14.137 | 3137.9 | 3491.4 | 6.3602 | 8.621 | 3038.5 | 3366.1 | 6.0531 |
| 650 | 23.45 | 3298.9 | 3697.6 | 6.7522 | 15.433 | 3251.6 | 3637.4 | 6.5229 | 9.619 | 3170.8 | 3536.3 | 6.2427 |
| 700 | 25.09 | 3401.2 | 3827.7 | 6.8894 | 16.646 | 3361.3 | 3777.5 | 6.6707 | 10.527 | 3294.1 | 3694.1 | 6.4093 |

**Table B-3** $H_2O$—Saturated Solid and Saturated Vapor (SI Units)

| Temp. C $t$ | Press. kPa | Specific Volume | | Internal Energy | | Enthalpy | | Entropy | |
|---|---|---|---|---|---|---|---|---|---|
| | | Sat. Solid $10^3v_i$ | Sat. Vapor $v_g$ | Sat. Solid $u_i$ | Sat. Vapor $u_g$ | Sat. Solid $h_i$ | Sat. Vapor $h_g$ | Sat. Solid $s_i$ | Sat. Vapor $s_g$ |
| 0.01 | 0.6113 | 1.0908 | 206.1 | −333.40 | 2375.3 | −333.40 | 2501.4 | −1.221 | 9.156 |
| 0 | 0.6108 | 1.0908 | 206.3 | −333.43 | 2375.3 | −333.43 | 2501.3 | −1.221 | 9.157 |
| −4 | 0.4375 | 1.0901 | 283.8 | −341.78 | 2369.8 | −341.78 | 2494.0 | −1.253 | 9.283 |
| −10 | 0.2602 | 1.0891 | 466.7 | −354.09 | 2361.4 | −354.09 | 2482.9 | −1.299 | 9.481 |
| −14 | 0.1815 | 1.0884 | 658.8 | −362.15 | 2355.9 | −362.15 | 2475.5 | −1.331 | 9.619 |
| −18 | 0.1252 | 1.0878 | 940.5 | −370.10 | 2350.3 | −370.10 | 2468.1 | −1.362 | 9.762 |
| −24 | 0.0701 | 1.0868 | 1640.1 | −381.80 | 2342.0 | −381.80 | 2456.9 | −1.408 | 9.985 |
| −30 | 0.0381 | 1.0858 | 2943. | −393.23 | 2333.6 | −393.23 | 2445.8 | −1.455 | 10.221 |
| −34 | 0.0250 | 1.0851 | 4419. | −400.71 | 2328.0 | −400.71 | 2438.4 | −1.486 | 10.386 |
| −40 | 0.0129 | 1.0841 | 8354. | −411.70 | 2319.6 | −411.70 | 2427.2 | −1.532 | 10.644 |

**Table B–4** $H_2O$—Compressed (Subcooled) Liquid (SI Units)

| | Pressure, MPa (saturation temperature in parentheses) | | | | | | | | | | | |
|---|---|---|---|---|---|---|---|---|---|---|---|---|
| | **2.5** (223.99 C) | | | | **5.0** (263.99 C) | | | | **10.0** (311.06 C) | | | |
| **Temp. C** $t$ | $10^3v$ | $u$ | $h$ | $s$ | $10^3v$ | $u$ | $h$ | $s$ | $10^3v$ | $u$ | $h$ | $s$ |
| 20 | 1.0006 | 83.80 | 86.30 | 0.2961 | 0.9995 | 83.65 | 88.65 | 0.2956 | 0.9972 | 83.36 | 93.33 | 0.2945 |
| 60 | 1.0160 | 250.67 | 253.21 | 0.8298 | 1.0149 | 250.23 | 255.30 | 0.8285 | 1.0127 | 249.36 | 259.49 | 0.8258 |
| 120 | 1.0590 | 502.68 | 505.33 | 1.5255 | 1.0576 | 501.80 | 507.09 | 1.5233 | 1.0549 | 500.08 | 510.64 | 1.5189 |
| 200 | 1.555 | 849.9 | 852.8 | 2.3294 | 1.1530 | 848.1 | 853.9 | 2.3255 | 1.1480 | 844.5 | 856.0 | 2.3178 |
| 280 | — | — | — | — | — | — | — | — | 1.3216 | 1220.9 | 1234.1 | 3.0548 |

| | **17.5** (354.75 C) | | | | **22.09** (374.14 C) | | | | **40** (> critical pressure) | | | |
|---|---|---|---|---|---|---|---|---|---|---|---|---|
| **Temp. C** $t$ | $10^3v$ | $u$ | $h$ | $s$ | $10^3v$ | $u$ | $h$ | $s$ | $10^3v$ | $u$ | $h$ | $s$ |
| 20 | 0.9939 | 82.91 | 100.31 | 0.2928 | 0.9919 | 82.64 | 104.56 | 0.2918 | 0.9845 | 81.59 | 120.97 | 0.2874 |
| 80 | 1.0211 | 330.94 | 348.80 | 1.0640 | 1.0190 | 329.95 | 352.46 | 1.0611 | 1.0114 | 326.28 | 366.74 | 1.0500 |
| 160 | 1.0901 | 666.52 | 685.60 | 1.9232 | 1.0871 | 664.39 | 688.40 | 1.9181 | 1.0764 | 656.51 | 699.56 | 1.8991 |
| 260 | 1.2505 | 1111.5 | 1133.4 | 2.8517 | 1.2427 | 1106.1 | 1133.6 | 2.8412 | 1.2161 | 1087.3 | 1136.0 | 2.8044 |
| 380 | — | — | — | — | — | — | — | — | 1.6813 | 1709.6 | 1776.8 | 3.8813 |

**Figure B.1** Temperature–Entropy Chart for Water, SI Units

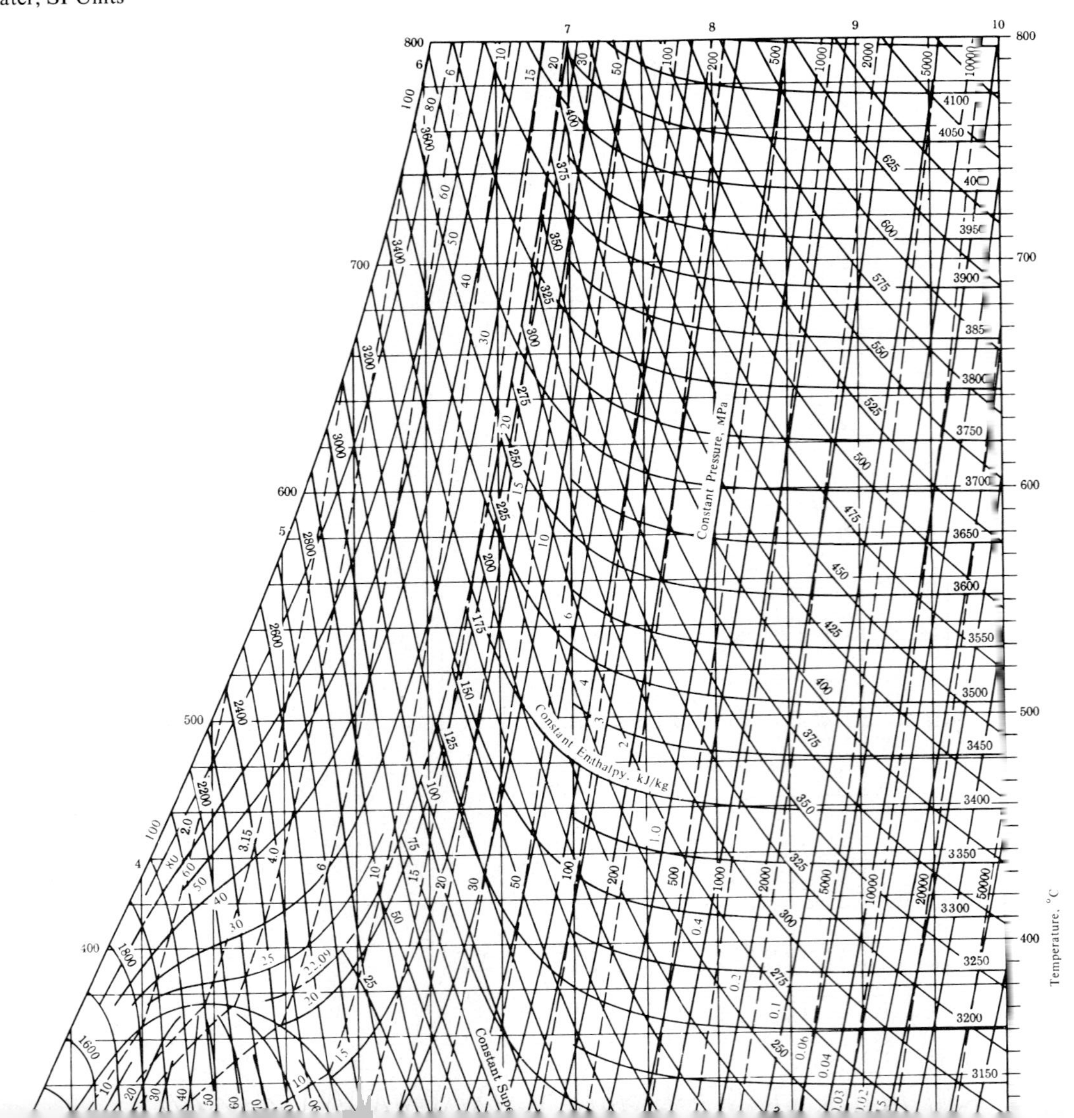

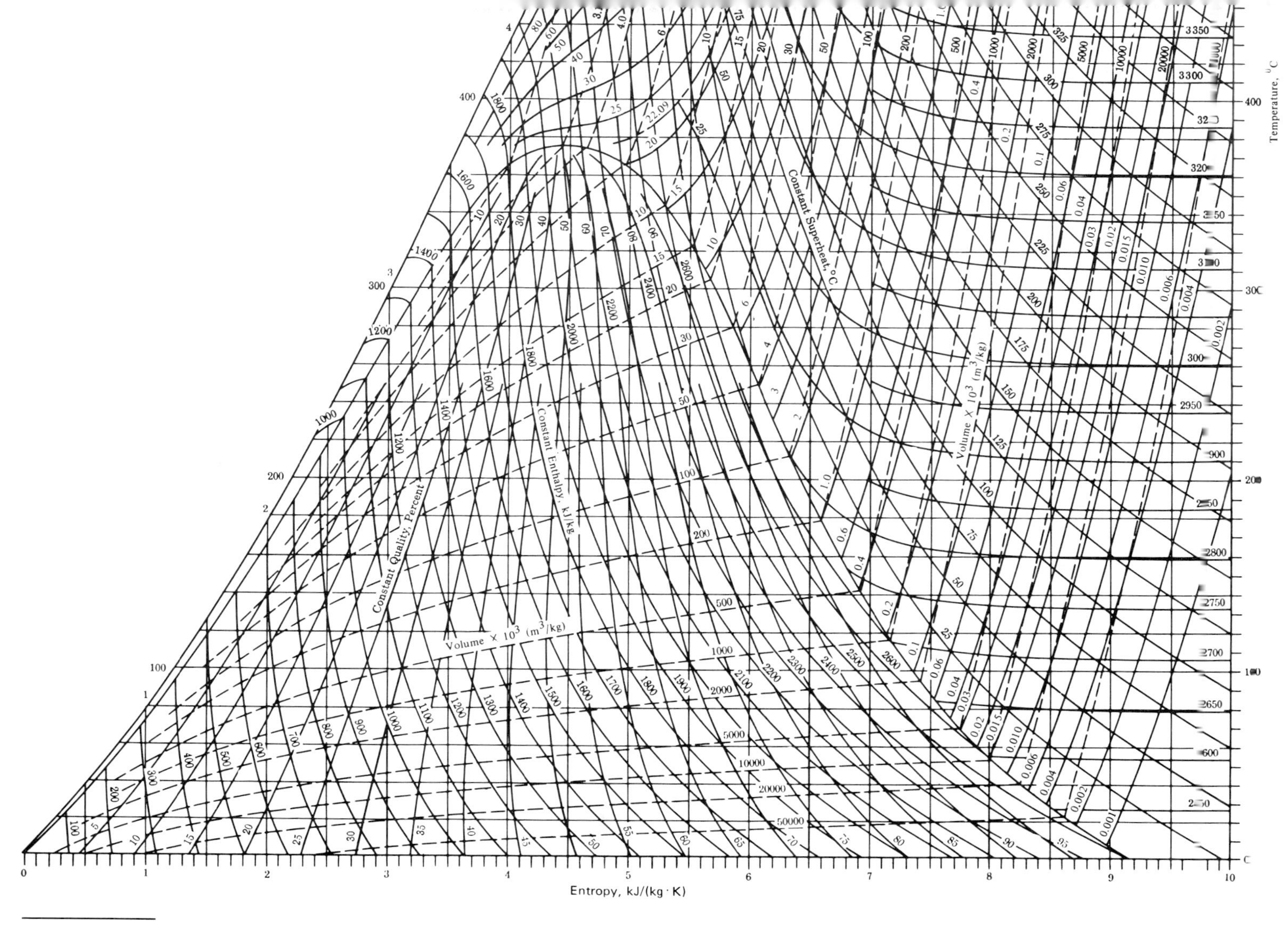


(Reprinted by permission from Keenan, Keyes, Hill, and Moore, *Steam Tables, SI Units.* Copyright 1969, 1974, John Wiley & Sons, Inc.)

**Figure B.2** Enthalpy–Entropy (Mollier) Chart for Water, SI Units

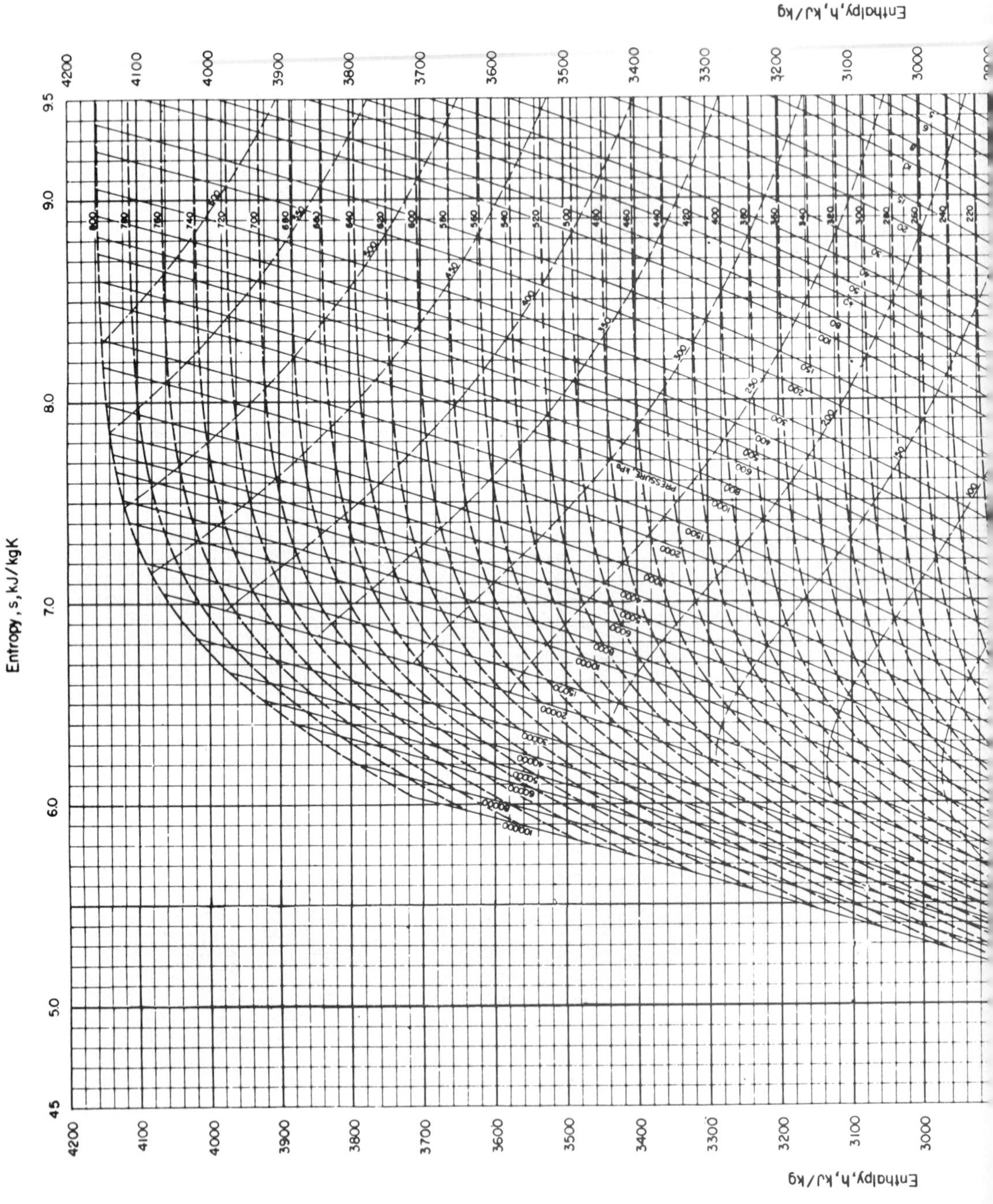

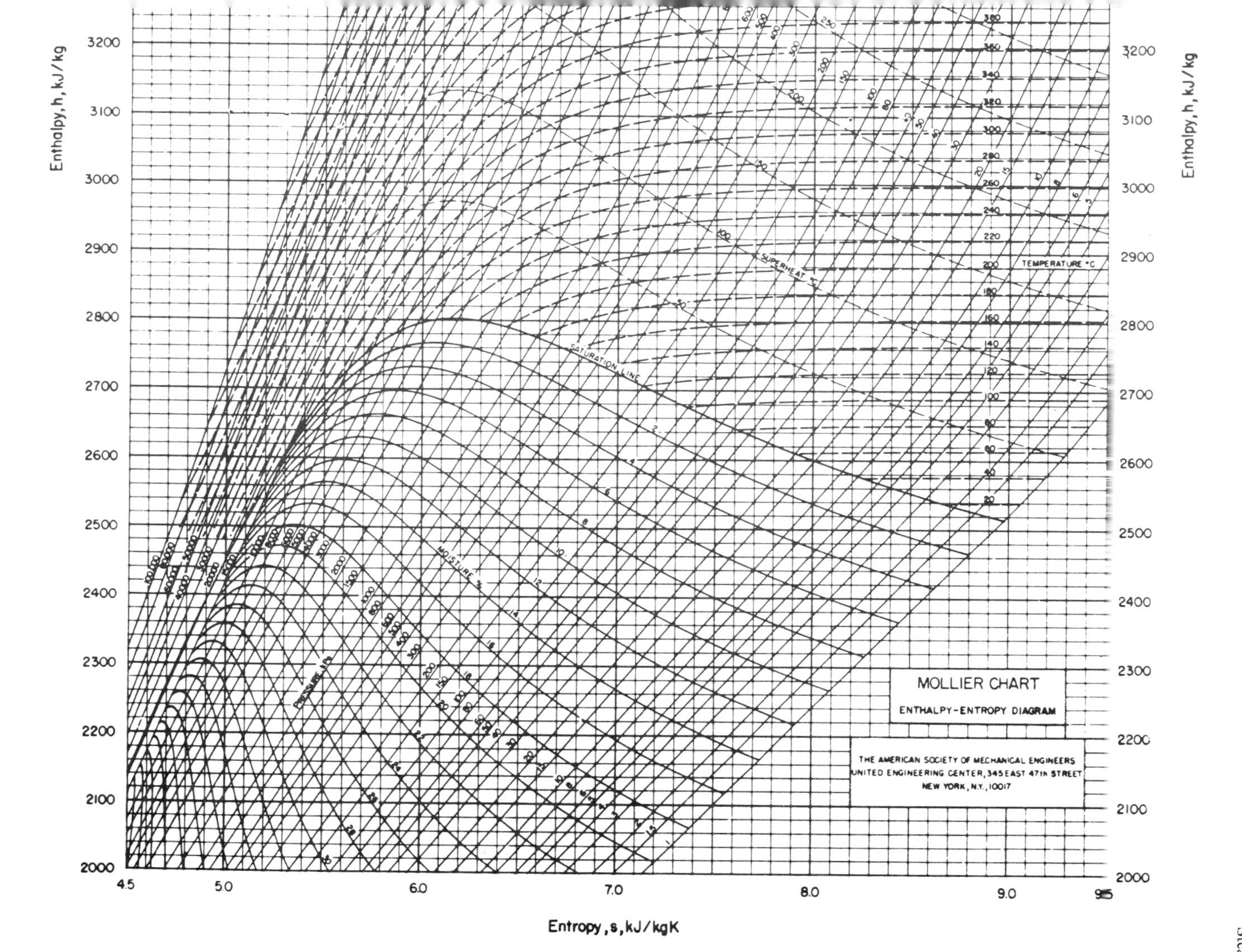


(Reprinted by permission from *ASME Steam Tables in SI (Metric) Units.* 1967. American Society of Mechanical Engineers, New York, NY)

# Appendix C

## Steam Tables (English Units)

### Symbols and Units

- $h$ specific enthalpy, Btu per lbm
- $p$ pressure, lbf/in$^2$ absolute
- $s$ specific entropy, Btu per lbm Rankine
- $t$ temperature, Fahrenheit
- $u$ specific internal energy, Btu per lbm
- $v$ specific volume, ft$^3$ per lbm

### Subscripts

- $f$ refers to a property of liquid in equilibrium with vapor
- $g$ refers to a property of vapor in equilibrium with liquid (above the triple point) or in equilibrium with solid (below the triple point)
- $i$ refers to a property of solid in equilibrium with vapor

**Table C–1** $H_2O$—Saturated Liquid and Saturated Vapor (English Units)

| Temp. F $t$ | Pressure lbf/in.² $p$ | Specific Volume: Sat. Liq. $v_f$ | Specific Volume: Sat. Vapor $v_g$ | Internal Energy: Sat. Liq. $u_f$ | Internal Energy: Sat. Vapor $u_g$ | Enthalpy: Sat. Liq. $h_f$ | Enthalpy: Sat. Vapor $h_g$ | Entropy: Sat Liq. $s_f$ | Entropy: Sat. Vapor $s_g$ |
|---|---|---|---|---|---|---|---|---|---|
| **32.018** | **0.08866** | 0.016022 | 3302 | 0.00 | 1021.2 | 0.01 | 1075.4 | 0.00000 | 2.1869 |
| **34** | 0.09601 | 0.016021 | 3062 | 1.99 | 1021.9 | 1.99 | 1076.3 | 0.00404 | 2.1799 |
| 35.02 | **0.10** | 0.016021 | 2946 | 3.02 | 1022.2 | 3.02 | 1076.7 | 0.00612 | 2.1764 |
| **40** | 0.12166 | 0.016020 | 2445 | 8.02 | 1023.9 | 8.02 | 1078.9 | 0.01617 | 2.1592 |
| 45.45 | **0.15** | 0.016022 | 2005 | 13.49 | 1025.7 | 13.49 | 1081.3 | 0.02706 | 2.1409 |
| **50** | 0.17803 | 0.016024 | 1704.2 | 18.06 | 1027.2 | 18.06 | 1083.3 | 0.03607 | 2.1259 |
| 53.15 | **0.20** | 0.016027 | 1526.3 | 21.22 | 1028.2 | 21.22 | 1084.7 | 0.04225 | 2.1158 |
| **60** | 0.2563 | 0.016035 | 1206.9 | 28.08 | 1030.4 | 28.08 | 1087.7 | 0.05555 | 2.0943 |
| 64.46 | **0.30** | 0.016041 | 1039.7 | 32.55 | 1031.9 | 32.56 | 1089.6 | 0.06411 | 2.0807 |
| **70** | 0.3632 | 0.016051 | 867.7 | 38.09 | 1033.7 | 38.09 | 1092.0 | 0.07463 | 2.0642 |
| 75.69 | **0.44** | 0.016063 | 723.8 | 43.78 | 1035.6 | 43.78 | 1094.5 | 0.08531 | 2.0478 |
| **80** | 0.5073 | 0.016073 | 632.8 | 48.08 | 1037.0 | 48.09 | 1096.4 | 0.09332 | 2.0356 |
| 85.19 | **0.60** | 0.016086 | 540.0 | 53.26 | 1038.7 | 53.27 | 1098.6 | 0.10287 | 2.0213 |
| **90** | 0.6988 | 0.016099 | 467.7 | 58.07 | 1040.2 | 58.07 | 1100.7 | 0.11165 | 2.0083 |
| 94.35 | **0.80** | 0.016112 | 411.7 | 62.41 | 1041.7 | 62.41 | 1102.6 | 0.11951 | 1.9968 |
| **100** | 0.9503 | 0.016130 | 350.0 | 68.04 | 1043.5 | 68.05 | 1105.0 | 0.12963 | 1.9822 |
| 101.70 | **1.0** | 0.016136 | 333.6 | 69.74 | 1044.0 | 69.74 | 1105.8 | 0.13266 | 1.9779 |
| **110** | 1.2763 | 0.016166 | 265.1 | 78.02 | 1046.7 | 78.02 | 1109.3 | 0.14730 | 1.9574 |
| 115.65 | **1.5** | 0.016187 | 227.7 | 83.65 | 1048.5 | 83.65 | 1111.7 | 0.15714 | 1.9438 |
| **120** | 1.6945 | 0.016205 | 203.0 | 87.99 | 1049.9 | 88.00 | 1113.5 | 0.16465 | 1.9336 |
| 126.04 | **2.0** | 0.016230 | 173.75 | 94.02 | 1051.8 | 94.02 | 1116.1 | 0.17499 | 1.9198 |
| **130** | 2.225 | 0.016247 | 157.17 | 97.97 | 1053.0 | 97.98 | 1117.8 | 0.18172 | 1.9109 |
| 141.43 | **3.0** | 0.016300 | 118.72 | 109.38 | 1056.6 | 109.39 | 1122.5 | 0.20089 | 1.8861 |
| **150** | 3.722 | 0.016343 | 96.99 | 117.95 | 1059.3 | 117.96 | 1126.1 | 0.21503 | 1.8684 |
| 170.03 | **6.0** | 0.016451 | 61.98 | 137.98 | 1065.4 | 138.00 | 1134.2 | 0.24736 | 1.8292 |
| **180.** | 7.515 | 0.016509 | 50.20 | 147.97 | 1068.3 | 147.99 | 1138.2 | 0.26311 | 1.8109 |
| 193.19 | **10.** | 0.016590 | 38.42 | 161.20 | 1072.2 | 161.23 | 1143.3 | 0.28358 | 1.7877 |
| **210** | 14.125 | 0.016702 | 27.82 | 178.10 | 1077.0 | 178.14 | 1149.7 | 0.30913 | 1.7599 |
| **211.99** | **14.696** | 0.016715 | 26.80 | 180.10 | 1077.6 | 180.15 | 1150.5 | 0.31212 | 1.7567 |
| 227.96 | **20.** | 0.016830 | 20.09 | 196.19 | 1082.0 | 196.26 | 1156.4 | 0.33580 | 1.7320 |
| **250** | 29.82 | 0.017001 | 13.826 | 218.49 | 1087.9 | 218.59 | 1164.2 | 0.36772 | 1.7001 |
| 267.26 | **40.** | 0.017146 | 10.501 | 236.03 | 1092.3 | 236.16 | 1170.0 | 0.39214 | 1.6767 |
| **290** | 57.53 | 0.017352 | 7.467 | 259.25 | 1097.7 | 259.44 | 1177.2 | 0.42360 | 1.6477 |
| 312.07 | **80.** | 0.017570 | 5.474 | 281.95 | 1102.6 | 282.21 | 1183.6 | 0.45344 | 1.6214 |
| 327.86 | **100.** | 0.017736 | 4.434 | 298.28 | 1105.8 | 298.61 | 1187.8 | 0.47439 | 1.6034 |
| **350** | 134.53 | 0.017988 | 3.346 | 321.35 | 1109.8 | 321.80 | 1193.1 | 0.50329 | 1.5793 |
| 373.13 | **180.** | 0.018273 | 2.533 | 345.68 | 1113.4 | 346.29 | 1197.8 | 0.53292 | 1.5553 |
| **400** | 247.1 | 0.018638 | 1.8661 | 374.27 | 1116.6 | 375.12 | 1202.0 | 0.56672 | 1.5284 |
| 417.43 | **300.** | 0.018896 | 1.5442 | 393.0 | 1118.2 | 394.1 | 1203.9 | 0.5883 | 1.5115 |
| **440** | 381.2 | 0.019260 | 1.2192 | 417.62 | 1119.3 | 418.98 | 1205.3 | 0.61605 | 1.4900 |
| 456.39 | **450.** | 0.019547 | 1.0326 | 435.7 | 1119.6 | 437.4 | 1205.6 | 0.6360 | 1.4746 |
| **475** | 539.3 | 0.019901 | .8594 | 456.6 | 1119.2 | 458.5 | 1204.9 | 0.6586 | 1.4571 |
| 486.33 | **600.** | 0.02013 | .7702 | 469.4 | 1118.6 | 471.7 | 1204.1 | 0.6723 | 1.4464 |
| **500** | 680.0 | 0.02043 | .6761 | 485.1 | 1117.4 | 487.7 | 1202.5 | 0.6888 | 1.4335 |
| 526.76 | **860.** | 0.02109 | .5264 | 516.6 | 1113.6 | 519.9 | 1197.4 | 0.7212 | 1.4080 |
| **550** | 1044.0 | 0.02175 | .4249 | 544.9 | 1108.6 | 549.1 | 1190.6 | 0.7497 | 1.3851 |
| 572.56 | **1250.** | 0.02250 | .3454 | 573.4 | 1101.7 | 578.6 | 1181.6 | 0.7778 | 1.3619 |
| 591.88 | **1450.** | 0.02326 | .2888 | 598.9 | 1093.9 | 605.1 | 1171.4 | 0.8024 | 1.3409 |
| **600** | 1541.0 | 0.02363 | .2677 | 609.9 | 1090.0 | 616.7 | 1166.4 | 0.8130 | 1.3317 |
| 636.00 | **2000.** | 0.02565 | .18813 | 662.4 | 1066.6 | 671.9 | 1136.3 | 0.8623 | 1.2861 |
| **650** | 2205. | 0.02673 | .16206 | 685.0 | 1053.7 | 695.9 | 1119.8 | 0.8831 | 1.2651 |
| 662.31 | **2400.** | 0.02791 | .14067 | 706.4 | 1039.3 | 718.8 | 1101.8 | 0.9028 | 1.2441 |
| **680.** | 2705. | 0.03032 | .11127 | 741.7 | 1011.0 | 756.9 | 1066.7 | 0.9350 | 1.2068 |
| 695.52 | **3000.** | 0.03431 | .08404 | 783.4 | 968.8 | 802.5 | 1015.5 | 0.9732 | 1.1575 |
| **705.44*** | **3203.6** | 0.05053 | .05053 | 872.6 | 872.6 | 902.5 | 902.5 | 1.0580 | 1.0580 |

*Critical point

**Table C–2** $H_2O$—Superheated Vapor (English Units)

| Temp. F $t$ | Pressure, lbf/in² absolute (saturation temperature in parentheses) | | | | | | | | | | | |
|---|---|---|---|---|---|---|---|---|---|---|---|---|
| | **0.2 (53.16 F)** | | | | **0.4 (72.84 F)** | | | | **0.8 (94.35 F)** | | | |
| | $v$ | $u$ | $h$ | $s$ | $v$ | $u$ | $h$ | $s$ | $u$ | $v$ | $h$ | $s$ |
| 60 | 1546.7 | 1030.5 | 1087.7 | 2.1217 | — | — | — | — | — | — | — | — |
| 70 | 1576.6 | 1033.8 | 1092.2 | 2.1302 | — | — | — | — | — | — | — | — |
| 80 | 1606.4 | 1037.2 | 1096.6 | 2.1385 | 802.7 | 1037.1 | 1096.5 | 2.0619 | — | — | — | — |
| 90 | 1636.3 | 1040.5 | 1101.1 | 2.1467 | 817.7 | 1040.4 | 1100.9 | 2.0701 | — | — | — | — |
| 100 | 1666.1 | 1043.9 | 1105.6 | 2.1548 | 832.6 | 1043.8 | 1105.4 | 2.0782 | 415.9 | 1043.6 | 1105.1 | 2.0014 |
| 120 | 1725.8 | 1050.6 | 1114.5 | 2.1705 | 862.5 | 1050.5 | 1114.4 | 2.0939 | 430.9 | 1050.3 | 1114.1 | 2.0171 |
| 150 | 1815.2 | 1060.8 | 1127.9 | 2.1931 | 907.3 | 1060.7 | 1127.8 | 2.1165 | 453.4 | 1060.5 | 1127.6 | 2.0399 |
| 200 | 1964.3 | 1077.7 | 1150.4 | 2.2285 | 981.9 | 1077.7 | 1150.4 | 2.1520 | 490.7 | 1077.6 | 1150.2 | 2.0755 |

| Temp. F $t$ | Pressure, lbf/in² absolute (saturation temperature in parentheses) | | | | | | | | | | | |
|---|---|---|---|---|---|---|---|---|---|---|---|---|
| | **1.0 (101.70 F)** | | | | **5.0 (162.21 F)** | | | | **10 (193.19 F)** | | | |
| | $v$ | $u$ | $h$ | $s$ | $v$ | $u$ | $h$ | $s$ | $u$ | $v$ | $h$ | $s$ |
| 120 | 344.6 | 1050.2 | 1114.0 | 1.9924 | — | — | — | — | — | — | — | — |
| 160 | 386.6 | 1063.8 | 1132.0 | 2.0225 | — | — | — | — | — | — | — | — |
| 200 | 392.5 | 1077.5 | 1150.1 | 2.0508 | 78.15 | 1076.3 | 1148.6 | 1.8715 | 38.85 | 1074.7 | 1146.6 | 1.7927 |
| 240 | 416.4 | 1091.2 | 1168.3 | 2.0775 | 83.00 | 1090.3 | 1167.1 | 1.8987 | 41.32 | 1089.0 | 1165.5 | 1.8205 |
| 300 | 452.3 | 1112.0 | 1195.7 | 2.1150 | 90.24 | 1111.3 | 1194.8 | 1.9367 | 44.99 | 1110.4 | 1193.7 | 1.8592 |
| 400 | 511.9 | 1147.0 | 1241.8 | 2.1720 | 102.24 | 1146.6 | 1241.2 | 1.9941 | 51.03 | 1146.1 | 1240.5 | 1.9171 |
| 500 | 571.5 | 1182.8 | 1288.5 | 2.2235 | 114.20 | 1182.5 | 1288.2 | 2.0458 | 57.04 | 1182.2 | 1287.7 | 1.9690 |
| 600 | 631.1 | 1219.3 | 1336.1 | 2.2706 | 126.15 | 1219.1 | 1335.8 | 2.0930 | 63.03 | 1218.9 | 1335.5 | 2.0164 |

| Temp. F $t$ | Pressure, lbf/in² absolute (saturation temperature in parentheses) | | | | | | | | | | | |
|---|---|---|---|---|---|---|---|---|---|---|---|---|
| | **14.696 (211.99 F)** | | | | **20 (227.96 F)** | | | | **40 (267.26 F)** | | | |
| | $v$ | $u$ | $h$ | $s$ | $v$ | $u$ | $h$ | $s$ | $v$ | $u$ | $h$ | $s$ |
| 220 | 27.15 | 1080.6 | 1154.4 | 1.7624 | — | — | — | — | — | — | — | — |
| 240 | 28.00 | 1087.9 | 1164.0 | 1.7764 | 20.475 | 1086.5 | 1162.3 | 1.7405 | — | — | — | — |
| 280 | 29.69 | 1102.4 | 1183.1 | 1.8030 | 21.73 | 1101.4 | 1181.8 | 1.7676 | 10.711 | 1097.3 | 1176.6 | 1.6857 |
| 320 | 31.36 | 1116.8 | 1202.1 | 1.8280 | 22.98 | 1116.0 | 1201.0 | 1.7930 | 11.360 | 1112.8 | 1196.6 | 1.7124 |
| 360 | 33.02 | 1131.2 | 1221.0 | 1.8516 | 24.21 | 1130.6 | 1220.1 | 1.8168 | 11.996 | 1128.0 | 1216.8 | 1.7373 |
| 400 | 34.67 | 1145.6 | 1239.9 | 1.8741 | 25.43 | 1145.1 | 1239.2 | 1.8395 | 12.623 | 1143.0 | 1236.4 | 1.7606 |
| 440 | 36.31 | 1160.1 | 1258.8 | 1.8956 | 26.64 | 1159.6 | 1258.2 | 1.8611 | 13.243 | 1157.8 | 1255.8 | 1.7828 |
| 500 | 38.77 | 1181.8 | 1287.3 | 1.9263 | 28.46 | 1181.5 | 1286.8 | 1.8919 | 14.164 | 1180.1 | 1284.9 | 1.8140 |
| 600 | 42.86 | 1218.6 | 1335.2 | 1.9737 | 31.47 | 1218.4 | 1334.8 | 1.9395 | 15.685 | 1217.3 | 1333.4 | 1.8621 |
| 700 | 46.93 | 1256.1 | 1383.8 | 2.0175 | 34.47 | 1255.9 | 1383.5 | 1.9834 | 17.196 | 1255.1 | 1382.4 | 1.9063 |

*Continued on next page*

**Table C–2—*Continued***

| Temp. F t | Pressure, lbf/in² absolute (saturation temperature in parentheses) | | | | | | | | | | | |
|---|---|---|---|---|---|---|---|---|---|---|---|---|
| | 80 (312.07 F) | | | | 100 (327.86 F) | | | | 150 (358.48 F) | | | |
| | v | u | h | s | v | u | h | s | v | u | h | s |
| 320 | 5.544 | 1106.0 | 1188.0 | 1.6271 | — | — | — | — | — | — | — | — |
| 340 | 5.717 | 1114.3 | 1199.0 | 1.6409 | 4.521 | 1111.1 | 1194.8 | 1.6121 | — | — | — | — |
| 360 | 5.886 | 1122.5 | 1209.7 | 1.6541 | 4.662 | 1119.7 | 1205.9 | 1.6259 | 3.024 | 1111.9 | 1195.9 | 1.5715 |
| 400 | 6.217 | 1138.5 | 1230.6 | 1.6790 | 4.934 | 1136.2 | 1227.5 | 1.6517 | 3.221 | 1130.1 | 1219.5 | 1.5997 |
| 450 | 6.621 | 1158.0 | 1256.0 | 1.7078 | 5.265 | 1156.2 | 1253.6 | 1.6812 | 3.455 | 1151.5 | 1247.4 | 1.6312 |
| 500 | 7.017 | 1177.2 | 1281.1 | 1.7346 | 5.587 | 1175.7 | 1279.1 | 1.7085 | 3.679 | 1171.9 | 1274.1 | 1.6598 |
| 600 | 7.794 | 1215.3 | 1330.7 | 1.7838 | 6.216 | 1214.2 | 1329.3 | 1.7582 | 4.111 | 1211.6 | 1325.7 | 1.7110 |
| 700 | 8.561 | 1253.6 | 1380.3 | 1.8285 | 6.834 | 1252.8 | 1379.2 | 1.8033 | 4.531 | 1250.8 | 1376.6 | 1.7568 |
| 800 | 9.321 | 1292.4 | 1430.4 | 1.8700 | 7.445 | 1291.8 | 1429.6 | 1.8449 | 4.944 | 1290.2 | 1427.5 | 1.7989 |
| 900 | 10.078 | 1332.0 | 1481.2 | 1.9087 | 8.053 | 1331.5 | 1480.5 | 1.8838 | 5.353 | 1330.2 | 1478.8 | 1.8381 |

| Temp. F t | Pressure, lbf/in² absolute (saturation temperature in parentheses) | | | | | | | | | | | |
|---|---|---|---|---|---|---|---|---|---|---|---|---|
| | 200 (381.86 F) | | | | 250 (401.04 F) | | | | 300 (417.43 F) | | | |
| | v | u | h | s | v | u | h | s | v | u | h | s |
| 400 | 2.361 | 1123.5 | 1210.8 | 1.5600 | — | — | — | — | — | — | — | — |
| 420 | 2.437 | 1132.9 | 1223.1 | 1.5741 | 1.9075 | 1126.5 | 1214.7 | 1.5420 | 1.5517 | 1119.6 | 1205.7 | 1.5136 |
| 440 | 2.511 | 1142.0 | 1234.9 | 1.5874 | 1.9709 | 1136.3 | 1227.5 | 1.5563 | 1.6086 | 1130.3 | 1219.6 | 1.5292 |
| 480 | 2.654 | 1159.5 | 1257.7 | 1.6122 | 2.092 | 1154.9 | 1251.7 | 1.5826 | 1.7154 | 1150.1 | 1245.3 | 1.5572 |
| 520 | 2.792 | 1176.3 | 1279.7 | 1.6351 | 2.207 | 1172.5 | 1274.6 | 1.6065 | 1.8156 | 1168.6 | 1269.4 | 1.5823 |
| 580 | 2.993 | 1200.8 | 1311.6 | 1.6667 | 2.372 | 1197.8 | 1307.6 | 1.6392 | 1.9580 | 1194.8 | 1303.5 | 1.6161 |
| 640 | 3.188 | 1224.9 | 1342.9 | 1.6959 | 2.532 | 1222.4 | 1339.6 | 1.6691 | 2.094 | 1220.0 | 1336.2 | 1.6467 |
| 700 | 3.379 | 1248.8 | 1373.8 | 1.7234 | 2.688 | 1246.7 | 1371.1 | 1.6970 | 2.227 | 1244.6 | 1368.3 | 1.6751 |
| 800 | 3.693 | 1288.6 | 1425.3 | 1.7660 | 2.943 | 1287.0 | 1423.2 | 1.7401 | 2.442 | 1285.4 | 1421.0 | 1.7187 |
| 900 | 4.003 | 1328.9 | 1477.1 | 1.8055 | 3.193 | 1327.6 | 1475.3 | 1.7799 | 2.653 | 1326.3 | 1473.6 | 1.7589 |

| Temp. F t | Pressure, lbf/in² absolute (saturation temperature in parentheses) | | | | | | | | | | | |
|---|---|---|---|---|---|---|---|---|---|---|---|---|
| | 400 (444.70 F) | | | | 500 (467.13 F) | | | | 600 (486.33 F) | | | |
| | v | u | h | s | v | u | h | s | v | u | h | s |
| 460 | 1.977 | 1128.5 | 1217.1 | 1.4984 | — | — | — | — | — | — | — | — |
| 480 | 1.2421 | 1139.6 | 1231.5 | 1.5139 | 0.9543 | 1127.7 | 1216.0 | 1.4759 | — | — | — | — |
| 500 | 1.2843 | 1150.1 | 1245.2 | 1.5282 | 0.9924 | 1139.7 | 1231.5 | 1.4923 | 0.7947 | 1128.0 | 1216.2 | 1.4592 |
| 550 | 1.3833 | 1174.6 | 1277.0 | 1.5605 | 1.0792 | 1167.7 | 1266.6 | 1.5279 | 0.8749 | 1158.2 | 1255.4 | 1.4990 |
| 600 | 1.4760 | 1197.3 | 1306.6 | 1.5892 | 1.1583 | 1191.1 | 1298.3 | 1.5585 | 0.9456 | 1184.5 | 1289.5 | 1.5320 |
| 650 | — | — | — | — | 1.2327 | 1214.0 | 1328.0 | 1.5860 | 1.0109 | 1208.6 | 1320.9 | 1.5609 |
| 700 | 1.6503 | 1240.4 | 1362.5 | 1.6397 | 1.3040 | 1236.0 | 1356.7 | 1.6112 | 1.0727 | 1231.5 | 1350.6 | 1.5872 |
| 800 | 1.8163 | 1282.1 | 1416.6 | 1.6844 | 1.4407 | 1278.8 | 1412.1 | 1.6571 | 1.1900 | 1275.4 | 1407.6 | 1.6343 |
| 900 | 1.9776 | 1323.7 | 1470.1 | 1.7252 | 1.5723 | 1321.0 | 1466.5 | 1.6987 | 1.3021 | 1318.4 | 1462.9 | 1.6766 |
| 1000 | 2.136 | 1365.5 | 1523.6 | 1.7632 | 1.7008 | 1363.3 | 1520.7 | 1.7371 | 1.4108 | 1361.2 | 1517.8 | 1.7155 |

*Continued on next page*

**Table C-2** *Continued*

| Temp. F $t$ | Pressure, lbf/in² absolute (saturation temperature in parentheses) 800 (518.36 F) $v$ | $u$ | $h$ | $s$ | 1000 (544.75 F) $v$ | $u$ | $h$ | $s$ | 1450 (591.88 F) $v$ | $u$ | $h$ | $s$ |
|---|---|---|---|---|---|---|---|---|---|---|---|---|
| 540 | 0.6015 | 1131.8 | 1220.8 | 1.4378 | — | — | — | — | — | — | — | — |
| 570 | 0.6415 | 1152.0 | 1247.0 | 1.4637 | 0.4795 | 1131.7 | 1220.5 | 1.4179 | — | — | — | — |
| 600 | 0.6776 | 1170.1 | 1270.4 | 1.4861 | 0.5140 | 1153.7 | 1248.8 | 1.4450 | 0.2992 | 1104.0 | 1184.3 | 1.3531 |
| 650 | 0.7324 | 1197.2 | 1305.6 | 1.5186 | 0.5637 | 1184.7 | 1289.1 | 1.4822 | 0.3493 | 1151.3 | 1245.0 | 1.4092 |
| 700 | 0.7829 | 1222.1 | 1338.0 | 1.5471 | 0.6080 | 1212.0 | 1324.6 | 1.5135 | 0.3882 | 1186.6 | 1290.7 | 1.4495 |
| 760 | 0.8399 | 1250.3 | 1374.6 | 1.5779 | 0.6569 | 1242.1 | 1363.7 | 1.5464 | 0.4282 | 1222.2 | 1337.1 | 1.4885 |
| 800 | 0.8764 | 1268.5 | 1398.2 | 1.5969 | 0.6878 | 1261.2 | 1388.5 | 1.5664 | 0.4525 | 1243.8 | 1365.3 | 1.5112 |
| 900 | 0.9640 | 1312.9 | 1455.6 | 1.6408 | 0.7610 | 1307.3 | 1448.1 | 1.6120 | 0.5085 | 1294.1 | 1430.5 | 1.5611 |
| 1000 | 1.0482 | 1356.7 | 1511.9 | 1.6807 | 0.8305 | 1352.2 | 1505.9 | 1.6530 | 0.5601 | 1341.6 | 1491.9 | 1.6047 |
| 1100 | 1.1300 | 1400.5 | 1567.8 | 1.7178 | 0.8976 | 1396.8 | 1562.9 | 1.6908 | 0.6090 | 1388.1 | 1551.5 | 1.6442 |

| Temp. F $t$ | Pressure, lbf/in² absolute (saturation temperature in parentheses) 2400 (662.31 F) $v$ | $u$ | $h$ | $s$ | 3200 (705.27 F) $v$ | $u$ | $h$ | $s$ | 5000 (> critical) $v$ | $u$ | $h$ | $s$ |
|---|---|---|---|---|---|---|---|---|---|---|---|---|
| 680 | 0.16420 | 1080.1 | 1153.0 | 1.2894 | — | — | — | — | 0.02535 | 690.6 | 714.1 | 0.8873 |
| 700 | 0.18261 | 1110.2 | 1191.3 | 1.3228 | — | — | — | — | 0.02676 | 721.8 | 746.6 | 0.9156 |
| 720 | 0.19745 | 1133.6 | 1221.3 | 1.3484 | 0.10226 | 1030.3 | 1090.8 | 1.2192 | 0.02867 | 756.5 | 783.0 | 0.9468 |
| 760 | 0.2218 | 1170.6 | 1269.1 | 1.3883 | 0.13672 | 1109.5 | 1190.4 | 1.3024 | 0.03657 | 849.6 | 883.4 | 1.0303 |
| 800 | 0.2422 | 1200.8 | 1308.4 | 1.4200 | 0.15865 | 1155.2 | 1249.1 | 1.3498 | 0.05932 | 987.2 | 1042.1 | 1.1583 |
| 850 | 0.2646 | 1233.7 | 1351.2 | 1.4534 | 0.18027 | 1198.3 | 1305.0 | 1.3934 | 0.08556 | 1092.7 | 1171.9 | 1.2596 |
| 900 | 0.2849 | 1263.4 | 1389.9 | 1.4824 | 0.19863 | 1234.2 | 1351.8 | 1.4285 | 0.10385 | 1155.1 | 1251.1 | 1.3190 |
| 1000 | 0.3215 | 1317.8 | 1460.6 | 1.5326 | 0.2302 | 1296.1 | 1432.4 | 1.4857 | 0.13120 | 1242.0 | 1363.4 | 1.3988 |
| 1040 | 0.3352 | 1338.5 | 1487.4 | 1.5507 | 0.2416 | 1318.9 | 1461.9 | 1.5057 | 0.14038 | 1270.7 | 1400.6 | 1.4240 |
| 1100 | 0.3549 | 1368.9 | 1526.5 | 1.5763 | 0.2578 | 1351.8 | 1504.5 | 1.5335 | 0.15302 | 1310.6 | 1452.2 | 1.4577 |

**Table C-3** $H_2O$—Saturated Solid and Saturated Vapor (English Units)

| Temp. F $t$ | Pressure Lbf/in.² $p$ | Specific Volume Sat. Solid $v_i$ | Sat. Vapor $v_g \times 10^{-3}$ | Internal Energy Sat. Solid $u_i$ | Sat. Vapor $u_g$ | Enthalpy Sat. Solid $h_i$ | Sat. Vapor $h_g$ | Entropy Sat. Solid $s_i$ | Sat. Vapor $s_g$ |
|---|---|---|---|---|---|---|---|---|---|
| 32.018 | 0.0887 | 0.01747 | 3.302 | −143.34 | 1021.2 | −143.34 | 1075.4 | −0.292 | 2.187 |
| 32 | 0.0886 | 0.01747 | 3.305 | −143.35 | 1021.2 | −143.35 | 1075.4 | −0.292 | 2.187 |
| 30 | 0.0808 | 0.01747 | 3.607 | −144.35 | 1020.5 | −144.35 | 1074.5 | −0.294 | 2.195 |
| 25 | 0.0641 | 0.01746 | 4.506 | −146.84 | 1018.9 | −146.84 | 1072.3 | −0.299 | 2.216 |
| 20 | 0.0505 | 0.01745 | 5.655 | −149.31 | 1017.2 | −149.31 | 1070.1 | −0.304 | 2.238 |
| 15 | 0.0396 | 0.01745 | 7.13 | −151.75 | 1015.5 | −151.75 | 1067.9 | −0.309 | 2.260 |
| 10 | 0.0309 | 0.01744 | 9.04 | −154.17 | 1013.9 | −154.17 | 1065.7 | −0.314 | 2.283 |
| 5 | 0.0240 | 0.01743 | 11.52 | −156.56 | 1012.2 | −156.56 | 1063.5 | −0.320 | 2.306 |
| 0 | 0.0185 | 0.01743 | 14.77 | −158.93 | 1010.6 | −158.93 | 1061.2 | −0.325 | 2.330 |
| −5 | 0.0142 | 0.01742 | 19.03 | −161.27 | 1008.9 | −161.27 | 1059.0 | −0.330 | 2.354 |
| −10 | 0.0109 | 0.01741 | 24.66 | −163.59 | 1007.3 | −163.59 | 1056.8 | −0.335 | 2.379 |
| −15 | 0.0082 | 0.01740 | 32.2 | −165.89 | 1005.6 | −165.89 | 1054.6 | −0.340 | 2.405 |
| −20 | 0.0062 | 0.01740 | 42.2 | −168.16 | 1003.9 | −168.16 | 1052.4 | −0.345 | 2.431 |
| −30 | 0.0035 | 0.01738 | 74.1 | −172.63 | 1000.6 | −172.63 | 1048.0 | −0.356 | 2.485 |
| −40 | 0.0019 | 0.01737 | 133.8 | −177.00 | 997.3 | −177.00 | 1043.6 | −0.366 | 2.542 |

**Table C–4** $H_2O$—Compressed (Subcooled) Liquid (English Units)

| | Pressure, lbf/in.$^2$ abs. (saturation temperature in parentheses) | | | | | | | | | | | |
|---|---|---|---|---|---|---|---|---|---|---|---|---|
| | 500 (467.13 F) | | | | 1000 (544.75 F) | | | | 1500 (596.39 F) | | | |
| Temp. F *t* | *v* | *u* | *h* | *s* | *v* | *u* | *h* | *s* | *v* | *u* | *h* | *s* |
| 100 | 0.016106 | 67.87 | 69.36 | 0.12932 | 0.016082 | 67.70 | 70.68 | 0.12901 | 0.016058 | 67.53 | 71.99 | 0.12870 |
| 200 | 0.016608 | 167.65 | 169.19 | 0.29341 | 0.016580 | 167.26 | 170.32 | 0.29281 | 0.016554 | 166.87 | 171.46 | 0.29221 |
| 300 | 0.017416 | 268.92 | 270.53 | 0.43641 | 0.017379 | 268.24 | 271.46 | 0.43552 | 0.017343 | 267.58 | 272.39 | 0.43463 |
| 400 | 0.018608 | 373.68 | 375.40 | 0.56604 | 0.018550 | 372.55 | 375.98 | 0.56472 | 0.018493 | 371.45 | 376.59 | 0.56343 |
| 500 | — | — | — | — | 0.02036 | 483.8 | 487.5 | 0.6874 | 0.02024 | 481.8 | 487.4 | 0.6853 |

| | 2000 (636.00 F) | | | | 2500 (668.31 F) | | | | 3000 (695.52 F) | | | |
|---|---|---|---|---|---|---|---|---|---|---|---|---|
| Temp. F *t* | *v* | *u* | *h* | *s* | *v* | *u* | *h* | *s* | *v* | *u* | *h* | *s* |
| 100 | 0.016034 | 67.37 | 73.30 | 0.12839 | 0.016010 | 67.20 | 74.61 | 0.12808 | 0.015987 | 67.04 | 75.91 | 0.12777 |
| 200 | 0.016527 | 166.49 | 172.60 | 0.29162 | 0.016501 | 166.11 | 173.75 | 0.29104 | 0.016476 | 165.74 | 174.89 | 0.29046 |
| 350 | 0.017822 | 318.15 | 324.74 | 0.49929 | 0.017780 | 317.33 | 325.56 | 0.49826 | 0.017739 | 316.53 | 326.38 | 0.49725 |
| 500 | 0.02014 | 479.8 | 487.3 | 0.6832 | 0.02004 | 478.0 | 487.3 | 0.6813 | 0.019944 | 476.2 | 487.3 | 0.6794 |
| 660 | — | — | — | — | 0.02729 | 698.4 | 711.0 | 0.8954 | 0.02629 | 687.6 | 702.2 | 0.8853 |

| | 3204 (critical pressure) (705.44 F) | | | | 5000 (> critical pressure) | | | | 10000 (> critical pressure) | | | |
|---|---|---|---|---|---|---|---|---|---|---|---|---|
| Temp. F *t* | *v* | *u* | *h* | *s* | *v* | *u* | *h* | *s* | *v* | *u* | *h* | *s* |
| 100 | 0.015978 | 66.97 | 76.45 | 0.12764 | 0.015897 | 66.40 | 81.11 | 0.12651 | 0.015684 | 64.92 | 93.94 | 0.12333 |
| 200 | 0.016465 | 165.59 | 175.36 | 0.29022 | 0.016376 | 164.32 | 179.47 | 0.28818 | 0.016145 | 161.07 | 190.94 | 0.28277 |
| 350 | 0.017722 | 316.21 | 326.71 | 0.49684 | 0.017583 | 313.28 | 329.75 | 0.49334 | 0.017241 | 306.73 | 338.63 | 0.48442 |
| 500 | 0.019906 | 475.5 | 487.3 | 0.6786 | 0.019603 | 469.8 | 487.9 | 0.6724 | 0.018938 | 456.8 | 491.8 | 0.6579 |
| 700 | 0.03323 | 779.8 | 799.5 | 0.9695 | 0.02676 | 721.8 | 746.6 | 0.9156 | 0.02321 | 674.3 | 717.3 | 0.8707 |

**Figure C.1** Temperature–Entropy Chart for Water, English Units

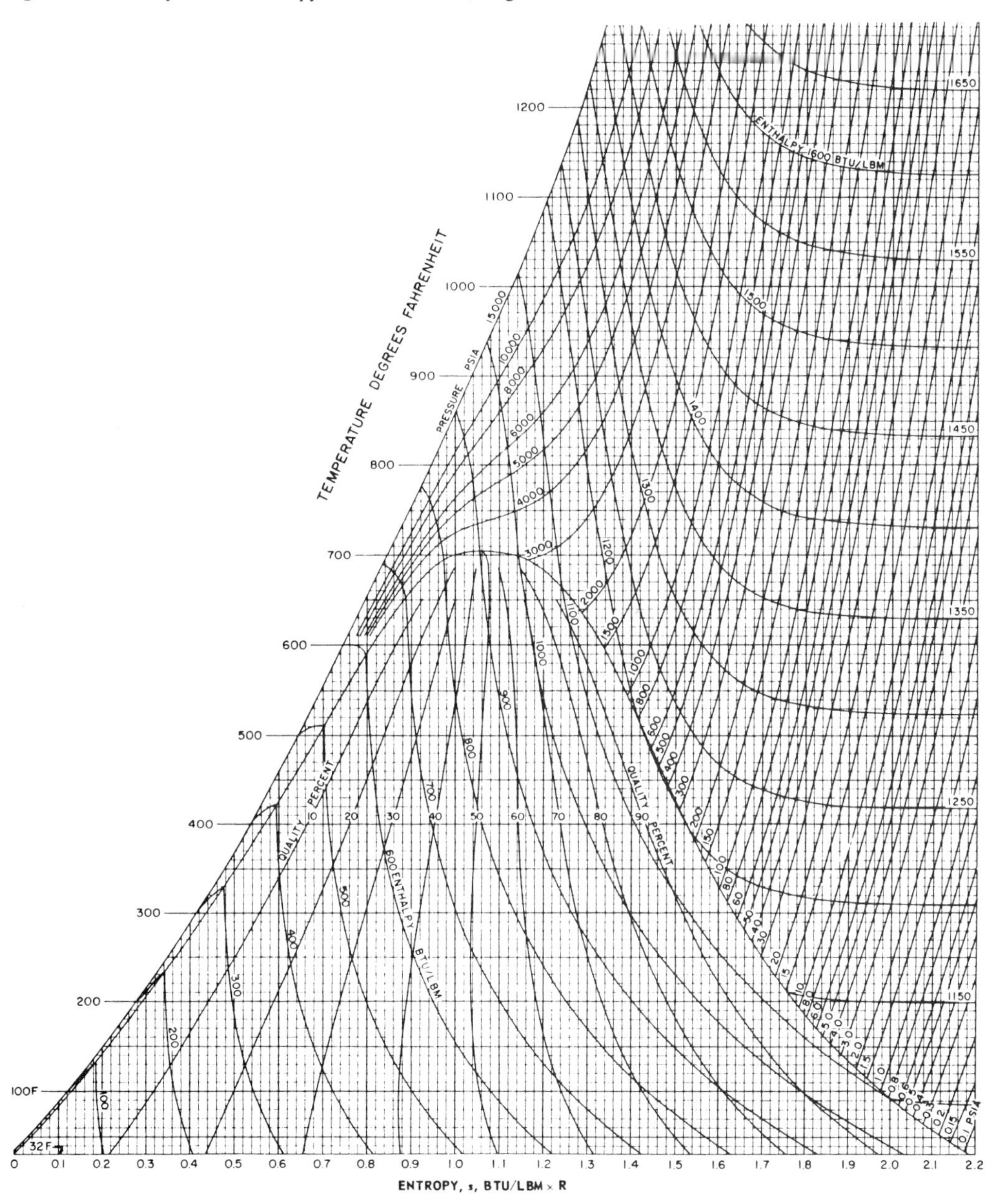


(Reprinted by permission from *ASME Steam Tables,* 3d ed. American Society of Mechanical Engineers, New York, NY)

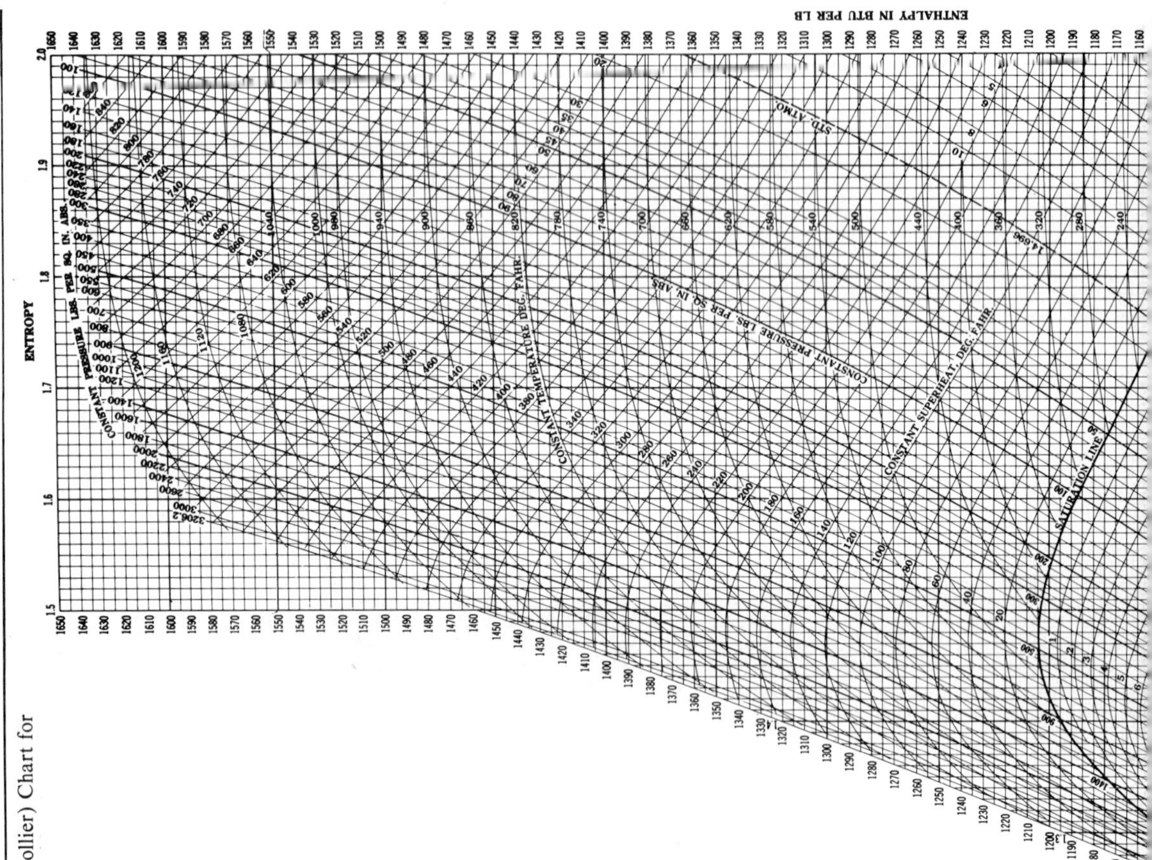


**Figure C.2** Enthalpy–Entropy (Mollier) Chart for Water, English Units

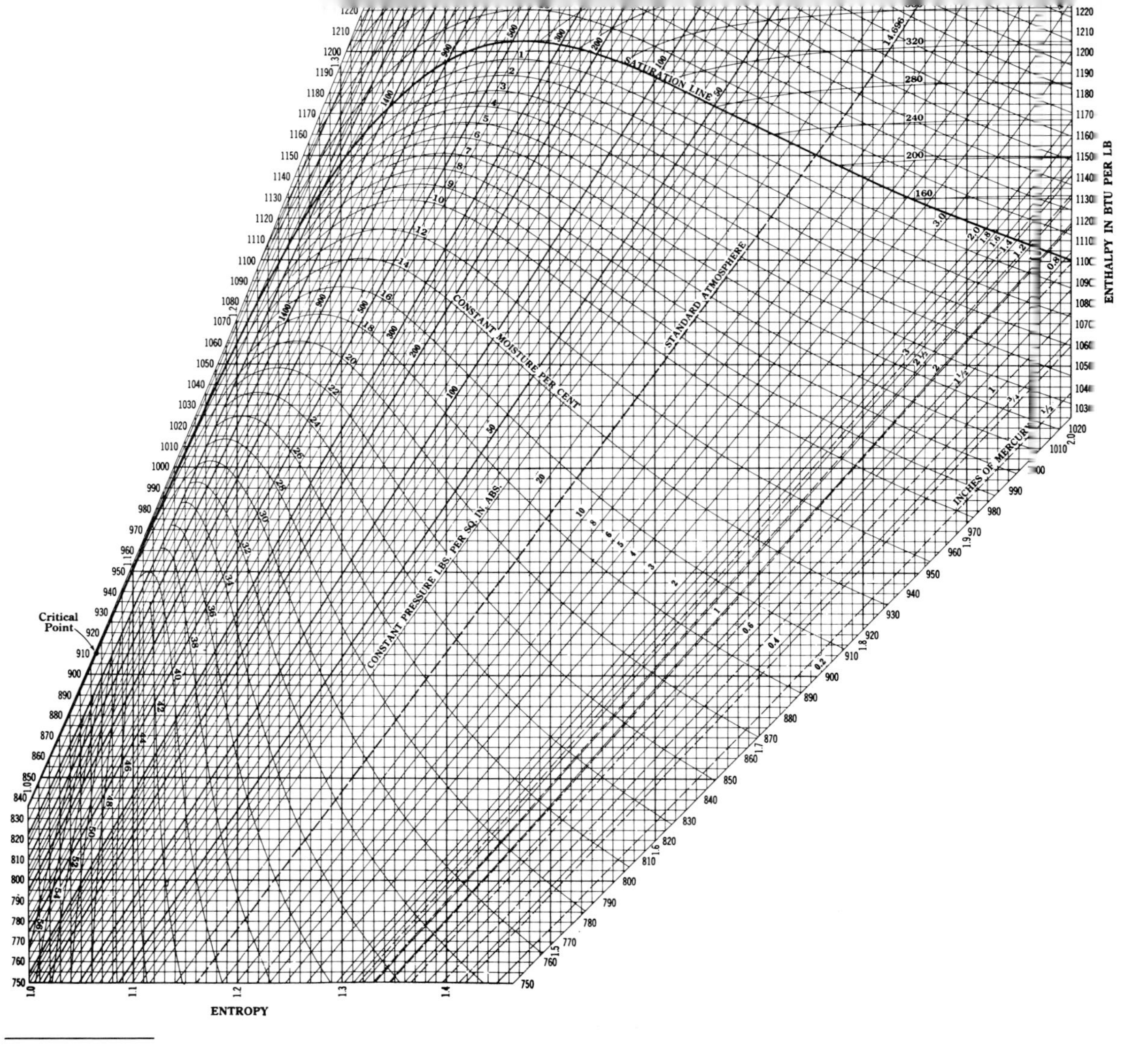


(Reprinted by permission from *Steam Tables*. Copyright 1940. Combustion Engineering, Inc., Windsor, CN)

# Appendix D

## Thermodynamic Properties of Ammonia (Refrigerant 717) English Units

Critical pressure—1657 psia
Critical temperature—271.4 F

**Table D–1** Saturated Ammonia Temperature Table

| Temperature, F | Pressure, psia | Specific Volume $ft^3/lbm$ | | Enthalpy, B/lbm | | Entropy, B/lbm R | | Temperature, F |
|---|---|---|---|---|---|---|---|---|
| | | $v_f$ | $v_g$ | $h_f$ | $h_g$ | $s_f$ | $s_g$ | |
| −60 | 5.55 | 0.02278 | 44.73 | −21.2 | 589.6 | −0.0517 | 1.4769 | −60 |
| −50 | 7.67 | 0.02299 | 33.08 | −10.6 | 593.7 | −0.0256 | 1.4497 | −50 |
| −40 | 10.41 | 0.02322 | 24.86 | 0.0 | 597.6 | 0.0000 | 1.4242 | −40 |
| −30 | 13.90 | 0.02345 | 18.97 | 10.7 | 601.4 | 0.0250 | 1.4001 | −30 |
| −20 | 18.30 | 0.02369 | 14.68 | 21.4 | 605.0 | 0.0497 | 1.3774 | −20 |
| −10 | 23.74 | 0.02393 | 11.50 | 32.1 | 608.5 | 0.0738 | 1.3558 | −10 |
| 0 | 30.42 | 0.02419 | 9.116 | 42.9 | 611.8 | 0.0975 | 1.3352 | 0 |
| 5 | 34.27 | 0.02432 | 8.150 | 48.3 | 613.3 | 0.1092 | 1.3253 | 5 |
| 10 | 38.51 | 0.02446 | 7.304 | 53.8 | 614.9 | 0.1208 | 1.3157 | 10 |
| 20 | 48.21 | 0.02474 | 5.910 | 64.7 | 617.8 | 0.1437 | 1.2969 | 20 |
| 30 | 59.74 | 0.02503 | 4.825 | 75.7 | 620.5 | 0.1663 | 1.2790 | 30 |
| 40 | 73.32 | 0.02533 | 3.971 | 86.8 | 623.0 | 0.1885 | 1.2618 | 40 |
| 50 | 89.19 | 0.02564 | 3.294 | 97.9 | 625.2 | 0.2105 | 1.2453 | 50 |
| 60 | 107.6 | 0.02597 | 2.751 | 109.2 | 627.3 | 0.2322 | 1.2294 | 60 |
| 70 | 128.8 | 0.02632 | 2.312 | 120.5 | 629.1 | 0.2537 | 1.2140 | 70 |
| 80 | 153.0 | 0.02668 | 1.955 | 132.0 | 630.7 | 0.2749 | 1.1991 | 80 |
| 90 | 180.6 | 0.02707 | 1.661 | 143.5 | 632.0 | 0.2958 | 1.1846 | 90 |
| 100 | 211.9 | 0.02747 | 1.419 | 155.2 | 633.0 | 0.3166 | 1.1705 | 100 |
| 110 | 247.0 | 0.02790 | 1.217 | 167.0 | 633.7 | 0.3372 | 1.1566 | 110 |
| 120 | 286.4 | 0.02836 | 1.047 | 179.0 | 634.0 | 0.3576 | 1.1427 | 120 |

Data for Appendix D were abstracted from National Bureau of Standards Circular No. 142 and Circular No. 472

Additional reference: Haar, L., and J. S. Gallagher. Thermodynamic Properties of Ammonia. *J. Phys. and Chem. Reference Data,* vol. 7, no. 3, 1978.

**Table D–2** Saturated Ammonia Pressure Table

| Pressure, psia | Temperature, F | Specific Volume, ft³/lbm $v_f$ | $v_g$ | Enthalpy, B/lbm $h_f$ | $h_g$ | Entropy, B/lbm R $s_f$ | $s_g$ | Pressure, psia |
|---|---|---|---|---|---|---|---|---|
| 5.00 | −63.11 | 0.02271 | 49.31 | −24.5 | 588.3 | −0.0599 | 1.4857 | 5.00 |
| 10.0 | −41.34 | 0.02319 | 25.81 | −1.4 | 597.1 | −0.0034 | 1.4276 | 10.0 |
| 15.0 | −27.29 | 0.02350 | 17.67 | 13.6 | 602.4 | 0.0318 | 1.3938 | 15.0 |
| 20.0 | −16.64 | 0.02378 | 13.50 | 25.0 | 606.2 | 0.0578 | 1.3700 | 20.0 |
| 25.0 | −7.96 | 0.02398 | 10.96 | 34.3 | 609.1 | 0.0787 | 1.3515 | 25.0 |
| 30.0 | −0.57 | 0.02418 | 9.236 | 42.3 | 611.6 | 0.0962 | 1.3364 | 30.0 |
| 40.0 | 11.66 | 0.02451 | 7.047 | 55.6 | 615.4 | 0.1246 | 1.3125 | 40.0 |
| 50.0 | 21.67 | 0.02479 | 5.710 | 66.5 | 618.2 | 0.1475 | 1.2939 | 50.0 |
| 60.0 | 30.21 | 0.02504 | 4.805 | 75.9 | 620.5 | 0.1668 | 1.2787 | 60.0 |
| 70.0 | 37.69 | 0.02526 | 4.151 | 84.5 | 622.4 | 0.1834 | 1.2657 | 70.0 |
| 80.0 | 44.40 | 0.02546 | 3.655 | 91.7 | 624.0 | 0.1982 | 1.2545 | 80.0 |
| 100.0 | 56.05 | 0.02584 | 2.952 | 104.7 | 626.5 | 0.2237 | 1.2356 | 100.0 |
| 120.0 | 66.02 | 0.02618 | 2.476 | 116.0 | 628.4 | 0.2452 | 1.2201 | 120.0 |
| 140.0 | 74.79 | 0.02649 | 2.132 | 126.0 | 629.9 | 0.2638 | 1.2068 | 140.0 |
| 160.0 | 82.61 | 0.02678 | 1.875 | 135.0 | 631.1 | 0.2804 | 1.1953 | 160.0 |
| 200.0 | 96.34 | 0.02732 | 1.502 | 150.9 | 632.7 | 0.3090 | 1.1756 | 200.0 |
| 220.0 | 102.42 | 0.02758 | 1.367 | 158.0 | 633.2 | 0.3216 | 1.1671 | 220.0 |
| 250.0 | 110.8 | 0.02792 | 1.202 | 168.0 | 633.8 | 0.3388 | 1.1555 | 250.0 |
| 275.0 | 117.22 | 0.02823 | 1.091 | 175.6 | 634.0 | 0.3519 | 1.1466 | 275.0 |
| 300.0 | 123.21 | 0.02851 | 0.999 | 182.9 | 634.0 | 0.3642 | 1.1383 | 300.0 |

**Table D–3** Superheated Ammonia Vapor
$v$ = specific volume, ft³/lbm; $h$ = enthalpy, B/lbm; $s$ = entropy, B/lbm R

| Temp. F. | Pressure, psia (saturation temperature in parentheses) 15 (−27.29 F) $v$ | $h$ | $s$ | 20 (−16.64 F) $v$ | $h$ | $s$ | 25 (−7.96 F) $v$ | $h$ | $s$ | Temp. F |
|---|---|---|---|---|---|---|---|---|---|---|
| −20 | 18.01 | 606.4 | 1.4031 | — | — | — | — | — | — | −20 |
| −10 | 18.47 | 611.9 | 1.4154 | 13.74 | 610.0 | 1.3784 | — | — | — | −10 |
| 0 | 18.92 | 617.2 | 1.4272 | 14.09 | 615.5 | 1.3907 | 11.19 | 613.8 | 1.3616 | 0 |
| 10 | 19.37 | 622.5 | 1.4386 | 14.44 | 621.0 | 1.4025 | 11.47 | 619.4 | 1.3738 | 10 |
| 20 | 19.82 | 627.8 | 1.4497 | 14.78 | 626.4 | 1.4138 | 11.75 | 625.0 | 1.3855 | 20 |
| 30 | 20.26 | 633.0 | 1.4604 | 15.11 | 631.7 | 1.4248 | 12.03 | 630.4 | 1.3967 | 30 |
| 40 | 20.70 | 638.2 | 1.4709 | 15.45 | 637.0 | 1.4356 | 12.30 | 635.8 | 1.4077 | 40 |
| 50 | 21.14 | 643.4 | 1.4812 | 15.78 | 642.3 | 1.4460 | 12.57 | 641.2 | 1.4183 | 50 |
| 60 | 21.58 | 648.5 | 1.4912 | 16.12 | 647.5 | 1.4562 | 12.84 | 646.5 | 1.4287 | 60 |
| 70 | 22.01 | 653.7 | 1.5011 | 16.45 | 652.8 | 1.4662 | 13.11 | 651.8 | 1.4388 | 70 |
| 80 | 22.44 | 658.9 | 1.5108 | 16.78 | 658.0 | 1.4760 | 13.37 | 657.1 | 1.4487 | 80 |
| 90 | 22.88 | 664.0 | 1.5203 | 17.10 | 663.2 | 1.4856 | 13.64 | 662.4 | 1.4584 | 90 |
| 100 | 23.31 | 669.2 | 1.5296 | 17.43 | 668.5 | 1.4950 | 13.90 | 667.7 | 1.4679 | 100 |
| 120 | 24.17 | 679.6 | 1.5478 | 18.08 | 678.9 | 1.5133 | 14.43 | 678.2 | 1.4864 | 120 |
| 140 | 25.03 | 690.0 | 1.5655 | 18.73 | 689.4 | 1.5312 | 14.95 | 688.8 | 1.5043 | 140 |
| 160 | 25.88 | 700.5 | 1.5827 | 19.37 | 700.0 | 1.5485 | 15.47 | 699.4 | 1.5217 | 160 |
| 180 | 26.74 | 711.1 | 1.5995 | 20.02 | 710.6 | 1.5653 | 15.99 | 710.1 | 1.5387 | 180 |
| 200 | 27.59 | 721.7 | 1.6158 | 20.66 | 721.2 | 1.5817 | 16.50 | 720.8 | 1.5552 | 200 |
| 240 | — | — | — | 21.94 | 742.8 | 1.6135 | 17.53 | 742.5 | 1.5870 | 240 |
| 260 | — | — | — | — | — | — | 18.04 | 753.4 | 1.6025 | 260 |

*Continued on next page*

**Table D-3**—*Continued*

| Temp. F | Pressure, psia (saturation temperature in parentheses) | | | | | | | | | Temp. F |
|---|---|---|---|---|---|---|---|---|---|---|
| | 30 (−0.57 F) | | | 40 (11.66 F) | | | 50 (21.67 F) | | | |
| | *v* | *h* | *s* | *v* | *h* | *s* | *v* | *h* | *s* | |
| 10 | 9.492 | 617.8 | 1.3497 | — | — | — | — | — | — | 10 |
| 20 | 9.731 | 623.5 | 1.3618 | 7.203 | 620.4 | 1.3231 | — | — | — | 20 |
| 30 | 9.966 | 629.1 | 1.3733 | 7.387 | 626.3 | 1.3353 | 5.838 | 623.4 | 1.3046 | 30 |
| 40 | 10.20 | 634.6 | 1.3845 | 7.568 | 632.1 | 1.3470 | 5.988 | 629.5 | 1.3169 | 40 |
| 50 | 10.43 | 640.1 | 1.3953 | 7.746 | 637.8 | 1.3583 | 6.135 | 635.4 | 1.3286 | 50 |
| 60 | 10.65 | 645.5 | 1.4059 | 7.922 | 643.4 | 1.3692 | 6.280 | 641.2 | 1.3399 | 60 |
| 70 | 10.88 | 650.9 | 1.4161 | 8.096 | 648.9 | 1.3797 | 6.423 | 646.9 | 1.3508 | 70 |
| 80 | 11.10 | 656.2 | 1.4261 | 8.268 | 654.4 | 1.3900 | 6.564 | 652.6 | 1.3613 | 80 |
| 90 | 11.33 | 661.6 | 1.4359 | 8.439 | 659.9 | 1.4000 | 6.704 | 658.2 | 1.3716 | 90 |
| 100 | 11.55 | 666.9 | 1.4456 | 8.609 | 665.3 | 1.4098 | 6.842 | 663.7 | 1.3816 | 100 |
| 110 | 11.77 | 672.2 | 1.4550 | 8.777 | 670.7 | 1.4194 | 6.980 | 669.2 | 1.3914 | 110 |
| 120 | 11.99 | 677.5 | 1.4642 | 8.945 | 676.1 | 1.4288 | 7.117 | 674.7 | 1.4009 | 120 |
| 130 | 12.21 | 682.9 | 1.4733 | 9.112 | 681.5 | 1.4381 | 7.252 | 680.2 | 1.4103 | 130 |
| 140 | 12.43 | 688.2 | 1.4823 | 9.278 | 686.9 | 1.4471 | 7.387 | 685.7 | 1.4195 | 140 |
| 160 | 12.87 | 698.8 | 1.4998 | 9.609 | 697.7 | 1.4648 | 7.655 | 696.6 | 1.4374 | 160 |
| 180 | 13.30 | 709.6 | 1.5168 | 9.938 | 708.5 | 1.4820 | 7.921 | 707.5 | 1.4548 | 180 |
| 200 | 13.73 | 720.3 | 1.5334 | 10.27 | 719.4 | 1.4987 | 8.185 | 718.5 | 1.4716 | 200 |
| 220 | 14.16 | 731.1 | 1.5495 | 10.59 | 730.3 | 1.5150 | 8.448 | 729.4 | 1.4880 | 220 |
| 260 | 15.02 | 753.0 | 1.5808 | 11.24 | 752.3 | 1.5465 | 8.970 | 751.6 | 1.5197 | 260 |
| 300 | — | — | — | 11.88 | 774.6 | 1.5766 | 9.489 | 774.0 | 1.5500 | 300 |

| Temp. F | Pressure, psia (saturation temperature in parentheses) | | | | | | | | | Temp. F |
|---|---|---|---|---|---|---|---|---|---|---|
| | 60 (30.21 F) | | | 80 (44.40 F) | | | 100 (56.05 F) | | | |
| | *v* | *h* | *s* | *v* | *h* | *s* | *v* | *h* | *s* | |
| 40 | 4.933 | 626.8 | 1.2913 | — | — | — | — | — | — | 40 |
| 50 | 5.060 | 632.9 | 1.3035 | 3.712 | 627.7 | 1.2619 | — | — | — | 50 |
| 60 | 5.184 | 639.0 | 1.3152 | 3.812 | 634.3 | 1.2745 | 2.985 | 629.3 | 1.2409 | 60 |
| 70 | 5.307 | 644.9 | 1.3265 | 3.909 | 640.6 | 1.2866 | 3.068 | 636.0 | 1.2539 | 70 |
| 80 | 5.428 | 650.7 | 1.3373 | 4.005 | 646.7 | 1.2981 | 3.149 | 642.6 | 1.2661 | 80 |
| 90 | 5.547 | 656.4 | 1.3479 | 4.098 | 652.8 | 1.3092 | 3.227 | 649.0 | 1.2778 | 90 |
| 100 | 5.665 | 662.1 | 1.3581 | 4.190 | 658.7 | 1.3199 | 3.304 | 655.2 | 1.2891 | 100 |
| 110 | 5.781 | 667.7 | 1.3681 | 4.281 | 664.6 | 1.3303 | 3.380 | 661.3 | 1.2999 | 110 |
| 120 | 5.897 | 673.3 | 1.3778 | 4.371 | 670.4 | 1.3404 | 3.454 | 667.3 | 1.3104 | 120 |
| 130 | 6.012 | 678.9 | 1.3873 | 4.460 | 676.1 | 1.3502 | 3.527 | 673.3 | 1.3206 | 130 |
| 140 | 6.126 | 684.4 | 1.3966 | 4.548 | 681.8 | 1.3598 | 3.600 | 679.2 | 1.3305 | 140 |
| 150 | 6.239 | 689.9 | 1.4058 | 4.635 | 687.5 | 1.3692 | 3.672 | 685.0 | 1.3401 | 150 |
| 160 | 6.352 | 695.5 | 1.4148 | 4.722 | 693.2 | 1.3784 | 3.743 | 690.8 | 1.3495 | 160 |
| 180 | 6.576 | 706.5 | 1.4323 | 4.893 | 704.4 | 1.3963 | 3.883 | 702.3 | 1.3678 | 180 |
| 200 | 6.798 | 717.5 | 1.4493 | 5.063 | 715.6 | 1.4136 | 4.021 | 713.7 | 1.3854 | 200 |
| 220 | 7.019 | 728.6 | 1.4658 | 5.231 | 726.9 | 1.4304 | 4.158 | 725.1 | 1.4024 | 220 |
| 240 | 7.238 | 739.7 | 1.4819 | 5.398 | 738.1 | 1.4467 | 4.294 | 736.5 | 1.4190 | 240 |
| 260 | 7.457 | 750.9 | 1.4976 | 5.565 | 749.4 | 1.4626 | 4.428 | 747.9 | 1.4350 | 260 |
| 280 | 7.675 | 762.1 | 1.5130 | 5.730 | 760.7 | 1.4781 | 4.562 | 759.4 | 1.4507 | 280 |
| 300 | 7.892 | 773.3 | 1.5281 | 5.894 | 772.1 | 1.4933 | 4.695 | 770.8 | 1.4660 | 300 |

*Continued on next page*

**Table D-3**—*Continued*

| Temp. F | Pressure, psia (saturation temperature in parentheses) 120 (66.02 F) | | | 140 (74.79 F) | | | 170 (86.29 F) | | | Temp. F |
|---|---|---|---|---|---|---|---|---|---|---|
| | *v* | *h* | *s* | *v* | *h* | *s* | *v* | *h* | *s* | |
| 70 | 2.505 | 631.3 | 1.2255 | — | — | — | — | — | — | 70 |
| 80 | 2.576 | 638.3 | 1.2386 | 2.166 | 633.8 | 1.2140 | — | — | — | 80 |
| 90 | 2.645 | 645.0 | 1.2510 | 2.228 | 640.9 | 1.2272 | 1.784 | 634.4 | 1.1952 | 90 |
| 100 | 2.712 | 651.6 | 1.2628 | 2.288 | 647.8 | 1.2396 | 1.837 | 641.9 | 1.2087 | 100 |
| 110 | 2.778 | 658.0 | 1.2741 | 2.347 | 654.5 | 1.2515 | 1.889 | 649.1 | 1.2215 | 110 |
| 120 | 2.842 | 664.2 | 1.2850 | 2.404 | 661.1 | 1.2628 | 1.939 | 656.1 | 1.2336 | 120 |
| 130 | 2.905 | 670.4 | 1.2956 | 2.460 | 667.4 | 1.2738 | 1.988 | 662.8 | 1.2452 | 130 |
| 140 | 2.967 | 676.4 | 1.3058 | 2.515 | 673.7 | 1.2843 | 2.035 | 669.5 | 1.2563 | 140 |
| 150 | 3.029 | 682.5 | 1.3157 | 2.569 | 679.9 | 1.2945 | 2.081 | 675.9 | 1.2669 | 150 |
| 160 | 3.089 | 688.4 | 1.3254 | 2.622 | 686.0 | 1.3045 | 2.217 | 682.3 | 1.2773 | 160 |
| 170 | 3.149 | 694.3 | 1.3348 | 2.675 | 692.0 | 1.3141 | 2.172 | 688.5 | 1.2873 | 170 |
| 180 | 3.209 | 700.2 | 1.3441 | 2.727 | 698.0 | 1.3236 | 2.216 | 694.7 | 1.2971 | 180 |
| 200 | 3.326 | 711.8 | 1.3620 | 2.830 | 709.9 | 1.3418 | 2.303 | 706.9 | 1.3159 | 200 |
| 220 | 3.442 | 723.4 | 1.3793 | 2.931 | 721.6 | 1.3594 | 2.389 | 719.0 | 1.3338 | 220 |
| 240 | 3.557 | 734.9 | 1.3960 | 3.030 | 733.3 | 1.3763 | 2.473 | 730.9 | 1.3512 | 240 |
| 260 | 3.671 | 746.5 | 1.4123 | 3.129 | 745.0 | 1.3928 | 2.555 | 742.8 | 1.3679 | 260 |
| 280 | 3.783 | 758.0 | 1.4281 | 3.227 | 756.7 | 1.4099 | 2.637 | 754.6 | 1.3841 | 280 |
| 300 | 3.89 | 769.6 | 1.4435 | 3.323 | 768.3 | 1.4243 | 2.718 | 766.4 | 1.3999 | 300 |
| 320 | — | — | — | 3.420 | 780.0 | 1.4395 | 2.798 | 778.3 | 1.4153 | 320 |
| 340 | — | — | — | — | — | — | 2.878 | 790.1 | 1.4303 | 340 |

| Temp. F | Pressure, psia (saturation temperature in parentheses) 200 (96.34 F) | | | 240 (108.09 F) | | | 280 (118.45 F) | | | Temp. F |
|---|---|---|---|---|---|---|---|---|---|---|
| | *v* | *h* | *s* | *v* | *h* | *s* | *v* | *h* | *s* | |
| 100 | 1.520 | 635.6 | 1.1809 | — | — | — | — | — | — | 100 |
| 110 | 1.567 | 643.4 | 1.1947 | 1.261 | 635.5 | 1.1621 | — | — | — | 110 |
| 120 | 1.612 | 650.9 | 1.2077 | 1.302 | 643.5 | 1.1764 | 1.078 | 635.4 | 1.1473 | 120 |
| 130 | 1.656 | 658.1 | 1.2200 | 1.342 | 651.3 | 1.1898 | 1.115 | 644.0 | 1.1621 | 130 |
| 140 | 1.698 | 665.0 | 1.2317 | 1.380 | 658.8 | 1.2025 | 1.151 | 652.2 | 1.1759 | 140 |
| 150 | 1.740 | 671.8 | 1.242 | 1.416 | 666.1 | 1.2145 | 1.184 | 660.1 | 1.1888 | 150 |
| 160 | 1.780 | 678.4 | 1.2537 | 1.452 | 673.1 | 1.2259 | 1.217 | 667.6 | 1.2011 | 160 |
| 170 | 1.820 | 684.9 | 1.2641 | 1.487 | 680.0 | 1.2369 | 1.249 | 674.9 | 1.2127 | 170 |
| 180 | 1.859 | 691.3 | 1.2742 | 1.521 | 686.7 | 1.2475 | 1.279 | 681.9 | 1.2239 | 180 |
| 200 | 1.935 | 703.9 | 1.2935 | 1.587 | 699.8 | 1.2677 | 1.339 | 695.6 | 1.2449 | 200 |
| 220 | 2.009 | 716.3 | 1.3120 | 1.651 | 712.6 | 1.2867 | 1.396 | 708.8 | 1.2647 | 220 |
| 240 | 2.082 | 728.4 | 1.3296 | 1.714 | 725.1 | 1.3049 | 1.451 | 721.8 | 1.2834 | 240 |
| 260 | 2.154 | 740.5 | 1.3467 | 1.775 | 737.5 | 1.3224 | 1.505 | 734.4 | 1.3013 | 260 |
| 280 | 2.225 | 752.5 | 1.3631 | 1.835 | 749.8 | 1.3392 | 1.558 | 747.0 | 1.3184 | 280 |
| 300 | 2.295 | 764.5 | 1.3791 | 1.894 | 762.0 | 1.3554 | 1.610 | 759.4 | 1.3350 | 300 |
| 320 | 2.364 | 776.5 | 1.3947 | 1.954 | 774.1 | 1.3712 | 1.661 | 771.7 | 1.3511 | 320 |
| 340 | 2.432 | 788.5 | 1.4099 | 2.012 | 786.3 | 1.3866 | 1.712 | 784.0 | 1.3667 | 340 |
| 360 | 2.500 | 800.5 | 1.4247 | 2.069 | 798.4 | 1.4016 | 1.762 | 796.3 | 1.3819 | 360 |
| 380 | 2.568 | 812.5 | 1.4392 | 2.126 | 810.6 | 1.4163 | 1.811 | 808.7 | 1.3967 | 380 |
| 400 | — | — | — | — | — | — | 1.861 | 821.0 | 1.4112 | 400 |

# Appendix E

## Thermodynamic Properties of Freon-12 (F-12) Dichlorodifluoromethane English Units

**Table E–1** Saturation Pressure-Saturation Temperature, Freon-12

| Pressure, psia | Temperature, F | Pressure, psia | Temperature, F |
|---|---|---|---|
| 0.20 | −144.9 | 90.00 | 73.79 |
| 0.40 | −130.7 | 100.00 | 80.76 |
| 0.60 | −121.6 | 120.00 | 93.29 |
| 0.80 | −114.8 | 140.00 | 104.35 |
| 1.00 | −109.2 | 160.00 | 114.30 |
| 2.00 | −90.71 | 180.00 | 123.38 |
| 3.00 | −78.76 | 200.00 | 131.74 |
| 4.00 | −69.72 | 220.00 | 139.51 |
| 5.00 | −62.35 | 240.00 | 146.77 |
| 6.00 | −56.08 | 260.00 | 153.60 |
| 8.00 | −45.71 | 280.00 | 160.06 |
| 10.00 | −37.23 | 300.00 | 166.18 |
| 14.696 | −21.62 | 320.00 | 172.01 |
| 15.00 | −20.75 | 340.00 | 177.57 |
| 20.00 | −8.13 | 360.00 | 182.90 |
| 25.00 | 2.23 | 380.00 | 188.02 |
| 30.00 | 11.11 | 400.00 | 192.93 |
| 35.00 | 18.92 | 420.00 | 197.67 |
| 40.00 | 25.93 | 440.00 | 202.25 |
| 50.00 | 38.15 | 460.00 | 206.67 |
| 60.00 | 48.64 | 500.00 | 215.10 |
| 70.00 | 57.90 | 550.00 | 224.95 |
| 80.00 | 66.21 | | |

Critical state 596.9 psia, 233.6 F

Tables in Appendix E courtesy of E. I. DuPont de Nemours & Company

**Table E–2** Saturation State Properties, Freon-12

| Temperature, F | Pressure, psia | Specific Volume, ft³/lbm $v_f$ | $v_g$ | Enthalpy, B/lbm $h_f$ | $h_g$ | Entropy, B/lbm R $s_f$ | $s_g$ | Temperature, F |
|---|---|---|---|---|---|---|---|---|
| −150 | 0.15359 | 0.0095822 | 178.65 | −22.697 | 60.837 | −0.062619 | 0.20711 | −150 |
| −140 | 0.25623 | 0.0096579 | 110.46 | −20.652 | 61.896 | −0.056123 | 0.20208 | −140 |
| −130 | 0.41224 | 0.0097359 | 70.730 | −18.609 | 62.968 | −0.049830 | 0.19760 | −130 |
| −120 | 0.64190 | 0.0098163 | 46.741 | −16.565 | 64.052 | −0.043723 | 0.19359 | −120 |
| −110 | 0.97034 | 0.0098992 | 31.777 | −14.518 | 65.145 | −0.037786 | 0.19002 | −110 |
| −100 | 1.4280 | 0.0099847 | 22.164 | −12.466 | 66.248 | −0.032005 | 0.18683 | −100 |
| −90 | 2.0509 | 0.010073 | 15.821 | −10.409 | 67.355 | −0.026367 | 0.18398 | −90 |
| −80 | 2.8807 | 0.010164 | 11.533 | −8.3451 | 68.467 | −0.020862 | 0.18143 | −80 |
| −70 | 3.9651 | 0.010259 | 8.5687 | −6.2730 | 69.580 | −0.015481 | 0.17916 | −70 |
| −60 | 5.3575 | 0.010357 | 6.4774 | −4.1919 | 70.693 | −0.010214 | 0.17714 | −60 |
| −50 | 7.1168 | 0.010459 | 4.9742 | −2.1011 | 71.805 | −0.005056 | 0.17533 | −50 |
| −40 | 9.3076 | 0.010564 | 3.8750 | 0 | 72.913 | 0 | 0.17373 | −40 |
| −30 | 11.999 | 0.010674 | 3.0585 | 2.1120 | 74.015 | 0.004961 | 0.17229 | −30 |
| −25 | 13.556 | 0.010730 | 2.7295 | 3.1724 | 74.563 | 0.007407 | 0.17164 | −25 |
| −20 | 15.267 | 0.010788 | 2.4429 | 4.2357 | 75.110 | 0.009831 | 0.17102 | −20 |
| −15 | 17.141 | 0.010846 | 2.1924 | 5.3020 | 75.654 | 0.012234 | 0.17043 | −15 |
| −10 | 19.189 | 0.010906 | 1.9727 | 6.3716 | 76.169 | 0.014617 | 0.16989 | −10 |
| −5 | 21.422 | 0.010968 | 1.7794 | 7.4444 | 76.735 | 0.016979 | 0.16937 | −5 |
| 0 | 23.849 | 0.011030 | 1.6089 | 8.5207 | 77.271 | 0.019323 | 0.16888 | 0 |
| 5 | 26.483 | 0.011094 | 1.4580 | 9.6005 | 77.805 | 0.021647 | 0.16842 | 5 |
| 10 | 29.335 | 0.011160 | 1.3241 | 10.684 | 78.335 | 0.023954 | 0.16798 | 10 |
| 15 | 32.415 | 0.011227 | 1.2050 | 11.771 | 78.861 | 0.026243 | 0.16758 | 15 |
| 20 | 35.736 | 0.011296 | 1.0988 | 12.863 | 79.385 | 0.028515 | 0.16719 | 20 |
| 25 | 39.310 | 0.011366 | 1.0039 | 13.958 | 79.904 | 0.030772 | 0.16683 | 25 |
| 30 | 43.148 | 0.011438 | 0.91880 | 15.058 | 80.419 | 0.033013 | 0.16648 | 30 |
| 35 | 47.263 | 0.011512 | 0.84237 | 16.163 | 80.930 | 0.035240 | 0.16616 | 35 |
| 40 | 51.667 | 0.011588 | 0.77357 | 17.273 | 81.436 | 0.037453 | 0.16586 | 40 |
| 45 | 56.373 | 0.011666 | 0.71149 | 18.387 | 81.937 | 0.039652 | 0.16557 | 45 |
| 50 | 61.394 | 0.011746 | 0.65537 | 19.507 | 82.433 | 0.041839 | 0.16530 | 50 |
| 60 | 72.433 | 0.011913 | 0.55839 | 21.766 | 83.409 | 0.046180 | 0.16479 | 60 |
| 70 | 84.888 | 0.012089 | 0.47818 | 24.050 | 84.359 | 0.050482 | 0.16434 | 70 |
| 80 | 98.870 | 0.012277 | 0.41135 | 26.365 | 85.282 | 0.054751 | 0.16392 | 80 |
| 90 | 114.49 | 0.012478 | 0.35529 | 28.713 | 86.174 | 0.058997 | 0.16353 | 90 |
| 100 | 131.86 | 0.012693 | 0.30794 | 31.100 | 87.029 | 0.063227 | 0.16315 | 100 |
| 110 | 151.86 | 0.012924 | 0.26769 | 33.531 | 87.844 | 0.067451 | 0.16279 | 110 |
| 120 | 172.35 | 0.013174 | 0.23326 | 36.013 | 88.610 | 0.071680 | 0.16241 | 120 |
| 130 | 195.71 | 0.013447 | 0.20364 | 38.553 | 89.321 | 0.075927 | 0.16202 | 130 |
| 140 | 221.32 | 0.013746 | 0.17799 | 41.162 | 89.967 | 0.080205 | 0.16159 | 140 |
| 150 | 249.31 | 0.014078 | 0.15564 | 43.850 | 90.534 | 0.084531 | 0.16110 | 150 |
| 160 | 279.82 | 0.014449 | 0.13604 | 46.633 | 91.006 | 0.088927 | 0.16053 | 160 |
| 180 | 349.00 | 0.015360 | 0.10330 | 52.562 | 91.561 | 0.098039 | 0.15900 | 180 |
| 200 | 430.09 | 0.016659 | 0.076728 | 59.203 | 91.278 | 0.10789 | 0.15651 | 200 |
| 212 | 485.01 | 0.017830 | 0.062517 | 63.764 | 90.337 | 0.11448 | 0.15404 | 212 |
| 220 | 524.43 | 0.018986 | 0.053140 | 67.246 | 89.036 | 0.11943 | 0.15149 | 220 |
| 230 | 577.03 | 0.021854 | 0.039435 | 72.893 | 85.122 | 0.12739 | 0.14512 | 230 |
| 233.6* | 596.9 | 0.02870 | 0.02870 | 78.86 | 78.86 | 0.1359 | 0.1359 | 233.6* |

*Critical state

Freon and F-12 are registered trademarks of E. I. du Pont de Nemours & Co., Inc., Data for Appendix E were abstracted from Table T-12, Freon Products Division, E. I. du Pont de Nemours & Co., Inc., by permission.

**Table E-3** Superheated Vapor, Freon-12

$v$ = specific volume, ft³/lbm; $h$ = enthalpy, B/lbm; $s$ = entropy, B/lbm R

| | Pressure, psia (saturation temperature in parenthesis) | | | | | | | | | | | | |
|---|---|---|---|---|---|---|---|---|---|---|---|---|---|
| | 5 (−62.35 F) | | | 10 (37.23 F) | | | 15 (−20.75 F) | | | 20 (−8.13 F) | | | |
| Temp. F | $v$ | $h$ | $s$ | $v$ | $h$ | $s$ | $v$ | $h$ | $s$ | $v$ | $h$ | $s$ | Temp. F |
| 0 | 8.0611 | 78.582 | 0.19663 | 3.6945 | 78.246 | 0.18471 | 2.6201 | 77.902 | 0.17751 | 1.9390 | 77.550 | 0.17222 | 0 |
| 20 | 8.4265 | 81.309 | 0.20244 | 4.1691 | 81.014 | 0.19061 | 2.7494 | 80.712 | 0.18349 | 2.0391 | 80.403 | 0.17829 | 20 |
| 40 | 8.7903 | 84.090 | 0.20812 | 4.3556 | 83.828 | 0.19635 | 2.8770 | 83.561 | 0.18931 | 2.1373 | 83.289 | 0.18419 | 40 |
| 60 | 9.1528 | 86.922 | 0.21367 | 4.5408 | 86.689 | 0.20197 | 3.0031 | 86.451 | 0.19498 | 2.2340 | 86.210 | 0.18992 | 60 |
| 80 | 9.5142 | 89.806 | 0.21912 | 4.7248 | 89.596 | 0.20746 | 3.1281 | 89.383 | 0.20051 | 2.3295 | 89.168 | 0.19550 | 80 |
| 100 | 9.8747 | 92.738 | 0.22445 | 4.9079 | 92.548 | 0.21283 | 3.2521 | 92.357 | 0.20593 | 2.4241 | 92.164 | 0.20095 | 100 |
| 120 | 10.234 | 95.717 | 0.22968 | 5.0903 | 95.546 | 0.21809 | 3.3754 | 95.373 | 0.21122 | 2.5179 | 95.198 | 0.20628 | 120 |
| 140 | 10.594 | 98.743 | 0.23481 | 5.2720 | 98.586 | 0.22325 | 3.4981 | 98.429 | 0.21640 | 2.6110 | 98.270 | 0.21149 | 140 |
| 160 | 10.952 | 101.812 | 0.23985 | 5.4533 | 101.669 | 0.22830 | 3.6202 | 101.525 | 0.22148 | 2.7036 | 101.380 | 0.21659 | 160 |
| 180 | 11.311 | 104.925 | 0.24479 | 5.6341 | 104.793 | 0.23326 | 3.7419 | 104.661 | 0.22646 | 2.7957 | 104.528 | 0.22159 | 180 |
| 200 | 11.668 | 108.079 | 0.24964 | 5.8145 | 107.957 | 0.23813 | 3.8632 | 107.835 | 0.23135 | 2.8874 | 107.712 | 0.22649 | 200 |
| 220 | 12.026 | 111.272 | 0.25441 | 5.9946 | 111.159 | 0.24291 | 3.9841 | 111.046 | 0.23614 | 2.9789 | 110.932 | 0.23130 | 220 |
| 240 | — | — | — | 6.1745 | 114.398 | 0.24761 | 4.1049 | 114.292 | 0.24085 | 3.0700 | 114.186 | 0.23602 | 240 |
| 260 | — | — | — | — | — | — | 4.2254 | 117.574 | 0.24547 | 3.1609 | 117.475 | 0.24065 | 260 |
| 280 | — | — | — | — | — | — | — | — | — | 3.2517 | 120.796 | 0.24520 | 280 |

| | Pressure, psia (saturation temperature in parenthesis) | | | | | | | | | | | | |
|---|---|---|---|---|---|---|---|---|---|---|---|---|---|
| | 25 (2.23 F) | | | 30 (11.11 F) | | | 35 (18.92 F) | | | 40 (25.93 F) | | | |
| Temp. F | $v$ | $h$ | $s$ | $v$ | $h$ | $s$ | $v$ | $h$ | $s$ | $v$ | $h$ | $s$ | Temp. F |
| 20 | 1.6125 | 80.088 | 0.17414 | 1.3278 | 79.765 | 0.17065 | 1.1240 | 79.434 | 0.16761 | — | — | — | 20 |
| 40 | 1.6932 | 83.012 | 0.18012 | 1.3969 | 82.730 | 0.17671 | 1.1850 | 82.442 | 0.17375 | 1.0258 | 82.148 | 0.17112 | 40 |
| 60 | 1.7723 | 85.965 | 0.18591 | 1.4644 | 85.716 | 0.18257 | 1.2442 | 85.463 | 0.17968 | 1.0789 | 85.206 | 0.17712 | 60 |
| 80 | 1.8502 | 88.950 | 0.19155 | 1.5306 | 88.729 | 0.18826 | 1.3021 | 88.504 | 0.18542 | 1.1306 | 88.277 | 0.18292 | 80 |
| 100 | 1.9271 | 91.968 | 0.19704 | 1.5957 | 91.770 | 0.19379 | 1.3589 | 91.570 | 0.19100 | 1.1812 | 91.367 | 0.18854 | 100 |
| 120 | 2.0032 | 95.021 | 0.20240 | 1.6600 | 94.843 | 0.19918 | 1.4148 | 94.663 | 0.19643 | 1.2309 | 94.480 | 0.19401 | 120 |
| 140 | 2.0786 | 98.110 | 0.20763 | 1.7237 | 97.948 | 0.20445 | 1.4701 | 97.785 | 0.20172 | 1.2798 | 97.620 | 0.19933 | 140 |
| 160 | 2.1535 | 101.234 | 0.21276 | 1.7868 | 101.086 | 0.20960 | 1.5248 | 100.938 | 0.20689 | 1.3282 | 100.788 | 0.20453 | 160 |
| 180 | 2.2279 | 104.393 | 0.21778 | 1.8494 | 104.258 | 0.21463 | 1.5789 | 104.122 | 0.21195 | 1.3761 | 103.985 | 0.20961 | 180 |
| 200 | 2.3019 | 107.588 | 0.22269 | 1.9116 | 107.464 | 0.21957 | 1.6327 | 107.338 | 0.21690 | 1.4236 | 107.212 | 0.21457 | 200 |
| 220 | 2.3756 | 110.817 | 0.22752 | 1.9735 | 110.702 | 0.22440 | 1.6832 | 110.586 | 0.22175 | 1.4707 | 110.469 | 0.21944 | 220 |
| 240 | 2.4491 | 114.080 | 0.23225 | 2.0351 | 113.973 | 0.22915 | 1.7394 | 113.865 | 0.22651 | 1.5176 | 113.757 | 0.22420 | 240 |
| 260 | 2.52233 | 117.375 | 0.23689 | 2.0965 | 117.275 | 0.23380 | 1.7923 | 117.175 | 0.23117 | 1.5642 | 117.074 | 0.22888 | 260 |
| 280 | 2.5953 | 120.703 | 0.24145 | 2.1576 | 120.609 | 0.23837 | 1.8450 | 120.515 | 0.23575 | 1.6106 | 120.421 | 0.23347 | 280 |
| 300 | 2.6681 | 124.061 | 0.24593 | 2.2186 | 123.973 | 0.24286 | 1.8976 | 123.885 | 0.24024 | 1.6568 | 123.796 | 0.23797 | 300 |

| | Pressure, psia (saturation temperatures in parentheses) | | | | | | | | | | | | |
|---|---|---|---|---|---|---|---|---|---|---|---|---|---|
| | 50 (38.15 F) | | | 60 (48.64 F) | | | 70 (57.90 F) | | | 80 (66.21 F) | | | |
| Temp. F | $v$ | $h$ | $s$ | $v$ | $h$ | $s$ | $v$ | $h$ | $s$ | $v$ | $h$ | $s$ | Temp. F |
| 40 | 0.80248 | 81.540 | 0.16655 | — | — | — | — | — | — | — | — | — | 40 |
| 60 | 0.84713 | 84.676 | 0.17271 | 0.69210 | 84.126 | 0.16892 | 0.58088 | 83.552 | 0.16556 | — | — | — | 60 |
| 80 | 0.89025 | 87.811 | 0.17862 | 0.72964 | 87.330 | 0.17497 | 0.61458 | 86.832 | 0.17175 | 0.52795 | 86.316 | 0.16885 | 80 |
| 100 | 0.93216 | 90.953 | 0.18434 | 0.76588 | 90.528 | 0.18079 | 0.64685 | 90.091 | 0.17768 | 0.55734 | 89.640 | 0.17489 | 100 |
| 120 | 0.97313 | 94.110 | 0.18988 | 0.80110 | 93.731 | 0.18641 | 0.67803 | 93.343 | 0.18339 | 0.58556 | 92.945 | 0.18070 | 120 |
| 140 | 1.0133 | 97.286 | 0.19527 | 0.83551 | 96.945 | 0.19186 | 0.70836 | 96.597 | 0.18891 | 0.61286 | 96.242 | 0.18629 | 140 |
| 160 | 1.0529 | 100.485 | 0.20051 | 0.86928 | 100.176 | 0.19716 | 0.73800 | 99.862 | 0.19427 | 0.63943 | 99.542 | 0.19170 | 160 |
| 180 | 1.0920 | 103.708 | 0.20563 | 0.90252 | 103.427 | 0.20233 | 0.76708 | 103.141 | 0.19948 | 0.66543 | 102.851 | 0.19696 | 180 |
| 200 | 1.1307 | 106.958 | 0.21064 | 0.93531 | 106.700 | 0.20736 | 0.79571 | 106.439 | 0.20455 | 0.69095 | 106.174 | 0.20207 | 200 |
| 220 | 1.1690 | 110.235 | 0.21553 | 0.96775 | 109.997 | 0.21229 | 0.82397 | 109.756 | 0.20951 | 0.71609 | 109.513 | 0.20706 | 220 |
| 240 | 1.2070 | 113.539 | 0.22032 | 0.99988 | 113.319 | 0.21710 | 0.85191 | 113.096 | 0.21435 | 0.74090 | 112.872 | 0.21193 | 240 |
| 260 | 1.2447 | 116.871 | 0.22502 | 1.0318 | 116.666 | 0.22182 | 0.87959 | 116.459 | 0.21909 | 0.76544 | 116.251 | 0.21669 | 260 |
| 280 | 1.2823 | 120.231 | 0.22962 | 1.0634 | 120.039 | 0.22644 | 0.90705 | 119.846 | 0.22373 | 0.78975 | 119.652 | 0.22135 | 280 |
| 320 | 1.3569 | 127.032 | 0.23857 | 1.1262 | 126.863 | 0.23543 | 0.96142 | 126.693 | 0.23274 | 0.83781 | 126.521 | 0.23039 | 320 |
| 360 | — | — | — | — | — | — | — | — | — | 0.88529 | 133.482 | 0.23910 | 360 |

*Continued on next page*

## Table E-3—*Continued*

| Temp. F | Pressure, psia (saturation temperatures in parentheses) 90 (73.79 F) | | | 100 (80.76 F) | | | 125 (96.17 F) | | | 150 (109.45 F) | | | Temp. F |
|---|---|---|---|---|---|---|---|---|---|---|---|---|---|
| | *v* | *h* | *s* | *v* | *h* | *s* | *v* | *h* | *s* | *v* | *h* | *s* | |
| 100 | 0.48749 | 89.175 | 0.17234 | 0.43138 | 88.694 | 0.16996 | 0.32943 | 87.407 | 0.16455 | — | — | — | 100 |
| 120 | 0.51346 | 92.536 | 0.17824 | 0.45562 | 92.116 | 0.17597 | 0.35086 | 91.008 | 0.17087 | 0.28007 | 89.800 | 0.16629 | 120 |
| 140 | 0.53845 | 95.879 | 0.18391 | 0.47881 | 95.507 | 0.18172 | 0.37098 | 94.537 | 0.17686 | 0.29845 | 93.498 | 0.17256 | 140 |
| 160 | 0.56268 | 99.216 | 0.18938 | 0.50118 | 98.884 | 0.18726 | 0.39015 | 98.023 | 0.18258 | 0.31566 | 97.112 | 0.17849 | 160 |
| 180 | 0.58629 | 102.557 | 0.19469 | 0.52291 | 102.257 | 0.19262 | 0.40857 | 101.484 | 0.18807 | 0.33200 | 100.675 | 0.18415 | 180 |
| 200 | 0.60941 | 105.905 | 0.19984 | 0.54413 | 105.633 | 0.19782 | 0.42642 | 104.934 | 0.19338 | 0.34769 | 104.206 | 0.18958 | 200 |
| 220 | 0.63213 | 109.267 | 0.20486 | 0.56492 | 109.018 | 0.20287 | 0.44380 | 108.380 | 0.19835 | 0.36285 | 107.720 | 0.19483 | 220 |
| 240 | 0.65451 | 112.644 | 0.20976 | 0.58538 | 112.415 | 0.20780 | 0.46081 | 111.829 | 0.20353 | 0.37761 | 111.226 | 0.19992 | 240 |
| 260 | 0.67662 | 116.040 | 0.21455 | 0.60554 | 115.828 | 0.21261 | 0.47750 | 115.287 | 0.20840 | 0.39203 | 114.732 | 0.20485 | 260 |
| 280 | 0.69849 | 119.456 | 0.21923 | 0.62546 | 119.258 | 0.21731 | 0.49394 | 118.756 | 0.21316 | 0.40617 | 118.242 | 0.20967 | 280 |
| 300 | 0.72016 | 122.892 | 0.22381 | 0.64518 | 122.707 | 0.22191 | 0.51016 | 122.238 | 0.21780 | 0.42008 | 121.761 | 0.21436 | 300 |
| 320 | 0.74166 | 126.349 | 0.22830 | 0.66472 | 126.176 | 0.22641 | 0.52619 | 125.737 | 0.22235 | 0.43379 | 125.290 | 0.21894 | 320 |
| 340 | 0.76301 | 129.828 | 0.23271 | 0.68411 | 129.665 | 0.23083 | 0.54207 | 129.252 | 0.22680 | 0.44753 | 128.833 | 0.22343 | 340 |
| 360 | 0.78423 | 133.329 | 0.23703 | 0.70338 | 133.174 | 0.23517 | 0.55781 | 132.785 | 0.23116 | 0.46074 | 132.390 | 0.22782 | 360 |
| 400 | — | — | — | — | — | — | — | — | — | 0.48719 | 139.552 | 0.23635 | 400 |

| Temp. F | Pressure, psia (saturation temperatures in parentheses) 175 (121.18 F) | | | 200 (131.74 F) | | | 250 (150.24 F) | | | 300 (166.18 F) | | | Temp. F |
|---|---|---|---|---|---|---|---|---|---|---|---|---|---|
| | *v* | *h* | *s* | *v* | *h* | *s* | *v* | *h* | *s* | *v* | *h* | *s* | |
| 140 | 0.24595 | 92.373 | 0.16859 | 0.20579 | 91.137 | 0.16180 | — | — | — | — | — | — | 140 |
| 160 | 0.26198 | 96.142 | 0.17478 | 0.22121 | 95.100 | 0.17130 | 0.16249 | 92.717 | 0.16462 | — | — | — | 160 |
| 180 | 0.27697 | 99.823 | 0.18062 | 0.23535 | 98.921 | 0.17737 | 0.17605 | 96.925 | 0.17130 | 0.13482 | 94.556 | 0.16537 | 180 |
| 200 | 0.29120 | 103.447 | 0.18620 | 0.24860 | 102.652 | 0.18311 | 0.18824 | 100.930 | 0.17747 | 0.14697 | 98.975 | 0.17217 | 200 |
| 220 | 0.30485 | 107.036 | 0.19156 | 0.26117 | 106.325 | 0.18860 | 0.19952 | 104.809 | 0.18326 | 0.15774 | 103.136 | 0.17838 | 220 |
| 240 | 0.31804 | 110.605 | 0.19674 | 0.27323 | 109.962 | 0.19387 | 0.21014 | 108.607 | 0.18877 | 0.16761 | 107.140 | 0.18419 | 240 |
| 260 | 0.33087 | 114.162 | 0.20175 | 0.28489 | 113.576 | 0.19896 | 0.22027 | 112.351 | 0.19404 | 0.17685 | 111.043 | 0.18969 | 260 |
| 280 | 0.34339 | 117.717 | 0.20662 | 0.29623 | 117.178 | 0.20390 | 0.23001 | 116.060 | 0.19913 | 0.18562 | 114.879 | 0.19495 | 280 |
| 300 | 0.35567 | 121.273 | 0.21137 | 0.30730 | 120.775 | 0.20870 | 0.23944 | 119.747 | 0.20405 | 0.19402 | 118.670 | 0.20000 | 300 |
| 320 | 0.36773 | 124.835 | 0.21599 | 0.31815 | 124.373 | 0.21337 | 0.24862 | 123.420 | 0.20882 | 0.20214 | 122.430 | 0.20489 | 320 |
| 340 | 0.37963 | 128.407 | 0.22052 | 0.32881 | 127.974 | 0.21793 | 0.25759 | 127.088 | 0.21346 | 0.21002 | 126.171 | 0.20963 | 340 |
| 360 | 0.39137 | 131.989 | 0.22494 | 0.33932 | 131.583 | 0.22239 | 0.26639 | 130.754 | 0.21799 | 0.21770 | 129.900 | 0.21423 | 360 |
| 380 | 0.40298 | 135.585 | 0.22927 | 0.34969 | 135.202 | 0.22675 | 0.27504 | 134.423 | 0.22241 | 0.22522 | 133.624 | 0.21872 | 380 |
| 400 | 0.41448 | 139.194 | 0.23352 | 0.35994 | 138.832 | 0.23102 | 0.28356 | 138.097 | 0.22674 | 0.23260 | 137.346 | 0.22310 | 400 |
| 420 | — | — | — | 0.37010 | 142.475 | 0.23521 | 0.29197 | 141.780 | 0.23097 | 0.23987 | 141.071 | 0.22739 | 420 |

| Temp. F | Pressure, psia (saturation temperatures in parentheses) 360 (182.90 F) | | | 420 (197.67 F) | | | 500 (215.10 F) | | | 600 (Super-critical) | | | Temp. F |
|---|---|---|---|---|---|---|---|---|---|---|---|---|---|
| | *v* | *h* | *s* | *v* | *h* | *s* | *v* | *h* | *s* | *v* | *h* | *s* | |
| 200 | 0.11081 | 96.143 | 0.16572 | 0.081569 | 92.191 | 0.15812 | — | — | — | — | — | — | 200 |
| 220 | 0.12189 | 100.838 | 0.17273 | 0.094795 | 98.037 | 0.16685 | 0.064207 | 92.399 | 0.15683 | — | — | — | 220 |
| 240 | 0.13152 | 105.189 | 0.17904 | 0.10490 | 102.949 | 0.17398 | 0.077620 | 99.218 | 0.16672 | 0.047488 | 91.024 | 0.15335 | 240 |
| 260 | 0.14023 | 109.340 | 0.18489 | 0.11356 | 107.452 | 0.18032 | 0.087054 | 104.526 | 0.17421 | 0.061922 | 99.741 | 0.16566 | 260 |
| 280 | 0.14833 | 113.365 | 0.19041 | 0.12135 | 111.722 | 0.18618 | 0.094923 | 109.277 | 0.18072 | 0.070859 | 105.637 | 0.17374 | 280 |
| 300 | 0.15596 | 117.304 | 0.19567 | 0.12855 | 115.846 | 0.19168 | 0.10190 | 113.729 | 0.18666 | 0.078059 | 110.729 | 0.18053 | 300 |
| 320 | 0.16325 | 121.185 | 0.20071 | 0.13532 | 119.872 | 0.19691 | 0.10829 | 117.997 | 0.19221 | 0.084333 | 115.420 | 0.18663 | 320 |
| 340 | 0.17026 | 125.026 | 0.20557 | 0.14176 | 123.830 | 0.20192 | 0.11426 | 122.143 | 0.19746 | 0.090017 | 119.871 | 0.19227 | 340 |
| 360 | 0.17705 | 128.840 | 0.21028 | 0.14795 | 127.740 | 0.20675 | 0.11992 | 126.205 | 0.20247 | 0.095289 | 124.167 | 0.19757 | 360 |
| 380 | 0.18366 | 132.637 | 0.21486 | 0.15392 | 131.618 | 0.21142 | 0.12533 | 130.207 | 0.20730 | 0.10025 | 128.355 | 0.20262 | 380 |
| 400 | 0.19011 | 136.422 | 0.21931 | 0.15973 | 135.473 | 0.21596 | 0.13054 | 134.166 | 0.21196 | 0.10498 | 132.466 | 0.20746 | 400 |
| 420 | 0.19643 | 140.202 | 0.22366 | 0.16540 | 139.313 | 0.22038 | 0.13559 | 138.096 | 0.21648 | 0.10952 | 136.523 | 0.21213 | 420 |
| 440 | 0.20265 | 143.980 | 0.22791 | 0.17094 | 143.144 | 0.22468 | 0.14051 | 142.004 | 0.22087 | 0.11391 | 140.539 | 0.21664 | 440 |
| 460 | 0.20876 | 147.760 | 0.23206 | 0.17638 | 145.970 | 0.22889 | 0.14531 | 145.898 | 0.22515 | 0.11816 | 144.526 | 0.22102 | 460 |
| 500 | — | — | — | — | — | — | 0.15464 | 153.662 | 0.23341 | 0.12637 | 152.444 | 0.22945 | 500 |

# Appendix F

## Psychrometric Charts

Reproduced with the permission of Carrier Corporation

Figure F.1

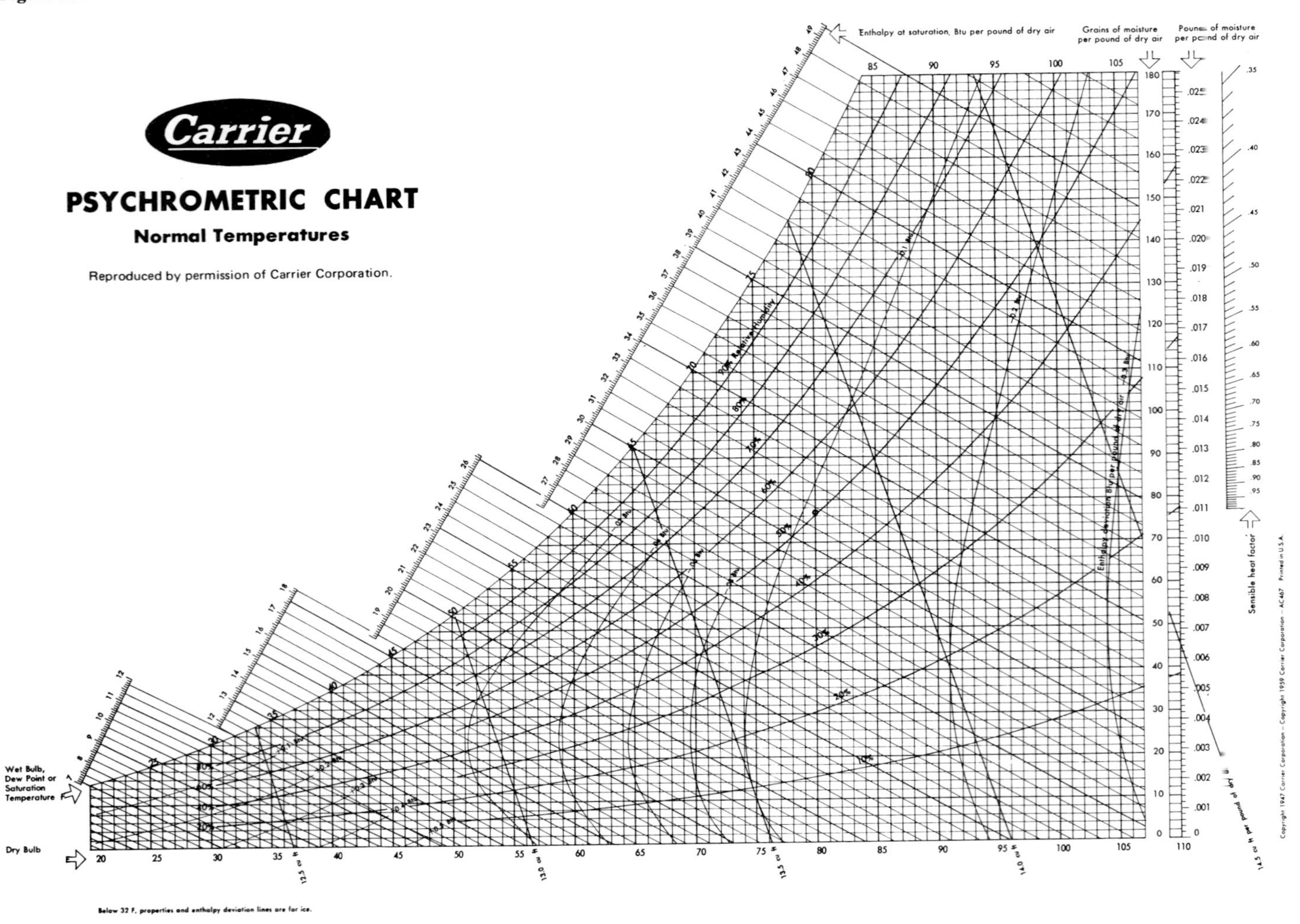
Carrier
PSYCHROMETRIC CHART
Normal Temperatures
Reproduced by permission of Carrier Corporation.
Wet Bulb, Dew Point or Saturation Temperature
Dry Bulb
Enthalpy at saturation, Btu per pound of dry air
Grains of moisture per pound of dry air
Pounds of moisture per pound of dry air
Sensible heat factor
Enthalpy deviation Btu per pound of dry air
Relative Humidity
Below 32 F, properties and enthalpy deviation lines are for ice.

Figure F.2

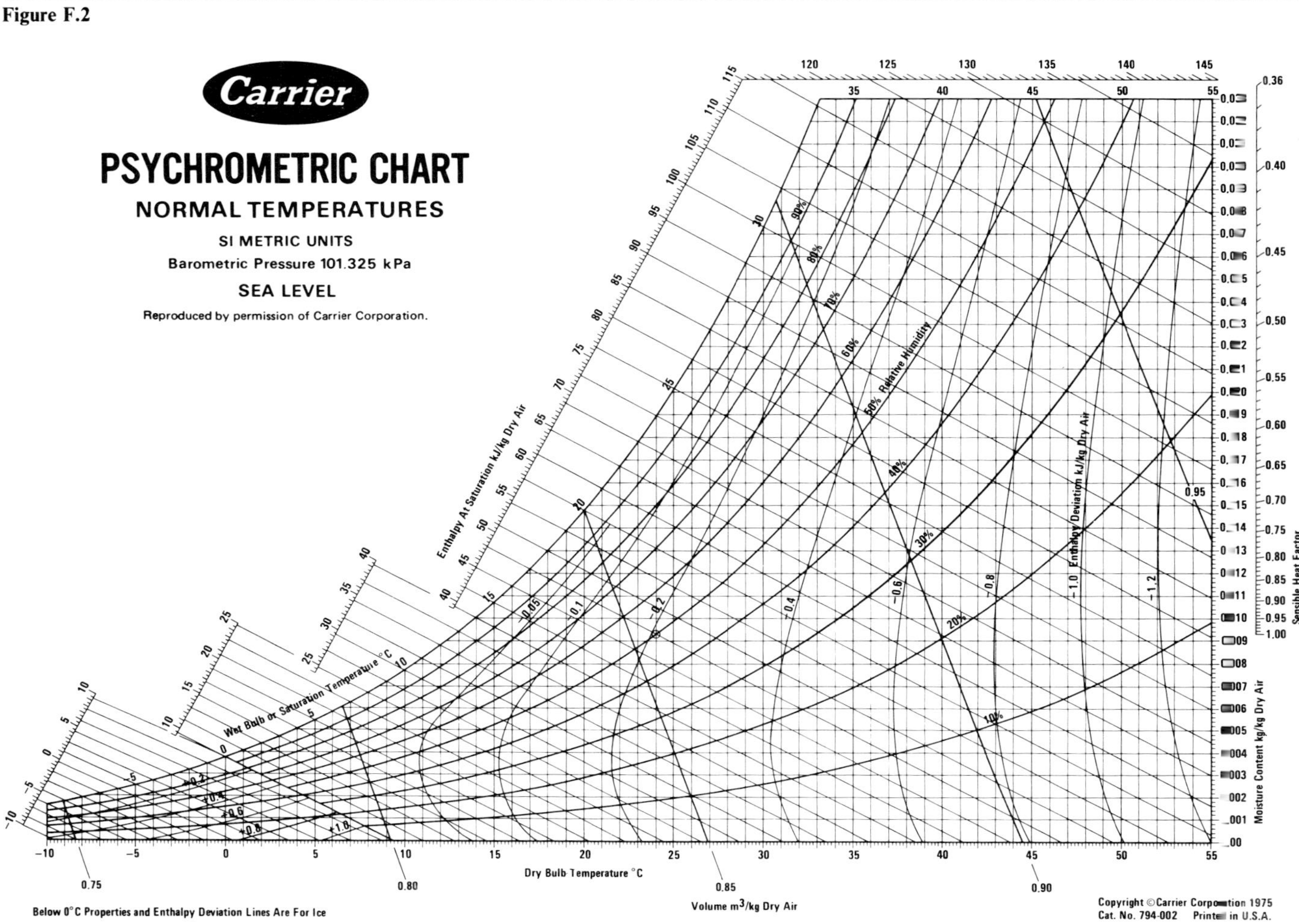
Carrier
PSYCHROMETRIC CHART
NORMAL TEMPERATURES
SI METRIC UNITS
Barometric Pressure 101.325 kPa
SEA LEVEL
Reproduced by permission of Carrier Corporation.
Dry Bulb Temperature °C
Volume m³/kg Dry Air
Enthalpy At Saturation kJ/kg Dry Air
Wet Bulb or Saturation Temperature °C
Relative Humidity
Enthalpy Deviation kJ/kg Dry Air
Moisture Content kg/kg Dry Air
Sensible Heat Factor
Below 0°C Properties and Enthalpy Deviation Lines Are For Ice
Copyright © Carrier Corporation 1975
Cat. No. 794-002 Printed in U.S.A.

# Appendix G

## Average Specific Heats for Several Gases

Average molal specific heats $\overline{C}_{p_a}$ between 537 R (298 K) and indicated temperature. See eq. (5.11).

| | | | | | |
|---|---|---|---|---|---|
| G–1. | Oxygen, $O_2$ | G–5. | Carbon monoxide, CO | G–9. | Ethane, $C_2H_6$ |
| G–2. | Nitrogen, $N_2$ | G–6. | Carbon dioxide, $CO_2$ | G–10. | Propane, $C_3H_8$ |
| G–3. | Hydrogen, $H_2$ | G–7. | Sulfur dioxide, $SO_2$ | G–11. | n-Butane, $C_4H_{10}$ |
| G–4. | Water vapor, $H_2O$ | G–8. | Methane, $CH_4$ | | |

Values for Tables G–1 through G–6 are calculated from equations in Sweigert, R. L., and M. W. Beardsley. Empirical Specific Heat Equations Based on Spectroscopic Data. State Engr. Exp. Sta. Bulletin No. 2, 1938. Table G–7 calculated values from Bryant, W. M. D. Empirical Molecular Heat Equations from Spectroscopic Data. *Ind. & Engr. Chem.*, vol. 25, p. 820, 1933. Tables G–8 through G–11 calculated from Fallon, J. F., & K. M. Watson. *Nat. Petrol. News,* Tech. Section, June 9, 1944. See 12.3.

**Table G–1** Average Molal Specific Heats for **Oxygen,** $O_2$, Between 537 R (298 K) and Listed Temperatures. Molecular Weight 32.00

| $T$, R | $\overline{C}_{p_a}$, $\frac{B}{lb\ mol\ R}$ | $T$, R | $\overline{C}_{p_a}$, $\frac{B}{lb\ mol\ R}$ |
|---|---|---|---|
| 500 | 6.912 | 2500 | 8.013 |
| 550 | 6.952 | 2600 | 8.046 |
| 600 | 6.993 | 2700 | 8.079 |
| 650 | 7.033 | 2800 | 8.110 |
| 700 | 7.072 | 2900 | 8.141 |
| 750 | 7.111 | 3000 | 8.170 |
| 800 | 7.149 | 3100 | 8.199 |
| 850 | 7.186 | 3200 | 8.227 |
| 900 | 7.222 | 3300 | 8.253 |
| 950 | 7.257 | 3400 | 8.279 |
| 1000 | 7.292 | 3500 | 8.305 |
| 1100 | 7.358 | 3600 | 8.329 |
| 1200 | 7.420 | 3700 | 8.353 |
| 1300 | 7.480 | 3800 | 8.377 |
| 1400 | 7.536 | 3900 | 8.399 |
| 1500 | 7.590 | 4000 | 8.422 |
| 1600 | 7.641 | 4100 | 8.443 |
| 1700 | 7.690 | 4200 | 8.464 |
| 1800 | 7.736 | 4300 | 8.485 |
| 1900 | 7.781 | 4400 | 8.505 |
| 2000 | 7.823 | 4500 | 8.524 |
| 2100 | 7.864 | 4600 | 8.543 |
| 2200 | 7.904 | 4700 | 8.562 |
| 2300 | 7.941 | 4800 | 8.580 |
| 2400 | 7.978 | 4900 | 8.598 |
| | | 5000 | 8.616 |

| $T$, K | $\overline{C}_{p_a}$, $\frac{kJ}{kmol\ K}$ | $T$, K | $\overline{C}_{p_a}$, $\frac{kJ}{kmol\ K}$ |
|---|---|---|---|
| 275 | 28.92 | 1450 | 33.70 |
| 300 | 29.07 | 1500 | 33.82 |
| 325 | 29.22 | 1550 | 33.94 |
| 350 | 29.37 | 1600 | 34.06 |
| 375 | 29.52 | 1650 | 34.17 |
| 400 | 29.67 | 1700 | 34.28 |
| 425 | 29.82 | 1750 | 34.38 |
| 450 | 29.96 | 1800 | 34.49 |
| 475 | 30.10 | 1850 | 34.59 |
| 500 | 30.23 | 1900 | 34.68 |
| 550 | 30.50 | 1950 | 34.78 |
| 600 | 30.75 | 2000 | 34.87 |
| 650 | 30.99 | 2050 | 34.96 |
| 700 | 31.22 | 2100 | 35.05 |
| 750 | 31.43 | 2150 | 35.14 |
| 800 | 31.64 | 2200 | 35.22 |
| 850 | 31.84 | 2250 | 35.30 |
| 900 | 32.03 | 2300 | 35.38 |
| 950 | 32.21 | 2350 | 35.46 |
| 1000 | 32.39 | 2400 | 35.54 |
| 1050 | 32.56 | 2450 | 35.61 |
| 1100 | 32.72 | 2500 | 35.69 |
| 1150 | 32.87 | 2550 | 35.76 |
| 1200 | 33.02 | 2600 | 35.83 |
| 1250 | 33.17 | 2650 | 35.90 |
| 1300 | 33.31 | 2700 | 35.97 |
| 1350 | 33.44 | 2750 | 36.03 |
| 1400 | 33.57 | | |

**Table G–2** Average Molal Specific Heats for **Nitrogen,** $N_2$, Between 537 R (298 K) and Listed Temperatures. Molecular Weight 28.01.

| $T$, R | $\overline{C}_{p_a}$, $\frac{\text{B}}{\text{lb mol R}}$ | $T$, R | $\overline{C}_{p_a}$, $\frac{\text{B}}{\text{lb mol R}}$ |
|---|---|---|---|
| 500 | 7.095 | 2500 | 7.632 |
| 550 | 7.013 | 2600 | 7.648 |
| 600 | 6.960 | 2700 | 7.679 |
| 650 | 6.929 | 2800 | 7.709 |
| 700 | 6.913 | 2900 | 7.738 |
| 750 | 6.908 | 3000 | 7.766 |
| 800 | 6.911 | 3100 | 7.793 |
| 850 | 6.920 | 3200 | 7.819 |
| 900 | 6.934 | 3300 | 7.844 |
| 950 | 6.951 | 3400 | 7.869 |
| 1000 | 6.970 | 3500 | 7.892 |
| 1100 | 7.014 | 3600 | 7.915 |
| 1200 | 7.062 | 3700 | 7.936 |
| 1300 | 7.111 | 3800 | 7.958 |
| 1400 | 7.160 | 3900 | 7.978 |
| 1500 | 7.209 | 4000 | 7.998 |
| 1600 | 7.256 | 4100 | 8.017 |
| 1700 | 7.302 | 4200 | 8.036 |
| 1800 | 7.347 | 4300 | 8.054 |
| 1900 | 7.390 | 4400 | 8.072 |
| 2000 | 7.431 | 4500 | 8.089 |
| 2100 | 7.471 | 4600 | 8.105 |
| 2200 | 7.509 | 4700 | 8.121 |
| 2300 | 7.546 | 4800 | 8.137 |
| 2400 | 7.581 | 4900 | 8.152 |
| | | 5000 | 8.167 |

| $T$, K | $\overline{C}_{p_a}$, $\frac{\text{kJ}}{\text{kmol K}}$ | $T$, K | $\overline{C}_{p_a}$, $\frac{\text{kJ}}{\text{kmol K}}$ |
|---|---|---|---|
| 275 | 29.75 | 1450 | 32.03 |
| 300 | 29.41 | 1500 | 32.15 |
| 325 | 29.19 | 1550 | 32.26 |
| 350 | 29.05 | 1600 | 32.37 |
| 375 | 28.97 | 1650 | 32.48 |
| 400 | 28.93 | 1700 | 32.58 |
| 425 | 28.92 | 1750 | 32.68 |
| 450 | 28.94 | 1800 | 32.78 |
| 475 | 28.97 | 1850 | 32.87 |
| 500 | 29.03 | 1900 | 32.96 |
| 550 | 29.16 | 1950 | 33.05 |
| 600 | 29.33 | 2000 | 33.13 |
| 650 | 29.50 | 2050 | 33.22 |
| 700 | 29.69 | 2100 | 33.30 |
| 750 | 29.87 | 2150 | 33.37 |
| 800 | 30.06 | 2200 | 33.45 |
| 850 | 30.24 | 2250 | 33.52 |
| 900 | 30.42 | 2300 | 33.59 |
| 950 | 30.59 | 2350 | 33.66 |
| 1000 | 30.76 | 2400 | 33.73 |
| 1050 | 30.92 | 2450 | 33.80 |
| 1100 | 31.08 | 2500 | 33.86 |
| 1150 | 31.23 | 2550 | 33.92 |
| 1200 | 31.37 | 2600 | 33.99 |
| 1250 | 31.51 | 2650 | 34.05 |
| 1300 | 31.65 | 2700 | 34.10 |
| 1350 | 31.78 | 2750 | 34.16 |
| 1400 | 31.91 | | |

**Table G–3** Average Molal Specific Heats for **Hydrogen,** $H_2$, Between 537 R (298 K) and Listed Temperatures. Molecular Weight 2.02.

| $T$, R | $\overline{C}_{p_a}$, $\frac{\text{B}}{\text{lb mol R}}$ | $T$, R | $\overline{C}_{p_a}$, $\frac{\text{B}}{\text{lb mol R}}$ |
|---|---|---|---|
| 500 | 6.938 | 2500 | 7.184 |
| 550 | 6.932 | 2600 | 7.206 |
| 600 | 6.928 | 2700 | 7.228 |
| 650 | 6.925 | 2800 | 7.250 |
| 700 | 6.923 | 2900 | 7.273 |
| 750 | 6.923 | 3000 | 7.295 |
| 800 | 6.924 | 3100 | 7.318 |
| 850 | 6.925 | 3200 | 7.342 |
| 900 | 6.928 | 3300 | 7.365 |
| 950 | 6.931 | 3400 | 7.389 |
| 1000 | 6.934 | 3500 | 7.413 |
| 1100 | 6.943 | 3600 | 7.437 |
| 1200 | 6.954 | 3700 | 7.461 |
| 1300 | 6.966 | 3800 | 7.485 |
| 1400 | 6.980 | 3900 | 7.509 |
| 1500 | 6.995 | 4000 | 7.534 |
| 1600 | 7.011 | 4100 | 7.558 |
| 1700 | 7.028 | 4200 | 7.582 |
| 1800 | 7.045 | 4300 | 7.605 |
| 1900 | 7.063 | 4400 | 7.627 |
| 2000 | 7.083 | 4500 | 7.648 |
| 2100 | 7.102 | 4600 | 7.670 |
| 2200 | 7.122 | 4700 | 7.690 |
| 2300 | 7.142 | 4800 | 7.710 |
| 2400 | 7.163 | 4900 | 7.730 |
| | | 5000 | 7.749 |

| $T$, K | $\overline{C}_{p_a}$, $\frac{\text{kJ}}{\text{kmol K}}$ | $T$, K | $\overline{C}_{p_a}$, $\frac{\text{kJ}}{\text{kmol K}}$ |
|---|---|---|---|
| 275 | 29.05 | 1450 | 30.18 |
| 300 | 29.02 | 1500 | 30.26 |
| 325 | 29.01 | 1550 | 30.34 |
| 350 | 28.99 | 1600 | 30.43 |
| 375 | 28.99 | 1650 | 30.51 |
| 400 | 28.98 | 1700 | 30.60 |
| 425 | 28.98 | 1750 | 30.69 |
| 450 | 28.99 | 1800 | 30.77 |
| 475 | 28.99 | 1850 | 30.86 |
| 500 | 29.00 | 1900 | 30.95 |
| 550 | 29.03 | 1950 | 31.04 |
| 600 | 29.06 | 2000 | 31.13 |
| 650 | 29.10 | 2050 | 31.22 |
| 700 | 29.14 | 2100 | 31.31 |
| 750 | 29.19 | 2150 | 31.41 |
| 800 | 29.24 | 2200 | 31.50 |
| 850 | 29.30 | 2250 | 31.59 |
| 900 | 29.36 | 2300 | 31.69 |
| 950 | 29.43 | 2350 | 31.78 |
| 1000 | 29.49 | 2400 | 31.87 |
| 1050 | 29.56 | 2450 | 31.95 |
| 1100 | 29.63 | 2500 | 32.02 |
| 1150 | 29.71 | 2550 | 32.10 |
| 1200 | 29.78 | 2600 | 32.18 |
| 1250 | 29.86 | 2650 | 32.25 |
| 1300 | 29.93 | 2700 | 32.32 |
| 1350 | 30.01 | 2750 | 32.39 |
| 1400 | 30.09 | | |

**Table G–4** Average Molal Specific Heats for **Water Vapor,** $H_2O$, at Low Pressure Between 537 R (298 K) and Listed Temperatures. Molecular Weight 18.02.

| $T$, R | $\overline{C}_{p_a}$, $\frac{\text{B}}{\text{lb mol R}}$ | $T$, R | $\overline{C}_{p_a}$, $\frac{\text{B}}{\text{lb mol R}}$ |
|---|---|---|---|
| 500 | 8.108 | 2500 | 9.419 |
| 550 | 8.053 | 2600 | 9.495 |
| 600 | 8.018 | 2700 | 9.568 |
| 650 | 8.002 | 2800 | 9.641 |
| 700 | 7.999 | 2900 | 9.711 |
| 750 | 8.007 | 3000 | 9.780 |
| 800 | 8.024 | 3100 | 9.848 |
| 850 | 8.047 | 3200 | 9.914 |
| 900 | 8.074 | 3300 | 9.978 |
| 950 | 8.106 | 3400 | 10.041 |
| 1000 | 8.142 | 3500 | 10.103 |
| 1100 | 8.219 | 3600 | 10.163 |
| 1200 | 8.304 | 3700 | 10.222 |
| 1300 | 8.391 | 3800 | 10.280 |
| 1400 | 8.481 | 3900 | 10.337 |
| 1500 | 8.572 | 4000 | 10.392 |
| 1600 | 8.663 | 4100 | 10.447 |
| 1700 | 8.752 | 4200 | 10.500 |
| 1800 | 8.841 | 4300 | 10.552 |
| 1900 | 8.929 | 4400 | 10.604 |
| 2000 | 9.015 | 4500 | 10.654 |
| 2100 | 9.099 | 4600 | 10.703 |
| 2200 | 9.182 | 4700 | 10.752 |
| 2300 | 9.263 | 4800 | 10.799 |
| 2400 | 9.342 | 4900 | 10.846 |
| | | 5000 | 10.892 |

| $T$, K | $\overline{C}_{p_a}$, $\frac{\text{kJ}}{\text{kmol K}}$ | $T$, K | $\overline{C}_{p_a}$, $\frac{\text{kJ}}{\text{kmol K}}$ |
|---|---|---|---|
| 275 | 33.98 | 1450 | 39.78 |
| 300 | 33.75 | 1500 | 40.06 |
| 325 | 33.60 | 1550 | 40.33 |
| 350 | 33.52 | 1600 | 40.60 |
| 375 | 33.49 | 1650 | 40.86 |
| 400 | 33.50 | 1700 | 41.12 |
| 425 | 33.54 | 1750 | 41.37 |
| 450 | 33.61 | 1800 | 41.61 |
| 475 | 33.70 | 1850 | 41.85 |
| 500 | 33.80 | 1900 | 42.09 |
| 550 | 34.05 | 1950 | 42.32 |
| 600 | 34.34 | 2000 | 42.55 |
| 650 | 34.65 | 2050 | 42.77 |
| 700 | 34.98 | 2100 | 42.99 |
| 750 | 35.32 | 2150 | 43.21 |
| 800 | 35.66 | 2200 | 43.42 |
| 850 | 36.00 | 2250 | 43.62 |
| 900 | 36.34 | 2300 | 43.83 |
| 950 | 36.68 | 2350 | 44.03 |
| 1000 | 37.01 | 2400 | 44.22 |
| 1050 | 37.34 | 2450 | 44.41 |
| 1100 | 37.67 | 2500 | 44.60 |
| 1150 | 37.99 | 2550 | 44.79 |
| 1200 | 38.30 | 2600 | 44.97 |
| 1250 | 38.61 | 2650 | 45.15 |
| 1300 | 38.91 | 2700 | 45.33 |
| 1350 | 39.21 | 2750 | 45.50 |
| 1400 | 39.50 | | |

**Table G–5** Average Molal Specific Heats for **Carbon Monoxide,** CO, Between 537 R (298 K) and Listed Temperatures. Molecular Weight 28.01.

| $T$, R | $\overline{C}_{p_a}$, $\frac{\text{B}}{\text{lb mol R}}$ | $T$, R | $\overline{C}_{p_a}$, $\frac{\text{B}}{\text{lb mol R}}$ |
|---|---|---|---|
| 500 | 7.097 | 2500 | 7.679 |
| 550 | 7.029 | 2600 | 7.711 |
| 600 | 6.988 | 2700 | 7.742 |
| 650 | 6.971 | 2800 | 7.771 |
| 700 | 6.956 | 2900 | 7.799 |
| 750 | 6.957 | 3000 | 7.826 |
| 800 | 6.964 | 3100 | 7.852 |
| 850 | 6.977 | 3200 | 7.878 |
| 900 | 6.994 | 3300 | 7.902 |
| 950 | 7.013 | 3400 | 7.925 |
| 1000 | 7.034 | 3500 | 7.948 |
| 1100 | 7.081 | 3600 | 7.970 |
| 1200 | 7.130 | 3700 | 7.991 |
| 1300 | 7.180 | 3800 | 8.011 |
| 1400 | 7.230 | 3900 | 8.031 |
| 1500 | 7.279 | 4000 | 8.050 |
| 1600 | 7.326 | 4100 | 8.069 |
| 1700 | 7.372 | 4200 | 8.087 |
| 1800 | 7.416 | 4300 | 8.105 |
| 1900 | 7.459 | 4400 | 8.121 |
| 2000 | 7.499 | 4500 | 8.138 |
| 2100 | 7.538 | 4600 | 8.154 |
| 2200 | 7.576 | 4700 | 8.170 |
| 2300 | 7.612 | 4800 | 8.185 |
| 2400 | 7.646 | 4900 | 8.199 |
| | | 5000 | 8.214 |

| $T$, K | $\overline{C}_{p_a}$, $\frac{\text{kJ}}{\text{kmol K}}$ | $T$, K | $\overline{C}_{p_a}$, $\frac{\text{kJ}}{\text{kmol K}}$ |
|---|---|---|---|
| 275 | 29.75 | 1450 | 32.29 |
| 300 | 29.47 | 1500 | 32.41 |
| 325 | 29.30 | 1550 | 32.52 |
| 350 | 29.19 | 1600 | 32.63 |
| 375 | 29.13 | 1650 | 32.73 |
| 400 | 29.12 | 1700 | 32.83 |
| 425 | 29.13 | 1750 | 32.93 |
| 450 | 29.16 | 1800 | 33.02 |
| 475 | 29.21 | 1850 | 33.11 |
| 500 | 29.28 | 1900 | 33.20 |
| 550 | 29.43 | 1950 | 33.28 |
| 600 | 29.60 | 2000 | 33.36 |
| 650 | 29.79 | 2050 | 33.44 |
| 700 | 29.98 | 2100 | 33.52 |
| 750 | 30.16 | 2150 | 33.60 |
| 800 | 30.35 | 2200 | 33.67 |
| 850 | 30.53 | 2250 | 33.74 |
| 900 | 30.71 | 2300 | 33.81 |
| 950 | 30.88 | 2350 | 33.88 |
| 1000 | 31.05 | 2400 | 33.94 |
| 1050 | 31.21 | 2450 | 34.01 |
| 1100 | 31.36 | 2500 | 34.07 |
| 1150 | 31.51 | 2550 | 34.13 |
| 1200 | 31.65 | 2600 | 34.19 |
| 1250 | 31.79 | 2650 | 34.25 |
| 1300 | 31.92 | 2700 | 34.30 |
| 1350 | 32.05 | 2750 | 34.36 |
| 1400 | 32.18 | | |

**Table G–6** Average Molal Specific Heats for **Carbon Dioxide,** $CO_2$, Between 537 R (298 K) and Listed Temperatures. Molecular Weight 44.10.

| $T$, R | $\overline{C}_{p_a}$, $\frac{\text{B}}{\text{lb mol R}}$ | $T$, R | $\overline{C}_{p_a}$, $\frac{\text{B}}{\text{lb mol R}}$ |
|---|---|---|---|
| 500 | 8.852 | 3100 | 12.58 |
| 550 | 8.958 | 3200 | 12.64 |
| 600 | 9.078 | 3300 | 12.70 |
| 650 | 9.203 | 3400 | 12.76 |
| 700 | 9.331 | 3500 | 12.81 |
| 750 | 9.459 | 3600 | 12.87 |
| 800 | 9.584 | 3700 | 12.92 |
| 850 | 9.708 | 3800 | 12.97 |
| 900 | 9.827 | 3900 | 13.02 |
| 950 | 9.944 | 4000 | 13.06 |
| 1000 | 10.05 | 4100 | 13.11 |
| 1100 | 10.27 | 4200 | 13.15 |
| 1200 | 10.46 | 4300 | 13.20 |
| 1300 | 10.65 | 4400 | 13.24 |
| 1400 | 10.82 | 4500 | 13.28 |
| 1500 | 10.98 | 4600 | 13.31 |
| 1600 | 11.13 | 4700 | 13.35 |
| 1700 | 11.27 | 4800 | 13.39 |
| 1800 | 11.40 | 4900 | 13.42 |
| 1900 | 11.52 | 5000 | 13.46 |
| 2000 | 11.64 | | |
| 2100 | 11.75 | | |
| 2200 | 11.85 | | |
| 2300 | 11.95 | | |
| 2400 | 12.04 | | |
| 2500 | 12.13 | | |
| 2600 | 12.21 | | |
| 2700 | 12.29 | | |
| 2800 | 12.37 | | |
| 2900 | 12.44 | | |
| 3000 | 12.51 | | |

| $T$, K | $\overline{C}_{p_a}$, $\frac{\text{kJ}}{\text{kmol K}}$ | $T$, K | $\overline{C}_{p_a}$, $\frac{\text{kJ}}{\text{kmol K}}$ |
|---|---|---|---|
| 275 | 37.02 | 1450 | 51.18 |
| 300 | 37.41 | 1500 | 51.48 |
| 325 | 37.85 | 1550 | 51.77 |
| 350 | 38.32 | 1600 | 52.04 |
| 375 | 38.80 | 1650 | 52.31 |
| 400 | 39.28 | 1700 | 52.56 |
| 425 | 39.76 | 1750 | 52.80 |
| 450 | 40.23 | 1800 | 53.04 |
| 475 | 40.69 | 1850 | 53.26 |
| 500 | 41.14 | 1900 | 53.48 |
| 550 | 42.01 | 1950 | 53.69 |
| 600 | 42.82 | 2000 | 53.89 |
| 650 | 43.58 | 2050 | 54.09 |
| 700 | 44.30 | 2100 | 54.28 |
| 750 | 44.97 | 2150 | 54.46 |
| 800 | 45.59 | 2200 | 54.64 |
| 850 | 46.18 | 2250 | 54.81 |
| 900 | 46.73 | 2300 | 54.98 |
| 950 | 47.26 | 2350 | 55.14 |
| 1000 | 47.75 | 2400 | 55.30 |
| 1050 | 48.21 | 2450 | 55.45 |
| 1100 | 48.65 | 2500 | 55.60 |
| 1150 | 49.07 | 2550 | 55.75 |
| 1200 | 49.47 | 2600 | 55.89 |
| 1250 | 49.84 | 2650 | 56.02 |
| 1300 | 50.20 | 2700 | 56.15 |
| 1350 | 50.54 | 2750 | 56.28 |
| 1400 | 50.87 | | |

**Table G–7** Average Molal Specific Heats for **Sulfur Dioxide,** $SO_2$, Between 537 R (298 K) and Listed Temperatures. Molecular Weight 64.06.

| $T$, R | $\overline{C}_{p_a}$, $\frac{\text{B}}{\text{lb mol R}}$ | $T$, K | $\overline{C}_{p_a}$, $\frac{\text{kJ}}{\text{kmol K}}$ |
|---|---|---|---|
| 500 | 9.513 | 275 | 39.78 |
| 550 | 9.622 | 300 | 40.19 |
| 600 | 9.727 | 325 | 40.59 |
| 650 | 9.832 | 350 | 40.99 |
| 700 | 9.934 | 375 | 41.37 |
| 750 | 10.03 | 400 | 41.76 |
| 800 | 10.13 | 425 | 42.13 |
| 850 | 10.22 | 450 | 42.50 |
| 900 | 10.32 | 475 | 42.86 |
| 950 | 10.41 | 500 | 43.22 |
| 1000 | 10.50 | 550 | 43.91 |
| 1100 | 10.68 | 600 | 44.57 |
| 1200 | 10.84 | 650 | 45.21 |
| 1300 | 11.00 | 700 | 45.82 |
| 1400 | 11.16 | 750 | 46.41 |
| 1500 | 11.30 | 800 | 46.97 |
| 1600 | 11.44 | 850 | 47.50 |
| 1700 | 11.56 | 900 | 48.00 |
| 1800 | 11.68 | 950 | 48.48 |
| 1900 | 11.80 | 1000 | 48.94 |
| 2000 | 11.90 | 1050 | 49.36 |
| 2100 | 12.00 | 1100 | 49.76 |
| 2200 | 12.09 | 1150 | 50.14 |
| 2300 | 12.17 | 1200 | 50.48 |
| 2400 | 12.24 | 1250 | 50.80 |
| 2500 | 12.31 | 1300 | 51.10 |
| 2600 | 12.37 | 1350 | 51.37 |
| 2700 | 12.42 | 1400 | 51.61 |
| | | 1450 | 51.82 |
| | | 1500 | 52.01 |

**Table G–8** Average Molal Specific Heats for **Methane,** $CH_4$, Between 537 R (298 K) and Listed Temperatures. Molecular Weight 16.04.

| $T$, R | $\bar{C}_{p_a}$, $\frac{B}{lb\ mol\ R}$ |
|---|---|
| 500 | 8.214 |
| 550 | 8.429 |
| 600 | 8.640 |
| 650 | 8.849 |
| 700 | 9.057 |
| 750 | 9.262 |
| 800 | 9.465 |
| 850 | 9.667 |
| 900 | 9.866 |
| 950 | 10.06 |
| 1000 | 10.25 |
| 1050 | 10.45 |
| 1100 | 10.64 |
| 1150 | 10.82 |
| 1200 | 11.01 |
| 1250 | 11.19 |
| 1300 | 11.38 |
| 1350 | 11.56 |
| 1400 | 11.73 |
| 1450 | 11.91 |
| 1500 | 12.08 |
| 1550 | 12.25 |
| 1600 | 12.42 |
| 1650 | 12.59 |
| 1700 | 12.75 |
| 1750 | 12.92 |
| 1800 | 13.08 |
| 1850 | 13.24 |

| $T$, K | $\bar{C}_{p_a}$, $\frac{kJ}{kmol\ K}$ |
|---|---|
| 275 | 34.30 |
| 300 | 35.10 |
| 325 | 35.90 |
| 350 | 36.70 |
| 375 | 37.48 |
| 400 | 38.26 |
| 425 | 39.03 |
| 450 | 39.80 |
| 475 | 40.55 |
| 500 | 41.30 |
| 525 | 42.04 |
| 550 | 42.78 |
| 575 | 43.51 |
| 600 | 44.23 |
| 625 | 44.94 |
| 650 | 45.65 |
| 675 | 46.34 |
| 700 | 47.04 |
| 725 | 47.72 |
| 750 | 48.40 |
| 775 | 49.07 |
| 800 | 49.73 |
| 825 | 50.38 |
| 850 | 51.03 |
| 875 | 51.67 |
| 900 | 52.31 |
| 925 | 52.93 |
| 950 | 53.55 |
| 975 | 54.16 |
| 1000 | 54.77 |

**Table G–9** Average Molal Specific Heats for **Ethane,** $C_2H_6$, Between 537 R (298 K) and Listed Temperatures. Molecular Weight 30.07.

| $T$, R | $\bar{C}_{p_a}$, $\frac{B}{lb\ mol\ R}$ |
|---|---|
| 500 | 12.28 |
| 550 | 12.75 |
| 600 | 13.21 |
| 650 | 13.67 |
| 700 | 14.11 |
| 750 | 14.55 |
| 800 | 14.99 |
| 850 | 15.41 |
| 900 | 15.83 |
| 950 | 16.24 |
| 1000 | 16.65 |
| 1050 | 17.04 |
| 1100 | 17.43 |
| 1150 | 17.82 |
| 1200 | 18.19 |
| 1250 | 18.56 |
| 1300 | 18.92 |
| 1350 | 19.28 |
| 1400 | 19.62 |
| 1450 | 19.96 |
| 1500 | 20.30 |
| 1550 | 20.62 |
| 1600 | 20.94 |
| 1650 | 21.25 |
| 1700 | 21.56 |
| 1750 | 21.85 |
| 1800 | 22.15 |
| 1850 | 22.43 |

| $T$, K | $\bar{C}_{p_a}$, $\frac{kJ}{kmol\ K}$ |
|---|---|
| 275 | 51.24 |
| 300 | 53.01 |
| 325 | 54.75 |
| 350 | 56.48 |
| 375 | 58.17 |
| 400 | 59.84 |
| 425 | 61.49 |
| 450 | 63.12 |
| 475 | 64.72 |
| 500 | 66.29 |
| 525 | 67.84 |
| 550 | 69.37 |
| 575 | 70.87 |
| 600 | 72.35 |
| 625 | 73.81 |
| 650 | 75.24 |
| 675 | 76.65 |
| 700 | 78.03 |
| 725 | 79.39 |
| 750 | 80.72 |
| 775 | 82.03 |
| 800 | 83.32 |
| 825 | 84.58 |
| 850 | 85.82 |
| 875 | 87.03 |
| 900 | 88.22 |
| 925 | 89.38 |
| 950 | 90.53 |
| 975 | 91.64 |
| 1000 | 92.74 |

**Table G–10** Average Molal Specific Heats for **Propane,** $C_3H_8$, Between 537 R (298 K) and Listed Temperatures. Molecular Weight 44.10.

| $T$, R | $\overline{C}_{p_a}$, $\frac{\text{B}}{\text{lb mol R}}$ | $T$, K | $\overline{C}_{p_a}$, $\frac{\text{kJ}}{\text{kmol K}}$ |
|---|---|---|---|
| 500 | 17.17 | 275 | 71.61 |
| 550 | 17.89 | 300 | 74.31 |
| 600 | 18.59 | 325 | 76.96 |
| 650 | 19.28 | 350 | 79.58 |
| 700 | 19.96 | 375 | 82.16 |
| 750 | 20.63 | 400 | 84.71 |
| 800 | 21.28 | 425 | 87.21 |
| 850 | 21.93 | 450 | 89.67 |
| 900 | 22.56 | 475 | 92.09 |
| 950 | 23.18 | 500 | 94.48 |
| 1000 | 23.79 | 525 | 96.82 |
| 1050 | 24.39 | 550 | 99.12 |
| 1100 | 24.98 | 575 | 101.3 |
| 1150 | 25.55 | 600 | 103.6 |
| 1200 | 26.12 | 625 | 105.8 |
| 1250 | 26.67 | 650 | 107.9 |
| 1300 | 27.21 | 675 | 110.0 |
| 1350 | 27.74 | 700 | 112.1 |
| 1400 | 28.25 | 725 | 114.1 |
| 1450 | 28.77 | 750 | 116.1 |
| 1500 | 29.25 | 775 | 118.0 |
| 1550 | 29.73 | 800 | 120.0 |
| 1600 | 30.20 | 825 | 121.8 |
| 1650 | 30.67 | 850 | 123.7 |
| 1700 | 31.11 | 875 | 125.5 |
| 1750 | 31.55 | 900 | 127.2 |
| 1800 | 31.97 | 925 | 128.9 |
| 1850 | 32.38 | 950 | 130.6 |
| | | 975 | 132.2 |
| | | 1000 | 133.8 |

**Table G–11** Average Molal Specific Heats for **n-Butane,** $C_4H_{10}$, Between 537 R (298 K) and Listed Temperatures. Molecular Weight 58.12.

| $T$, R | $\overline{C}_{p_a}$, $\frac{\text{B}}{\text{lb mol R}}$ | $T$, K | $\overline{C}_{p_a}$, $\frac{\text{kJ}}{\text{kmol K}}$ |
|---|---|---|---|
| 500 | 23.41 | 275 | 97.65 |
| 550 | 24.32 | 300 | 101.0 |
| 600 | 25.20 | 325 | 104.4 |
| 650 | 26.07 | 350 | 107.7 |
| 700 | 26.93 | 375 | 110.9 |
| 750 | 27.77 | 400 | 114.1 |
| 800 | 28.60 | 425 | 117.3 |
| 850 | 29.41 | 450 | 120.4 |
| 900 | 30.21 | 475 | 123.5 |
| 950 | 30.99 | 500 | 126.5 |
| 1000 | 31.76 | 525 | 129.4 |
| 1050 | 32.52 | 550 | 132.3 |
| 1100 | 33.26 | 575 | 135.2 |
| 1150 | 33.98 | 600 | 138.0 |
| 1200 | 34.69 | 625 | 140.7 |
| 1250 | 35.39 | 650 | 143.4 |
| 1300 | 36.07 | 675 | 146.1 |
| 1350 | 36.73 | 700 | 148.7 |
| 1400 | 37.38 | 725 | 151.3 |
| 1450 | 38.02 | 750 | 153.8 |
| 1500 | 38.64 | 775 | 156.2 |
| 1550 | 39.25 | 800 | 158.6 |
| 1600 | 39.84 | 825 | 161.0 |
| 1650 | 40.42 | 850 | 163.3 |
| 1700 | 40.98 | 875 | 165.6 |
| 1750 | 41.53 | 900 | 167.8 |
| 1800 | 42.07 | 925 | 169.9 |
| 1850 | 42.58 | 950 | 172.0 |
| | | 975 | 174.1 |
| | | 1000 | 176.1 |

# Answers to Selected Problems

## Chapter 2

2.2 35.75 lbf

2.5 2.1 cm

2.6 60.8 ft

2.12 281.4 psia

2.17 55 psia

2.20 1.910 $m^3$/kg

2.24 414.20 degrees

2.26 (a) 531.4 degrees; (b) 135.6 degrees Z

2.27 248.8 $ft^2$

2.30 (a) 123.5 m/s; (b) 63,200 kg/hr

2.36 64.5 psia

## Chapter 3

3.1 −40.1 kJ, 2.1 kJ, 2.1 kJ

3.5 System energy is constant, bar energy increases, gas energy decreases.

3.7 (a) 63,395 ft lbf; (b) 182.1 B; (c) 91.5 B

3.11 853 kJ

3.19 0.776 kg/s

3.20 (a) 73.5 HP; (b) −1764 B/min

3.26 (a) 170.5 B/lbm; (b) 40,200 HP; (c) 6.13 $ft^2$

3.29 183,400 kW

3.32 $3.32/day

3.35 Heat pump $0.40/hr, resistance heat $0.85/hr

## Chapter 4

4.8 No—pressure is below triple-point pressure

4.12 1.231 lbm liquid, 0.769 lbm vapor, 4.98 $ft^3$ vapor

4.15 358.48 F, 580 F

4.16 (a) 66.98 psia; (b) 66.98 psia; (c) 22.0 psia

4.17 (a) 0.5400 $ft^3$/lbm; (b) 0.02024 $ft^3$/lbm

4.27 (a) 0.99874 kg liquid/kg mix, 0.00126 kg vapor/kg mix; (b) 683 kJ

4.28 $q$ = 262.1 B/lbm, enthalpy change = 262.1 B/lbm

4.32 13.5 percent

4.34 3974 lbm/hr

4.35 426 gal/min

## Chapter 5

5.2 1.543 $m^3$, 1068 kPa

5.4 (a) 50.75 psia or 36.46 psig; (b) 0.01653 lbm air

5.6 128 $ft^3$/min

5.10 470 kPa

5.14 734 R

5.17 80.4 kW

5.20 (a) 151 F; (b) 3.34 $ft^2$

5.22 −218 kJ

5.28 772 K, 1039 K

5.33 (a) −77.2 B; (b) −85.7 B

5.36 0.0101 $ft^2$

5.45 $q = -37$ kJ/kg, $w = -212$ kJ/kg

5.48 $\beta_R = 2.79$, $\beta_P = 3.79$

5.51 (a) 18.93 $ft^3$; (b) 18.46 $ft^3$; (c) 18.11 $ft^3$

## Chapter 6

6.6 Yes

6.9 (a) 40.3 percent; (b) 40.3 percent

6.12 3.04

6.17 350 kJ/min, 350 kJ/min

6.20 0.562 kW

6.22 Yes

6.24 $3.12/day

6.26 141 kJ

## Chapter 7

7.1 0, 0.04769 B/(lbm R), −0.04769 B/(lbm R), 0

7.5 0.254 kJ/K

7.10 (a) $t_2 = 24$ C, $p_2 = 350$ kPa; (b) 0.199 kJ/(kg K)

7.12 (a) 327.86 F, 1.4856 B/(lbm R), 3.971 $ft^3$/lbm (b) 219.56 B/lbm, 0.36702 B/(lbm R), 0.016972 $ft^3$/lbm (c) 66.98 psia, 1114.6 B/lbm, 6.006 $ft^3$/lbm (d) 1000 F, 1493.5 B/lbm, 1.6094 B/(lbm R) (e) 0.0142 psia, 387.36 B/lbm, 8555.7 $ft^3$/lbm (f) 100 F, 0.12963 B/(lbm R), 0.016130 $ft^3$/lbm

7.14 198.53 kPa, −134.5 kJ/kg

7.21 (a) 0.0284 B/(lbm R); (b) 352 B/lbm

7.22 (a) 0.0218 B/(lbm R); (b) 181 B/lbm

7.25 −0.100 B/R

7.38 1251 ft/sec, 11.79 psia, 640 R

## Chapter 8

8.1 6.47 B

8.3 13.17 kJ/kg

8.6 −83.3 kJ

8.11 75.4 B/lbm $H_2O$

8.14 0.198 B, 2.79 B

8.18 429.3 B/lbm, 500 B/lbm, equal

8.20 (a) 60.2 kJ/kg; (b) 30.79 kJ/kg; (c) 86.6 kJ/kg

8.23 13.2 kJ/kg

8.28 (a) 78 percent; (b) 37.3 B/lbm

8.30 629 kJ/kg steam

8.36 53 percent

## Chapter 9

9.1 827.1 B/lbm, 538.6 B/lbm, 34.9 percent

9.8 (a) 32.8 percent; (b) 33.7 percent; (c) 35.0 percent

9.9 480 B/lbm, 505 B/lbm

9.12 (a) 37.4 percent; (b) 36.2 percent

9.14 38.0 percent

9.15 37.3 percent

9.20 39.6 percent, 217,600 lbm/hr, 165,000 lbm/hr, 373 tons/day, 114.1 kW, 22,840 gal/min

9.23 Plan A: 38,740 lbm/hr; Plan B 44,380 lbm/hr, 1705 kW

9.32 (a) 5.75; (b) 0.820 HP/ton; (c) 3.35 $ft^3$/min ton

9.39 Refrigerator: 16.6 B/lbm; Freezer, 33.6 B/lbm; Work = −20.7 B/lbm

9.44 $0.25, $1.06, $4.10

## Chapter 10

10.1 6.86

10.5 (a) 142.5 HP; (b) 56.5 percent; (c) 0.225 lbm/HP hr

10.7 (a) 21.1; (b) 2.30; (c) 352 B/lbm

10.11 Otto: 7.97, 56.4 percent; Diesel: 18.70, 64.9 percent

10.14 (a) 200 B/lbm; (b) 394 B/lbm; (c) 1328 R; (d) 50.8 percent

10.16 (a) 329 B/lbm; (b) 60.7 percent

10.18 (a) 140 B/lbm; (b) 373 B/lbm; (c) 1492 R; (d) 37.5 percent

10.20 (a) 74.75 psia; (b) 3490 ft/sec; (c) 185 lbm/sec

10.23 (a) $27.68 \times 10^6$ B/hr, $8.41 \times 10^6$ B/hr; (b) 56.4 percent

## Chapter 11

11.2 (a) $N_2$: 0.1004 lb mol, $H_2$: 0.2685 lb mol, $O_2$: 0.1475 lb mol; (b) $N_2$: 4.023 psia, $H_2$: 10.76 psia, $O_2$: 5.91 psia; (c) 143.7 $ft^3$

11.4 (a) 70.24 lbm/min; (b) $N_2$: 0.3163, $O_2$: 0.6837; (c) 851 $ft^3$/min

11.8 $N_2$: 26.94 kPa, $H_2$: 55.88 kPa, $CO_2$: 21.44 kPa, 0.735 kg/$m^3$; 17.23 kg/kmol

11.11 (a) $CO_2$: 0.108, He: 0.622; (b) 32.28 kJ/(kmol K), 23.96 kJ/(kmol K); (c) 2.188 $m^3$

11.14 208 F

11.18 (a) 520.8 B/lbm mix; (b) 0.2987 B/(lbm mix R); (c) He: 0.092 B/(lbm mix R), $N_2$: 0.2067 B/(lbm mix R)

11.23 792 kg/$m^2$s

11.25 (a) 0.249 lbm; (b) $4.0 \times 10^{-3}$ $ft^3$; (c) 0.178 psia; (d) 948 $ft^3$

11.26 (a) 63 F; (b) 60 percent; (c) 0.073 lbm/$ft^3$

11.28 45.4 percent

11.32 1.61 percent

11.35 (a) 63.5 F; (b) 0.00455 lbm $H_2O$/lbm da; (c) 27 percent

11.37 (a) 0.01062 lbm $H_2O$/lbm da; (b) 58 F; (c) 36.2 percent; (d) 14.38 $ft^3$/lbm; (e) 68.82 lbm da

11.39 (a) 26.7 lbm/hr; (b) 86,460 B/hr

11.43 (a) 1204 B/min; (b) 605 B/min

11.45 (a) 1092 lbm/hr; (b) 28,500 B/min; 142.3 tons; (d) 152 HP

11.47 (a) $3.716 \times 10^7$ lbm/hr; (b) $6.076 \times 10^5$ lbm/hr; (c) $2.40 \times 10^8$ $ft^3$/hr

## Chapter 12

12.1 (a) 9.84 $ft^3$ da/$ft^3$ fuel, 17.3 lbm da/lbm fuel; (b) 24.6 $ft^3$ da/$ft^3$ fuel, 15.69 lbm da/lbm fuel; (c) 10.57 $ft^3$ da/$ft^3$ fuel, 17.1 lbm da/lbm fuel

12.4 (a) 11,190 $ft^3$/min; (b) 19,500 $ft^3$/min; (c) 137 F

12.7 (a) 19.4 percent; (b) 63.0 lbm $H_2O$/lb mol fuel; (c) Input, lbm: fuel, 36.6; dry air, 600.0; air supply $H_2O$; 10.7; total, 647.3: Output, lbm: $CO_2$, 85.85; CO, 8.37; $O_2$, 29.15; $N_2$, 461.6; $H_2O$, 63.0; total, 648

12.10 −383,029 B/lb mol, −1,420,386 B/lb mol, −121,749 B/lb mol

12.12 (a) −121,750 B/lb mol; (b) −121,750 B/lb mol

12.22 −687,830 kJ/kmol

12.26 −318,500 B/lb mol

12.27 1340 R, 1220 R

12.30 201 percent

12.32 (a) 4780 R; (b) 3780 R

12.37 5190 R

## Chapter 13

13.1 (a) 121.5 psia, 740 R; (b) 76.8 psia, 740 R; (c) 0.0313 B/(lbm R)

13.3 1266.6 B/lbm, 146 psia, 488 F

13.6 (a) 1076 R; (b) 15,680 HP

13.9 (a) 592 R, 18.06 psia; (b) 4.36 $ft^2$; (c) 7464 lbf opposite to the flow direction

13.10 (a) 1395 K, 125 kPa; (b) 398.5 K, 199.2 kPa

13.16 (a) 740 R, 121.6 kPa; (b) 2.06

13.22 (a) 319 K; (b) 215 kPa; (c) 393 m/s; (d) 1.119

13.25 (a) 2.75 lbm/sec, 1.0; (b) 0.903 lbm/sec, 0.98

13.29 3.74

13.32 0.00433 $ft^2$, 0.00601 $ft^2$

13.34 0.00650 $ft^2$

13.36 94.8 percent

13.39 (a) either; (b) converging-diverging; (c) converging-diverging; (d) converging-diverging; (e) converging

## Chapter 14

14.2 (a) −70.5 B/lbm; (b) −93.9 B/lbm; (c) −89.4 B/lbm

14.5 14,150 HP

14.6 (a) 202 HP, −4270 B/min; (b) 194 HP, −3330 B/min

14.8 (a) 191 HP; (b) −778 B/min, −696 B/min; (c) 330 $ft^3$/min

14.11 1760 HP

14.12 0.59 $ft^3$

14.15 (a) 3.39 $ft^3$/(ton min); (b) 1.22 HP/ton

14.16 (a) 38.6 HP; (b) 26.1 tons; (c) 6560 B/min

14.18 58 percent

14.23 (a) 1.66 HP

14.27 (a) 589 ft/sec; (b) 38.9 degrees

# Index